Puy

185

20

Springer Series in Solid-State Sciences

Volume 194

The Springer Series in Solid-State Sciences consists of fundamental scientific books prepared by leading researchers in the field. They strive to communicate, in a systematic and comprehensive way, the basic principles as well as new developments in theoretical and experimental solid-state physics.

More information about this series at http://www.springer.com/series/682

Michael Kneissl · Andreas Knorr ·
Stephan Reitzenstein · Axel Hoffmann
Editors

Semiconductor Nanophotonics

Materials, Models, and Devices

Springer

Editors
Michael Kneissl
Institute of Solid State Physics
Technische Universität Berlin
Berlin, Germany

Andreas Knorr
Institute of Theoretical Physics
Technische Universität Berlin
Berlin, Germany

Stephan Reitzenstein
Institute of Solid State Physics
Technische Universität Berlin
Berlin, Germany

Axel Hoffmann
Institute of Solid State Physics
Technische Universität Berlin
Berlin, Germany

ISSN 0171-1873 ISSN 2197-4179 (electronic)
Springer Series in Solid-State Sciences
ISBN 978-3-030-35655-2 ISBN 978-3-030-35656-9 (eBook)
https://doi.org/10.1007/978-3-030-35656-9

This Springer imprint is published by the registered company Springer Nature Switzerland AG
The registered company address is: Gewerbestrasse 11, 6330 Cham, Switzerland

Preface

This book is a compendium of twelve years of research within the collaborative research centre "Semiconductor Nanophotonics" (CRC 787) which was funded by the German Research Foundation (DFG) between 2008 and 2019. Although all CRC 787 projects contributed to the book, it is not meant to be mere project report. In addition to summarizing key research results of CRC 787, the book is intended to provide a broad overview of the state-of-the-art in the development of semiconductor nanostructures and nanophotonic devices. Consequently, all book chapters were organized along with key scientific topics with contributions from different projects interwoven within thematic chapters. It covers epitaxial growth and the structural characteristics of group III-arsenide and III-nitride semiconductor materials and nanostructures, the description of fundamental optic, electronic, and vibronic properties of nanomaterials as well as the design and realization of a wide range of semiconductor nanophotonic devices.

After a brief introduction into basic features and applications of semiconductor nanophotonic devices in Chap. 1, the Chap. 2 covers the epitaxial growth as well as structural, electronic, and optical properties of InAs submonolayer quantum dots and their application in semiconductor optical amplifiers and lasers. This discussion is complemented by a review of site-controlled nucleation of InGaAs quantum dots using the buried stressor growth mode in Chap. 3, which concludes with a demonstration of single-photon emission from an electrically driven quantum light source based on a site-controlled quantum dot. The ultrafast carrier and photon dynamics in semiconductors quantum dots and nanophotonic devices are presented in Chap. 4 including the dynamics of optical gain spectra and their application in optical amplifiers, laser diodes, and multi-section mode-locked laser devices. The Chap. 5 focuses on the description of a series of advanced nanoscale characterization tools, encompassing scanning transmission electron microscopy cathodoluminescence (STEM-CL), tip-enhanced Raman spectroscopy (TERS), microphotoluminescence (μPL), high-resolution X-ray diffraction (XRD), and cross-sectional scanning tunnelling microscopy and spectroscopy (STM/STS). This chapter also covers the growth of III-nitride nanowires and nanorods as well as analysis of their structural and optical properties. It is followed by Chap. 6

presenting a microscopic description of light emission from quantum dots and coupled quantum dot—microcavity structures with respect to their correlated photon emission statistics. It provides a comprehensive examination of individual electronic excitations such as excitons, the fundamental Coulomb interaction between the electronic states as well as their coupling to other quasiparticles such as phonons and photons. Furthermore, phenomena such as photon entanglement and intraband spectroscopy are reviewed. Building on these fundamental descriptions of light emission in the quantum regime of light–matter interaction, Chap. 7 presents basic concepts and numerical methods for the self-consistent modelling and multi-dimensional simulation of a wide range of semiconductor nanophotonic devices. Recent advances in device-scale modelling of electrically driven quantum-dot-based single-photon sources and laser diodes are described creating new numeric tools in the design and development of future integrated light sources and quantum devices. In Chap. 8, deterministic fabrication technologies for quantum-dot-based sources of single photons and entangled-photon pairs for applications in optical quantum communication systems are discussed. This includes the controlled integration of quantum dots into microlenses for enhanced photon-extraction efficiency, the deterministic fabrication of on-chip integrated quantum circuits with high functionality as well as cutting-edge quantum-optical measurements. Building on this, Chap. 9 describes quantum networks based on single-photon emitters and covers topics like frequency conversion of quantum light, single-photon storage and quantum repeaters, quantum key distribution protocols, and the demonstration of a free-space optical link as a test bed for a future quantum network. Chapter 10 reviews vertical-cavity surface-emitting laser diodes (VCSELs), another critical nanophotonic device with a wide range of applications in optical data communication systems, photonic-electronic integrated circuits as well as sensing and tracking systems. Different VCSEL designs including high contrast gratings and VCSEL arrays, as well as various key fabrication technologies will be discussed, and the performance characteristics of VCSEL devices with record ultra-high modulation bandwidths and energy efficiencies will be presented. Silicon photonic interconnect technologies based on VCSELs emitting at tele- and datacom wavelengths are examined in Chap. 11. Incoherent, as well as coherent VCSEL-based transmission links, e.g. using quadrature phase-shift keying (QPSK), are presented and analysed. The final two chapters are dedicated to group III-nitride materials and devices. In Chap. 12, microcavities with InGaN quantum wells or GaN quantum dots as active medium are explored as building blocks for electrically driven surface-emitting lasers and room-temperature single-photon emitters. Key components like distributed Bragg mirrors in the visible to deep UV spectral range as well as the structural, electronic, and optical properties of GaN quantum dots and microcavities will be discussed. Finally, Chap. 13 explores the realization of current-injection AlGaN-based laser diodes emitting in the deep ultraviolet spectral range. This includes the fabrication of low defect density AlN templates, the pseudomorphic growth of AlGaN laser heterostructures, advanced fabrication technologies for ohmic contact formation and tunnel injection, and the investigation of optical gain and losses in deep UV lasers.

We like to express our appreciation to all the co-authors for their important contributions to the extensive chapters of this book. This book represents a truly joint effort with a total of 67 co-authors contributing to its content including all principal investigators of the third phase of CRC 787 as well as a number of postdoctoral researchers and Ph.D. students. We would also like to thank the German Research Foundation, in particular, Dr. Michael Mößle and Mrs. Brit Redöhl, as well as all the members of the DFG grant committee for their continued support over this twelve-year journey. Last but not least, a special "Thank you" goes to Thomas Kure, scientific secretary of the CRC 787, for his dedicated effort and organizational skills that were crucial in order to bring all contributions together in a timely, coherent, and smooth manner.

Berlin, Germany

Michael Kneissl
Andreas Knorr
Stephan Reitzenstein
Axel Hoffmann

Contents

Contributors

Uwe Bandelow Weierstraß-Institut für Angewandte Analysis und Stochastik, Berlin, Germany

Oliver Benson Institute of Physics, Humboldt-Universität zu Berlin, Berlin, Germany

Christoph Berger Institut für Physik, Otto-von-Guericke-Universität Magdeburg, Magdeburg, Germany

Frank Bertram Institut für Physik, Otto-von-Guericke-Universität Magdeburg, Magdeburg, Germany

Fabian Böhm Institut für Optik und Atomare Physik, Technische Universität Berlin, Berlin, Germany

Samir Bounouar Institute of Solid State Physics, Technische Universität Berlin, Berlin, Germany

Sven Burger Zuse-Institut Berlin, Berlin, Germany

Alexander Carmele Institut für Theoretische Physik, Nichtlineare Optik und Quantenelektronik, Technische Universität Berlin, Berlin, Germany

Jürgen Christen Institut für Physik, Otto-von-Guericke-Universität Magdeburg, Magdeburg, Germany

Armin Dadgar Institut für Physik, Otto-von-Guericke-Universität Magdeburg, Magdeburg, Germany

Mario Dähne Institute of Solid State Physics, Technische Universität Berlin, Berlin, Germany

Holger Eisele Institut für Festkörperphysik, Technische Universität Berlin, Berlin, Germany

Ronald Freund Heinrich Hertz Institute, Fraunhofer Institute for Telecommunications, Berlin, Germany

Ludwig A. Th. Greif Institut für Festkörperphysik, Technische Universität Berlin, Berlin, Germany

Tobias Heindel Institute of Solid State Physics, Technische Universität Berlin, Berlin, Germany

Bastian Herzog Institute of Optics and Atomic Physics, Technische Universität Berlin, Berlin, Germany

Axel Hoffmann Institut für Festkörperphysik, Technische Universität Berlin, Berlin, Germany

Theresa Höhne Zuse-Institut Berlin, Berlin, Germany

Stefan T. Jagsch Institut für Festkörperphysik, Technische Universität Berlin, Berlin, Germany

Lina Jaurigue Institut für Theoretische Physik, Technische Universität Berlin, Berlin, Germany

Markus Kantner Weierstraß-Institut für Angewandte Analysis und Stochastik, Berlin, Germany

Michael Kneissl Institute of Solid State Physics, Technische Universität Berlin, Berlin, Germany;
Ferdinand-Braun-Institut, Leibniz-Institut für Höchstfrequenztechnik, Berlin, Germany

Andreas Knorr Institut für Theoretische Physik, Nichtlineare Optik und Quantenelektronik, Technische Universität Berlin, Berlin, Germany

Mirco Kolarczik Institut für Optik und Atomare Physik, Technische Universität Berlin, Berlin, Germany

Thomas Koprucki Weierstraß-Institut für Angewandte Analysis und Stochastik, Berlin, Germany

Tim Kroh Institute of Physics, Humboldt-Universität zu Berlin, Berlin, Germany

Christian Kuhn Institute of Solid State Physics, Technische Universität Berlin, Berlin, Germany

Sandra C. Kuhn Institut für Theoretische Physik, Nichtlineare Optik und Quantenelektronik, Technische Universität Berlin, Berlin, Germany

Michael Lehmann Institute of Optics and Atomic Physics, Technische Universität Berlin, Berlin, Germany

Andrea Lenz Institute of Solid State Physics, Technische Universität Berlin, Berlin, Germany

Benjamin Lingnau Institute of Theoretical Physics, Technische Universität Berlin, Berlin, Germany

James A. Lott Center of Nanophotonics, Institute of Solid State Physics, Technical University Berlin, Berlin, Germany

Kathy Lüdge Institute of Theoretical Physics, Technische Universität Berlin, Berlin, Germany;
Institut für Theoretische Physik, Technische Universität Berlin, Berlin, Germany

Janina Maultzsch Department of Physics, Friedrich-Alexander-Universität Erlangen-Nürnberg, Erlangen, Germany

Stefan Meinecke Institut für Theoretische Physik, Technische Universität Berlin, Berlin, Germany

Sebastian Metzner Institut für Physik, Otto-von-Guericke-Universität Magdeburg, Magdeburg, Germany

Alexander Mielke Weierstraß-Institut für Angewandte Analysis und Stochastik, Berlin, Germany;
Institut für Mathematik, Humboldt-Universität zu Berlin, Berlin, Germany

Chris Müller Institute of Physics, Humboldt-Universität zu Berlin, Berlin, Germany

Marcus Müller Institut für Physik, Otto-von-Guericke-Universität Magdeburg, Magdeburg, Germany

Tore Niermann Institute of Optics and Atomic Physics, Technische Universität Berlin, Berlin, Germany

Nina Owschimikow Institute of Optics and Atomic Physics, Technische Universität Berlin, Berlin, Germany

Nicolas Perlot Heinrich Hertz Institute, Fraunhofer Institute for Telecommunications, Berlin, Germany

Alexander Pimenov Weierstraß-Institut für Angewandte Analysis und Stochastik, Berlin, Germany

Udo W. Pohl Institute of Solid State Physics, Technische Universität Berlin, Berlin, Germany

Emanuele Poliani Institut für Festkörperphysik, Technische Universität Berlin, Berlin, Germany

Stephan Reitzenstein Institute of Solid State Physics, Technische Universität Berlin, Berlin, Germany

Marten Richter Institut für Theoretische Physik, Nichtlineare Optik und Quantenelektronik, Technische Universität Berlin, Berlin, Germany

Jasper Rödiger Heinrich Hertz Institute, Fraunhofer Institute for Telecommunications, Berlin, Germany

Sven Rodt Institute of Solid State Physics, Technische Universität Berlin, Berlin, Germany

Andrei Schliwa Institute of Solid State Physics, Technische Universität Berlin, Berlin, Germany

Gordon Schmidt Institut für Physik, Otto-von-Guericke-Universität Magdeburg, Magdeburg, Germany

Frank Schmidt Zuse-Institut Berlin, Berlin, Germany

Philipp-Immanuel Schneider JCMwave GmbH, Berlin, Germany

Pascal M. Seiler Technische Universität Berlin, Institut für HF- und HL-Systemtechnologien, Berlin, Germany

André Strittmatter Institut für Physik, Otto-von-Guericke-Universität Magdeburg, Magdeburg, Germany

Luca Sulmoni Institute of Solid State Physics, Technische Universität Berlin, Berlin, Germany

Bernd Tillack Technische Universität Berlin, Institut für HF- und HL-Systemtechnologien, Berlin, Germany;
IHP, Leibniz Institut für innovative Mikroelektronik, Frankfurt, Germany

Günther Tränkle Ferdinand-Braun-Institut, Leibniz-Institut für Höchstfrequenztechnik, Berlin, Germany

Peter Veit Institut für Physik, Otto-von-Guericke-Universität Magdeburg, Magdeburg, Germany

Andrei Vladimirov Weierstraß-Institut für Angewandte Analysis und Stochastik, Berlin, Germany

Markus R. Wagner Institut für Festkörperphysik, Technische Universität Berlin, Berlin, Germany

Tim Wernicke Institute of Solid State Physics, Technische Universität Berlin, Berlin, Germany

Markus Weyers Ferdinand-Braun-Institut, Leibniz-Institut für Höchstfrequenztechnik, Berlin, Germany

Ulrike Woggon Institut für Optik und Atomare Physik, Technische Universität Berlin, Berlin, Germany

Matthias Wolfrum Weierstraß-Institut für Angewandte Analysis und Stochastik, Berlin, Germany

Hans-Jürgen Wünsche Weierstraß-Institut für Angewandte Analysis und Stochastik, Berlin, Germany;
Ferdinand-Braun-Institut, Leibniz-Institut für Höchstfrequenztechnik, Berlin, Germany

Lars Zimmermann Technische Universität Berlin, Institut für HF- und HL-Systemtechnologien, Berlin, Germany;
IHP, Leibniz Institut für innovative Mikroelektronik, Frankfurt, Germany

Lin Zschiedrich JCMwave GmbH, Berlin, Germany

Chapter 1
A Short Introduction to Semiconductor Nanophotonics

Michael Kneissl

Abstract Semiconductor nanophotonic devices, confining electronic excitations and light on a nanometer spatial scale, could provide valuable solutions to many challenges that society is facing. One example are energy efficient high-speed vertical cavity surface emitting lasers (VCSELs) for applications in multi-terabus systems to curb the rapidly increasing power consumption of the global internet traffic. Another relates to cyber security and the development of key components for quantum cryptography, like qubit and entangled photon emitters operating at high qubit rates. Nanophotonic technologies also have significant economic impact with a wide range of applications ranging from materials processing and 3D printing to medical diagnostics and sensing. In this book, some of the key features of nanophotonic devices will be introduced and emphasize the strong interaction between development of nanomaterials, key advances in the performance of nanophotonic devices, like VCSELs and non-classical light emitters and their theoretical description of the electronic and optical properties on an nano-, micro- and macroscopic scale. We will present highlights of nanophotonic device development and illustrate synergies between different device designs and fabrication technologies and how this may be exploited to create a tool box for future generation integrated photonic circuits and quantum communication networks.

1.1 Nanophotonics and Internet Traffic

Applied research is often triggered by global challenges that society is facing and the search for technological solutions that can help solve these problems. This also applies to research in the field of semiconductor nanophotonics, which is the subject of this book. Explaining research in relationship to solving these enormous challenges provides also an excellent conduit for explaining the importance of these research activities to a wider audience and even to the general public which is often curious,

---

M. Kneissl (✉)
Institute of Solid State Physics, Technische Universität Berlin, Berlin, Germany
e-mail: kneissl@physik.tu-berlin.de

Ferdinand-Braun-Institut, Leibniz-Institut für Höchstfrequenztechnik, Berlin, Germany

M. Kneissl et al. (eds.), *Semiconductor Nanophotonics*, Springer Series
in Solid-State Sciences 194, https://doi.org/10.1007/978-3-030-35656-9_1

but not necessarily technology savvy. In the following we will present three examples that illustrate the challenges that our increasingly interconnected world is facing and some of the solutions that developments in the area of semiconductor nanophotonics may provide.

A first example is global warming and the consequences of climate change, which are among the most prominent and critical challenges that humankind is currently facing. In order to reduce greenhouse gas emissions many efforts are geared towards developing alternative renewable energy sources, like photovoltaics, hydroelectric and wind power. However, much less attention is focused on curbing future energy consumption, and even if, the focus is mostly devoted to making transportation and buildings more energy-efficient. A fact that is much less known is that the power consumption of the global internet could increase to more than 20% of total electric power by 2030 [1, 2]. This dramatic increase is driven by the tremendous growth of global internet traffic that has already entered the zettabyte era in 2016. One zettabyte corresponds to 10^{21} bytes of data per year and is equivalent to transferring the information stored on 35 Mio. DVDs every hour. Figure 1.1 shows a forecast of the global internet protocol (IP) based data traffic published by the company CISCO in 2018 [3]. Already in 2019 it is expected that 200 exabytes of data will to be transferred each month corresponding to 2.4×10^{21} bytes of data for the entire year. Since most of the data is transported via optical fiber communication

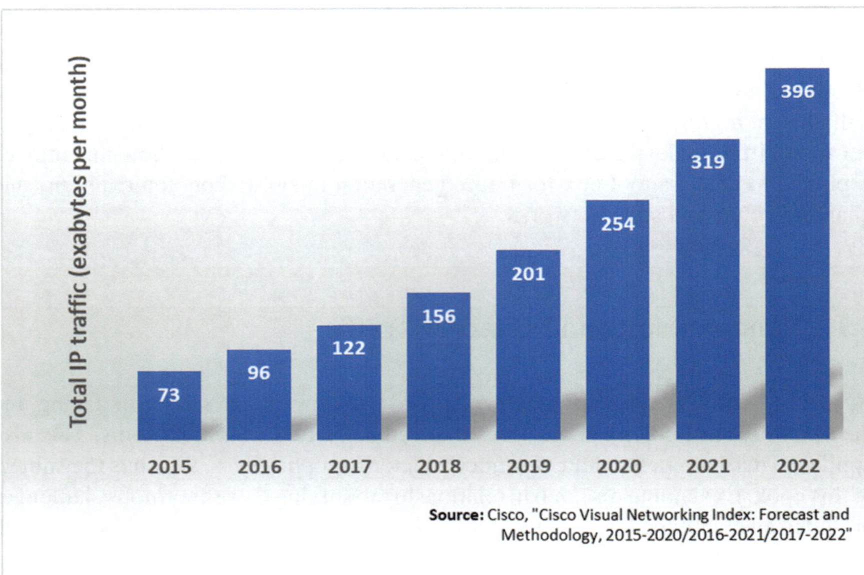

Fig. 1.1 Forecast of the global IP traffic [3]. Already in 2019, it is expected that 200 exabytes of data are being transferred globally each month. Please note that one exabyte corresponds to 10^{18} bytes of data and that five exabytes are roughly equal to a transcript of all the words ever spoken by human beings

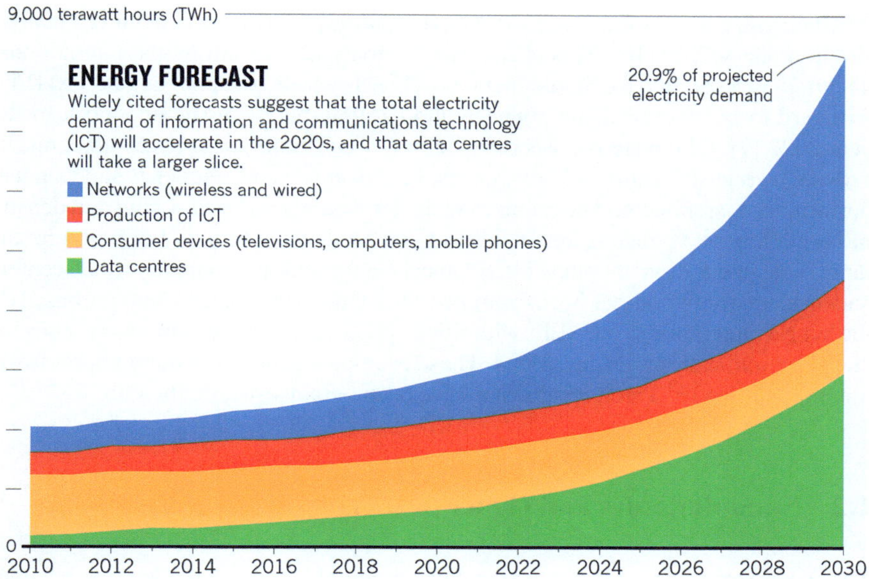

Fig. 1.2 Forecast of the electric power consumption for information and communication technologies (ICT) [2]

systems, it becomes obvious that the development of ultrafast and highly efficient nanophotonic devices is critical to keep up with the increasing demand. In order to transport, route, retrieve, and search such large volumes of data requires not only increasingly faster laser diodes and modulators, but also significantly more energy-efficient devices. A recent study by Andres Andrae shows [4] that the global power consumption from the usage of information and communication technologies (ICT) is expected to increase from 8.2% of global electricity usage in 2015 to 20.9% in 2030 [2, 4], corresponding to an incredible 8.100 TWh of energy (Fig. 1.2). In order to put this number in perspective, the world's energy consumption for transportation of passengers by automobiles, planes, buses, and trains in 2012 was only twice that value [5]. This means that the amount of energy we consume to transport and route data will soon rival that energy consumption for transporting people unless great efforts are put into developing more energy efficient ICT technologies.

The challenges discussed above were directly addressed by research activities in the collaborative research center 787 "Semiconductor Nanophotonics": For example, the fundamental limits of energy efficiency and high frequency operation of ultra-small vertical cavity surface emitting lasers (VCSELs) for applications in multi-terabus systems have been explored. In order to compare the energy consumption of different VCSELs the energy-to-data rate is used as one of the key performance parameters. The energy-to-data rate can be calculated from the ratio of the electric power consumed by the VCSEL, i.e. operating current times drive voltage, and the bit rate at error-free operation. In the course of the CRC 787 data transmission at

25 Gb/s with a record-low energy dissipation of only 56 fJ/bit was achieved using a single-mode VCSEL [6]. This is close to the goals set forward by the International Technology Roadmap for Semiconductors (ITRS) whose roadmap targets 100 fJ/bit for board-to-board optical interconnects and 10 fJ/bit for on-chip interconnects by the year 2022 [7]. Of course besides reduced power consumption optical interconnects provide a series of additional advantages, e.g., distance and frequency independent performance, architectural advantages with significantly reduced wiring and density of interconnects, voltage isolation, and a more predictable timing. Furthermore, the large scale integration of optical interconnects in future supercomputers, data centers and local area networks has been examined and a silicon photonic I/O engine based on the hybrid integration of VCSELs with silicon photonic integrated circuits for chip-to-chip communication was developed. These examples underscore quite impressively the impact of the research results that have been obtained within the CRC 787.

1.2 Nanophotonics and Cyber Security

The second example is also related to data communication and the internet but concerns cyber security. In a 2017 public opinion survey regarding cyber security in the 28 European Union countries, 87% of the respondents see cybercrime as an important challenge to the internal security of the EU [8]. "*Respondents express high levels of concern about the security of their online transactions*" *and "73% of Internet users are concerned that their online personal information could not be kept secure by websites. A rising majority of respondents are concerned about experiencing or being victims of cybercrimes, ..., with the largest proportions of respondents expressing concern about discovering malicious software on their device (69%), identity theft (69%) and bank card and online banking fraud (66%)*" [8]. Hereby the "*misuse of personal data and the security of online payments continue to be the most significant concerns of Internet users*" [8]. However, cyber security requires multifaceted approaches that are integrated within the software and hardware of the devices. One key element of cyber security concerns the safe transfer of data between different parties and this is already being applied to secure E-Mails exchanges, safely stored data, which provide secure authentication systems, and to establish virtual private networks (VPN). This is typically implemented by employing cryptography, i.e. encoding the data using encryption algorithms and encryption keys. While public key cryptography is already widely used, these technologies are not considered 100% safe since increasing computing power and the development of complex deciphering algorithms may allow these encryption systems to be cracked. Therefore, the development of quantum key distribution (QKD) systems which use quantum states, or qubits, and the laws of quantum mechanics to ensure a 100% secure transfer of encryption keys. One example of such a quantum key distributions system is shown in Fig. 1.3, which utilizes the so-called "BB84" protocol originally proposed by Bennett et al. [9]. In this example a transmitter called "Alice" sends a sequence of photons to the recipient named "Bob" through a quantum channel such as an optical fiber or

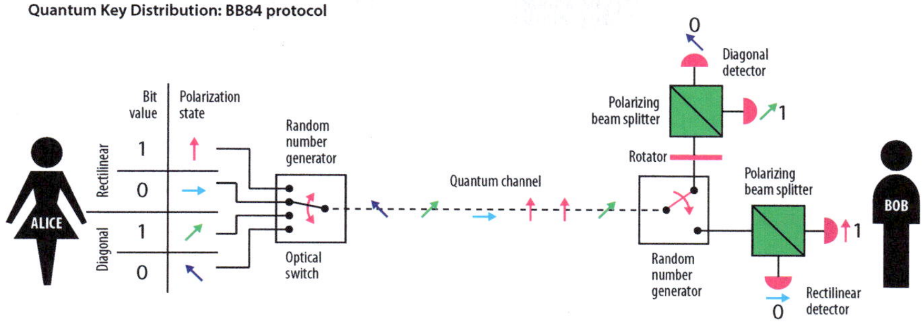

Quantum transmission & detection	ALICE sends photons	↖	↗	→	↑	↑	↗	↖	↑
	ALICE's random bits	0	1	0	1	1	1	0	1
	BOB's detection events	↑	↗	↖	↑	↗	↗	↖	↖
	BOB's detected bit values	1	1	0	1	1	1	0	0
Public discussion (i.e., sifting)	BOB tells ALICE the basis choices he made	Rect	Diag	Diag	Rect	Diag	Diag	Diag	Diag
	ALICE tells BOB which bits to keep		✔		✔		✔	✔	
	ALICE and BOB's shared sifted key	–	1	–	1	–	1	0	–

Fig. 1.3 Schematic example showing quantum key distribution rough a quantum channel between two parties named "Alice" and "Bob" (Reprint permission by the National Security Agency) [10]

via free-space. Hereby the information is encoded in non-orthogonal quantum states or qubits, e.g., single photons with diagonal or rectilinear polarization directions. Each of these photons represent a bit of information with the logical value of "1" corresponding to photons with 0° or 45° polarization and the value of "0" corresponding to photons with −45° or 90° polarization. In the first step Alice transmits a random string of qubits together with a random sequence of polarization bases. When Bob receives these photons he decides independently and at random whether to measure the photons diagonal (i.e. −45° or +45°) or rectilinear (i.e. 0° or 90°) polarization. From this measurements Bob creates a string of bit values represented by "1" and "0". In a subsequent step Bob and Alice exchange information on a public communication channel which of the photons were successfully received by Bob and measured with the correct basis. These remaining bits now constitute the shared sifted key, and consequently QKD protocols enable the secure distribution of encoding keys which in turn can then be used to encrypt messages that are exchanged between two parties.

In this field of quantum communication, one of the challenges addressed by the research work in CRC 787 was the development of key components for quantum cryptography. This refers in particular to the realization of electrically driven quantum key distribution systems that are based on qubit and entangled photon emitters operating at high qubit rates as well as their implementation in real information networks. Progress towards that goal included the realization of deterministic single quantum dot emitters for the generation of on-demand frequency-locked indistinguishable photons and the generation of time-bin entangled photon pairs. As part of this effort a stand-alone and compact quantum dot based single photon source [11] as well as the deterministic integration of quantum dots into an on-chip multimode interference beam splitters [12] have recently been demonstrated. We have also made great progress in the implementation of a free-space optical link and demonstrated the transmission of quantum information via single and entangled photons. One important step on this path was the successful demonstration of coherent frequency conversion in order to generate single photons in the 1.55 μm telecom wavelength band [13]. Another highlight was the demonstration of a free space optical link including the design of QKD antennas with optical tracking for long-term measurements [14]. Progress has also been made in the quest for room temperature operation of electrically pumped UV single photon emitters based on GaN quantum dots embedded in microcavities. Single photon emission from clustered GaN/AlN quantum dots produced via growth interruption has been successfully demonstrated [15]. Overall, the devices and network system developed within the CRC 787 show quite impressively the impact the nanophotonics has in this area. Of course further progress is needed before these system can be implemented into real-world applications, but these first successful steps show the way to a more secure future.

1.3 Economic Impact of Nanophotonics

Last, but not least, nanophotonic technologies also have a significant economic impact. This can be seen from a number of indicators and clearly outlined in the recently published "Strategic Roadmap 2021–2027" by the European Technology Platform Photonics21 [16]. The report entitled "Europe's age of light! How photonics will power growth and innovation" strongly emphasizes the importance of photonics technologies for European economies. "*Today, the European Photonics industry, comprised of mainly SMEs, is fast-growing and thriving: there are an estimated 5000 companies that have created more than 300,000 highly skilled jobs in this sector alone with an annual turnover in excess of €60 billion. With a compound annual growth rate (CAGR) of 6.2%, the European photonics industry is growing four times faster than the European GDP. Europe is one of the leading players in the global photonics market, ranking only second to China*" [16]. The study identifies key application areas in photonics, in particular information and communication technologies, lighting and displays, industrial manufacturing, life sciences and health, as well as security, metrology and sensors. Especially for the last decade, the development of high power and high brilliance IR lasers as well as deep UV laser diodes (UV-LEDs) has provided new light sources for a wide range of applications. As can be seen from Fig. 1.4 there are a wide range of applications for lasers, in particular in the area of materials processing, communication, medical diagnostics, phototherapy, sensing, 3D printing, lithography, and optical storage.

Developments for tools used in many of these applications can be also found within the Collaborative Research Centre 787 (CRC 787). For instance, in the application segment of materials processing great progress has been made in the development of high power and high efficiency infrared lasers, where GaAs-based stacked laser diode bars reach multi kW output power levels in quasi cw mode operation [18]. These high-power lasers are used for materials processing, e.g., laser cutting and welding or as pump sources for even higher power solid-state and fiber lasers systems. Higher power levels have also been reached for surface emitting laser. High power VCSEL arrays could enable new applications in the area of low-cost LiDAR systems for autonomous driving automobiles and free-space data communications. For example, 7-element 980 nm VCSEL array with peak output power of 50 mW and a 3 dB bandwidth of more than 25 GHz have been realized [19]. Besides higher power levels, accessing new wavelength regimes, in particular in the ultraviolet spectral range, was one of the goals of the CRC 787. Based on the relatively unexplored AlGaN materials system, significant progress has been made in the realization of deep UV laser diodes. Optically pumped lasing from AlGaN multiple quantum well heterostructures has been demonstrated with record short emission wavelength of 237 nm [20]. Also progress towards the realization of current-injection UV laser diodes has been made. High efficiency UV-LEDs, the precursors for demonstrating laser diodes, have been demonstrated with output power levels of more than 30 mW continuous-wave and more than 100 mW in pulsed operation. Finally, we also were able to push the limits of UV-LED emission down to wavelength of 217 nm [21].

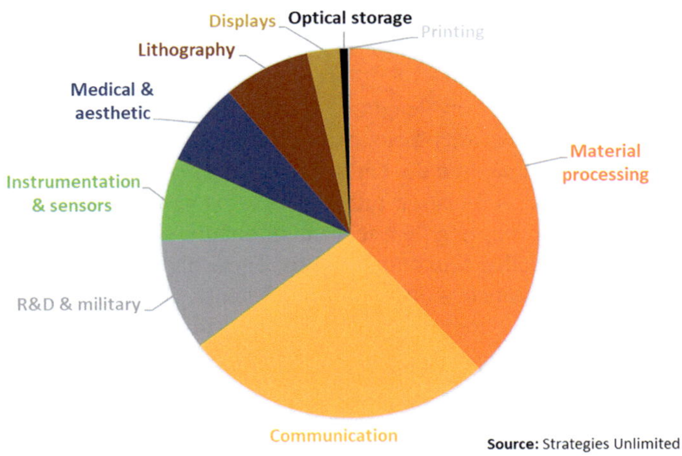

Fig. 1.4 Market share of different laser application segments [17]

1.4 Semiconductor Nanophotonics

According to the Merriam-Webster dictionary [22], "photonics is a branch of physics that deals with the properties and applications of photons", in particular the generation, detection, and manipulation of photons. The prefix "nano" stems from the Greek word "nanos" meaning "dwarf," refers to the scale of the objects under investigation, in fact something very small, i.e. structures on a nanometer (~10^{-9} m) scale. The small dimensions of the structure has several immediate consequences in particular effecting the electronic and optical properties of nanophotonic devices. First of all, nanometer-scale objects require a quantum mechanical description of their respective electronic properties. It also allows the manipulation of these properties by simple changing the size or shape of the nanostructures. One frequently used example is the change of the density of states with the reduction of the dimension of the objects, starting with two-dimensional quantum systems like quantum wells and going down all the way to zero-dimensional objects, i.e. quantum dots, which exhibit an atom-like electronic structure. Smallness also effects the interaction of light with matter and depending on the strength of the interaction results in different quantum electrodynamic effects. For example, when an emitter is placed within an optical cavity the interaction in the weak coupling regime can result in the so-called Purcell effect which alters the spontaneous emission rate. If the emitter is in resonance with a high Q-cavity this can lead to an enhancement of spontaneous emission which can be described by the Purcell factor F_{P} [23]

$$F_{\mathrm{P}} = \frac{3}{4\pi^2}\left(\frac{\lambda}{n}\right)^3 \frac{Q}{V}$$

where λ corresponds to the wavelength, n to the refractive index within the cavity, and Q and V represent the quality factor and mode volume of the cavity, respectively. Therefore, by placing a quantum dot emitter into a high-Q cavity with a small mode volume the radiative recombination rate of single photon emitters can be greatly enhanced. In the strong coupling regime excitons and the optical cavity modes form new quasiparticles, so-called exciton-polaritons. These phenomena can be employed to demonstrate the spontaneous emission of coherent light from exciton-polariton Bose-Einstein condensates and "threshold-less" polariton lasing [24–26].

All in all, nanophotonics is one the fastest growing fields in optoelectronics covering a wide range of research areas, including plasmonics, photonic crystals, metamaterials, quantum optics, nanostructures like quantum dots, nanolasers, and quantum emitters. Research in nanophotonics also covers a wide range of materials including group III–V semiconductors, silicon, diamond, graphene and more recently 2D materials like transition metal dichalcogenides (TMDC). The research activities within our collaborative research center "Semiconductor Nanophotonics" (CRC 787) have been focusing on the three most relevant families of semiconducting materials for light emitting devices, i.e. group III-arsenides, III-phosphides, and III-nitrides. These material systems cover a wide variety of emission wavelengths ranging from the deep ultraviolet to the visible and near-infrared spectrum. Progress in the development of group III–V nanostructures was key to the realization of on-demand single-photon sources, ultrafast amplifiers, highly energy-efficient VCSELs, and UV lasers. A number of advances have been realized in the growth of III–V nanostructures including a novel buried-stressor approach that allows the controlled positioning of InGaAs quantum dots and has led to the demonstration of electrically operated single-photon devices with almost pure single-photon characteristic and resolution-limited linewidth. Furthermore, desorption induced GaN quantum dots have been realized by metalorganic vapor phase epitaxy (MOVPE) on (0001) AlN [27] exhibiting single photon emission characteristics [15].

All advances in nanomaterials and devices were guided by the theoretical and numerical modeling ranging from the description of fundamental optic, electronic, and vibronic properties of nanomaterials to the simulation of entire nanophotonic devices. Key to understanding and advancing device performance requires the development of theoretical and numerical models ranging from the description of fundamental optic, electronic, and vibronic properties of nanomaterials to the simulation of entire nanophotonic devices, bridging the length scales from nano- to micrometers [28]. In order to enable a comprehensive description of a variety of nanophotonic devices a hierarchy of physical phenomena was investigated ranging from fully quantized light-matter interaction and fundamental quantum optics [29] to microscopically motivated rate equation models and semi-classical descriptions of fully functional, complex devices [30]. Utilizing this modular approach we were able to connect fundamental microscopic properties of quantum dots and novel gain materials (e.g., sub-monolayer quantum dots) to the performance characteristics of nanophotonic devices, like VCSELs, single photon emitters, semiconductor optical amplifiers, and UV laser diodes.

Although the eco system of semiconductor nanophotonic devices is quite diverse, one can find a lot of commonalities in the device structures and device fabrication technologies. These are illustrated in Fig. 1.5 displaying schematic cross-sections of three in principle very different nanophotonic devices, including an InGaAs multi-quantum-well (MQW) VCSEL, InGaAs single photon emitter (SPE), and an InGaN MQW microcavity. Even though the materials involved to fabricate these devices are quite different, these device structures exhibit a lot of similarities. One common feature is the use of distributed Bragg reflectors (DBR) as highly reflective mirrors, based on different semiconducting (e.g., AlAs/GaAs, AlInN/GaN) or dielectric materials (SiO_2/Ta_2O_5). In all cases III–V nanostructures, i.e. InGaAs or InGaN multiple quantum well or quantum dots are employed as optically active components. Current confinement is provided by lateral aluminum oxide apertures or SiNx combine with an indium tin oxide (ITO) current spreading layer. These synergies in technologies and design as well as the development of hierarchical models provides some confidence that a better understanding of one nanophotonic device will also spur advances in many others. Eventually we will be able to develop a comprehensive nanophotonic toolbox that allows the design of complex nanophotonic devices as well as the prediction of their electro-optic characteristics, similar to what has been already

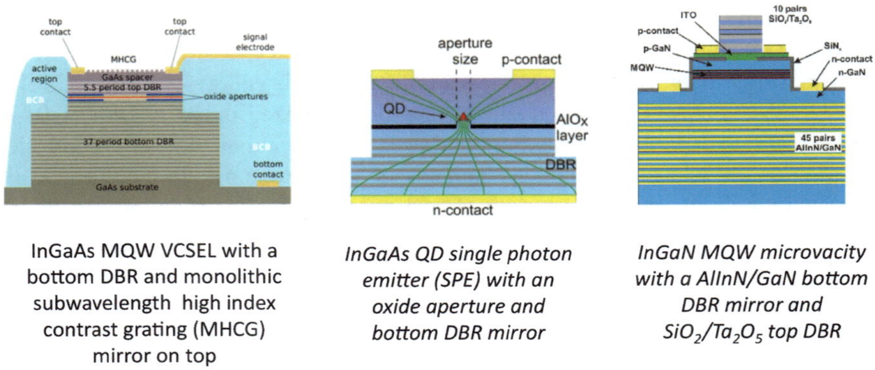

Fig. 1.5 Schematic cross-sections of different nanophotonic devices, including InGaAs MQW VCSELs, InGaAs SPE, and an InGaN MQW microcavity [31, 32]

established in the silicon electronic industry. Through these advances we should be able to enter the next stage of development going from individual nanophotonic devices to complex systems, like integrated photonic circuits (IPC) and quantum communication networks.

References

1. A.S.G. Andrae, T. Edler, Challenges **6**, 117 (2015)
2. N. Jones, The information factories. Nature **561**, 163 (2018)
3. Cisco Visual Networking Index: Forecast and Trends, 2017–2022 (2018). Retrieved from www.cisco.com/c/m/en_us/solutions/service-provider/vni-forecast-highlights.html. 20 March 2019
4. A. Andrae, Total consumer power consumption forecast. Retrieved from www.researchgate.net/publication/320225452_Total_Consumer_Power_Consumption_Forecast. 24 May 2019
5. U.S. Energy Information Administration. Retrieved from https://www.eia.gov/outlooks/ieo/pdf/transportation.pdf. 20 March 2019
6. P. Moser, J.A. Lott, D. Bimberg, IEEE J. Sel. Top. QE **19**, 1702212 (2013)
7. D.A.B. Miller, Proc. IEEE **97**, 1166 (2009)
8. *Special Eurobarometer 464a Report—Europeans' Attitudes Towards Cyber Security* (European Union, 2017). https://doi.org/10.2837/82418
9. H. Bennett et al., Quantum cryptography: public key distribution and coin tossing, in *Proceedings of the IEEE ICCSSP*, vol. 174 (1984)
10. Forecasting single-photon detector technology, The Next Wave, vol. 20, No. 1 (2013)
11. A. Schlehahn et al., A stand-alone fiber-coupled single-photon source. Sci. Rep. **8**, 1340 (2018)
12. P. Schnauber et al., Nano Lett. **18**, 2336 (2018)
13. Kroh et al., Quant. Sci. Tech. **2**, 034007 (2017)
14. N. Perlot et al., in *Photonics Networks, 19th ITG-Symposium. VDE* (2018)
15. G. Schmidt et al., Appl. Phys. Lett. **106**, 252101 (2015)
16. Retrieved 18 May 2019 from www.photonics21.org
17. Image retrieved 9 May 2019 from LaserFocusWorld magazine. https://www.laserfocusworld.com/articles/print/volume-55/issue-01/features/what-goes-up.html
18. K. Posilovic et al., Electron. Lett. **48**, 1419 (2012)

19. N. Haghighi, P. Moser, J.A. Lott, IEEE J. Sel. Top. Quantum Electron. (2019). https://doi.org/10.1109/jstqe.2019.2922843
20. M. Martens, Dissertation (2018)
21. F. Mehnke et al., Appl. Phys. Express **12**, 012008 (2019)
22. Retrieved 18 May 2019 from www.merriam-webster.com
23. E.M. Purcell, Spontaneous emission probabilities at radio frequencies. Phys. Rev. **69**, 681 (1946)
24. A. Imamoglu, R.J. Ram, S. Pau, Y. Yamamoto, Non equilibrium condensates and lasers without inversion: exciton-polariton lasers. Phys. Rev. A **53**, 4250 (1996)
25. J. Kasprzak et al., Bose-Einstein condensation of exciton polaritons. Nature **443**, 409 (2006)
26. C. Schneider, A. Rahimi-Iman, N.Y. Kim, J. Fischer, I.G. Savenko, M. Amthor, M. Lermer, A. Wolf, L. Worschech, V.D. Kulakovskii, I.A. Shelykh, M. Kamp, S. Reitzenstein, A. Forchel, Y. Yamamoto, S. Höfling, Nature **497**, 348 (2013)
27. K. Bellmann, F. Tabataba-Vakili, T. Wernicke, A. Strittmatter, G. Callsen, A. Hoffmann, M. Kneissl, Phys. Status Solidi RRL **9**, 526 (2015)
28. M. Kantner, M. Mittnenzweig, T. Koprucki, Phys. Rev. B **96**, 205301 (2017)
29. S.C. Kuhn, A. Knorr, S. Reitzenstein, M. Richter, Opt. Express **24**, 25446 (2016)
30. S. Kreinberg, X. Porte, D. Schicke, B. Lingnau, C. Schneider, S. Höfling, I. Kanter, K. Lüdge, S. Reitzenstein, Nat. Commun. **10**, 1539 (2019)
31. M. Gębski, J.A. Lott, T. Czyszanowski, Opt. Express **27**, 7139 (2019)
32. W. Unrau et al., Appl. Phys. Lett. **101**, 211119 (2012)

Chapter 2
Submonolayer Quantum Dots

N. Owschimikow, B. Herzog, B. Lingnau, K. Lüdge, A. Lenz, H. Eisele, M. Dähne, T. Niermann, M. Lehmann, A. Schliwa, A. Strittmatter and U. W. Pohl

Abstract The cycled deposition of small InAs islands into a GaAs matrix leads to the formation of a tailored rough quantum well containing densely spaced In-rich agglomerations referred to as submonolayer quantum dots, which support an efficient exciton formation. Carrier localization properties of the submonolayer structures are further enhanced by alloying with antimony. In this chapter we address the growth, the structure, and the optical and optoelectronic properties of alloyed and unalloyed InAs submonolayer quantum dots and devices based on these structures. Based on structural and optical characterization, we find densities of localization centers exceeding those of self-assembled quantum dots by an order of magnitude. Submonolayer quantum dots show quantum-dot like ultrafast carrier dynamics, while at the same time providing a significantly larger modal gain, which reaches values known for InGaAs quantum-well structures. We develop a numerical model for the density of states and relevant scattering channels in the submonolayer potential landscape. Alloyed and unalloyed submonolayer quantum dots differ predominantly in the degree of hole localization, which is dramatically increased by the addition of antimony. We show that the alloyed submonolayer quantum dots support a heterodimensional confinement, from fully zero-dimensional to hetero-confinement with zero-dimensionally confined holes and electrons free in two dimensions.

N. Owschimikow (✉) · B. Herzog · T. Niermann · M. Lehmann
Institute of Optics and Atomic Physics, Technische Universität Berlin, 10623 Berlin, Germany
e-mail: nina.owschimikow@physik.tu-berlin.de

B. Lingnau · K. Lüdge
Institute of Theoretical Physics, Technische Universität Berlin, 10623 Berlin, Germany

A. Lenz · H. Eisele · M. Dähne · A. Schliwa · U. W. Pohl
Institute of Solid State Physics, Technische Universität Berlin, 10623 Berlin, Germany

A. Strittmatter
Institute of Physics, Otto-von Guericke Universität, Magdeburg, Germany

M. Kneissl et al. (eds.), *Semiconductor Nanophotonics*, Springer Series in Solid-State Sciences 194, https://doi.org/10.1007/978-3-030-35656-9_2

2.1 Carrier Localization in Quantum Dots

Quantum dots (QDs) represent a tailored artificial counterpart to atoms and have found applications in many advanced semiconductor devices [1–3]. Ensembles of QDs were also implemented to act as optical gain media in devices like semiconductor lasers and semiconductor optical amplifiers [4, 5]. These applications require a high optical gain obtained from a dense QD ensemble, usually a narrow gain spectrum to include the emission of most dots in the ensemble, and a high thermal stability.

2.1.1 Stranski-Krastanow and Submonolayer Quantum Dots

Dense ensembles of QDs are generally fabricated by employing the self-organized growth of Stranski-Krastanow islands, which are embedded into a semiconductor matrix with larger fundamental bandgap energy. Such self-assembly generates defect-free QDs of high optical quality; however, due to slight variations in the dot size, shape, and composition the energy levels of the individual dots show some variations, leading to an inhomogeneous broadening of transition energies within the dot ensemble. Moreover, the maximum dot density is limited to about 10^{10}–10^{11} cm^{-2} due to their typical base lengths of 7–26 nm [6–9] and a repulsive elastic dot-dot interaction [10, 11]. Another drawback is the inherent coupling of Stranski-Krastanow QDs (SK QDs) to a quantum well (QW) formed by the wetting layer, which generally remains after the two-dimensional to three-dimensional transition of the strained QD material, see Fig. 2.1a. This coupling can limit the thermal stability of carrier confinement due to a loss of carriers confined in the QD to states of the QW, which has a high density of electronic states. In addition, carrier capture after non-resonant optical or electrical excitation occurs predominantly via states of the wetting layer [12–14]. Such a process can impose limits to the dynamical operation in high-speed devices [15, 16].

Submonolayer quantum dots (SML QDs) were introduced as an alternative approach for QD formation [17]. The fabrication involves the deposition of less than 1 monolayer (ML) of, e.g., InAs on GaAs, capping it with a thin (few MLs)

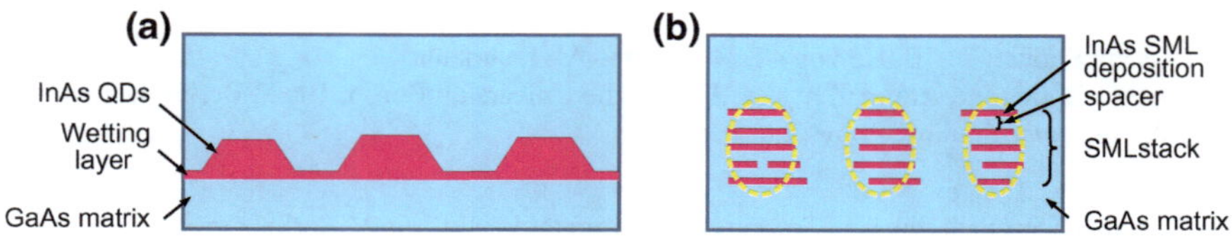

Fig. 2.1 **a** Schematic of QDs formed by a Stranski-Krastanow transition. **b** Concept of QDs formed by a cycled deposition of submonolayers of low bandgap material, covered by thin spacers of matrix material; *yellow ellipses* indicate the wavefunction envelopes of confined carriers

spacer layer of GaAs, and repeating the procedure several times to form a submonolayer (SML) stack. Since InAs has a larger lattice constant than GaAs, the thin GaAs spacer layer covering the only partially complete InAs ML is tensely strained where it covers InAs material. The related local increase in lattice constant provides favorable sites for the nucleation of InAs also in the subsequent layer, leading in the ideal case to a vertical alignment of the SML InAs material as illustrated in Fig. 2.1b. The vertical distance between the InAs depositions is much smaller than the wavefunction of a confined carrier; the wavefunction hence extends over one whole InAs stack, i.e., the InAs SML stack forms an InGaAs quantum dot in GaAs matrix. Such SML QDs do not develop a wetting layer, and their density is substantially higher ($\geq 10^{12}$ cm^{-2} [18, 19]) than that of SK QDs.

Another favorable property of an SML QD ensemble is the narrow luminescence linewidth. The full width at half maximum (FWHM) of typically 10 meV is similar to that of a QW and hence much smaller than that found for SK QDs. Figure 2.2 shows the photoluminescence (PL) spectra of a sample comprising three SML stacks (blue and cyan curves) and the PL of a sample with three SK QD layers (red and orange curves). The SML stacks were grown by a ten-fold repeated deposition of 0.5 ML InAs and 2.3 ML GaAs on GaAs(001), while the SK QDs were formed from 2.7 ML $In_{0.65}Ga_{0.35}As$ on GaAs(001). The PL was excited non-resonantly at room temperature and above as indicated at the spectra. We note the narrow spectra of the SML QDs and the broad emission of the SK QDs, the latter originating from the inhomogeneous broadening of the QD ensemble and contributions of occupied excited states. In addition, the SML QDs show a very intense peak luminescence. This feature and the high QD density lead to a very high optical gain, which was exploited for edge-emitting lasers [20] and vertical external cavity surface-emitting

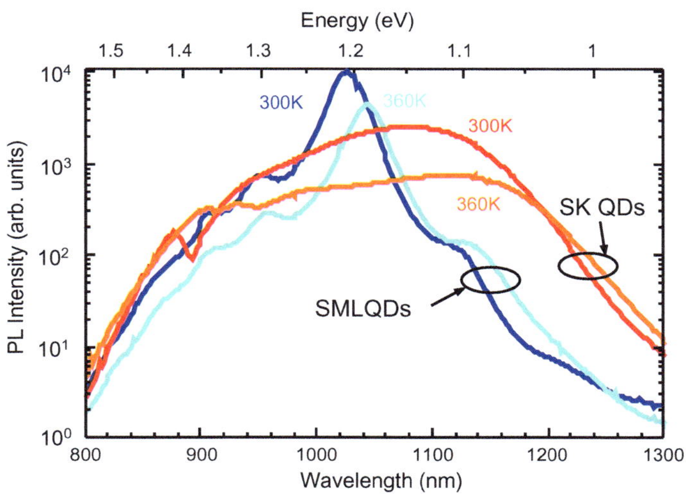

Fig. 2.2 Photoluminescence spectra of three ten-fold InAs/GaAs SML stacks (*blue and cyan curves*) and three-fold $In_{0.65}Ga_{0.35}As$/GaAs SK QD layers (*red and orange curves*)

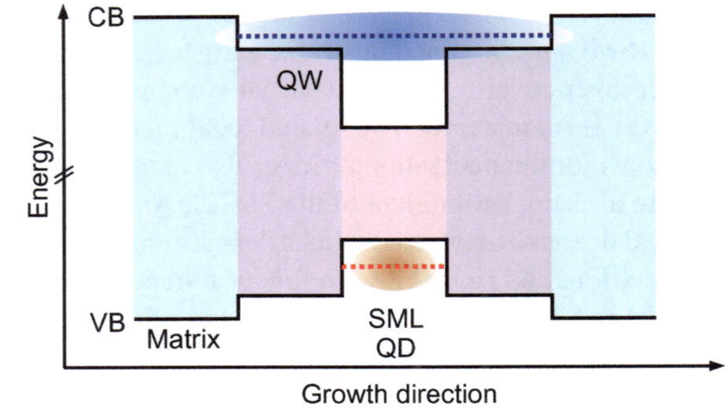

Fig. 2.3 Energy scheme of a SML QD with electron and hole confined states indicated (dotted lines). The hole is zero-dimensionally confined to the InAs SML stack, while the lowest confined state of the electron is bound by the QW defined by the cladding GaAs matrix. The ellipses indicate electron (blue) and hole (red) wavefunctions

lasers (VECSELs) [21, 22]; for a vertical cavity surface-emitting laser (VCSEL) high-speed operation at frequencies above 20 GHz was demonstrated [23, 24].

The origin of narrow luminescence and gain spectra is not obvious, because structural investigations reported in Sect. 2.3 show a rather inhomogeneous distribution of InAs in the SML structures. The apparent contradiction could be resolved by an investigation of the Bohr radii of carriers confined in the SML QDs: The magneto-optical measurements provided clear evidence for a heterodimensional confinement (see crossed excitons in Sect. 4.2.3). The light electrons extend laterally over several InAs SML stacks and are hence two-dimensionally confined by the cladding GaAs barriers, while the heavier holes are zero-dimensionally confined in the individual InAs stacks (see Fig. 2.3) [25]. The study showed two characteristic length scales for the disorder in the SML samples, sensed by excitons: The first refers to low magnetic fields (including zero field) with a large Bohr radius where the exciton *averages* over the highly nonuniform SML stacks, yielding a line broadening given by large-scale fluctuations. At high magnetic field, in contrast, the exciton is squeezed to an extent where it probes smaller-scale fluctuations produced by monolayer fluctuations of the QW, which is defined by the cladding GaAs barriers. In both cases the disorder-induced line broadening is substantially smaller than that observed for an ensemble of SK QDs.

In this chapter we first address the electronic structure, the fabrication, the spatial structure, and the optical properties of InGaAs SML QDs as well as the effect of an alloying with antimony. The observed carrier dynamics and the dependence on alloying-induced localization is discussed in detail, and the properties of laser and semiconductor optical amplifiers based on SML QDs are presented.

2.1.2 Electronic Structure of InAs Submonolayer Quantum Dots

The electronic properties of InAs SML QDs in GaAs can be tailored by various means: (1) the total amount of In within an SML stack, (2) the ratio of incorporated indium within the SML QDs relative to the In within the surrounding QW, and (3) the addition of antimony during SML growth. Thereby energies, localization properties, and resulting decay dynamics of the charge carriers can be adjusted. In the following we investigate the evolving electronic properties by employing eight-band $\mathbf{k} \cdot \mathbf{p}$ theory, emphasizing particularly the localization properties of the involved charge carriers. The SML QDs are modelled according to the procedure depicted in Fig. 2.4. First, a three-dimensional model of the SML structure is set up on a mesoscopic level. Next, the inhomogeneous strain distribution is calculated based on a continuum mechanic approach, and the thereby arising piezoelectric field is determined. In a subsequent step, single-particle energies and wavefunctions of electrons and holes are calculated. Finally, a Hartree self-consistent cycle is employed to account for Coulomb interaction using the results from the previous step as start vectors.

Heterodimensional Carrier Confinement

The heterodimensional carrier confinement illustrated in Fig. 2.3 was concluded from magneto-optical measurements reported in [25]: Low-temperature PL and magneto-PL demonstrate strong vertical confinement by the GaAs barriers and only weak lateral confinement by the InGaAs QW, yielding two-dimensional excitons. In contrast, high-temperature (400 K) magneto-PL reveals excited states that fit a Fock-Darwin

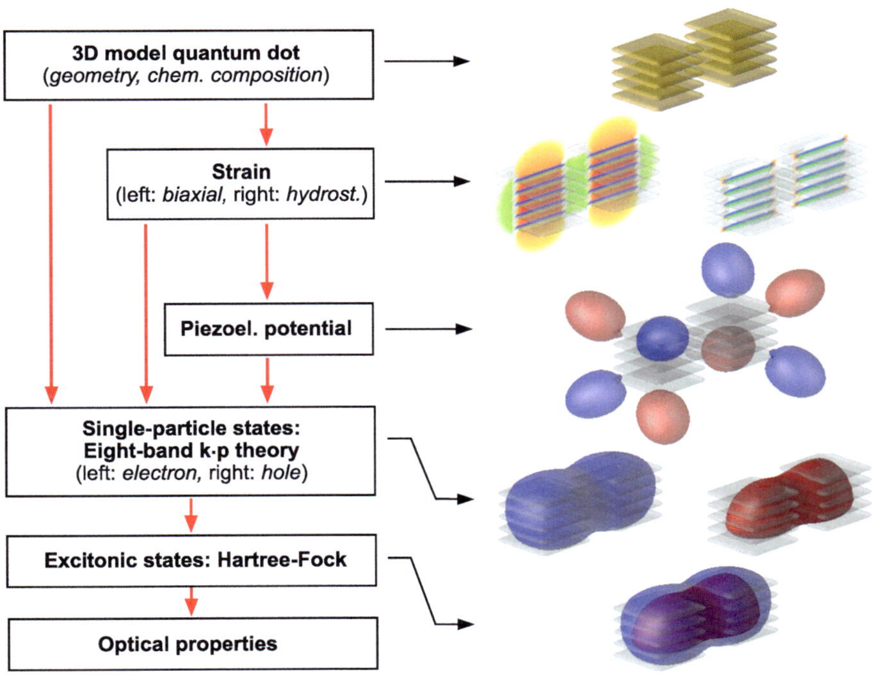

Fig. 2.4 Schematic of the modeling procedure applied for calculating electron and hole states of an SML QD

spectrum, characteristic of a zero-dimensional system in a magnetic field [26]. This apparent paradox can be resolved by the heterodimensional nature of the system: The light electrons extend laterally over several In-rich agglomerations and see only the InGaAs QW, while the heavier holes are confined within the In-rich regions—the InAs SML QDs.

The conditions for the formation of heterodimensionality in InAs SML QDs was studied in [25] by using an effective-mass approach: Fig. 2.5, left, shows the extent of the heterodimensional phase (green, two-dimensional electrons and zero-dimensional holes) as function of indium content in QD (x) and QW (y). For small values of y most of the phase space comprises mono-dimensional zero-dimensionally localized electrons and holes, which relates to the classical case of SK QDs. As the In content y in the QW increases, the heterodimensional phase becomes dominant.

To judge the effects of lateral coupling and the Coulomb interaction between electrons and holes, we performed a more sophisticated calculation of a periodic two-dimensional array of QDs with 5 nm diameter in an $In_{0.25}Ga_{0.75}As$ QW using eight-band $\mathbf{k} \cdot \mathbf{p}$ theory [27] in conjunction with a self-consistent Hartree approach. Figure 2.5 (right) shows the results of such a calculation. In the plane of the QW the In-rich agglomerations are modeled by an In content varying as $\cos^2$ functions along the two orthogonal directions in the lateral plane of the sample, superposed by a broad Gaussian variation in In content over all the dots, such that the peak In content in the center dot is 0.5 and 0.75 in Fig. 2.5b, c, respectively. The variation of In content in different dots mimics their morphological diversity while ensuring that the exciton is centered at the QD in the center. The results are largely consistent with the effective mass calculations and the experimental data. The electron is delocalized over several QDs, but eventually gets localized at high In content ($x \cong 0.6$). Localization of the holes is delayed by lateral coupling, but the hole is consistently much more localized

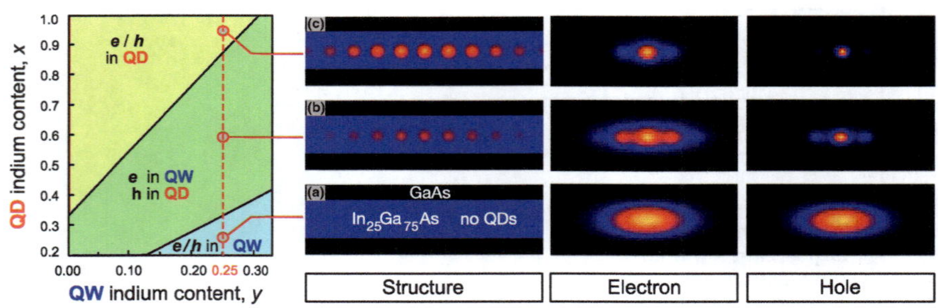

Fig. 2.5 (Left) Effective mass calculation of electron e and hole h confinement for spherical $In_xGa_{1-x}As$ QDs with 5 nm diameter in a 13 nm thick In_yGa_{1-y} As QW, with varying In content (reproduced from [25]). The *solid lines* are guides to the eye indicating the approximate positions of the phase boundaries. (Right) Color contour plots of electron and hole wavefunctions in a 14 nm thick $In_{0.25}$ $Ga_{0.75}As$ QW with **a** no QDs, **b** an array of QDs with $x \leq 0.5$, and **c** an array of QDs with $x \leq 0.75$. The In composition is maximal in the center of each dot and follows a Gaussian profile across the array. The left-hand panel shows the morphology of the structure, while the middle and right-hand panels show the electron and hole wavefunctions, respectively, calculated by $\mathbf{k} \cdot \mathbf{p}$ theory. Reprinted after [25] © APS (2016)

than the electron. The Coulomb interaction enhances electron localization, but this is not sufficient to eliminate the heterodimensional behavior.

Strong Charge-Carrier Localization due to Incorporation of Antimony

The electronic confinement of carriers in InAs SML QDs can be influenced by adding antimony during growth. The effect of Sb alloying on the wavefunctions of electrons and holes was evaluated using eight-band $\mathbf{k} \cdot \mathbf{p}$ simulations [28] for three different Sb incorporation schemes:

(A) Sb is incorporated into the QDs (Fig. 2.6b)
(B) Even distribution of Sb throughout the SML stack (not shown)
(C) Incorporation of Sb in between the QDs (Fig. 2.6c).

The basic structure for the simulation consists of an 8.5 nm thick InGaAs QW with 15% In content. Spherical In-rich agglomerations with a $\cos^2$ composition profile, a maximum local In content of 35%, and a spacing of 13.6 nm are embedded into this QW.

First of all, the incorporation of Sb into the SML stack leads to a redshift of the PL emission in all three simulated cases, the strongest being the shift of 140 meV in case (B). To achieve a similar redshift for the two other cases, a peak Sb content of more than 20% is necessary.

Electron and hole wavefunctions depend quite differently on the Sb incorporation, as the schematic band alignment in Fig. 2.7 demonstrates: Holes localize at the Sb-rich regions, while the electrons localize at the In-rich regions and delocalize from Sb-rich regions (Fig. 2.8). The spatial distribution of the wavefunctions also influences the electric dipole moment in the semiconductor structure. Their overlap defines the Huang-Rhys factor [29], which is also a measure for the coupling to LO phonons. A better overlap of the wavefunctions leads to a smaller Huang-Rhys factor. Figure 2.7 shows the band structures for the three different assumptions on

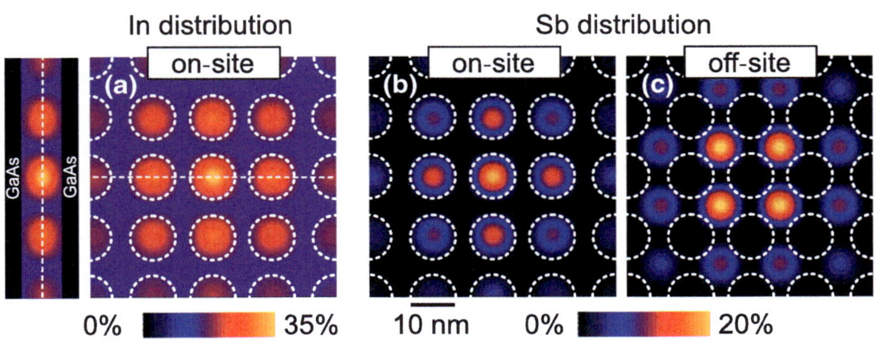

Fig. 2.6 Indium and antimony distributions assumed for the eight-band k · p simulations. The *dashed white circles* indicate the positions of the QDs as defined by In-rich regions: **a** In distribution, the maximum local In content is 35%, a sigmoidal In distribution is used. **b** Sb distributions for the cases (**a**) and (**c**), the maximum local Sb content is 20%. Reprinted after [28], © APS (2015)

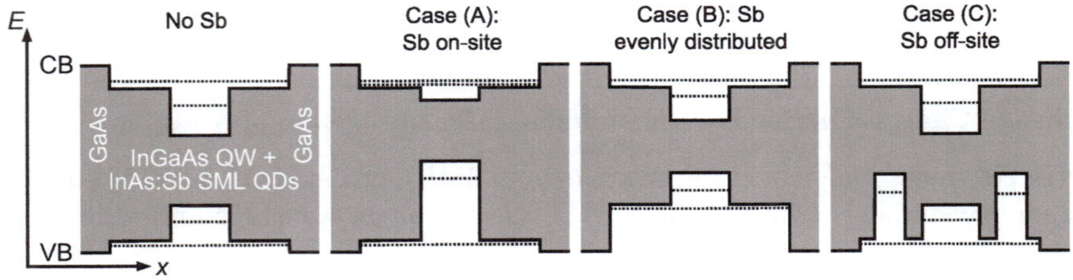

Fig. 2.7 Schematic band alignment for the different cases of Sb distribution; *dotted lines* indicate the energy levels of the QW and the embedded QDs. Reprinted after [28], © APS (2015)

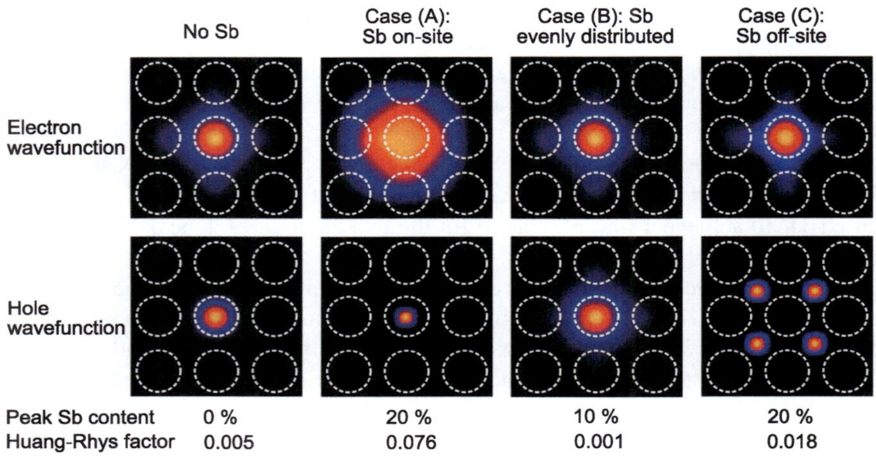

Fig. 2.8 Electron and hole localization for the different cases of Sb distribution. Huang-Rhys factors for electron-hole pairs are shown below. Reprinted after [28], © APS (2015)

Sb incorporation together with the case without Sb. In case (A), the Sb incorporated into the QDs leads to a stronger localization of the hole wavefunction, whereas the electron wavefunction delocalizes (see Figs. 2.7 and 2.8). This results in a strongly reduced electron-hole overlap, leading to a 15-fold increase of the Huang-Rhys factor from 0.005 without Sb to 0.076 for a maximum local Sb content of 20%. An even distribution of the Sb throughout the SML stack as in case (B) only leads to a slightly stronger delocalization of the hole wavefunction (Fig. 2.8), resulting in a better overlap and consequently a smaller Huang-Rhys factor of only 0.001 for 10% Sb content. Case (C), in which Sb is incorporated in between the QDs, leads to a localization of the hole wavefunction at the Sb-rich regions (Fig. 2.8), yielding a strong delocalization from the QDs; thereby the Huang-Rhys factor increases to 0.018 for 20% Sb content due to the reduced overlap. This case yields the lowest overall strain in the semiconductor structure.

The PL spectra shown below in Fig. 2.10a indicate a strong increase of phonon coupling upon introduction of Sb into the SML QDs. This clearly excludes case (B) from the experimentally observed effect of Sb alloying. A detailed evaluation of the PL spectra performed in [28] yields a coupling strength indicating case (A) to be the most likely, i.e., a preferred incorporation of antimony in the SML QDs. The findings from the structural investigations (Sect. 2.3) also show that Sb is found in or directly at the sides of the In-rich agglomerations, leading to conditions corresponding to case (A).

2.2 Epitaxy of Submonolayer Quantum Dots

The fabrication of SML QDs requires a deposition control with a precision of only a fraction of a single monolayer within each deposition cycle. A strained material of submonolayer thickness is forced by energy minimization to form islands of certain size corresponding to an equilibrium state between elastic strain energy and surface energy of the islands. Given a highly strained material combination, such as InAs layers on bulk GaAs(001) surfaces, the lateral island size may become as small as a few nm. The size and shape distribution of the islands within such a SML deposition can largely be preserved by capping with a strain-free material; usually the SML structure is buried by the host substrate material.

A single SML island of a lower bandgap semiconductor embedded into a higher bandgap host material may be too thin to efficiently confine the electron and hole wavefunctions for providing three-dimensionally confined excitonic QD states. Therefore, a cycled deposition technique is applied, stacking several vertically separated SML sheets on top of each other to provide vertically extended low-bandgap inclusions; these lead eventually to zero-dimensional excitonic confinement, i.e. QD states, within an optically direct semiconductor material. This technique is referred to as epitaxial submonolayer growth, or SML growth for short.

2.2.1 InAs/GaAs Submonolayers

In the following we restrict ourselves to the SML growth within the InAs/GaAs(001) material system. Submonolayers were realized first by molecular beam epitaxy [30], but we focus here on metalorganic vapor-phase epitaxy applied to fabricate the investigated structures. The fabrication of SML has its roots in ultrathin QW layers whose emission properties are dominated by thickness fluctuations at the interfaces with the surrounding matrix material. Gerard et al. [31] observed excitonic emission with small linewidth from 1 ML thin InAs/GaAs QWs, which was confirmed by further investigations of ultrathin InAs insertions into GaAs [32–36].

InAs/GaAs SML growth does not differ much from the conventional QW growth regime, except for the care that is taken to realize precisely deposited SML stacks of the low-bandgap material. Low substrate temperature and a low V/III ratio favor a reduced migration length of In atoms adsorbed on a GaAs(001) surface; such conditions should produce a high density of small InAs islands on the surface. For metalorganic vapor-phase epitaxy, a minimum substrate temperature is required for precursor decomposition; this sets a lower limit of about 450–500 °C in the case of InAs/GaAs growth. At such low temperatures the use of tertiarybutyl-arsine (TBAs) as precursor for arsenic is preferred. Another technical detail to be mentioned is the switching speed of mechanical valves, being on the order of 0.3 s. Low growth rates on the order of 0.1 ML/s (about 0.3 nm/s) are hence important for reproducible growth.

In a series of growth experiments of InAs/GaAs SML stacks, the substrate temperature, the V/III ratio, and the growth procedure were varied. These parameters did not significantly change the optical properties measured by integral and spatially resolved PL. More sophisticated measurements shown in subsequent sections reveal, however, that both structural and electronic properties of these SML structures differ from those of ordinary QW layers. On the one hand, their structure is indeed composed of In-rich regions with sizes in the nm range, being vertically separated by the GaAs spacer layers but not as well defined as intended: A vertical segregation of In atoms across the GaAs spacer layers is observed as discussed in Sect. 2.3. On the other hand, the wavefunctions of holes and electrons are confined differently in these SML QDs as pointed out in Sect. 2.1.2.

2.2.2 *InAs/GaAs Submonolayers with Antimony*

When antimony is added during the deposition cycle of InAs/GaAs SML structures, the electronic properties are substantially changed. Integral and spatially resolved luminescence measurements confirm the formation of QD states similar to those observed for Stranski-Krastanow QDs [28]. We applied a flush of triethyl-antimony (TESb) prior to each InAs deposition (Fig. 2.9) to employ the surfactant behavior of Sb for further reducing the In adatom mobility and for suppressing the vertical In segregation. A combined analysis of the Sb content by X-ray diffraction and X-ray fluorescence is summarized in Fig. 2.9 [37]. As shown, a significant part of antimony atoms gets incorporated into each sheet of the SML stack and from comparison of optical data to theoretical modelling a prevailing incorporation within the InAs agglomerations is concluded. No significant change in lateral or vertical In distribution is found, and the size and density of the agglomerations are rather constant. The way of Sb incorporation can be described by a Langmuir-type model assuming fast and slow reaction components [37].

The optical properties gradually change with increasing Sb incorporation. As can be seen in the PL measurements displayed in Fig. 2.10, an increased broadening of the emission lines occurs. On the long-wavelength side of the emission a second peak of lower intensity appears. It can be attributed to LO-phonon emission, which was not observed in the pure InAs/GaAs SML structures. This finding is related

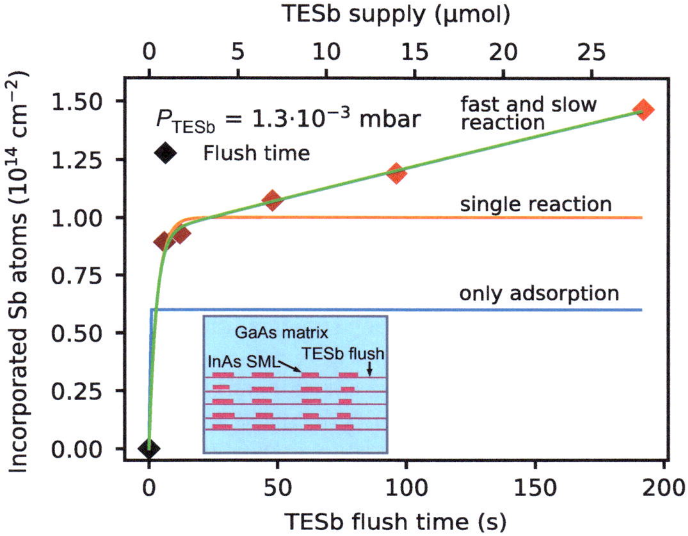

Fig. 2.9 Temporal evolution of the Sb incorporation per SML cycle at a constant TESb flush level. The *solid lines* show simulation results based on Langmuir-type adsorption models using fast and slow reaction components. Inset: Schematic of InAs:Sb SMLs in GaAs; antimony is supplied via single TESb flushes prior to the InAs growth. Reprinted after [28], © APS (2015)

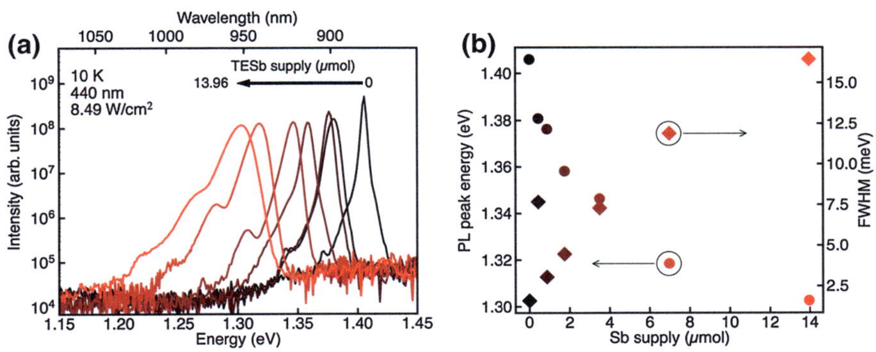

Fig. 2.10 Photoluminescence (PL) spectra at low temperatures **a** and peak positions with FWHM values, **b** as extracted from the PL data. With increasing Sb supply, a redshift of the PL, a linewidth broadening, and an increased intensity of LO-phonon replica are observed. Reprinted after [28], © APS (2015)

to an increased exciton-phonon coupling, pointing to an increased electric dipole moment within the In-rich agglomerations with incorporated Sb. The modelling of the electronic states discussed in Sect. 2.1.2 yields an enlargement of the electric dipole moment due to the stronger spatial separation of the electron and hole wavefunctions.

Pure electronic coupling between individual vertically stacked InAs:Sb depositions can be concluded from the close spacing of the layers verified by X-ray reflectometry measurements. Figure 2.11 shows results of a sample series of SML stacks with increasing GaAs spacer thickness between the InAs:Sb depositions. The perfection of the periodicity within the SML stack manifests itself as superlattice fringes appearing at larger angular positions due to the very short superlattice period. With decreasing spacer thickness, the characteristic features shift towards larger incidence angles. By simulation of the obtained reflectometry curves the observed decline of the measured intensity can be attributed to the general decrease of the specular intensity with increasing incidence angle. Any additional other reason for the decrease in intensity, which would indicate an increasing structural imperfection of the superlattice interfaces, can be ruled out.

Electronically, the onset of carrier localization can be investigated by temperature-dependent photoluminescence spectra. SML stacks with Sb and a spacer thickness below 1.6 MLs exhibit at low temperature a deviation from a purely temperature induced bandgap shrinking. This behavior, commonly referred to as an S-shape in the temperature-dependent peak position, can be modelled by a broadening of the potential landscape which allows for 3D carrier localization [37]. A more direct proof of the QD-like emission properties is found in cathodoluminescence spectroscopy data where individual, spectrally sharp lines are found in an otherwise broad ensemble spectrum at high spatial resolution [28].

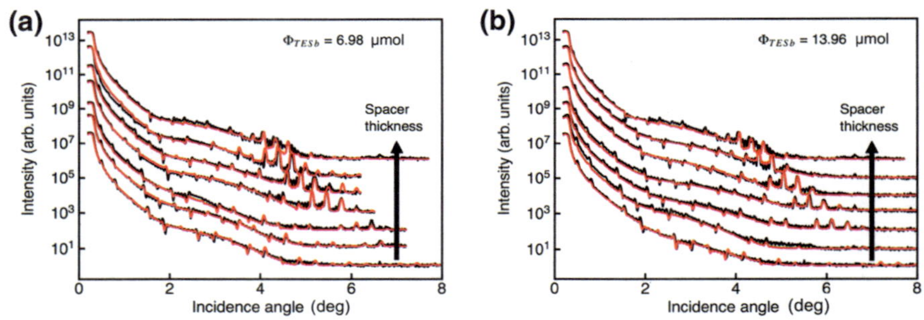

Fig. 2.11 X-ray reflectometry measurements (black) and respective simulations (red) on two sample series with stacks of InAs:Sb/GaAs SMLs and increasing GaAs spacer thickness between the InAs:Sb SML layers. **a** TESb supply of 6.98 μmol, GaAs spacer varied from 1.6 to 2.9 ML, **b** 13.96 μmol TESb supply, spacer varied from 1.6 to 2.7 ML. Characteristic fringes due to internal SML interfaces appear at angles between 4° and 7°. The X-ray data were recorded with Cu-K_{α} radiation. Reprinted after [37], © Elsevier (2018)

2.3 Atomic Structure of Submonolayer Quantum Dots

Growth is a highly non-linear process, so that the resulting atomic arrangement of heterostructural depositions within a device-like structure does not necessarily follow the growth recipe. Due to precursor decomposition, physisorption, chemisorption, surface kinetics and surface diffusion, re-evaporation, strain, surface reconstructions, and especially due to segregation during capping [38], typically the embedded material in a double heterostructure gets re-arranged and mostly intermixed by the capping layer material [39]. These effects apply also for SML depositions, even if the embedded and strained material amount is quite low. Due to quantum effects, the resulting electronic states and thereby the optoelectronic properties of the SML QDs strongly depend on the local geometric arrangement of the constituent materials (Sect. 2.1.2). Thus, it is necessary to study the actual atomic structure in detail in order to obtain information for optimizing the growth on the one hand as well as for the interpretation of the physical effects on the other hand. This applies to SML QDs in the same way as it does for SK QDs.

2.3.1 Methods for Structural Analysis

In order to study the atomic arrangement of different materials within a three-dimensional heterostructure—an SML QD or an SK QD—there are quite a couple of analytic methods available. They all have advantages and disadvantages for answering particular structural questions, while none of the methods is able to monitor every aspect of atomic arrangement. An overview of the preferential methods for several aspects is given in Table 2.1. Typically, X-ray diffraction (XRD) is an efficient method to monitor the properties of ensembles of SML QDs, especially a possible geometric ordering, such as, e.g., the spacing between several SML depositions by the capping material (see Fig. 2.11). The different (scanning) transmission-electron microscopy [(S)TEM] imaging modes are very versatile for many aspects of the actual SML structure. Especially chemical analysis in STEM by high-angular annular dark field analysis (HAADF) of the Z-contrast or the energy-dispersive X-ray spectroscopy (EDX) achieves nowadays an atomically resolved contrast of the material. Nevertheless, (S)TEM typically images the three-dimensional material along atomic columns of the thickness of the specimen, resulting in a *two*-dimensional data set, see Fig. 2.12. Top-view scanning-tunneling microscopy (STM) and atomic-force microscopy (AFM) can be easily applied at growth surfaces before capping. Cross-sectional scanning tunneling microscopy (XSTM) mainly provides atomically resolved information from a single cross-sectional surface of the capped structure. Atom-probe tomography (APT) would be an ideal method for several structural aspects, but unfortunately for III–V semiconductor materials the geometric resolution of the single atomic species is limited to several nm, since mainly clusters of

Table 2.1 Overview on methods to be applied on special structural aspects of nanostructures like SK QDs or SML QDs in crystalline semiconductor heterostructures

Structural analysis method	XRD ensemble	(S)TEM top/side	STM/AFM top	XSTM side	APT 3-d
Density of QDs or SMLs	X	**X**	**X**	(x)	(x)
Lateral ordering of QDs or SMLs	**X**	**X**	**X**	–	–
Vertical stacking of QDs or SMLs	**X**	**X**	–	X	X
Height of QDs or SMLs	X	X	**X**	X	X
Lateral size of QDs or SMLs	X	X	**X**/(x)	X	X
Shape of QDs or SMLs uncapped	(x)	X	**X**/(x)	–	(x)
Shape of QDs or SMLs capped	(x)	X	–	X	(x)
Atomic stoichiometry/local arrangement of atoms	X	X	(x)/ –	**X**	(x)
Structure of a wetting layer (WL)	X	X	(x)/ –	**X**	(x)
Online growth monitoring	**X**	–	X/–	–	–
Electric fields in the structure	–	**X**	–	(x)	–

X: should be used preferentially, X: will give substantial statements, (x): can give good hints or weaker statements, –: no result can be obtained

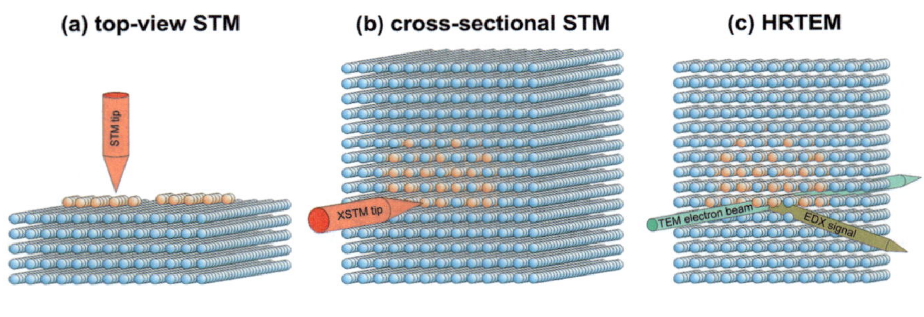

Fig. 2.12 Methods for structural analysis of nanostructures. *Blue and red dots* represent structural units of GaAs and InAs, respectively

the group-V material (especially for P, As, and Sb, less for N) get ablated, demonstrating that surface decomposition and re-arrangement processes occur during the measurement [40]. All of the analytic measurement methods require typically additional simulations of the crystal structure by numerical methods in order to interpret the measured data (for details see [41, 42]).

2.3.2 Analysis of InAs Submonolayer Depositions

For a comprehensive analysis of the SMLs and especially for the structural analysis of single QD-like structures as they are used in devices, and in addition to the X-ray characterization discussed in Sect. 2.2, we applied (S)TEM and XSTM. Details of sample growth, sample preparation, and technical aspects of the (S)TEM and XSTM analyses are given in [28, 41, 43]. The central questions are whether the less than 1 ML deposited InAs within an SML cycle is distributed statistically within a single layer, or if and how it gets re-distributed during further growth and capping by GaAs. It is of particular interest if locally In-rich material is formed, which is able to act as a zero-dimensional QD-like structure for optoelectronic applications. Further on, there is an important question about the material distribution upon Sb supply, which has a strong influence on the optical emission characteristics: Optical transitions may remain locally direct due to a type-1 band alignment, or the Sb atoms may be locally separated from the In atoms, resulting rather in a type-2 band alignment of the electronic states.

In order to explore first how SML growth works on the atomic scale, a couple of samples were studied containing single 0.5 ML thick InAs depositions separated by 16 ML GaAs spacer layers. Figure 2.13a shows a [010] zone-axis high-resolution TEM micrograph. The [001] growth direction is aligned vertically in the image. From this image the local In content in the sample was calculated by an analysis of the {002} Fourier coefficients: As it is visible in Fig. 2.13b, the five In-containing layers are all thicker than 1 ML, which amounts to 0.283 nm for unstrained GaAs or 0.325 nm for the coherently strained $In_{0.5}Ga_{0.5}As$ within a GaAs matrix. Furthermore, the local In content within the layers does not exceed 20% and hence does not reach the nominal value of 50%. Thus, In is not found within one single monolayer as it was nominally

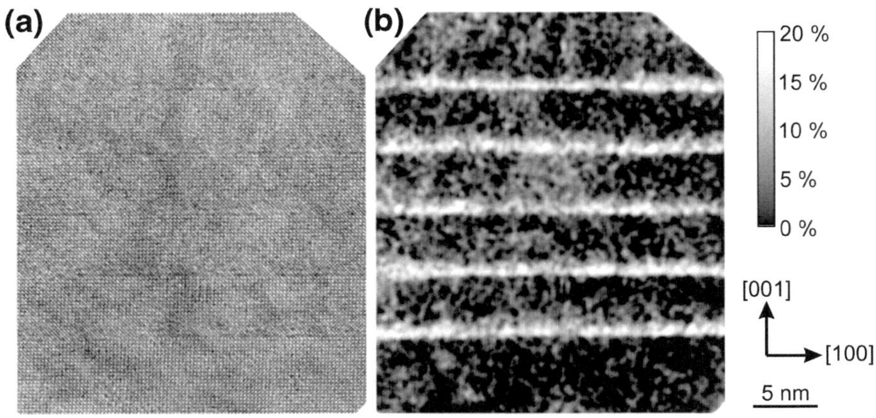

Fig. 2.13 **a** High resolution TEM micrograph of five InAs SML depositions with 0.5 ML nominal thickness, separated by 16 ML thick GaAs spacers. **b** Composition map evaluated from (**a**), where the *gray scale* indicates the In composition *x*. Reprinted after [41], © AIP publishing (2012)

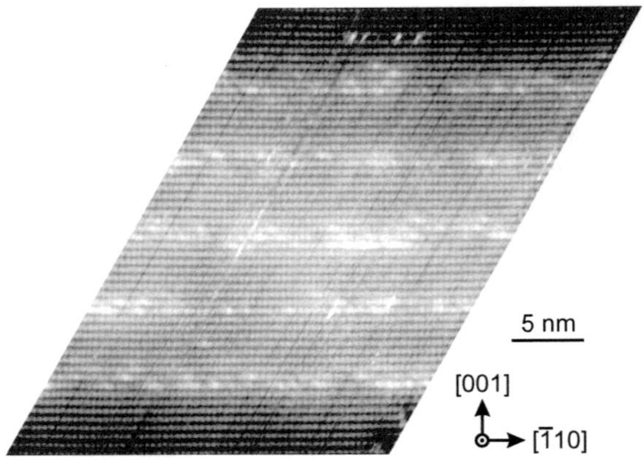

Fig. 2.14 Close-view filled-state XSTM image of five SMLs of 0.5 ML InAs, alternating with 16 ML GaAs, taken at a sample voltage of $V_T = -2.5$ V and a tunneling current of $I_T = 60$ pA. In-rich material appears as bright spots. Reprinted after [43], © Japan Society of Applied Physics (2010)

deposited, but it is strongly redistributed vertically. A closer inspection of the local In content reveals that the lower interface of the In-containing layers is always sharper than the upper one.

Figure 2.14 shows the corresponding XSTM image of the same sample. Again, it is visible that the brighter appearing In atoms form a sharp interface to the underlying GaAs layer, while they are more distributed along growth direction, resulting in a rougher upper interface. Moreover, it is clearly visible that laterally separated In-rich material is formed.

A model explaining these findings is the vertical segregation process, as e.g. described by Muraki et al. [44]. During overgrowth of the InAs by the GaAs spacer material, some of the strained In atoms in the second topmost layer release a part of their strain by exchanging their crystal site with Ga atoms from the topmost layer. When this process occurs with a certain probability (the so-called segregation coefficient r) for each In atom, it can be described for the present $N = 5$ deposited InAs layers as

$$c(n) = \sum_{k=1}^{N=5} c_{0,k}\theta(n - l_k)r^{(n-l_k)}(1 - r), \tag{2.1}$$

where $c(n)$ is the In concentration found in the atomic layer n, θ is the Heaviside step-function, and l_k is the number of the layer in which the kth In deposition occurred. $c_{0,k}$ is the initial In concentration in the kth deposition, which is left as a free fit parameter for all depositions in order to accommodate for local fluctuations in the measured samples.

Figure 2.15 shows the local In concentration, averaged perpendicular to the growth direction, as determined by both methods. The TEM data shown in Fig. 2.15a are derived from the composition map in Fig. 2.13b. The XSTM data are derived by analyzing in the image (Fig. 2.14) the local lattice parameter along growth direction,

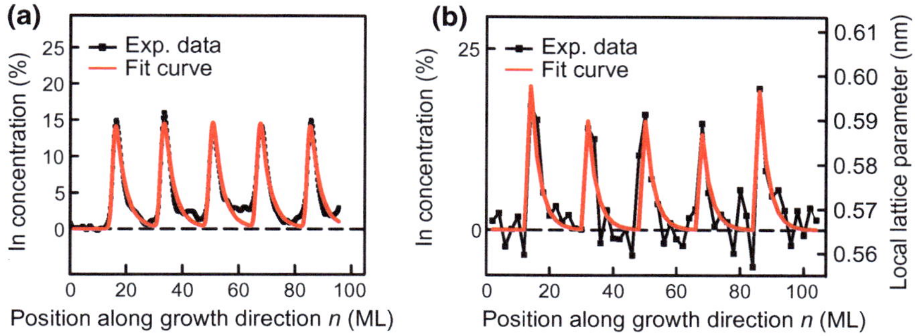

Fig. 2.15 Laterally averaged composition profiles along the growth direction. **a** Profile derived from the TEM data shown in Fig. 2.13. Reprinted after [41], © AIP publishing (2012). **b** Analysis of the local lattice parameter along the growth direction, determined from the XSTM image in Fig. 2.14. *Black dots and connecting lines* represent measured data, and *red solid lines* are composition profiles obtained from a fit to the segregation model. Reprinted after [43], © Japan Society of Applied Physics (2010)

which is related to the In concentration due to strain [45]. For both methods the segregation model fits well with the measured data, exhibiting very similar segregation coefficients of $r = 0.75 \pm 0.06$ in the case of TEM and $r = 0.71 \pm 0.04$ in the case of XSTM. The initially deposited amount of In was found to be 0.76 ± 0.07 ML in the case of TEM and $\sim$0.56 ML in the case of XSTM. These values probably differ since they were locally measured at two different regions of the same sample. They furthermore differ from the nominally deposited amount of 0.5 ML InAs because of the same reason. Summarizing these results, the incorporated In is found to segregate strongly along growth direction with a segregation coefficient around 0.73, and the maximum local In concentration in the layer, in which the deposition occurred, is limited to about 20% instead of the nominal 50%.

Typically, SML structures used for applications in devices are separated by much thinner GaAs spacers, i.e., in the range of 1.5–4.0 nm. Therefore, also samples with thinner spacer layers were studied. Figure 2.16a shows an XSTM image of a five-fold SML stack being separated by only 4 ML of GaAs. In Fig. 2.16b the respective determination of the In concentration is shown. Due to the thinner GaAs spacer layers the In content does not reach the value of almost zero between subsequent InAs depositions any more. The segregation coefficient is found to be $r = 0.75 \pm 0.07$, agreeing well with the values obtained for the thicker spacer layers. In addition to the segregation along growth direction, we observe that the segregated In is incorporated into the structure mainly above In-rich regions in previous layers, leading to the formation of columns of In-rich material. Such a formation of laterally separated In-rich material was observed also by TEM for a ten-fold SML deposition with 1.9 ML GaAs spacers, which was grown on top of an InGaAs/GaAs seed layer with SK QDs, as shown in Fig. 2.17.

In order to summarize the experimental findings on the structural characterization we note that the InAs material is not found in single monolayers, as would be

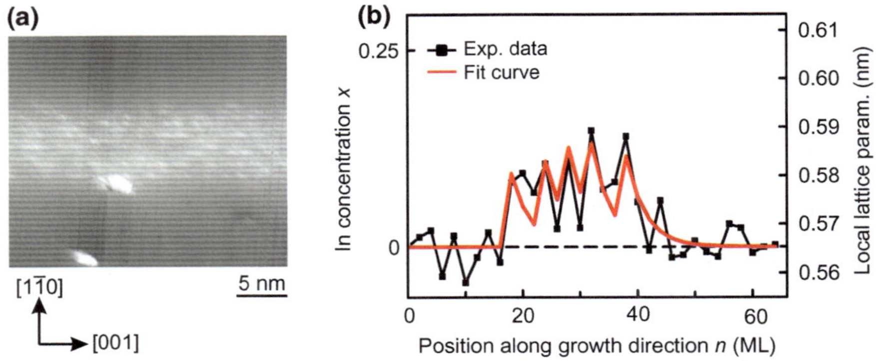

Fig. 2.16 **a** Close-view filled-state XSTM image of a stack of five SMLs of 0.5 ML InAs alternating with 4 ML GaAs, taken at $V_T = -2.4$ V and $I_T = 60$ pA. **b** Analysis of the local lattice parameter along growth direction, averaged perpendicular across the entire XSTM image in (**a**). The measured data are drawn as *back squares*, while the fitted In concentration is shown as *solid red line*. Reprinted after [19], © AIP (2011)

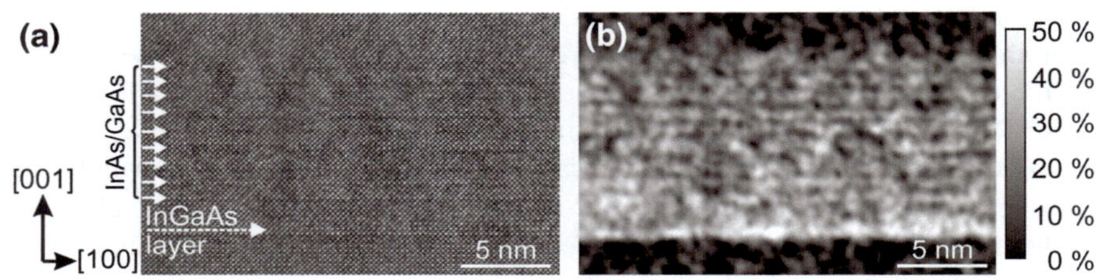

Fig. 2.17 **a** HRTEM micrograph of a sample with 10 InAs layers of 0.6 ML thickness with 1.9 ML GaAs spacers, grown on top of 2.6 ML $In_{0.73}Ga_{0.27}As$ and 2.6 ML GaAs, **b** composition map evaluated from (**a**). Reprinted after [41], © AIP publishing (2012)

expected from the nominal growth sequence. The concept of SML QDs illustrated in Fig. 2.1b and redrawn in more detail in Fig. 2.18a must hence be considered as oversimplified. Instead, these In-rich agglomerations extend vertically over several atomic layers and form columnar structures in the case of thin spacer layers, as depicted in Fig. 2.18b. As pointed out in Sect. 2.1.2 the lateral In-rich agglomerations act as quantum dots with a strong zero-dimensional confinement for the holes and a two-dimensional confinement for the electrons, which interact with the holes by Coulomb interaction. This conclusion was also supported by a line-shape analysis of PL spectra of the SMLs with narrow spacers [19]. The emission line can be described by a superposition of a symmetrical Gaussian part representing the zero-dimensionally localized contribution with a Boltzmann tail on the high-energy side due to the two-dimensional contribution. The different confinement of the electrons and holes are related to the structural finding of (columnar) In-rich agglomerations (zero-dimensional) and the complete layer of the SML stack (two-dimensional) [25].

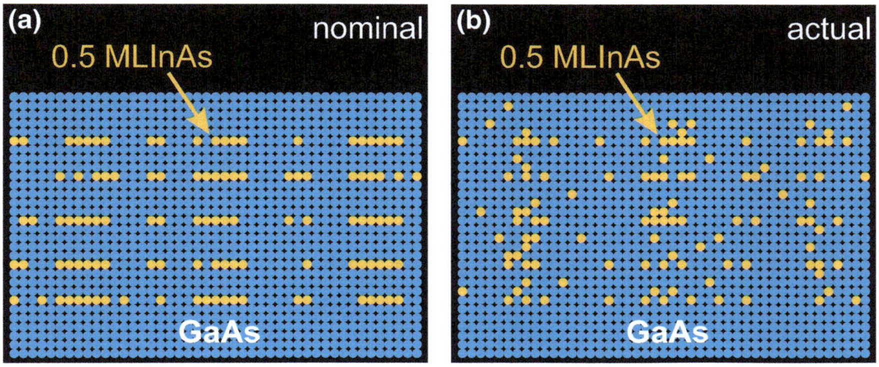

Fig. 2.18 Sketch of the distribution of In within the SML structure. In atoms are drawn in yellow and Ga atoms in blue. **a** Nominal SML deposition, leading to monolayer high islands. **b** Material arrangement observed experimentally, including segregation along the [001] growth direction and lateral surface diffusion, resulting in a vertical correlation of the In-rich agglomerations. Reprinted after [19], © AIP (2011)

2.3.3 Analysis of InAs Submonolayer Depositions with Antimony

The electronic properties of the InAs SMLs are substantially changed upon the addition of Sb as discussed in Sect. 2.1.2 and evidenced in Fig. 2.10. We applied (S)TEM and XSTM to analyze how the different supplied materials are arranged within the resulting SML structure. We first studied the chemically sensitive EDX signal in an STEM experiment of an SML structure consisting of 7 layers of 0.6 ML InAs with previous Sb exposure, separated by 2.6 ML thick GaAs spacers. A HAADF image and EDX images of the contributing chemical components are presented in Fig. 2.19a. The respective concentration profiles of the different elements can be found in Fig. 2.19b. The Sb signal is quite weak due to the small fraction of Sb atoms. We find that Sb is distributed across the entire SML stack, similar to the distribution obtained for the In atoms. Both elements, In and Sb, show also similar segregation along growth direction. The thickness of the complete SML stack is found to be 3.5 nm, which is significantly less than the nominally deposited 5.7 nm. The segregation coefficient is found to be $r = 0.68 \pm 0.05$; this value is similar to the value of 0.66 derived from XRD measurements [28]. We note that this segregation coefficient is significantly smaller than that found in SML structures without Sb supply [41, 43]. So even if the Sb is solely supplied by a flush before the individual SML depositions, Sb atoms are incorporated across the complete layer of the cycled deposition and influence the growth behavior considerably.

In order to study the local arrangement of the incorporated Sb atoms with respect to the In-rich agglomerations, we performed XSTM measurements. Using the chemical selectivity of bias-dependent imaging [46], it is possible to distinguish between the group-III and group-V sublattice of the zincblende structure in XSTM measurements

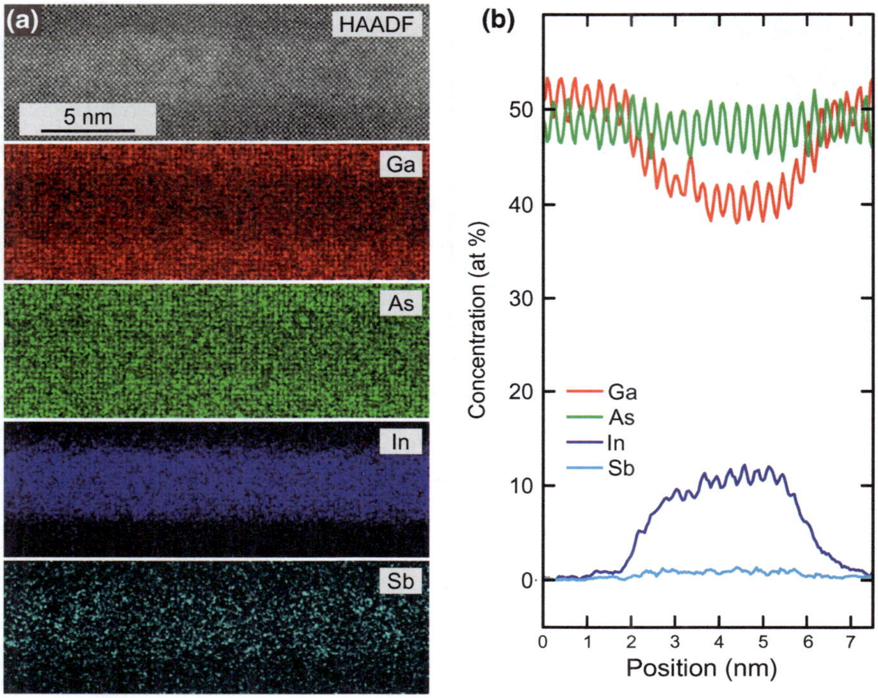

Fig. 2.19 **a** High-angular annular dark field analysis (HAADF) image of an InAs:Sb/GaAs stack together with the chemically selective STEM-EDX images, yielding the spatial distributions of In, Ga, As, and Sb atoms. **b** The concentration profiles of the respective components along the [001] growth direction, averaged perpendicular to the growth direction in the respective images in (**a**)

[47]. At positive sample bias mainly the group-III sublattice is imaged, allowing to distinguish between darker Ga atoms and brighter In atoms, while at negative sample bias mainly the group-V sublattice is detected, with darker As atoms and brighter Sb atoms [42]. Figure 2.20 shows XSTM results of the same sample region taken in (a) at positive sample bias and in (b) at negative sample bias. In Fig. 2.20a the bright areas indicate In-rich material. In the case of Sb supply, the extensions of the In-rich regions are slightly smaller as compared with those without Sb. This indicates that the presence of Sb slightly reduces the lateral In surface diffusion. In Fig. 2.20b we marked single Sb atoms, which can be detected within the layers. From a first inspection it is already obvious that the Sb atoms are not present underneath the InAs layer, where they were deposited during growth, but they are segregated in growth direction even more than the In atoms. Sb is hence found rather in the upper regions of the In-rich agglomerations. This is a typical behavior when Sb is involved in the growth on GaAs, where Sb acts as a surfactant, since the Sb dimers are less strained on the GaAs(001) surface as compared with the As dimers. Thus, in the

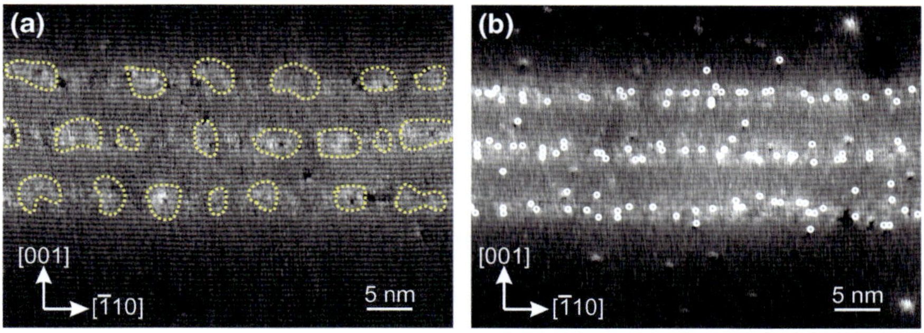

Fig. 2.20 XSTM images of three InAs:Sb SML stacks, each consisting of 4 layers of 0.44 ML InAs separated by 3.22 ML thick spacers. **a** Empty-state image emphasizing the group-III sublattice, taken at $V_T = +2.4$ V and $I_T = 40$ pA, showing the bright In-rich regions (yellow dotted lines) in the stacked SMLs. **b** Filled-state image, taken at $V_T = -1.8$ V and $I_T = 40$ pA, emphasizing the group-V sublattice with single, bright Sb atoms marked by white solid circles. The images are not taken at exactly the same position of the sample. Reprinted after [28], © APS (2015)

case of quaternary InGaAsSb growth on GaAs, the strained Sb segregates more to the growth surface than the similarly strained In [43, 48, 49].

In Fig. 2.21a negative bias image with the identified Sb atoms is shown, which we superposed by contours of the identified In-rich agglomerations, as derived from a positive bias image taken exactly at the same position of the sample surface. Now it becomes obvious that the Sb atoms are mainly found within the upper regions of the In-rich agglomerations or at their sides, but definitely not at their lower interface, where the deposition occurred, and also not within the GaAs regions in between the agglomerations. This result is an important input for the calculations of the optical transitions and the Huang Rhys factor (Sect. 2.1.2).

Fig. 2.21 Merging of the identified In-rich agglomerations (*yellow*) with the single Sb atoms (*white circles*) from the same imaging area of a four-fold stacked InAs:Sb/GaAs SML structure. The underlying filled-state XSTM image was taken at $V_T = -2.2$ V and $I_T = 40$ pA

Summarizing the results of the structural characterization of the InAs:Sb SML QDs we found that Sb leaves the In-rich regions within the GaAs matrix almost unchanged. The Sb atoms segregate even stronger than the In atoms, leading to an incorporation mostly at the upper sides of the In-rich regions.

2.4 Optical and Excitonic Properties

2.4.1 InAs Submonolayer Quantum-Dot Ensembles

In this section we focus on the optical and electronic properties of SML QDs, and on the excitonic properties and the carrier dynamics that can be derived from the optical emission. Essential features of the InAs SML QD luminescence are illustrated in Fig. 2.2 and contrasted with the luminescence of typical SK QDs. SML QDs display a narrower and simultaneously brighter luminescence than SK QDs emitting in the same wavelength range. SML QDs also do not show saturation effects typical for SK QDs due to the limited number of localization centers to be occupied by excitons [50]. The luminescence differs, however, also from the spectrum emitted by a QW; this indicates a different and peculiar density of states (DOS) of the optically active excitons in SML QD systems compared to SK QDs and QWs. Figure 2.22 shows as solid lines the electroluminescence of a 0.5 mm long SML QD-based semiconductor optical amplifier (SOA) at different injection currents ranging from 0.8 to 3.3 times the transparency current. Simulated electroluminescence curves with several test functions for the effective DOS of excitons are shown as dashed lines [51]. The Gaussian and step-like test functions represent the densities of states displayed for an inhomogeneously broadened SK QD ensemble and a two-dimensional QW, respectively. While the SML QD luminescence is too much skewed towards the blue to be accounted for by a SK-QD-like Gaussian-like DOS, a step-like two-dimensional DOS dramatically overestimates the blue side of the emission. The best fit is obtained by implementing an extended Gaussian-like DOS described by

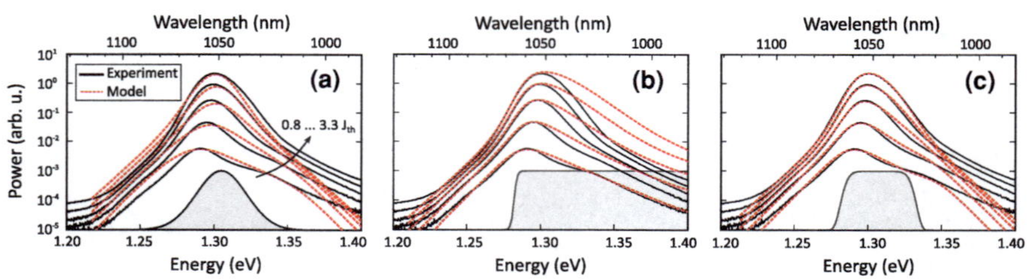

Fig. 2.22 Measured (solid lines) and simulated (dashed lines) electroluminescence spectra of an SOA based on InAs SML QDs in the active region for different injection currents. For the simulations **a** a Gaussian, **b** step-like, and **c** an extended Gaussian-like effective density of states were assumed for the excitons. Reprinted after [51], © APS (2016)

$$\mathcal{D}^k \propto \exp\left[-4\ln 2\left(\frac{\hbar\omega^k_{\mathrm{SML}} - E_{\mathrm{SML}}}{\Delta E^{\mathrm{inh}}_{\mathrm{SML}}}\right)^8\right], \tag{2.2}$$

where E_{SML} is the SML central transition energy, $\Delta E^{\mathrm{inh}}_{\mathrm{SML}}$ the inhomogeneous broadening (in this case 30 meV), and k an index that runs over SML QD exciton subensembles $\hbar\omega^k_{\mathrm{SML}}$.

Qualitatively this shape of the DOS can be understood to originate from the interplay between Coulomb forces and kinetic energy influencing the two-dimensional delocalized electrons, as addressed in Sect. 2.1.2. In optically active excitons, the long-range Coulomb interaction enforces a correlation between the localized carrier and its delocalized conjugate, in the present case being the holes and the electrons, respectively. This makes it available for optical recombination up to a limiting kinetic energy of the two-dimensionally delocalized electrons, above which such a process is prevented. The comparatively strong zero-dimensional localization of holes does not permit a diffusion of the exciton as a whole entity. A question of immediate importance for possible applications is the influence of unbound carriers on the optical properties and performance of devices based on SMLs.

Experimental Pump-Probe Method

The issue discussed above can be addressed using heterodyne-detected pump-probe spectroscopy, a technique which delivers simultaneously the changes in optical gain and refractive index induced in the investigated sample. The technique is illustrated in Fig. 2.23. Optical pump and probe pulses with a relative variable time delay Δt are coupled into the SOA waveguide and detected behind the waveguide with a spectrally integrating detector, e.g., a fast diode. The waveguide geometry enforces a collinear and co-polarized propagation of both pump and probe pulses. To make both pulses separable, the probe pulse is split into a probe and a reference (local oscillator) part prior to entering the sample waveguide, and a frequency shift is imposed on the probe part by an acousto-optic frequency shifter. When the light transmitted through the sample and the local oscillator are combined on the integrating detector, this frequency shift causes a temporal beating pattern in the detected intensity.

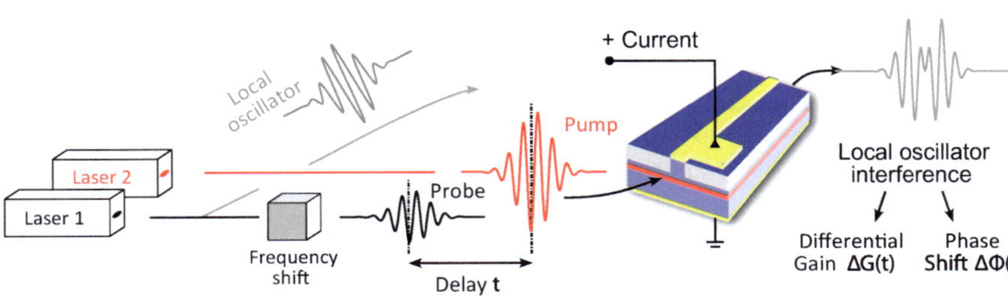

Fig. 2.23 Schematic of a heterodyne-detected pump-probe experiment. Two femtosecond pulses are coupled collinearly into the SML SOA waveguide to excite and analyze the electronic system with high temporal resolution. The probe pulse carries a marker frequency to enable a separation of pump and probe pulses in heterodyne detection

The beat frequency is ideally chosen in the radio-frequency range, where it can be detected by a lock-in detector. Since low intensities are desirable for the probe pulse in such experiments, balanced detection is used to suppress the common mode noise originating from the pump-probe laser source and the emitted luminescence of the sample, which also hits the same detector. It is thus possible to use pump and probe pulses in resonance with optical transitions in the active medium. Spurious effects created by the cascading down of hot carriers from the barrier material, if short pump wavelengths are used, can thus be avoided. The interferometric detection scheme of this type of experiment yields information about the pump-induced change in optical gain $\Delta G(t)$ via the amplitude of the signal and the pump-induced change in the refractive index via probe-reference phase shift $\Delta\Phi(t)$ [52, 53]. Thus, edge-emitting electro-optic devices can be studied under operating conditions. A more detailed description of the experimental method can be found in Sect. 4.2.1.

Dynamical Properties of InAs Submonolayer Quantum Dots

Representative pump-probe traces of $\Delta G(t)$ and $\Delta\Phi(t)$ for the device used in Fig. 2.22 are shown in Fig. 2.24a, b, respectively [51, 52]. The traces were taken with pump and probe pulses tuned into resonance with the peak of the electroluminescence and with a temporal FWHM of 300 fs; this width approximately limits the time resolution of the experiment. Time zero is defined as full temporal overlap of pump and probe pulses. A positive change in $\Delta G(t)$ indicates optical pumping by the pump pulse with a consecutive increase in gain. A negative change signifies the depletion of an inverted transition by the pump pulse, and a consecutive increased absorption of the probe pulse. At the transparency current J_{tr}, or at a semi-inverted transition, both effects cancel. A logarithmic scale of the time axis helps to highlight the different dynamic processes and their respective time range.

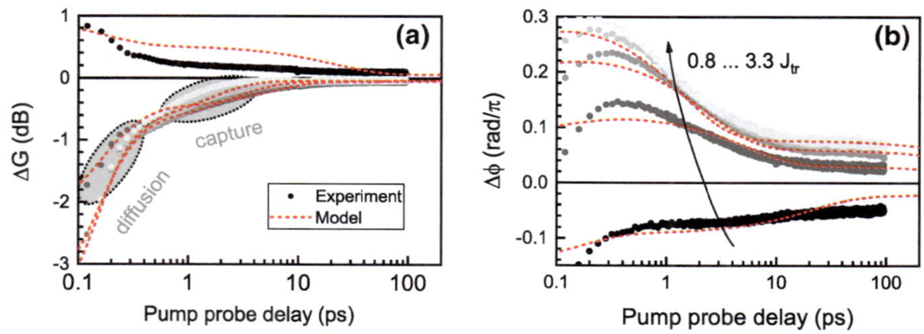

Fig. 2.24 Pump-probe traces with pump and probe resonant to the peak of the emission of the SML QD SOA of Fig. 2.22 for injection currents ranging from 0.3 to 3.3 times the transparency current J_{tr}. The dots are experimental data for different injection currents, the dashed lines represent the results of numerical simulations. Time intervals of diffusion and carrier capture are marked as gray areas in the differential gain in (**a**). The differential phase traces in (**b**) show a large phase shift up to 0.3 π at high injection current

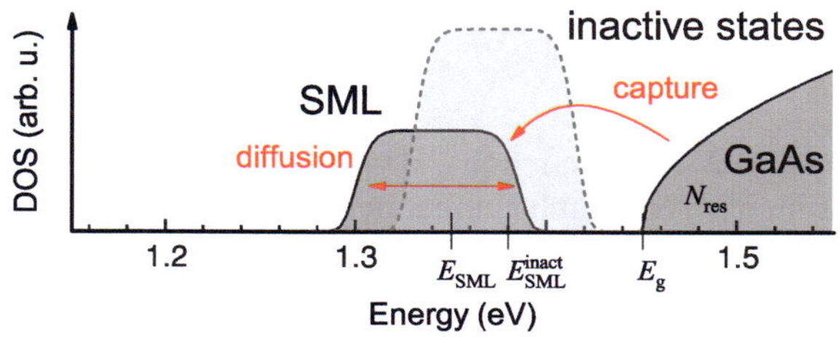

Fig. 2.25 Density of states and carrier relaxation pathways for SML QDs. Inactive states in terms of free carriers form an optically inactive, but efficient carrier reservoir. Reprinted after [51], © APS (2016)

The key observation to be drawn from the $\Delta G(t)$ time trace is the appearance of a sub-picosecond recovery process with very large amplitude, which is absent in similar experiments on self-assembled SK QDs. The speed of this recovery process requires that the refilling carriers diffuse from the immediate vicinity of the depleted transition. The process is hence a carrier-diffusion process within the SML QD layer. The second stage of gain recovery on the few-picosecond time scale shows a peculiar current-dependence that gives clues of the dimensionality of the reservoir feeding it. Unlike in SK QDs, the carrier population of SML QDs is recovered immediately from a three-dimensional reservoir, in this case GaAs, without the necessity of a two-dimensional intermediate state (the wetting layer of SK QDs), leading to a one order of magnitude faster gain recovery for SML QDs [51]. The direct capture from the GaAs reservoir hints at less strict momentum selection rules compared to the carrier capture into SK QDs [54].

Via causality and the Kramers-Kronig relations, every optical transition is accompanied by a change in the refractive index. It shows a zero-crossing at the energy of the resonance and in its wings decays slower than the Lorentzian function describing the change in gain upon crossing a resonance. Ideally, for excitons with resonances distributed evenly around a central energy, the net refractive index change, which an optical pulse tuned to this central energy experiences, is equal to zero. It was very early observed that this is not the case for semiconductor optoelectronic devices, since only a net refractive index change accompanying an optical transition can explain the unusual linewidth broadening in semiconductor lasers [55]. This effect is essentially due to the existence of a reservoir of carriers on the blue side of the observed transition, and hence asymmetric in energy relative to the emitting states in semiconductor devices. It accounts for a small yet measurable phase shift that the probe laser pulse experiences when passing, e.g., an SK QD-based device following a pump pulse. In contrast to SK QDs, SML-QD-based devices display a much larger phase response following an optical perturbation. A phase effect of this magnitude can only be accounted for by assuming the existence of an *additional* reservoir of free carriers, which do not contribute to the changes in optical gain, but nevertheless give rise to a refractive index change via off-resonant transitions into higher lying levels.

Modelling of InAs Submonolayer Quantum Dot Dynamics

These conclusions are supported in a numerical model for gain and phase recovery, which is condensed in Fig. 2.25 [50, 51]. The three-dimensional reservoir is coupled to the SMLs according to

$$\frac{\partial}{\partial t}N_{\text{res}} = J - \frac{N_{\text{res}}}{\tau_{\text{res}}} - 2N^{\text{SML}}\sum_k \mathcal{D}^k_{\text{SML}}\tilde{R}^{\text{cap}}_{\text{SML}}(k), \tag{2.3}$$

where N_{res} is the number density of carriers in the reservoir with a recombination time τ_{res}, N_{SML} the number density of localization center, and $\tilde{R}^{\text{cap}}_{\text{SML}}(k)$ the capture rate into the kth SML QD subensemble. The recovery of the occupation probability ρ^k_{SML} of the kth SML QD subensemble contains a diffusive process, the capture from the three-dimensional reservoir, and a decay process due to stimulated emission:

$$\frac{\partial}{\partial t}\rho^k_{\text{SML}} = -\frac{\left(\rho^k_{\text{SML}}\right)^2}{\tau_{\text{SML}}} + \tilde{R}^{\text{cap}}_{\text{SML}}(k) + \tilde{R}^{\text{rel}}_{\text{SML}}(k) - R^{k,\text{stim}}_{\text{SML}} \tag{2.4}$$

with capture rates

$$\tilde{R}^{\text{cap}}_{\text{SML}}(k) = R^{\text{cap}}_{\text{SML}}(J)\left[F\left(\hbar\omega^k_{\text{SML}}, \mu, T\right) - \rho^k_{\text{SML}}\right], \tag{2.5}$$

where $F(E, \mu, T)$ is the quasi-Fermi function, E is the transition energy, μ is the quasi-Fermi level and T is the carrier temperature. We need to assume a current-dependent capture rate $R^{\text{cap}}_{\text{SML}}(J) = 60\,\text{ns}^{-1}(J/J_{\text{tr}})^2$, and relaxation rates

$$\begin{aligned}\tilde{R}^{\text{rel}}_{\text{SML}}(k) = \sum_{j\neq k} f^j \Big\{ & R^{\text{rel}}_{\text{SML}}\left(1 - \rho^k_{SML}\right)\rho^j_{\text{SML}} \\ & - R^{\text{rel}}_{\text{SML}}\rho^k_{\text{SML}}\left(1 - \rho^j_{\text{SML}}\right)\mathrm{e}^{-\left(\hbar\omega^j_{\text{SML}} - \hbar\omega^k_{\text{SML}}\right)/(k_B T)}\Big\}.\end{aligned} \tag{2.6}$$

Table 2.2 summarizes the numerical values used in the simulations. The comparison between experimental and numerical results shown in Fig. 2.24 demonstrates that the fast part of the gain recovery is excellently described by a capture of carriers from neighboring localization centers with a rate of 15 ps^{-1}. Although the optical gain is large, most of the localization centers must be assumed as optically inactive at the probe wavelength to account for the highly efficient gain recovery below 1 ps. This model of SML QD-reservoir coupling describes all essential effects of the dynamics in InAs SMLs. When antimony is introduced into the SML QDs, their optical and dynamical properties are substantially changed.

2.4.2 InAs:Sb Submonolayer Quantum-Dot Ensembles

One of the fundamental ideas behind alloying SMLs with antimony is to shift the optical emission of the SML system towards the red and hence towards the technologically important window around 1060 nm. In the case of SML QDs, apart from an

Table 2.2 Parameters used for growth and simulation of the InAs SML QD SOA

Description	Symbol	Value
SML thicknesses InAs/GaAs		6 × (0.4 ML/1.6 ML)
Bulk lifetime	τ_{res}	2 ns
SML QD lifetime	τ_{SML}	0.2 ns
SML QD transition energy	$E_{SML}\ (J = 0)$	1.313 eV
SML QD ensemble inhom. width	ΔE_{SML}^{inhom}	30 meV
SML QD areal density	N_{SML}	$3.3 \times 10^{11}\ cm^{-2}$
Capture rate into SML QDs	$R_{SML}^{cap}\ (J)$	$60\ ns^{-1} \times (J/J_{tr})^2$
Diffusive SML QD scattering rate	R_{SML}^{rel}	$15\ ps^{-1}$
Bandgap energy	$E_g\ (J = 0)$	1.38 eV
Inactive SML QD energy	$E_{SML}^{inact}\ (J = 0)$	1.35 eV

overall redshift, alloying with antimony gives rise to an asymmetric emission with a pronounced red wing and an overall broadening also of the main luminescence peak. This is visualized in Fig. 2.26, which compares the sub-threshold electroluminescence of semiconductor lasers based on unalloyed SML QDs in the left panel and Sb-alloyed SML QDs in the right panel. Both devices are designed for laser emission in the 1060 nm wavelength range [56]. At low injection currents, the emission of the

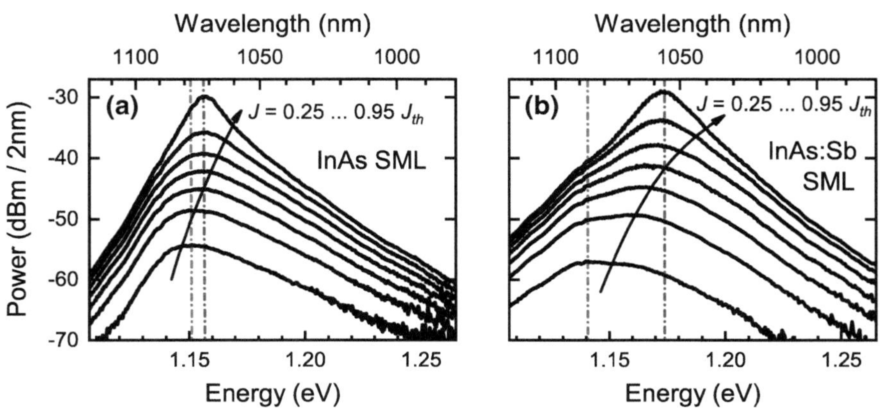

Fig. 2.26 Electroluminescence spectra of semiconductor lasers based on **a** InAs SML QDs and **b** SML QDs alloyed with Sb. Dash-dotted lines mark the evolution of the lasing mode. Reprinted after [56] © IEEE (2019)

InAs:Sb SML QD based laser is broad with a peak at around 1080 nm. At higher currents, a second peak displaying essentially SML QD-like features as discussed above is observed in the blue part of the luminescence spectrum. The initially more red-centered emission develops into a shoulder of the dominating blue emission peak. A detailed study of the static and dynamic optical properties of InAs:Sb SML QDs including the origin of the emission in the red wing of the spectrum leads to the conclusion that they contain two distinct distributions of localization centers with different effective dimensionality [56].

A schematic representation of the DOS and the scattering channels describing the emission and gain recovery observed in InAs:Sb SML QDs is displayed in Fig. 2.27. At higher energies the DOS is identical to the DOS assumed for the unalloyed samples containing active and inactive SML states (see Fig. 2.25). The double peak structure of the emission is accounted for by introducing an additional red-shifted Gaussian-distributed DOS for excitons. As opposed to the SML states, a lateral diffusion is *not* allowed within the low-energetic Gaussian-distributed ensemble. This agrees with results described in Sect. 2.1.2, demonstrating increased localization of carriers upon introduction of Sb. Carriers are captured as well from the extended active and inactive SML QDs as from the GaAs reservoir. These additional localization centers thus have properties very similar to SK QDs with zero-dimensional confinement.

For a quantitative description, the equation for the occupation probability of the delocalized SML QD states ρ^k_{SML} [see also (2.4)] is modified by an additional loss channel into the localized SML QD states, identified by the index SML:Sb:

$$\frac{\partial}{\partial t}\rho^k_{\mathrm{SML}} = -\frac{\rho^k_{\mathrm{SML}}}{\tau_{\mathrm{SML}}} + \tilde{R}^{\mathrm{cap}}_{\mathrm{SML}}(k) + \tilde{R}^{\mathrm{rel}}_{\mathrm{SML}}(k) - R^{k,\mathrm{stim}}_{\mathrm{SML}} - \sum_j \mathcal{D}^j_{\mathrm{SML:Sb}} \tilde{R}^{\mathrm{cap}}_{\mathrm{SML:Sb}}(\Delta E_j). \tag{2.7}$$

τ_{SML} is the SML QD carrier lifetime and R^{stim} the stimulated emission rate. The numerical values for the parameters related to the delocalized SML subensemble can be found in Table 2.2.

The occupation probability of the *localized* SML QD states evolves according to

$$\frac{\partial}{\partial t}\rho^k_{\mathrm{SML:Sb}} = -\frac{\rho^k_{\mathrm{SML:Sb}}}{\tau_{\mathrm{SML:Sb}}} - R^{k,\mathrm{stim}}_{\mathrm{SML:Sb}} + \sum_j \mathcal{D}^j_{\mathrm{SML}} \tilde{R}^{\mathrm{cap}}_{\mathrm{SML:Sb}}(\Delta E_j). \tag{2.8}$$

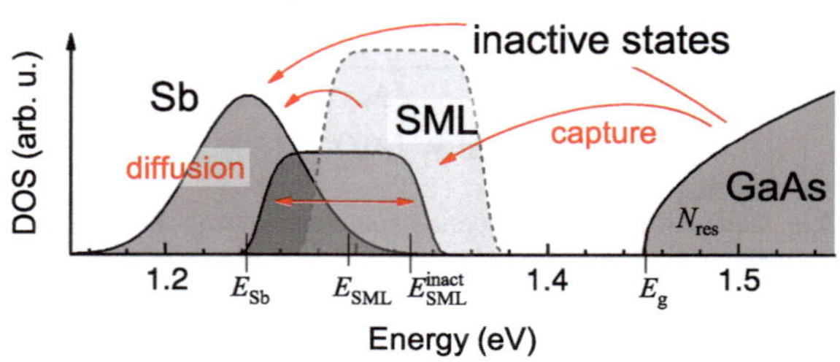

Fig. 2.27 Density of states and carrier relaxation path-ways for SML QDs alloyed with antimony. The alloy induces an additional low-energy gaussian carrier distribution. Reprinted after [56], © IEEE (2019)

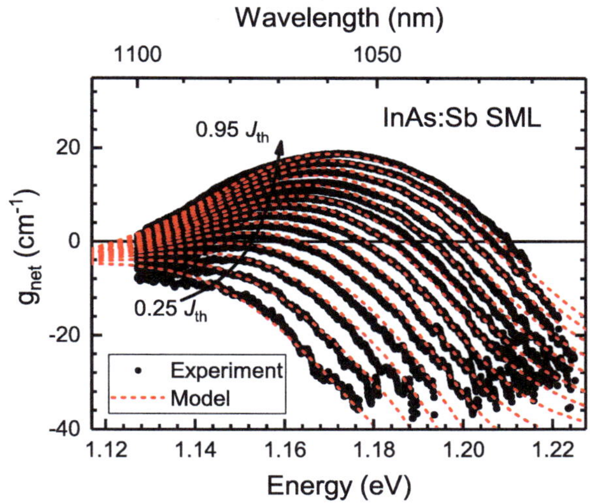

Fig. 2.28 Measured and modeled gain spectra of the InAs:Sb SML laser for injection currents ranging from 0.25 to 0.95 times the threshold current. Reprinted after [56], © IEEE (2019)

From the occupation of these states and the combined DOS D_m where m iterates over the SML and the SML:Sb states, the modal gain $g(E)$, can be derived:

$$g(E) = \frac{\mathcal{G}(E)}{v_g} m \int \left[2\rho^m(E') - 1\right]\mathcal{D}_m(E')\mathcal{L}(E' - E)dE' - \alpha_{\text{int}}. \tag{2.9}$$

Here $\mathcal{G}(E)$ is the gain coefficient, v_g the group velocity, α_{int} the internal losses and $\mathcal{L}(E)$ the Lorentzian lineshape of the optical transition according to:

$$\mathcal{L}(E) = \frac{T_2}{\pi\hbar}\text{sech}\left(\frac{T_2 E}{\hbar}\right). \tag{2.10}$$

This model provides an excellent description of the full modal gain of an InAs:Sb SML laser as shown in Fig. 2.28. The set of parameters describing the parameters specifying the dynamics of the InAs:Sb SMLs and the relevant parameters for the calculation of the gain spectra are collected in Table 2.3.

2.5 Devices Based on Submonolayer Quantum Dots

2.5.1 Gain and Efficiency

Apart from the fast gain recovery after depletion of the active region, InAs SML QDs and InAs:Sb SML QDs both display a large optical gain due to the combined effects of a large number density of localization centers and efficient carrier capture. This is qualitatively obvious from the lack of saturation in the electroluminescence with

Table 2.3 Parameters used for growth and simulation of the InAs:Sb SML QDs

Description	Symbol	Value
SML thicknesses InAs/GaAs		8 × (0.83 ML/1.59 ML)
Sb content per cycle		7 μmol
InAs:Sb SML QD lifetime	$\tau_{\mathrm{SML:Sb}}$	1 ns
InAs SML QD inhomogeneous width	$\Delta E_{\mathrm{SML}}^{\mathrm{inhom}}$	50 meV
InAs:Sb SML QD inhomogeneous width	$\Delta E_{\mathrm{SML:Sb}}^{\mathrm{inhom}}$	65 meV
InAs SML QD areal density	N_{SML}	$3.5 \times 10^{11}\ \mathrm{cm}^{-2}$
InAs:Sb SML QD areal density	N_{Sb}	$0.7 \times 10^{11}\ \mathrm{cm}^{-2}$
Capture rate into InAs SML QD states	$R_{\mathrm{SML}}^{\mathrm{cap}}\ (J)$	$75\ \mathrm{ns}^{-1} \times (J/J_{\mathrm{tr}})^2$
Diffusive InAs SML QD scattering rate	$R_{\mathrm{SML}}^{\mathrm{rel}}$	$15\ \mathrm{ps}^{-1}$
Capture rate into InAs:Sb SML QD states	$R_{\mathrm{SML:Sb}}^{\mathrm{cap}}$	$1\ \mathrm{ps}^{-1}$
Dephasing time	$T_2\ (J = 0)$	150 fs
Gain coefficient	G	$3 \times 10^{-4}\ \mathrm{eV\ m^2\ s^{-1}}$

increasing injection current, quite unlike the emission of SK-QDs, which saturates fast due to a limited DOS in the QD ground state.

Figure 2.29a shows the net gain g_{net} versus energy obtained for the InAs and the InAs:Sb SML QD SOAs by the Hakki-Paoli method [57]. It relies on the extraction of the gain from the peak-to-valley ratio r of the Fabry-Pérot cavity modes below

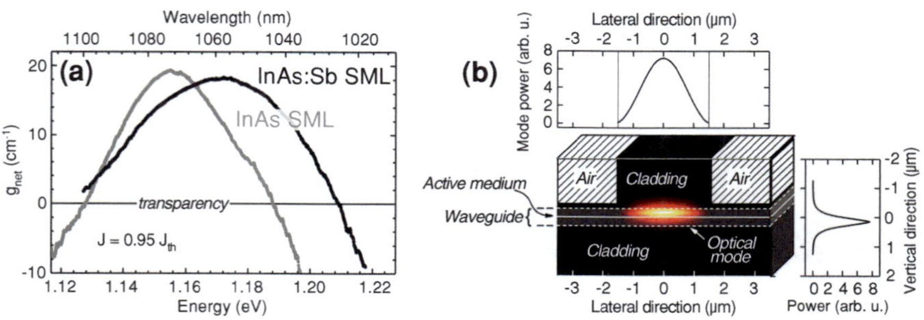

Fig. 2.29 **a** Net modal gain of an InAs/GaAs and an InAs:Sb SML based semiconductor laser with a 0.5 mm cavity at 0.95 times the respective threshold current. **b** Modeled optical field distribution for determination of the mode confinement factor **Γ**. Reprinted after [56], © IEEE (2019)

threshold:

$$g_{\mathrm{net}} = \Gamma g_{\mathrm{mat}} - \alpha_{\mathrm{mirr}} = \frac{1}{L}\ln\left(\frac{\sqrt{r}-1}{\sqrt{r}+1}\right) - \frac{1}{L}\ln(R), \tag{2.11}$$

using the facet reflectivity R and the cavity length L. Further contributions to g_{net} are the material gain g_{mat}, the confinement factor Γ of the optical mode, the outcoupling losses α_{mirr}, and the internal losses α_{int}. The relatively low values in the graph are due to the fact that all the devices contain only a single layer of QDs in the active region. The confinement factor quantifies the overlap of the optical mode with the active region. The optical field distribution in Fig. 2.29b is calculated with a one-dimensional mode solver using the refractive index profile of the waveguide [58]. Estimating a thickness of the active material from the TEM and XSTM measurements (see Sect. 2.3.1 of this chapter) of about 5.7 and 3.5 nm for InAs SML QDs and InAs:Sb SML QDs, respectively, a very large material gain of 1100 and 1700 cm^{-1} is obtained. Extrapolating from the luminescence of both samples, it is reasonable to expect a more broadband gain for the InAs:Sb SML QD system. This is indeed the case, however, the delocalized ensemble produces a much larger gain than the localized fraction. Thus, in Fig. 2.29a the gain spectrum of the InAs:Sb SML QDs has a higher intensity on the blue side rather than on the red as compared to the one of the pure InAs SML QDs.

Consequently, lasing starts on the part of the spectrum, at which the delocalized part of the DOS is located (Fig. 2.26b). The light-current characteristics of lasers based on InAs SML QDs and InAs:Sb SML QDs is compared in Fig. 2.30. While the threshold current is dramatically different for both systems, from the laser emission a very similar external quantum efficiency η_{ext} of about 33% per facet is deduced. Though the localized states must be saturated first and lead to a high threshold current, they do not deteriorate the slope of the light-current characteristics at lasing.

Rather than as lasers, the devices can be operated as photodetectors when driving at reverse bias. InAs SML QDs used in infrared photodetectors have shown enhanced peak responsivity and bandwidth as compared to SK QDs [59–61]. The InAs:Sb SML QDs contribute on the red side of the spectrum, further extending the response

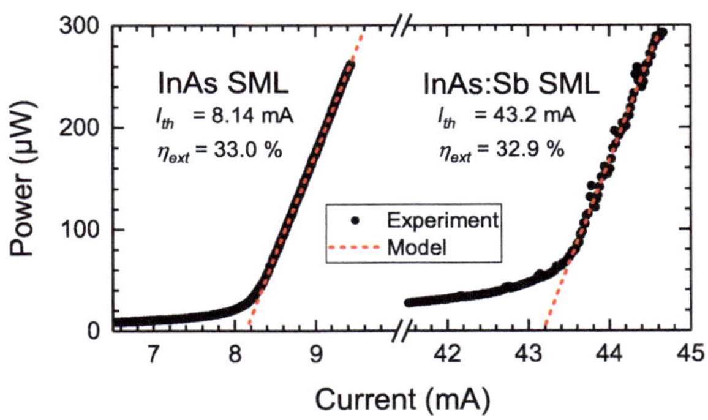

Fig. 2.30 Light-current characteristics for InAs and InAs:Sb SML QD SOAs. Dots mark experimental data; dashed lines represent a linear fit for the determination of the threshold current and slope efficiency. Reprinted after [56], © IEEE (2019)

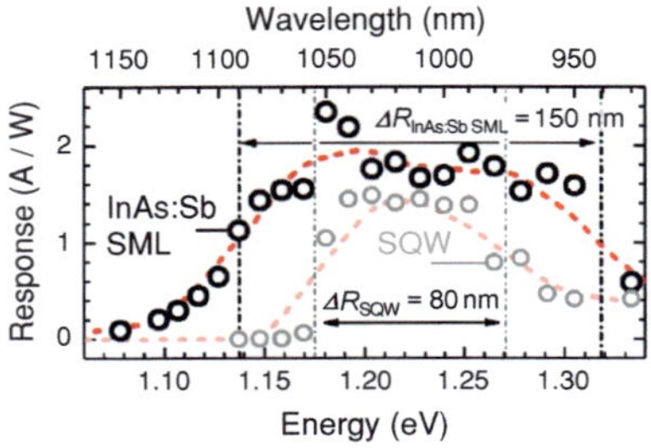

Fig. 2.31 Spectral response of a InAs:Sb SML SOA in comparison to a single-QW SOA at a bias of −1 V and 10 nm excitation bandwidth. Dots mark experimental data; dashed lines represent an interpolation to estimate the spectral response. Reprinted after [56], © IEEE (2019)

bandwidth [56]. A comparison of InAs:Sb SML QDs to a single-QW-based device (8.3 nm $In_{0.23}GaAs$ SQW) with transition energies centered in the same wavelength range is shown in Fig. 2.31. The response R relates the average induced photocurrent $I(\lambda)$ after an optical excitation with the power P_{in} of the excitation according to $I(\lambda) = R(\lambda) \times P_{in}$. The bandwidth of the InAs:Sb SML QD detector covers about 150 nm, or 180 meV, twice as much as the SQW-based detector.

2.5.2 Amplitude-Phase Coupling

The amplitude-phase coupling in optoelectronic devices is quantified in terms of the so-called α parameter, relating the charge carrier-induced changes in refractive index $n(N)$ and gain $g(N)$, scaled by the mode wavelength λ [55]. The large phase response of SML structures also implies a large α parameter in these systems [51, 62]. It is notoriously difficult to quantify α exactly, as it strongly depends on a wide range of parameters; different methods are known to yield differing numbers, even if applied to the same device [20, 63, 64]. It is, however, possible to draw qualitative conclusions in case the parameter is extracted by the same method from identically prepared devices. It is possible to extract α from gain ΔG and phase $\Delta\Phi$ traces recorded in a heterodyne detected pump-probe experiment [50], using the following relation which connects the observed gain and phase changes with changes in the carrier population N in the active region:

$$\alpha = -\frac{4\pi}{\lambda}\frac{\mathrm{d}n}{\mathrm{d}N} / \frac{\mathrm{d}g}{\mathrm{d}N} = -20\log_{10}(e)\frac{\mathrm{d}\Phi(\mathrm{rad})}{\mathrm{d}G(\mathrm{d}B)}. \tag{2.12}$$

Figure 2.32a shows α parameters determined in this way for the InAs:Sb SML SOA with the emission spectrum shown in Fig. 2.26b. The delay-time dependence of the pump-probe traces gives rise to a delay-time dependence equally in α, which we label for this reason α_{dyn}. The curves start at single-digit values, rise thereafter

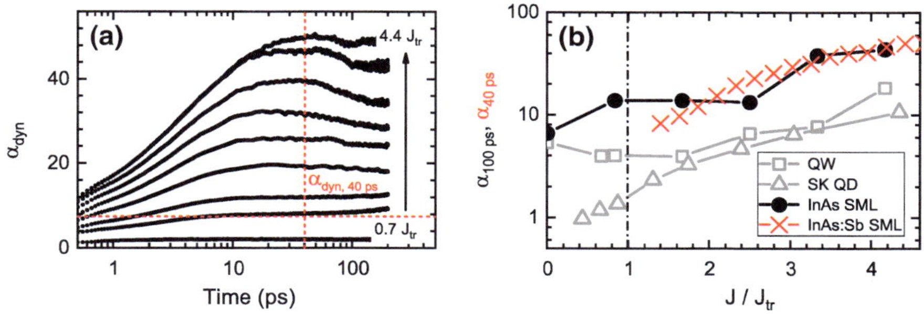

Fig. 2.32 Dynamical α parameter for SOAs based on QWs, SK QDs, InAs SML QDs and InAs:Sb SML QDs extracted from the ratio of phase and amplitude changes in heterodyne-detected pump-probe experiments. **a** Temporal evolution for different injection currents with saturation after 40 ps. **b** Saturation values for SOAs based on different gain media. Both types of SML QDs show similar results, which are significantly increased as compared to the MQW and SK QD-based devices. Reprinted after [56], © IEEE (2019)

until saturation is reached at delay times larger than about 40 ps. The reason for these dynamics is a desynchronized behavior of the active QDs and the carrier reservoir. The perturbation created by the stimulating optical pulse does not immediately propagate to the carrier reservoir and hence creates a delayed response. The application-relevant saturation value of $\alpha_{\rm dyn}$ is plotted in Fig. 2.32b for a range of identically processed SOAs based on a QW, on self-assembled SK QDs, on InAs SML QDs, and on InAs:Sb SML QDs in the active region; all SOAs are designed to emit in the same wavelength range [50]. The index citing a picosecond value indicates the pump-probe delay time, for which the equation was evaluated, since different nanostructures have differing relaxation rates. The equilibrium values of $\alpha_{\rm dyn}$ are plotted in the figure versus the reduced injection current $J/J_{\rm tr}$.

The lowest $\alpha_{\rm dyn}$ is displayed by the device based on SK QDs. The peculiarity of this structure is the relatively large energy separation of active states and carrier reservoir. The carrier reservoir in this case is formed by a QW embedding the SK QDs (dots-in-a-well). It is thus not surprising that for large currents, $\alpha_{\rm dyn}$ of SOAs with either SK QDs or a QW are nearly coinciding. The much larger $\alpha_{\rm dyn}$ (note the logarithmic scale) observed in the InAs SML QDs and InAs:Sb SML QDs stems from the spectral proximity of the carrier reservoir. The speed of gain recovery hence increases at the expense of a large amplitude-phase coupling over a large range of wavelengths in devices based on SML QDs. Note that the dynamical α parameter of InAs SML QDs and InAs:Sb SML QDs is of the same order, leading to the conclusion that the fast optical and excitonic properties are governed by the delocalized SML QD states present in both systems.

A complementary approach to determine the α parameter for an unperturbed free running device is the method of Henning and Collins [65], in which the refractive index change with injection current (i.e., with carrier population) is extracted from the shift of the comb of longitudinal Fabry-Pérot resonator modes modulating the

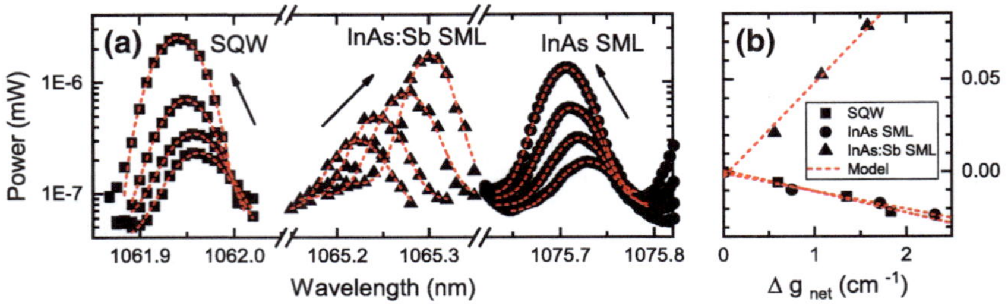

Fig. 2.33 **a** Shift of longitudinal modes for single-QW, InAs:Sb SML QD, and InAs SML QD-based semiconductor lasers with increasing injection currents (arrows). **b** Wavelength shift of the mode peak versus mode gain change. Symbols mark experimental data; dashed lines in (**a**) represent peak fits to determine λ_i and linear fits in (*b*) to calculate α_{stat}

emission spectrum. Since the parameter strongly depends on operating conditions, we label the Henning-Collings α parameter as α_{stat}. It is given by

$$\alpha_{\text{stat}} = \frac{2}{\delta\lambda}\frac{\mathrm{d}\lambda_i}{\mathrm{d}g_{\text{net},i}}, \tag{2.13}$$

for which i signifies the number of the measured longitudinal mode and $\delta\lambda$ the mode spacing. The wavelength shift $\Delta\lambda_{\text{i}}$ of a mode as a function of net gain growth Δg_{net} with increasing injection current for the SOAs containing a single QW, InAs SML QDs or InAs:Sb SML QD SOAs are shown in Fig. 2.33a, b. The absolute α_{stat} values for single QW and InAs SML QD lasers of 1.4 ± 0.1 and 1.3 ± 0.1, respectively, are almost identical; for the InAs:Sb SML QD laser the absolute value of α_{stat} of 6.5 ± 0.3 is larger. Interestingly, it displays the opposite sign as compared with the other devices with the modes red-shifting with increasing current. This exotic property may be very beneficial for applications, as it implies that the waveguide of index-guiding devices will improve with increasing current.

2.6 Conclusion and Perspectives

In this chapter we have shown that InAs submonolayer quantum dots represent a semiconductor nanostructure with mixed zero- and two-dimensionally confined charge carriers, thus uniting properties of excitons confined in semiconductor quantum dots and quantum wells. InAs SML QDs are formed via cycled deposition of fractions of InAs monolayers into GaAs host material forming vertically extended In-rich agglomerations. These areas act as localization centers for holes, while electrons in these structures remain only loosely bound within the area of the entire SML stack by Coulomb interaction. Such conditions are referred to as heteroconfinement and give rise to a density of states for excitons with a sharp limit at the blue side, originating from the limited capacity of localized holes to bind high-energy electrons. Optoelectronic devices based on SML QDs combine a large material gain comparable to quantum wells with a fast gain recovery as typical for quantum dots. However, the unbound electrons create a large phase response to an optical perturbation, and thus a large refractive index change. The applicability in telecommunications is therefore limited to phase-sensitive signal processing, e.g., cross phase modulation rather than linear amplification. At typically less than 1000 nm emission wavelength, however, the main application fields are light emitting devices, like VCSELs and VECSELs, possibly as mode-locking devices. Devices based on SML QDs are also exceptionally prone to chaotic dynamics upon optical feedback.

The alloying with antimony, which is incorporated in the In-rich agglomerations, pushes the emission of SML QDs from the wavelength range shorter than 1000 nm into the technologically relevant 1060 nm spectral range. Apart from a general red-shift, the emission of InAs:Sb SML QDs has a much larger bandwidth than the emission of InAs SML QDs or quantum wells. The emerging band structure of this system leads to enhanced hole confinement and therewith decreased exciton Bohr radius. The localization centers at lower energy in these InAs:Sb SML QDs display true zero-dimensional QD properties and they coexist with states at higher energy showing SML-like properties. Since Sb induces a (slight) shrinking of the In-rich agglomerations during the growth process, InAs:Sb SML QDs have an exceptionally large material gain, which could be beneficial for an application as active material for VCSELs. The broadband emission renders the usually applied chirped growth to enlarge the bandwidth unnecessary. In the wavelength window around 1060 nm broadband sources are also used for optical coherence tomography in material science and medical imaging.

Acknowledgements We thank Anna Carlsson, Thermo Fischer Scientific, for EDX measurements (Fig. 2.19). We are grateful to David Quandt and Jan-Hindrik Schulze for growing samples, Konstantin Zak, Jonas Becker, and Zeno Dimer for structural characterization, and Mirco Kolarczik and Ulrike Woggon for fruitful discussions and support with the time-resolved experiments.

References

1. Z.M. Wang (ed.), *Quantum Dot Devices* (Springer, New York, 2012)
2. D.J. Mowbray, M.S. Skolnick, New physics and devices based on self-assembled semiconductor quantum dots. J. Phys. D: Appl. Phys. **38**, 2059 (2005)
3. W.W. Chow, F. Jahnke, On the physics of semiconductor quantum dots for applications in lasers and quantum optics. Prog. Quant. Electron. **37**, 109 (2013)
4. F. Hopfer, M. Kuntz, M. Lämmlin, G. Fiol, N.N. Ledentsov, A.R. Kovsh, S.S. Mikrin, I. Kaiander, V. Haisler, A. Lochmann, A. Mutig, C. Schubert, A. Umbach, V. M. Ustinov, U.W. Pohl, D. Bimberg, Quantum dot photonics: edge emitter amplifier, and VCSEL. Proc. SPIE **7009**, 700902 (2008)
5. D. Bimberg, U.W. Pohl, Quantum dots: promises and accomplishments. Mater. Today **14**, 388 (2011)

6. Q. Xie, A. Madhukar, P. Chen, N. Kabayashi, Vertically self-organized InAs quantum box islands on GaAs(100). Phys. Rev. Lett. **75**, 2542 (1995)
7. H. Eisele, O. Flebbe, T. Kalka, C. Preinesberger, F. Heinrichsdorff, A. Krost, D. Bimberg, M. Dähne-Prietsch, Cross-sectional scanning-tunneling microscopy of stacked InAs quantum dots. Appl. Phys. Lett. **75**, 106 (1999)
8. J.H. Blokland, M. Bozkurt, J.M. Ulloa, D. Reuter, A.D. Wieck, P.M. Koenraad, P.C.M. Christianen, J.C. Maan, Ellipsoidal InAs quantum dots observed by cross-sectional scanning tunneling microscopy. Appl. Phys. Lett. **94**, 023107 (2009)
9. S. Gaan, G. He, R.M. Feenstra, J. Walker, E. Towe, Size, shape, composition, and electronic properties of InAs/GaAs quantum dots by scanning tunneling microscopy and spectroscopy. J. Appl. Phys. **108**, 114315 (2010)
10. M. Schmidbauer, M. Hanke, R. Köhler, Asymmetric correlation function describing the positional ordering of liquid-phase-epitaxy Si–Ge nanoscale islands. Phys. Rev. B **71**, 115323 (2005)
11. V.A. Shchukin, N.N. Ledentsov, P.S. Kop'ev, D. Bimberg, Spontaneous ordering of arrays of coherent strained islands. Phys. Rev. Lett. **75**, 2968 (1995)
12. R. Heitz, M. Veit, N.N. Ledentsov, A. Hoffmann, D. Bimberg, V.M. Ustinov, P.S. Alferov, Z.I. Kopev, Energy relaxation by multiphonon processes in InAs/GaAs quantum dots. Phys. Rev. B **56**, 10435 (1997)
13. T.R. Nielsen, P. Gartner, F. Jahnke, Many-body theory of carrier capture and relaxation in semiconductor quantum-dot lasers. Phys. Rev. B **69**, 235314 (2004)
14. T. Markussen, P. Kristensen, B. Tromborg, T.W. Berg, J. Mork, Influence of wetting-layer wave functions on phonon-mediated carrier capture into self-assembled quantum dots. Phys. Rev. B **74**, 195342 (2006)
15. T.W. Berg, S. Bischoff, I. Magnusdottir, J. Mork, Ultrafast gain recovery and modulation limitations in self-assembled quantum-dot devices. IEEE Photon. Technol. Lett. **13**, 541 (2001)
16. J. Gomis-Bresco, S. Dommers, V.V. Temnov, U. Woggon, M. Laemmlin, D. Bimberg, E. Malic, M. Richter, E. Schöll, A. Knorr, Impact of Coulomb scattering on the ultrafast gain recovery in InGaAs quantum dots. Phys. Rev. Lett. **101**, 256803 (2008)
17. S.S. Mikhrin, A.E. Zhukov, A.R. Kovsh, N.A. Maleev, V.M. Ustinov, Y.M. Shernyakov, I.P. Soshnikov, D.A. Livshits, I.S. Tarasov, D.A. Bedarev, B.V. Volovik, M.V. Maximov, A.F. Tsatsul'nikov, N.N. Ledentsov, P.S. Kop'ev, D. Bimberg, Z.I. Alferov, 0.94 μm diode lasers based on Stranski-Krastanow and sub-monolayer quantum dots. Semicond. Sci. Technol. **15**, 1061 (2000)
18. Z. Xu, D. Birkedal, J.M. Hvam, Z. Zhao, Y. Liu, K. Yang, A. Kanjilal, J. Sadowski, Structure and optical anisotropy of vertically correlated submonolayer InAs/GaAs quantum dots. Appl. Phys. Lett. **82**, 3859 (2003)
19. A. Lenz, H. Eisele, J. Becker, J.-H. Schulze, T.D. Germann, F. Luckert, K. Pötschke, E. Lenz, L. Ivanova, A. Strittmatter, D. Bimberg, U.W. Pohl, M. Dähne, Atomic structure and optical properties of submonolayer InAs depositions in GaAs. J. Vac. Sci. Technol., B **29**, 04D104 (2011)
20. Z. Xu, D. Birkedal, M. Juhl, J.M. Hvam, Submonolayer InGaAs/GaAs quantum dot lasers with high modal gain and zero-linewidth enhancement factor. Appl. Phys. Lett. **85**, 3259 (2004)
21. T.D. Germann, A. Strittmatter, J. Pohl, U.W. Pohl, D. Bimberg, J. Rautiainen, M. Guina O.G. Okhotnikov, High-power semiconductor disk laser based on InAs/GaAs submonolayer quantum dots. Appl. Phys. Lett. 92, 101123 (2008)
22. U.W. Pohl, D. Bimberg, *Semiconductor Disk Lasers Based on Quantum Dots*, ed by O.G. Okhotnikov (Wiley-VCH, Weinheim, 2010), pp. 187–211
23. F. Hopfer, G. Fiol, M. Kuntz, V.A. Shchukin, V.A. Haisler, T. Warming, E. Stock, S.S. Mikhrin, I.L. Krestnikov, D.A., Livshits, A.R. Kovsh, C. Bornholdt, A. Lenz, H. Eisele, M. Dähne, N.N. Ledentsov, D. Bimberg, 20 Gb/s 85 °C error-free operation of VCSELs based on submonolayer deposition of quantum dots. IEEE J. Sel. Top. Quant. Electron. **13**, 1302 (2007)
24. N.N. Ledentsov, D. Bimberg, F. Hopfer, A. Mutig, V.A. Shchukin, A.V. Savel'ev, G. Fiol, E. Stock, H. Eisele, M. Dähne, D. Gerthsen, U. Fischer, D. Litvinov, A. Rosenauer, S.S. Mikhrin,

A.R. Kovsh, N.D. Zakharov, P. Werner, Submonolayer quantum dots for high speed surface emitting lasers. Nanoscale Res. Lett. **2**, 417 (2007)

25. S. Harrison, M.P. Young, P.D. Hodgson, R.J. Young, M. Hayne, L. Danos, A. Schliwa, A. Strittmatter, A. Lenz, H. Eisele, U.W. Pohl, D. Bimberg, Heterodimensional charge-carrier confinement in stacked submonolayer InAs in GaAs. Phys. Rev. B **93**, 085302 (2016)
26. M. Hayne, B. Bansal, High-field magneto-photoluminescence of semiconductor nanostructures. Luminescence **19**, 179 (2012)
27. O. Stier, M. Grundmann, D. Bimberg, Electronic and optical properties of strained quantum dots modeled by 8-band k·p theory. Phys. Rev. B **59**, 5688 (1999)
28. D. Quandt, J.-H. Schulze, A. Schliwa, Z. Diemer, C. Prohl, A. Lenz, H. Eisele, A. Strittmatter, U.W. Pohl, M. Gschrey, S. Rodt, S. Reitzenstein, D. Bimberg, M. Lehmann, M. Weyland, Strong carrier localization in InAs/GaAs sub-monolayer stacks prepared by Sb-assisted metalorganic vapor-phase epitaxy. Phys. Rev. B **91**, 235418 (2015)
29. K. Huang, A. Rhys, Theory of light absorption and non-radiative transitions in F-centres. Proc. Royal Soc. A **204**, 406 (1950)
30. P.D. Wang, N.N. Ledentsov, C.M. Sotomayor Torres, P.S. Kop'ev, V.M. Ustinov, Optical characterization of submonolayer and monolayer InAs structures grown in a GaAs matrix on (100) and high-index surfaces. Appl. Phys. Lett. **64**, 1526 (1994)
31. J.M. Marzin, J.Y. Gerard, High quality ultrathin InAs/GaAs quantum wells grown by standard and low-temperature modulated-fluxes molecular beam epitaxy. Appl. Phys. Lett. **53**, 568 (1988)
32. O. Brandt, R. Cingolani, H. Lage, G. Scamarcio, L. Tapfer, K. Ploog, Radiative decay of excitonic states in bulklike GaAs with a periodic array of InAs lattice planes. Phys. Rev. B **42**, 11396(R) (1990)
33. O. Brandt, L. Tapfer, R. Cingolani, K. Ploog, M. Hohenstein, F. Phillipp, Structural and optical properties of (100) InAs single-monolayer quantum wells in bulklike GaAs grown by molecular-beam epitaxy. Phys. Rev. B **41**, 12599 (1990)
34. J.H. Lee, K.Y. Hsieh, R.M. Kolbas, Photoluminescence and stimulated emission from monolayer-thick pseudomorphic InAs single-quantum-well heterostructures. Phys. Rev. B **41**, 7678 (1990)
35. O. Brandt, H. Lage, K. Ploog, Excitons in InAs/GaAs submonolayer quantum wells. Phys. Rev. B **43**, 14285 (1991)
36. O. Brandt, H. Lage, G.C.L. Rocca, A. Heberle, K. Ploog, Role of broken translational invariance for the optical response of excitons in lowdimensional semiconductors. Surf. Sci. **267**, 319 (1992)
37. D. Quandt, J. Bläsing, A. Strittmatter, Analysis of InAsSb/GaAs submonolayer stacks. J. Cryst Growth **1**, 494 (2018)
38. H. Eisele, A. Lenz, R. Heitz, R. Timm, M. Dähne, Y. Temko, T. Suzuki, K. Jacobi, Change of InAs/GaAs quantum dot shape and composition during capping. J. Appl. Phys. **104**, 124301 (2008)
39. H. Eisele, P. Ebert, N. Liu, A.L. Holmes Jr., C.-K. Shih, Reverse mass transport during capping of $In_{0.5}Ga_{0.5}As$/GaAs quantum dots. Appl. Phys. Lett. **101**, 233107 (2012)
40. M. Müller, B. Gault, G.D.W. Smith, C.R.M. Grovenor, Accuracy of pulsed laser atom probe tomography for compound semiconductor analysis. J. Phys.: Conf. Ser. **326**, 12031 (2011)
41. T. Niermann, F. Kießling, M. Lehmann, J.-H. Schulze, T.D. Germann, K. Pötschke, A. Strittmatter, U.W. Pohl, Atomic structure of closely stacked InAs submonolayer depositions in GaAs. J. Appl. Phys. **112**, 083505 (2012)
42. H. Eisele, *Cross-Sectional Scanning Tunneling Microscopy of InAs/GaAs Quantum Dots* (Wissenschaft & Technik Verlag, Berlin, 2002)
43. A. Lenz, H. Eisele, J. Becker, L. Ivanova, E. Lenz, F. Luckert, K. Pötschke, A. Strittmatter, U.W. Pohl, D. Bimberg, M. Dähne, Atomic structure of buried InAs sub-monolayer depositions in GaAs. Appl. Phys. Express **3**, 105602 (2010)
44. K. Muraki, S. Fukatsu, Y. Shiraki, Surface segregation of In atoms during molecular beam epitaxy and its influence on the energy levels in InGaAs/GaAs quantum wells. Appl. Phys. Lett. **61**, 557 (1992)

45. H. Eisele, A. Lenz, C. Hennig, R. Timm, M. Ternes, M. Dähne, Atomic structure of InAs and InGaAs quantum dots determined by cross-sectional scanning tunneling microscopy. J. Cryst. Growth **248**, 322 (2003)
46. R.M. Feenstra, J.A. Stroscio, J. Tersoff, A.P. Fein, Atom-selective imaging of the GaAs(110) surface. Phys. Rev. Lett. **58**, 1192 (1987)
47. R. Timm, H. Eisele, A. Lenz, T.-Y. Kim, F. Streicher, K. Pötschke, U.W. Pohl, D. Bimberg, M. Dähne, Structure of InAs/GaAs quantum dots grown with Sb surfactant. Physica E **32**, 25 (2006)
48. V. Haxha, R. Garg, M.A. Migliorato, W. Drouzas, J.M. Ulloa, P.M. Koenraad, M.J. Steer, H.Y. Liu, M. Hopkinson, D.J. Mowbray, The use of Abell-Tersoff potentials in atomistic simulations of InGaAsSb/GaAs. Opt. Quant. Electron. **40**, 1143 (2008)
49. R. Timm, H. Eisele, A. Lenz, L. Ivanova, G. Balakrishnan, D. Huffaker, M. Dähne, Self-organized formation of GaSb/GaAs quantum rings. Phys. Rev. Lett. **101**, 256101 (2008)
50. B. Herzog, B. Lingnau, M. Kolarczik, Y. Kaptan, D. Bimberg, A. Maaßdorf, U.W. Pohl, R. Rosales, J.-H. Schulze, A. Strittmatter, M. Weyers, U. Woggon, K. Lüdge, N. Owschimikow, Strong amplitude-phase coupling in submonolayer-quantum-dots. Appl. Phys. Lett. **109**, 201102 (2016)
51. B. Lingnau, K. Lüdge, B. Herzog, M. Kolarczik, Y. Kaptan, U. Woggon, N. Owschimikow, Ultrafast gain recovery and large nonlinear optical response in submonolayer quantum dots. Phys. Rev. B **94**, 014305 (2016)
52. B. Herzog, N. Owschimikow, J.-H. Schulze, R. Rosales, Y. Kaptan, M. Kolarczik, T. Switaiski, A. Strittmatter, D. Bimberg, U.W. Pohl, U. Woggon, Fast gain and phase recovery of semiconductor optical amplifiers based on submonolayer quantum dots. Appl. Phys. Lett. **107**, 201102 (2015)
53. M. Kolarczik, N. Owschimikow, B. Herzog, F. Buchholz, Y. Kaptan, U. Woggon, Exciton dynamics probe the energy structure of a quantum dot-in-a-well system: the role of Coulomb attraction and dimensionality. Phys. Rev. B **91**, 235310 (2015)
54. A. Wilms, D. Breddermann, P. Mathé, Theory of direct capture from two- and three-dimensional reservoirs to quantum dot states. Phys. Status Solidi C **9**, 1278 (2012)
55. C.H. Henry, Theory of the linewidth of semiconductor lasers. IEEE J. Quantum Electron. **18**, 259 (1982)
56. B. Herzog, B. Lingnau, M. Kolarczik, S. Helmrich, A.W. Achtstein, K. Thommes, F. Alhussein, D. Quandt, A. Strittmatter, U.W. Pohl, O. Brox, M. Weyers, U. Woggon, K. Lüdge, N. Owschimikow, Broadband light-emitting devices based on heteroconfined excitons in InGa(Sb)As submonolayer quantum dots. IEEE J. Select. Top. Quantum Electron. **25**, 1900310 (2019)
57. B.W. Hakki, T.L. Paoli, Gain spectra in GaAs double-heterostructure injection lasers. J. Appl. Phys. **46**, 1299 (1975)
58. K. Vynck, M. Burresi, F. Riboli, D.S. Wiersma, Photon management in two-dimensional disordered media. Nat. Mater. **11**, 1017 (2012)
59. D.Z.-Y. Ting, S.V. Bandara, S.D. Gunapala, J.M. Mumolo, S.A. Keo, C.J. Hill, J.K. Liu, E.R. Blazejewski, S.B. Rafol, Y.-C. Chang, Submonolayer quantum dot infrared photodetector. Appl. Phys. Lett. **94**, 111107 (2009)
60. J.O. Kim, S. Sengupta, A.V. Barve, Y.D. Sharma, S. Adhikary, S.J. Lee, S.K. Noh, M.S. Allen, J.W. Allen, S. Chakrabarti, S. Krishna, Multi-stack InAs/InGaAs sub-monolayer quantum dots infrared photodetectors. Appl. Phys. Lett. **102**, 011131 (2013)
61. H. Ghadi, S. Sengupta, S. Shetty, A. Manohar, A. Balgarkashi, S. Chakrabarti, N.B. Pendyala, S.L. Prajapati, A. Kumar, Comparison of three design architectures for quantum dot infrared photodetectors: InGaAs-capped dots; dots-in-a-well; and submonolayer quantum dots. IEEE Transact. Nanotechnol. **14**, 603 (2015)
62. C.G.E. Alfieri, D. Waldburger, J. Nürnberg, M. Golling, L. Jaurigue, K. Lüdge, U. Keller, Mode-locking instabilities for high-gain semiconductor disk lasers based on active submonolayer quantum dots. Phys. Rev. Appl. **92**, 044015 (2018)

63. B. Lingnau, B. Herzog, M. Kolarczik, U. Woggon, K. Lüdge, N. Owschimikow, Dynamic phase response and amplitude-phase coupling of self-assembled semiconductor quantum dots. Appl. Phys. Lett. **110**, 241102 (2017)
64. M. Osinski, J. Buus, Linewidth broadening factor in semiconductor lasers—an overview. IEEE J. Quant. Electron. **23**, 9 (1987)
65. I.D. Henning, J.V. Collins, Measurements of the semiconductor linewidth broadening factor. Electron. Lett. **19**, 927 (1983)

Chapter 3
Stressor-Induced Site Control of Quantum Dots for Single-Photon Sources

U. W. Pohl, A. Strittmatter, A. Schliwa, M. Lehmann, T. Niermann, T. Heindel, S. Reitzenstein, M. Kantner, U. Bandelow, T. Koprucki and H.-J. Wünsche

Abstract The strain field of selectively oxidized AlO_x current apertures in an AlGaAs/GaAs mesa is utilized to define the nucleation site of InGaAs/GaAs quantum dots. A design is developed that allows for the self-aligned growth of single quantum dots in the center of a circular mesa. Measurements of the strain tensor applying transmission-electron holography yield excellent agreement with the calculated strain field. Single-dot spectroscopy of site-controlled dots proves narrow excitonic linewidth virtually free of spectral diffusion due to quantum-dot growth in a defect-free matrix. Implementation of such dots in an electrically driven *pin* structure yields single-dot electroluminescence. Single-photon emission with excellent purity is proved for this device using a Hanbury Brown and Twiss setup. The injection efficiency of the initial *pin* design is affected by a substantial lateral current spreading close to the oxide aperture. Applying 3D carrier-transport simulation a *ppn* doping profile is developed achieving a substantial improvement of the current injection.

3.1 Site-Controlled Nucleation of Quantum Dots

The introduction of quantum structures into semiconductor physics and technology has generated novel electronic and optoelectronic devices with unique properties. The electronic density of states in such nanometer-sized structures gets discrete, allowing customization of desired properties [1]. Quantum dots (QDs) represent

U. W. Pohl (✉) · A. Schliwa · T. Heindel · S. Reitzenstein
Institute of Solid State Physics, Technische Universität Berlin, Berlin, Germany
e-mail: pohl@physik.tu-berlin.de

A. Strittmatter
Institute of Physics, Otto-von Guericke Universität, Magdeburg, Germany

M. Lehmann · T. Niermann
Institute of Optics and Atomic Physics, Technische Universität Berlin, Berlin, Germany

M. Kantner · U. Bandelow · T. Koprucki · H.-J. Wünsche
Weierstraß-Institut für Angewandte Analysis und Stochastik, Berlin, Germany

M. Kneissl et al. (eds.), *Semiconductor Nanophotonics*, Springer Series in Solid-State Sciences 194, https://doi.org/10.1007/978-3-030-35656-9_3

the limit of quantization with carrier confinement in all spatial directions, and their implementation into device structures led to devices with remarkable performance [2].

The fabrication of nanoscale QDs makes high technological demands. While dot structures for transport applications can be realized by a planar lithographic process defining potential wells in a two-dimensional electron gas [3], the fabrication of QDs for optical applications by etching a two-dimensional quantum well produces defective structures of poor quality [4]. Today QDs of excellent performance are made by a self-assembled process employing the fundamental Stranski-Krastanow growth mode [5, 6] of lattice-mismatched semiconductors. In this approach, a thin strained two-dimensional layer spontaneously forms three-dimensional islands to lower the elastic strain energy; QDs are created when these islands are embedded into a higher-bandgap semiconductor.

Recent developments in quantum-information processing and quantum cryptography created needs for on-demand sources of single photons. A single quantum dot in a diode structure is an appropriate realization of this requirement. Since Stranski-Krastanow growth leads to random nucleation sites of the islands, means to accurately align the QD with respect to the surrounding device structure are needed. A successful approach locates the device to the random position of a QD [7, 8], see also Chap. 8. A more widely applied and upscalable technique defines the location of the QD nucleation-site by patterning the substrate with nanometer-sized pits; at these sites the critical thickness for island formation is reduced and single QDs can be generated in a perfectly ordered array [9, 10]. A drawback of this technique is the vicinity of the QD to an etched surface, leading to spectral diffusion induced by random charging of nearby defects [11, 12]. Another restraint is the limited maximum distance between nucleation sites, above which additional parasitic nucleation occurs.

The novel approach reported in this chapter employs the strain of a stressor to define favorable sites for QD nucleation. The effect of strain on QD nucleation is well known from the mutual dot-dot correlation in QD ensembles [13]. All forms of vertical and lateral correlations in nucleation and growth of self-assembled QDs are related to the elastic properties of the matrix material [14]; the interaction is particularly pronounced for a large elastic anisotropy of the matrix and was nicely demonstrated in vertically anticorrelated multilayer QD growth of CdSe/ZnSe [15], InAs/GaAs [16], and IV–VI [17] semiconductors.

The device structure for an electrically operated single-photon source based on a single quantum dot resembles the structure of a vertical-cavity surface-emitting laser (VCSEL), as illustrated later in Fig. 3.22a. The active region is embedded in an optical cavity, and the emitted light passes a ring contact on the top surface. A VCSEL generally employs a non-conducting aperture to confine the vertical current flow to the active region; in AlGaAs-based VCSELs this aperture consists of aluminum oxide, which introduces a large amount of strain into the device structure [18]. The basic concept of our novel approach is to employ this strain to control the site for the most favorable nucleation of Stranski-Krastanow islands. A successful implementation of this concept combines the correct location of the QD in the device structure with an efficient electrical injection due to the effect of the current confinement via the oxide aperture.

Application of the concept requires a slight modification of the VCSEL scheme, where two oxide apertures cladding the active region or a single aperture on top of the active region are used. In the projected single-photon source the aperture must be located *below* the QD layer to induce the strain profile required for the site-controlled QD growth. A simulation of the strain in such a device structure clarifies whether the strain induced by the aperture is sufficiently strong at the location of the QD layer to control the nucleation of Stranski-Krastanow islands.

3.2 Simulation of Strain

Strain has a strong impact on the confinement of carriers in a quantum dot with a magnitude comparable to that of the band offset between QD and matrix material [19]. The approaches of continuum elasticity [20] and valence force-field [21] are hence widely applied to calculate the strain in mismatched epitaxial structures and the effect on electronic properties.

3.2.1 Model for Strain Simulation

We apply the continuum-elasticity model assuming a linear relation between stress σ and strain ε according to Hooke's law

$$\sigma_{ij} = \sum_{kl} C_{ijkl}\,\varepsilon_{kl}, \tag{3.1}$$

where $i, \ldots, l = 1, 2, 3$. The elastic strain energy becomes minimum in equilibrium; this yields the condition for the components i of the stress tensor

$$\frac{\partial \sigma_{i1}}{\partial x_1} + \frac{\partial \sigma_{i2}}{\partial x_2} + \frac{\partial \sigma_{i3}}{\partial x_3} = 0. \tag{3.2}$$

For strain simulation we apply the finite difference method to minimize the elastic strain energy under the constraint of (3.2); in the computation we use the iterative conjugated-gradient approximation [22]. At interfaces, proper boundary conditions are obtained from the condition of a continuous stress tensor.

The simulated strain-inducing aperture structure is described in Sect. 3.3.1. It basically consists of a circular GaAs mesa with a buried AlAs layer, the outer ring of which is oxidized to an AlO_x aperture; this buried layer is clad by $Al_{0.90}Ga_{0.10}As$ and graded $Al_xGa_{1-x}As$ layers. Such a layer sequence is simulated by a central cylindrical 41 nm thick AlAs layer of 430 nm diameter with a surrounding AlO_x ring

Table 3.1 Materials parameters used in the strain simulation

Parameter	GaAs	AlAs	AlO_x
a_0 (Å)	5.6503	5.66	5.26153
C_{11} (GPa)	118.8	120.2	33.37498
C_{12} (GPa)	53.8	57.0	12.34927
C_{44} (GPa)	59.4	58.9	10.73786

of equal thickness, both symmetrically clad by 47 nm thick $Al_{0.80}Ga_{0.20}As$ layers; the experimentally implemented graded interface between these layers is neglected, since results are not significantly affected by thickness variations within ~20%. For the measurement of the strain components pointed out in Sect. 3.4, a thin lamella was cut out from the mesa to obtain transparency for electron microscopy; since the cut-out introduces some relaxation of strain, we also simulate the strain in this lamella assuming a lamella thickness of 240 nm.

Materials parameters for GaAs and AlAs are well known [23] and listed in Table 3.1. The elastic constants for the amorphous AlO_x, however, are scarce; we use quantities reported from an evaluation of indentation measurements assumed to be most reliable [24]. The amorphous AlO_x layer is simulated by an *isotropic* cubic material; this was performed by putting (in Voigt notation) $C_{44} = \frac{1}{2}(C_{11} - C_{12})$.

3.2.2 Strain in a Mesa and in a Lamella Structure

Results of the strain simulation are given in Fig. 3.1. Tensile and compressive strain are marked by red and blue colors, respectively. Strain maxima appear near the edges of the AlO_x aperture, which is drawn in white. The strain maxima of the lateral strain component ε_{xx} merge to a single tensile maximum at larger distances above the aperture center. The other lateral component ε_{yy} does hardly show such merging; still the strain-induced relative change of the area of the surface unit cell, expressed by the biaxial strain $\left(\varepsilon_{xx} + \varepsilon_{yy}\right)$ and shown in Fig. 3.2, does exhibit a pronounced tensile maximum at larger distances above the aperture.

The maximum of tensile strain on the GaAs mesa surface is important for defining a favorable nucleation site of single InGaAs QDs above the AlO_x aperture. Compressively strained QDs like InGaAs/GaAs QDs will preferentially nucleate at surface sites with enlarged surface unit-cells, i.e., tensely strained GaAs. These sites are above the aperture *edges* for *thin* strained GaAs layers on top of the aperture, and above the aperture *center* for *thicker* GaAs layers. The same applies respectively for large and small aperture diameters at fixed GaAs layer thickness.

The results for the lamella shown in Fig. 3.1 (right column) show a largely similar strain as found for the entire mesa, with a generally smaller strain magnitude. An apparent qualitative difference is found in the lateral component ε_{yy}; a tensile strain is calculated in the aperture center of the lamella, while a compressive strain results

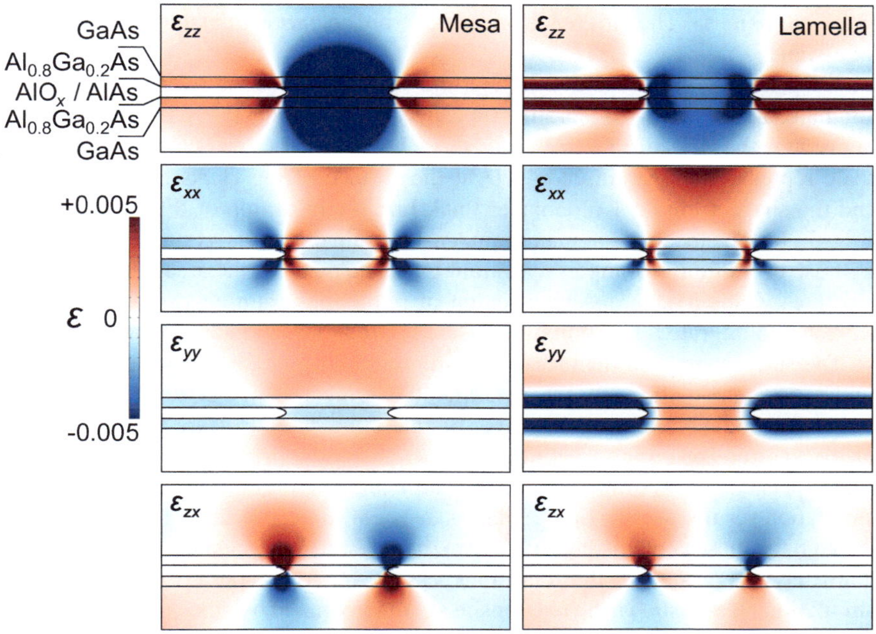

Fig. 3.1 Strain distribution (cross-section) simulated for a GaAs circular mesa (**left column**) and a lamella cut-out from the mesa center (**right column**), both strained by a buried oxide aperture. *White areas* represent the AlO_x aperture formed from the selectively oxidized part of an AlAs layer. Adapted with permission from [25], © 2015 American Physical Society

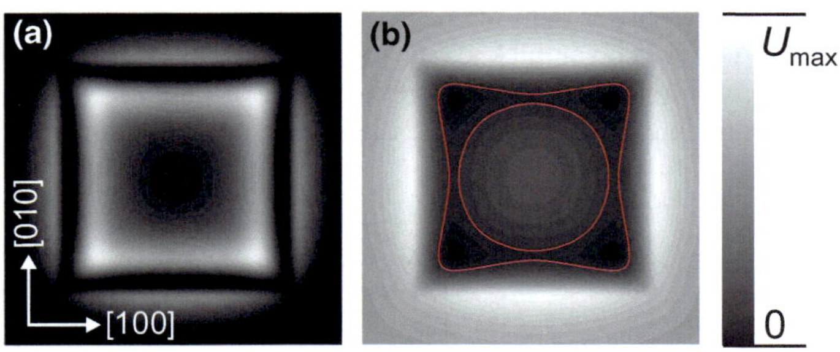

Fig. 3.2 Strain-energy distribution (top view) simulated for a GaAs mesa strained by a buried quadratic oxide aperture. **a** Strain energy within GaAs at the surface of the mesa indicating high tensile strain at the aperture edges with maxima at the corners, **b** strain energy in a thin InAs layer on top of the strained GaAs mesa, showing minimum strain energy at locations of maximum strain energy of the GaAs underneath. *Red lines* delimitate the area of minimum InAs strain energy where island nucleation is expected. Adapted with permission from [26], © 2012 John Wiley and Sons

in the entire mesa. The difference originates from an elastic relaxation of the lamella side-walls.

A top view on the area above the aperture of an entire mesa is given in Fig. 3.2. The simulated aperture has a quadratic shape, because the processing parameters applied to produce the AlO_x from AlAs leads to such a geometry, see Sect. 3.3.1. The gray scale represents the strain energy U at the surface of the GaAs mesa (Fig. 3.2a) and the strain within a 1 monolayer thick InAs layer on top of the strained GaAs (Fig. 3.2b). The locations of maximum tensile strain on the GaAs surface correspond to those of minimum compressive strain in the InAs layer; the area bounded by the red lines yields a square-shaped region of minimum strain energy where InAs is expected to migrate for forming InAs islands, with maximum density at the corners.

The surface strain of the mesa structure was simulated for various parameters to find optimum conditions for the experimental verification of a controlled QD nucleation. The thickness of the GaAs layer on top of the AlO_x aperture and the diameter of the aperture opening turned out to be of particular importance for achieving a single-peaked tensile strain maximum of sufficient magnitude [26]. Figure 3.3 shows simulation results for a varied thickness d_{GaAs} of the GaAs layer for a vertical cut through the mesa center. A mesa of 3 μm diameter and a circular aperture of 0.5 μm diameter were assumed; the total thickness of the AlO_x aperture is 130 nm. The simulation clearly shows the two tensile strain maxima above the aperture edges for thin GaAs layers (30 and 50 nm) evolving to a single maximum for thicker layers. The calculation yields an optimum thickness on the order of 100 nm and about 2% enlargement of the surface unit cell for the given parameters; such a magnitude is very promising for defining a favorable nucleation position of InGaAs islands on top of such a layer.

Another critical parameter is the diameter of the aperture. Results of the respective variation for a fixed GaAs layer thickness on top of the aperture and otherwise similar parameters are given in Fig. 3.4. They resemble those of Fig. 3.3, with a thin GaAs thickness corresponding to a large aperture diameter. The simulation indicates that a

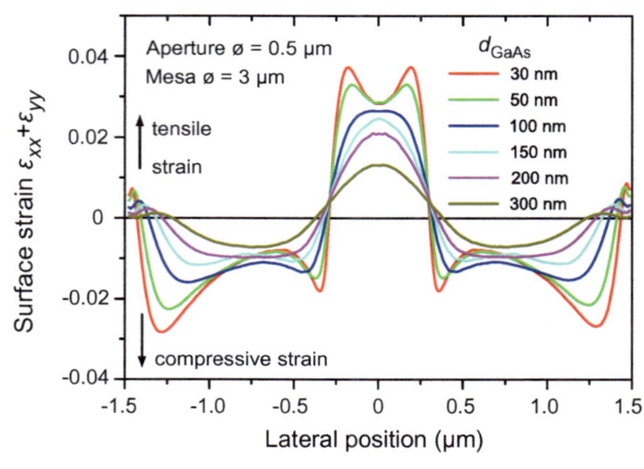

Fig. 3.3 Simulated strain $(\varepsilon_{xx} + \varepsilon_{yy})$ at the mesa surface depending on the thickness d_{GaAs} of the GaAs layer on top of the AlO_x aperture. For structural parameters given in the figure the tensile strain merges to a single maximum above the aperture center for 100 nm surface—aperture separation. Adapted with permission from [26], © 2012 John Wiley and Sons

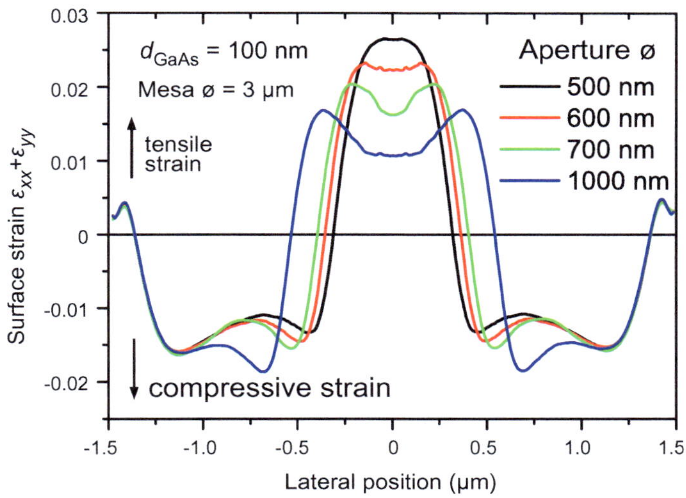

Fig. 3.4 Surface strain across a circular mesa of 3 µm diameter with various aperture diameters. Adapted with permission from [26], © 2012 John Wiley and Sons

small aperture is important for obtaining a single tensile strain maximum above the aperture center.

Figures 3.3 and 3.4 both show a further maximum of tensile strain near the outer edge of the mesa, i.e., at large absolute lateral position. This feature originates from strain relaxation of the mesa side-walls and is also experimentally observed in the nucleation density of InGaAs islands. In a device this parasitic nucleation is not detrimental, because such QDs are not excited for proper current injection (Sect. 3.6); furthermore, these QDs are covered by the ring-shaped top-contact.

Further parameter variations involve the mesa diameter, the aperture thickness, and the shape of the aperture. These parameters prove to be of minor significance for the strain distribution on the surface near the aperture. The effect of a varied lateral shape of the aperture is shown in Fig. 3.5. The simulation yields only a small influence, basically affecting the maximum and minimum magnitude of the strain.

3.3 Nucleation Control by a Buried Aperture Stressor

The simulation results reported above demonstrated that a buried aperture introduces significant strain into a GaAs top layer. Oxide apertures are produced in $Al_xGa_{1-x}As$-based VCSELs by a selective oxidation of layers with a high Al composition [27]. The oxidation process of an AlAs layer performed in a wet ambient is described by the reaction

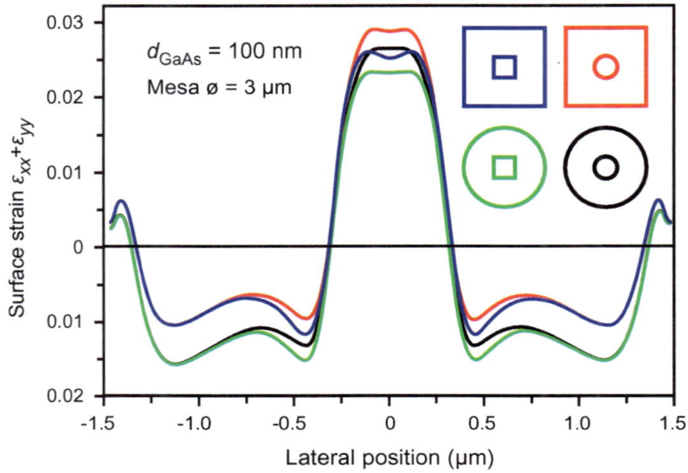

Fig. 3.5 Surface strain for various lateral shapes of the mesa and the apertures as indicated in the insets; their outer forms indicate mesa shapes, inner shapes represent apertures

$$2\,AlAs + 6\,H_2O(g) \rightarrow Al_2O_3 + As_2O_3 + 6\,H_2; \tag{3.3}$$

the reaction is exothermic, yielding at 420 °C a change of Gibbs free energy by $\Delta G = -473$ kJ/mol. Replacing AlAs with GaAs produces an *endothermic* reaction with $\Delta G = +10$ kJ/mol; the process is hence highly selective, and the oxidation rate can be adjusted by the Ga composition in high-Al content layers. The intermediate product As_2O_3 is reduced to As and further to volatile AsH_3 in presence of hydrogen, leaving a stable and largely amorphous AlO_x layer with defect-free interfaces to unoxidized GaAs.

The oxidation is accompanied by a substantial shrinkage of the initial $Al_xGa_{1-x}As$ layer, with a linear magnitude of ~13% for AlAs and ~7% for $Al_{0.92}Ga_{0.08}As$. This was found to introduce a strong compressive stress in the GaAs layers (on the order of mid 10^7 Pa) of a GaAs/AlGaAs distributed Bragg reflector of a VCSEL [18]. The simulation of a single buried AlAs/AlO_x aperture layer (Sect. 3.2.2) yields a tensile strain in GaAs of sufficient magnitude to favor a controlled InGaAs/GaAs island nucleation. The experimental feasibility of such a nucleation was therefore studied using appropriate test structures.

3.3.1 Development of a Buried-Stressor Design

Since the oxidation process introduces a large stress acting particularly at the GaAs/AlO_x interface, a study of the stability of the GaAs layer on top of the aperture is an important issue. To reduce this stress linearly graded 10 nm thick $Al_xGa_{1-x}As$

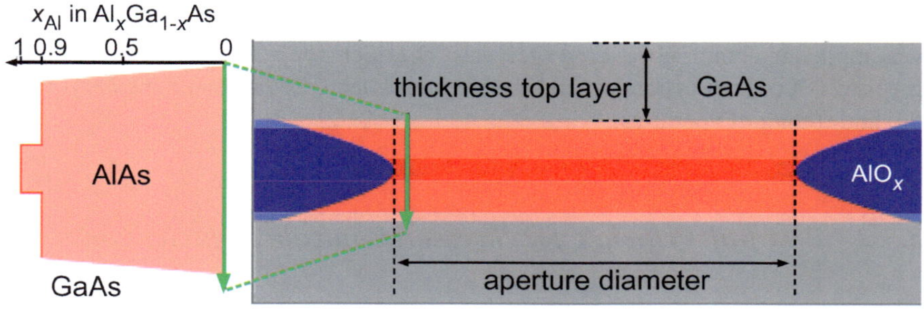

Fig. 3.6 Cross section of the aperture with concentration profile of the Al composition in the $Al_xGa_{1-x}As$ layers and $Al_{0.90}Ga_{0.10}As$ layers cladding the AlAs layer, which is subjected to selective oxidation

layers with x ranging between 0.1 and 0.9 and additional $Al_{0.90}Ga_{0.10}As$ layers with a total thickness of (10 + 40) nm were inserted between AlAs and GaAs, see Fig. 3.6. Applying these precautions a peeling off of the GaAs top layer could reliably be avoided.

Parameters for the selective wet oxidation of AlAs were studied using large apertures. High processing temperatures and low H_2O partial pressure generally favor an isotropic, diffusion-limited oxidation, while low temperature and higher H_2O partial pressure lead to an anisotropic, reaction-limited process [27]. Results for two different parameter sets given in Fig. 3.7 clearly illustrate the effect of the limiting processes. The micrographs show an optical top view on the circular mesa with bright outer areas of oxides and darker central areas of unoxidized AlAs. The mesa processed at 500 °C shown in Fig. 3.7a has a circular aperture formed by an isotropic process, while a process performed at lower temperature and higher H_2O partial pressure yields a faster oxidation along ⟨100⟩ directions with respect to ⟨110⟩ directions and hence a more quadratic aperture.

GaAs structures for stressor-induced InGaAs island nucleation comprised 300 nm high mesa with diameters between 22.0 and 24.8 μm (200 nm steps), defined by standard photolithography and dry etching. For nucleation studies and device fabrication a low oxidation temperature of 420 °C with flows of 50 ml/min water vapor and

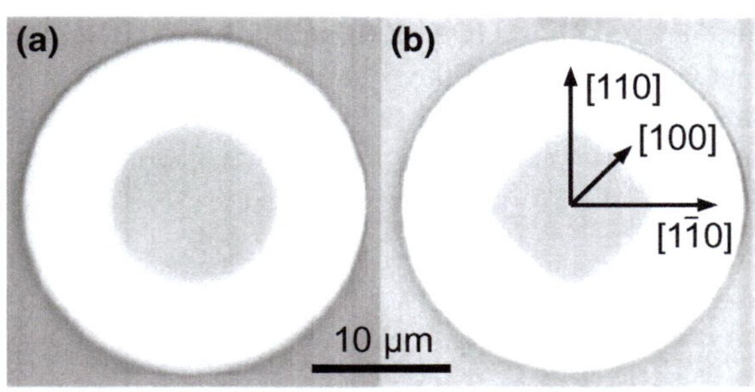

Fig. 3.7 Optical micrographs (top view) of buried $Al_{0.98}Ga_{0.02}As$ layers after **a** isotropic oxidation at 500 °C and **b** anisotropic oxidation at 420 °C

3 slpm nitrogen carrier gas were used. In situ optical control of the oxidation front was applied to obtain sub-micron apertures. The microscopy image allows for a 1 μm precision of oxidation control, and the step-wise decreased mesa diameters provide the fine-tuning of apertures.

3.3.2 Proof-of-Principle for Stressor-Controlled Nucleation

Using mesa fabricated by the process described above, a single ~2 monolayers thick $In_{0.65}Ga_{0.35}As$ layer was deposited in a second epitaxial growth on top of the upper strained GaAs mesa layer for forming site-controlled islands. This second epitaxy is initiated by a thermal surface-oxide desorption at 720 °C, followed by growth of 50 nm homoepitaxial GaAs grown at 685 °C; the InGaAs QD layer and the GaAs cap layer are subsequently grown at 485 °C.

The thickness of the InGaAs layer is carefully calibrated on planar and reference mesa samples. The layer thickness is adjusted close to the 2D-3D Stranski-Krastanow transition, yielding an island density near or below 10^8 cm^{-2} on planar samples. This requirement proved important to obtain a good selectivity between islands nucleated above the aperture and parasitic nucleation in surrounding areas. An average island density of mid 10^9 cm^{-2} maintains the nucleation-site control, but yields additional homogeneous nucleation outside the aperture region (in Fig. 3.4 the area beyond ±0.5 μm) [26]. Such additional nucleation is largely suppressed at low 10^8 cm^{-2} average island density, translating to the requirement of a thickness control of InGaAs deposition better than 0.1 monolayers.

A proof of principle for nucleation-site control is provided by atomic force micrographs taken from InGaAs islands on strained GaAs mesa covered by a 6 nm thin GaAs cap. An experimental result for the site control of island nucleation is shown in Fig. 3.8. The oxide aperture in this mesa structure is buried 170 nm beneath the GaAs surface. On this strained surface a 2.4 monolayers thick $In_{0.65}Ga_{0.35}As$ layer was grown, yielding the displayed distribution of islands. The comparison with the simulated strain shown in Fig. 3.8b demonstrates the close correlation of surface strain and nucleation sites. The locations of islands reproduce many details of the calculated strain distribution as indicated by the arrows. The tensile strain maximum above the aperture edge induces the highest density of islands, while no islands nucleate in the compressive strain minimum. We also note the absolute density maxima above the corners of the square aperture, in agreement with the simulated absolute tensile strain maxima at the GaAs surface shown in Fig. 3.2a. At these corners the first islands appear, when a very low amount of InGaAs is deposited [26].

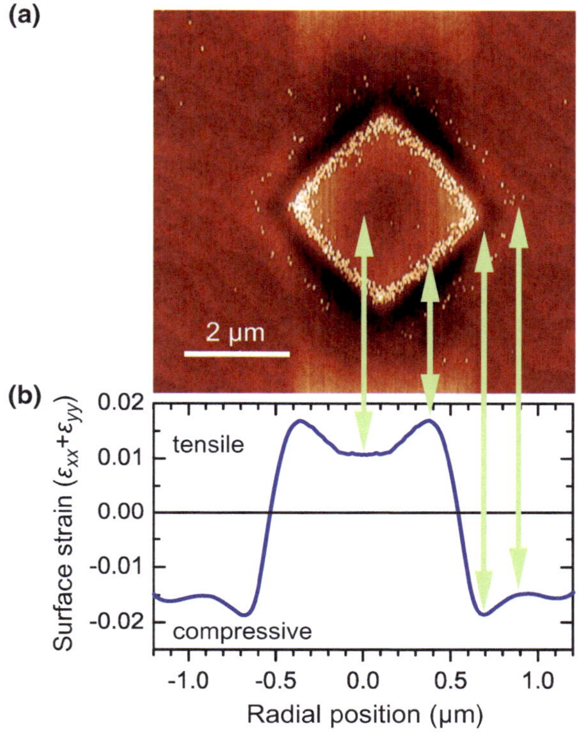

Fig. 3.8 **a** Atomic-force micrograph of InGaAs Stranski-Krastanow islands (bright spots) nucleated on a strained GaAs surface with a buried aperture underneath. **b** Simulated strain ($\varepsilon_{xx} + \varepsilon_{yy}$) on the GaAs surface along the direction through the center of the aperture. Double arrows indicate areas of corresponding strain in panels (**a**) and (**b**). Adapted with permission from [25], © 2015 American Physical Society

3.3.3 *Site-Control of Single Quantum Dots*

Luminescence studies proved that the site-controlled islands presented above form quantum dots of excellent optical quality after deposition of a GaAs cap layer [26, 28, 29]. The strain simulations shown in Fig. 3.4 predict a single tensile strain maximum for small apertures with an opening well below 1 μm. Decreasing the aperture size will hence focus the optimum nucleation site above the aperture center. The experimental verification of this trend is depicted in Fig. 3.9. The diameter of the aperture was gradually decreased by decreasing the mesa diameter from 14.7 μm to 14.0 μm and 13.2 μm. At the large aperture (Fig. 3.9a) islands nucleate above the square-shaped aperture boundaries with a density of 10^{10} cm^{-2}, with some residual nucleation outside on the order of 10^8 cm^{-2}. Above the submicron aperture a cluster of islands nucleates above the center of the aperture as shown in Fig. 3.9b. Two single islands are observed above the further reduced aperture diameter as displayed in Fig. 3.9c.

In the next step, single Stransi-Krastanow islands like those shown in Fig. 3.9c are covered with GaAs matrix material to form quantum dots. The optical properties of such QDs were investigated using micro-photoluminescence. The left inset in

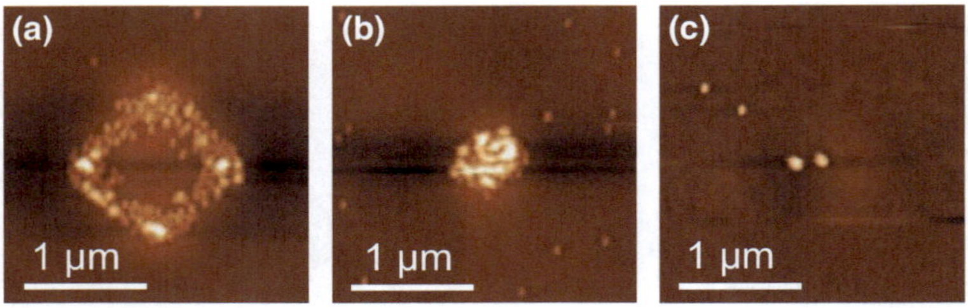

Fig. 3.9 Atomic-force micrographs of InGaAs islands nucleated above small apertures. **a** aperture diameter 1 μm, **b** submicron aperture, **c** aperture diameter several 100 nm. Adapted with permission from [26], © 2012 John Wiley and Sons

Fig. 3.10 depicts the low-temperature luminescence of an entire mesa with a submicron aperture. We observe a bright luminescence at the outer edge of the mesa, originating from a high QD density due to the outer tensile strain maximum visible in Figs. 3.3 and 3.4. The bright single QD in the center of the mesa is measured in the main panel of Fig. 3.10 and the right inset. The excitonic emission is a sharp line close to the resolution limit of the setup; for resonant excitation of similar QDs linewidths below 10 μeV are observed [29]. The narrow linewidth is an indication for the high quality of such site-controlled QDs.

Site-controlled QDs often suffer from line broadening due to a vicinity to processed interfaces [9, 12]. Defects introduced by processing lead to carrier traps, which produce fluctuating electrical fields at the location of a QD due to random charging and decharging [11]. This in turn induces random shifts of the transition energy on a short time scale by a quantum-confined Stark effect and thereby to line broadening. A series of luminescence spectra of our site-controlled QDs, each recorded for 100 ms, did not give an indication for spectral diffusion or blinking. We attribute this

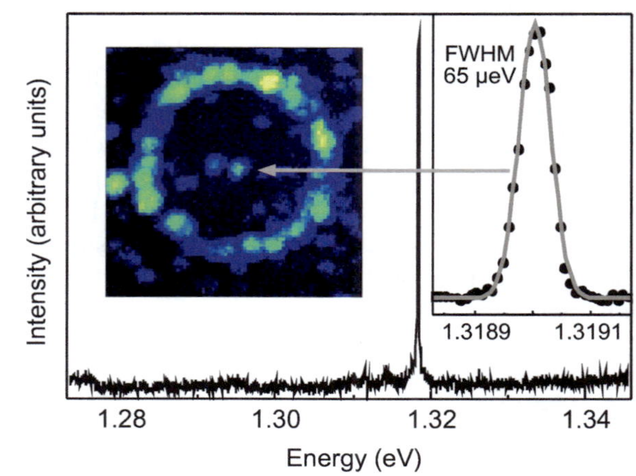

Fig. 3.10 Micro-photoluminescence of site-controlled quantum dots nucleated on a mesa with a buried sub-micron aperture, recorded at $T = 10$ K. The left inset shows a spatially resolved image with color-coded intensities, the right inset is a local spectrum of the central quantum dot; from [30]

feature to a large distance of over 100 nm to the aperture and a basically defect-free unprocessed environment of the quantum dot.

3.4 Strain Measurement Applying Electron Holography

The strain fields calculated in Sect. 3.2 depend on the validity of the applied material parameters and the correctness of the geometric model. Especially the edges of the AlO_x aperture in the buried stressor delineated in Fig. 3.6 are locations with a high density of deformation energy; they have a strong influence on the overall strain field and must hence be modeled correctly. Furthermore, the elastic constants describing the AlO_x material are not well known.

Electron holography in the transmission electron microscope provides a means to measure components of the strain tensor in a specimen. Thus the validity of the strain calculation and the growth as well as the oxidation process can be proved.

3.4.1 Reconstruction of the Strain Tensor

The components of the strain tensor are connected to the displacement field $\vec{u}$ via

$$\varepsilon_{ij} = \frac{1}{2}\left(\frac{\partial u_i}{\partial x_j} + \frac{\partial u_j}{\partial x_i}\right). \tag{3.4}$$

Transmission-electron microscopy is an imaging technique operating in transmission; the components of the displacement field perpendicular to the electron's direction can hence principally be measured by lattice imaging, using the crystal lattice as a natural reference of the displacement field. However, the displacement components must be measured with picometer precision, while on the other hand the entire field must be measured with a field of view ranging several hundred nanometers. High-resolution imaging as measurement technique can therefore not be applied; instead, the measurement requires a combination of diffraction (sufficient sensitive to small strain but non-local) and imaging techniques (local but not sufficient precise to small displacements). Dark-field electron holography provides such a combination of techniques.

According to Laue's law the scattered wave for a crystalline material is a superposition of plane waves (beams) with diffraction vectors corresponding to the material's reciprocal lattice vectors $\vec{g}$. Thus, the full exit wavefunction $\psi(\vec{r})$ on the back side of the specimen can be conveniently written as a Fourier series with complex amplitudes $\psi_{\vec{g}}$ [31]:

$$\psi(\vec{r}) = \sum_{\vec{g}} \psi_{\vec{g}} \exp\left(i\, 2\pi \left(\vec{g} + \vec{k}\right) \cdot \vec{r}\right), \tag{3.5}$$

where $\vec{k}$ is the wave vector of the incident wave. For a lateral displacement field $\vec{u}(\vec{r})$ with $\vec{k} \cdot \vec{u} = 0$, which varies on length scales larger than the lattice spacing, it can be assumed that the specimen's potential and thus the exit wave is shifting rigidly. These displacements can be described by envelope functions of the Fourier coefficients:

$$\psi'(\vec{r}) = \psi(\vec{r} + \vec{u}(\vec{r})) = \sum_{\vec{g}} \psi'_{\vec{g}}(\vec{r}) \exp\left(i\, 2\pi \left(\vec{g} + \vec{k}\right) \cdot \vec{r}\right), \tag{3.6}$$

with

$$\psi'_{\vec{g}}(\vec{r}) = \psi_{\vec{g}} \exp(i\, 2\pi \vec{g} \cdot \vec{u}(\vec{r})). \tag{3.7}$$

In comparison to the diffracted beams $\psi_{\vec{g}}$ of an unstrained area, the diffracted beams $\psi'_{\vec{g}}(\vec{r})$ of strained areas have the displacement field encoded in an additional position-dependent phase modulation. This phase modulation originating from the displacement field is called the *geometric phase* [32]. Due to the typically small angles involved in electron scattering, we can furthermore assume that the information obtained from a point of the back side of the specimen originates only from a very small atom column of a few nanometer radius around this point in the specimen (*column approximation*).

So far, we assumed a constant displacement field in beam direction. When the variations of the lateral components of the displacement field in beam direction (defined here in TEM geometry as y direction) are small, i.e. $\partial u/\partial y \ll 1$, the complex amplitude of the diffracted beam in two-beam conditions can be approximated as weighted integral over the displacement field [33]:

$$\psi'_{\vec{g}}(x, z) = \psi_{\vec{g}} \exp\left(i\, 2\pi \int_0^t f_{\vec{g},t}(y) \vec{g} \cdot \vec{u}(x, y, z) \mathrm{d}y \right), \tag{3.8}$$

where t is the overall specimen thickness. The weighting function $f_{\vec{g},t}(y)$ depends on the extinction length of the reflection and on the thickness.

The weighting functions for $\{400\}$ and $\{202\}$ reflections of a 275 nm thick GaAs specimen are depicted in Fig. 3.11. The weighting function for $\{400\}$ reflections is close to zero for ~10 nm at the top and at the bottom of the lamella. Thus, geometric phases measured from $\{400\}$ reflections are less influenced by surface relaxations. This is in contrast to $\{202\}$ reflections, which are more strongly influenced by surface effects for this thickness.

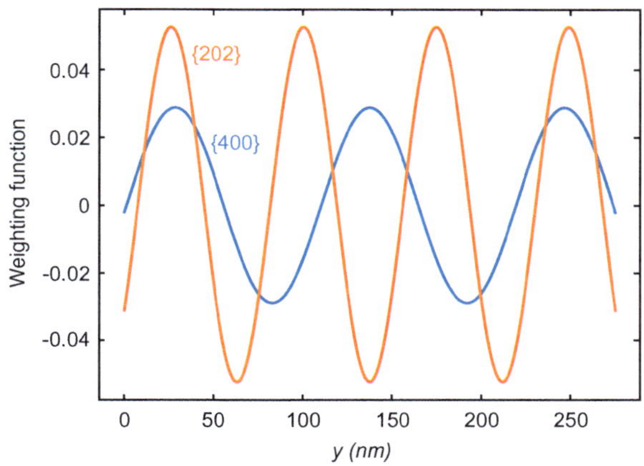

Fig. 3.11 Weighting function for {400} and {202} reflections across a 275 nm thick lamella of GaAs

3.4.2 Phase Analysis of Dark-Field Electron Holography

Dark-field electron holography (DFEH) allows spatially resolved measurements of the diffracted beam phases [34]. Figure 3.12 shows the setup used for these measurements: The objective lens maps the specimen plane to the intermediate image plane. Since the specimen is crystalline, the diffracted beams occur only in discrete directions according to Laue's law. Independent of where they emanate from the specimen, all diffracted rays belonging to the same diffracted beam, i.e., having the same Bragg angle, will intersect the corresponding diffraction spot in the back focal plane of the objective lens. Here the objective aperture selects a single diffracted beam; the phase of this beam is measured. An electron-optical biprism located between back focal plane and image plane interferes beams from two areas of the specimen.

Typically, the specimen is oriented with respect to the setup such that one of the two overlapping areas contains only unstrained homogeneous material (zero geometric phase). This area serves as a reference. The other area contains the specimen region, where the strain is measured. This allows obtaining directly the geometric phase from the phase modulation of the interference pattern.

In the image plane a cosine-like interference pattern, the hologram, is observed (see Fig. 3.13). The period of this cosine-like pattern increases with increasing angle of overlap between the beams, which are deflected by the biprism. For a specimen area with tensile strain the diffracted beam is scattered under a smaller angle and results in a larger overlap angle in the hologram (see dotted lines in Fig. 3.12). The phase modulation of the interference pattern directly corresponds to the phase modulation of the beam as given by (3.7). When like in (3.7) the geometric phase is proportional to the displacement field, the strain is readily calculated from the derivatives of the phase.

It must be noted that also other influences than the geometric phase may affect the measured phase. The geometric phase introduced in (3.7) is a modulation of the complex amplitude $\psi_{\vec{g}}$ of the beam emanating from unstrained material. However,

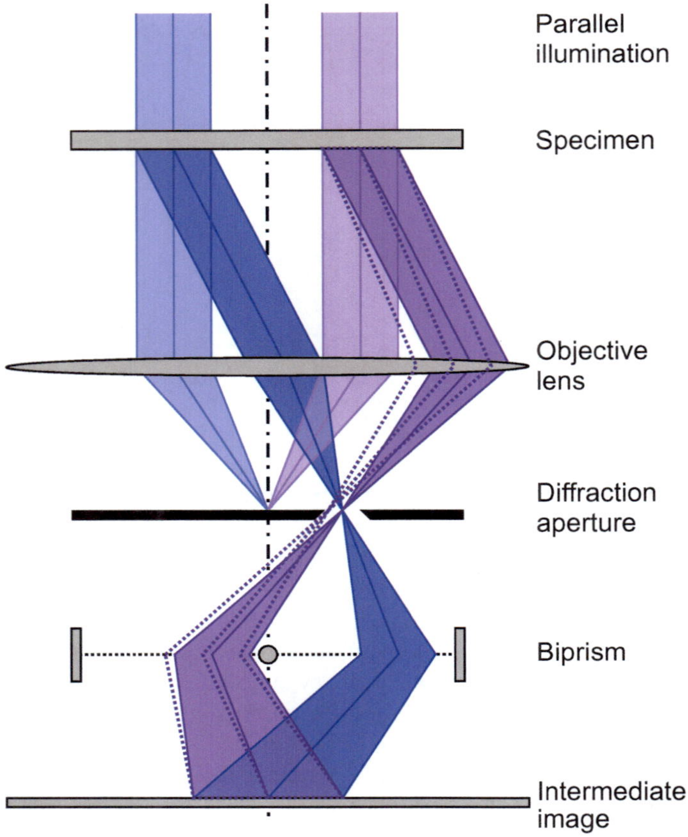

Fig. 3.12 Ray path of the dark-field electron holography technique. Exemplarily the rays of the two unstrained specimen areas, which are interfered in the image plane, are shown in different *colors*. The *dotted lines* on the right-hand side show the rays diffracted from tensely strained material

this complex amplitude $\psi_{\vec{g}}$ also depends on the material itself and on the specimen thickness. Changes in the material like a transition from AlAs to GaAs as found in the stressor structures will cause additional jumps in the measured phases. The influence of the specimen thickness can be mitigated by preparation of lamellas of constant thickness by means of focused-ion-beam preparation and choosing reflections with longer extinction length (e.g., {111} reflections in the GaAs system should be avoided).

The precision of the measured phases is proportional to the modulus of the diffracted beams. In order to have a strong diffracted beam, the specimen is oriented in two-beam excitation conditions for the selected beam. Furthermore, sample preparation intends to meet thicknesses close to ½, 1½, 2½, … times the extinction length, where the diffracted beams are highly excited and also the weighting function in (3.8) is smoother. While in principle any two non-collinear reflections will

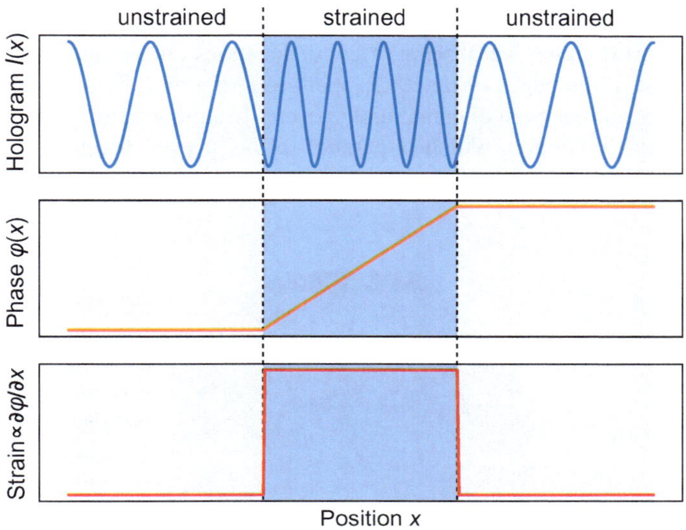

Fig. 3.13 Evaluation of dark-field electron holograms. Depicted are the hologram (interference pattern), the phase, and the derived strain originating from the interfering beams of strained and unstrained specimen areas. In the interference pattern of the strained area the phase is changed

suffice for the reconstruction of the displacement field, this condition makes a set of reflections of equivalent crystallographic planes preferable.

3.4.3 Strain Analysis in a Lamella of a GaAs Mesa

The measurements were performed using the FEI Titan 80-300 Berlin Holography Special microscope operated at 300 kV. Experimentally, the dark-field holography was performed in centered dark-field mode (instead of a shifted aperture like in Fig. 3.12), allowing the diffracted beam to be oriented along the optical axis of the microscope; this avoids optical aberrations in the measurement. A double biprism setup with the first biprism being located in an image plane and the second biprism in the drop shadow of the first one was used to avoid diffraction artifacts caused by the biprism itself [35, 36]. The microscope was operated in Lorentz mode in order to decrease the magnification in front of the biprisms and to allow for a wider field of view. Furthermore, 30 holograms of 10 s exposure were averaged with adjustment for specimen and biprism drifts to achieve a sufficient high signal to noise ratio [37]. Phase variations caused by image distortions of the camera and the projective system of the microscope were removed, employing the usual technique of empty holograms.

A bright-field image as well as dark-field electron holograms of the investigated stressor structure are shown in Fig. 3.14. The studied AlAs/AlO_x layer is described in detail in Sect. 3.3.1. 130 nm above the AlAs/AlO_x structure an InGaAs QD layer is visible. Furthermore, we observe a few dislocations emanating from the highly strained areas at the edges of the AlO_x aperture in the dark-field holograms. Both dark-field holograms were obtained under strong-beam conditions; Fig. 3.14b uses the $\vec{g} = (004)$ reflection, which is parallel to the growth direction (z-direction). Figure 3.14d was obtained using the $\vec{g} = (400)$ reflection, which is parallel to the

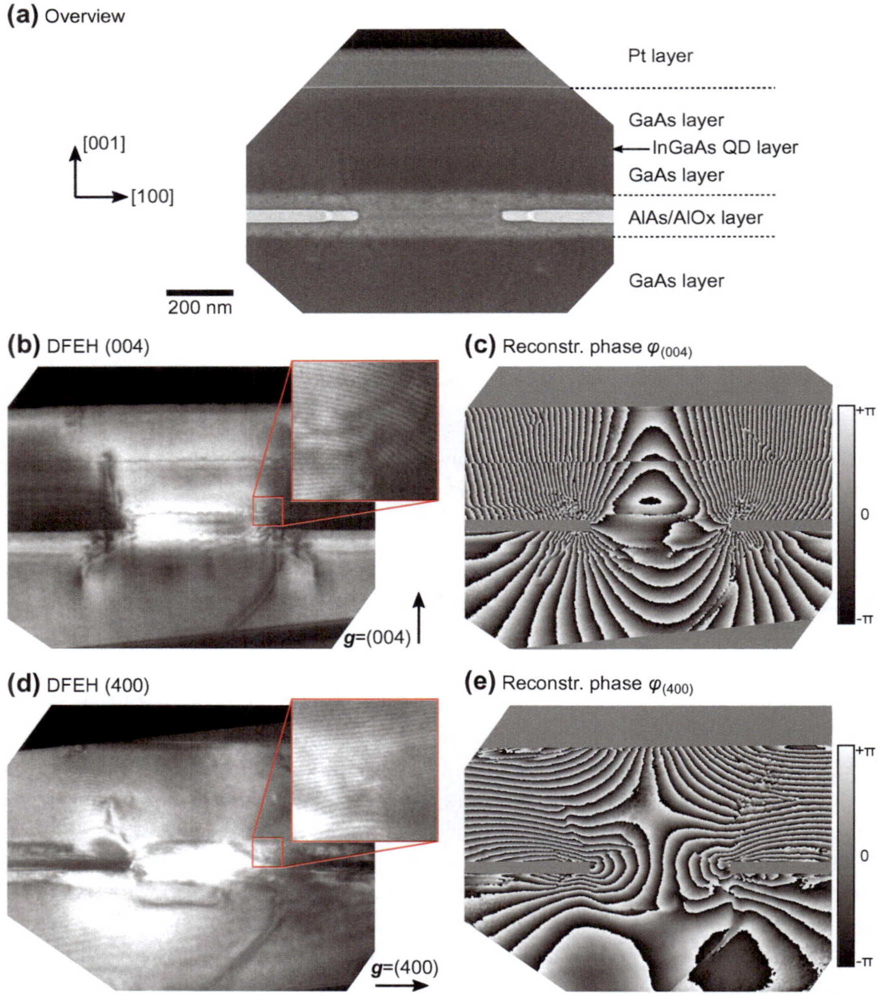

Fig. 3.14 **a** Bright field image of the investigated stressor structure; **b** dark-field electron hologram using the (004) reflection; **c** phase reconstructed from the (004) hologram; **d** dark-field electron hologram using the (400) reflection; **e** reconstructed phase from the (400) hologram. The *scale bar in panel* (**a**) applies to all subfigures

in-plane direction, i.e. the x-direction. The details of the holograms show nicely the bending and change in density of the interference fringes. The main direction and distance of the fringes reflects the overlap of the two interfering partial waves, with the reference area taken from the substrate below the structure.

The phase reconstructed from the holograms in Fig. 3.14b and d are shown in Fig. 3.14c and e. The phase is not well defined in areas of the image, were the amplitude of the reflected beam is close to zero or outside the area of the interference pattern. Thus, these areas are masked by gray patches to guide the eye. The general behavior of the dark-field phases showing the component of the displacement field parallel to the reflection as described by (3.7) can be nicely seen at the InGaAs QD layer. A pseudomorphically grown InGaAs layer, even if it is only a few nanometers thick, causes a step in the displacement field in growth directions of a few ten picometers; this appears also as a step in the reconstructed phases. This step appears as expected only in the phase of the (004) reflection in growth direction, but not in the phase of the (400) reflection in in-plane direction, were the lattice constants are matched.

Since the phases $\varphi_{(400)}$ and $\varphi_{(004)}$ are proportional to the displacements in their directions, the components of the strain field are easily calculated from their numerical derivatives using (3.4) [32]. In addition, the rotation in the xz-plane is given by the antisymmetric part of the displacement's Jacobian:

$$\omega_{xz} = \frac{1}{2}\left(\frac{\partial u_x}{\partial z} - \frac{\partial u_z}{\partial x}\right). \tag{3.9}$$

The evaluated components are shown in Fig. 3.15. At the heterointerfaces between AlAs/GaAs as well as in the InGaAs layer artifacts are visible, which result from the change of the structure factor ($\psi_{\vec{g}}$ in (3.7)). Further artifacts are visible at the dislocations, where their strong strain field is not spatially sampled with sufficient resolution.

The experimentally measured strain can be compared with the calculated strain shown in Fig. 3.1. For this comparison, the geometry of the investigated lamella must be considered: the full mesa has a diameter of several micrometers, which is too thick to be transparent for the electron beam. Thus, the DFEH investigations were performed using a 275 ± 5 nm thick lamella, cut from the central part of the mesa. The extinction length is 109 nm for the $\{400\}$ reflection, such that the reflections were still strong enough.

However, strongly reducing the specimen size in one dimension allows for strain relaxation at the lamella surfaces. These relaxation effects cause variations of the displacement field in beam direction. Since the weighting function for $\{004\}$ reflections at a thickness of 275 nm as shown in Fig. 3.11 exhibits a weight of close to zero at the surfaces, the measured dark-field phases mainly represent the strain state at the center of the lamella and not at the surfaces.

For comparison, the calculated strain in the center of a relaxed lamella can be found in the right column of Fig. 3.1. This calculation agrees remarkably well with the experiment. The critical merging of the tensile lateral ε_{xx} component above

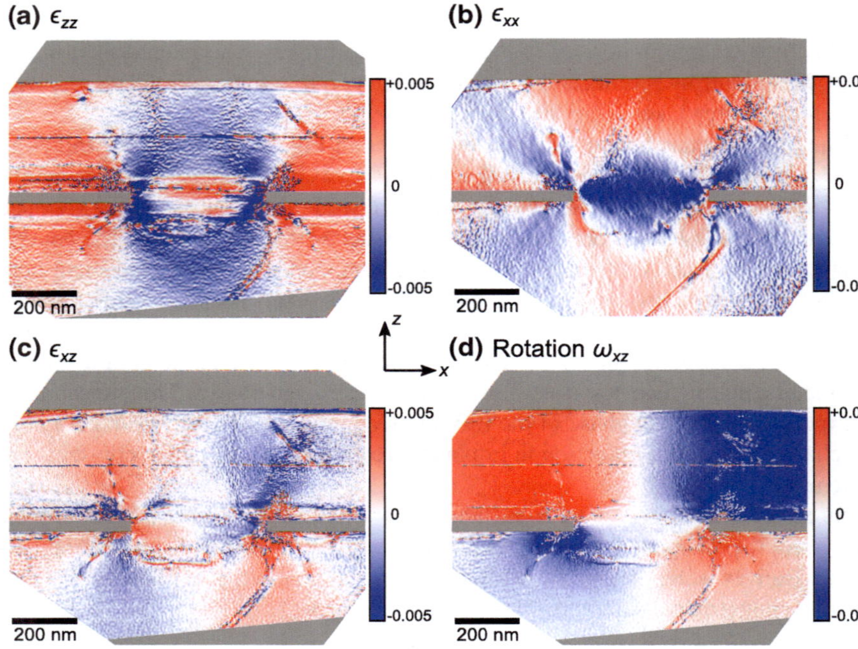

Fig. 3.15 Components of the strain tensor and rotation tensor in the *xz*-plane. Please note the change in the color scale for the rotation component. Adapted with permission from [25], © 2015 American Physical Society

the center of the aperture is observed in Fig. 3.15b. The excellent agreement of experimental and simulated data prove the validity of the assumptions presumed in the calculation [25].

3.5 Single-Photon Source Based on Stressor-Induced Site Control of Quantum Dots

The fabrication of an electrically operated single-photon source based on a single semiconductor QD faces a few challenges such as operation at cryogenic temperatures, selection of a QD, tolerance of the optical properties for the necessary processing, and efficient current injection into ideally just one single QD. Randomly distributed QDs grown by the Stranski-Krastanow mode can be selected *post-growth* by optical or structural means in combination with selective etching [7, 8; see also Chap. 8]. In order to isolate single QDs for integration into μm-scale devices it is beneficial—if not mandatory—to reduce the QD density to the low 10^8 QD/cm^2 range. The average distance between QDs is then on the order of 1 μm and mesa etching for selection becomes feasible. The larger the inter-dot separations are the less delicate becomes the processing, because the mesa size can be significantly enlarged. Very low QD densities down to 10^6 QD/cm^2 within the Stranski-Krastanow regime can be achieved by controlling the initial material deposition prior to the 2D-3D transition [38]. At such low densities, however, finding suitable but randomly distributed QDs for post-growth selection is an increasingly time-consuming effort which motivates deterministic ways of QD positioning. Still the reduction of the QD density to the 10^8 QD/cm^2 range remains a prerequisite also for the application of deterministic concepts.

3.5.1 Development of an Electroluminescence Quantum-Dot Diode

Inherent to any kind of an electrically operated device based on deterministic QD positioning is a *two-step growth process*, wherein the first growth step concerns the fabrication of a suitable surface topology regarding morphology or strain. By applying the stressor-induced approach an elegant way of creating a strain topology at the growth surface is realized. The *first growth step* concludes with layers providing the stressor structure outlined in Sect. 3.3.1; the complete layer sequence may also include monolithically integrated Bragg mirrors, sacrificial layers, and the like. For electrical operation, doping of these layers is required. Since growth chemistry and physics of III–V semiconductors favor a structure of *pn*-junctions with *p*-type layers grown on top of *n*-type layers, most layers grown during the first step are doped *n*-type.

After the first growth step, the templates undergo mesa processing and oxidation to form the buried stressor structure described in Sect. 3.3. The *second growth step* begins with the removal of the native surface oxide and growth of a thin GaAs buffer layer. Since the range of the stressor field of an oxidized GaAs/AlAs/GaAs aperture structure is reasonably far, the growth of a 50–100 nm thick buffer does not significantly affect the positioning effect. Subsequently, In(Ga)As Stranski-Krastanow islands are grown and embedded into GaAs similar to the procedure applied for conventional quantum-dot devices. All layers forming the intrinsic region of the *pin* diode structure are undoped. Finally, a *p*-doped contact structure is grown completing the growth sequence of step 2.

Due to anisotropic growth rates along the GaAs [110] and [$\bar{1}$10] directions, the edges of the mesa become a source of roughness that causes problems for subsequent processing steps. Fortunately, the buried stressor approach allows for applying mesa sizes larger than 10 μm that enable re-shaping of the mesa etches by selective dry etching. Also mesa structures with specific facet orientations along favorable in-plane directions can be designed to largely suppress the edge-induced roughness.

It is furthermore important to electrically insulate the side facets by a dielectric such as SiN_x in order to avoid short-circuit of the *pn* junction upon metallization of the p-contact. The *pin*-diode device is completed by an n-contact metallization of the substrate backside, see Fig. 3.22a.

3.5.2 Operation Characteristics of a Single-Photon Source

Based on the developments presented in Sect. 3.5.1, electrically driven devices were fabricated to demonstrate proof-of-principle operation. For this purpose, electro-optical characterization of buried-stressor based single-photon LEDs was performed using a high-resolution micro-electroluminescence (μEL) setup at low temperatures (10–50 K). Tested devices were mounted onto chip carriers and connected electrically via wire-bonds. Prior to the optical characterization, the current-voltage dependence was measured, revealing clear diode behavior with an onset voltage in forward direction of 1.4 V and a low saturation current <0.1 nA in reverse direction (not shown).

Figure 3.16a shows μEL spectra of an exemplary device in a 10 meV wide spectral window centered at 1.285 meV (964.9 nm) at a temperature of 12.5 K. Spectra are recorded for injection currents from 0.01 to 0.60 μA with the microscope objective being spatially located right above the aperture region. The selected device has a mesa diameter of 24.6 μm and an estimated aperture size of 0.2–0.4 μm, estimated from the adjusted oxidation rate and time during fabrication. At low injection current (0.01 μA) an emission-line doublet labelled X in the figure is visible at 962.4 nm originating from a single charge-neutral exciton (X). Increasing the injection current, further emission lines appear at lower and higher wavelengths, which result from the

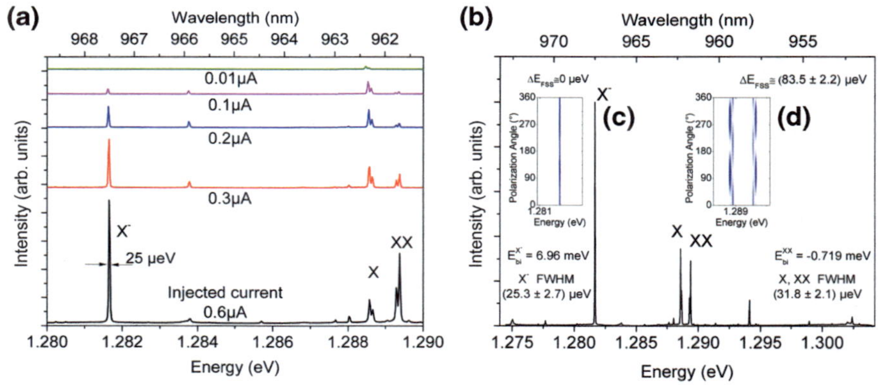

Fig. 3.16 **a** Micro-electroluminescence spectra for various injection currents. Emission lines from exciton (X), bi-exciton (XX), and charged exciton (X^-) transitions are indicated; **b** high resolution spectrum at 0.6 μA current injection level with annotations of linewidths and binding energies. Insets **c** and **d** show the polarization dependence of the X^-, X, and XX lines. Reprinted from [39] with the permission from AIP Publishing, © 2012 AIP

radiative recombination of multi-particle states of the same QD. The biexciton state (XX) at the high energy side is identified due to its super-linear intensity increase with rising injection current as well as due to its doublet structure. At the low-energy side of the spectrum, a particularly bright emission of the negatively charged trion state (X^-) is observed, assigned due to its binding energy [40]. The observation of such a clean emission spectrum under electrical pumping is possible due to the very low QD density directly above the aperture region. Moreover, all lines exhibit narrow spectral linewidths below 50 μeV, with best values of 25 μeV (resolution limited), which represents a major progress as compared to previous reports on site-controlled QD growth [9, 41–46]. Such low spectral linewidths confirm, that the QD is embedded in a semiconductor matrix with high crystalline quality and low defect densities. Emission of the electrically pumped QD is further analyzed employing polarization-resolved measurements. Figure 3.16b shows a high-resolution μEL spectrum of the same QD as in panel (a) at an injection current of 0.6 μA in a wider spectral range for a given polarization.

The insets (c) and (d) present spectra of the emission of the trion state X^- and the XX-X radiative cascade as a function of the polarization angle in a contour plot. While a fine-structure splitting (FSS) of the exciton and biexciton state is clearly visible, the trion does not exhibit any polarization splitting as expected. A closer look at Fig. 3.16d also reveals the typical polarization dependence of the radiative XX-X cascade, at which the horizontally and vertically polarized components of both states show the largest and smallest energy separation, respectively. Quantitatively analyzing the spectra reveals a relatively large FSS of (84 ± 2) μeV for the observed XX-X cascade and an antibinding XX state with a binding energy of -0.72 meV. Both values are in very good agreement with a statistical analysis of QDs for which a correlation of QD size and shape to ground-state optical transition levels was established [40]. The exciton emission-energy and biexciton binding energy suggest, that the QD under study has a size of about 3 monolayers in growth direction and a base length of approximately 10 nm. The binding energy of the X^- state is 6.96 meV, which also agrees well with the estimated size and shape of the QD.

Due to its clean and bright emission spectrum, this QD LED is suitable for non-classical light generation under electrical pumping, as demonstrated in the following. Photon-statistics measurements on the X^- emission was performed using a Hanbury Brown and Twiss (HBT)-type setup based on a 50:50 multimode fiber beam-splitter and two silicon-based single-photon counting modules with a timing resolution of 0.8 ns. Figure 3.17 displays the measured second-order photon auto-correlation $g^{(2)}(\tau)$ together with a fit of the experimental data. Clear photon anti-bunching with $g^{(2)}(0) = 0.04$ is obtained at an injection current of 0.32 μA. Remarkably, single-photon emission with such a high purity is among the best reported by now for site-controlled QDs [47]. For electrically operated QD-based single-photon emitters, the result compares well to best values of $g^{(2)}(0) = 0.02$ measured on devices with stochastically distributed QDs [48]. Moreover, future work combining our device concept with suitable microcavity structures or microlenses reported in Chap. 8 will enable the development of deterministic single-photon sources with high extraction efficiency.

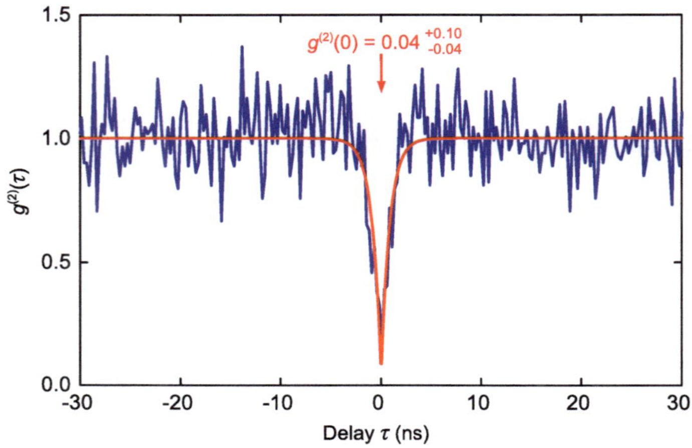

Fig. 3.17 Measured photon auto-correlation obtained from the X^- emission of a single-photon LED based on the buried-stressor approach. The red line is a fit to the experimental data revealing close-to-ideal single-photon emission

3.5.3 Development of a Resonant-Cavity Structure

With the lateral position of single QDs being controlled by the aperture of the buried stressor structure as presented in Sect. 3.3.3, the quest for resonant-cavity designs employing monolithically integrated distributed Bragg reflectors (DBRs) for surface-emission of single-photons is open. Such resonant microcavity structures are needed for an efficient directed outcoupling of the light and the utilization of the Purcell-effect, i.e., the enhancement of the spontaneous emission rate by the reduction of the optical mode volume. The microcavity is formed by two mirrors surrounding the emission center; they must be separated by an integer multiple of half the emission wavelength. The resonance condition for the spectrally narrow QD emission and the microcavity dip is exceedingly difficult to meet. With regard to the cavity design the inherently required interruption of the growth process for processing of the stressor structure and the subsequent regrowth for the QD deposition are serious obstacles. Since the electron and hole wavefunctions within a quantum dot depend very sensitively on material stoichiometry, geometrical size, and shape, the adjustment of QD growth parameters for the precise control of a targeted emission wavelength is impossible.

The obstacle may be overcome by employing the tuning potential of the resonance structure. The Stranski-Krastanow growth regime of InGaAs/GaAs allows for reproducing QDs emitting within a relatively narrow wavelength window of about 30 nm. The emission wavelength of a QD depends furthermore on temperature due to bandgap variation. This enables an externally controlled wavelength tuning to the optical resonance of the cavity by sample heating or cooling, because the cavity resonance shifts at a much lower rate. Still the operation temperature of InGaAs/GaAs

QDs should be as low as possible in order to obtain single photons of high statistical purity. Given the unavoidable randomness of the expected emission wavelength, the spectral position of the cavity resonance should be designed towards slightly longer wavelengths. The QD emission may then tuned into the cavity resonance by some sample heating.

A further means for resonance tuning is the design of the upper DBR mirror. Either a dielectric stack of quarter-wavelength layer pairs or a monolithically integrated, epitaxially grown DBR can be used. Amorphous dielectric mirrors are easy to fabricate by sputtering techniques, and their local optical properties are not much affected by the surface morphology of the mesa template. They also allow for an optical characterization of the stressor-controlled QD in the center of the mesa prior to deposition; this enables an adjustment of the cavity resonance to the detected emission wavelength. The monolithic integration of an epitaxially grown DBR is much more delicate as the growth becomes very sensitive to any surface corrugation.

We have investigated the epitaxial growth of AlGaAs/GaAs and InGaP/GaAs DBR mirrors on deeply (~4 μm) and shallowly (<1 μm) etched mesa structures containing a buried stressor for QD positioning, but without any dielectric surface mask. Thereby growth rates can be reliably inferred from calibration growth runs on planar surfaces, as there will be no enhancement factor due to an excess amount of species. This is of particular importance for the control of a low-density of QDs ($\leq 10^8$ cm^{-2}) to ensure selective growth at the stressor aperture.

Conformal layer growth is observed at the low temperatures required after QD formation, see Fig. 3.18. For deeply etched mesas, the growth direction of the DBR structure at the mesa side-walls is perpendicular to the wafer surface. On shallowly etched mesas, the orientation of the DBR layers remains parallel to the wafer surface. Since the mesa structure separates the growth fronts at different crystallographic side facets, large and irregularly shaped growth fronts occur along the [110] and [$\bar{1}$10] directions during coalescence. These defective areas increase in size with the number of layer pairs.

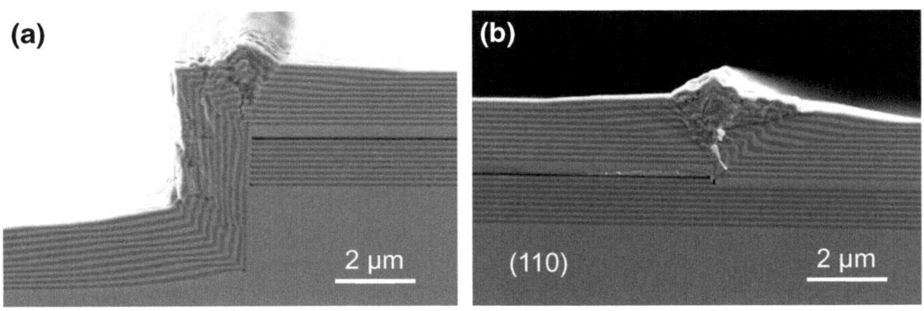

Fig. 3.18 Cross-section scanning electron micrographs of AlGaAs/GaAs DBRs grown on **a** deeply and **b** shallowly etched stressor templates with bottom AlGaAs/GaAs DBR beneath the aperture (*black layer*). Irregularly shaped growth fronts evolve upon coalescence of the growth fronts at the top edges of the mesa

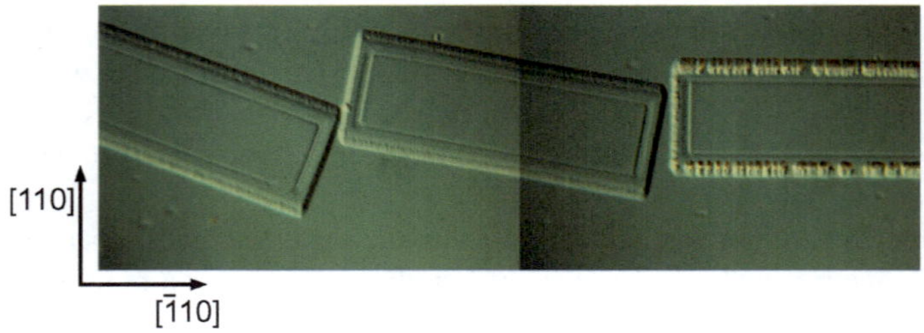

Fig. 3.19 Nomarski micrograph of rectangular mesa structures after overgrowth with InGaP/GaAs, showing the orientation dependence of the sidewall roughness. The *right mesa* is exactly oriented along the [$\bar{1}$10] direction. The *middle and left mesa* are rotated by 5° and 10°, respectively

To large extent, growth of the AlGaAs layers enhances the generation and expansion of these defects because of the high sticking coefficient of Al. Better smoothness of sidewalls and less pronounced defect areas are obtained with DBR mirrors made of InGaP/GaAs layers. Nomarski micrographs of rectangular mesa structures after overgrowth with InGaP/GaAs layers are given in Fig. 3.19. Prior to the deposition of the top DBR the mesa have been laterally oxidized; this leads to a surface undulation as seen in the image as an inner frame-like pattern. The right mesa is exactly oriented with the long side along the [$\bar{1}$10] direction. The middle and left mesa are rotated by 5 and 10 degrees, respectively. Off-orientation of the side facets from the [110] and [$\bar{1}$10] directions enables smooth overgrowth with InGaP/GaAs DBR structures.

The problem of defect generation upon overgrowth can hence be solved by geometrical shaping of the mesas. As shown in Fig. 3.19, the crystallographic orientation of the mesa edges is decisive for the smoothness of the overgrowth. If no facet is oriented along the [110] or [$\bar{1}$10] direction, smooth overgrowth with InGaP/GaAs DBR mirror structures is realized. A combination of shallowly etched, rectangular mesas and facet orientation along the [100] or [010] direction suppresses the formation of irregular growth fronts.

Smooth overgrowth of shallowly etched mesas with laterally oxidized stressor structures is possible with InGaP/GaAs-DBR structures. Figure 3.20 shows that separate GaAs layers on the mesa and on the etched bottom with (111) facet are formed at the beginning of the overgrowth with GaAs. With the first InGaP layer, the gap between those layers is filled. The exposure of a (111) facet is connected with high growth rates for InGaP layers and reduced growth rates for GaAs layers. The (001)-oriented InGaP and GaAs layers close to the edges gain in thickness along an approximately 200 nm wide region. This region shifts gradually towards the mesa center according to the angle of the (111) facet. However, the coalescence region continuously smoothens, and for typical mesa with several tens of microns lateral dimensions such a coalescence region is well under control. The smoothness of the overgrowth allows for subsequent photolithography and dry etching to remove the coalescence region, and to newly define the device area.

Having solved the overgrowth issues, the final challenge for a resonant-cavity single-photon source is the control of the detuning of the QD emission wavelength

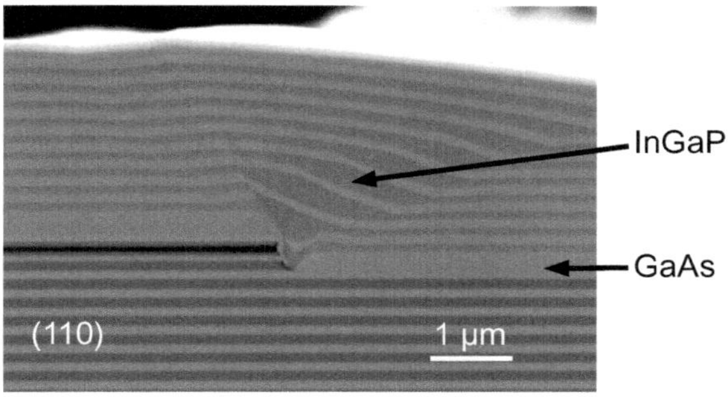

Fig. 3.20 Cross-section SEM image of an overgrown 8.5-fold InGaP/GaAs DBR structure on a shallowly etched mesa with laterally oxidized stressor structure (*black layer*). GaAs layers show *bright contrast* while InGaP layers appear *gray*. The bottom mirror consist of GaAs/AlGaAs, where AlGaAs layers appear *dark gray*. The coalescence region monotonously straighten with increasing number of DBR pairs. The growth-rate overshoot region shifts towards the center by about 1 μm

on the stressor structure with respect to the cavity resonance. Around a wavelength of 1000 nm, room-temperature measurements of the cavity resonance-wavelength must be reduced by about 6 nm to account for low-temperature operation below 50 K. InGaAs/GaAs QDs show hardly any sharp spectral line at room-temperature; repeated low-temperature, highly spatially and spectrally resolved luminescence measurements are hence necessary to tune the QD emission towards the short-wavelength side of the cavity resonance. With the buried stressor approach, the time-consuming effort to find the target QD is greatly reduced. Post-growth annealing is an option to shift QD emission towards shorter wavelengths by as much as 100 nm [49]; this provides a means to bring detuned QDs towards the long-wavelength side into resonance with the cavity.

An electrically operated resonant-cavity single-photon source at an operation temperature of 20 K is shown in Fig. 3.21. A single QD is successfully positioned by the buried stressor in the center of the device. Within an area of $10 \times 10\ \mu\text{m}^2$ only the electrically excited emission of the central quantum dot is dominant, exhibiting a narrow linewidth of only 77 μeV at a drive current of 150 nA. The cavity resonance is at 922 nm, while the QD emission occurs at 931 nm. Although a cavity with a relatively wide spectral resonance of about 2 nm was employed, a resonance effect is not observed in this device. Time-resolved electroluminescence measurements do accordingly not yield any lifetime reduction of the spontaneous emission. Still there remains no growth or fabrication issue for realizing such devices with full functionality.

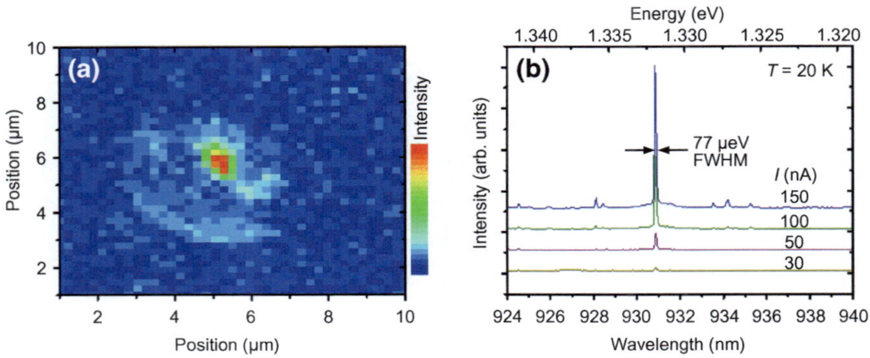

Fig. 3.21 **a** Spatially-resolved micro-electroluminescence of a resonant-cavity single-photon source. Efficient and selective excitation of the positioned QD is demonstrated. **b** Electroluminescence spectrum of the central site-controlled quantum dot for various injection currents at $T = 20$ K

3.6 Realization of an Efficient Current Injection into a Single Quantum Dot

The efficient electrical pumping of sub-micrometer sized domains is essential to achieve deterministic control over the carrier population within optically active nanostructures such as QDs. Based on the experience with the injection-current confinement via selectively oxidized AlO_x layers in VCSELs [50], the device design shown in Fig. 3.22a is expected to provide a very efficient carrier injection into the

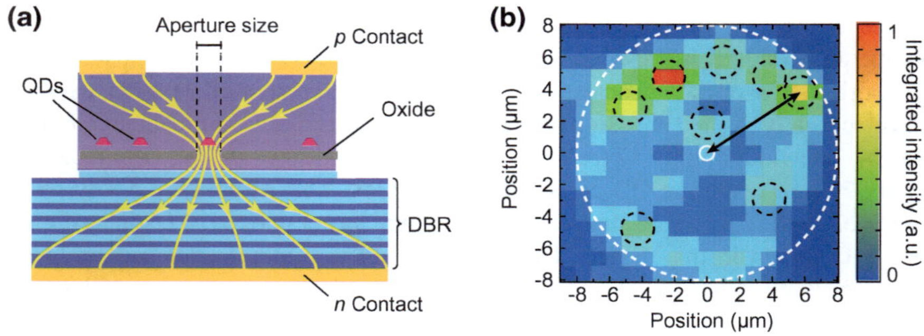

Fig. 3.22 **a** Schematic cross section of the *pin* device. The QD positions are indicated by capped *red triangles* and the naively expected current flow is shown by *yellow arrows*. **b** Electroluminescence map of a 16 μm diameter mesa featuring several parasitic QDs recorded at low DC injection (10 nA) with at a lateral resolution of 1 μm. The *white solid circle* indicates the aperture (~800 nm diameter), the *white dashed circle* indicates the mesa edge. The *dashed black circles* highlight the positions of several parasitic QDs, the *black arrow* corresponds to a distance of 6.7 μm between a QD and the aperture center. Adapted from [51]

single QD or few central QDs grown by the buried stressor approach. The experimentally recorded electroluminescence map depicted in Fig. 3.22b, however, reveals merely a minor effect of the aperture on the lateral current distribution: The emission measured directly above the aperture shows a weak luminescence of a few (about 2–4) QDs, whose intensity is about 30 times weaker than the emission originating from parasitic QDs in the outer part of the mesa [51]. This measurement indicates an insufficient current confinement by the oxidized aperture, since parasitic QDs located several micrometers away from the mesa center can be electrically excited. Conversely, the observation suggests a rapid lateral current spreading above the oxide layer, which clearly contradicts the naively expected current path indicated by the yellow arrows in Fig. 3.22a. In the following, we will outline how the current confining property of the oxidized aperture breaks down at low injection currents and cryogenic temperatures, which are the typical operation conditions of single-photon emitting diodes. Finally, a modification of the doping profile is suggested, which allows for the selective electrical excitation of the central QD(s) above the aperture.

3.6.1 Modeling of the Current Flow in the Device

As the lateral carrier-density distributions and the current flow in the device are not directly accessible by experiment, the mechanism underlying the lateral current spreading must be investigated theoretically by means of numerical simulations. An efficient description of the carrier transport on macroscopic scales is provided by the van Roosbroeck system [52, 53], which describes the transport and recombination dynamics of electrons and holes in their self-consistently generated electric field.

In order to focus on the lateral current spreading above the aperture, some details of the experimental structure are simplified in the simulation. The highly p-doped cap layer is replaced by an (ideal) ohmic contact boundary-condition, neglecting the small potential drop across the layer. The same is done with the entire bottom DBR section. Thereby, the computational domain is restricted to the interesting region ranging from 200 nm below to 270 nm above the aperture such that the current flow between the aperture and the QD layer is decoupled from the complex phenomena in the top and bottom layers of the device. As the impact of a few, isolated QDs on the macroscopic carrier transport is considered to be small on large scales [51], the QD layer is neglected in the simulations. Moreover, the detailed multi-layer $Al_xGa_{1-x}As/AlAs$ sandwich structure in the aperture region of the real device is replaced by a single insulating layer with an aperture radius of 0.5 μm and a thickness of 20 nm. By exploiting the rotational symmetry of the mesa, it is sufficient to analyze the current flow in the 2D cross section shown in Fig. 3.23a. Surface recombination at the semiconductor-oxide interface is neglected for the sake of simplicity. In order to include the effects of the cryogenic operation temperature, the lattice temperature is set to $T = 30$ K throughout this section. Further parameters used in the calculation can be found in [51]. The numerical simulations are carried out using a finite volume Scharfetter–Gummel method [54]. Note that the cryogenic

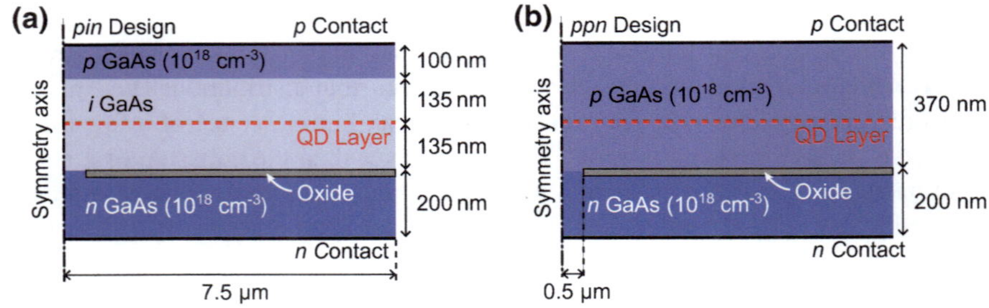

Fig. 3.23 Cross sections of the two simulated device designs. **a** The *pin* design contains a 270 nm thick intrinsic layer between the *n*- and *p*-doped regions. The QD layer is embedded in the center of the intrinsic domain (*dashed red line*). The thickness of the oxide layer is set to 20 nm, the radius of the aperture is chosen as 0.5 µm. Except for the oxide, the material is GaAs throughout. **b** In the *ppn* design the *p*-type doping is applied down to the oxide, replacing the former intrinsic layer

operation temperature poses a particular challenge to the numerical routines [55]. Details on the van Roosbroeck system and the numerical methods can be found in Sect. 7.2.1.

First, the experimentally indicated lateral current spreading phenomena shall be reproduced by the simulation. Figure 3.24a shows the electron density n_{QD} at the QD layer in the *pin* design at stationary operation. The electron density is very sensitive to the applied bias and rapidly increases as the bias is ramped up. Remarkably, in the *low-injection regime*, the electron density is nearly constant at the QD layer (blue curve in Fig. 3.24a at $V = 1.46$ V). Since the capture rate to the QDs is assumed to be proportional to the bulk carrier density [56, 57]—which is virtually independent

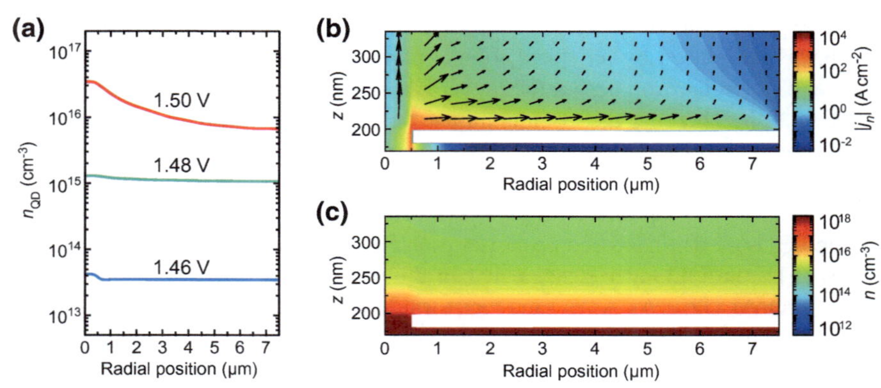

Fig. 3.24 Current flow and carrier-density distributions in the *pin* design. **a** Radial electron-density distribution at the QD layer for different applied voltages. **b** Electron current-density distribution between the oxide (*white*) and the QD layer (*top border*) at $V = 1.48$ V. The magnitude of the current density is color-coded and the arrows point into the direction of electron flow. **c** Electron-density distribution in the same domain as (**b**). Adapted from [51]

of the radial position—the occupation of the QDs in the low-injection regime is expected to be homogeneous within the whole QD layer. This agrees well with the experimentally observed luminescence of parasitic QDs far away from the aperture, see Fig. 3.22b. The underlying mechanism is shown in Fig. 3.24b, which reveals how the electrons flow around the aperture and then diffuse outwards within a thin layer directly above the oxide. In this layer, the electron density reaches nearly 10^{18} cm^{-3} and is practically independent of the radial position, see Fig. 3.24c. Therefore, the quasi-homogeneous and highly conductive thin layer acts like an equipotential plane. As a consequence, also the electron distribution everywhere above it (and thus the injection into the QDs) is homogeneous along the radial direction.

At first glance, the current spreading might be caused by the high electron mobility in the intrinsic region. Hence, a reversal of the doping profile should help to reduce the current spreading since the mobility of holes is much smaller. However, the simulation of the corresponding *nip*-design showed almost no improvement. Consequently, the carrier mobility does not sufficiently explain the lateral current spreading in the low-injection regime. Another possible reason is found in the thin, highly conductive layer above the oxide, where the electron density exceeds the density of holes by several orders of magnitude. As a result, the recombination rate R is very small and the electron lifetime $\tau_n = n/R$ becomes extremely large. The simulation yields lifetimes on the order of milliseconds, see Fig. 3.25a. Therefore, the electron diffusion length $L_n = \sqrt{D_n \tau_n}$ reaches the order of millimeters, which exceeds the mesa radius by two orders of magnitude, such that the electrons can easily occupy the whole mesa area. Due to the freeze-out of minority carriers, this mechanism is particularly enhanced at cryogenic temperatures.

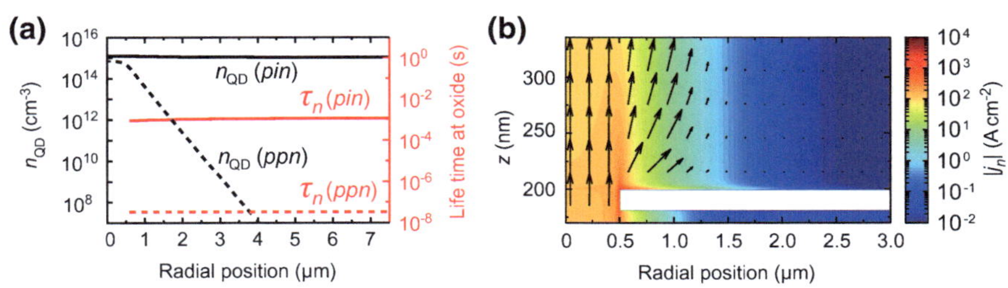

Fig. 3.25 **a** Radial electron-density distributions n_{QD} at the QD layer (*black*) and electron lifetime τ_n directly above the oxide (*red*). *Solid lines* correspond to the *pin* design, *dashed lines* are for the *ppn* design. **b** Electron-current density distribution between the oxide (*white*) and the QD layer (*top border*) in the *ppn* design. The arrows indicate the direction of electron flow, the magnitude of the current density is color-coded. Adapted from [51]

3.6.2 *Current Confinement in* pin *and* ppn *Designs*

The basic conclusion of the previous section is that the lateral current spreading above oxidized apertures is inevitable in the low-injection regime as long as the carrier-recombination rate above the oxide remains small. Hence, doping of the active region containing the QD layer will likely be helpful to suppress the lateral current spreading. In the following, the conventional *pin* design will be compared with the *ppn* design shown in Fig. 3.23b, where only the doping profile has been altered.

3.6.2.1 Stationary Operation

Due to the availability of holes as recombination partners in the active region of the *ppn* design, the electron lifetime drops by several orders of magnitude, see Fig. 3.25a, such that the current spreading is significantly reduced because of the decreased diffusion length. Accordingly, the radial electron-density distribution in the QD layer drops exponentially with increasing radius, such that the excitation of parasitic QDs far away from the center is strongly suppressed. This coincides with the current flow in the *ppn* design shown in Fig. 3.25b, which indicates a narrowly confined excitation of the central region above the aperture only. In conclusion, the selective excitation of the central QD(s) can be achieved even in the low-injection regime by a sufficiently strong *p*-type doping in the active region (here $N_A = 10^{18}$ cm^{-3}).

3.6.2.2 Pulsed Operation

For practical applications, the generation of single photons on demand in response to an externally triggered excitation is essential. In this section, the performance of the *pin* and the *ppn* design is compared under pulsed electrical excitation. In particular, we show that the improved current confinement of the *ppn* design persists in the transient case. In the following, the device response to short voltage pulses ($\Delta t_{\text{pulse}} = 100$ ps) with an amplitude ΔV superimposed on a constant bias V_0 with a repetition frequency of 1 GHz (see inset of Fig. 3.26b) is investigated. For each design, the corresponding DC bias V_0 is chosen such that the electron density in the QD layer is slightly below 10^{12} cm^{-3}. At this low carrier density, the capture rate to the QDs is very small, such that no excitations in between subsequent pulses are expected (see [51] for details). The pulse amplitude ΔV is chosen such that the bulk carrier density in the QD layer reaches $n_{\text{QD}} \approx 10^{16}$ cm^{-3} during the excitation.[1]

The time evolution of the radial electron-density distribution in the QD layer in response to a periodically pulsed contact voltage is shown in Fig. 3.26. In both designs, the carrier density follows the pulses with a short delay of only a few

[1]We assume that this carrier density corresponds to a capture rate of around 10 ns^{-1} during the excitation pulse time Δt_{pulse}; see [51] for details.

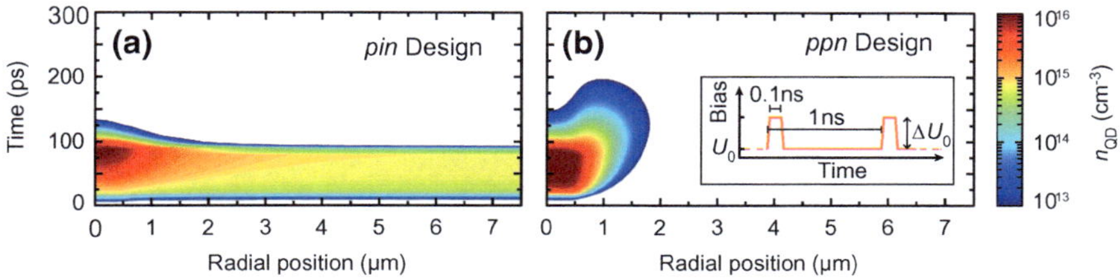

Fig. 3.26 Time evolution of the radial electron-density distribution in the QD layer in response to a periodically pulsed contact voltage for the **a** *pin* design ($V_0 = 1.43$ V, $\Delta V = 70$ mV) and the **b** *ppn* design ($V_0 = 1.48$ V, $\Delta V = 30$ mV). The inset shows the voltage-pulse train for transient excitation. Adapted from [51]

picoseconds. Therefore, the internal time scale of the carrier transport processes to pump the QDs is sufficiently fast to allow for very high repetition rates beyond the GHz scale. However, external effects like parasitic capacitances and inductivity of bond wires, which are neglected here, will lead to limitations in real devices. The radial electron-density distribution differs drastically between both designs. In the *pin* design, the lateral current spreading quickly leads to a nearly homogeneous electron-density distribution in the QD layer until $n_{\mathrm{QD}} \approx 10^{15}$ cm^{-3} is reached in the whole plane. Close to the end of the excitation pulse, a peak density of $n_{\mathrm{QD}} \approx 10^{16}$ cm^{-3} is reached only above the aperture, while the electron density above the oxide saturates at $n_{\mathrm{QD}} \approx 10^{15}$ cm^{-3}. Hence, just as in the stationary case (cf. Fig. 3.24a), a small drop in the radial electron-density distribution can be observed. The pulse response of the *ppn* design depicted in Fig. 3.26b shows a strongly localized excitation restricted to the central region of the mesa. The radial range in which the electron capture to the QDs is significantly different from zero is restricted to $r < 1.5$ μm. In conclusion, both under stationary and pulsed operation, the simulation indicates a distinctly more efficient current injection into the central QD(s) using the *ppn* design.

3.6.3 Demonstration of a ppn *QD Diode with Efficient Current Confinement*

The theoretical concept developed in the previous section was experimentally verified. Diode structures with oxide apertures and a site-controlled QD layer were grown, using identical designs apart from the doping profile. Results for one of these test structures with the QDs embedded in an intrinsic GaAs region is shown in the upper row of Fig. 3.27. The series shows photos of the electroluminescence of this *pin* device, recorded by a digital camera using an IR filter. When the injection current is slowly turned on, the emission appears first at the *outer* part of the mesa where the QD density is particularly high (see also Fig. 3.10). This behavior agrees well with the predictions of the modeling presented in Sect. 3.6.1. On the other hand, when the QDs are embedded in a *p*-doped region according to the *ppn* design, the spatial

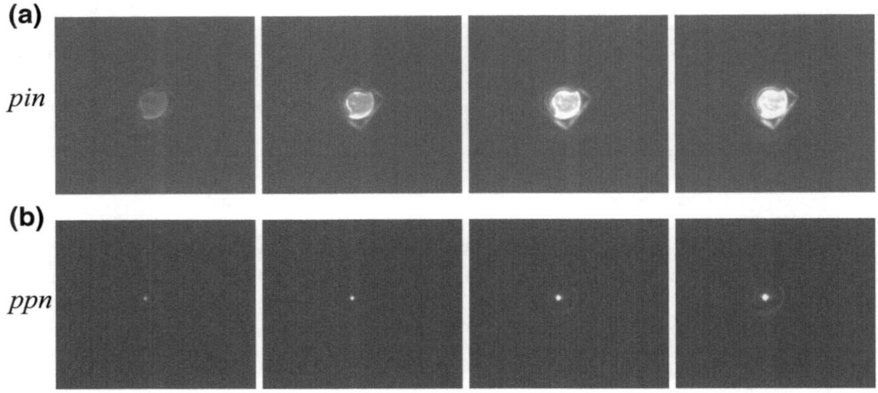

Fig. 3.27 Comparison of the electroluminescence of devices fabricated with doping either **a** according to the *pin* design or **b** according to the *ppn* design. The injection current is increased from left to right

emission characteristics are different. We see in the lower row of Fig. 3.27 that now the emission appears first at the *inner* part of the mesa, providing a clear proof of the current confinement and efficient injection in the *ppn* device.

3.7 Conclusion and Perspectives

We have presented a new method to define the location for the nucleation of semiconductor quantum dots. The discussed example employs the strain field produced by an oxide aperture in a circular mesa to locate a quantum dot in the current path through the device. The simulated strain profile was verified by electron holography, and growth studies proved the alignment of QD nucleation sites at locations of tensile strain maxima. Luminescence studies showed exciton emission from site-controlled QDs with narrow linewidth; this indicates strongly reduced spectral diffusion, which often originate from process-induced defects observed for other positioning methods. The quality of the quantum dots which are site-controlled by our approach are similar to those of freely grown quantum dots without any site control. Implementing our approach into a diode with single site-controlled QDs lead to the demonstration of an electrically operating single-photon source. The presented device has a resolution-limited linewidth and delivers almost perfectly pure single photons. An efficient current injection was achieved by developing an appropriate doping profile for the device structure.

The presented application focused on the nucleation of single quantum dots, using basically small circular apertures formed by the selective oxidation. Further developments may take advantage of the versatility of the described method. The aperture may be formed to virtually any shape and size: vertically at the aperture edge by a sequence of appropriately alloyed $Al_{1-x}Ga_xAs$ layers with varied Ga content to control the oxidation speed; laterally by defining mesa of appropriate shapes and dimensions.

By using an aperture of larger diameter also *ensembles* of quantum dots can be confined to a predefined area in the device. This may be interesting for the fabrication of microlasers. In an appropriate design the active quantum dots are located in the maximum of the optical mode, and losses can be minimized. This applies for various kinds of such devices. In a micro-pillar laser or a superradiant LED the quantum dots should be located in the center like those shown in Fig. 3.9b. By contrast in a whispering-gallery laser the quantum dots should be located at the outer edge like those shown in Fig. 3.9a.

Applications for site control are not restricted to the confinement of nucleation sites on a spot or a ring-like structure. By defining elongated mesa also linearly aligned quantum-dot ensembles can be created. Such an approach is attractive for optically active waveguide structures. This implies edge-emitting devices as well as photo detectors. There are various benefits of using the presented buried-stressor approach for such devices. The active low-bandgap medium is embedded vertically and laterally in a high-bandgap matrix, by a self-aligned bottom-up approach—without post growth processing and effects of related defects. The active material can be adapted to the optical mode, thereby inducing efficient index guiding and low losses.

The mentioned examples indicate that the presented buried-stressor approach provides a technology, which may serve as a toolbox to implement a large variety of structures which are based on site-controlled semiconductor quantum dots—both individual dots and ensembles confined to a predefined area.

Acknowledgements We would like to thank Jan-Hindrik Schulze for growing samples and Felix Kießling for structural characterization.

References

1. U.W. Pohl, Low-dimensional semiconductors, in *Handbook of Materials Data*, 2nd edn., ed. by H. Warlimont, W. Martienssen (Springer Nature, Cham, Switzerland, 2018), pp. 1077–1100
2. D. Bimberg, U.W. Pohl, Quantum dots: promises and accomplishments. Mater. Today **14**, 388 (2011)
3. M. Ciorga, A.S. Sachrajda, P. Hawrylak, C. Gould, P. Zawadzki, S. Jullian, Y. Feng, Z. Wasilewski, Addition spectrum of a lateral dot from Coulomb and spin-blockade spectroscopy. Phys. Rev. B **61**, R16315 (2000)
4. A. Forchel, H. Leier, B.E. Maile, R. Germann, Fabrication and optical spectroscopy of ultra small III-V compound semiconductor structures, in *Festkörperprobleme/Advances in Solid State Physics*, vol. 28, ed. by U. Rössler (Vieweg, Braunschweig, 1988), pp. 99–119
5. I.N. Stranski, L. Krastanow, Zur Theorie der orientierten Ausscheidung von Ionenkristallen aufeinander. Monatsh. Chem. **71**, 351 (1937). (On the theory of oriented deposition of ionic crystals upon each other, in German)
6. U.W. Pohl, *Epitaxy of Semiconductors* (Springer, Berlin, 2013)

7. K. Hennessy, A. Badolato, M. Winger, D. Gerace, M. Atatüre, S. Gulde, S. Fält, E.L. Hu, A. Imamoğlu, Quantum nature of a strongly coupled single quantum dot–cavity system. Nature **445**, 896 (2007)
8. M. Gschrey, R. Schmidt, J.-H. Schulze, A. Strittmatter, S. Rodt, S. Reitzenstein, Resolution and alignment accuracy of low-temperature *in situ* electron beam lithography for nanophotonic device fabrication. J. Vacuum Sci. Technol. B **33**, 021603 (2015)
9. P. Atkinson, O.G. Schmidt, S.P. Bremner, D.A. Ritchie, Formation and ordering of epitaxial quantum dots. C. R. Phys. **9**, 788 (2008)
10. C. Schneider, M. Strauß, T. Sünner, A. Huggenberger, D. Wiener, S. Reitzenstein, M. Kamp, S. Höfling, A. Forchel, Lithographic alignment to site-controlled quantum dots for device integration. Appl. Phys. Lett. **92**, 183101 (2008)
11. V. Türck, S. Rodt, O. Stier, R. Heitz, R. Engelhardt, U.W. Pohl, D. Bimberg, R. Steingrüber, Effect of random field fluctuations on excitonic transitions of individual CdSe quantum dots. Phys. Rev. B **61**, 9944 (2000)
12. C. Schneider, A. Huggenberger, T. Sünner, T. Heindel, M. Strauß, S. Göpfert, P. Weinmann, S. Reitzenstein, L. Worschech, M. Kamp, Single site-controlled In(Ga)As/GaAs quantum dots: growth, properties and device integration. Nanotechnology **20**, 434012 (2009)
13. O.G. Schmidt (ed.), *Lateral Alignment of Epitaxial Quantum Dots* (Springer, Berlin, 2007)
14. V. Holý, G. Springholz, M. Pinczolits, G. Bauer, Strain induced vertical and lateral correlations in quantum dot superlattices. Phys. Rev. Lett. **83**, 356 (1999)
15. I.L. Krestnikov, M. Straßburg, M. Caesar, A. Hoffmann, U.W. Pohl, D. Bimberg, N.N. Ledentsov, P.S. Kop'ev, Zh.I. Alferov, D. Litvinov, A. Rosenauer, D. Gerthsen, Control of the electronic properties of CdSe submonolayer superlattices via vertical correlation of quantum dots. Phys. Rev. B **60**, 8695 (1999)
16. X.-D. Wang, N. Liu, C.K. Shih, S. Govindaraju, A.L. Holmes Jr., Spatial correlation-anticorrelation in strain-driven self-assembled InGaAs quantum dots. Appl. Phys. Lett. **85**, 1356 (2004)
17. G. Springholz, M. Pinczolits, P. Mayer, V. Holy, G. Bauer, H.H. Kang, L. Salamanca-Riba, Tuning of vertical and lateral correlations in self-organized PbSe/$Pb_{1-x}Eu_xTe$ quantum dot superlattices. Phys. Rev. Lett. **84**, 4669 (2000)
18. V.A. Haisler, F. Hopfer, R.L. Sellin, A. Lochmann, K. Fleischer, N. Esser, W. Richter, N.N. Ledentsov, D. Bimberg, C. Möller, N. Grote, Micro-Raman studies of vertical-cavity surface-emitting lasers with Al_xO_y/GaAs distributed Bragg reflectors. Appl. Phys. Lett. **81**, 2544 (2002)
19. F. Guffarth, R. Heitz, A. Schliwa, O. Stier, N.N. Ledentsov, A.R. Kovsh, V.M. Ustinov, D. Bimberg, Strain engineering of self-organized InAs quantum dots. Phys. Rev. B **64**, 085305 (2001)
20. J.F. Nye, *Physical Properties of Crystals* (Claredon Press, Oxford, 1972)
21. E.O. Kane, Phonon spectra of diamond and zinc-blende semiconductors. Phys. Rev. B **31**, 7865 (1985)
22. W. Press, B.P. Flannery, S.A. Teukolsky, W.T. Vetterling, *Numerical Recipes in C* (Cambridge University Press, Cambridge, 1988)
23. I. Vurgaftman, J.R. Meyer, L.R. Ram Mohan, Band parameters for III/V compound semiconductors and their alloys. J. Appl. Phys. **89**, 5815 (2001)
24. B. Zhou, B.C. Prorok, A new paradigm in thin film indentation. J. Mater. Res. **25**, 1671 (2010)
25. F. Kießling, T. Niermann, M. Lehmann, J.-H. Schulze, A. Strittmatter, A. Schliwa, U.W. Pohl, Strain field of a buried oxide aperture. Phys. Rev. B **91**, 075306 (2015)
26. A. Strittmatter, A. Holzbecher, A. Schliwa, J.-H. Schulze, D. Quandt, T.D. Germann, A. Dreismann, O. Hitzemann, E. Stock, I. Ostapenko, S. Rodt, W. Unrau, U.W. Pohl, A. Hoffmann, D. Bimberg, V. Haisler, Site-controlled quantum dot growth on buried oxide stressor layers. Phys. Status Solidi A **209**, 2411 (2012)
27. K.D. Choquette, K.M. Geib, C.I.H. Ashby, R.D. Twesten, O. Blum, H.Q. Hou, D.M. Follstaedt, B.E. Hammons, D. Mathes, R. Hull, Advances in selective wet oxidation of AlGaAs alloys. IEEE J. Sel. Top. Quantum Electron. **3**, 916 (1997)

28. A. Strittmatter, A. Schliwa, J.H. Schulze, T.D. Germann, A. Dreismann, O. Hitzemann, E. Stock, I. Ostapenko, S. Rodt, W. Unrau, U.W. Pohl, V. Haisler, A. Hoffmann, D. Bimberg, Lateral positioning of InGaAs quantum dots using a buried stressor. Appl. Phys. Lett. **100**, 093111 (2012)
29. M. Strauß, A. Kaganskiy, R. Voigt, P. Schnauber, J.-H. Schulze, S. Rodt, A. Strittmatter, S. Reitzenstein, Resonance fluorescence of a site-controlled quantum dot realized by the buried-stressor growth technique. Appl. Phys. Lett. **110**, 111101 (2017)
30. O. Hitzemann, E. Stock, A. Strittmatter, A. Schliwa, J.-H. Schulze, T.D. Germann, D. Quandt, W. Unrau, U.W. Pohl, A. Hoffmann, D. Bimberg, V. Haisler, Optical excitation channels of a single site-controlled quantum dot. Paper HL 74.8 Presented at Spring Meeting of the German Physical Society, University of Regensburg, Regensburg, 10–15 Mar 2013
31. M. De Graef, *Introduction to Conventional Transmission Electron Microscopy* (Cambridge University Press, Cambridge, 2003)
32. M. Hÿtch, E. Snoeck, R. Kilaas, Quantitative measurement of displacement and strain fields from HREM micrographs. Ultramicroscopy **74**, 131 (1998)
33. A. Lubk, E. Javon, N. Cherkashin, S. Reboh, C. Gatel, M. Hÿtch, Dynamic scattering theory for dark-field electron holography of 3D strain fields. Ultramicroscopy **42**, 136 (2014)
34. M. Hÿtch, F. Houdellier, F. Hüe, E. Snoeck, Nanoscale holographic interferometry for strain measurements in electronic devices. Nature **453**, 1086 (2008)
35. K. Harada, A. Tonomura, Y. Togawa, T. Akashi, T. Matsuda, Double-biprism electron interferometry. Appl. Phys. Lett. **84**, 3229 (2004)
36. F. Genz, T. Niermann, B. Buijsse, B. Freitag, M. Lehmann, Advanced double-biprism holography with atomic resolution. Ultramicroscopy **147**, 33 (2014)
37. T. Niermann, M. Lehmann, Averaging scheme for atomic resolution off-axis electron holograms. Micron **63**, 28 (2014)
38. D. Richter, R. Hafenbrak, K.D. Jöns, W.-M. Schulz, M. Eichfelder, M. Heldmaier, R. Roßbach, M. Jetter, P. Michler, Low density MOVPE grown InGaAs QDs exhibiting ultra-narrow single exciton linewidths. Nanotechnology **21**, 125606 (2010)
39. W. Unrau, D. Quandt, J.-H. Schulze, T. Heindel, T.D. Germann, O. Hitzemann, A. Strittmatter, S. Reitzenstein, U.W. Pohl, D. Bimberg, Electrically driven single photon source based on a site-controlled quantum dot with self-aligned current injection. Appl. Phys. Lett. **101**, 211119 (2012)
40. S. Rodt, A. Schliwa, K. Pötschke, F. Guffarth, D. Bimberg, Correlation of structural and few-particle properties of self-organized InAs/GaAs quantum dots. Phys. Rev. B **71**, 155325 (2005)
41. C. Taylor, E. Marega, E.A. Stach, G. Salamo, L. Hussey, M. Munoz, A. Malshe, Directed self-assembly of quantum structures by nanomechanical stamping using probe tips. Nanotechnology **19**, 015301 (2008)
42. J. Skiba-Szymanska, A. Jamil, I. Farrer, M.B. Ward, C.A. Nicoll, D.J.P. Ellis, J.P. Griffiths, D. Anderson, G.A.C. Jones, D.A. Ritchie, A.J. Shields, Narrow emission linewidths of positioned InAs quantum dots grown on pre-patterned GaAs (100) substrates. Nanotechnology **22**, 065302 (2011)
43. M. Felici, P. Gallo, A. Mohan, B. Dwir, A. Rudra, E. Kapon, Site-controlled InGaAs quantum dots with tunable emission energy. Small **5**, 938 (2009)
44. A. Huggenberger, S. Heckelmann, C. Schneider, S. Höfling, S. Reitzenstein, L. Worschech, M. Kamp, A. Forchel, Narrow spectral linewidth from single site-controlled In (Ga) As quantum dots with high uniformity. Appl. Phys. Lett. **98**, 131104 (2011)
45. M. Mehta, D. Reuter, A. Melnikov, A.D. Wieck, S.M. de Vasconcellos, T. Baumgarten, A. Zrenner, C. Meier, Intentionally positioned self-assembled InAs quantum dots in an electroluminescent *p–i–n* junction diode. Phys. E **42**, 2749 (2010)
46. C. Schneider, T. Heindel, A. Huggenberger, T.A. Niederstrasser, S. Reitzenstein, A. Forchel, S. Höfling, M. Kamp, Microcavity enhanced single photon emission from an electrically driven site-controlled quantum dot. Appl. Phys. Lett. **100**, 091108 (2012)
47. K.D. Jöns, P. Atkinson, M. Müller, M. Heldmaier, S.M. Ulrich, O.G. Schmidt, P. Michler, Triggered indistinguishable single photons with narrow line widths from site-controlled quantum dots. Nano Lett. **13**, 126 (2013)

48. R.B. Patel, A.J. Bennett, K. Cooper, P. Atkinson, C.A. Nicoll, D.A. Ritchie, A.J. Shields, Quantum interference of electrically generated single photons from a quantum dot. Nanotechnology **21**, 274011 (2010)
49. F. Heinrichsdorff, M. Grundmann, O. Stier, A. Krost, D. Bimberg, Influence of In/Ga intermixing on the optical properties of InGaAs/GaAs quantum dots. J. Crystal Growth **195**, 540 (1998)
50. R. Michalzik, VCSEL fundamentals, in *VCSELs—Fundamentals, Technology and Applications of Vertical-Cavity Surface-Emitting Lasers*, ed. by R. Michalzik. Springer Series in Optical Sciences, vol. 166 (Springer, Berlin, Heidelberg, 2013), pp. 19–75
51. M. Kantner, U. Bandelow, T. Koprucki, J.-H. Schulze, A. Strittmatter, H.-J. Wünsche, Efficient current injection into single quantum dots through oxide-confined *p-n*-diodes. IEEE Trans. Electron. Dev. **63**, 2036 (2016)
52. S. Selberherr, *Analysis and Simulation of Semiconductor Devices* (Springer, Vienna, 1984)
53. W.W. van Roosbroeck, Theory of the flow of electrons and holes in germanium and other semiconductors. Bell Syst. Tech. J. **29**, 560 (1950)
54. P. Farrell, N. Rotundo, D.H. Doan, M. Kantner, J. Fuhrmann, T. Koprucki, Drift-diffusion models, in *Handbook of Optoelectronic Device Modeling and Simulation*, ed. by J. Piprek. Lasers, Modulators, Photodetectors, Solar Cells, and Numerical Methods, vol. 2 (CRC Press, Taylor & Francis, Boca Raton, 2017), pp. 733–771
55. M. Kantner, T. Koprucki, Numerical simulation of carrier transport in semiconductor devices at cryogenic temperatures. Opt. Quantum Electron. **48**, 543 (2016)
56. A. Wilms, P. Mathé, F. Schulze, T. Koprucki, A. Knorr, U. Bandelow, Influence of the carrier reservoir dimensionality on electron-electron scattering in quantum dot materials. Phys. Rev. B **88**, 235421 (2013)
57. T.R. Nielsen, P. Gartner, F. Jahnke, Many-body theory of carrier capture and relaxation in semiconductor quantum-dot lasers. Phys. Rev. B **69**, 235314 (2004)

Chapter 4
Coherent and Incoherent Dynamics in Quantum Dots and Nanophotonic Devices

M. Kolarczik, F. Böhm, U. Woggon, N. Owschimikow, A. Pimenov, M. Wolfrum, A. Vladimirov, S. Meinecke, B. Lingnau, L. Jaurigue and K. Lüdge

Abstract The interest in coherent and incoherent dynamics in novel semiconductor gain media and nanophotonic devices is driven by the wish to understand the optical gain spectrally, dynamically, and energetically for applications in optical amplifiers, lasers or specially designed multi-section devices. This chapter is devoted to the investigation of carrier dynamics inside nanostructured gain media as well as to the dynamics of the resulting light output. It is structured into two parts. The first part deals with the characterization of ultrafast and complex carrier dynamics via the optical response of the gain medium with pump-probe methods, two-color four-wave mixing setups and quantum-state tomography. We discuss the optical nonlinearities resulting from light-matter coupling and charge carrier interactions using microscopically motivated rate-equation models. In the second part, nanostructured mode-locked lasers are investigated, with a focus on analytic insights about the regularity of the pulsed light emission. A method for efficiently predicting the timing fluctuations is presented and used to optimize the device properties. Finally, one specific design of a mode-locked laser with tapered gain section is discussed which draws the attention to alternative ways of producing very stable and high intensity laser pulses.

S. Meinecke · B. Lingnau · L. Jaurigue · K. Lüdge (✉)
Institut für Theoretische Physik, Technische Universität Berlin,
Hardenbergstraße 36, 10623 Berlin, Germany
e-mail: luedge@tu-berlin.de

M. Kolarczik · F. Böhm · U. Woggon · N. Owschimikow
Institut für Optik und Atomare Physik, Technische Universität Berlin,
Hardenbergstraße 36, 10623 Berlin, Germany
e-mail: nina.owschimikow@physik.tu-berlin.de

A. Pimenov · M. Wolfrum · A. Vladimirov
Weierstraß-Institut für Angewandte Analysis und Stochastik,
Mohrenstraße 39, 10117 Berlin, Germany
e-mail: vladimir@wias-berlin.de

M. Kneissl et al. (eds.), *Semiconductor Nanophotonics*, Springer Series in Solid-State Sciences 194, https://doi.org/10.1007/978-3-030-35656-9_4

4.1 Introduction

In optoelectronic devices with semiconductor quantum dots as the active medium a wealth of dynamical phenomena on vastly different timescales leave their imprints on the output of the device. Some of these effects are desired, like modelocking to produce ultrashort broadband laser pulses, others are detrimental to the performance, like excessive laser linewidth broadening. Exploring the performance limits and implementing novel technologies demands a fundamental understanding of the underlying physics. In every case, the mutual interaction of carrier population of the active region with the electric field in the device cavity is governing the dynamical behavior of the device, and will be tackled in this chapter from an experimental as well as a theoretical perspective.

Quantum dots form an inherently broadband, thermally stable and fast gain medium which is particularly suited for the generation of ultrashort pulses. Depending on the timescale investigated, they can appear as nearly ideal two-level systems, or display a complex energy level structure with even mixed-dimensional carrier confinement. This transition covers the picosecond timescale, which is approached by modern ultrafast data transmission techniques. In the short time limit, in the regime of coherent energy exchange between electrical field and carriers, these even have to be treated as wave functions with amplitude and phase. Quantum-optical processes like Rabi oscillations leave distinct patterns on the electrical field coupling to the carriers. The few-picosecond time range represents the spreading of the initial perturbation caused by absorption or emission of photons through the energy levels of the device up to the carrier reservoir. It gives rise to a dynamically varying amplitude phase coupling, as the initial response of gain and refractive index is decoupled due to the cascading nature of the gain recovery process. Data sequences transmitted or manipulated by a device may experience a varying amplification and chirp.

Understanding and controlling these processes can be very helpful for the design of devices that generate ultra short pulses, i.e. mode-locked lasers. In addition to controlling the carrier dynamics, a crucial role is also played by controlling the cavity modes via the cavity design, cavity sectioning and facet reflections.

The chapter is structured into two parts. Section 4.2 focuses on different methods to characterize the ultrafast carrier dynamics. Complementary setups ranging from pump-probe methods, four-wave mixing to quantum-state tomography will be used to measure the optical response of different semiconductor gain materials. It will be shown that the optical nonlinearity resulting from light-matter coupling and complex charge-carrier interactions, e.g. crossed excitons, have to be taken into account in order to interpret and, thus, to correctly model the nonlinear optical response. The second part, Sect. 4.3, deals with mode-locked laser devices that consist of these materials. In the focus of the discussion is the regularity of the pulsating light emission. Here, an analytic method is presented that helps to efficiently characterize the timing fluctuations without extensive stochastic simulations and that is therefore very helpful for optimizing those devices. One specific design of a mode-locked laser with tapered gain section will also be discussed to draw the attention to alternative ways of building integrated devices with very stable and high intensity laser pulses.

4.2 Ultrafast Carrier Dynamics in Semiconductors with Reduced Dimensionality

Zero-dimensional (0D) localization centers for carriers assist the radiative recombination of electrons and holes in semiconductors [1]. The concept is well established in a wide range of applications, ranging from the exploration of fundamental issues of light-matter interaction to the use in opto-electronic devices [2–4]. Frequently, the 0D centers are embedded in a solid state matrix, which supplies excitation and scattering channels via its characteristic energy structure. A classical example for 0D confinement are quantum dots (QDs) grown by the Stranski-Krastanov (SK) method, where narrow-bandgap material is deposited on a large-bandgap matrix until it self-assembles into lens-shaped agglomerations upon reaching a critical thickness [5]. SK-QDs are the work-horse of QD-based opto-electronics, as they provide a clean, spectrally well-separated ground state (GS) transition and a fast charge-carrier recovery after depletion [6–8]. A drawback of this system, however, is that due to the growth method the areal density that can be achieved is limited. For the In(Ga)As material system it is on the order of a few $10^{10}\,cm^{-2}$, which implies a massively reduced density of states (DOS) and achievable optical gain per unit length in comparison e.g. to the two-dimensional (2D) quantum wells (QWs) [9].

For the InAs/GaAs material system, the growth of submonolayer (SML)-based localization centers as In-rich agglomerations in a GaAs matrix promises to combine a large DOS with the favorable properties of QDs [10–13]. Consequently, these devices are expected to provide high optical gain and ultrafast gain recovery. Recently, we have shown that gain and phase recovery in an SML-based opto-electronic device are indeed as fast as observed in SK-QDs, making the system an attractive choice as active medium in opto-electronics [14]. Due to the fast exciton dissociation within the SMLs, they are strong candidates for the application in photodetector and solar cell devices, where the SML medium can lead to a fast and efficient device operation [15, 16]. The rather strong phase response of SML devices makes them favorable candidates for nonlinear optical applications employing self or cross-phase-modulation.

We will investigate the properties of these different nanostructured materials and focus especially on the carrier dynamics with various methods, i.e. pump-probe spectroscopy (Sect. 4.2.1), four wave mixing analysis (Sect. 4.2.2), two color heterodyne techniques (Sect. 4.2.3), and quantum state tomogrophy (Sect. 4.2.4).

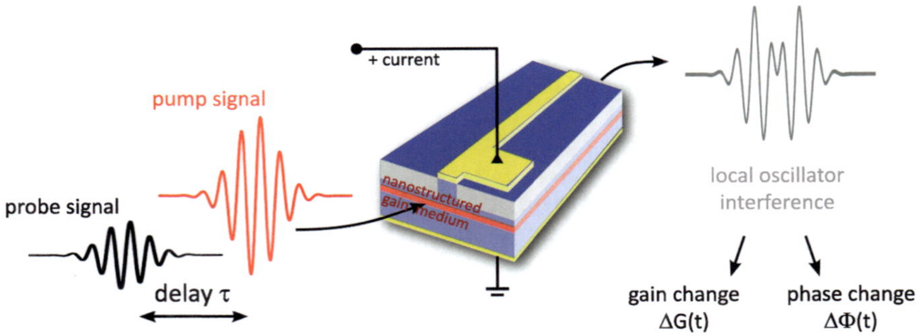

Fig. 4.1 Illustration of the heterodyne pump-probe experiment. By interference of the probe pulse with a local oscillator pulse the pump-induced changes of gain ΔG (population) and phase $\Delta\varphi$ (refractive index) can be resolved. Reprinted with permission from [18]. Copyright (2019) by the American Physical Society

4.2.1 Ultrafast Gain and Phase Recovery Dynamics

In this section we discuss the gain and phase dynamics using the ultrafast heterodyne-detected pump-probe experiment [17], as shown in Fig. 4.1, and compare the results to simulations obtained with microscopically motivated rate equations. The internal charge-carrier dynamics of an optically active material determines its physical properties and device performance to a large extent. Utilizing the heterodyne pump-probe method, it is possible to quantify the dynamic timescales of the involved optical transitions and gain an insight into the charge-carrier dynamics of the investigated material.

The experiment utilizes an interferometric approach to access carrier and phase dynamics in the active region of a waveguide, where pump and probe pulses are transmitted collinear and co-polarized. By interference with a local oscillator pulse on a photodetector, a complex signal $S(t)$ is recorded, from which the differential changes of amplitude and phase of the probe pulse are retrieved. The differential intensity gain is defined as $\Delta G = 20 \times \log(\mathrm{Re}(S)/\mathrm{Re}(S_0))$, and the differential phase as $\Delta\Phi = \mathrm{Im}(S) - \mathrm{Im}(S_0)$, where S_0 corresponds to the signal of the undisturbed system (probe pulse only), and S to the signal after a pump pulse has been applied. The experiment is described in greater detail in [19]. Differential gain and phase recovery for the SML-QDs have been published in [14].

Two different devices will be compared with respect to their recovery properties. One is a semiconductor optical amplifier (SOA) structure containing InAs SK-QDs in a in a dot-in-a-well (DWELL) structure (see sketch in Fig. 4.2a). The QD GS is emitting at 1280 nm which can be seen by the electroluminescence spectra in Fig. 4.2a3 [19]. The QW emission is located at 1160 nm [20]. This corresponds to a separation in energy of about 100 meV between the QW reservoir and dot states, ensuring good 0D confinement in the QD GS. Note that for a comparison of carrier dynamics, this relative separation of localization center and continuum

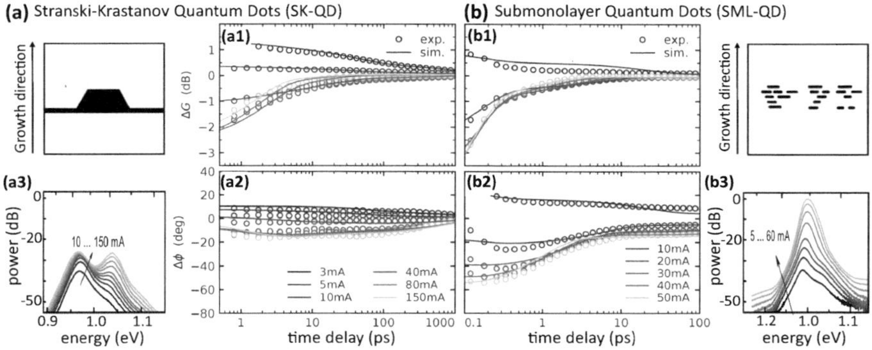

Fig. 4.2 Stranski-Krastanov **a** and submonolayer quantum dots, **b** compared via their respective electroluminescence spectra (**a3**, **b3**), as well as their dynamic gain (**a2**, **b2**) and phase recovery curves (**a1**, **b1**) after perturbation by an optical pump pulse centered at zero time delay. Shown are the results for increasing pump currents (dark to light curves), with the transparency currents being approximately 6 mA for the SK-QDs and 15 mA for the SML-QDs. The symbols show the measured recovery obtained from the pump-probe setup with the simulation results superimposed as solid lines. Excitation wavelength is 1290 nm in (**a**), and 960 nm in (**b**)

is important, and not the absolute emission energy. For the second device we use a SOA based on InAs/GaAs SML-QDs grown by metal organic chemical vapor phase epitaxy as described in [18]. A schematic of the active region containing the SMLs is shown in Fig. 4.2b. The resulting localization centers provide a relatively shallow confinement, as also illustrated in the energy band diagram in Fig. 4.5c, in particular the electron wavefunction is delocalized over many InAs agglomerations, while the hole wavefunction remains more localized [21]. The electroluminescene (EL) obtained from the device is centered at 960 nm at low injection current and shows a slight blueshift for higher currents (Fig. 4.2b3). The strong growth of the luminescence (note the logarithmic power axis) promises a high optical gain.

The optical response of the medium is determined not only by the electronic states resonant with the desired optical wavelength, but by all optically active transitions. From the Kramers-Kronig relations it follows that the medium's refractive index is constituted of the contributions from all non-resonant transitions, and consequently dynamic changes of the refractive index are inevitable in all semiconductor media. A varying index of refraction imprints a change in the optical phase, leading to nonlinear modification and chirp of a propagating optical signal [22]. In a semiconductor medium the different electronic states are usually coupled via Coulomb- or phonon-scattering processes [23], which thus lead to a coupling of the gain and refractive index dynamics. This amplitude-phase coupling [24–26] is commonly described by an α-factor, and quantifies the nonlinear index response of a given active medium. Depending on the application, a small α, i.e. a small dynamic index change, is generally preferred for linear amplification of signals. A high α can be utilized in nonlinear applications such as self or cross-phase modulation [27], or generation of deterministic optical chaos in semiconductor lasers [28, 29].

The measured dynamic gain and phase recovery of the Stranski-Krastanov (SK) and submonolayer (SML) quantum dot materials are shown in Fig. 4.2a, b. Both material systems exhibit an ultrafast gain recovery, revealing an efficient carrier refilling mechanism of the localized states (see panels a1, b1). For SK-QDs, typical recovery times of a few ps are found, which become faster with stronger electrical pumping of the QDs. The SML QDs surpass the gain recovery speed, revealing a sub-ps gain recovery, due to a highly efficient lateral coupling of charge carriers in the SML layers [18]. The dynamic phase response reveals important differences between the materials (compare Fig. 4.2a2, b2). SK-QDs show only a small change in the optical phase, owing to the strongly localized ground and excited states, which reduce the influence of all other off-resonant optical transitions [18, 30, 31]. The main contribution to the phase change comes from the surrounding carrier reservoir states, with a much longer lifetime on a ns scale, leading to a much longer lasting perturbation of the refractive index after an initial perturbation of the medium [32]. The submonolayer QD material, in contrast, shows a stronger amplitude-phase coupling, with a threefold larger phase shift compared to the SK-QDs. From numerical simulations (lines in Fig. 4.2) this strong nonlinear index change can be attributed to the energetically broad SML transitions, as well as a charge carrier reservoir formed by the SML layer itself, facilitating the ultrafast gain recovery but also inducing the observed strong phase shift [18].

The charge-carrier dynamics in the SK-QD structure is determined to a large extent by the charge-carrier capture from the surrounding two-dimensional QW carrier reservoir into the inhomogeneously broadened QD states, and the subsequent carrier relaxation processes from energetically higher excited states (ES) to the QD ground state (GS) [33–35]. A sketch of the energy states of the SK-QDs can be seen later on in Fig. 4.18b. QD materials differ in their inhomogeneous broadening and electronic confinement energy, which in turn influence their carrier dynamics and optical properties [36]. These parameters can be influenced, to an extent, by the choice of active material composition and growth parameters. The impact of the particular structure of the SK-QDs on the gain and phase recoveries can be investigated by performing rigorous parameter studies in numerical simulations. The dependence of the dynamic recovery timescales are shown in Fig. 4.3. The data have been obtained from numeric integration of microscopically motivated rate equations similar to those used later on for the mode-locked laser in Sect. 4.3.4. The full model can be found in [36]. We vary the QD confinement energy of the GS between 55 and 220 meV below the QW band edge. The QD inhomogeneous broadening FWHM is varied between 0 meV, corresponding to perfect homogeneity, and 80 meV. Evaluating the gain and phase recovery simulations shown in Fig. 4.2, we can extract characteristic times, e.g. the time needed for a 90% recovery from the dynamic simulations after perturbation of the medium by a strong optical pulse.

Figure 4.3a depicts this gain recovery time. The gain recovery generally improves with increasing QD confinement, due to the decoupling of the QD states from the slow reservoir recovery and reduced escape rates of carriers into the reservoir. This is due to the decreasing dependence of the quasi-equilibrium QD occupation on changes of the quasi-Fermi level in the reservoir. For high inhomogeneous broadening, spectral hole

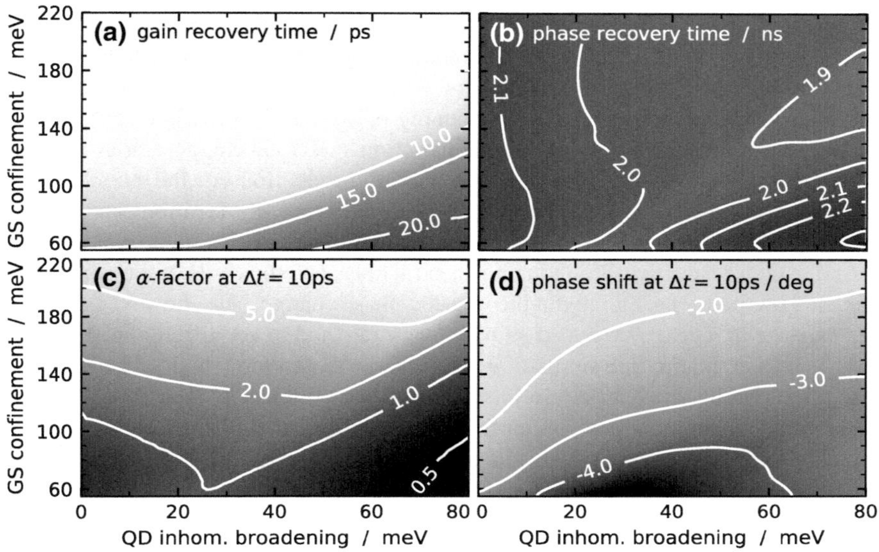

Fig. 4.3 Recovery timescales and quantities after optical pulse amplification in dependence of the QD inhomogeneous broadening FWHM and the QD confinement. **a** Gain recovery time, defined as the time after the pulse when the optical gain has recovered by 90%. **b** Phase recovery time. **c** α-factor at a time delay of $\Delta t = 10\,\text{ps}$ after the pulse amplification. **d** Phase shift at a time delay of $\Delta t = 10\,\text{ps}$ and a pump current of 10 times the transparency pump current. Modified from [36] with the permission of AIP Publishing

burning becomes significant, which slows down the gain recovery. Since no direct coupling between different QD subgroups is possible, the spectral hole recovers via carrier capture from the reservoir on a slow timescale, whereas for narrower QD distributions the carrier variation is more uniform and the recovery is dominated by the faster relaxation processes. The phase recovery time in Fig. 4.3b, on the other hand, shows a much smaller dependence on the QD structure, and occurs on the slow reservoir recovery time over the complete parameter range, due to the dominating role of the reservoir transitions in the phase recovery. Figure 4.3c shows the resulting α-factor at 10 ps after the optical excitation, defined from the gain change and phase change via [37]

$$\alpha(t) := -\frac{20}{\ln 10}\frac{\Delta\phi(t)/\text{rad}}{\Delta G(t)/\text{dB}}\,. \tag{4.1}$$

The α-factor evaluated at $\Delta t = 10\,\text{ps}$ exhibits a clear increase with stronger QD confinement. The reason for this is again the decoupling of the gain and phase recovery with increasing level separation, which improves the gain recovery but leaves the phase recovery nearly unchanged. The low α-factor observed for high inhomogeneous broadening must therefore be treated with caution, as it is a consequence of slow gain recovery, as shown in panel (a), and not an indication of smaller optical phase change. A direct evaluation of the phase shift at $\Delta t = 10\,\text{ps}$ is therefore shown in Fig. 4.3d. Here, the absolute phase change can instead be seen to decrease slightly with increasing GS confinement. The phase change exhibits a local maximum around an inhomogeneous broadening of around 30 meV. For narrower QD distributions, the higher QD differential gain limits the total charge-carrier variation and thus the induced phase change. For broad QD distributions, on the other hand, the lower energy part of the ES spectrum overlaps with the GS transitions and become resonant with the pump pulse. The carrier depletion in the ES is subsequently transferred to their respective GS, inducing spectral holes on either side of the pump pulse wavelength, partly balancing out the induced phase change. This, however, comes at the expense of a greatly reduced gain recovery rate and lower differential gain.

The measurements and simulation results show that in semiconductor media gain and refractive index changes are intricately linked. While the choice and tailoring of material parameters can influence both quantities, the type of active material dictates some fundamental dynamic properties. Depending on the application, not all QD materials are equally suited. In the following, we consider the phenomenon of ultrafast coherent optical nonlinearities in SK QDs—where both the ultrafast charge carrier dynamics as well as the long-lived polarization lifetime in SK-QDs are exploited.

4.2.2 Ultrafast Coherent Optical Nonlinearities

The sub-picosecond charge-carrier dynamics in self-organized semiconductor QDs not only enhance the optical signal amplification bandwidth, but also enable applications utilizing the ultrafast carrier dynamics itself. The propagation of a strong optical pulse through the QD medium generally perturbs the carrier distribution by spectral hole burning or total carrier depletion, followed by the ultrafast, nonlinear recovery process. This nonlinearity in turn generates a strongly nonlinear optical susceptibility in the active medium. Such optical nonlinearities are commonly employed for all-optical wavelength conversion schemes, such as nonlinear wave mixing [38–42], cross-gain modulation [43–45], or for the generation of optical supercontinua and frequency combs [46–48]. The ultrafast charge-carrier dynamics in SK-QDs has previously been exploited in ultra-high bandwidth four-wave mixing (FWM) applications [49–51]. In addition to the ultrafast carrier dynamics, optical amplifiers based on InAs QDs (QDSOAs) were shown to possess long polarization dephasing times [52, 53], making it possible to drive the QDs in the strong coupling regime in high-Q optical cavities [54–56], and to excite Rabi oscillations by amplification of strong optical pulses [57, 58]. These Rabi oscillations lead to a splitting of the optical spectrum by the Rabi frequency, which is determined by the optical field amplitude [59, 60]. The excitation of Rabi oscillations can thus be viewed as a high-order optical nonlinearity. The interplay between the nonlinear FWM processes and the induced Rabi oscillations in QDSOAs can invoke substantial differences in the optical response when the Rabi frequency is close to the desired frequency conversion bandwidth.

To maximize the nonlinear FWM performance, semiconductor optical amplifiers must be operated within the nonlinear amplification regime. Here, the optical signal is strong enough to perturb the charge-carrier distribution, while the optical gain remains sufficient to produce nonlinear mixing products of appreciable power. Increasing the optical input power further saturates the optical gain of the amplifier and generally deteriorates the FWM performance. We define the FWM conversion efficiency, η_{FWM}, for two injected optical cw signals, the pump and the probe field with a frequency detuning Δ, as

$$\eta_{\text{FWM}} := \frac{P_{\text{out}}(-\Delta)}{P_{\text{in}}(+\Delta)}, \tag{4.2}$$

which yields the power ratio of the newly generated signal at a detuning $-\Delta$ with respect to the injected data signal. In Fig. 4.4a, we show the FWM conversion efficiency in a QDSOA in dependence of the optical power of the input beams from numerical simulations [61]. The FWM efficiency is calculated by keeping the pump beam centered at a wavelength of $\lambda = 1.3\,\mu\text{m}$, close to the GS transition frequency, and varying the pump-probe detuning Δ. The conversion efficiency curves exhibit a characteristic cone-like shape, showing a maximum in η_{FWM} for small $|\Delta|$ which decreases with higher detuning frequency. We observe a nearly symmetric shape with respect to the sign of the detuning Δ, which is a typical feature in QDSOAs [44] and can be explained by the small amplitude-phase coupling in QD active media, i.e., small dynamic change of the refractive index in the active medium as discussed and plotted in Fig. 4.3c. The QDSOA allows for efficient four-wave mixing with $\eta_{\text{FWM}} > -30\,\text{dB}$ within a detuning range of $\pm 1\,\text{THz}$ at an input optical power of 0 dBm. With increasing optical power the conversion efficiency can be seen to

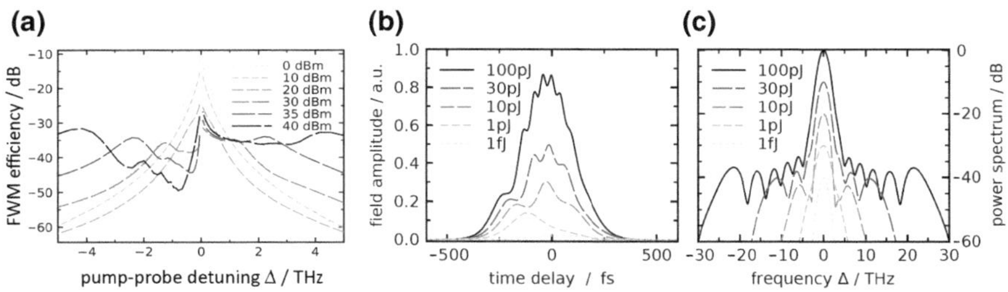

Fig. 4.4 **a** Four-wave mixing conversion efficiency η_{FWM} in dependence of the pump-probe detuning Δ, for increasing optical input power. The pump and probe field amplitudes were chosen to be equal. The deformation of the FWM efficiency curves at high input power is due to resonant excitation of Rabi oscillations. **b** Electric field amplitude profiles (left) and optical power spectra (right) of optical pulses after propagation through a SK-QD optical amplifier for different input pulse energies. The spectra in the right plot are shifted by 10 dB with respect to each other for better readability

decrease, as the device is driven further into gain saturation. An optimal FWM efficiency is found for an input power of 0 dBm, which is strong enough to induce a nonlinear response of the active medium while still maintaining considerable optical gain.

At even higher optical input power, additional features can be seen. A shoulder in the FWM efficiency curve appears, which develops into pronounced peaks moving towards higher detuning with increasing optical power. The peak structure greatly enhances the FWM efficiency at high detuning frequencies, reaching up to -32 dB at $\Delta = \pm 4$ THz for an input power of 40 dBm, increasing the FWM bandwidth fourfold in comparison to the weaker optical input. The observed peaks in the conversion efficiency are caused by the resonant excitation of Rabi oscillations and their harmonics by the beating frequency of the injected pump and probe signals. The dependence of this induced efficiency optimum on the field amplitude might enable high-bandwidth conversion over several THz: For any desired conversion frequency the optical power should be chosen such that its Rabi frequency matches the pump-probe detuning to ensure high conversion efficiency.

While the SK-QD gain medium allows harnessing Rabi-oscillations for optical wavelength conversion under cw excitation, this is possible only for optical input power above 25 dBm. The cw optical powers required for the Rabi-enhanced FWM efficiency in the QD active medium exceed 1W of continuous optical excitation, which in many cases is unwanted or unfeasible. Ultrashort optical pulses, on the other hand, reach high peak intensities while maintaining a reasonably low average optical power. The nonlinear pulse propagation of such pulses in QDSOAs has been shown to induce Rabi oscillations even at room temperature [57], as soon as the input pulse area approaches 2π [62]. Experimental results of the pulse shaping of a strong optical pulse with an FWHM of 230 fs after propagation through a 1.5 mm long QDSOA have been obtained using the previously introduced heterodyne pump-probe method. Both below and above transparency the output pulse develops a characteristic change in the pulse shape, with a pronounced dip at the trailing edge of the pulse (positive time delay), accompanied by a significant change in optical phase. Simulated pulse propagation, using a traveling wave Maxwell-Bloch model [57], reproduces this change in pulse shape, linking it to a partial Rabi flop induced by the strong optical pulse. The experimental setup however limits the temporal resolution to the pulse width of the pump pulse, as only a convolution of the propagated and reference pulse is measured. From the numerical simulations, we can extract the actual shape of the electric field amplitude. We simulate amplification of pulses with energies between 1 fJ and 100 pJ. Generation of fs optical pulses in this power range has recently been achieved using high-power mode-locked lasers [63, 64]. The time traces of the pulses and pulse spectra resulting from our modeling are shown in Fig. 4.4b, c [61]. The weakest pulses with an energy of 1 fJ can be seen to retain their Gaussian shape both in time as well as frequency space, being linearly amplified. With increasing pulse energy a strong shaping of the pulse profile appears, leading to a pronounced shoulder at the leading pulse edge which evolves into a modulation on top of the pulse. These induced Rabi oscillations create several additional frequency components in the optical pulse spectrum at either side of the central peak. With increasing pulse amplitude the new spectral components span a larger frequency range, corresponding to the faster modulation of the output pulse shape. The optical spectrum shown in Fig. 4.4c exhibits an intricate fringe pattern, which can be interpreted as an interference pattern of the field emitted by the QD medium at different times during the passage of the strong optical pulse [65]. We remark that this spectral pattern, explained in [65] for a single two-level system, is retained even for the inhomogeneously broadened gain medium considered here. For the highest pulse energy of 100 pJ, the optical spectrum is extended over a frequency spanning almost 50 THz with a power spectral density of roughly −40 dB relative to the center frequency. Our results show that QDSOAs driven deep into the saturated gain regime could be used for the creation of ultra-broadband optical frequency combs by employing optically induced Rabi oscillations.

4.2.3 Crossed Excitons

In the context of submonolayer (SML) structures (for more details about these structures please also see Chap. 1 of this book), it is necessary and well–accepted to assume a heterodimensional confinement with 0D electron states and 2D hole states. This heteroconfinement does not hinder efficient light emission from SML structures as was discussed in the last sections. In the case of QDs, this type of confinement is usually not considered. In DWELL structures, however, a similar behavior can be expected at least for shallow QD confinement. Bound 0D-2D states have previously been discussed and been referred to as *crossed excitons* [66, 67]. In Fig. 4.5, the intermediate scenario between SK-QDs and SMLs is illustrated: For the SK-QD, the lowest energy interband transition is the direct 0D exciton (Fig. 4.5a). A crossed exciton in the same potential landscape will have a higher transition energy (Fig. 4.5b). This state is similar to the SML exciton (Fig. 4.5c) where this case represents the lowest transition energy.

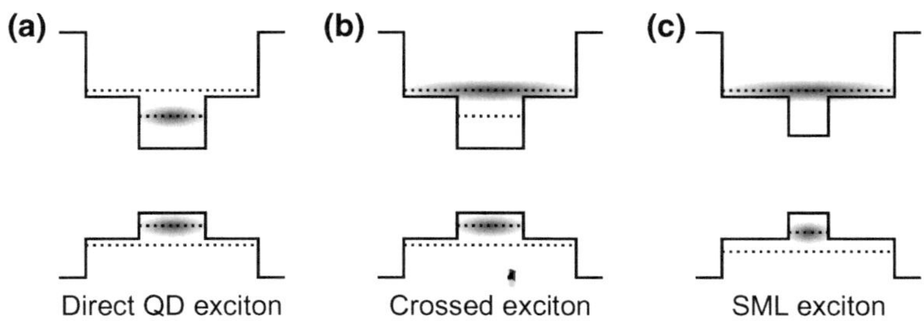

Fig. 4.5 Crossed excitons can be considered as an intermediate case of the QD and SML confinement: **a** In QDs, usually only homoconfined 0D excitons are considered. **b** In contrast, a crossed exciton is a heteroconfined state bound to the QD, which is not the lowest energy transition. **c** In SMLs, this type of 0D-2D excitons represents the lowest bound interband state

The presence of crossed excitons in the coupled QD/QW (or QD/bulk) system would have implications on two levels: Firstly, additional optical transitions are expected with transition energies in the intermediate range between the respective QD and QW (bulk) transition energies. Secondly, crossed excitons can be considered as additional localized states, which will have also an impact on capture and escape processes as well as the inter-dot diffusion. By two-color pump-probe spectroscopy on QDs embedded in the active region of an SOA, we identify the contributions of these additional bound states and provide evidence for the existence of both, QD/QW and QD/bulk crossed excitons.

A first experimental series focuses on the optical excitability of QD/bulk crossed excitons [20]. Both, pump and probe pulse, are shaped to a spectral width of 15 nm, while the amplified spontaneous emission indicates an inhomogeneous broadening of the QD GS ensemble of 30 nm. The probe central wavelength is kept at 1280 nm, corresponding to the QD GS, throughout this series. The pump pulse central wavelength is scanned in a range from 1200 nm (0.968 eV) to 1060 nm (1.170 eV) in steps of 10 nm. The experiment probes the immediate response of the QD GS to an excitation in an energy range where no bound states are predicted by a model based on a linear superposition of QD and QW states. As an additional degree of freedom, the injection current of the SOA is tuned from 0 to 200 mA.

Each of the resulting large number of pump-probe curves (see Fig. 4.7) is reduced to two parameters: The immediate response at 0 ps is determined by a linear fit of the gain dynamics right after the excitation (1–5 ps), the slope of the fit indicates a subsequent feeding or depletion of the QD GS. As a first result, this method yields the exact spectral position of the QW band edge. It is determined by the cross-over from a falling to a rising slope, corresponding to pumping above and below the band edge of the reservoir, respectively.

When displayed over the pump central wavelength, the immediate GS response reveals a remarkable pattern: The excitation is minimal in the range of the QW band edge, but shows distinct peaks below and above this energy. The measured pattern is shown in Fig. 4.6, left. The peak below the QW band edge coincides with the luminescence from the QD ES. However, the structured response above the band edge is not expected by the traditional model. The peaks can, however, be attributed to crossed excitons formed of a QD electron and a bulk hole. The fact, that the crossed excitons and the QD GS transition share this electron, explains the immediacy of the response.

In contrast, QD/QW crossed excitons are expected at energies between GS and QW. They therefore overlap with the inhomogeneously broadened QD states and are obscured by the direct excitons, in particular in the spectral proximity of the QW at the ES luminescence peak. The spectrally most isolated QD/QW crossed exciton has the lowest energy and is formed by a QD electron and a QW hole. It is located about 30 meV above the respective GS transition, which is a sufficient separation to address these energy ranges in a two-color pump-probe experiment. We investigate the coupling of these states in a second set of pump-probe experiments [19]: The probed GS subensemble is chosen at the low energy flank of the GS luminescence. The corresponding crossed exciton, which is addressed by the pump pulse, is therefore

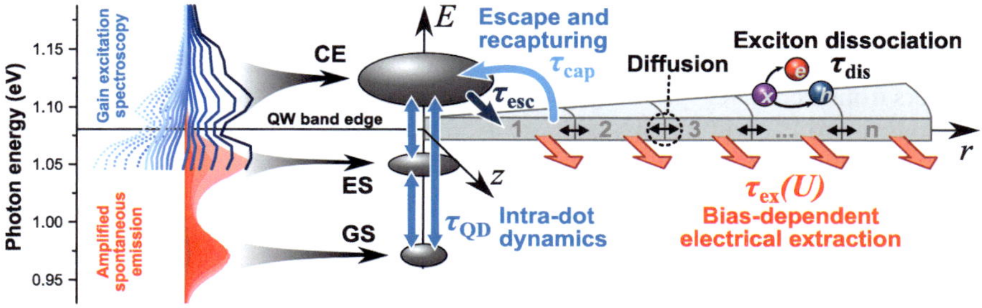

Fig. 4.6 Left: Gain excitation spectroscopy reveals the existence of optical transitions above the QW band edge energy that can be attributed to QD/bulk crossed excitons. These transitions are directly coupled to the QD GS transition. Right: Diffusion dynamics can be described by a rate equation model consisting of QD ground state (GS), excited state (ES), a generalized crossed exciton state (CE), and a 2D continuum. The continuum is described in a 1D representation of concentric circles that allow carrier diffusion by means of a random walk

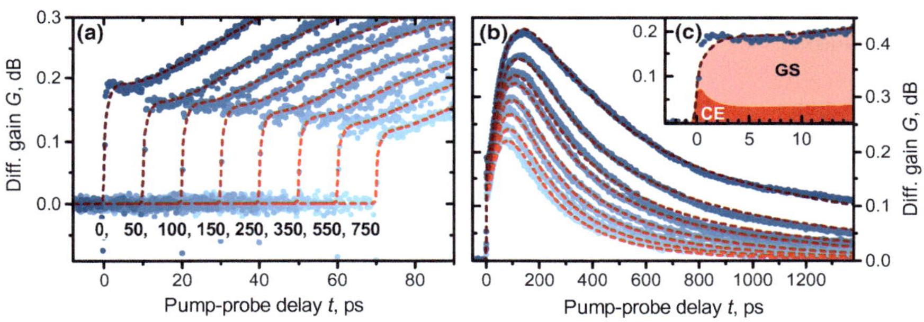

Fig. 4.7 Diffusion dynamics observed in two-color pump-probe experiments (blue scatter) and explained by a linear rate equation model (red dashed) under variation of the extraction potential (bold numerals, mV). **a** An immediate response is observed, although pump and probe pulse are spectrally separated. **b** Inter-dot diffusion takes place on the timescale of 100 ps before carrier decay becomes the predominant process. **c** The model shows that the crossed exciton is the excitation channel that rapidly feeds the GS direct exciton transition. Reprinted from [19]

located at the high energy flank, but clearly below the ES luminescence. Like in the QD/bulk crossed exciton case, the two-color pump-probe curves show an immediate onset.

This immediate onset on the timescale of 500 fs is shown in Fig. 4.7a cannot be explained by inter-dot diffusion, although the pump pulse does also address higher energy direct excitons in smaller QDs. The diffusion process takes place on the timescale of 100 ps and can be observed subsequently in Fig. 4.7b: The gain curves rise up to a maximum at 130 ps, before the decay channels start to dominate the dynamics. Two prerequisites allow the observation of diffusion dynamics in this measurement series: Firstly, it is performed without electrical carrier injection, so only optically created excitons are observed. Secondly, the spatial isolation of the QDs ensures that the predominant number of excitons is also spatially isolated.

To gain deeper insight into this diffusion dynamics, the SOA bias is varied below the carrier injection threshold. In the unbiased case, the diode band bending extracts carriers from the active region. This band bending can be fully compensated, resulting in a longer period of diffusion. Additionally, the linearity of the response to the pump pulse is monitored by simultaneous multi-power measurements.

Intra-dot dynamics and diffusion in the QW can be modeled by a linear rate equation system [19], which is also possible due to the spatial isolation. The model scheme, which incorporates an additional crossed exciton state in the QD, is shown in Fig. 4.6 right. This additional crossed exciton state acts as a gate between the QD and the QW. This implies that capture and escape processes occur stepwise, with one single carrier entering (leaving) the QD first and forming a crossed exciton. The diffusion is emulated by a random walk in a one-dimensional representation of the QW. The rate equation scheme in Fig. 4.6 is solved nine times, for three carrier types (excitons, electrons, holes) and three transition energy subensembles, respectively. The model system yields excellent agreement with the total of 178 pump-probe curves in a resonant and an off-resonant series. It also reveals the contributions of direct and crossed excitons to the immediate onset in Fig. 4.7c. The increased degree of localization by the crossed exciton state is necessary to describe the resonant and off-resonant series by a single set of parameters.

A coherent interaction of crossed and direct exciton is already implied by the comparison of Fig. 4.5a, b. In this example, both excitons share the electron as a single carrier. Two-dimensional coherent spectroscopy is an established tool to investigate these coherent couplings [68]. It requires the phase-resolved observation of the four-wave mixing (FWM) signal of two pulses under variation of two delays. In this particular case, however, the relevant states spread over an energy range of 200 meV and the SOA restricts the experiment to a collinear geometry. This energy range can be addressed by modern fiber laser systems with supercontinuum generation, and the collinearity is one of the original purposes of the heterodyne approach. We extended the heterodyne pump-probe setup by implementing multiband-detection [69] including the FWM band [17], phase-resolution following the FROSCH concept [57], and time referencing based on simultaneous multi-power measurements [19]. The resulting setup, referred to as *Supercontinuum based Two-dimensional Observation of Radiation Coherences by Heterodyning* (STORCH), is displayed in Fig. 4.8a.

In Fig. 4.8b, the amplitude of a STORCH spectrum of the unbiased SOA is shown. Along the diagonal, the direct GS transition contributes with a strong signal. Above this area, a wide range of off-diagonal signal is observed, indicating a coherent coupling to other states. This signal can be attributed to the QD/QW crossed excitons that have already been addressed in the two-color pump-probe experiments, causing the immediate onset in Fig. 4.7a. The STORCH spectrum now allows a much more detailed insight into the coupled system. It also provides important information like the dephasing time T_2, which is in this case in excellent agreement with literature values [70]. The unparalleled spectral width and the successful extraction of first data makes STORCH a promising tool for the future investigation of coupled nanostructures.

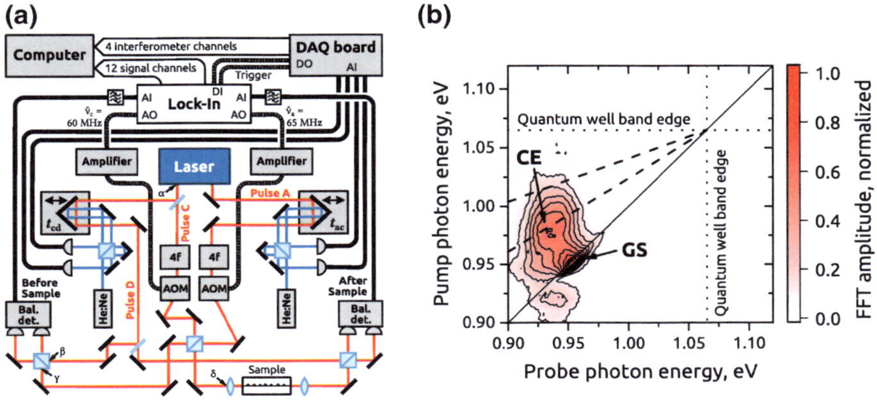

Fig. 4.8 **a** The STORCH setup for two-dimensional coherent spectroscopy. **b** A STORCH spectrum of the unbiased SOA. The off-diagonal coupling signal can be attributed to QD/QW crossed excitons, while the diagonal signal is dominated by direct excitons

4.2.4 Quantum State Tomography

In this section, we discuss an alternative experimental method that is able to extract the carrier dynamics directly from the statistics of the emitted light. It is thus possible to investigate the system without perturbing it by a prior pulse [71]. As discussed in Sect. 4.2.1, the knowledge of both phase and amplitude dynamics is crucial for our understanding of quantum-confined semiconductor nanostructures. Quantum state tomography (QST) is a technique to characterize quantum optical properties of a light field and is particularly useful when dealing with macroscopic or mesoscopic QD ensembles, from which the emission of non-classical light is expected [72, 73]. Here QST is applied to analyse the output of a semiconductor optical amplifier based on InAs QDs as gain material. For the input and the output state the Wigner function, density matrix and intensity autocorrelations are obtained by a balanced detection scheme in a self-heterodyning setup [71, 74]. The full quantum state is reconstructed from the quantum statistical measurement of the field fluctuations [75, 76] which can also be applied to pump-probe experiments.

The Wigner function W of an optical field is a well-known example for the most complete representation of the quantum state of light, i.e. the quantum-optical analogue of its optical phase space. It is defined as

$$W(X_1, X_2) = \int_{-\infty}^{\infty} \left\langle X_1 - \frac{x}{2} |\rho| X_1 + \frac{x}{2} \right\rangle e^{iX_2 x} \mathrm{d}x$$

where X_1, X_2 are the quadrature components, and ρ is the density matrix operator of the light field. The vacuum state, for example, is characterized by Gaussian-

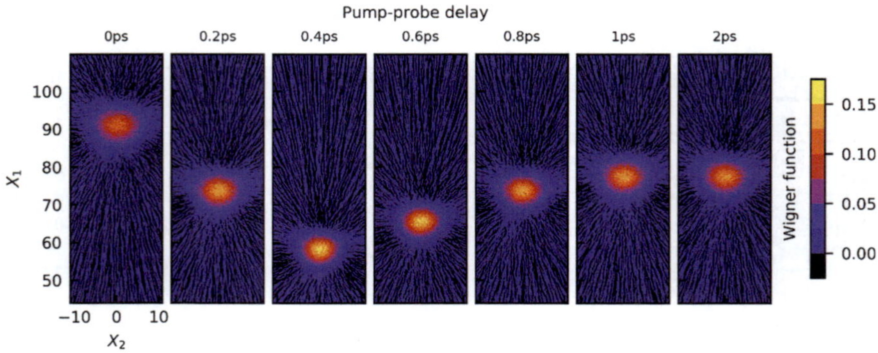

Fig. 4.9 Wigner functions observed in pump-probe quantum state tomography. Shown is the amplified probe state (QD GS) under the influence of the QD ES pump with the delay varied. The injection current is 80 mA above gain threshold

distributed fluctuations around the origin of the X_1, X_2-plane with minimal extension. The coherent state resembles a displaced vacuum state, which rotates around the origin. Though the Wigner function cannot be measured directly, it is possible to reconstruct it from projections of the electromagnetic field of interest onto a well-defined reference field, mapped for all angles of the reference field [71, 77].

The combination of QST with a pump-probe approach allows us to directly analyse the state inside the amplifier at a picosecond timescale. Using both electrical and optical carrier injection we measure the photon statistics as a function of pump-probe delay. The detection sensitivity is at a level where the fluctuations of the vacuum field can be clearly measured. Exemplary Wigner functions constructed for different pump-probe delay times are shown in Fig. 4.9. Evident in the pump delays series of Fig. 4.9, with the pump resonant to the QD ES and the probe resonant to the GS, we witness in the Wigner function a clear decrease in amplification of the signal, as well as a decrease in the noise of both quadratures. The qualitative influence of pumping the ES is to deplete both the gain and noise of the amplified probe which is resonant to the GS. In traditional pump-probe experiments, at these ultrafast timescales, coherent dynamics and population dynamics can merge, and are hard to quantify independently. However, with the QST setup, we can separate those effects, which is a clear advantage of the method and allows one to time-resolve the thermal-to-coherent evolution as a laser pulse is travelling through a semiconductor optical amplifier.

A complementary, more robust approach to data analysis is to extract the photon number distribution from the raw data [74]. The diagonal elements of the photon number distribution can be sampled without phase information, but it is also possible to recover the full photon density matrix. The photon density matrix in the basis of Fock-states is recovered from a convolution of field fluctuations and pattern functions, as has been pioneered in [75, 78].

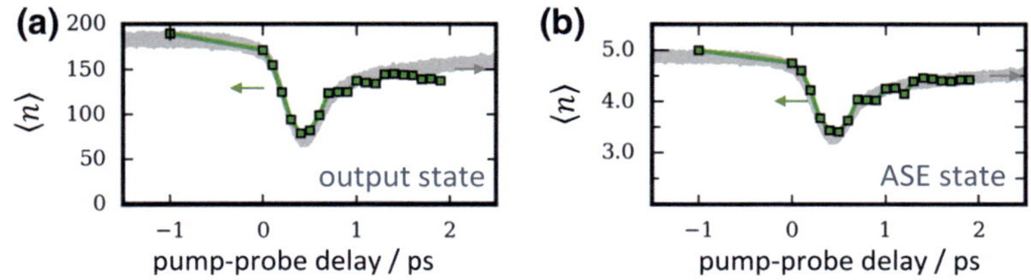

Fig. 4.10 Mean number of photons (green symbols) for an output state after amplification (**a**) and for the amplified spontaneous emission (ASE state) as a function of pump-probe delay (**b**). The measured differential gain is shown as a gray curve (arb. units) [74]

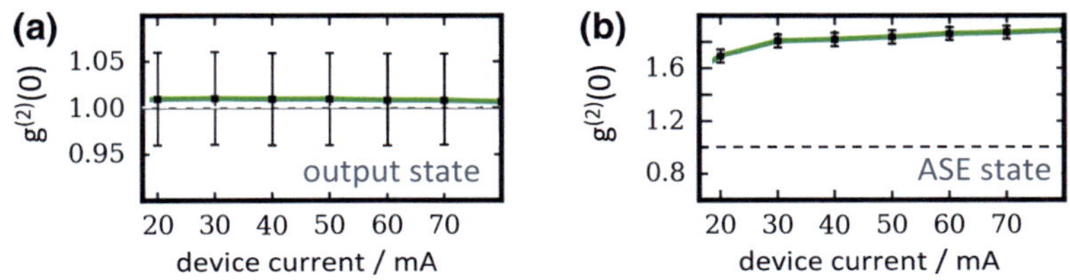

Fig. 4.11 Intensity autocorrelation function $g^{(2)}(0)$ as a function of the pump current for **a** a pulse transmitted through the SK-QD amplifier at the ground state energy and **b** the amplified spontaneous emission exiting the amplifier

The interferometric detection achieves a high sensitivity in the few photon limit and allows one to correctly reconstruct the photon statistics of states with a very low number of photons. Figure 4.10 shows the measured mean photon number both for an amplified state, i.e. after the light passed the pumped amplifier (Fig. 4.10a), and for the pure amplified noise that exits the unpumped device. As can be seen in the latter case (Fig. 4.10b), our technique reaches a resolution of $\approx$3 photons, making it a complimentary option to standard Hanbury Brown-Twiss experiments in the characterization of few-emitter systems. From the pump-probe data shown in Fig. 4.10, second order autocorrelation functions $g^{(2)}(0)$ have been calculated with high precision and shown in Fig. 4.11a as a function of the pump current.

The intensity autocorrelation function $g^{(2)}$ is a powerful tool to characterize light as it displays different specific values depending on the photon statistics. The most classical quantum state of light is the coherent state of the laser where $g^{(2)}(\tau) = 1$ independently of the time delay τ between two detected photons [79]. A $g^{(2)}(0) = 2$ is commonly associated with a thermal photon statistics, also referred to as bunched light, while a value of $g^{(2)}(0) < 1$ is called anti-bunching and a signature of non-classical light emission (see Chap. 7 of this book for more details on single photon sources). $g^{(2)}(0)$ can be obtained from the photon number distribution via

$$g^{(2)}(0) = 1 + \frac{\langle n^2 \rangle - \langle n \rangle}{\langle n \rangle^2} \tag{4.3}$$

where the moments of the photon number distribution can be extracted from the data according to

$$\langle n\rangle = \langle\langle X_1^2\rangle\rangle_{X_1,\theta} - \frac{1}{2}, \quad \text{and} \quad \langle n^2\rangle = \frac{2}{3}\langle\langle X_1^4\rangle\rangle_{X_1,\theta} + \frac{1}{2}. \tag{4.4}$$

By using a pulsed reference, the intensity autocorrelation can be determined with spectral and temporal resolution only given by the reference pulse.

In Fig. 4.11a it can be seen that the statistics of the amplified pulse (output state) is close to the coherent limit of $g^{(2)}(0) = 1$. However, caused by the contribution of the amplified spontaneous emission, we observe a slight deviation to values of $g^{(2)}(0) = 1.01$. Even though in this particular experiment the number of photons originating from the pulse was up to 60 times larger than the number of photons originating from the thermal background, the latter make an observable contribution to the statistics. The intensity autocorrelation $g^{(2)}(0)$ derived without a probe pulse, i.e. for the pure amplified spontaneous emission, is shown in Fig. 4.11b. It shows values approaching $g^{(2)}(0) \approx 2$ as expected for thermal radiation.

The presented QST method can also easily be extended to include the study of other material systems and other device types. For laser systems, it might be interesting to study the quantum mechanical effects in small cavities and their influence on the emission characteristics. Special interest lies in the transformation from super- to sub-Poissonian statistics at the laser threshold, as effects of sub- and superradiance have been predicted in this regime [73]. Furthermore, it could be of high interest to study time dependent effects such as the evolution of the photon statistics at the laser turn on.

4.3 Multisection Mode-Locked Semiconductor Lasers

After focussing on different methods to characterize the carrier dynamics in nanostructured materials, this section will now focus on optical devices. Combining the active materials described in the last section into a two section device as sketched in Fig. 4.12a, it is possible to synchronize the longitudinal cavity modes within the gain spectrum. This so called passive mode-locking (ML) leads to the emission of regular pulse trains at a repetition rate that is given by the cavity roundtrip time, which is on the order of tenth of GHz for integrated devices. Using semiconductor materials, these mode-locked lasers are easy to fabricate and thus in principle inexpensive sources of short optical pulses and promising candidates in various applications, such as data communication, optical clocking, high speed optical sampling, all-optical clock recovery, and microscopy among other applications [80–83]. Furthermore, the mode locked regime is associated with the so-called optical frequency combs in the frequency domain, consisting of a series of discrete, equally spaced frequency lines. Optical frequency combs are used in numerous scientific and practical applications, including high precision spectroscopy, metrology, medicine, optical clocking [84].

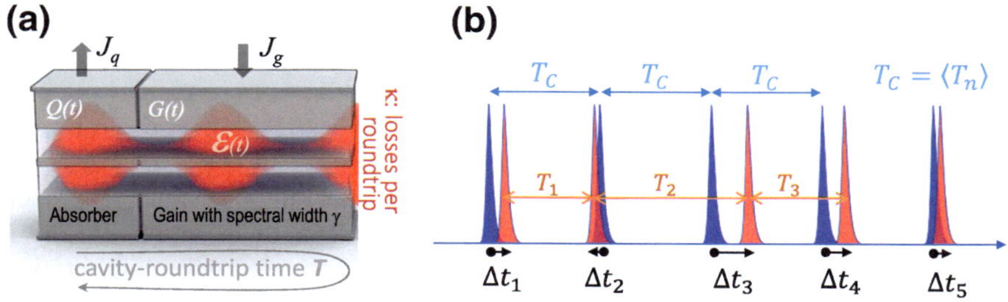

Fig. 4.12 **a** Mode-locked semiconductor laser with gain and absorber sections. **b** Sketch of the varying pulse positions Δt_N of the emitted pulse train (red pulses) as deviations from an ideal pulse train (blue) with mean repetition period T_C

However, in the presence of noise the pulse train emitted by such lasers deviates from that of the perfectly periodic one (see Fig. 4.12b for an illustration). In semiconductors the spontaneous emission rate is rather high and thus large fluctuation of the temporal pulse positions, also called pulse timing jitter, occur [85].

In the course of this section we will use a delay differential equation (DDE) model proposed by [86] and present the different dynamic regimes in Sect. 4.3.1. The model effectively describes the emission of a mode-locked laser but still allows to find analytic results. In Sect. 4.3.2 we will focus on the description of the timing fluctuations of the pulse positions (timing jitter) which stem from spontaneous emission noise and degrade the performance of the ML laser. A lot of effort has been made to reduce the timing jitter in these lasers and therefore we will discuss two methods of timing jitter reduction in Sect. 4.3.3 before we present an alternative tapered multi-section design that is capable of producing high power pulses with small timing jitter in Sect. 4.3.4.

4.3.1 Delay Differential Equation Modeling

A standard theoretical approach to model mode-locked semiconductor lasers is based on the application of the so-called travelling wave equations governing the temporal and spatial evolution of complex electric envelopes of the two counterpropagating waves in the laser cavity and carrier density in the laser sections [87, 88]. This set of partial differential equations must be supplied with proper boundary conditions on the laser facets and can be solved numerically with the help of specially designed software. Another traditional method for modelling mode-locked lasers is based on the so-called Haus master equations [89]. Haus master equations are much simpler than the traveling wave model and under certain simplifying assumptions yield an analytical expression for a ML pulse in the form of hyperbolic secant [89]. However, the derivation of the Haus master equation relies on small gain and loss approximations, which are hardly valid for semiconductor lasers. This, in particular, explains why the master equation fails to reproduce the asymmetry of the mode-locked pulses emitted by semiconductor lasers. Furthermore, the form of the Haus master equations depends on the ratios of the gain and absorber relaxation rates to the cavity decay rate: the so called slow absorber [90] and fast absorber [91] limits. To go beyond the limitations of the Haus theory, the delay differential equation (DDE) model of a mode-locked laser was proposed in [86, 92, 93] using the lumped element approach. Similar to the Haus master equations this model assumes unidirectional lasing in a ring cavity. However, the DDE ML model is free from the small gain and loss approximation and is valid for any values of the gain and absorber relaxation times (for more details on the relation between DDE model and Haus master equations see [94]).

In the following we use this DDE ML model. It governs time evolution of slowly varying electric field amplitude $\mathscr{E}$ at the entrance of the absorber section, the saturable gain G and the saturable loss Q, which are determined by the integrals of the carrier density over the amplifying and absorber laser sections, respectively. It reads:

$$\dot{\mathscr{E}}(t) = -\gamma \mathscr{E}(t) + \gamma R(t-T)\, e^{-i\Delta\Omega T} \mathscr{E}(t-T) + D\xi(t) \tag{4.5}$$

$$\dot{G}(t) = J_g - \gamma_g G(t) - e^{-Q(t)}\left(e^{G(t)} - 1\right)|\mathscr{E}(t)|^2, \tag{4.6}$$

$$\dot{Q}(t) = J_q - \gamma_q Q(t) - r_s e^{-Q(t)}\left(e^{Q(t)} - 1\right)|\mathscr{E}(t)|^2, \tag{4.7}$$

where the amplification and losses of the electric field during one roundtrip in the cavity are given by the factor

$$R(t) \equiv \sqrt{\kappa}\, e^{\frac{1}{2}((1-i\alpha_g)G(t)-(1-i\alpha_q)Q(t))}. \tag{4.8}$$

The pump parameter is described by J_g while J_q describes the unsaturated absorption (see also Fig. 4.12a for a visualization of the device under study). The carrier lifetimes in the gain and absorber sections are given by $1/\gamma_g$ and $1/\gamma_q$, respectively. The factor r_s is the ratio of the saturation intensities in the gain and absorber sections. The cold cavity round trip time is defined as $T \equiv v/L$, where L is the length of the ring cavity. The spectral bandwidth of the output radiation is limited either by the finite width of the amplification line or by a special spectral filtering section incorporating a Bragg reflection element [95]. Spectral filtering of laser radiation is modelled by a linear Lorentzian-shaped filter function of width γ. The parameter $\Delta\Omega$ describes the shift between the reference frequency and the central frequency of the spectral filter. Internal and out-coupling losses per roundtrip are taken into account by the attenuation factor κ. The linewidth enhancement factors (α-factor) in the gain and absorber sections are denoted α_g and α_q, respectively. As discussed in Sect. 4.2, the α-factor can dynamically vary and strongly depend on the material. They have been set to zero for simplicity in the following sections. Some results on the impact of α can be found here [96–98] although a systematic study of its effect is still subject of future work. The last term in (4.5) models spontaneous emission noise with strength D and a complex Gaussian white noise source $\xi(t) = \xi_1(t) + i\xi_2(t)$ where real and imaginary noise contributions are uncorrelated, $\langle \xi_i(t)\xi_j(t')\rangle = \delta_{ij}\delta(t-t')$

Table 4.1 Parameter values used in numerical simulations, unless stated otherwise

Symbol	Value	Symbol	Value	Symbol	Value	Symbol	Value
T	25 ps	γ	2.66 ps^{-1}	κ	0.1		
γ_g	1 ns^{-1}	γ_q	75 ns^{-1}	α_g	0	α_q	0
J_g	0.12 ps^{-1}	J_q	0.3 ps^{-1}	r_s	25.0	$\Delta\Omega$	0

with zero mean $\langle \xi_i(t) \rangle = 0$. The parameter values used in numerical simulations are summarized in Table 4.1. They are typical for monolithic semiconductor lasers.

In the absence of noise the dynamics of the DDE mode-locking model and some of its modifications was studied in [86, 88, 92, 93, 96, 99–103]. It was demonstrated that this model captures all the main dynamical regimes exhibited by passively ML monolithic semiconductor lasers, including Q-switching, Q-switched mode-locking, fundamental and harmonic ML regimes, and irregular pulsing in nice agreement with experimental results.

Figures 4.13a summarizes the different dynamic regimes found by numeric integration in the parameter plane spanned by pump current J_g and unsaturated absorption J_q with the different colors referring to fundamental mode-locking (white), first- and second order harmonic mode locking (light and darker blue), constant wave (cw) or off states (grey) and non-periodic dynamics (black). The bifurcations bordering the different regions can be found via path continuation with DDE-Biftool and are also plotted in Fig. 4.13a; grey, blue and green lines indicate the positions of Hopf-bifurcations of cw solutions, Torus (T)-bifurcation of the periodic solution, and saddle-node (SN) bifurcations of limit cycles. It is noted that the T and SN lines border the region of stable mode-locking operation (see also [96, 104, 105]), even though other solutions as e.g. harmonic mode-locking exist multi-stable to this solution (see white and light blue regions in Fig. 4.13a).

In the regime of fundamental mode-locking (white in Fig. 4.13a) the interplay of saturable gain G and saturable absorption Q leads to a fixed phase relation, i.e phase locking, of a certain number of cavity modes resulting in a generation of a sequence of short pulses with the repetition period close to the cavity round trip time T. A rough estimation of the number of modes that can be involved in the mode-locking process is given by the ratio of the bandwidth of the spectral filter γ and the free spectral range of the empty cavity $1/T$. In Fig. 4.13a1, a2 time traces of the absolute laser intensity $|\mathscr{E}|^2$ while emitting stable mode locking pulses are shown (blue lines). Further, the saturable gain G (green line), the total loss Q_{tot} (red line), which is the sum of saturable and non-resonant losses, and the net gain $\mathscr{G} \equiv G - Q + \ln \kappa$ (black line) which the field amplitude experiences during one cavity round trip are plotted. The pulse repetition rate, i.e. the period of the oscillation T_0 is close to 40 GHz for the 1 mm device modelled here. Due to the interaction with the gain medium the period T_0 does not exactly equal the roundtrip time.

During mode-locked operation the dynamics can be divided into a slow motion between two subsequent pulses (slow stage) and a fast excursion in the phase space

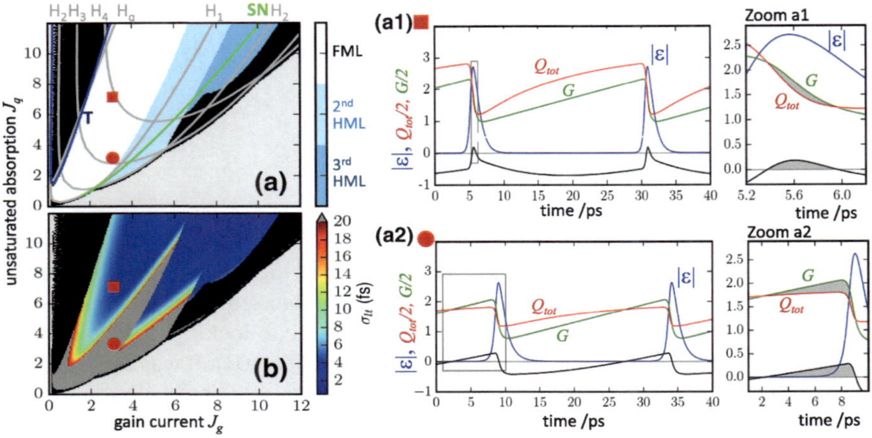

Fig. 4.13 Mode-locked laser dynamics **a** operation regimes found for changing pump current J_g and unsaturated absorption J_q; white, light blue, and blue colored regions are fundamental mode-locking (FML), second and third order harmonic mode locking (HML). The light grey regions indicate constant wave or no emission. Blacks indicates non-periodic dynamics. Grey, blue and green lines indicate the positions of the Hopf, torus and saddle-node bifurcations as labelled atop of the figure. **a1**, **a2** Time series of the absolute value of the amplitude $|\mathscr{E}|$ (blue), gain $G_{\text{red}} = G/2$ (green), loss $Q_{\text{red}} = (Q + |\ln \kappa|)/2$ (red) and net gain $\mathscr{G} = G_{\text{red}} - Q_{\text{red}}$ (black) for two different parameter choices indicated with the red square in circle in (a), respectively. **b** Semi-analytically calculated long term jitter for the dynamic regimes found in (**a**). Other parameters are as in Table 4.1

during the pulse (fast stage). In each of the two stages approximate analytical solutions for the dynamical system (4.5)–(4.7) were found and a 3D map describing the transformation of the pulse energy, saturable gain, and saturable loss after a cavity round trip was derived in [86]. Using this map the stability domain of the ML regime is was roughly estimated by using the net gain stability criterion proposed in [106] (see [86] and [99] for more details concerning the validity of the net gain stability criterion and a generalization of the approach to allow an estimation of the pulse width). According to this criterion, the pulse emission is stable if the net gain $\mathscr{G}$ is negative during the whole slow stage where the amplitude of the electric field is close to zero. The two mode-locked solutions shown in Fig. 4.13a1, a2 have been obtained for the parameter values indicated with the red filled square and circle in Fig. 4.13a, respectively. They do not both satisfy the net gain stability criterion as can be nicely seen in the zoom of the dynamics plotted in the right column where the grey areas show the regions with positive net gain ($\mathscr{G} > 0$ long before the pulse in (a2)). Physically, the net gain stability criterion means that small perturbations of the electric field introduced for example due to spontaneous emission noise, decay from one cavity round trip to the next. As we discuss later on in this section, the stability of the pulses shown in Fig. 4.13a2 are indeed much more susceptible to noise and vary their positions in time much more (Fig. 4.13b).

It is noted that recently modifications of the DDE model were applied to study also the dynamics of tapered mode-locked lasers [107–109], the formation of localized states [110] and optical bullets in wide-aperture lasers [111] and of bound pulse trains in arrays of coupled mode-locked lasers [112].

4.3.2 Timing Jitter Calculation

Due to the absence of the reference clock and presence of the time-shift and the phase-shift symmetries in the passively mode-locked laser, the spikes in the time trace are not fluctuating near fixed temporal positions but rather exhibit a random walk. One can adapt the *von der Linde* method [113] to characterize the timing jitter as in [114]. There, for each noise realization the timing fluctuations $\{\Delta t\} \equiv \{\Delta t_l\}_{l=1}^N$ are extracted from the time series, as sketched in Fig. 4.12b, as a set of timing deviations of the pulse train from a clock $T_C = \langle T_n \rangle$, i.e., $\Delta t_l \equiv t_l - lT_C$, where t_l is the sum of the first l inter-spike intervals and the repetition period T_C is obtained by averaging over all pulse separations in all time series obtained for different noise realizations. By applying a discrete Fourier transform to each sequence of pulse position deviations $\{\Delta t\}_j$ and then averaging over simulations with different noise realizations, the timing spectral density and with it the phase noise spectrum can be calculated. Thus, the root-mean-square (rms) timing jitter that follows from this *von der Linde* method is obtained by integrating the phase noise spectrum $\mathscr{L}(\nu)$ over the frequency range from ν_{low} to ν_{high}

$$\sigma_{\text{rms}}(\nu_{\text{low}}, \nu_{\text{high}}) \equiv \frac{T_C}{2\pi}\sqrt{\int_{\nu_{\text{low}}}^{\nu_{\text{high}}} 2\mathscr{L}(\nu)d\nu}, \tag{4.9}$$

where ν_{low} and ν_{high} are the minimal and the maximal frequency offset from the carrier frequency, i.e., from the mean repetition frequency $1/T_C$.

Instead of the spectral approach described above, which is not very efficient in numerical simulations, one can also estimate the so-called *long-term* timing jitter σ_{lt}. For that we average the timing fluctuations Δt_j (see also Fig. 4.12b for their definition) over all noise realizations and obtain the estimate of the timing jitter σ_{lt} via the standard deviation of the distribution

$$\sigma_{lt}^2 = \frac{\langle |\Delta t_N|^2 \rangle}{N}. \tag{4.10}$$

In order for this approach to be valid, the integration time has to be long enough for all short time correlations to decay. For more details on the comparison between the different numerical methods for jitter calculations see [97, 98]. The different numeric estimations of timing jitter in ML semiconductor lasers has been widely used for travelling wave [114, 115] and DDE [97, 116–118] models, but due to the time-consuming computation they have always been limited to relatively small parameter regions. Thus an analytic approach is needed to deepen the understanding.

A first analytic approach to study the influence of noise on soliton mode-locking regime was proposed by Haus and Mecozzi using a master equation approach [119]. Later this technique was extended by taking into account the finite carrier density relaxation rate in semiconductor lasers [85]. The master equation has secant-shaped ML pulses as a solution, and a small perturbation of this state can be studied using the linearized equation of motion. The perturbed pulse is described by four parameters: the perturbations of the pulse amplitude, phase, frequency, and timing. Using the orthogonality of the solutions of the linearized equation to the solutions of the adjoint homogeneous linear system, coupled first order differential equations of motion, driven by noise, can be written out. However, due to multiple simplifying assumptions underlying the Haus master equation, this approach is not directly applicable to the analysis of semiconductor laser devices, which operate far away from the soliton mode-locked regime. A modification of the Haus and Mecozzi approach in application to semiconductor lasers was proposed in [120] and then used to study the effect of nonlinear phenomena such as bifurcations and bistability on timing jitter. The semi-analytic results were found to be in good qualitative agreement with experimental data. Furthermore, a generalisation of this semi-analytical method to study passively ML lasers with multiple delayed feedback was proposed in [96, 121] and used to derive an analytic formula for the timing jitter for resonant feedback delay lengths as we show later in this section.

4.3.2.1 Semi-analytic Perturbative Jitter Calculation

We will now describe the perturbative approach to estimate timing jitter of a passively ML laser using the DDE model equations (4.5)–(4.7) and the assumption of weak enough noise $D \ll 1$, in more detail. In the absence of noise ($D = 0$) the model equations (4.5)–(4.7) are invariant with respect to time and phase shifts: $t \to t + t_0$ and $A \to A \exp i\phi_0$ with arbitrary t_0 and ϕ_0. This means that the linear stability analysis of a solution with periodic laser intensity corresponding to a passively ML regime yields at least two zero eigenvalues related to time and phase shift symmetries. Therefore, one can expect that the inclusion of small noise into the model equations will result in a random walk along the neutrally stable directions leading to a pulse timing and phase diffusion. In particular, the diffusion related to the time shift symmetry is responsible for a large timing jitter in a passively ML laser while the phase shift symmetry contributes to a large phase diffusion in these lasers.

Let us consider a ML solution of the unperturbed system (4.5)–(4.7) with $D = 0$ in the form $E = E_0(t) \exp(i\omega_0 t - i\omega t)$, $G = G_0(t)$ and $Q = Q_0(t)$ with T_0-periodic functions of time E_0, G_0, and Q_0, i.e. a solution similar to the one plotted in Fig. 4.13a1. Here T_0 is the pulse repetition period of the ML regime and ω_0 has the meaning of its central frequency, thus the ML solution can be represented as a T_0-periodic vector $\psi_0 = (\mathrm{Re}\mathcal{E}_0\, \mathrm{Im}\mathcal{E}_0, G_0, Q_0)^{\mathrm{T}}$. One should note, however, that due to the rotational symmetry, there is a family of such solutions $\Gamma_\varphi \cdot \psi_0 =$

$(\mathrm{Re}(e^{i\varphi}\mathcal{E}_0), \mathrm{Im}(e^{i\varphi}\mathcal{E}_0), G_0, Q_0)^{\mathrm{T}}$ parametrized by arbitrary constant phase φ, where Γ_φ denotes the corresponding matrix of rotation of the $\mathcal{E}_0$ plane. Since the amplitude of the noise is assumed to be reasonably small, $D \ll 1$, and we restrict our analysis to the situation when the solutions remain at a distance of order D from the torus filled by stable periodic solutions $\Gamma_\varphi \cdot \psi_0(t+\theta)$ at all times (that is, the probability of a large fluctuation of the solution is assumed to be negligible during the typical time interval of system observation). Here θ denotes a shift along the time coordinate. Under this assumption, the noise results in a slow diffusion of the time-shift θ of the solution, as well as a slow diffusion of the angular variable φ. Furthermore, one expects that the variance of the time-shift θ and of the variable φ increases linearly with time, that is $\langle\theta - \bar{\theta}\rangle^2 \propto t$, which expresses a simple diffusion process [122]. We will use the coefficient of proportionality in this relationship as a measure of the timing jitter.

The time-shift θ of a solution can be defined in several ways [123], which, in practice, lead to equivalent or close results when applied for the evaluation of the time-shift diffusion rate. In particular, the definition of the asymptotic time-shift is based on the fact that every solution $\psi(t)$ of the unperturbed system (4.5)–(4.7) with $D = 0$ converges to a periodic solution $\Gamma_\varphi \cdot \psi_0(t+\theta)$ in the limit $t \to \infty$ where the constant θ, called the asymptotic time-shift, and the angle φ are specific to the initial state of the solution $\psi(t)$. Recall that initial states of the DDE system (4.5)–(4.7) are functions defined on the interval $[-T, 0]$. The asymptotic time-shift θ and the angle φ remain constant along the trajectories of the unperturbed system. However, in the perturbed system, the asymptotic time-shift θ and the angular variable φ evolve as functions of the evolving state $\psi(t+r)$ $(r \in [-T, 0])$.

As the dynamics are restricted to a small neighborhood of the limit cycle ψ_0 (and its rotations $\Gamma_\varphi \cdot \psi_0$), the evolution of the time-shift can be deduced from the linearization of system (4.5)–(4.7), around this cycle ($\psi(t) = \psi_0(t) + \delta\psi(t)$):

$$\frac{\mathrm{d}}{\mathrm{d}t}\delta\psi(t) = A(t)\delta\psi(t) + B_m(t-T)\delta\psi(t-T) + Dw(t), \tag{4.11}$$

where A and B_m are T_0-periodic Jacobi matrices of the linearization and, $Dw(t) = D(\xi_1(t), \xi_2(t), 0, 0)^{\mathrm{T}}$ is the small noise term. Noise results in a slow diffusion of the variables θ and φ along the neutral periodic eigenmodes of the linearized unperturbed system (4.11) (for $D = 0$) with the variance proportional to time. As it was already mentioned, there are two such neutral modes,

$$\delta\psi_\theta(t) = \frac{\mathrm{d}}{\mathrm{d}t}\left(\mathrm{Re}\mathcal{E}_0, \mathrm{Im}\mathcal{E}_0, G_0, Q_0\right)^T, \quad \delta\psi_\varphi(t) = \left(-\mathrm{Im}\mathcal{E}_0, \mathrm{Re}\mathcal{E}_0, 0, 0\right)^T, \tag{4.12}$$

which correspond to the time-shift and phase-shift symmetries of the unperturbed $(D = 0)$ nonlinear system (4.5)–(4.7), respectively; all the other Floquet modes are exponentially decaying. Two neutral modes $\delta\psi_\theta^\dagger(t)$ and $\delta\psi_\varphi^\dagger(t)$ of the adjoint linear system for a row vector $\delta\psi^\dagger(t) = (\delta\psi_1^\dagger, \delta\psi_2^\dagger, \delta\psi_3^\dagger, \delta\psi_4^\dagger)$

$$\frac{\mathrm{d}}{\mathrm{d}t}\delta\psi^{\dagger}(t) + \delta\psi^{\dagger}(t)A(t) + \delta\psi^{\dagger}(t+T)B_m(t) = 0 \tag{4.13}$$

can be used for calculating the projections of noise onto the eigendirections $\delta\psi_\theta$ and $\delta\psi_\varphi$ with the help of the following definition for a bilinear form [124, 125]

$$\left[\delta\psi^{\dagger}, \delta\psi\right](t) = \delta\psi^{\dagger}(t)\delta\psi(t) + \int_{-T}^{0} \delta\psi^{\dagger}(t+r+T)B_m(t+r)\delta\psi(t+r)dr. \tag{4.14}$$

Using the perturbation expansion with respect to the small parameter D, and adapting the asymptotic analysis from [126], we obtain first equations for the noise-driven slow evolution of the time-shift θ and the angular variable φ of solutions to (4.5)–(4.7) [121] which can then be transformed into a Fokker-Planck equation for the joint probability density $p(t, \theta, \varphi)$ of the stochastic process. Since, for $D \ll 1$, the probability density function $p(t, \theta, \varphi)$ changes slowly, the Fokker-Planck equation can be averaged over the period T_0 of the functions $\delta\psi_i^{\dagger}(t+\theta)$, resulting in the diffusion equation with constant coefficients [127]. The diffusion coefficient

$$\bar{d}_{11} = \frac{D^2}{T_0}\int_0^{T_0} \left\{ \left[\delta\psi_{\theta,1}^{\dagger}(s)\right]^2 + \left[\delta\psi_{\theta,2}^{\dagger}(s)\right]^2 \right\}\mathrm{d}s \tag{4.15}$$

approximates the rate of diffusion of the time-shift θ which, finally, gives us the pulse timing jitter calculated over a long time interval $n\tilde{T}_0$ (with $n \gg 1$ and average period $\tilde{T}_0 \approx T_0$) normalized by the number of round-trips n, as the product of the diffusion rate by the period

$$\sigma_{\mathrm{var}}^2 = \bar{d}_{11}T_0 = D^2\int_0^{T_0} \left\{ \left[\delta\psi_{\theta,1}^{\dagger}(s)\right]^2 + \left[\delta\psi_{\theta,2}^{\dagger}(s)\right]^2 \right\}\mathrm{d}s. \tag{4.16}$$

This value is approximately equal to the variance of $\theta(n\tilde{T}_0)$ divided by $n \gg 1$. We note that in the case when the number of roundtrips $n \geq 1$ is not sufficiently large in numerical simulations, the calculated timing jitter can deviate considerably from that given by (4.16), since in this case the numerically calculated value can be affected by amplitude noise, or, in other words, stable eigendirections can play a role as well.

Quantitative comparison of the obtained formula with the numerically calculated long term jitter shows very good agreement as long as the dynamics is far from bifurcation points [96, 121]. The impact of those instabilities on the timing jitter in comparison to experiments with QD lasers can also be found in [120]. In our case of the solitary mode-locked laser, the timing jitter calculated with the above method is plotted as a color code in Fig. 4.13b. The general trends shown there are that the timing jitter is lower for higher unsaturated absorption values J_q and that the harmonically mode-locked solutions (HML) have a slightly lower timing jitter than the fundamentally mode-locked solution. The latter can be understood by considering that at higher pump currents there is more energy in the system and therefore the noise has a smaller influence. Please note also that not all regions with mode-locked dynamics (white regions in Fig. 4.13a) show small timing jitter which can be understood by the stability criterion mentioned above. Consequently, parameter regions, where the mode-locked pulses show positive net gain windows outside the pulse, are unstable in its pulse positions (e.g. time series shown in Fig. 4.13a2 with parameters marked as a circle in Fig. 4.13b).

The asymptotic expression (4.16) can also be applied for estimation of the timing jitter in a ML laser with coherent optical feedback [121] where it yields an explicit dependence of the timing jitter on the feedback delay time in the particular case of resonant feedback. This will be shown in the next section after we shortly summarize different methods for timing jitter reduction.

4.3.3 Reducing Timing Jitter by Optical Perturbations

Since passively mode-locked lasers are subject to very high timing and optical phase jitter due to the absence of the reference clock, the most efficient way to reduce the timing jitter is by introducing of weak forcing into the system. Two possible ways are sketched in Fig. 4.14. Using extensions of the DDE model one can study the following approaches:

- **Single-mode optical injection** into the slave mode-locked laser can suppress phase jitter of the optical field. This method can be described in the DDE model by adding a CW injection into the field equation [126]

$$\dot{\mathscr{E}}(t) = -\gamma \mathscr{E}(t) + \gamma R(t-T)e^{-i\Delta\Omega T}\mathscr{E}(t-T) + \eta e^{i\Delta\omega_0 t},$$

  where $\Delta\omega_0$ is the frequency difference of the injected light and the ML laser frequency and η is its injection strength.

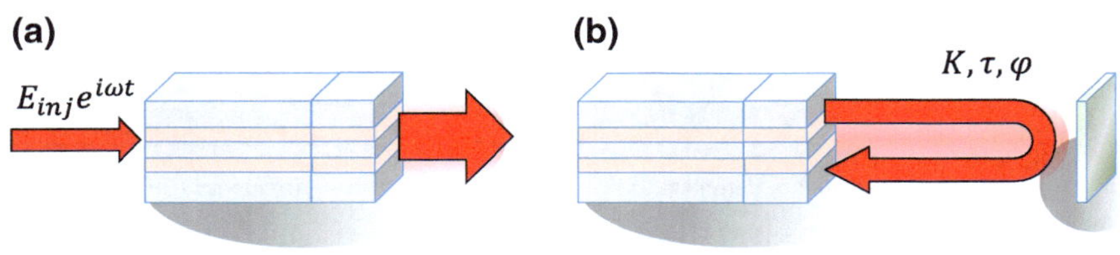

Fig. 4.14 Sketch of a mode-locked semiconductor laser subjected to optical injection (**a**) or optical feedback (**b**) to reduce the timing jitter

- **Hybrid mode-locking** technique can be applied by periodic modulation of the reverse absorber voltage to suppress the timing jitter [128]. This method can be described in the DDE model by modifying the equation for the absorber:

$$\dot{Q}(t) = (1 + \eta f(t))\left(J_q - \gamma_q Q(t)\right) - r_s e^{-Q(t)}\left(e^{Q(t)} - 1\right)|\mathscr{E}(t)|^2,$$

where $f(t)$ is a periodic modulation and η is the modulation strength.
- **Two-mode optical injection** can suppress both phase and timing jitter, and it can be modelled similarly to single-mode injection by adding another CW injection with a second frequency ω_1 to the field equation, see [129] for details.

In all these techniques the mode-locked laser is locked (synchronized) to the weak external forcing similarly to actively mode-locked laser thus suppressing the random walk of the spike temporal positions by destroying corresponding symmetries of the passive system and eliminating long-term timing jitter so that pulses fluctuate near fixed temporal positions. Therefore, the most interesting question to study here is the locking conditions. In particular, one can study the so-called locking range, i. e. the interval of frequencies of the periodic forcing to which the slave mode-locked laser can lock, and compare to the experiments (see Fig. 4.15). Moreover, one can perform a perturbation analysis of the mode-locked solution of the DDE model in the limit of weak forcing $\eta \ll 1$ similarly to the case of weak noise in the section below to see that the locking range is proportional to the forcing strength η.

- **Optical feedback from one or more external cavities** is another interesting method to reduce the timing jitter. It has been successfully implemented [131–134]. For the modelling in the framework of the described delay model, the field equation with optical feedback takes the form

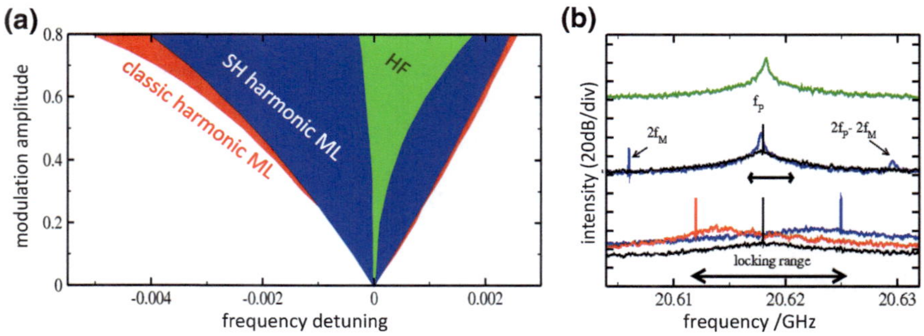

Fig. 4.15 Hybrid mode-locking: **a** simulated domains of stable mode-locking in the plane of frequency detuning and modulation amplitude depicted as red, blue, and green cones placed on top of each other in the corresponding order, which represent modulations with different frequencies: $1/T_C$ (classic), $1/(2T_C)$ (SH), and $2/T_C$ (HF), respectively. **b** Measured electrical spectra of passively and hybrid ML laser. ©2019 IEEE. Reprinted, with permission, from [130]

$$\dot{\mathscr{E}}(t) = -\gamma\mathscr{E}(t) + \gamma R(t-T)e^{-i\Delta\Omega T}\mathscr{E}(t-T) + D\xi(t)$$
$$+ \gamma \sum_{m=1}^{M}\sum_{l=1}^{\infty} K_{m,l}e^{-ilC_m}R(t-T-l\tau_m)e^{-i\Delta\Omega(T+l\tau_m)}\mathscr{E}(t-T-l\tau_m). \tag{4.17}$$

The optical feedback is described by the sum in (4.17). Here l is the number of round-trips in the external cavity, $K_{m,l}$ is the round-trip dependent feedback strength of the mth feedback cavity and C_m is the phase shift that accumulates over one round-trip in the external cavity. Below we consider feedback contributions only from light that has made one round-trip in the external cavities ($K_{m,1} = K_m$). The basic effect, that feedback can have a stabilizing effect on the dynamics of the system has already been experimentally suggested in [135]. Numerical investigations for one delay were performed in [116] and later on also shown experimentally [136–139].

From an application point of view the last method needs less effort for realization. However, optical feedback does not destroy the symmetries of the passively mode-locked laser, hence long-term timing jitter will persist, and we will thus study the delayed model in more detail to understand the mechanism of timing jitter reduction and the conditions and limits of this method.

4.3.3.1 Effect of Optical Feedback on Pulse Timing Jitter

From experimental investigations it was found that the longer the feedback delay time τ is in a certain interval of delays, the smaller is the timing jitter and the phase noise. This can be intuitively understood by noting that in the case of a resonant feedback, i.e. when the delay time $\tau = qT_0$ is a multiple of the pulse repetition period T_0, the feedback introduces correlations between ML pulses separated by q periods T_0. Therefore, for large delay the pulse positions are correlated over longer time spans (for larger integer numbers q) and the long term timing jitter and phase noise must decrease.

A modification of the perturbative approach of Sect. 4.3.2 that allows one to describe the effect of delayed optical feedback on the noise characteristics of passively ML laser was developed in [121]. In this case the (4.16) for the timing jitter estimation still remains valid. However, instead of (4.13) the adjoint linear system taking into account the feedback term in (4.17) should be used for calculation of the neutral adjoint eigenvectors $\delta\psi_{\theta,\varphi}$. For more details, see [121], where these eigenvectors were calculated numerically. Resonant feedback applied to the fundamental ML regime does not significantly affect the dynamical behavior of the system and the laser output remains periodic. When the feedback delay time is tuned from one resonance to the next, bifurcations can occur and the dynamical behavior can change. This is described in detail in [97, 116]. The agreement between the perturbative approach and numerical simulations is thus only good within the locking regions, i.e. the regions

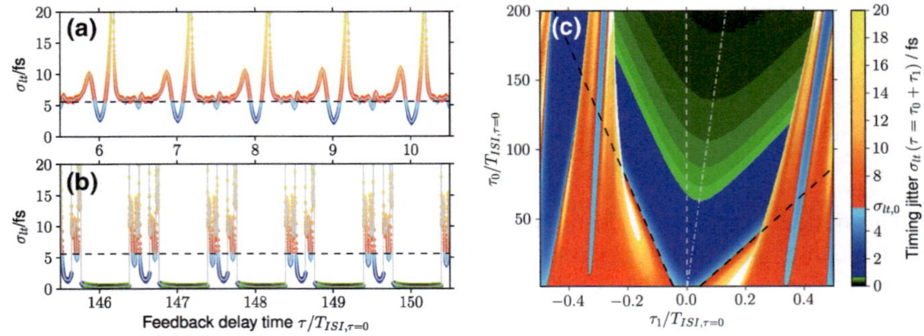

Fig. 4.16 **a**, **b** Timing jitter σ_{lt} in dependence of the feedback cavity delay time. The color code indicates the timing jitter according to the color bar given in subplot (**c**). The black dashed line indicates the timing jitter of the solitary laser. **c** Timing jitter σ_{lt} in dependence of the feedback cavity delay time, where $\tau = \tau_0 + \tau_1$ for any given point. The horizontal axis spans one T_0 and is centered on an exact main resonance. The vertical axis indicates the number of the main resonance. The timing jitter is indicated by the color code and $\sigma_{lt,0}$ is the timing jitter of the solitary laser. Parameters: $K = 0.1$, $D = 0.2$, $\alpha_g = 0$, $\alpha_q = 0$, others as in Table 4.1. Reprinted with permission from [121]. Copyright (2019) by the American Physical Society

where the laser still emits periodic pulse trains. At the edges, a sharp increase in the timing jitter is observed associated with bifurcation points of the deterministic DDE system, and large discrepancy between numeric and semi-analytic results can occur.

The decrease of the timing jitter with increased resonant feedback delay time τ is illustrated in Fig. 4.16, where the subplots (a) and (b) correspond to a relatively short and a long delay time τ, respectively. It is seen that for longer delay times depicted in Fig. 4.16b the timing jitter reduction is greater and the regions with strong timing jitter suppression (green in Fig. 4.16) widen. This can also nicely be seen in the color plot in Fig. 4.16c. The horizontal axis in this plot covers the interval of delays equal to the pulse repetition period T_0 and is centered at the exact main resonances $\tau = qT_0$, while the vertical axis gives the number q of the main resonance. Blue and green (red) colors in Fig. 4.16c indicate a reduction (increase) in the timing jitter with respect to that in the solitary laser without feedback, $K = 0$. In particular, within the green region the timing jitter is reduced by a factor of 10 or greater. It is seen from Fig. 4.16c that the timing jitter decreases with the resonance number q and for $q \lessapprox 50$ the width of the frequency-pulling regions characterized by reduced timing jitter, increases approximately linearly with the number q. At larger q a bistability between the main and higher order resonances appears [116]. Interestingly, the minimal jitter is typically not obtained at the exact resonance but along the slightly tiled dashed white line in Fig. 4.16c. A clear advantage of the semi-analytic method is its applicability for long delay times. A pure numeric evaluation of the stochastic dynamics would be harder due to very long integration times.

In order to quantify the decrease in the timing jitter with increasing delay time τ at the main resonances, we have plotted the numerically determined timing jitter in Fig. 4.17a with the green dots. For this resonant situation, $\tau = qT_0$, an analytic

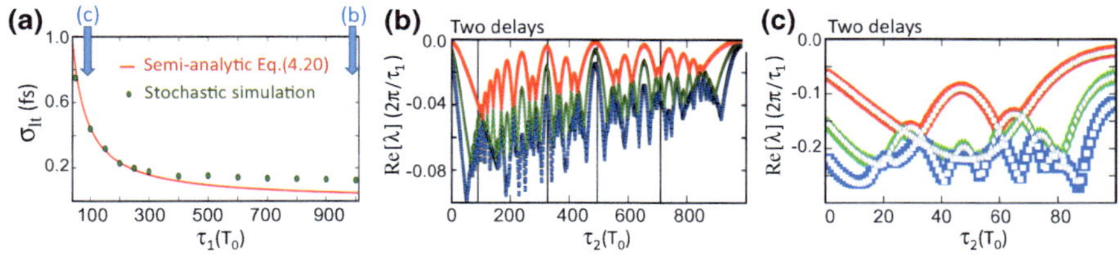

Fig. 4.17 **a** Comparison between numerically (green dots) and analytically (red line) determined timing jitter as a function of the delay time τ_1. **b**, **c** Lines show the three largest Floquet multipliers of the ML dynamics (red line represents the largest Floquet multiplier) for increasing value of a second delay τ_2 when the first delay time is $\tau_1 = 1000$ (**b**) and $\tau_1 = 100$ (**c**)

expression for the timing jitter estimation was derived in [121] with the help of the perturbative approach described in Sect. 4.3.2 and plotted as the red line in Fig. 4.17a. As can be seen there, the agreement is good as long as the delay stays smaller than $\tau < 200T_0$.

Let us first derive the equation before we discuss its limits. At the resonances the solutions to the DDE model are identical for all q, and the periodicity is the same as that of the laser with zero delay (instantaneous) feedback, i.e. T_0. Therefore, (4.14) can be expressed as

$$\left[\delta\psi^\dagger, \delta\psi\right] = \left[\delta\psi^\dagger, \delta\psi\right]^{\tau=0} + Kq \int\limits_{-T_0}^{0} \left(\delta\psi^\dagger\left(t+r+T\right)\right)^{\mathrm{T}} B_1\left(t+r\right)\delta\psi\left(t+r\right)\mathrm{d}r, \tag{4.18}$$

where $\left[\delta\psi^\dagger, \delta\psi\right]^{\tau=0}$ is the bilinear form (4.14) for $\tau = 0$ ($q = 0$). Equation (4.16) can thus be expressed as

$$\sigma_{lt}^{\tau=qT_0} = \frac{\sigma_{lt}^{\tau=0}\left(K\right)}{1 + Kq\mathscr{F}\left(K\right)}, \tag{4.19}$$

$$\mathscr{F}\left(K\right) = \frac{\int_{-T_0}^{0} \delta\psi_{0t}^{\dagger *}\left(t+r+T\right) B_1\left(t+r\right)\delta\psi_1^*\left(t+r\right)\mathrm{d}r}{\left[\delta\psi_1^{\dagger *}, \delta\psi_1^*\right]^{\tau=0}},$$

where $\sigma_{lt}^{\tau=0}\left(K\right)$ is the timing jitter for $\tau = 0$. Moreover, fitting the minimum jitter curve for various feedback strengths we find that the relation

$$\sigma_{lt}^{\min} \approx \frac{\sigma_{lt}^{\tau=0}\left(K\right)}{1 + Kq}, \tag{4.20}$$

holds well for low feedback strengths.

Several physical insights can be gained from (4.19) and (4.20). First, the decrease in the timing jitter, with increasing delay length, is directly related to the increase in the length of the history of the solution. The influence on the dynamics of the system, of a perturbation, at one time point is smaller if the solution space is larger. This can be understood by considering the definition of the asymptotic time shift. If the solution to the homogeneous system is perturbed at one point in time, then in the large time limit the solution will once again converge to the solution of the homogeneous system, but with a certain time shift with respect to the unperturbed solution. If the solution space is larger, the resulting time shift will be smaller. Another way to think of this is to consider that the pulse positions are correlated over the history of the solution. Thus, if the pulse positions are correlated over longer time spans, via increased feedback delay lengths, the timing jitter becomes smaller. The timing jitter reduction, therefore, has nothing to do with the stability of the system, as one might have expected based on studies on feedback stabilization. Second, in contrast to previous works on timing jitter reduction, these results highlight the importance of the pulse shape, and hence the gain and absorber dynamics. The pulse shape enters in the integral for $\mathscr{F}(K)$, as this influences the overlap of $\delta\psi_{0t}^{\dagger}(t+T)$ and $B_1(t)\delta\psi_1(t)$. This overlap is maximal when $T_0 = T$, and the deviation of the period from the cold cavity round-trip time is intrinsically linked to the asymmetry of the pulses. Increasing the width of the Lorentzian filter reduces the interspike interval time, as $T_0 \approx T + \gamma^{-1}$ [86], this results in an increase in $\mathscr{F}(K)$, and hence improved timing jitter reduction at the exact main resonance. The fact that the minimum timing jitter does not coincide with the exact main resonances shows that pulse reshaping also affects the timing jitter.

Recalling the comparison between the semi-analytic results and the numeric values for the timing jitter shown in Fig. 4.17a, we now understand that the analytic value is just a lower bound. For the long delay length the assumption of strongly negative Floquet multipliers is not justified as the delay induces an undamping of certain modes. A solution to this problem is a second delay line. Choosing a non-commensurate ratio between the length of both can stabilize the unstable directions and consequently reduce the timing jitter. A plot of the Floquet multipliers for the case of two delay lines as a function of the second delay time τ_2 are shown in Fig. 4.17b, c for $\tau_1 = 100$ and $\tau_1 = 1000$, respectively. Looking at the red line, which gives the value for the most unstable mode, it is obvious that the stability can increase by a factor of more than 10 if a short cavity is added to the long one (Fig. 4.17b). Adding the second cavity to a short delay (Fig. 4.17c) also has an effect but it is considerably smaller since the delay has not destabilized the delay induced modes that strong in the first place. Interestingly the dependence of the stability on τ_1 and τ_2 is very generic for oscillating systems and can be described by a simple characteristic equation (see [140] for details).

4.3.4 Tapered Multi-section Mode-Locked Laser

As we have seen so far, passively mode-locked semiconductor lasers are able to produce short optical pulses with high repetition rates. For applications in high-precision metrology [46, 141] and high-capacity optical interconnects [81, 142],

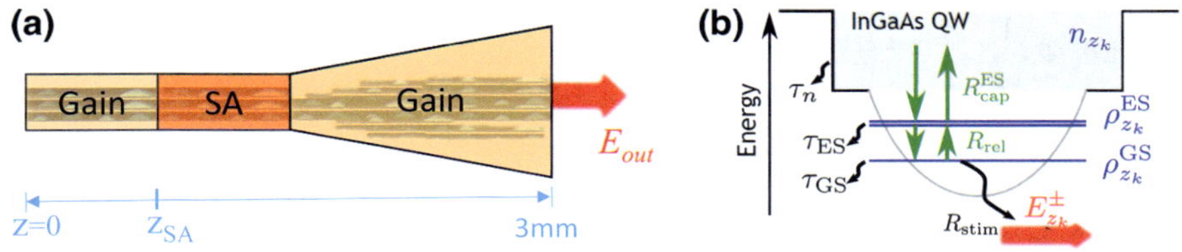

Fig. 4.18 **a** Sketch of the multi-section mode-locked laser with tapered gain sections and an absorber (SA) that starts in between the gains sections at z_{SA}, **b** energy band structure of the gain material based upon Stranski-Krastanov quantum dots (SK-QDs) as used for the rate-equation modeling

where pulse trains with low amplitude and timing jitter are required, the approaches described so far (hybrid mode-locking, optical injection and optical feedback) allow for the reduction of the timing jitter, but add further electronics and optics and thereby also additional costs. New concepts of monolithic mode-locked semiconductor laser geometries for improved pulse stability have therefore become of major interest. One of those ideas utilizing a tapered gain section will be discussed now.

Following the ideas in [143] we simulate a monolithically integrated three section tapered mode-locked QD semiconductor laser with two gain sections (labeled Gain in Fig. 4.18a) and a saturable absorber (labeled SA in Fig. 4.18a). In contrast to the pure DDE modeling described in the last section, we use a multi-section traveling-wave like approach where all the details of the gain and phase dynamics of the QD gain material can be included. Here, we describe a 3 mm long device (13.24 GHz fundamental repetition frequency) that consists of two gain sections; one with a taper angle, and an absorber section in the middle that starts at z_{SA}. The left(right) facet is high(anti)-reflection coated. The active region contains 10 layers of InAs QDs embedded in GaAs quantum wells, which are electrically pumped. The band structure used for the rate equation modeling of the carrier dynamics is depicted in Fig. 4.18b.

4.3.4.1 Modeling Spatio-temporal Intra-cavity Dynamics

Our aim is to study the impact of the device geometry on the pulse train stability, e.g. changes in the emission dynamics induced by a changing angle of the tapered section or a different position of the absorber. Thus, we require our numerical model to be capable of describing the device specific spatio-temporal evolution of the electric field, but also to be computationally efficient to perform parameter studies and obtain sufficient pulse train statistics. The field propagation is described in the slowly-varying envelope and rotating wave approximation, which lead to first order partial differential equations for the forward (+) and backward (−) moving traveling-wave envelope function $E^{\pm}(z, t)$ [144]

$$\left(\pm\partial_z + \frac{1}{v_g}\partial_t\right) E^{\pm}(z,t) = \frac{g_{\mathrm{GS}}}{2}(2\rho^{\mathrm{GS}} - 1)G^{\pm}_{\mathrm{GS}} - i\frac{\delta\omega^{\mathrm{ES}}}{2}\left(2\rho^{\mathrm{ES}} - 1\right)E^{\pm} \quad (4.21)$$

where the right hand side summarizes the source term which was derived from the Maxwell-Bloch equations under the assumption that only the QD ground state contributes to the amplification and absorption of the field while the QD excited state only causes refractive index shifts and that the charge carrier dynamics are slow compared to the polarization. Integration along the characteristic curve of (4.21) yields a time delayed algebraic propagation scheme [145, 146], which is used for the numeric integration as it allows for a much coarser discretization than the direct integration of the PDE (see [143] for details). In the above equation $\rho^{\mathrm{GS,ES}}$ denote the QD ground and excited state occupation numbers, g_{GS} the differential gain and $\delta\omega^{\mathrm{ES}}$ the amplitude-phase coupling coefficient. The dynamical variable $G^{\pm}$ takes a role similar to the polarization by filtering the optical field. The corresponding equation reads

$$\frac{\mathrm{d}}{\mathrm{d}t}G^{\pm}_{\mathrm{GS}} = \frac{1}{T_2}(E^{\pm} - G^{\pm}_{\mathrm{GS}}) + i\Delta\omega_{\mathrm{GS}}G^{\pm}_{\mathrm{GS}} + \sqrt{D\rho^{\mathrm{GS}}}\xi(t) \quad (4.22)$$

where T_2 denotes the effective polarization dephasing time of the inhomogeneously broadened QD assemble, $\Delta\omega_{\mathrm{GS}}$ the detuning from the optical center frequency, and the last term models spontaneous emission noise via a delta correlated Gaussian white noise term $\xi(t)$ and a noise strength D.

The charge-carrier model is excitonic and includes the occupation numbers of the QD ground and excited state and the charge-carrier density in the surrounding quantum well. Their dynamics at each spatial coordinate are described by a set of coupled rate equations [36] which was also used for simulating the gain recovery dynamics in Sect. 4.2.1.

$$\frac{\mathrm{d}}{\mathrm{d}t}n = -\frac{n}{\tau^n} + J - 4N^{\mathrm{QD}}R^{\mathrm{ES}}_{\mathrm{cap}} \quad (4.23)$$

$$\frac{\mathrm{d}}{\mathrm{d}t}\rho^{\mathrm{ES}} = -\frac{\rho^{\mathrm{ES}}}{\tau^{\mathrm{ES}}} + R^{\mathrm{ES}}_{\mathrm{cap}} - \frac{1}{2}R_{\mathrm{rel}} \quad (4.24)$$

$$\frac{\mathrm{d}}{\mathrm{d}t}\rho^{\mathrm{GS}} = -\frac{\rho^{\mathrm{GS}}}{\tau^{\mathrm{GS}}} + R_{\mathrm{rel}} - g^{\mathrm{GS}}\eta(2\rho^{\mathrm{GS}} - 1)\mathrm{Re}\left(G^{+}_{\mathrm{GS}}E^{+^*} + G^{-}_{\mathrm{GS}}E^{-^*}\right) \quad (4.25)$$

with the pump current density J, the characteristic carrier lifetimes $\tau_{\mathrm{GS,ES,n}}$, the factor 4 for spin and ES degeneracy, the net carrier capture from the wetting layer $R^{\mathrm{ES}}_{\mathrm{cap}}$ and the net intra-dot carrier relaxation R_{rel} [35, 147, 148]. The net scattering rates include Pauli-blocking terms and a Boltzmann-factor with the energy difference of the respective levels to account for a detailed balance between the in and out-scattering processes.

The last term in (4.25) is the stimulated emission with the photon to field conversion factor $\eta = \varepsilon_b v_g h^{QW}/(4\hbar\Gamma N^{QD})$. The impact of the reverse bias that is applied to the absorber section is modeled via an exponential dependency of the QD excited state lifetime on the applied voltage [149–151]. To account for the tapered geometry of the long gain section, we not only change the width of the active region but also adapt the non-resonant wave-guide losses to mimic the results from beam propagation calculations [108, 152]. To compare with experiments, the pump current P is related to the pump current density J via $P = JA_G a_L e/\eta$ where A_G is the area of the active region, a_L the number of QD layers, e the electron charge and η the injection effiency, which is assumed to be 30%, and the out-coupled power is calculated according to $P_{out} = 2\hbar\omega a_L w_0 N^{QD}\eta\,|E_{out}|^2$ where the out-coupled electric field at the right facet of the tapered section is given by $|E_{out}|^2 = (1-\kappa_R)\,|E(z=l)|^2$. The relevant figures of merit can then be calculated either directly from time series or from numerically computed auto-correlation functions. Amplitude and timing jitter are calculated as the standard deviation of the pulse maxima and inter-pulse intervals as was introduced in (4.10).

4.3.4.2 Optimum Saturable Absorber Position

To study the influence of the saturable absorber (SA) position within the multi-section cavity, we perform scans of the pump current P as a function of the SA position. The resulting dynamics are shown in Fig. 4.19a, where the observed mode-locking states are depicted color-coded. For a SA in the middle of the tapered device (position $z_{SA} = 0.7$ mm) we observe a sequence of Q-switched mode-locking (QSML, pink regions), fundamental mode-locking (FML, blue regions), unstable fundamental mode-locking (uFML, light blue regions) and third order harmonic mode-locking (HML, orange regions). Shifting the absorber closer to the left facet (below $z_{SA} = 0.5$ mm), we find emission regimes, which exhibit two pulses circulating in the cavity, but contrary to HML the pulses have different inter-pulse spacings. We therefore refer to them as asymmetric two-pulse states (A2P, dark green regions in Fig. 4.19a). Similar pulse emission was also found in a two-section quantum well based mode locked laser [146] and in V-shaped external cavities [153]. Below $z_{SA} \approx 0.15$ mm, i.e. for a SA close to the left facet, unstable asymmetric two-pulse states (uA2P, light green regions) are observed. Looking at the full parameter scan of Fig. 4.19a, we see that the lasing threshold remains constant under spatial shifts of the absorber section, while the extend of the FM region increases if the absorber is moved closer to the facet.

For a better visualization we illustrate the different mode-locking states in Fig. 4.20 as pseudo space-time plots of the emitted light. Here the time-series are sliced into pieces with the length of the cold cavity round-trip time and stacked on top of each other to create a color-coded 2D-map of the pulse evolution. Figure 4.20a shows Q-switched mode-locking build of broad pulses with inter-pulse spacings of about 25 ps and a slow envelope with a period of 5–10 μs. FML is represented by a narrow line in the space-time plot in Fig. 4.20b. It is tilted to the right because

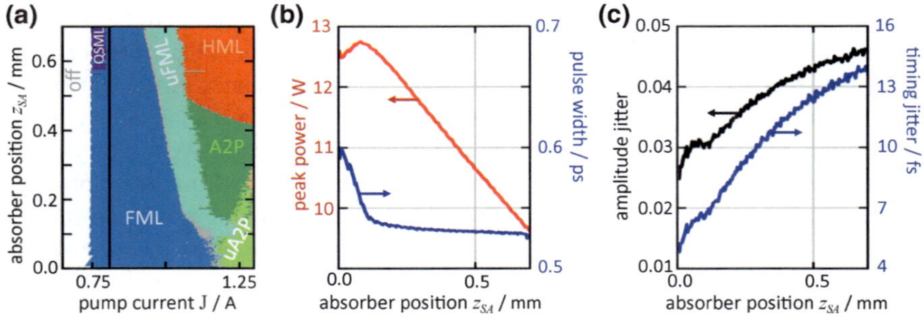

Fig. 4.19 Influence of the saturable absorber position z_{SA}. **a** Dynamic regimes of the tapered laser where Q-switched mode-locking (QSML), fundamental mode-locking (FML), unstable FML (uFML), third-order harmonic mode-locking (HML), asymmetric two-pulse mode-locking (A2P) and unstable A2P (uA2P) are observed. The black vertical line in (**a**) indicates a 1D cut, for which peak power (red) and pulse width (blue) are shown in (**b**) and amplitude (black) and timing jitter (blue) are shown in (**c**)

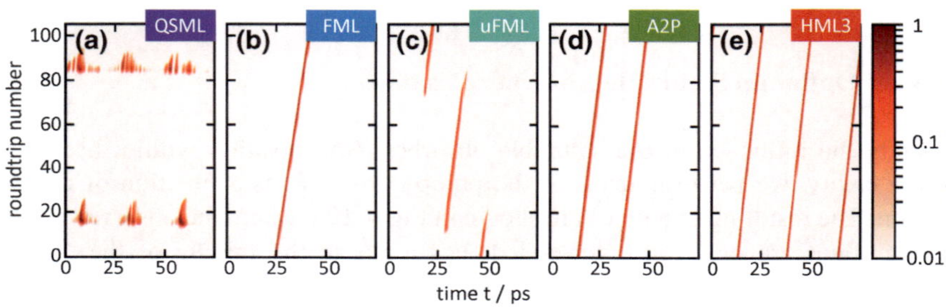

Fig. 4.20 Simulated pseudo space-time plots showing **a** Q-switched mode-locking (QSML, $P = 0.75\,\text{A}$), **b** fundamental mode-locking (FML, $P = 0.81\,\text{A}$), **c** unstable fundamental mode-locking (uFML, $P = 0.95\,\text{A}$), **d** asymmetric two-pulse mode-locking (A2P, $P = 1.18\,\text{A}$) and **e** third order harmonic mode-locking (HML3, $P = 1.08\,\text{A}$). Time is cut into slices with the length of the cold cavity round-trip time $T_0 = 75\,\text{ps}$ and the optical power is color coded and normalized to its respective maximum

the pulse period is slightly longer than the cold-cavity round-trip time. The unstable FML pulse train is shown in Fig. 4.20c, here noise induced perturbations create a competing pulse train that ends up taking the gain from the original pulse train. Asymmetric two-pulse mode-locking (A2P) in Fig. 4.20d shows two stable pulses with an inter-pulse spacing of ≈21 and ≈54 ps. Third order HML is depicted in Fig. 4.20e showing three pulses with an equidistant inter-pulse spacings within the cavity. For it to exist the absorber has to be in the middle of the device.

To examine the influence of the SA position on the pulse performance, we plot the peak power (red) and pulse width (blue) in Fig. 4.19c and the amplitude (black) and timing jitter (blue) in Fig. 4.19d along the black line of constant pump current in 4.19a. Decreasing the SA starting position, the peak power increases from 9 W to 13 W, while the pulse width remains almost constant. Further moving the SA starting position closer to the left facet, however, leads to a increase of the pulse width from 530 to 600 fs, which causes a rollover of the peak power. This behavior is caused by the involved time scales of the laser: For a SA starting position closer to the cavity back facet, the absorber has less time to recover before the reflected pulses reenters and thus less photons are absorbed leading to a greater peak power. Up to $z_{SA} = 0.1$ mm, the pulse shortening in the straight gain sections occurs to the left and to the right of the absorber section. For $z_{SA} < 0.1$ mm, however, the right moving pulse that leaves the absorber is to weak to fully saturate the gain section and therefore the pulse shortening mechanism is reduced in its impact, leading to a broadening of the pulses at the out-coupling facet. Amplitude and timing jitter both improve when shifting the saturable absorber towards the back-facet. We attribute this improvement again to the reduced absorber recovery time, which reduces the deterioration of the jitter for the right moving pulse within the absorber section.

Based on these simulations, we can thus predict an optimum performance for a saturable absorber starting position slightly shifted away from the facet. We believe that this approach can be used to optimize other devices and thus has a large potential of pushing the achievable performance even further. Efficient numerical simulations as presented in this work are hugely beneficial in this quest as they can precisely guide the future development of mode-locked semiconductor lasers.

4.4 Conclusion

Coherent and incoherent dynamics of charge carriers inside semiconductor gain material as well as of light travelling through an optical device leads to complex behaviour on different timescales, ranging from fs to ns, that can be exploited for various applications in data communication or for short pulse generation. Developing an analytic understanding is crucial in order to be able to optimize and create new functionalities. e.g. in the field of optical data transmission and reliable optical clocks. We presented sophisticated methods for characterizing the internal charge carrier dynamics in nanostructured materials (Sect. 4.2) ranging from pump-probe spectroscopy (Sect. 4.2.1) to quantum-state tomography (Sect. 4.2.4) and derived semi-classical models that are able to quantitatively describe the light matter interaction within different nanostructured devices and verified them with experimental results.

Further, we concentrated on the optical emission of mode-locked lasers (Sect. 4.3) and developed analytic methods for predicting their noise properties without the need for extensive stochastic integration (Sect. 4.3.2). In Sect. 4.3.4 we present a multi-section delay differential model that is able to effectively trace the light propagation while taking into account all microscopic details of the nanostructured gain and absorber media. Using this model we are able to suggest a multi-section design for noise reduced integrated mode-locked devices.

References

1. M. Grundmann, O. Stier, D. Bimberg, Phys. Rev. B **52**, 11969 (1995). https://doi.org/10.1103/physrevb.52.11969
2. Y. Todorov, A.M. Andrews, R. Colombelli, S. De Liberato, C. Ciuti, P. Klang, G. Strasser, C. Sirtori, Phys. Rev. Lett. **105**, 196402 (2010). https://doi.org/10.1103/physrevlett.105.196402
3. F. Jahnke, *Quantum Optics with Semiconductor Nanostructures* (Woodhead, 2012)
4. M. Grundmann, *Nano-Optoelectronics: Concepts, Physics and Devices* (Springer, Berlin, 2002). https://doi.org/10.1007/978-3-642-56149-8
5. P.M. Petroff, *Epitaxial Growth and Electronic Structure of Self-Assembled Quantum Dots* (Springer, 2003)
6. M.S. Skolnick, D.J. Mowbray, Annu. Rev. Mater. Res. **34**(1), 181 (2004). https://doi.org/10.1146/annurev.matsci.34.082103.133534
7. T.W. Berg, J. Mørk, J.M. Hvam, New J. Phys. **6**, 178 (2004)
8. J. Gomis-Bresco, S. Dommers, V.V. Temnov, U. Woggon, M. Lämmlin, D. Bimberg, E. Malić, M. Richter, E. Schöll, A. Knorr, Phys. Rev. Lett. **101**(25), 256803 (2008). https://doi.org/10.1103/physrevlett.101.256803
9. N.A. Cherkashin, M.V. Maksimov, A.G. Makarov, V.A. Shchukin, V.M. Ustinov, N.V. Lukovskaya, Y.G. Musikhin, G.E. Cirlin, N.A. Bert, Z.I. Alferov, N.N. Ledentsov, D. Bimberg, Semiconductors **37**(7), 861 (2003). https://doi.org/10.1134/1.1592865
10. S. Krishna, D. Zhu, J. Xu, K.K. Linder, O. Qasaimeh, P. Bhattacharya, D.L. Huffaker, J. Appl. Phys. **86**(11), 6135 (1999). https://doi.org/10.1063/1.371664
11. T. Niermann, F. Kiessling, M. Lehmann, J.H. Schulze, T.D. Germann, K. Pötschke, A. Strittmatter, U.W. Pohl, J. Appl. Phys. **112**(8), 083505 (2012). https://doi.org/10.1063/1.4758301
12. D. Arsenijević, C.Y. Liu, A. Payusov, M. Stubenrauch, D. Bimberg, I.E.E.E. Photon, Technol. Lett. **24**(11), 906 (2012). https://doi.org/10.1109/lpt.2012.2190396
13. P. Qiao, C.Y. Lu, D. Bimberg, S.L. Chuang, Opt. Express **21**(25), 30336 (2013). https://doi.org/10.1364/oe.21.030336
14. B. Herzog, N. Owschimikow, J.H. Schulze, R. Rosales, Y. Kaptan, M. Kolarczik, T. Switaiski, A. Strittmatter, D. Bimberg, U.W. Pohl, U. Woggon, Appl. Phys. Lett. **107**, 201102 (2015). https://doi.org/10.1063/1.4935792
15. D.Z.Y. Ting, S.V. Bandara, S.D. Gunapala, J.M. Mumolo, S.A. Keo, C.J. Hill, J.K. Liu, E.R. Blazejewski, S.B. Rafol, Y.C. Chang, Appl. Phys. Lett. **94**(11), 111107 (2009). https://doi.org/10.1063/1.3095812
16. P. Lam, J. Wu, M. Tang, Q. Jiang, S. Hatch, R. Beanland, J. Wilson, R. Allison, H. Liu, Sol. Energy Mater. Sol. Cells **126**, 83 (2014). https://doi.org/10.1016/j.solmat.2014.03.046
17. P. Borri, W. Langbein, J. Mørk, J.M. Hvam, Opt. Commun. **169**(1), 317 (1999). https://doi.org/10.1016/s0030-4018(99)00391-0
18. B. Lingnau, K. Lüdge, B. Herzog, M. Kolarczik, Y. Kaptan, U. Woggon, N. Owschimikow, Phys. Rev. B **94**(1), 014305 (2016). https://doi.org/10.1103/physrevb.94.014305
19. M. Kolarczik, N. Owschimikow, B. Herzog, F. Buchholz, Y. Kaptan, U. Woggon, Phys. Rev. B **91**, 235310 (2015). https://doi.org/10.1103/physrevb.91.235310
20. N. Owschimikow, M. Kolarczik, Y. Kaptan, N.B. Grosse, U. Woggon, Appl. Phys. Lett. **105**(10), 101108 (2014). https://doi.org/10.1063/1.4895558
21. S. Harrison, M.P. Young, P.D. Hodgson, R.J. Young, M. Hayne, L. Danos, A. Schliwa, A. Strittmatter, A. Lenz, U.W. Pohl, D. Bimberg, Phys. Rev. B **93**, 085302 (2016). https://doi.org/10.1103/physrevb.93.085302
22. M.W. Fleming, A. Mooradian, Appl. Phys. Lett. **38**, 511 (1981). https://doi.org/10.1063/1.92434
23. H. Haug, S.W. Koch, *Quantum Theory of the Optical and Electronic Properties of Semiconductors*, 3rd edn. (World Scientific, Singapore, 1993)
24. C.H. Henry, IEEE J. Quantum Electron. **18**(2), 259 (1982). https://doi.org/10.1109/jqe.1982.1071522

25. A.V. Uskov, E.P. O'Reilly, D. McPeake, D. Bimberg, G. Huyet, N.N. Ledentsov, Appl. Phys. Lett. **84**(2), 272 (2004). https://doi.org/10.1063/1.1639933
26. B. Lingnau, W.W. Chow, K. Lüdge, Opt. Express **22**(5), 4867 (2014). https://doi.org/10.1364/oe.22.004867
27. G.P. Agrawal, N.A. Olsson, IEEE J. Quantum Electron. **25**(11), 2297 (1989). https://doi.org/10.1109/3.42059
28. T. Erneux, P. Glorieux, *Laser Dynamics* (Cambridge University Press, UK, 2010). https://doi.org/10.1017/cbo9780511776908
29. M.C. Soriano, J. García-Ojalvo, C.R. Mirasso, I. Fischer, Rev. Mod. Phys. **85**, 421 (2013). https://doi.org/10.1103/revmodphys.85.421
30. S. Melnik, G. Huyet, A.V. Uskov, Opt. Express **14**(7), 2950 (2006). https://doi.org/10.1364/oe.14.002950
31. B. Lingnau, K. Lüdge, W.W. Chow, E. Schöll, Phys. Rev. E **86**(6), 065201(R) (2012). https://doi.org/10.1103/physreve.86.065201
32. J. Kim, C. Meuer, D. Bimberg, G. Eisenstein, Appl. Phys. Lett. **94**(4), 041112 (2009). https://doi.org/10.1063/1.3073715
33. E. Malić, K.J. Ahn, M.J.P. Bormann, P. Hövel, E. Schöll, A. Knorr, M. Kuntz, D. Bimberg, Appl. Phys. Lett. **89**, 101107 (2006). https://doi.org/10.1063/1.2346224
34. K. Lüdge, E. Schöll, IEEE J. Quantum Electron. **45**(11), 1396 (2009). https://doi.org/10.1109/jqe.2009.2028159
35. N. Majer, K. Lüdge, E. Schöll, Phys. Rev. B **82**, 235301 (2010). https://doi.org/10.1103/physrevb.82.235301
36. B. Lingnau, B. Herzog, M. Kolarczik, U. Woggon, K. Lüdge, N. Owschimikow, Appl. Phys. Lett. **110**, 241102 (2017). https://doi.org/10.1063/1.4985705
37. S. Schneider, P. Borri, W. Langbein, U. Woggon, R.L. Sellin, D. Ouyang, D. Bimberg, IEEE J. Quantum Electron. **40**(10), 1423 (2004). https://doi.org/10.1109/jqe.2004.834779
38. J.M.H. Elmirghani, H.T. Mouftah, IEEE Commun. Mag. **38**(3), 86 (2000). https://doi.org/10.1109/35.825645
39. S.J.B. Yoo, J. Light. Technol. **14**(6), 955 (1996). https://doi.org/10.1109/50.511595
40. S. Diez, C. Schmidt, R. Ludwig, H.G. Weber, K. Obermann, S. Kindt, I. Koltchanov, K. Petermann, IEEE J. Sel. Top. Quantum Electron. **3**(5), 1131 (1997). https://doi.org/10.1109/2944.658587
41. C. Politi, D. Klonidis, M.J. O'Mahony, J. Lightwave Technol. **24**(3), 1203 (2006). https://doi.org/10.1109/jlt.2005.863289
42. H. Schmeckebier, C. Meuer, D. Arsenijević, G. Fiol, C. Schmidt-Langhorst, C. Schubert, G. Eisenstein, D. Bimberg, IEEE Photon Technol. Lett. **24**(13), 1163 (2012). https://doi.org/10.1109/lpt.2012.2198635
43. J. Kim, M. Laemmlin, C. Meuer, D. Bimberg, G. Eisenstein, IEEE J. Quantum Electron. **45**(3), 240 (2009). https://doi.org/10.1109/jqe.2008.2010881
44. C. Meuer, H. Schmeckebier, G. Fiol, D. Arsenijević, J. Kim, G. Eisenstein, D. Bimberg, IEEE Photonics J. **2**(2), 141 (2010). https://doi.org/10.1109/jphot.2010.2044568
45. G. Contestabile, A. Maruta, S. Sekiguchi, K. Morito, M. Sugawara, K. Kitayama, IEEE J. Quantum Electron. **46**(12), 1696 (2010). https://doi.org/10.1109/jqe.2010.2060714
46. T. Udem, R. Holzwarth, T.W. Hänsch, Nature **416**, 233 (2002). https://doi.org/10.1038/416233a
47. J. Ye, H. Schnatz, L. Hollberg, IEEE J. Sel. Top. Quantum Electron. **9**(4), 1041 (2003). https://doi.org/10.1109/jstqe.2003.819109
48. A. Schliesser, N. Picque, T.W. Hänsch, Nat. Photonics **6**(7), 440 (2012). https://doi.org/10.1038/nphoton.2012.142
49. T. Akiyama, H. Kuwatsuka, N. Hatori, Y. Nakata, H. Ebe, M. Sugawara, IEEE Photon Technol. Lett. **14**(8), 1139 (2002). https://doi.org/10.1109/lpt.2002.1021995
50. D. Nielsen, S.L. Chuang, Phys. Rev. B **81**, 035305 (2010). https://doi.org/10.1103/physrevb.81.035305

51. G. Contestabile, A. Maruta, K. Kitayama, IEEE J. Quantum Electron. **50**(5), 379 (2014). https://doi.org/10.1109/jqe.2014.2313063
52. D. Birkedal, K. Leosson, J.M. Hvam, Phys. Rev. Lett. **87**, 227401 (2001). https://doi.org/10.1103/physrevlett.87.227401
53. P. Borri, W. Langbein, S. Schneider, U. Woggon, R.L. Sellin, D. Ouyang, D. Bimberg, Phys. Rev. Lett. **87**(15), 157401 (2001). https://doi.org/10.1103/physrevlett.87.157401
54. J.P. Reithmaier, G. Sek, A. Löffler, C. Hofmann, S. Kuhn, S. Reitzenstein, L.V. Keldysh, V.D. Kulakovskii, T.L. Reinecke, A. Forchel, Nature **432**, 197 (2004). https://doi.org/10.1038/nature02969
55. T. Yoshie, A. Scherer, J. Hendrickson, G. Khitrova, H.M. Gibbs, G. Rupper, C. Ell, O.B. Shchekin, D.G. Deppe, Nature **432**, 200 (2004). https://doi.org/10.1038/nature03119
56. G. Khitrova, H.M. Gibbs, M. Kira, S.W. Koch, A. Scherer, Nat. Phys. **2**, 81 (2006). https://doi.org/10.1038/nphys227
57. M. Kolarczik, N. Owschimikow, J. Korn, B. Lingnau, Y. Kaptan, D. Bimberg, E. Schöll, K. Lüdge, U. Woggon, Nat. Commun. **4**, 2953 (2013). https://doi.org/10.1038/ncomms3953
58. A. Capua, O. Karni, G. Eisenstein, J.P. Reithmaier, Phys. Rev. B **90**, 045305 (2014). https://doi.org/10.1103/physrevb.90.045305
59. L. Allen, J.H. Eberly, *Optical Resonance and Two-Level Atoms* (Dover, New York, 1975)
60. M.O. Scully, M.S. Zubairy, *Quantum Optics* (Cambridge University Press, 1997). https://doi.org/10.1017/cbo9780511813993
61. B. Lingnau, K. Lüdge, Opt. Quantum Electron. **50**(2), 111 (2018). https://doi.org/10.1007/s11082-018-1380-9
62. S.L. McCall, E.L. Hahn, Phys. Rev. **183**(2), 457 (1969). https://doi.org/10.1103/physrev.183.457
63. M. Hoffmann, O.D. Sieber, V.J. Wittwer, I.L. Krestnikov, D.A. Livshits, Y. Barbarin, T. Südmeyer, U. Keller, Opt. Express **19**(9), 8108 (2011). https://doi.org/10.1364/oe.19.008108
64. M. Scheller, T.L. Wang, B. Kunert, W. Stolz, S.W. Koch, J.V. Moloney, Electron. Lett. **48**(10), 588 (2012). https://doi.org/10.1049/el.2012.0749
65. A. Moelbjerg, P. Kaer, M. Lorke, J. Mørk, Phys. Rev. Lett. **108**, 017401 (2012). https://doi.org/10.1103/physrevlett.108.017401
66. G. Visimberga, G. Rainò, A. Salhi, V. Tasco, M.T. Todaro, L. Martiradonna, M. De Giorgi, A. Passaseo, R. Cingolani, M. De Vittorio, Appl. Phys. Lett. **93**(15), 151112 (2008). https://doi.org/10.1063/1.3000381
67. C.I. Shih, C.H. Lin, S.C. Lin, T.C. Lin, K.W. Sun, O. Voskoboynikov, C.P. Lee, Y.W. Suen, Nanoscale Res. Lett. **6**(1), 409 (2011). https://doi.org/10.1186/1556-276x-6-409
68. S.T. Cundiff, Opt. Express **16**(7), 4639 (2008). https://doi.org/10.1364/oe.16.004639
69. M. Kolarczik, C. Ulbrich, P. Geiregat, Y. Zhu, L.K. Sagar, A. Singh, B. Herzog, A. Achtstein, X. Li, D. van Thourhout, Z. Hens, N. Owschimikow, U. Woggon, APL Photonics **3**(1), 016101 (2018). https://doi.org/10.1063/1.5005490
70. P. Borri, S. Schneider, W. Langbein, U. Woggon, A.E. Zhukov, V.M. Ustinov, N.N. Ledentsov, Z.I. Alferov, D. Ouyang, D. Bimberg, Appl. Phys. Lett. **79**, 2633 (2001). https://doi.org/10.1063/1.1411986
71. N.B. Grosse, N. Owschimikow, R. Aust, B. Lingnau, A. Koltchanov, M. Kolarczik, K. Lüdge, U. Woggon, Opt. Express **22**(26), 32520 (2014). https://doi.org/10.1364/oe.22.032520
72. V.V. Temnov, U. Woggon, Phys. Rev. Lett. **95**, 243602 (2005). https://doi.org/10.1103/physrevlett.95.243602
73. H.A.M. Leymann, A. Foerster, F. Jahnke, J. Wiersig, C. Gies, Phys. Rev. Appl. **4**, 044018 (2015). https://doi.org/10.1103/physrevapplied.4.044018
74. F. Böhm, N.B. Grosse, M. Kolarczik, B. Herzog, A. Achtstein, N. Owschimikow, U. Woggon, Proc. SPIE **10359**, 1035907 (2017). https://doi.org/10.1117/12.2273855
75. D.T. Smithey, M. Beck, M.G. Raymer, A. Faridani, Phys. Rev. Lett. **70**, 1244 (1993). https://doi.org/10.1103/physrevlett.70.1244
76. M. Munroe, D. Boggavarapu, M.E. Anderson, M.G. Raymer, Phys. Rev. A **52**, 924(R) (1995). https://doi.org/10.1103/physreva.52.r924

77. J. Bertrand, P. Bertrand, Found. Phys. **17**(4), 397 (1987). https://doi.org/10.1007/bf00733376
78. U. Leonhardt, M. Munroe, T. Kiss, T. Richter, M.G. Raymer, Optics Commun. **127**(1), 144 (1996). https://doi.org/10.1016/0030-4018(96)00061-2
79. W.W. Chow, S. Reitzenstein, Appl. Phys. Rev. **5**(4), 041302 (2018). https://doi.org/10.1063/1.5045580
80. E.A. Avrutin, J.H. Marsh, E.L. Portnoi, IEE Proc. Optoelectron. **147**(4), 251 (2000). https://doi.org/10.1049/ip-opt:20000282
81. E.U. Rafailov, M.A. Cataluna, W. Sibbett, Nat. Photonics **1**(7), 395 (2007). https://doi.org/10.1038/nphoton.2007.120
82. J. Piprek, in *Handbook of Optoelectronic Device Modeling and Simulation: Fundamentals, Materials, Nanostructures, LEDs, and Amplifiers*. Series in Optics and Optoelectronics, vol. 1 (CRC Press, 2017). https://doi.org/10.1201/9781315152301
83. B. Ravoori, A.B. Cohen, J. Sun, A.E. Motter, T.E. Murphy, R. Roy, Phys. Rev. Lett. **107**, 034102 (2011). https://doi.org/10.1103/physrevlett.107.034102
84. J. Ye, S.T. Cundiff, *Femtosecond Optical Frequency Comb: Principle, Operation, and Applications* (Springer, New York, 2005). https://doi.org/10.1007/b102450
85. L.A. Jiang, M.E. Grein, H.A. Haus, E.P. Ippen, IEEE J. Sel. Top. Quantum Electron. **7**(2), 159 (2001). https://doi.org/10.1109/2944.954125
86. A.G. Vladimirov, D.V. Turaev, Phys. Rev. A **72**(3), 033808 (2005). https://doi.org/10.1103/physreva.72.033808
87. B. Tromborg, H.E. Lassen, H. Olesen, IEEE J. Quantum Electron. **30**(4), 939 (1994). https://doi.org/10.1109/3.291365
88. U. Bandelow, M. Radziunas, A.G. Vladimirov, B. Hüttl, R. Kaiser, Opt. Quantum Electron. **38**, 495 (2006). https://doi.org/10.1007/s11082-006-0045-2
89. H.A. Haus, IEEE, J. Sel. Top. Quantum Electron. **6**(6), 1173 (2000). https://doi.org/10.1109/2944.902165
90. H.A. Haus, IEEE J. Quantum Electron. **11**(9), 736 (1975). https://doi.org/10.1109/jqe.1975.1068922
91. H.A. Haus, J. Appl. Phys. **46**(7), 3049 (1975). https://doi.org/10.1063/1.321997
92. A.G. Vladimirov, D.V. Turaev, G. Kozyreff, Opt. Lett. **29**(11), 1221 (2004). https://doi.org/10.1364/ol.29.001221
93. A.G. Vladimirov, D.V. Turaev, Radiophys. Quantum Electron. **47**, 769 (2004). https://doi.org/10.1007/s11141-005-0015-8
94. T. Kolokolnikov, M. Nizette, T. Erneux, N. Joly, S. Bielawski, Physica D **219**, 13 (2006)
95. R. Kaiser, B. Hüttl, IEEE, J. Sel. Top. Quantum Electron. **13**(1), 125 (2007). https://doi.org/10.1109/jstqe.2006.885330
96. L.C. Jaurigue, *Passively Mode-Locked Semiconductor Lasers: Dynamics and Stochastic Properties in the Presence of Optical Feedback*. Springer Thesis edn. (Springer, Cham, 2017)
97. C. Otto, L.C. Jaurigue, E. Schöll, K. Lüdge, IEEE Photonics J. **6**(5), 1501814 (2014). https://doi.org/10.1109/jphot.2014.2352934
98. C. Otto, *Dynamics of Quantum Dot Lasers—Effects of Optical Feedback and External Optical Injection*. Springer Theses (Springer, Heidelberg, 2014). https://doi.org/10.1007/978-3-319-03786-8
99. D. Rachinskii, A.G. Vladimirov, U. Bandelow, B. Hüttl, R. Kaiser, J. Opt. Soc. Am. B **23**(4), 663 (2006). https://doi.org/10.1364/josab.23.000663
100. E.A. Viktorov, P. Mandel, A.G. Vladimirov, U. Bandelow, Appl. Phys. Lett. **88**(20), 201102 (2006). https://doi.org/10.1063/1.2203937
101. M. Nizette, D. Rachinskii, A.G. Vladimirov, M. Wolfrum, Physica D **218**(1), 95 (2006). https://doi.org/10.1016/j.physd.2006.04.013
102. E.A. Viktorov, P. Mandel, M. Kuntz, G. Fiol, D. Bimberg, A.G. Vladimirov, M. Wolfrum, Appl. Phys. Lett. **91**(23), 231116 (2007). https://doi.org/10.1063/1.2822808
103. A.G. Vladimirov, U. Bandelow, G. Fiol, D. Arsenijević, M. Kleinert, D. Bimberg, A.S. Pimenov, D. Rachinskii, J. Opt. Soc. Am. B **27**(10), 2102 (2010). https://doi.org/10.1364/josab.27.002102

104. L.C. Jaurigue, B. Krauskopf, K. Lüdge, Chaos **27**(11), 114301 (2017). https://doi.org/10.1063/1.5006743
105. A.G. Vladimirov, D. Rachinskii, M. Wolfrum, in *Nonlinear Laser Dynamics—From Quantum Dots to Cryptography*, ed. by K. Lüdge, Reviews in Nonlinear Dynamics and Complexity (Wiley-VCH, Weinheim, 2011), Chap. 8, pp. 183–213
106. G. New, IEEE J. Quantum Electron. **10**(2), 115 (1974). https://doi.org/10.1109/jqe.1974.1145781
107. M. Rossetti, X. Tianhong, P. Bardella, I. Montrosset, IEEE J. Quantum Electron. **47**(11), 1404 (2011). https://doi.org/10.1109/jqe.2011.2167131
108. M. Rossetti, T. Xu, P. Bardella, I. Montrosset, Phys. Status Solidi C **9**(2), 286 (2011). https://doi.org/10.1002/pssc.201100243
109. T. Xu, J. Cao, I. Montrosset, Nanoscale Res. Lett **10**(1), 26 (2015). https://doi.org/10.1186/s11671-014-0720-3
110. M. Marconi, J. Javaloyes, S. Balle, M. Giudici, Phys. Rev. Lett. **112**, 223901 (2014). https://doi.org/10.1103/physrevlett.112.223901
111. A.S. Pimenov, J. Javaloyes, S.V. Gurevich, A.G. Vladimirov, Philos. Trans. Royal Soc. A **376**, 2124 (2018). https://doi.org/10.1098/rsta.2017.0372
112. D. Puzyrev, A.G. Vladimirov, A.S. Pimenov, S.V. Gurevich, S. Yanchuk, Phys. Rev. Lett. **119**, 163901 (2017). https://doi.org/10.1103/physrevlett.119.163901
113. D. von der Linde, Appl. Phys. B **39**(4), 201 (1986). https://doi.org/10.1007/bf00697487
114. J. Mulet, J. Mørk, IEEE J. Quantum Electron. **42**(3), 249 (2006). https://doi.org/10.1109/jqe.2006.869808
115. B. Zhu, I.H. White, R.V. Penty, A. Wonfor, E. Lach, H.D. Summers, IEEE J. Quantum Electron. **33**(7), 1216 (1997). https://doi.org/10.1109/3.594887
116. C. Otto, K. Lüdge, A.G. Vladimirov, M. Wolfrum, E. Schöll, New J. Phys. **14**, 113033 (2012). https://doi.org/10.1088/1367-2630/14/11/113033
117. C. Simos, H. Simos, C. Mesaritakis, A. Kapsalis, D. Syvridis, Opt. Commun. **313**, 248 (2014). https://doi.org/10.1016/j.optcom.2013.10.034
118. L.C. Jaurigue, O. Nikiforov, E. Schöll, S. Breuer, K. Lüdge, Phys. Rev. E **93**(2), 022205 (2016). https://doi.org/10.1103/physreve.93.022205
119. H.A. Haus, A. Mecozzi, IEEE J. Quantum Electron. **29**(3), 983 (1993). https://doi.org/10.1109/3.206583
120. A.S. Pimenov, T. Habruseva, D. Rachinskii, S.P. Hegarty, G. Huyet, A.G. Vladimirov, Opt. Lett. **39**(24), 6815 (2014). https://doi.org/10.1364/ol.39.006815
121. L.C. Jaurigue, A.S. Pimenov, D. Rachinskii, E. Schöll, K. Lüdge, A.G. Vladimirov, Phys. Rev. A **92**(5), 053807 (2015). https://doi.org/10.1103/physreva.92.053807
122. A. Daffertshofer, Phys. Rev. E **58**(1), 327 (1998). https://doi.org/10.1103/physreve.58.327
123. L. Callenbach, P. Hänggi, S.J. Linz, Phys. Rev. E **65**, 051110 (2002). https://doi.org/10.1103/physreve.65.051110
124. J.K. Hale, *Theory of Functional Differential Equations* (Springer, New York, 1977). https://doi.org/10.1007/978-1-4612-9892-2
125. S. Guo, J. Wu, *Bifurcation Theory of Functional Differential Equations* (Springer, New York, 2013). https://doi.org/10.1007/978-1-4614-6992-6
126. N. Rebrova, G. Huyet, D. Rachinskii, A.G. Vladimirov, Phys. Rev. E **83**(6), 066202 (2011). https://doi.org/10.1103/physreve.83.066202
127. M.S. Krol, SIAM, J. Appl. Math. **51**(6), 1622 (1991). https://doi.org/10.1137/0151083
128. G. Fiol, D. Arsenijević, D. Bimberg, A.G. Vladimirov, M. Wolfrum, E.A. Viktorov, P. Mandel, Appl. Phys. Lett. **96**(1), 011104 (2010). https://doi.org/10.1063/1.3279136
129. R.M. Arkhipov, T. Habruseva, A.S. Pimenov, M. Radziunas, S.P. Hegarty, G. Huyet, A.G. Vladimirov, J. Opt. Soc. Am. B **33**(3), 351 (2016). https://doi.org/10.1364/josab.33.000351
130. R.M. Arkhipov, A.S. Pimenov, M. Radziunas, D. Rachinskii, A.G. Vladimirov, D. Arsenijevi, H. Schmeckebier, D. Bimberg, IEEE J. Sel. Top. Quantum Electron. **19**(4), 1100208 (2013). https://doi.org/10.1109/jstqe.2012.2228633

131. S. Breuer, W. Elsäßer, J.G. McInerney, K. Yvind, J. Pozo, E.A.J.M. Bente, M. Yousefi, A. Villafranca, N. Vogiatzis, J. Rorison, IEEE J. Quantum Electron. **46**(2), 150 (2010). https://doi.org/10.1109/jqe.2009.2033255
132. O. Nikiforov, L.C. Jaurigue, L. Drzewietzki, K. Lüdge, S. Breuer, Opt. Express **24**(13), 14301 (2016). https://doi.org/10.1364/oe.24.014301
133. H. Asghar, W. Wei, P. Kumar, E. Sooudi, J.G. McInerney, Opt. Express **26**(4), 4581 (2018). https://doi.org/10.1364/oe.26.004581
134. H. Asghar, J.G. McInerney, Opt. Lett. **42**(18), 3714 (2017). https://doi.org/10.1364/ol.42.003714
135. O. Solgaard, K.Y. Lau, IEEE Photon Technol. Lett. **5**(11), 1264 (1993)
136. C.Y. Lin, F. Grillot, Y. Li, R. Raghunathan, L.F. Lester, Opt. Express **18**(21), 21932 (2010). https://doi.org/10.1364/oe.18.021932
137. C.Y. Lin, F. Grillot, N.A. Naderi, Y. Li, L.F. Lester, Appl. Phys. Lett. **96**(5), 051118 (2010). https://doi.org/10.1063/1.3299714
138. M. Haji, L. Hou, A.E. Kelly, J. Akbar, J.H. Marsh, J.M. Arnold, C.N. Ironside, Opt. Express **20**(3), 3268 (2012). https://doi.org/10.1364/oe.20.003268
139. D. Arsenijevi, M. Kleinert, D. Bimberg, Appl. Phys. Lett. **103**(23), 231101 (2013). https://doi.org/10.1063/1.4837716
140. L.C. Jaurigue, E. Schöll, K. Lüdge, Phys. Rev. Lett. **117**, 154101 (2016). https://doi.org/10.1103/physrevlett.117.154101
141. U. Keller, Nature **424**, 831 (2003). https://doi.org/10.1038/nature01938
142. D. Bimberg, G. Fiol, M. Kuntz, C. Meuer, M. Lämmlin, N.N. Ledentsov, A.R. Kovsh, Phys. Status Solidi A **203**(14), 3523 (2006). https://doi.org/10.1002/pssa.200622488
143. S. Meinecke, L. Drzewietzki, C. Weber, B. Lingnau, S. Breuer, K. Lüdge, Sci. Rep. **9**, 1783 (2019). https://doi.org/10.1038/s41598-018-38183-1
144. G.H.M. van Tartwijk, G.P. Agrawal, Prog. Quantum Electron. **22**(2), 43 (1998). https://doi.org/10.1016/s0079-6727(98)00008-1
145. J. Javaloyes, S. Balle, Opt. Express **20**(8), 8496 (2012). https://doi.org/10.1364/oe.20.008496
146. A.G. Vladimirov, A.S. Pimenov, D. Rachinskii, IEEE J. Quantum Electron. **45**(5), 462 (2009). https://doi.org/10.1109/jqe.2009.2013363
147. K. Lüdge, in *Nonlinear Laser Dynamics—From Quantum Dots to Cryptography*, ed. by K. Lüdge (WILEY-VCH Weinheim, Weinheim, 2012), Chap. 1, pp. 3–34
148. T.R. Nielsen, P. Gartner, F. Jahnke, Phys. Rev. B **69**, 235314 (2004). https://doi.org/10.1103/physrevb.69.235314
149. D.B. Malins, A. Gomez-Iglesias, S.J. White, W. Sibbett, A. Miller, E.U. Rafailov, Appl. Phys. Lett. **89**(17), 171111 (2006). https://doi.org/10.1063/1.2369818
150. E.A. Viktorov, T. Erneux, P. Mandel, T. Piwonski, G. Madden, J. Pulka, G. Huyet, J. Houlihan, Appl. Phys. Lett. **94**(26), 263502 (2009). https://doi.org/10.1063/1.3159838
151. M. Wegert, D. Schwochert, E. Schöll, K. Lüdge, Opt. Quantum Electron. **46**, 1337 (2014). https://doi.org/10.1007/s11082-014-9878-2
152. T. Xu, P. Bardella, M. Rossetti, I. Montrosset, IET Optoelectron. **6**(2), 110 (2012). https://doi.org/10.1049/iet-opt.2011.0056
153. D. Waldburger, C.G.E. Alfieri, S.M. Link, S. Meinecke, L.C. Jaurigue, K. Lüdge, U. Keller, Opt. Express **26**(17), 21872 (2018). https://doi.org/10.1364/oe.26.021872

Chapter 5
Optical and Structural Properties of Nitride Based Nanostructures

Frank Bertram, Christoph Berger, Jürgen Christen, Holger Eisele, Ludwig A. Th. Greif, Axel Hoffmann, Janina Maultzsch, Marcus Müller, Emanuele Poliani, Gordon Schmidt, Peter Veit and Markus R. Wagner

Abstract Advanced characterization methods with nanoscale resolution are powerful tools in order to overcome the continuing challenges in the optimization of nitride semiconductor nanostructures for more efficient nanophotonic devices in the UV and green spectral range. This chapter is devoted to the study of optical, electronic, and structural properties of these nitride based nanostructures. In the first part, we discuss several state-of-the-art nanoscale characterization techniques including scanning transmission electron microscopy cathodoluminescence (STEM-CL), tip-enhanced Raman spectroscopy (TERS), micro-photoluminescence (μPL), X-ray diffraction (XRD), and scanning tunneling microscopy and spectroscopy (STM/STS). This selection of complementary microscopic and spectroscopic techniques provides unique insights into a multitude of nanostructure properties such as charge carrier excitation, relaxation, diffusion, and recombination dynamics, vibrational and structural properties including strain, segregation, as well as clustering, and surface and interface morphology. In the second part, we apply and combine these techniques to obtain detailed information on nanoscale properties of nitride based micro-columns, quantum wires, and heterostructures. The study of these nitride nanostructures provides not only insight into device limitations, but also contributes to the fundamental understanding of structural and optical properties of III-nitride nanostructures.

F. Bertram · C. Berger · J. Christen · M. Müller · G. Schmidt · P. Veit
Institut für Physik, Otto-von-Guericke-Universität Magdeburg, 39106 Magdeburg, Germany
e-mail: frank.bertram@ovgu.de

H. Eisele · L. A. Th. Greif · A. Hoffmann (✉) · E. Poliani · M. R. Wagner
Institut für Festkörperphysik, Technische Universität Berlin, 10623 Berlin, Germany
e-mail: axel.hoffmann@tu-berlin.de

M. R. Wagner
e-mail: markus.wagner@tu-berlin.de

J. Maultzsch
Department of Physics, Friedrich-Alexander-Universität Erlangen-Nürnberg, 91058 Erlangen, Germany

M. Kneissl et al. (eds.), *Semiconductor Nanophotonics*, Springer Series in Solid-State Sciences 194, https://doi.org/10.1007/978-3-030-35656-9_5

5.1 Introduction

The continuing miniaturization of semiconductor structures down to the nanometer scale has moved various materials in the research focus for novel photonic applications [1–4]. Quantum confinement in one or more directions introduces size dependencies of several physical parameters such as the band gap energy and density of states [1] leading to novel effects such as e.g. drastically reduced carrier scattering [2]. Nitride based nanostructures constitute a particularly interesting group of materials as their wide bandgap tunability allows the design and fabrication of emitters and sensors from the visible to the deep UV spectral range. While technologically mature quantum heterostructures based on the ternary InGaN material dominate the solid state lighting and display applications with emission from the near UV to green spectral range, major challenges for nitrides still exist to achieve comparable performance in the shorter and longer wavelength regime. In the case of longer wavelength emitters, In-segregation and clustering imposes material limitations in which fundamental aspects of these processes are still insufficiently understood. In the UV spectral range, light extracting, doping and control over non-radiative recombination channels pose major challenges. In order to expand the spectral range of nitride-based nanostructures for optoelectronics it is therefore crucial to apply modern characterization methods to reveal the structural, optical, vibrational, and electronic properties of nitrides with few nanometer spatial resolution.

In the first part of this chapter, we give an overview of several modern tools for the characterization of nanostructures with bandgaps in the visible and UV spectral range. We provide a short overview of the most important working principles and capabilities of (scanning) transmission electron microscopy cathodoluminescence (TEM/STEM CL), tip enhanced Raman scattering (TERS), micro-optical and quantum optical luminescence spectroscopy, X-ray diffraction (XRD), and scanning tunneling microscopy and spectroscopy (STM/STS).

In the second part, we apply and combine these techniques to obtain detailed information on nanoscale properties of nitride based micro-columns, quantum wires, and heterostructures. Using TERS, we analyze how the coupling between the metallic tip and the (doped) sample influences the Raman signature depending on the carrier concentration. Such measurements are not only of importance for the investigation of localized phonons in, e.g. nanowires and quantum wires, but also elucidate the nature of the enhancement process itself. Based on our combination of ultraviolet enhanced μPL and CL, both providing the capability for ps time-resolved and correlation spectroscopy, we discuss the strong coupling between excitons and phonons in nitride nanostructures. The combined TERS and μPL study strongly benefits from the correlation with our unique nm-resolved low temperature (S)TEM-CL technique, as it complementarily provides structural, compositional, and electronic/vibronic properties. In addition, STM data show the quality of the material at the atomic level in correlation with μPL results. The close link between the sub-μm spatial and ps-time-resolved techniques of μPL (ps-μPL) and CL (ps-CL) in nanostructures yields information about the kinetics of individual phonons and excitons, as well as their interrelation. The individual strengths of μPL and ps-CL interlock perfectly as they combine energy tunability in continuous wave and short pulse operation with resonant as well as off-resonant excitation (μPL) and deep UV above bandgap excitation with flexible excitation pulse length and shape up to steady state excitation (ps-CL). Thus, the interplay of both methods gives access to complex capture, relaxation, as well as recombination dynamics over a wide spectral range.

5.2 Advanced Tools for Nanostructure Characterization

5.2.1 TEM/STEM-CL

A detailed understanding of complex semiconductor heterostructures and the physics of devices based on them requires a systematic determination and correlation of structural, chemical, electronic and optical properties in the nanometer range. Luminescence techniques belong to the most sensitive, non-destructive methods of semiconductor research. The combination of luminescence spectroscopy at liquid He (LHe) temperatures with the spatial resolution of a scanning transmission electron microscope (STEM) ($\delta x < 1$ nm at RT, $\delta x < 5$ nm at 10 K) as realized by the technique of low temperature scanning transmission electron microscopy cathodoluminescence microscopy (STEM-CL) provides a unique tool for the optical nano-characterization of semiconductors, heterostructures, and interfaces.

This section focuses on STEM-CL method and its realization. Even though a few groups are able to perform optical nano-characterizations with this technique at room temperature [5–17], the use at LHe temperatures is worldwide unique.[1] First, the general concept of STEM-CL characterization is introduced followed by a description of the applied experimental setup. The fundamental aspects of the different imaging modes as well as the beam broadening in STEM are described. Subsequently, the cathodoluminescence in STEM and their effects on electron transparent TEM foils is discussed. In addition to the current chapter further examples of STEM-CL applications are described in Chap. 12.

Scanning Transmission Electron Microscopy

As 1938 realized by von Ardenne [18, 19], in scanning transmission electron microscopy the sample is scanned by a strong focused primary electron beam, which is in contrast to the homogeneous illumination in conventional TEM. Transmitted electrons are sequentially detected point-by-point to form the image. This allows a simultaneous, spatially correlated acquisition of secondary signals, like cathodoluminescence. The relationship between conventional TEM and STEM finds its expression in the reciprocity principle, which means that the electron optical paths behave reciprocally by exchange of electron source and detector.

[1] Until the completion of this publication.

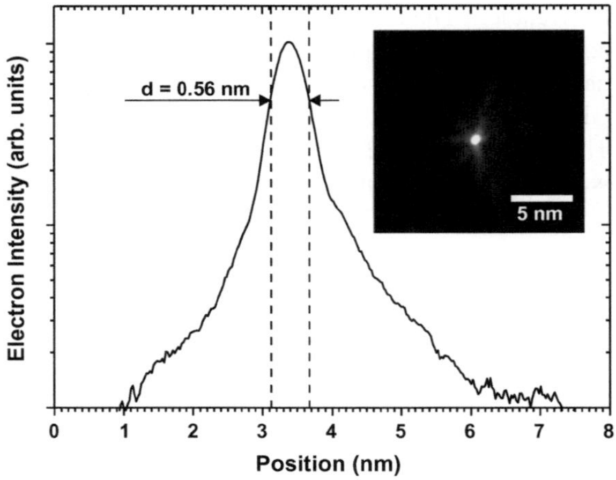

Fig. 5.1 Intensity profile of incident primary electron beam in STEM mode exhibits a 0.56 nm probe diameter, inset: image of two dimensional intensity profile of the probe [20]

Probe Diameter and Beam Broadening

The diameter of the incident primary electron probe should be as small as possible to enable highest spatial resolution in STEM. As an example, the lateral intensity profile of the electron probe is shown in Fig. 5.1. Here, the probe without sample interaction was imaged onto the CCD camera in nano-probe mode in a TecaniF20 for spot size of 11 μm and aperture of 30 μm with condenser lens 2. The full width at half maximum (FWHM) as criterion for the probe diameter was determined to 0.56 nm. The inset shows the two dimensional probe profile. In general, the probe diameter increases with increasing spot size and condenser aperture.

Regarding the determination of the spatial resolution in STEM-CL, the probe diameter as well as the scattering of primary electrons has to be considered. Due to the interaction of primary electrons with the sample, a broadening of the beam perpendicular to the beam direction occurs. This effect limits the resolution in the scanning mode of TEMs. The broadening of an infinitesimal wide electron beam can be described by elastic single scattering from Goldstein et al. [21]. The rewritten analytical expression with relativistic correction is [22]:

$$b_{\mathrm{STEM}} = 1.05 \cdot 10^5 \left(\frac{\varrho}{A}\right)^{1/2} \frac{Z}{E_{\mathrm{e}}} \frac{1 + E_{\mathrm{e}}/E_{\mathrm{e0}}}{1 + E_{\mathrm{e}}/2E_{\mathrm{e0}}} d_{\mathrm{foil}}^{3/2} \tag{5.1}$$

b_{STEM} beam broadening (cm)
ϱ mass density (g/cm^3)
A mean relative atomic mass (g/mol)
Z mean atomic number
E_{e} electron beam energy (eV)
E_{e0} electron rest energy (eV)
d_{foil} TEM foil thickness (cm)

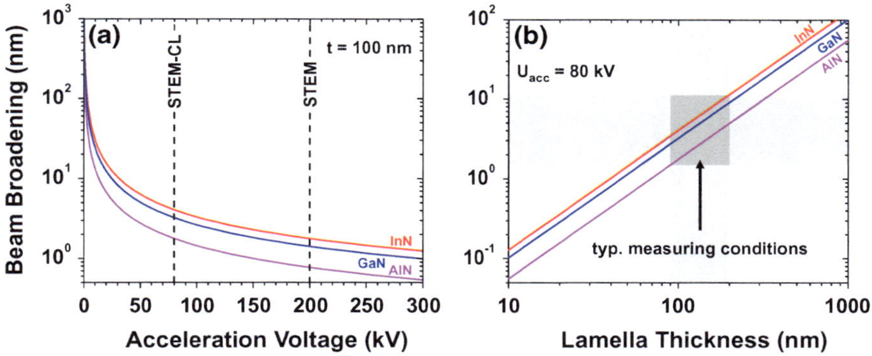

Fig. 5.2 Beam broadening in STEM mode after (5.1) as a function of **a** acceleration voltage for TEM foil thickness of 100 nm and of **b** foil thickness for 80 kV acceleration voltage in case of GaN, AlN and InN [20]

By reference of (5.1), a reduction of the beam broadening with increasing primary electron energy and decreasing foil thickness is considered. In Fig. 5.2a, the beam broadening in dependence of the acceleration voltage is plotted exemplarily for AlN, GaN and InN. Here, for STEM-CL characterization a reasonable foil thickness of 100 nm was assumed, lying within the typical range of 90–200 nm. In case of 80 kV acceleration voltage, which is used for all STEM-CL experiments, a beam broadening of 3.2 nm is obtained for GaN. STEM measurements at 200 kV yield a broadening of the electron beam of only 1.7 nm for GaN. The beam broadening as a function of foil thickness is represented for 80 kV in Fig. 5.2b.

Equation (5.1) represents a calculation of beam broadening only due to elastic single scattering events [21, 22]. In particular, for large foil thicknesses above the mean free path of elastically scattered electrons, multiple scattering can occur. Hence, the beam widening for an infinitesimal small primary electron beam is evaluated by Monte Carlo simulation and will be compared with the analytical calculated values. Figure 5.3 represents a Monte Carlo simulation for 100 nm thick GaN TEM foil at 80 kV acceleration voltage. The simulated histogram in Fig. 5.3 shows the frequency of the primary electrons as a function of their lateral positions after transmitting the sample. The beam broadening can be defined as the lateral width in which 95% of all transmitted primary electrons are scattered. Therefore, the beam broadening is 2.5 nm, which is slightly smaller than the calculated one after (5.1).

In conclusion, (5.1) seems to be a good approximation of the beam broadening for a typical GaN foil thickness of 100 nm. Reimer et al. verified for polystyrene spheres that the beam broadening calculated after (5.1) is the upper limit for a large range of foil thicknesses (see [22]). However, it should be noted that both, the analytical calculation in terms of (5.1) as well as the determination of the beam broadening by use of Monte Carlo simulation, do not account for the crystalline nature of a solid state.

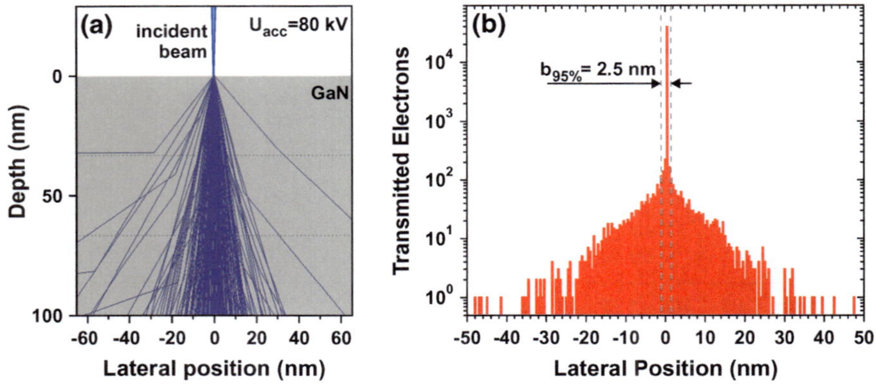

Fig. 5.3 Monte Carlo simulation of electron trajectories in 100 nm thick GaN TEM foil at 80 kV acceleration voltage: **a** cross-sectional view of electron trajectories, **b** histogram of transmitted primary electrons at foil exit; 95% of all primary electrons pass the GaN foil in a lateral width of 2.5 nm [20]

Bragg scattered electrons or channeling effects are not considered. Properly speaking, the beam broadening is only valid for amorphous samples. Nevertheless, these values give a good approximation of the beam broadening in crystalline structures.

Imaging Modes in STEM

In STEM, different detectors can be used for imaging. The most common ones are the high angle annular dark field (HAADF), annular dark field (ADF) and bright field (BF) detector. All of them integrate scattered electrons in a limited solid angle range. A schematic illustration of the STEM detector configuration is depicted in Fig. 5.4.

High angle scattered electrons are recorded by a HAADF detector ring above 50 mrad. Here, the contrast origin is mostly due to quasi-elastically, incoherently scattered electrons. The scattering at atoms/cores displaced from ideal lattice positions is most relevant in high angle regime. Since elastical scattering at a perfect crystal leads to intensities only in Bragg reflexes, once an atom/core is displaced from its lattice position, scattering intensities occur off the Bragg reflexes as a diffuse background. Such displacements can be caused by thermal motion thus this interaction is called thermal diffuse scattering. The thermal diffuse scattering is dominant for large scattering angles due to the reduced cross-section of Bragg scattered electrons barely influencing the HAADF signal. In accordance with Rutherford scattering at single atom cores, the intensity of the thermal diffuse scattering is approximately proportional to Z^2 for high angles giving the HAADF signal its synonym—Z contrast. The screening of the atom core's Coulomb potential in the solid-state results in a deviation from the bare Rutherford cross section leading to a Z^α dependence with α linked to the inner and outer detector angles. Furthermore, in HAADF the scattered secondary waves have no full phase relation anymore due to the derivation of the lattice periodicity caused by the displacement of atoms during thermal motion.

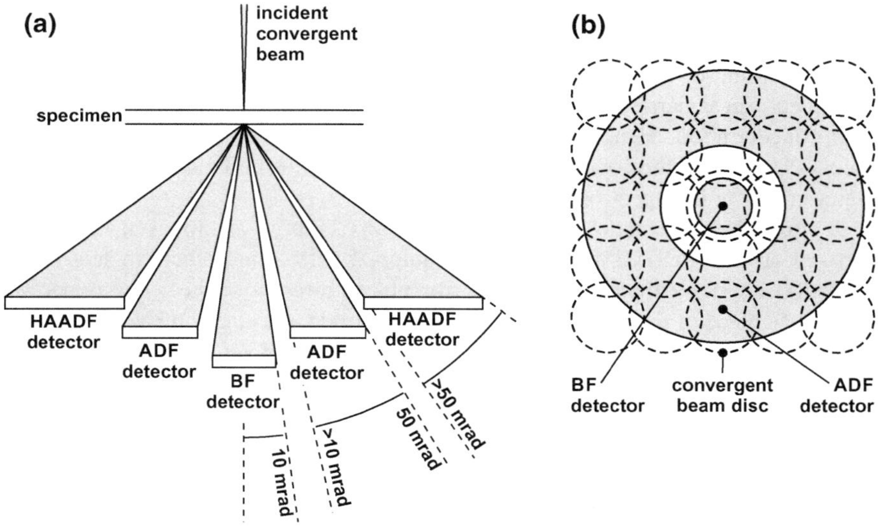

Fig. 5.4 Schematic illustration of **a** detector configuration in STEM and **b** plan-view on ADF and BF detector with diffraction discs, after [23, 24]. (**a**) Adapted by permission from Springer Nature: Springer Nature, How to 'See' Electrons by David B. Williams, C. Barry Carter, © Springer Science+Business Media, LLC 2009. (**b**) Adapted by permission from Springer Nature: Springer Nature, A Scan Through the History of STEM by Stephen J. Pennycook, © Springer Science+Business Media, LLC 2011

Consequently, the thermal diffuse scattering is incoherent and the HAADF signal can be regarded as convolution of probe and object giving an intuitive interpretation of STEM (HAADF) images. Moreover, a self-focusing effect of the primary electron beam occurs along the atomic rows due to the higher probability of presence of primary electrons in the vicinity of atom cores compared to the interstitial region [22]. This so-called channeling effect results in a higher HAADF intensity at the positions of atom rows [20]. Static structural changes, like crystal defects, are able to distort the lattice periodicity as well as the channeling effect, which is called dechanneling and can be used to observe such extended defects even in HAADF contrast [25].

Diffraction of the coherent convergent primary electron beam results in diffraction discs. In STEM imaging, the BF detector is located in a conjugated back-focal-plane, which means that the diffraction discs are imaged on the BF detector (see Fig. 5.4b). Hence, the direct unscattered beam disc and its superposition with the first order diffraction discs overlap with the BF detector integrating scattering events in the angle range from 0 to 10 mrad. Thus, in STEM (BF) images the diffraction contrast as well as mass thickness contrast contribute to contrast formation equivalent to the TEM (BF) imaging [24]. The ADF imaging in STEM is a superposition of diffraction and Z contrast. The diffraction discs of zero and higher-order Laue-zones are imaged on the detector ranging from 10 to 50 mrad.

Cathodoluminescence Spectroscopy in STEM

Cathodoluminescence results from the radiative recombination of electron beam generated excess carriers for which reason the CL intensity is a direct measure of their concentration. Hence, inelastic scattering events, leading to the creation of excess electrons in the conduction band and excess holes in the valence band, are a requirement for CL spectroscopy.

A simplified schematic image of our STEM-CL setup is shown in Fig. 5.5. It is based on a FEI (S)TEM Tecnai F20, equipped with a liquid helium stage (T = 10 K/300 K) and a light collecting mirror placed directly between the upper and lower pole piece of the objective lens. In STEM mode, the electron beam is focused into a spot and either kept at a single position or scanned over the region of interest over the sample. The emitted CL is collected by a retractable parabolic aluminum mirror and is focused onto the entrance slit of a grating monochromator MonoCL4 (Gatan). Using a Peltier-cooled GaAs(Cs) photomultiplier, panchromatic as well as spectrally resolved CL imaging is performed. Alternatively, a liquid-nitrogen cooled back-illuminated Si-CCD for parallel CL detection is used to record a complete CL spectrum at each pixel, enabling CL spectral imaging. Here, a 4-dimensional data set $I_{CL}(x, y, \lambda)$ is created while scanning over the area of interest. The set is used to

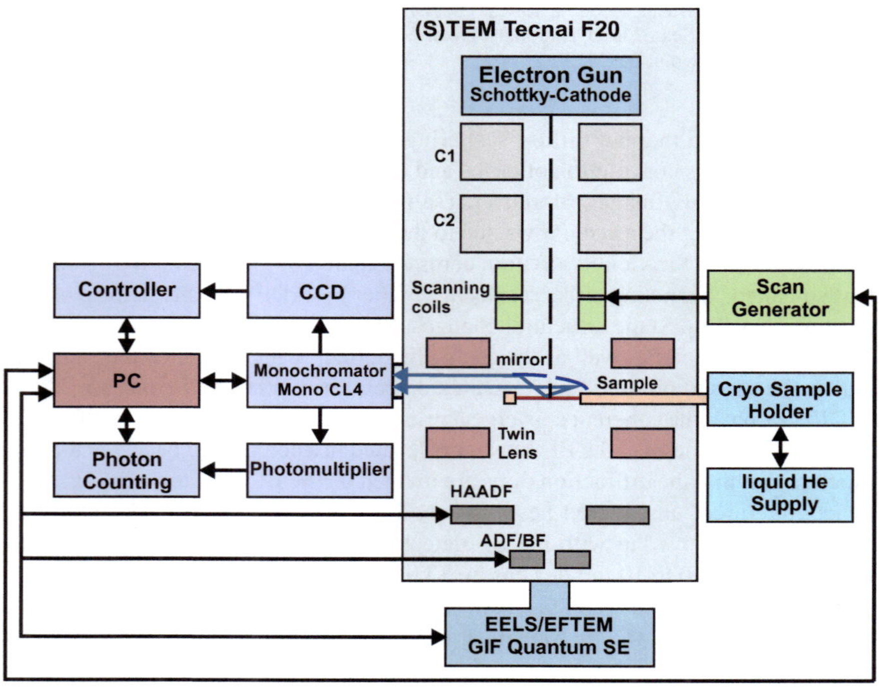

Fig. 5.5 Schematic STEM-CL setup for highly spatially resolved cathodoluminescence investigations and optional electron energy loss analysis

produce local spectra $I_{CL}(x_0, y_0, \lambda)$, sets of simultaneously recorded monochromatic CL images $\{I_{CL}(x_0, y_0, \lambda)\}$ (CLI), as well as CL peak wavelength images $\lambda_{Peak}(x, y)$ (CLWI), which map the peak wavelength of the local CL spectrum at each sampling point. Simultaneously to the detection of the CL-signal at each position, the electrons that are forward scattered into high as well as small solid angles are acquired by a (high-angle) annular dark-field detector (HA)ADF from Fischione (model 3000). In all STEM-CL experiments, we use an acceleration voltage of 80 kV to avoid sample damage and to prevent luminescence degradation under electron beam excitation.

Spatial Resolution

In CL microscopy, the spatial resolution is limited by two factors: the generation volume and the diffusion of generated carriers. Using electron transparent foils and high acceleration voltages, the excitation volume is drastically reduced in STEM-CL microscopy compared to cathodoluminescence in a scanning electron microscope due to the increased mean free paths for elastic and inelastic scattering events with increasing electron energy. This can be seen in a reduction of the beam broadening with increasing acceleration voltages and reduced thickness of TEM foil (see Sect. 5.2.1: probe diameter and beam broadening).

The excitation volume in STEM-CL microscopy is defined as the volume in which excess carriers are generated. Assuming a possible generation of excess carriers by scattered as well as unscattered primary electrons, the discussed probe diameter and the beam broadening can be used to calculate the excitation width. In the sense of a Gaussian probe profile, the lateral generation width is:

$$\Delta R = \sqrt{b_{\mathrm{STEM}}^2 + R_{\mathrm{probe}}^2}. \tag{5.2}$$

ΔR lateral excitation width
R_{probe} probe diameter

A GaN foil thickness of 100 nm, an acceleration voltage of 80 kV and a probe diameter of 0.56 nm results in a lateral generation width of 3.3 nm.

By detecting the cathodoluminescence globally (e.g. by a mirror), the diffusion is the second limiting factor for the spatial resolution, which decreases with increasing carrier diffusion length. Since a carrier concentration difference between generation volume and surrounding material, diffusion of excess carriers into the adjacent regions takes place. In case of a delta like excitation, the solution of the one dimensional diffusion differential equation is:

$$\Delta c(x) = G_0 \tau_{\mathrm{life}} \mathrm{e}^{-|x|/L_{\mathrm{diff}}}. \tag{5.3}$$

Δc density of generated carriers
G_0 generation rate
τ_{life} life time of carriers
L_{diff} diffusion length

In case of an extended excitation profiles, (5.3) has to be convoluted with the generation distribution. The derivation of the diffusion equation and its application to special problems in cathodoluminescence microscopy can be read in [26–29].

Exemplarily, the diffusion of excess carriers in STEM-CL microscopy is shown for a GaN/AlN quantum dot (QD) heterostructure grown by molecular beam epitaxy.[2] A metal organic vapor phase epitaxial grown AlN layer on top of a sapphire substrate serves as template for the following growth process. Subsequently, a stack of 10 GaN QD layers each embedded in 50 nm AlN barrier was grown under Stranski-Krastanow growth mode on top of the template. The AlN barriers with darker contrast and the QD layers with brighter contrast are clearly seen in STEM images of the cross section. Comparing the STEM images with the simultaneously recorded panchromatic CL intensity images at 16 K, a circular luminescence distribution is observable (see Fig. 5.6). However, not all QDs in ADF contrast are also optically active. Investigating the luminescence distribution of single QDs, a region of $175 \times 175\ \text{nm}^2$ was scanned. The intensity profile in Fig. 5.6c exhibit a monoexponential intensity slope in the direction of a single QD. Since the CL intensity is direct proportional to the excess carrier density, (5.3) is used as fit function. A diffusion length of 18 nm for generated carriers was extracted.

Excitation Conditions

Luminescence spectra may strongly depend on the excitation conditions due to e.g. screening of electrical fields, saturation of recombination channels, band filling, etc. with higher excitation density. In the case of low excitation density, the excess carrier density is lower than the majority carrier concentration and vice versa for the high excitation case. In the following, the excitation conditions in STEM-CL experiments are discussed.

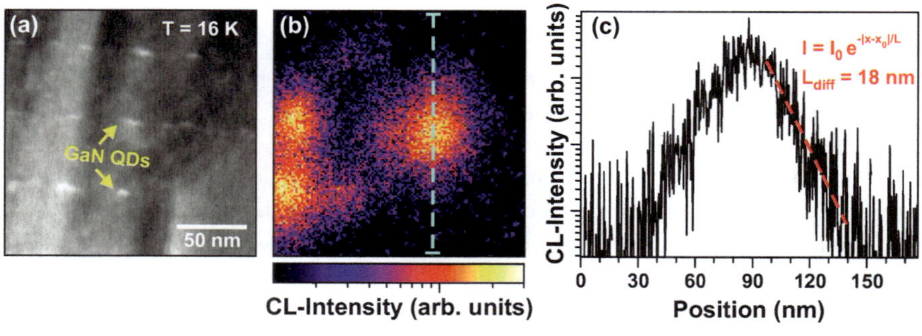

Fig. 5.6 LHe-STEM-CL investigation of a GaN/AlN QD stack ($T = 16$ K): **a** ADF image showing the QD stack with GaN QDs in light grey contrast, **b** simultaneously recorded panchromatic CL intensity image exhibiting circular intensity distribution around QDs, **c** linescan across an optical active QD with determined 18 nm diffusion length (marked in **b** with cyan line) [20]

[2]The sample was grown at CEA Grenoble at the Institut Nanosciences et Cryogénie in the group of Eva Monroy by Aparna Das.

The primary electron beam current is determined by a TEM sample holder with integrated Faraday cup. The variation of the spot size and condenser apertures leads to beam currents from 2 pA up to 4.6 nA. In typical STEM-CL experiments, a spot size of 6 is used resulting in beam currents ranging from 70 pA up to 680 pA. In contrast to CL spectroscopy in a SEM, where a single primary electron undergoes a cascade of inelastic scattering events, one primary electron experience solely a few or even no inelastic scattering event. The probability of j collisions for one electron is given by a Poisson statistic:

$$P_j = \frac{1}{j!}\left(\frac{d_{\text{foil}}}{l_{\text{mfp}}}\right)^j \exp\left\{\frac{-d_{\text{foil}}}{l_{\text{mfp}}}\right\}. \tag{5.4}$$

P_j probability for j inelastic scattering events
j number of inelastic scattering events
l_{mfp} mean free path for inelastic scattering

The mean free path is the mean distance between two consecutively occurring inelastic scattering events of a single primary electron. The relation $d_{\text{foil}}/l_{\text{mfp}}$ is the relative thickness and is equivalent to the expectation value. Hence, the mean number of inelastic scattering events is equal to the relative thickness. For a relative thickness of *one* (corresponds to 87 nm in GaN [30][3]) *only elastic* scattering events occur with a probability of 36.8% and *one single inelastic* scattering event has the same probability of 36.8%.

The electron hole generation energy is empirically calculated from the band edge energy of the semiconductor [31]:

$$E_{\text{i}} = 2.596E_{\text{g}} + 0.714. \tag{5.5}$$

E_{i} electron hole formation energy in eV
E_{g} band edge energy in eV

In GaN, the generation of excess carriers is possible for inelastic scattering processes with energy losses above 9.8 eV. The mean energy loss per inelastic scattering results from Poisson statistics and is [32]

$$\Delta E_{\text{inel}} = E_{\text{p}}\frac{d_{\text{foil}}}{l_{\text{mfp}}}, \tag{5.6}$$

E_{p} plasmon energy
ΔE_{inel} mean energy loss

whereas the low energy loss region is approximated to a series of plasmon peak replica. In experiments, intense plasmon peaks and their replicas in the low loss

[3] At 80 kV acceleration voltage and acceptance angle of 5 mrad.

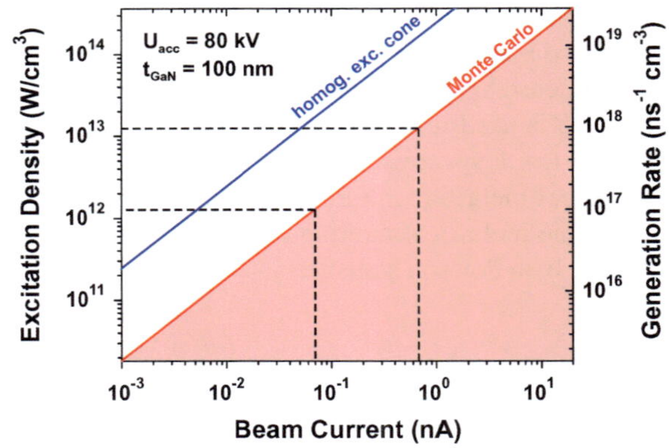

Fig. 5.7 Calculated excitation density and generation rate as a function of the primary electron beam current for a 100 nm thick GaN lamella in STEM-CL: red graph—from Monte-Carlo simulation; blue graph—model of homogeneous excited truncated cone [20]

region dominate the energy loss spectrum and the approximation seems to be justified. For a relative thickness of *one*, the mean energy loss accounts to 19.7 eV and exceeds the electron hole formation energy. Due to the above argumentations, it is reasonably assumed that every single primary electron generates one electron hole pair on average. This one to one conversion of primary electrons in excess carriers is the first approximation of the carrier generation. By approximating the generation volume to a truncated cone, the generation rate or excitation density can be estimated. In Fig. 5.7 the excitation density and generation rate as a function of the beam current for a 100 nm thick GaN lamella is plotted in case of the one-to-one conversion in the homogeneous excited truncated cone (blue graph in Fig. 5.7). For the typical STEM-CL beam currents, 1.7×10^{13}–1.6×10^{14} W/cm^3 and 1.3×10^{18}–1.3×10^{19} ns^{-1} cm^{-3} are obtained for excitation density and generation rate, respectively.

For a more precise estimation of the excitation density, Monte-Carlo simulations at a 100 nm thick GaN lamella at 80 kV acceleration voltage are carried out.[4] The resulting density distributions of the excess carriers are shown in Fig. 5.8 in cross-section and transversal cut of the generation volume. The carrier density follows from the number of scattering events with an energy loss exceeding the electron hole pair generation energy E_i. Since not every single scattering event results in generation of excess carriers, the given density can be regarded as the upper limit. In the cross section (Fig. 5.8a), the highest number of generated carriers per volume occurs in the area of the direct electron beam at $x = 0$. The density reduces in lateral direction drastically. The white marked region in transversal cut (Fig. 5.8b) contains 95% of the whole generated carriers with a mean carrier density of 11.4 nm^{-3}. The excitation density as well as the generation rate is calculated from this mean value and is plotted in Fig. 5.7 (red graph). For the typical STEM-CL beam currents, 1.3×10^{12}–1.3×10^{13} W/cm^3 and 1.0×10^{17}–9.9×10^{17} ns^{-1} cm^{-3} are obtained for excitation density and generation rate, respectively. For typical lifetime of 1 ns for the excess carriers, a carrier density above the background doping level and below

[4]5000 trajectories were simulated with the Monte-Carlo simulation software CASINO v2.48. An incident electron probe of 0.56 nm in diameter was chosen.

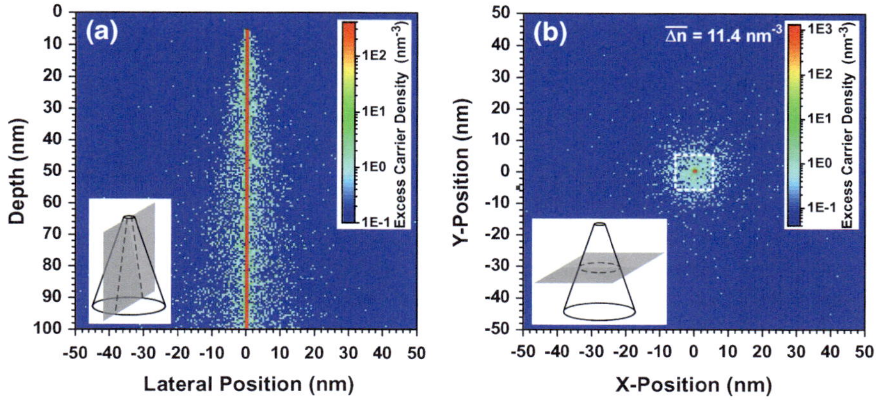

Fig. 5.8 Monte-Carlo simulations of the distribution of the maximum excess carrier density: **a** cross-section, **b** transversal section through the lamella center (white area with 11.4 nm^{-3} mean carrier density) [20]

the Mott density of GaN results from these calculations. It should be noted that for the estimation of the excitation density and generation rate from Monte-Carlo simulations as well as model of homogeneous excited truncated cone no further competing process to electron hole pair generation was considered. In particular, the generation of Auger electrons, plasmons, phonons and X-rays is not included. Hence, the given approximations of the excitation density and generation rate has to be regarded as upper limits.

5.2.2 Tip-Enhanced Raman Spectroscopy (TERS)

Raman spectroscopy is a versatile tool to investigate the structural, electronic, and vibrational properties of semiconductor nanostructures such as strain, doping, alloy composition, polymorphism, electron-phonon and phonon-plasmon coupling, and thermal conductivity [33–38]. The combination with μPL and CL spectroscopy therefore promises comprehensive insights into nanostructure properties. The optical diffraction limit of light, however, restricts the spatial resolution to approximately 500 nm. Furthermore, in many cases the Raman scattering cross section is low and requires a sufficiently large scattering volume, if no other enhancement effects play a role. As a result, very often only ensemble measurements of, e.g., many different quantum dots are feasible, which provide only averaged data.

Tip-enhanced Raman spectroscopy (TERS) can overcome these limitations and provide high spatial resolution combined with signal enhancement necessary for investigating objects much smaller than the excitation wavelength. Similar to surface-enhanced Raman scattering (SERS), it is based on the coupling of light with free electrons in metal surfaces, creating plasmons and plasmon polaritons which confine the light on the metal surface or, in the case of TERS, at the apex of a metallic tip [39]. The evanescent field, which is created by the coupling of plasmons with light, interacts with the sample in a range of few nanometers ($\approx$10 nm) away from the tip apex, making TERS sensitive to surfaces, very thin films, and nanostructures (see Fig. 5.9). The physical mechanism of the process differs from conventional Raman scattering due to the plasmonic interaction of the gold tip and is still subject to controversial discussions [40–44].

A typical TERS setup consists of an optical microscope, suitable for Raman spectroscopy (including lasers, monochromator, detector), and a scanning probe microscope, which can be an atomic force microscope (AFM) or a scanning tunnelling microscope (STM). Three configurations are common for coupling the optical microscope with the scanning probe microscope: back illumination, side illumination (c.f. Fig. 5.10) and top illumination. In the first configuration, the sample is approached from both sides: the scanning microscope at the top and the optical microscope from below. The excitation and detection are confocal through the objective, which limits the usage of this configuration to transparent substrates. The other two configurations, which excite and detect confocally through the objective located at the side or at the top of the sample, allow to perform TERS on non-transparent substrates. In order to exploit the plasmon-light coupling and then excite the evanescent field, the laser

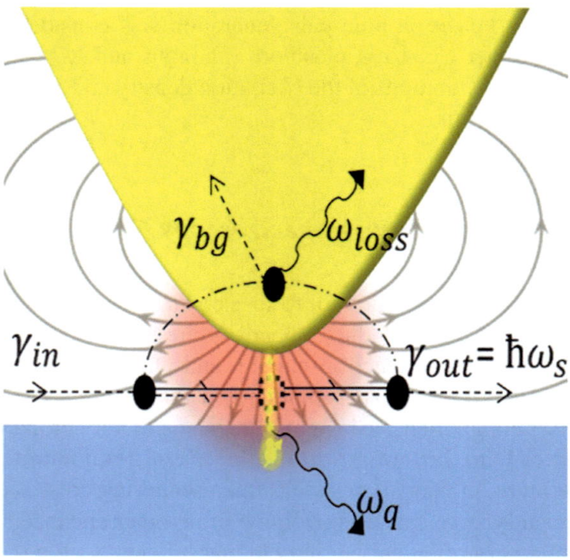

Fig. 5.9 Schematic representation of the tip-enhanced Raman scattering process. The symbols represent the relevant quasipartiles and interactions of the TERS process where γ_{in} are the photons of the excitation laser, γ_{out} are the inelastically scattered photons (Raman signal), γ_{bg} represent the photons of the broad plasmonic background generated by the laser excitation of the gold tip, ω_{loss} describes the energy loss by thermal phonons in the tip, and ω_{q} are the Raman active phonons of the sample. Reprinted with permission from [40]. Copyright (2017) American Chemical Society

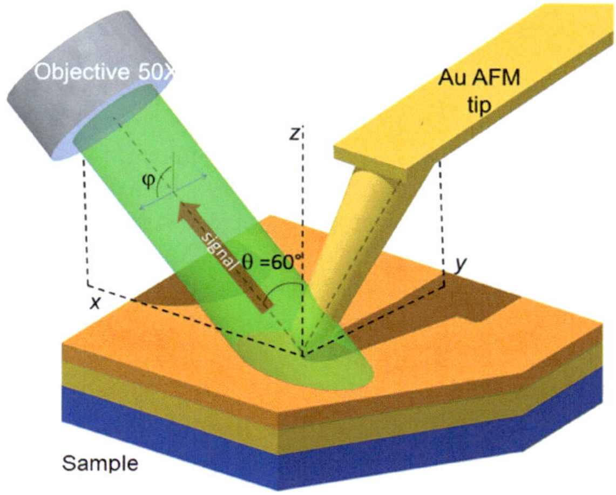

Fig. 5.10 Schematic illustration of tip enhanced Raman scattering (TERS) setup in side illumination configuration. The laserspot is focused by a long distance high NA microscope objective to overlap with the scanning AFM tip on the surface of the sample. Reprinted with permission from [40]. Copyright (2017) American Chemical Society

excitation energy must be close to the plasma frequency of the metallic tip. Typical materials used for the TERS tips are gold and silver. Plasmons of gold are in resonance with light in the green and red spectral range; plasmons of silver are in resonance with blue light. Several methods for tip fabrication are available; most common are sputtering of metals onto commercially available tips and electrochemical etching of bulk metal wires.

The application of TERS for the investigation of semiconductor nanostructures has appeared only recently in literature. Silicon was investigated with TERS by Saito et al. [45] and Lee et al. [46], who detected nanoscale strain of a silicon layer deposited on germanium with a spatial resolution of 25 nm and 20 nm, respectively in 2007. In the same year, Matsui et al. [47] observed a small enhancement of the Raman signal of a GaN layer deposited on sapphire. Other nanostructures studied by TERS include single nanocrystals of $BaTiO_3$ [48], bent InP nanowires [49], GaN nanowires [50], and GaAs [51]. TERS experiments on Ge and SiGe were reported by Ogawa in 2011 [52, 53], where the authors show high-resolution ($\approx$60 nm) Raman maps of a Ge nanowire and a Ge/Si quantum dot converted into maps of local properties like strain and chemical composition. Furthermore, Reparaz et al. [54] applied TERS to resolve local strain and an inhomogeneous distribution of Ge content in Ge nanowires. Poliani et al. demonstrated TERS on GaN nanostructures, i.e., on GaN nanorods as discussed in Sect. 5.3 [55], and on Ge-doped GaN epilayers [40]. In the latter case, the high doping levels achieved through Ge doping allowed to study the influence of carrier density in the sample on the TERS enhancement [40].

5.2.3 UV Optical and Quantum-Optical Characterization

The search for the limits of light-emitting semiconductor nanostructures (e.g. single-photon emitters, micro- and nanolasers) has led to cutting edge research at the crossroads between device physics and quantum-optics. Hence, a state-of-the-art characterization of such nanostructures requires the careful combination of photoluminescence (PL) and quantum-optical measurement techniques. To this end, accessing the statistical properties of the emission by means of second-order correlation measurements is a powerful tool to identify the underlying emission mechanism, e.g. a lasing transition in high-β nanolasers, single-photon or cascaded photon emission. The zero-delay time value, for example, is used to prove the onset of stimulated emission (power-dependent transition from $g^{(2)}(0) = 2$ to 1) and non-classical photon emission $\left(g^{(2)}(0) < 0.5\right)$.

A simplified sketch of the setup, optimized for the UV–visible spectral range, is provided in Fig. 5.11. The sample is placed in a helium-flow cryostat, facilitating temperature control in the range of 4–400 K. A custom Köhler illumination along

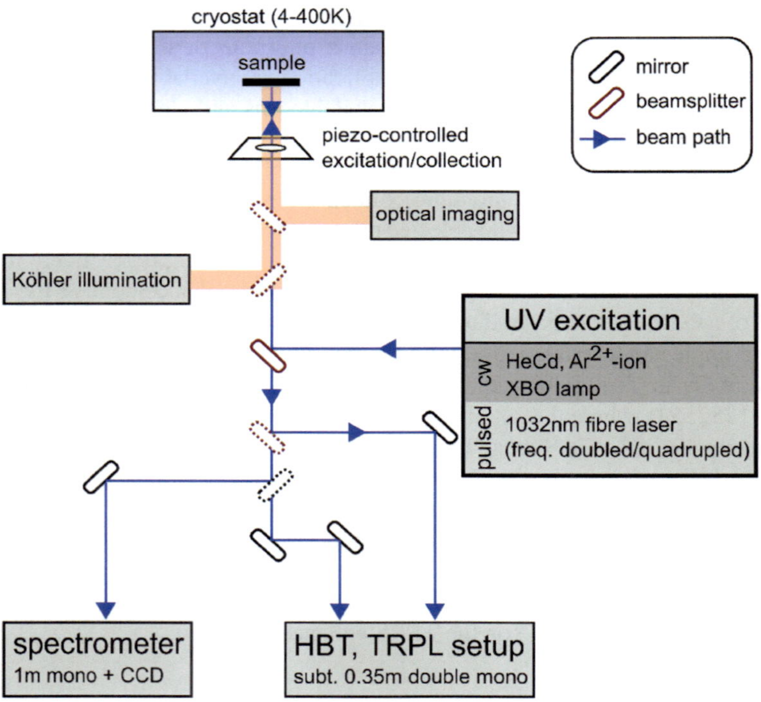

Fig. 5.11 Simplified schematic of the combined optical and quantum-optical characterization setup. UV excitation and signal collection take place in a piezo-controlled confocal arrangement. The detection capabilities comprise a UV-enhanced CCD setup, time-resolved μPL as well as a Hanbury-Brown and Twiss setup. Dashed contours represent flip-mounted optics

with a CCD camera are used for high contrast optical imaging of the sample's surface. The light from various available excitation sources shares a common beam path and is focused onto the sample via a UV objective in confocal arrangement. The objective is mounted on a closed-loop 3-axis piezo stage, which is used for high precision alignment and automated mapping of the sample. Excitation sources comprise a fiber-based pulsed laser (1032 nm, 516 nm, 258 nm) as well as continuous wave HeCd (325 nm, 442 nm) and Ar^{2+} gas lasers with emission lines between 458 and 528 nm for fundamental emission and between 229 and 264 nm for second harmonic generation. Additionally, a xenon short-arc (XBO) lamp equipped with a 0.22 m double monochromator is available for μPLE experiments in the UV–visible spectral range.

The bottom part of Fig. 5.11 indicates the detection side of the setup. Here, the luminescence is guided either through a single monochromator onto a UV-enhanced CCD array, or through a subtractive double monochromator onto single-photon counting detectors. The first detection path is used primarily for excitation power and polarization-dependent measurements. The second path is used for time-resolved μPL and intensity correlation measurements in Hanbury-Brown and Twiss configuration, investigating carrier relaxation/recombination dynamics and photon statistics, respectively. Long-time measurements on nanostructures (e.g. quantum dots or nanolasers) well in excess of 24 h are made possible by fully automated realignment routines using sample mapping. Several examples for the application of this technique are discussed for single GaN quantum dots in Chap. 12.

5.2.4 XRD

X-ray diffraction (XRD), especially high resolution X-ray diffraction (HRXRD) is a widely used and accepted standard technique for structural investigations of semiconductor layer systems and nanostructured materials. Since X-ray analysis is nondestructive, fast and automatable, X-ray measurements are often the first step of structural characterization. By performing X-ray diffraction on periodic structures like semiconductor crystals and superlattices, information on lattice constants, layer thicknesses and compositions can be obtained. Particularly HRXRD allows the investigation of deviations from an ideal crystal structure that can be caused by strain, defects or crystal mosaicity. Therefore, this technique is well-suited for characterization of single films as well as more sophisticated layer structures like multiple quantum wells (MQWs), field-effect transistors (FETs) or distributed Bragg reflectors (DBRs). In addition, XRD offers a variety of scanning geometries that can be applied to answer the most diverse questions about the structural properties of the samples. In the following, typical examples for more common types of XRD measurements, e.g. $\Theta/2\Theta$-scans, as well as some methods that are only rarely used in the scientific community are presented.

Figure 5.12 shows the $\Theta/2\Theta$-scan of an InGaN/GaN MQW heterostructure grown on a (0001) sapphire substrate. In this type of scan, the orientation of the scattering

Fig. 5.12 (0002) Θ/2Θ-scan of a GaN-based LED with a 5-pair InGaN/GaN MQW

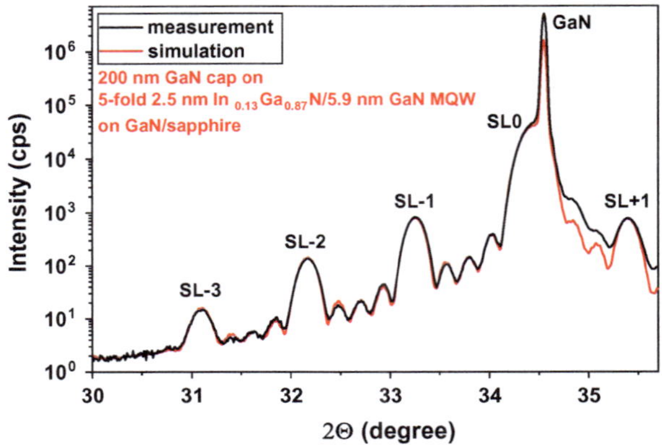

vector remains constant, i.e. perpendicular to the investigated lattice plane (in this case the *c*-plane), while incidence and excidence angle of the X-rays are varied simultaneously. Besides a dominant peak originating from the GaN buffer, several orders of superlattice (SL) peaks stemming from the periodic InGaN/GaN MQW are observed. While the higher intensity peaks with a larger angular distance are related to the sum of the thickness of one InGaN QW and one GaN barrier, the secondary maxima between the principal SL-peaks are determined by the total thickness of the 5-fold MQW structure. For such interference patterns a very good homogeneity in lateral and vertical direction with smooth interfaces between the layers is mandatory. With the help of a X-ray diffraction software, the thicknesses of the InGaN QWs and the GaN barriers, respectively as well as the In-concentration in the QWs can be determined. As a very good agreement between the measurement and the simulation is obvious, a high structural quality of the MQW-structure can be deduced.

In Fig. 5.13 a Θ/2Θ-scan of an AlGaN/GaN FET grown on a silicon (111) substrate is depicted. Thickness fringes from the AlGaN layer deposited on the thick GaN buffer are clearly visible. From the spacing between the fringes and the position of

Fig. 5.13 (0002) Θ/2Θ-scan of an AlGaN/GaN FET

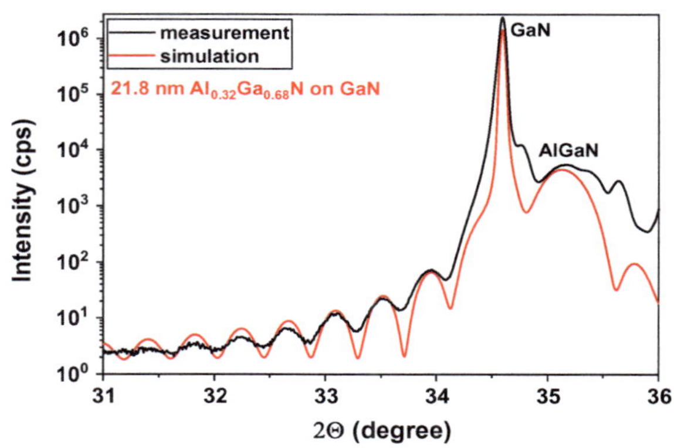

the AlGaN peak, the thickness of the AlGaN can be determined to 21.8 nm with an Al-content of 32%. The additional peaks on the right hand side of the GaN peak (which are not reproduced in the simulation) are caused by several AlGaN buffer layers with different composition that are initially grown on the Si-substrate and are neglected in the simulation.

The measurement of an AlN/GaN superlattice with GaN monolayers grown on c-plane sapphire is shown in Fig. 5.14. Several orders of superlattice peaks show up on both sides of the main peak of the AlN buffer. The period thickness amounts to 10.6 nm whereof 10.3 nm belong to the AlN and 0.3 nm to the GaN. Comparing the simulation that assumes ideal interfaces to the measurement, a reduced intensity of the higher order SL-peaks in the measurement is apparent. This indicates non-ideal AlN/GaN interfaces which might stem from fluctuating layer thicknesses or intermixing at the interfaces. Considering an interface roughness of 0.35 nm in the simulation, the observed behavior in the diffractogram is reproduced much better.

In contrast to the three previously shown measurements where the scattering vector is oriented perpendicular to the surface, the example in Fig. 5.15 shows a gracing incidence in-plane diffraction (GIID) map of a sample with GaN/AlN quantum disks. Here, the scattering vector lies parallel to the substrate and the X-rays are diffracted at the $(11\bar{2}0)$ planes. By this scanning geometry the lattice spacing within the growth plane and therefore the strain state of the layers can be evaluated. Using a small incidence angle below the critical value of total reflection, the X-rays are only able to penetrate the uppermost nanometers of the cap layer. In this case, the diffracted intensity stems only from a few monolayers from the sample surface. By increasing the incidence angle above the critical angle of total reflection, the penetration depth of the X-rays is drastically increased and also layers that are located deeper in the structure contribute to the diffraction signal. Using a small angle of incidence α_i of the X-rays around 0.1°, the scan mainly shows relaxed AlN that is located on the sample surface. By incremental increase of α_i the GaN peak becomes more intense and an additional contribution at around 58.7° is visible. This signal is caused by the first AlN layers grown on the GaN. These atomic layers are tensely strained and

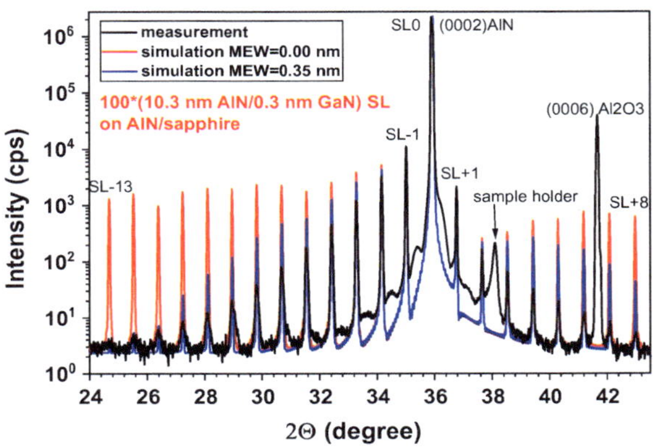

Fig. 5.14 (0002) Θ/2Θ-scan of a 100-pair AlN/GaN superlattice with GaN monolayer thickness

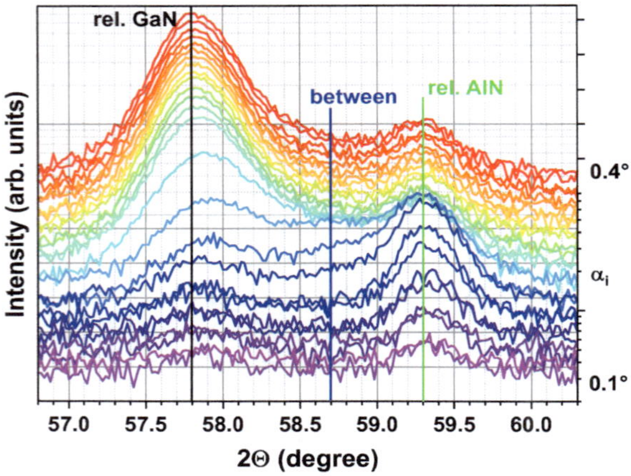

Fig. 5.15 Depth-dependent grazing incidence Θ/2Θ-scan around the (11$\bar{2}$0) reflection from $\alpha_i =$ 0.1° to 0.4°. Curves are shifted vertically for clarity

exhibit only a partial relaxation. Considering the GaN peak when measuring with a high penetration depth of the X-rays ($\alpha_i = 0.4°$), it is revealed that the deeper lying GaN layers are fully relaxed. However, at smaller α_i a shifting of the GaN peak towards higher angles is observed. This means, that the upper layers of GaN close to the AlN layer are compressively strained due to the smaller in-plane lattice constant of AlN. The example demonstrates that the method of GIID with a variation of the incidence angle α_i enables a depth-sensitive evaluation of the strain state of different layers.

Besides XRD, X-ray reflection (XRR) is a commonly applied method for the evaluation of the film thickness as well as surface- and interface-roughness. In comparison to diffraction measurements, also non-crystalline samples can be evaluated. Figure 5.16 illustrates the qualitatively investigation of the interface roughness between a GaN quantum-dot layer and an AlN cap layer by measuring the diffuse

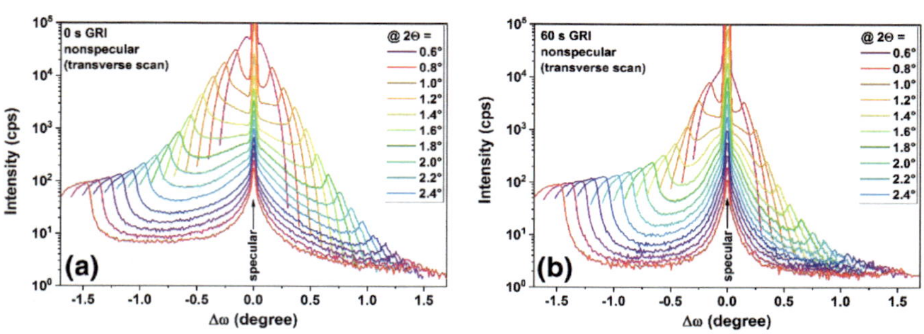

Fig. 5.16 Diffuse reflectivity scans of a GaN quantum dot layer embedded in AlN with **a** no growth interruption and with **b** 60 s growth interruption

scattering. This is done by performing ω-scans at different fixed 2Θ angles. The figure shows the diffuse reflection maps of a few monolayer thick GaN film embedded in AlN on sapphire. In Fig. 5.16a the AlN cap layer was grown immediately after the GaN layer and in (b) a growth interruption was induced, whereby the GaN layer was exposed only to hydrogen carrier gas for 60 s before capping with AlN. Transmission electron microscopy shows hexagonally-shaped GaN islands of 20 nm width and approximately 100 nm width for the sample without a growth interruption. Applying a growth interruption, desorption processes are induced leading to smaller and flatter GaN islands. Having a look at the diffuse scattered intensity on the left and right of the specular reflection peak, it is obvious that the diffuse scattering is significantly reduced in the sample with the 60 s GRI. Therefore, the XRR measurements directly indicate a reduced size of the nanoscopic objects and a GaN/AlN interface with a reduced roughness when applying a growth interruption.

The following example demonstrates that XRD can be used for determination of defect densities in crystalline structures as well. Figure 5.17 shows an extended scan of the reciprocal space of a GaN $(10\bar{1}3)$ layer grown on *m*-plane sapphire. This layer is characterized by a high density of stacking faults (SF) that are inducing SF streaks in the GaN (0001) plane along the GaN [0001] direction perpendicular to the stacking faults. The reciprocal space map is conducted around multiple lattice points reaching from GaN $(10\bar{1}2)$ to GaN $(10\bar{1}5)$. This is accomplished by performing ω-scans at different 2Θ angles. Since both axis need to be varied over a large angular region, such scans usually are very time-consuming. Using a customized X-ray setup with a rotating anode and a large area detector the measurement time can be drastically reduced. A pronounced stacking fault streak is observed between the GaN $(10\bar{1}1)$ peaks. The high intensity signal is directly caused by a strong diffuse X-ray scattering of the SFs. By simulation of the SF streak the density and the type of the SFs can be quantitatively analyzed.

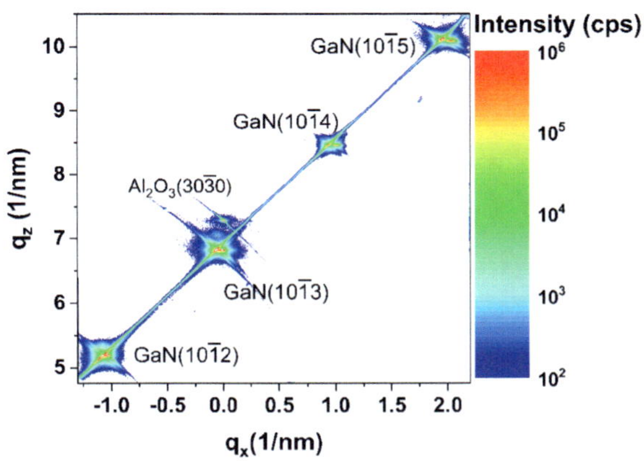

Fig. 5.17 Wide range reciprocal space map from $(10\bar{1}2)$ to $(10\bar{1}5)$ of $(10\bar{1}3)$ oriented GaN on *m*-plane sapphire

For a comprehensive analysis of the structural quality and the dislocation structure of epitaxial layers, measurements of multiple Bragg reflections in symmetric, asymmetric geometries are compulsory. Changing from one geometry to another is often elaborate or even different X-ray setups are necessary. Alternatively, peaks from multi-beam diffraction, so called umweganregungen, can be analyzed. These peaks are inherently caused by diffraction at lattice planes with different inclination angles and therefore contain information of the sample crystallinity in different directions. For investigation of multiple umweganregungen in one scan, a symmetric, specular geometry is used and the detector is positioned at the 2Θ angle of a forbidden reflection (i.e. GaN (0001)). Initially no diffraction intensity will be detected, but when the sample is rotated about the normal to the sample surface (ϕ angle), peaks caused by multiple-diffraction at two different lattice planes appear. Thereby, the sum of the corresponding Miller indices of the two involved planes [e.g. $(01\bar{1}1)/(0\bar{1}10)$] of one umweganregung equals to (0001). Figure 5.18 shows Renninger-maps of two GaN-based VCSEL structures with AlInN/GaN distributed Bragg reflectors. Since the diffraction pattern is reproduced after a ϕ-rotation of 30° due to the hexagonal symmetry of GaN, ϕ is only varied in this range. The 2Θ information is obtained by using a large area detector. A VCSEL structure with rough AlInN/GaN DBR interfaces, fluctuating layer thicknesses and high dislocation density is shown in Fig. 5.18a. A complex diffraction pattern with main contributions of approximately 9 pairs of lattice planes is revealed. However, from the intensity distribution along 2Θ direction no real information on the layer structure can be derived. In contrast, the scan of an improved structure with better periodicity, smoother interfaces and lower dislocation density (Fig. 5.18b) clearly shows SL-peaks along 2Θ comparable to standard symmetric Θ/2Θ-scans. Furthermore, the peaks from each umweganregung of the improved VCSEL-structure are less broadened in ϕ-direction hinting at a lower structural disorder and therefore a lower dislocation density as well.

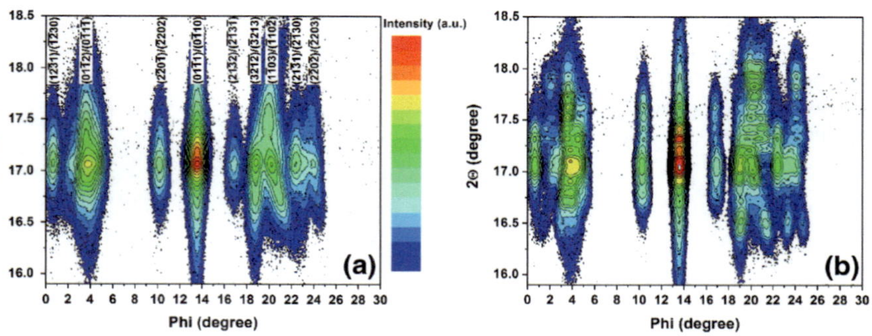

Fig. 5.18 Renninger ϕ-maps at (0001) reflection of two GaN-based VCSEL structures with **a** low and **b** high structural quality over 4 orders of magnitude on a logarithmic scale

5.2.5 Scanning Tunneling Microscopy and Spectroscopy (STM/STS)

Nanostructures, as they are analyzed in this Chapter, typically exhibit a high surface-to-volume ratio, turning the surface properties to be non-negligible as compared with bulk material. Especially group III-N nanowires typically formed by non-polar side facets, which may influence devices the electronic and opto-electronic properties of either core-shell-structures or embedded nano-discs. In order to study intrinsic surface properties with best resolution, i.e., atomically resolved, scanning tunneling microscopy and spectroscopy provides optimum conditions [56, 57]. While the polar [0001] and [000-1] surfaces were well investigated [58, 59], the intrinsic properties of group III/N non-polar surfaces were barely known. Especially, the scanning tunneling spectroscopy (STS) mode can be used in order to determine electronic properties, as e.g., Fermi level pinning, energetic alignment of surface states, charge transfer levels of dopants, etc. [60]. Further on, STM can be applied in the cross-sectional mode the so-called cross-sectional scanning tunneling microscopy and spectroscopy (XSTM/XSTS) in order to analyze geometric, stoichiometric, and local electronic properties of buried structures [61, 62]. Unfortunately, the group III-N materials exhibit a lot of defects, being detrimental for obtaining a flat clean cleavage surfaces, as they are necessary for such investigations.

One central problem of all (X)STM/STS investigations is the sample preparation: only clean, almost flat, and especially stoichiometric non-polar surfaces enables the determination of intrinsic material properties. In surface science typically applied procedures as sputtering and annealing sequences [63–66] lead to unclean, unflat, and non-stoichiometric surfaces. It looks like that the nitrogen atoms are disappearing into the vacuum during such process, turning to form nonstoichiometric surfaces. A competing surface preparation process is the in situ cleavage [67], which preserves at least the stoichiometry and the cleanness under UHV conditions, and typically exhibits a smaller or larger number of surface steps and terraces in between of them. Under certain growth conditions in a narrow parameter window also stoichiometric surface conditions can be reached via in-situ molecular beam epitaxy growth [68].

For the different group III-N materials the in situ cleavage is not trivial: GaN bulk material doesn't like to cleave along the non-polar (11-20) surface [67, 69], while along the competing (10-10) one it does [70, 71]. InN is not available as bulk material; here, the only possibility to investigate bulk properties are thick layers grown heteroepitaxially on foreign substrates, with as high crystal quality as possible, exhibiting only low defects concentrations [72]. The InN typically follows the cleavage surface direction of the underlying substrate material in accordance with the substrate-to-layer alignment: the (11-20) surface can be hence reached via a Si(111) substrate [73, 74], while the (10-10) surface is obtained by direct growth of InN on Ga(0001) [72].

As a result of the investigations one gets for the (10-10) surface an empty surface state, being energetically located 0.7–1.0 eV below the conduction band minimum within the fundamental band gap [75, 76]. It is not easy to access this state locally since it is unusually located about 0.24 nm more inside the surface as compared with the respective bulk state, leading to the name of a hidden state. Further on, this state can be re-filled from the conduction band only slowly, leading to a polarity dependent pinning of the Fermi level at the surface [77]. This situation results for *n*-type doped GaN in a very low non-radiative recombination rate of this in-gap surface state, for which the GaN(10-10) surface does not need to be protected by a wider band gap material to prevent from non-radiative surface recombination. In contrast, both non-polar InN surfaces, i.e. the (10-10) and the (11-20) surface, do not have their surface states within the fundamental band gap [72, 74]. As long as the surface is clean and stoichiometric, also both non-polar surfaces show no intrinsic electron accumulation. The Fermi level is energetically located within the fundamental band gap of InN, even at both, the (10-10) and the (11-20) surfaces. This finding in general enables effective *p*-type doping at non-polar surfaces as well as at hetero-interface as long as they are epitaxial and free of dislocations with internal dangling bonds.

5.3 Analysis of Nanostructure Growth in Nitrides

5.3.1 Growth of Nitride Based Nano- and Micro-columns

For LEDs, the growth of 3-dimensional GaN structures, like nano-wires or micro-columns, with a diameter in the range from nanometers up to several micrometers, offers an alternative concept in comparison to planar heterostructures. The growth of GaN nano-columns has been demonstrated for the first time by Yoshizawa et al. as well as Garcia et al. using MBE [78, 79]. The 3-dimensional structure of nano-columns brings substantial advantages. Due to the small column diameter and thus a small footage to the substrate, the growth on foreign substrates, like silicon or sapphire, can be carried out without the introduction of additional stress [80, 81]. Hersee et al. [82] was able to confirm, that threading dislocations bend to sidewalls at the bases of nano-columns resulting in a low density of extended defects. In addition, a core-shell growth of the active region on the non-polar side-facets of the columns reduces the effects of polarization fields (QCSE) as well as increases the active area of about one order of magnitude in comparison with planar structures, resulting in a lower current density of the final device [83].

Growth Modes of Nanowires

In principle, there are 3 different ways to grow nanowires (NWs):

- Self-assembled growth
- Selective area growth
- Top-down approach (selective etching)

The self-assembled growth can be carried out with the assistance of catalyst-assisted reactions in a vapor-liquid-solid (VLS) synthesis or without catalyst. In the

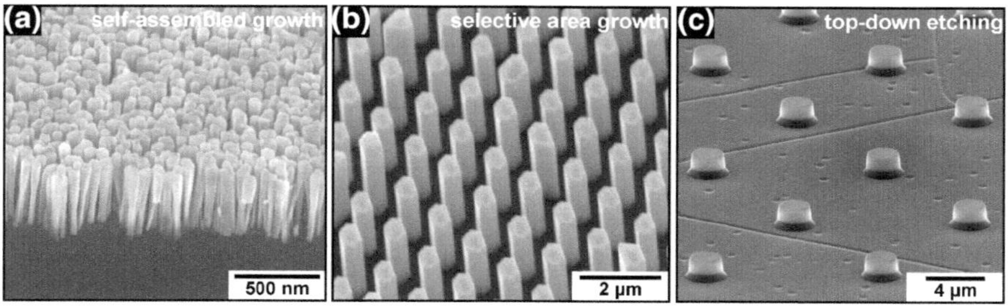

Fig. 5.19 Comparison of surface morphologies of **a** self-assembled growth of GaN nanowires, **b** selective growth, and **c** etched GaN micro-pillars [88]

catalyst-assisted VLS growth, metal drops (Au, Ni, Ta) act as catalyst deposited directly on a buffer layer or a substrate [84]. The catalyst creates a highly selective growth environment with growth occurring only at the catalyst location. The role of the metal is to form a liquid alloy droplet of relatively low freezing temperature. Material diffuses from the gas phase to liquid phase to become supersaturated [85–87]. The wire grows by precipitation of the catalyst from the liquid droplet. The catalyst-free growth takes place due to a systematic change of growth parameters. For example in MBE, N-rich conditions may lead to the growth of GaN columns [78, 79] (Fig. 5.19).

A problem with catalytic processes is that the metal catalyst will inevitably incorporate into the nanowire, which may be prohibitive for many semiconductor applications. Since the catalytic-assisted as well as in catalytic-free growth of NWs occurs in a self-assembled modus, there is a statistical variation in size and shape of the GaN wires as well as distance to each other [89]. Therefore, LED structures which are based on self-assembled grown GaN columns show structural inhomogeneities resulting in polychromatic emission [90] as well as a strong influence on electrical properties [91]. To address this issue, an alternative approach is used by means of selective dry-etching, the so-called top-down growth of nanowires [92, 93]. For this method, a mask is deposited on a planar structure and patterned by lithography. The layout of the mask determines the diameter and the pitch of the columns. Plasma-assisted, reactive ion etching leads to well-defined 3-dimensional structures (position and homogeneity). The main disadvantages are the surface damage as well as the rough, not facetted surfaces [94].

Selective Area Growth

For the first time *selective area growth* of nanowire was applied by Hersee et al. using pulsed metal organic vapor phase epitaxy [82]. For this method, the GaN sample is ex-situ patterned by a mask. Depending on the quality of the mask, the resume of GaN growth leads to coherent growth in the mask-free positions and suppressed growth on the mask. The required process steps are illustrated in Fig. 5.20. After the growth of a buffer layer, a mask is deposited on the sample. Dielectric layers, like

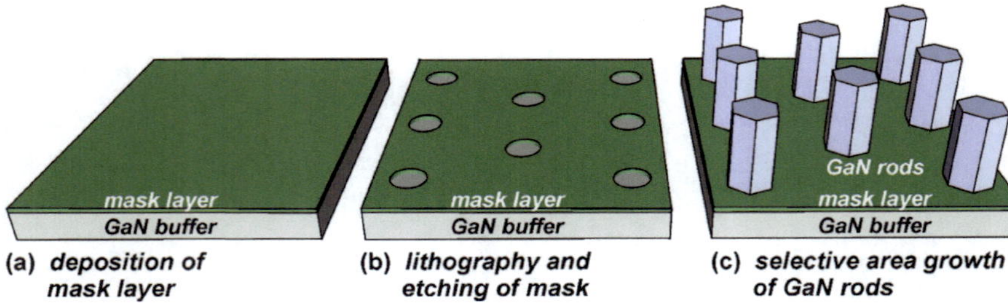

Fig. 5.20 Process steps for selective area growth of GaN [88]

SiO_2 or SiN, or metal films, like Mo or Ti, show a high growth selectivity to prevent parasitic GaN nucleation on the mask.

Subsequently, the mask is patterned by lithography and etching to expose the underlying GaN. Photolithography [82, 95], electron beam lithography [96, 97] as well as nano-imprint lithography [98, 99] are used to structure the mask. The spatial resolution of photolithography is limited down to 0.5 μm due to the optical diffraction rule of Abbe. In contrast, electron beam lithography is limited by the De-Broglie wavelength of electrons reaching a spatial resolution down to 10 nm [100, 101]. However, the long scanning time for exposing the photoresist restricts the area size. The nano-imprint lithography uses a structured stamp, which is pressed into the photoresist, to pattern the sample. The stamp itself can be produced by e-beam lithography or etching and the mask openings can reach a diameter down to nanometer scale. For high volume and low cost fabrication of patterned substrates, nano-imprint lithography is the more suitable choice [102]. After developing the photoresist, plasma assisted etching opens the mask at the intended positions. The layout of the mask controls the arrangement as well as the diameter of the columns.

During the subsequent epitaxy, the nucleation starts exclusively in the mask openings directly on the buffer due to the high selectively of GaN growth (Fig. 5.20c). Growth parameters have a direct influence on the morphology of GaN nanorods in MBE and MOVPE [103, 104]. The high growth temperature tends to enhance the Ga desorption and diffusion, and thus suppresses a parasitic nucleation on the mask itself. At a relatively low temperature, no selective area growth of the nanorods is observed. In addition, the V/III ratio was found to be very critical for the geometry of selective area growth of GaN nanorods. Low V/III ratio, e.g., less nitrogen flux favors selective nucleation with less parasitic nucleation by increasing the Ga diffusion length. Both the vertical and lateral growth rate of GaN increases with N_2 flux, evidencing that the selective area growth occurred under the nitrogen supply limitation scheme. Growth parameters, such as temperature, partial pressure, carrier gas, V/III ratio are adjusted to push the vertical growth and to suppress the lateral growth [105–107].

Increase of Active Area of a LED

In contrast to conventional planar LEDs, where the effective luminescent area is limited by the size of the substrate, the active area of LEDs on basis of coaxial InGaN/GaN micro-columns scales directly with the size of the shell area of the 3-dimensional structures. An effective increase of the luminescent area is achieved due to an increase of the aspect ratio of the columns. The gain in active area is calculated in this way:

$$F = \frac{A_{3D}}{A_{2D}} = \frac{8\pi}{\sqrt{3}} \cdot \left(\frac{r}{a}\right)^2 \cdot \left(\frac{h}{2r}\right) \cdot \left(\frac{k}{r}\right) \tag{5.7}$$

where r is the radius of the complete column, h is the height, k is the core radius, and a is the pitch of the columns to each other. The term $h/2r$ represents the aspect ratio and k/r expresses the ratio of core diameter to overall radius. The most important term for adjusting the gain is given by the aspect ratio $h/2r$.

In Fig. 5.21 the increase of active area depending on the aspect ratio of GaN micro-columns for different column densities is given. The ratio k/r was set to 0.9 in this case. With the assumption that the columns occupy 25% of the surface and the aspect ratio is given by $h/2r = 10$, the increase of active area in comparison to a planar LED is one order of magnitude. Two examples are given in Fig. 5.21 with high as well as low aspect ratio. In planar GaN based LEDs the so-called "efficiency droop" causes an decrease of quantum efficiency with increasing injection current density [108–110]. The dominant reason for the efficiency droop is a three-particle process by means of non-radiative Auger recombination [111], which strongly depends on

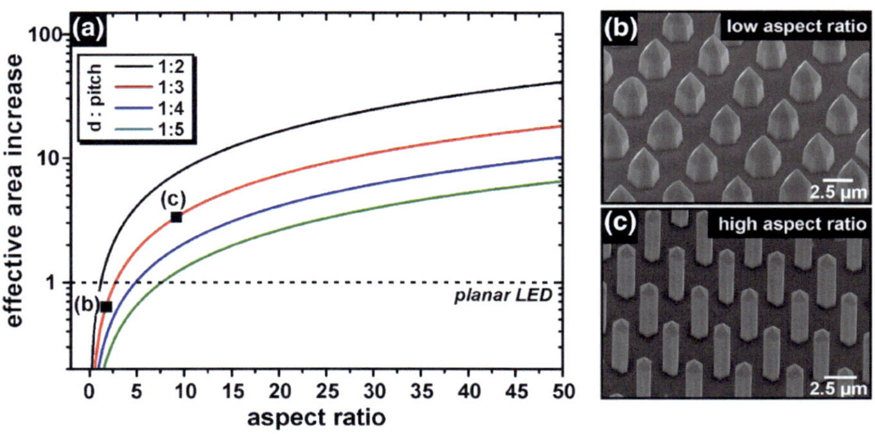

Fig. 5.21 **a** Increase of active area depending on the aspect ratio of GaN micro-columns for different column densities (pitches), examples for columns with **b** a low aspect ratio, and with **c** a high aspect ratio, the effective increase is marked in **a** as well as the planar LED [88]

the current density. The increased active area in micro-column based LEDs can be used to reduce the current density while maintaining the total current, resulting in a reduction of the efficiency droop.

5.4 Optical Analysis of Low-Dimensional Nitrides

5.4.1 Luminescence and Composition Inhomogeneities in InGaN/GaN Micro-columns

The samples were grown by MOVPE on a 6 μm thick optimized GaN/sapphire template. For the selective area growth, a SiO_2 mask was deposited and patterned by nano-imprint lithography and reactive ion etching resulting in hexagonally aligned, circular holes with a diameter of 3 μm. Subsequently, *n*-doped GaN micro-columns were grown by MOVPE under a temperature of $T = 1070$ °C, a nitrogen to gallium ratio of 100 and a Silane flow of 450 nmol/min. The high Silane flow favors the vertical growth of the GaN columns but leads to a relatively high free carrier concentration of 1×10^{20} cm^{-3} [112]. For high quality of the interface to the active shell layers as well as prevention the formation of a passivation layer, the SiH_4 concentration was reduced in the end of the GaN core growth by one order of magnitude.

The V/III ratio was increased by two orders of magnitude for the subsequent shell layers. First, an AlGaN buffer was deposited, followed by a *n*-doped GaN shell, an InGaN quantum well (QW) as well as an outer GaN layer. The growth of the InGaN QW and GaN barriers happened at 750 °C and 850 °C, respectively.

The selective area growth leads to homogeneously ordered array of micro-columns with density of 1.3×10^7 $columns/cm^2$ (visible in Fig. 5.22a). The core shell structures are terminated by nonpolar $\{1\bar{1}00\}$ side facets. At the apex, semipolar $\{1\bar{1}01\}$ facets form the tip of the pencil shaped column. The mean column height is 8.8 μm and the mean diameter is 1.1 μm. This leads to an aspect ratio of 8, which gives an increase of the active area of 3.5 in comparison to planar LED structures.

The column array was characterized by SEM-CL with an acceleration voltage of 5 kV, a beam current of about 500 pA and a temperature of 6 K. The spatially averaged CL spectrum is depicted in Fig. 5.22b. The near band edge (NBE) emission of GaN at 357 nm as well as the defect-related Yellow emission (YL) at 550 nm is obtainable. Both emission bands origin from the thick GaN buffer. The InGaN single quantum well (SQW) shows a broad, asymmetric luminescence between 410 and 490 nm, with a linewidth of 500 meV and a emission wavelength at 415 nm. Furthermore, the distribution of peak wavelengths (red histogram) in the spectral region of the InGaN SQW is broad as well as exhibits a wide shoulder to longer wavelengths. Already from the spectrally wide features in the histogram, we can conclude that local inhomogeneities of the InGaN SQW lead to the broadening of the integral spectrum. The reason for the broad distribution can be the local variation of the ensemble of micro-columns, i.e. differences from column to column and/or

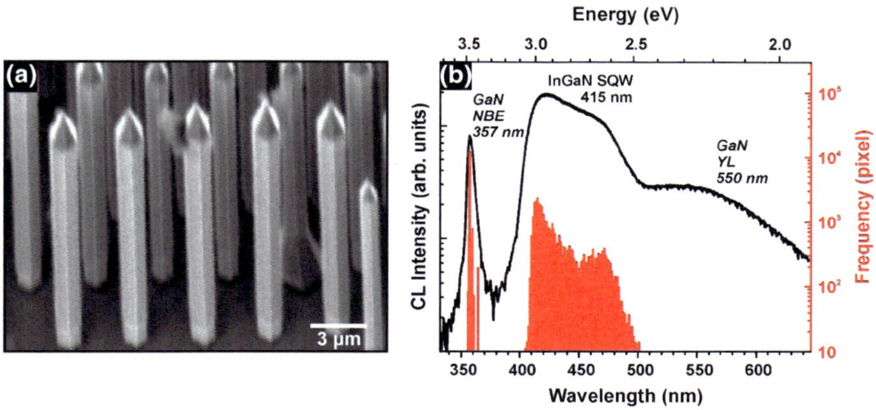

Fig. 5.22 **a** Bird's-eye view SEM image of the InGaN/GaN micro-columns and **b** spatially averaged CL spectrum at $T = 6$ K. The distribution of wavelength of the CL wavelength image Fig. 5.2b is given in red [88]

the composition as well as the thickness variation of the SQW within the column itself.

The CL intensity image in Fig. 5.23a proves that the InGaN SQW emission is exclusively emitted from the micro-columns. There is no parasitic deposition at the buffer layer. The columns show an uniform intensity distribution with respect to each other: the luminescence starts at a height of about 1 μm and increases along the $\{1\bar{1}00\}$ facets continuously over one order of magnitude up to the apex. The semipolar tip exhibits a drastic decrease of CL intensity. Here, we find a complex

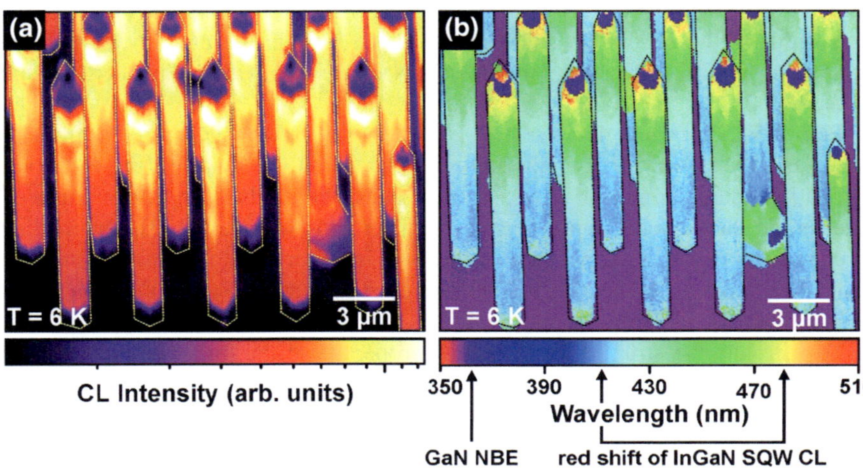

Fig. 5.23 **a** CL intensity image of the InGaN SQW luminescence from 400 to 500 nm and **b** corresponding CL wavelength image (CLWI) of the peak wavelength from 350 to 500 nm [88]

domain structure with a thinner SQW and different In content in comparison to the nonpolar facets.

Similar to the CL intensity, the wavelength variation along each column is uniform, too. For each individual micro-column, we find a continuous wavelength shift from shorter to longer wavelength starting at the base with 410 nm (bright blue contrast in CLWI Fig. 5.23b) and reaching 480 nm at the very top (yellow contrast in CLWI Fig. 5.23b). This shift in emission wavelength along the $\{1\bar{1}00\}$ facet of each column is the main reason for the broad linewidth of 500 meV in the integral spectrum.

Microscopic Correlation of Composition, Thickness and Emission of the Quantum Well

For a detailed structural characterization, the sample has been prepared for STEM measurements using focused ion beam preparation to get a homogeneous sample thickness over a whole wire. In Fig. 5.24a, a STEM image of a complete micro-column is shown in HAADF contrast. In the overview image, the GaN buffer in the lower part of the image shows a bright contrast while the directly following the SiO_2 mask exhibits a dark contrast. The STEM image proves that the GaN micro-column is coherently grown directly out of the hole of the SiO_2 mask. No extended defects are visible, the STEM image indicates a high material quality.

The InGaN shell starts to grow 1 μm above the column base. The diameter of the column increases from 1080 nm at the base to 1180 nm at the apex. Figure 5.24c, d show magnified HAADF maps of the active region. The InGaN SQW appears in brighter contrast, whereas the AlGaN layers appears in darker contrast in comparison to the GaN shell. All layers run parallel to the $\{1\bar{1}00\}$-facet.

The InGaN SQW increases in thickness along the wire by a factor of two. At the base, the InGaN shell has a width of 6.0 nm and near the apex a width of 13.1 nm. Figure 5.24b from the apex proves, that the growth rate of InGaN is drastically reduced at the semipolar $\{1\bar{1}01\}$-facet in comparison to the nonpolar $\{1\bar{1}00\}$-facet. The width of the semipolar InGaN shell at the apex is 6.8 nm. In addition, the brightness of the HAAD contrast of the InGaN shell changes along the micro-column (see Fig. 5.24).

Figure 5.25 shows highly resolved HAADF maps of the InGaN SQW from the upper and the lower part of the micro-column. With the help of "frozen-lattice" simulations and normalized HAADF-intensities, the medium In concentration is determined and depicted in profiles for the upper and lower part [113]. The medium In concentration in the upper part of the wire is about 13% and in the lower part about 11%. The HAADF signal evidences a clear increase of SQW thickness as well as In concentration along the micro-column.

To correlate the structural and optical properties of the micro-column we performed STEM-CL measurements at Helium temperatures ($T = 16$ K). The maps were taken under an acceleration voltage of 80 kV and a beam current of 130 pA. The STEM image (Fig. 5.26a) shows the GaN buffer in the lower part and the subsequently grown InGaN/GaN micro-column in bright field contrast. A contrast change from the lower to the upper part of the column is visible which is caused by the thickness gradient along the wedge shaped sample which is indicated by a black arrow in

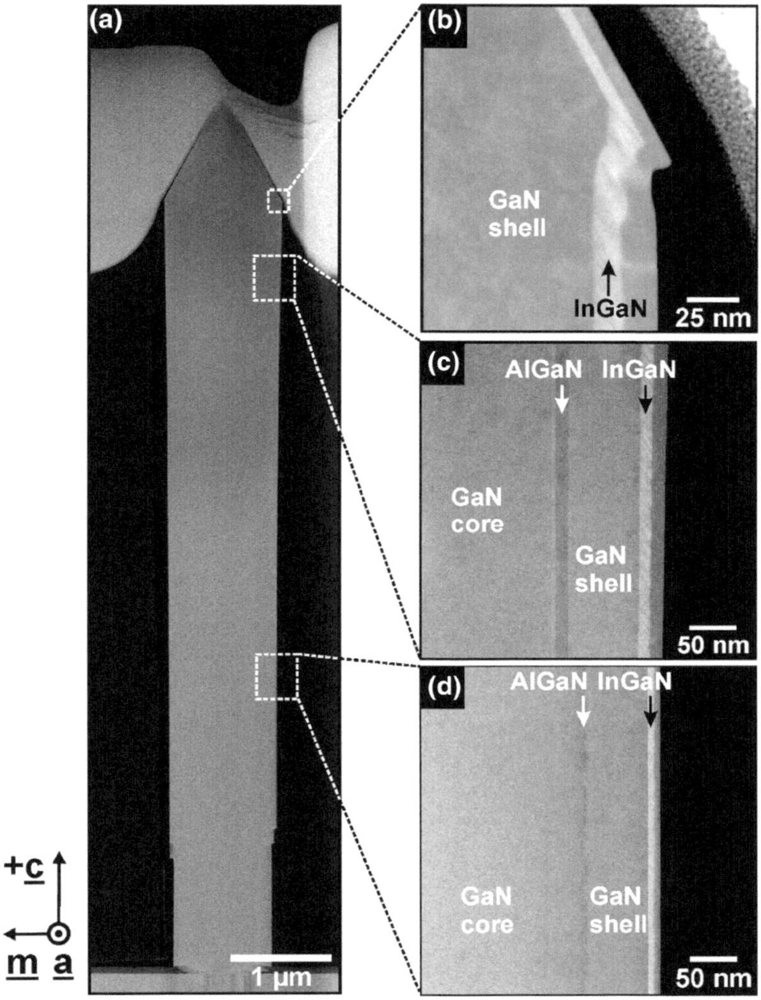

Fig. 5.24 Cross sectional STEM map of a InGaN/GaN micro-column in $[11\bar{2}0]$ projection. **a** Complete micro-column, magnified image of the core shell structure in the **b** upper part, **c** middle part, and **d** at the base of the micro-column. Reprinted with permission from [113]. Copyright {2016} American Chemical Society

the TEM image. This thickness gradient leads to a decrease of the CL intensity in the corresponding panchromatic CL intensity along the micro-column. In contrast, the SQW intensity increases from bottom to top (in agreement with the SEM-CL results).

The spectral evolution of the luminescence is visualized in the CL linescan running along the $\{1\bar{1}00\}$ facet in Fig. 5.26c. In the vertical range of $x = 0$–$0.25\ \mu$m, where the e-beam excites the GaN buffer, GaN NBE emission at 357 nm is observed. Starting

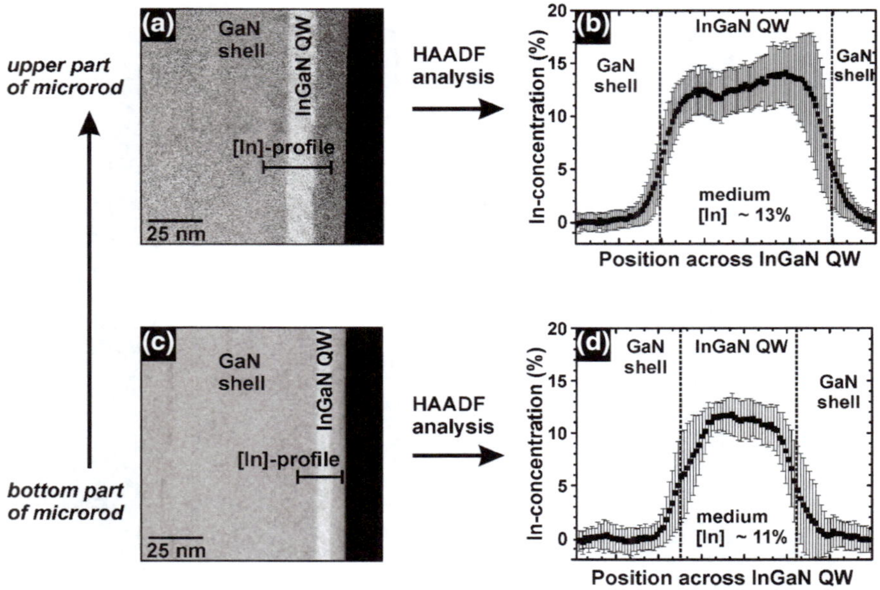

Fig. 5.25 *Z*-contrast analysis of the InGaN SQW: STEM-HAADF maps from the **a** upper part and the **c** lower part of the micro-column, together with the corresponding In-concentration profiles **b** and **d**. Reprinted with permission from [113]. Copyright {2016} American Chemical Society

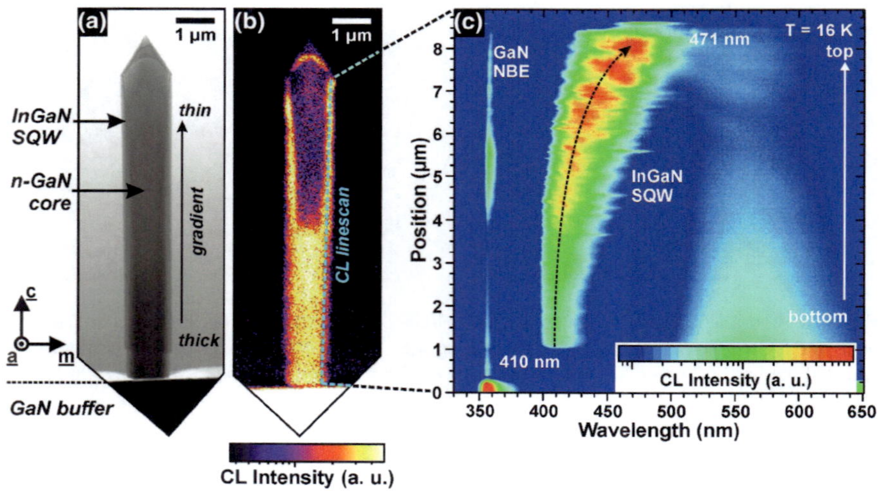

Fig. 5.26 Spatially resolved correlation of the structural and optical properties in cross section of the InGaN/GaN micro-column: **a** STEM map in bright field contrast, **b** panchromatic CL image and **c** CL linescan vertically along the nonpolar active region at $T = 16$ K. Reprinted with permission from [113]. Copyright {2016} American Chemical Society

at $x = 1.0$ μm, the InGaN SQW emission appears at spectral position of 410 nm. Only a slight shift in the InGaN emission (423 nm with shift of 50 meV) is visible up to $x = 5$ μm. The linewidth at $x = 5$ μm is 170 meV. Moving on to the apex of the micro-column, the spectral shift increases, reaching an emission wavelength of 471 nm (400 meV total energy shift). The linewidth increases up to 270 meV at the apex region.

A summary of the structural and optical evolution in growth direction of the SQW is given in Fig. 5.27. The SQW thickness and In concentration determined by HAAQF stays almost constant up to a height of about 5 μm (Fig. 5.27b). The shift to longer emission wavelength is in full agreement with those results (Fig. 5.27a). Beginning at a height of 5 μm, the SQW thickness increases significantly from 6.1 nm up to 13.1 nm. Simultaneously to the doubling of SQW thickness, the In content of the SQW increases from 11 to 13%. Keller et al. proved a strong link between a high growth rate and increase of In concentration in InGaN SQW on nonpolar facets [114]. Consequently, the SQW emission shifts to longer wavelengths reaching 471 nm showing a direct correlation of emission wavelength, composition as well as thickness of the SQW.

The reason for the change of thickness and composition of the quantum well could lay in inhomogeneous growth conditions along the nonpolar facet. There is a temperature gradient during the growth between basis and top of the micro-columns of about 0.1 °C determined by Schmitke et al. [115]. Another reason for the inhomogeneity of the SQW could be the reduced vapor phase diffusion during the MOVPE growth due to the high density of micro-columns. There is a detailed model based on calculations of Fang et al. [116] for the growth of 3-dimensional InGaN/GaN heterostructures pointing to the formation of a concentration gradient due to the reduced diffusion of the gas phase in the direction of the substrate. According to this work, a

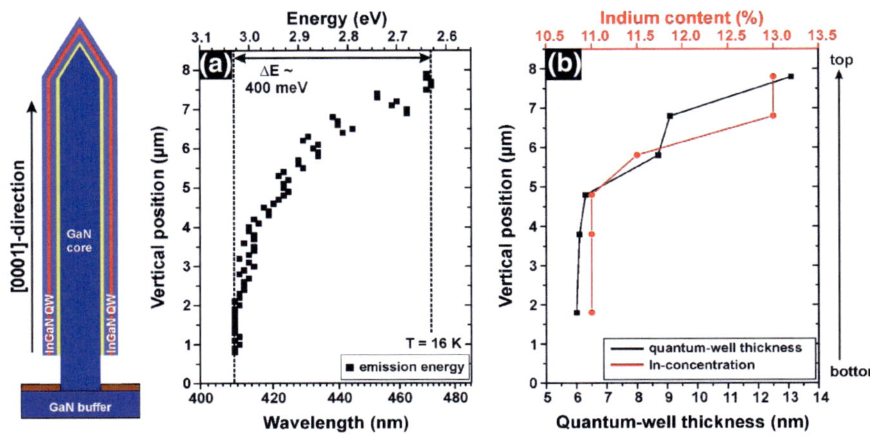

Fig. 5.27 Correlation of **a** InGaN SQW peak wavelength and **b** SQW thickness (black) and In composition (red) versus vertical position. Reprinted with permission from [113]. Copyright {2016} American Chemical Society

reduction of SQW thickness and In content in the lower part of the micro-columns is the result of reduced concentration of In- and Ga-precursors. Comparable results are reported by Gibbon et al. for the selective growth of GaInAsP structures [117] and Wunderer et al. for semipolar InGaN/GaN LEDs [118]. An increase of the growth temperature can lead to an improved diffusion in the gas phase, which can lead to a significant improvement of the homogeneity of the SQW emission along the micro-columns [119]. In addition, a decrease of the reactor pressure can increase the gas phase diffusion.

A closer look of the gradient in Fig. 5.26c shows a noncontiguous shift in wavelength as well as intensity. "Whispering gallery" modes, proved by Nobis and Grundmann in ZnO micro-columns [120], can be excluded as a reason since the hexagonal symmetry is lost in the TEM sample. The pattern could be caused by Fabry-Pérot thickness interferences at the wedge of the TEM sample, but detailed analysis exhibited no dependency on the sample thickness. The abrupt spectral changes of the SQW luminescence along the linescan are strong indications of bandgap fluctuations on sub-micrometer-scale. The following part of this section concentrates on the fluctuations of optical properties on nanometer scale.

Potential Fluctuations in the InGaN Quantum Well

In Fig. 5.28a the panchromatic intensity image of the upper part of the micro-column is depicted. To analyze the line shape, local spectra are taken (positions are marked in Fig. 5.28a) and shown in a waterfall plot in Fig. 5.28. The highest CL intensity yields from the nonpolar side facets. The core has one order of magnitude less intensity. The spatial distribution of the peak wavelength is shown in Fig. 5.28b. The Si-doped

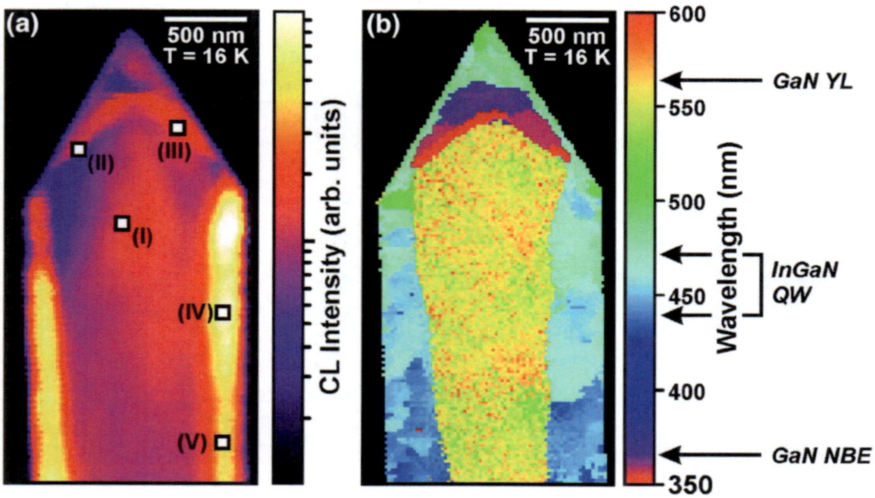

Fig. 5.28 STEM-CL mapping of the upper part of the InGaN/GaN micro-column taken at 16 K. **a** Panchromatic intensity image, **b** CL wavelength image. Reprinted with permission from [113]. Copyright {2016} American Chemical Society

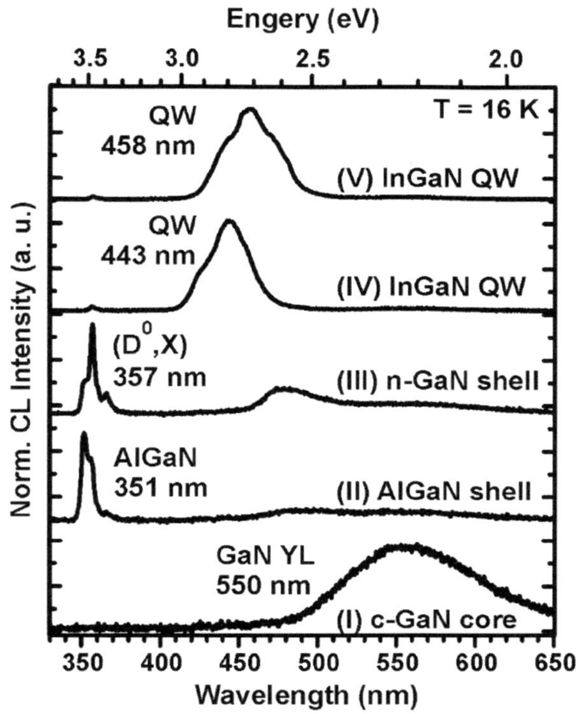

Fig. 5.29 Local spectra from different positions of the InGaN/GaN micro-column (local origin is marked in Fig. 5.7a) [88]

core is dominated by point-defect-related YL of GaN. The local spectrum (I) of Fig. 5.29 proves that no NBE emission of GaN is present here. Due to the strong Si-doping in this region, a high number of point defects are created during the 3-dimensional MOVPE growth. Most of the excess carriers recombine with the help of deep traps radiatively to emit yellow defect-emission or nonradiatively by means of Shockley-Read-Hall recombination.

In contrast, the GaN shell of the $\{1\bar{1}00\}$-facet shows NBE emission around 357 nm (local spectrum in Fig. 5.29). The intensity of YL is much weaker than in the inner core indicating fewer point defects. In addition, narrow emission from AlGaN is seen, in particular, in the complex domain structure at the apex.

The SQW emission is the dominant recombination channel at the nonpolar facet. The extension of this luminescence is determined by the diffusion of excess carriers from the GaN barriers into the InGaN SQW leading to a much broader appearance in the intensity and wavelength image. There is a slight red shift to longer wavelengths of the SQW emission in Fig. 5.27b starting in the lower part of the map at $\lambda = 440$ nm and reaching $\lambda = 470$ nm at the apex.

The semipolar facets at the apex show the complex impact of thickness change and In content of the SQW as well as polarization fields in different growth domains on the emission (Fig. 5.28).

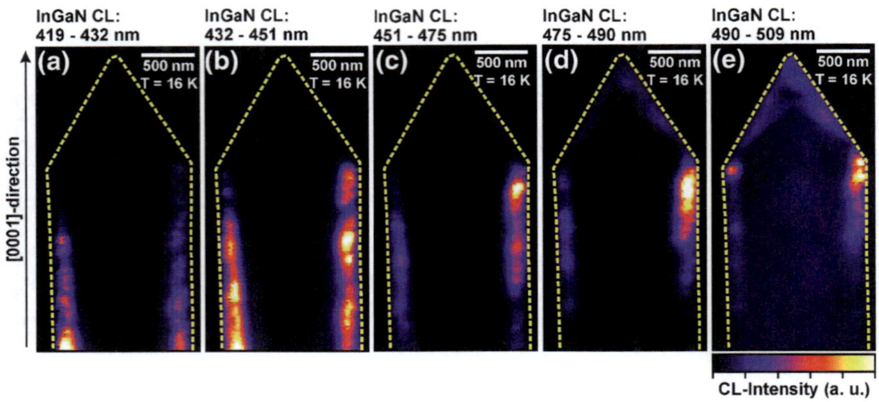

Fig. 5.30 Monochromatic CL intensity images of the InGaN SQW luminescence at $T = 16$ K in spectral region of **a** 419–432 nm, **b** 432–451 nm, **c** 451–475 nm, **d** 475–490 nm, and **e** 490–509 nm [88]

To characterize the emission fluctuations in detail, a set of monochromatic CL intensity maps taken at 16 K is displayed in Fig. 5.30. The InGaN luminescence begins at 419–423 nm in the lower part of the micro-column (Fig. 5.30a). With longer wavelengths, the emission shifts to higher positions along the micro-column. Additionally to the spectral shift along the nonpolar facet, we find short-ranging intensity fluctuations. The intensity in the monochromatic CL maps appears more and more inhomogeneous and spot-like for longer wavelength on a scale of under 200 nm.

It was already shown, that we do not find any extended defects, like dislocations or stacking faults, in the whole core shell structure. Non-radiative recombinations on extended defect can be excluded as reason for the local intensity fluctuation.

A reason for the inhomogeneous CL intensity distribution on nanometer scale can be local potential fluctuations. In general, potential fluctuations within a SQW happen of different length scales. Fluctuations on longer ranges caused by composition and thickness variations of the SQW as well as structural defects happen usually on micrometer scale. The spatial extension of those fluctuations is much larger than the diffusion length of excess carriers. In contrast, fluctuations on shorter scales happen in the nanometer range. Carriers are localized in the local potential minima of conduction and valance band. Chichibu et al. observed such behavior for the first time in planar films [121, 122]. The localization on such short scales can be assign to:

- Thickness variation of the quantum film [123–125]
- Formation of In-N-In complexes due to statistical composition fluctuations [126–129]
- In clusters within the quantum film [122, 130, 131].

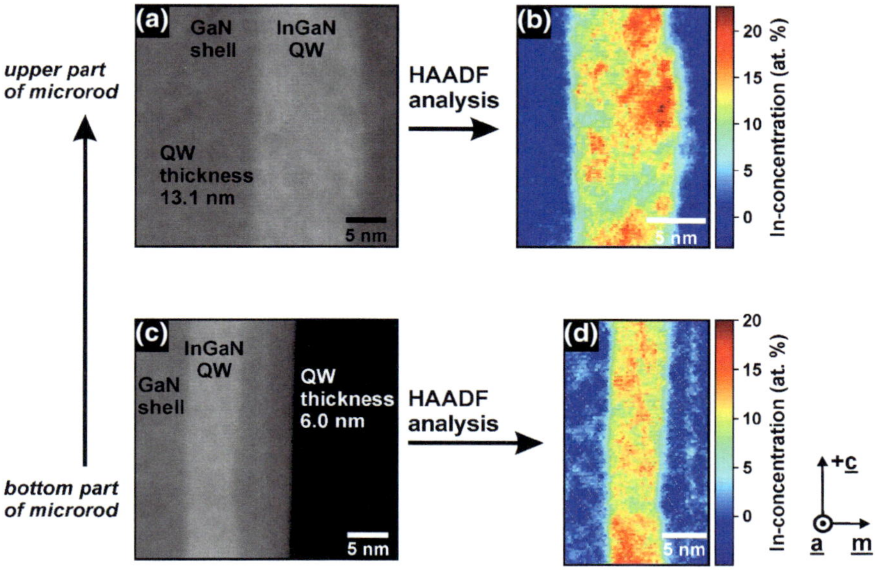

Fig. 5.31 HAADF mappings of the InGaN single quantum well from the **a** upper and **c** lower part of the micro-column, **b** and **d** depict the corresponding In-concentration. Reprinted with permission from [113]. Copyright {2016} American Chemical Society

Highly Spatially Resolved HAADF Characterization of the InGaN SQW

The local chemical composition as well as interface properties of the InGaN single quantum well are comprehensively investigated by highly resolved HAADF mappin.[5] The measurements were performed under an acceleration voltage of 300 kV with the zone axis in $[11\bar{2}0]$ direction. Figure 5.31 shows two HAADF images from the upper and lower part of the micro-column. The HAADF map gives direct access to the atomic number: The InGaN well appears brighter than the GaN matrix. In both images, the inner InGaN/GaN interface appears sharp, whereas the outer interface appears rougher.

To determine the local In-content of the InGaN SQW, the experimental, normalized HAADF intensities have been compared with frozen-lattice simulation [113]. The calculated In-content of the SQW is shown in Fig. 5.31b, d. Clearly, local inhomogeneities of the local composition is visible. Even more, the distribution of In within the quantum well strongly differs from a homogeneous, pure-statistical fluctuation. Figure 5.31b from the upper part of the micro-column yields a local variation of In of $[\mathrm{In}] = \pm 7\%$ from the mean amount of In of 13%. There are In-rich regions with $[\mathrm{In}] > 19\%$ as well as In-poor regions with $[\mathrm{In}] < 8\%$ in the nonpolar well. The spatial extension of such regions is in the range of 3–10 nm. In comparison, the In

---

[5]The presented HAADF measurements were performed by Florian Krause in the group of Prof. Andreas Rosenauer (University of Bremen).

concentration map from the lower part of the micro-column is more homogeneous. Nevertheless, also here we find variations of the In composition: In-rich regions with [In] > 14% and In-poor areas with 9%. Due to the thickness of the TEM sample of about 120 nm, we are not able to obtain the 3 dimensional geometric shape of the nanoscopic fluctuations.

To verify the findings, simulations of the HAADF intensity of an InGaN quantum well with statistical composition fluctuations as well as In-rich clusters are compared. The area for the simulation is 10 × 20 nm. In the simulation, parameter of the STEM experiment are inherited, like sample thickness of 120 nm and zone axis of $[11\bar{2}0]$ as well as an averaged In composition of 11%. In the color-coded maps of Fig. 5.32, the HAADF intensity is normalized to 11% (contrast of 1 = green). The left map with statistical fluctuations exhibits a relatively homogeneous distribution. The standard deviation yields $\sigma_{\text{statistical}} = 0.0014$. In contrast, the right map with In-rich clusters (spherical inclusions with 5 nm diameter) shows clearly an inhomogeneous distribution. Brighter HAADF intensity corresponds to higher In concentrations. The standard deviation is one order larger and yields $\sigma_{\text{cluster}} = 0.015$. This value is in agreement with the measurement of $\sigma_{\text{experimental data}} = 0.012$ in Fig. 5.31d. The simulation confirms the presents of In-rich cluster in the InGaN SQW even if the sample has a thickness of 120 nm.

An e-beam induced formation of In-rich clusters, as described by Smeeton et al. [132] or Li et al. [133], can be excluded. For the presented quantitative HAADF measurements, low electron dose (0.2 pA/nm^2) and low exposure time for the STEM map (40 s) was chosen, so that no alternation of the InGaN can occur [134, 135]. TEM characterizations of Bartel and Kieselowski [136] as well as Galtrey [124] prove, that optimized image conditions circumvent a damage of the InGaN well. The

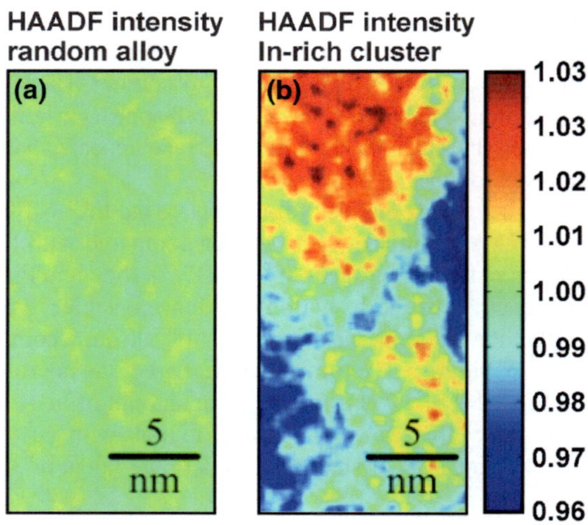

Fig. 5.32 Simulation of the HAADF-intensity for a InGaN quantum well with **a** statistical composition fluctuations and **b** In-rich clusters. Reprinted with permission from [113]. Copyright {2016} American Chemical Society

formation of In-rich clusters was evidenced by Humphrey et al. in a-planar InGaN wells [131] and Wu et al. in c-oriented InGaN wells [130].

Visualization of Nanoscopic Potential Fluctuations Using STEM-CL

Using CL spectroscopy directly performed in a STEM, the impact on potential fluctuations on the nanoscopic luminescence properties are visualized. The CL measurement was performed under an acceleration voltage of 80 kV with a beam current of 130 pA and a temperature of $T = 16$ K. Figure 5.33 shows a STEM-CL mapping of the active region over an area of 150×260 nm from the upper part of the InGaN/GaN micro-column. In the STEM bright field image (Fig. 5.12a), the InGaN SQW appears as dark contrast in the brighter GaN matrix. The quantum film is slightly tilted with respect to the electron beam leading to a broadening in the STEM image. The corresponding spatially averaged spectrum is depicted in Fig. 5.33c with an emission wavelength around $\lambda = 430$ nm and a line width of 250 meV. Exclusively, InGaN SQW emission is visible.

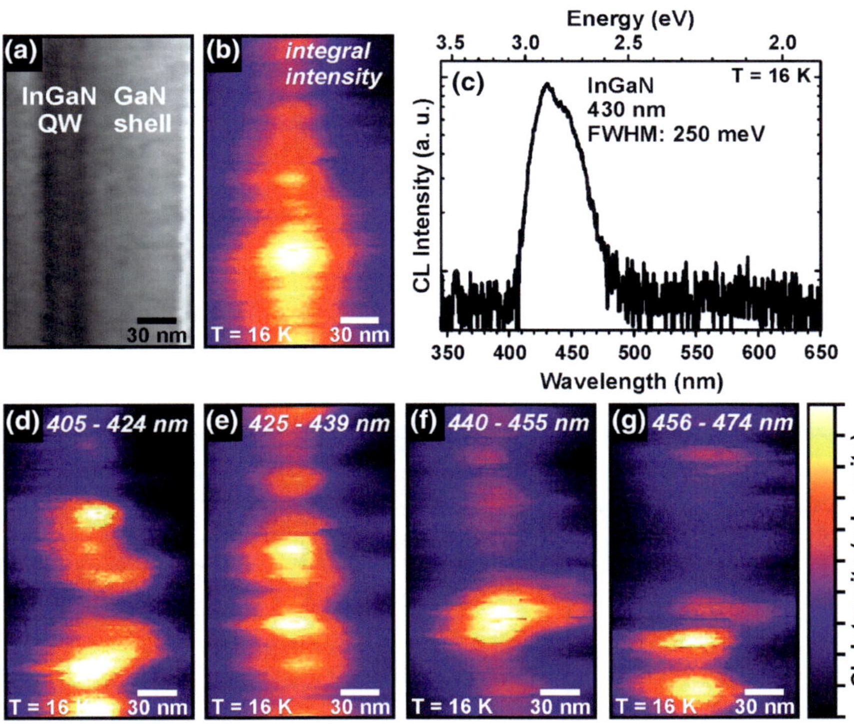

Fig. 5.33 Highly spatially resolved characterization of the InGaN SQW: **a** STEM bright field image, **b** panchromatic CL intensity image of the InGaN emission, **c** spatially averaged spectrum, **d–g** monochromatic CL intensity images of the InGaN emission. Reprinted with permission from [113]. Copyright {2016} American Chemical Society

The analysis of the panchromatic CL intensity distribution in Fig. 5.33b evidences the spatially inhomogeneous variation of the SQW luminescence. There are no extended defects visible in the HAADF image (Fig. 5.33a), so those can be excluded as reasons for the intensity variations. Corresponding highly-resolved monochromatic CL intensity maps give access to the spatial and spectral distribution of the SQW emission visible in Fig. 5.33d–g. Each image is scaled to its own maximum. For the shorter wavelength interval, the CL intensity map from 405 to 424 nm, spot-like areas with high intensity appear with an expansion of 30–50 nm. In comparison, the intensity maps of longer wavelengths 425–439, 440–455 and 456–474 nm appear with complementary contrast. The recombination centers vary in emission wavelength and intensity indicating strong localization. The spatial extension is larger than the expected interaction volume of the electron beam. For an acceleration voltage of $V_{\mathrm{acc}} = 80\,\mathrm{kV}$ and sample thickness of $t = 200$ nm a spatial resolution of about 10 nm is estimated. In addition, the features in the monochromatic intensity maps are larger than the size of the structural changes in the HAADF image (compositional In variation). Due to diffusion of excess carriers into local potential minima, the luminescence signal is widened in the images. This process can be used to determine the local minority diffusion length.

Local spectra taken with high spatial resolution are shown in Fig. 5.34. Narrow emission lines vary in the spectral rage from 429 to 458 nm. Several sharp lines with inhomogeneous broadening appear at the same local position indicating that more than one recombination center is active within the thickness of the TEM lamella. The

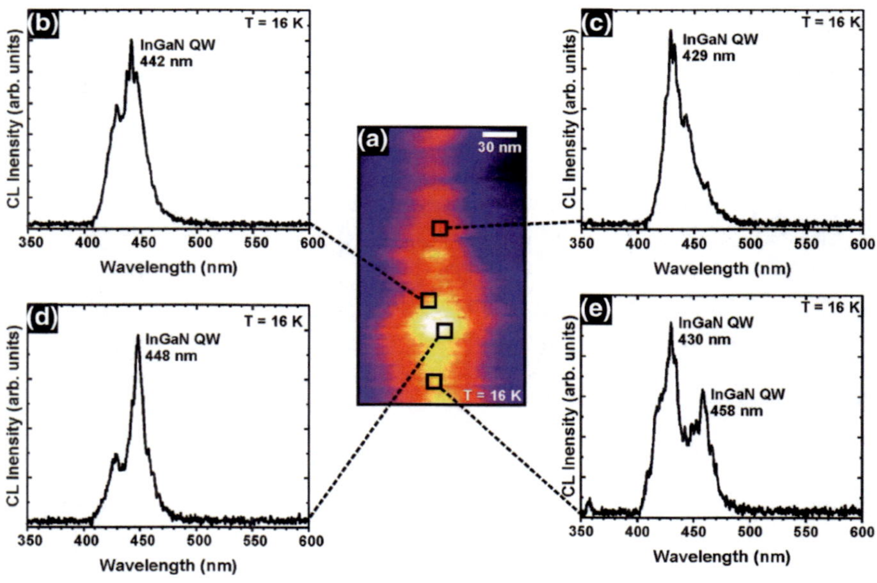

Fig. 5.34 **a** Highly spatially resolved CL intensity map of the InGaN emission, **b**–**e** local spectra taken at 16 K. Reprinted with permission from [113]. Copyright {2016} American Chemical Society

width of the emission line is around 15 meV, which is the spectral resolution of the measurement (150 l/mm grating). In comparison, μPL mappings of planar InGaN SQWs show line widths down to 0.8 meV [137].

At low temperatures, the kinetic energy of excess carriers is smaller than the barrier of the local potential fluctuation hampering the full relaxation of the carriers to the lowest energy. A "freeze-out" of excess carriers happens [138]. In result, localized carriers recombine from locally different potential minima with different emission energies. Due to a higher effective mass of holes in InGaN, the localization of the wavefunction of holes is stronger than for the electrons [126, 139]. Schulz et al. give a localization range of about 1–3 nm for holes in InGaN. In contrast, electrons average over a distance of 5–10 nm [126]. Since those length scales are in the range of the spatial resolution of the shown measurements, we are not able to differ between electron and hole localization.

When composition fluctuations reach the size of the exciton Bohr radius in InGaN (about 5 nm), quantum dot-like states are expected [137, 140]. Composition fluctuations as well as rough interfaces of the nonpolar InGaN SQW lead to spatial potential fluctuations. The STEM-CL measurements directly visualizes the inhomogeneous potential landscape on nanometer scale since the carriers are strongly localized.

5.4.2 InGaN/GaN Core-Shell Nanorods with Thick InGaN Shell

The InGaN/GaN core-shell nanorods discussed in this chapter are comparable to the previous section: instead of the thin quantum well heterostructure, a thick InGaN shell was deposited resulting in the formation of In-rich triangular nanoprisms at the edges of InGaN/GaN core-shell nanorods. Another unique feature is the radial cross section preparation of nanorods, which give additional structural and optical information.

The nanorods were grown by metal-organic vapor phase epitaxy using selective area growth approach. Patterned SiO_2 on GaN/sapphire substrate has been used as templates for the selective area growth. Using photolithography and reactive ion etching circular holes were opened into the SiO_2 mask. Subsequently, Si-doped GaN nanorods were grown. To achieve a high aspect ratio, the growth of the nanorods is performed in a two-step process: first, the vertical growth rate was increased by using a high Silane (SiH4) flow rate a low V/III ratio. In a second step, the SiH4 flow rate was decreased to reduce SiN_x passivation of the n-GaN core surface. An GaN spacer was deposited around the Si-doped GaN core. Finally, an InGaN shell layer with a nominal thickness of 30 nm was grown enfolding the GaN core structure. A schematic drawing of the core-shell nanorod heterostructure is depicted in Fig. 5.35a together with an SEM image of a single column in Fig. 5.35b.

The spatially averaged CL spectrum of the nanorod ensemble in Fig. 5.35d at $T = 6$ K is dominated by InGaN luminescence with two peaks at 393 and 480 nm. The

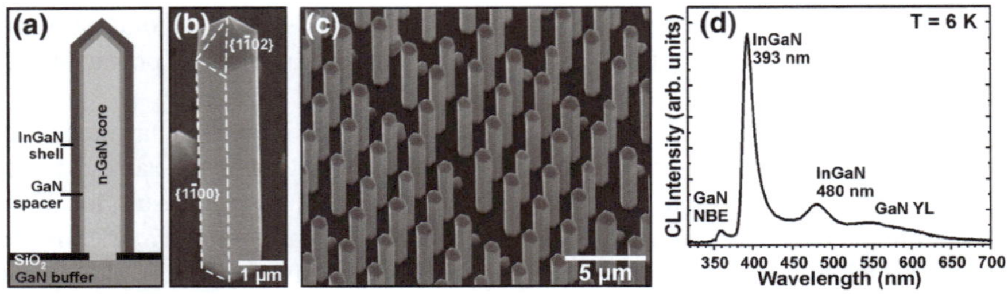

Fig. 5.35 **a** Schematic cross-section of the nanorod, **b** SEM image of a single nanorod, and **c** the as-grown nanorod array; **d** spatially averaged CL spectrum of nanorod ensemble measured at $T = 6$ K. Reprinted from [141]. Copyright (2018) licensed under CC BY 4.0

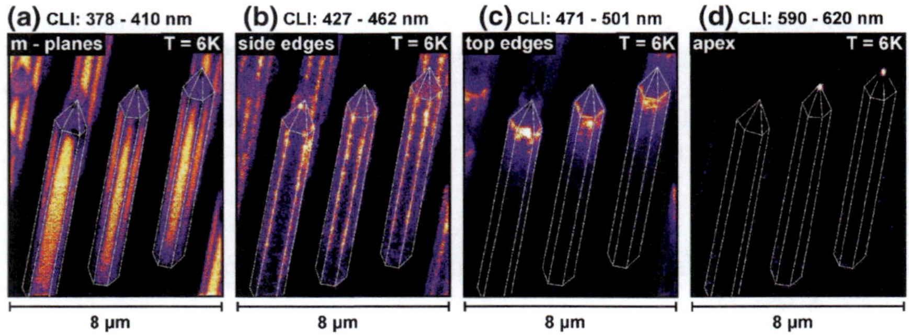

Fig. 5.36 Low temperature monochromatic images of the as-grwon nanorods. **a** 378–410 nm, **b** 427–462 nm, **c** 471–501 nm and **d** 590–620 nm measured at $T = 6$ K [88]

near band edge (NBE) emission of GaN as well as the yellow luminescence of GaN is obtained as well (Fig. 5.36).

Figure 5.37a shows the HAADF contrast of a single nanorod in radial cross-section perpendicular to the [0001]-direction. Dark regions correspond to GaN and brighter regions to InGaN with different [In] (brighter contrast corresponds to higher [In]).

The complete radial shell structure is resolved in the STEM map. Beginning with the nanorod center, the whole Si-doped GaN core appears homogeneously. STEM analysis verified the absence of extended defects like threading dislocations or stacking faults in the core. Enfolding this n-GaN core, a thin shell layer appears in brighter contrast. Finally, the GaN spacer decorates the *m*-plane core facets and the InGaN shell layer enfolds the complete GaN nanorod with sharp interfaces.

During the InGaN growth, nanoprisms with triangular cross-section are self-organized formed at the six vertices of the hexagonal cross-section. The brighter contrast of the triangles in the HAADF image evidences a higher In content in comparison to the *m*-plane shell layer. Figure 5.37b depicts a HAADF map of one of the hexagonal edges at higher magnification. At the edges of the GaN core, a {11-20}

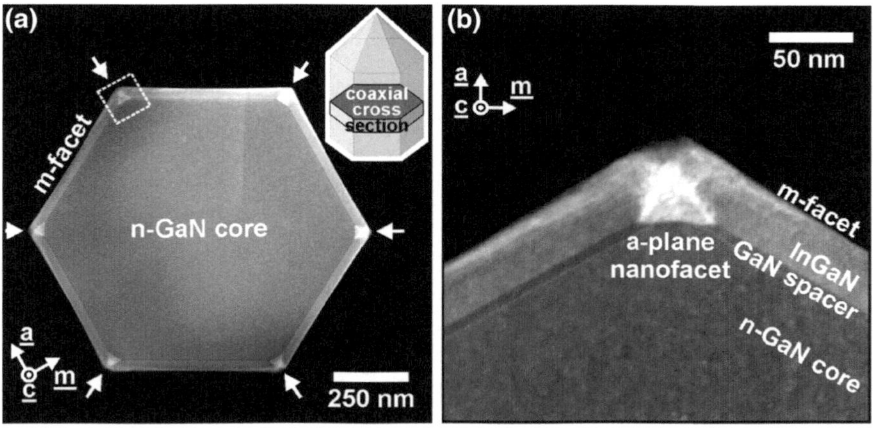

Fig. 5.37 **a** STEM-HAADF map of the radial cross-section (inset: schematic of nanorod section) of a nanorod in overview as well as **b** a corner in higher magnification. The whole layer structure is resolved: the n-GaN-core, the GaN spacer and the thick InGaN layer. A brighter HAADF contrast is found at the edges of the nanorod (marked by arrows in **a**) proving the formation of In-rich a triangular nanoprism. Reprinted from [141]. Copyright (2018) licensed under CC BY 4.0

a-plane has been formed. The formation of a-plane nanofacets at the edges has been previously reported for self-assembled GaN nanowires grown by MBE [142, 143]. Three crystallographic equivalent {11-20} facets forming an equilateral triangle with 60° internal angles, terminating the In-rich nanoprism.

Figure 5.38a–d show monochromatic CL images at low temperature ($T = 18$ K) recorded together with the corresponding HAADF map in Fig. 5.37a. Figure 5.38a, b focus on the GaN luminescence, visualizing the strong impact of the high Si doping of the inner core. Yellow luminescence (YL) only appears in the center part of the GaN core but not outside. Whereas intense band-band transition at 352 nm occurs in the outer region only. Due to a high free carrier concentration in the highly doped core leading to band filling and exciton screening, the (e, h) recombination dominates the near band edge emission.

In contrast, the YL is connected to point defects accompanied with deep levels incorporated during the growth with high Silane flow rate, which are denser in the inner core and reduce in density towards the outer region. The reason for this difference is that the central part of the core is grown vertically in c-direction whereas the outer part grows more or less laterally in semi-polar direction. This results in different densities of point defects which is visualized with the monochromatic intensity distributions of the (e, h) recombination and yellow defect luminescence.

Figure 5.38c, d shows two monochromatic CL images of the InGaN spectral region taken at 400 nm and at 500 nm. All six m-facets yield intense CL emission at the main InGaN peak over the whole thickness of the InGaN shell except the very corners. Exactly at these corners of the nanorod and exclusively there, the intensity of the 400 nm InGaN emission drops down and longer wavelength InGaN

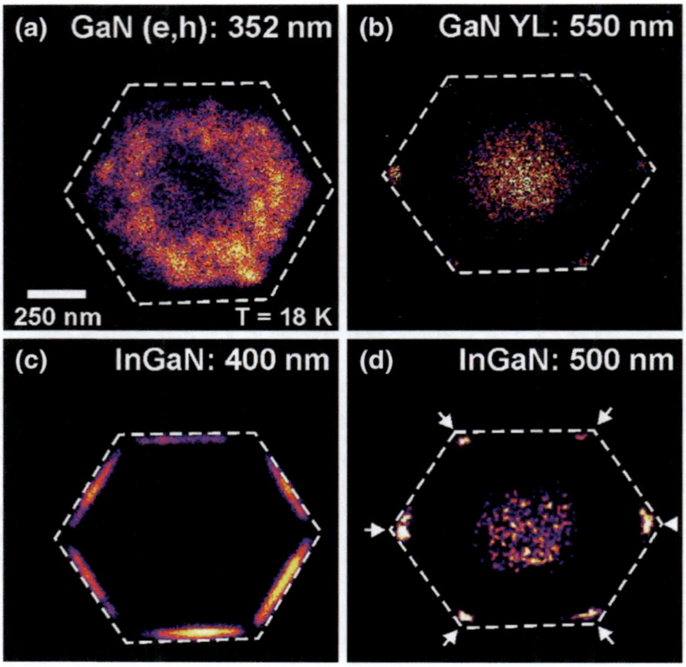

Fig. 5.38 Monochromatic CL intensity images taken at 18 K for **a** GaN, **e**, **h** emission at 352 nm, **b** GaN yellow luminescence (YL) at 550 nm, **c** *m*-plane InGaN at 400 nm, and **d** InGaN nanoprism at 500 nm, respectively. Reprinted from [141]. Copyright (2018) licensed under CC BY 4.0

CL (500 nm) appears. The formation of such In-rich InGaN nanoprisms and their redshifted luminescence in respect to the *m*-plane InGaN has been reported before by Griffiths et al. [143].

A spectrum linescan in *a*-direction was performed across an individual nanoprism (see Fig. 5.39). Starting in the GaN core a dominant GaN (e, h) emission can be observed (355 nm). Towards the surface 3 spatially and spectrally well separated InGaN luminescence contributions can be resolved. Emission at 417 nm is emitted close to the interface between the GaN core and the *a*-plane InGaN layer. This peak wavelength of 417 nm corresponds to a calculated In concentration of ~17%. Subsequently, a redshifted InGaN CL at 452 nm appears, followed by a weak cathodoluminescence at 500 nm close to the surface. The reduced intensity of the 500 nm emission can be attributed to the enhanced nonradiative recombination of generated carriers at the nearby surface as well as lower crystal quality. Different In incorporation can also be concluded from the HAADF image (see Fig. 5.38a), where we see a brightening of *Z*-contrast towards the surface. This is qualitatively in agreement with the redshifted InGaN emission in *a*-direction.

In summary, we investigated the In-incorporation in InGaN core-shell nanorods by low temperature cathodoluminescence performed in scanning transmission electron

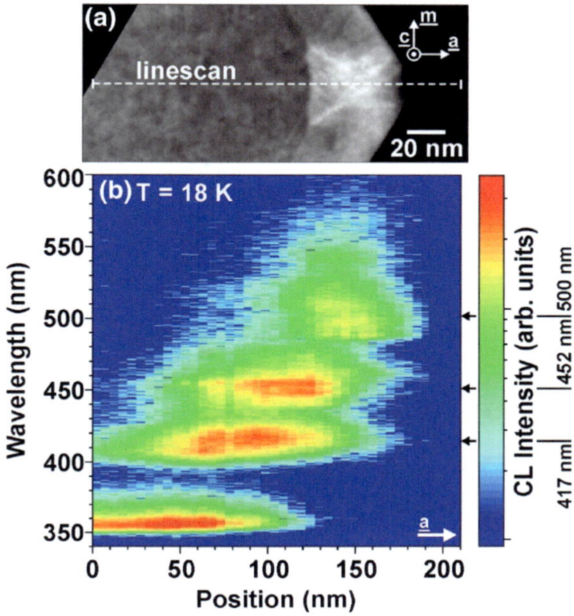

Fig. 5.39 **a** STEM-HAADF map of a nanoprism with **b** CL linescan along *a*-direction from the GaN core across the nanoprism (see dashed line in **a**), which shows 3 characteristic InGaN emission channels at 417 nm at the base of the triangle, 452 nm in the middle part and 500 nm at the top of the triangle. Reprinted from [141]. Copyright (2018) licensed under CC BY 4.0

microscopy. Self-organized In-rich triangular nanoprisms are formed at the edges of InGaN/GaN core-shell nanorods terminated by *a*-plane nanofacets and exhibit strongly redshifted luminescence in respect to the *m*-plane side walls. Inside the nanoprisms, three spatially separated InGaN emission contributions from 417 nm up to 500 nm peak wavelength towards the nanorod surface indicate the lattice-pulling effect. At least 4% higher In-concentration is concluded for the triangular shaped region compared to the *m*-plane sidewalls. Additionally, close to the surface of the *m*-plane InGaN domains of higher In-incorporation are resolved. This study clearly illustrates a self-organized complex In-accumulation for thick InGaN shell layers grown on GaN nanorod core.

5.4.3 Full InGaN/GaN LED Micro-column Structures

Based on the previous reported results, the microscopic luminescence characterization of InGaN/GaN core-shell-based full LEDs will be discussed in this section. Also in this sample, the growth starts with highly *n*-doped GaN micro-columns acting as inner cores of the structure. To achieve better interface quality between the core and the shell layers, an undoped GaN buffer as well as an AlGaN-shell with higher V-III-ratio was deposited. Subsequently, an *n*-doped GaN layer was grown (nominal concentration $N_D = 2 \times 10^{18}$ cm^{-3}), followed by a single InGaN quantum well and a nominal undoped GaN barrier. An electron blocking layer (EBL) consisting

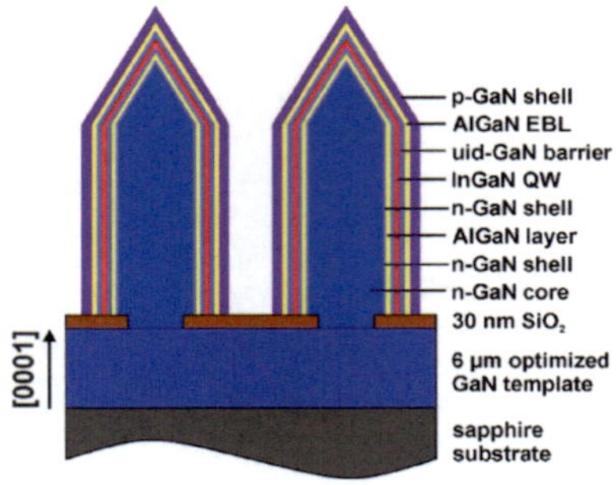

Fig. 5.40 Schematic sample setup of the InGaN/GaN micro-column full LED structure [88]

of AlGaN and the Mg-doped *p*-layer is completing the LED structure. A schematic drawing of the core-shell LED is visible in Fig. 5.40.

A STEM-mapping of cross-section of a complete LED column is depicted in Fig. 5.41. The bright field image shows, that the initial selective GaN-growth leads to the formation of a hexagonal nucleation pyramid with $\{1\bar{1}02\}$ side facets inside the SiO_2 mask holes (marked in the picture). Parallel to the growth direction, dark

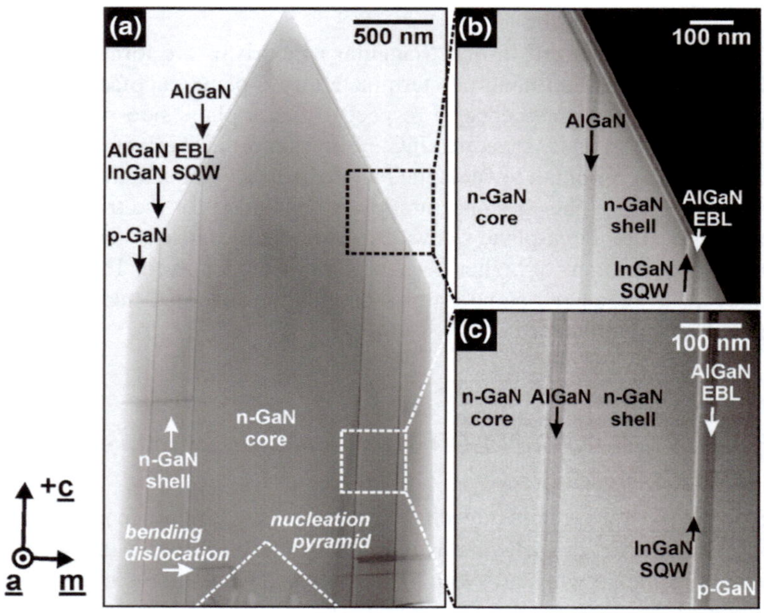

Fig. 5.41 **a** STEM cross-section of the whole micro-column LED in bright field contrast together with HAADF mappings of the shell layers from **b** semi-polar and **c** from non-polar facet [88]

dislocation lines are visible in the lower part of the nucleation pyramid. The dislocations tilt at the base of the column with a 90° angle to run laterally along the $[1\bar{1}00]$ direction resulting in a dislocation free upper part of the GaN core.

New defects are locally generated at the interface to the inner AlGaN layer. Those defects are visible as horizontally running (perpendicular to [0001]) dark contrasts, which terminate at the column surface. Under the chosen image conditions, partial dislocation and basal plane stacking faults (BSF) can be seen. Griffiths et al. [144] confirmed the formation of BSF in comparable structures. The growth under high Silane flow may lead to the creation of a Si-rich passivation layer as well as a roughening of the surface of the GaN core [145–147]. Both effects could create new BSF and partial dislocation at the shell interface. In addition, strain relaxation along the *c*-direction could lead to such defects [148, 149].

In Fig. 5.41b, c the complete layer structure of all shells is obtainable. The AlGaN-shells appear with darker contrast due to the smaller median *Z*, whereas the InGaN quantum well is brighter in comparison to GaN. All interfaces are smooth and abrupt. The QW thickness yields to (6 ± 1) nm. There is no indication for a thickness gradient along the *c*-direction. In contrast, the outer *p*-doped GaN shell increases its thickness from the base (270 ± 5) nm to the top (170 ± 5) nm along the *c*-direction just before the semi-polar facet starts. The reason for the thicker base could be the lower gas-phase diffusion of the carrier gas within the micro-columns [115]. The thickness gradient of the *p*-layer has negative impact on the current injection of the LED, because of the locally different current density.

Figure 5.42 depicts the locally averaged spectrum ($T = 14$ K) taken over the whole cross-section of the sample corresponding to the HAADF map Fig. 5.41. The spectrum shows intensive emission of the InGaN-QW at 435 nm with a FWHM

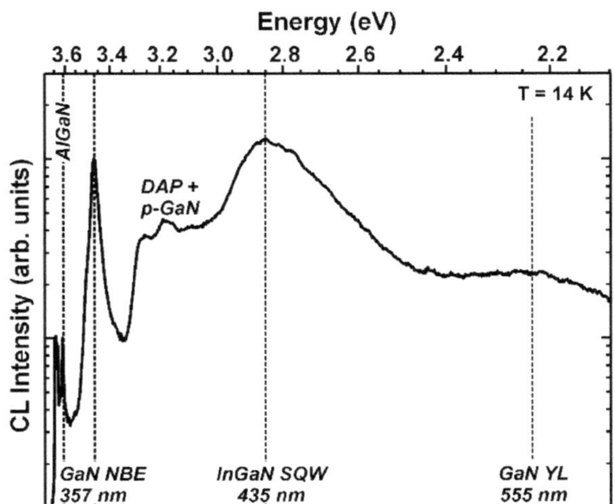

Fig. 5.42 Spatially averaged CL spectrum of a single micro-column in cross section ($T = 14$ K)

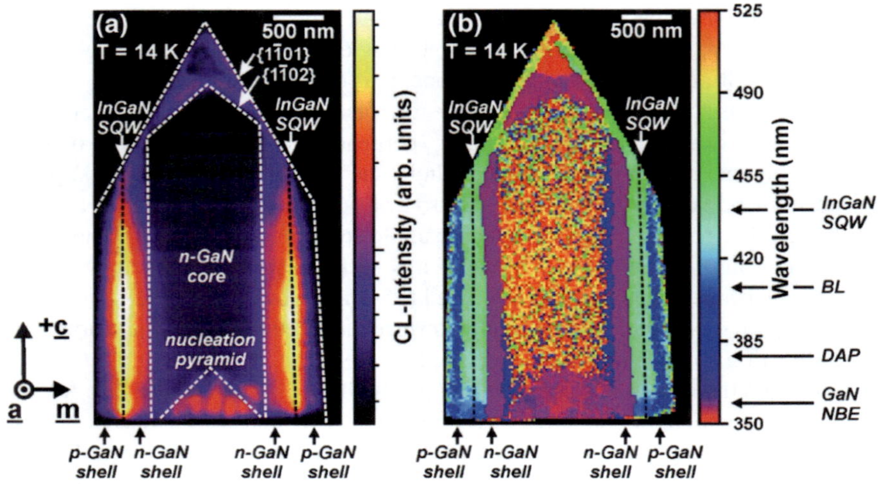

Fig. 5.43 STEM-CL mapping of micro-column LED in cross-section taken at $T = 14$ K: **a** panchromatic CL intensity image and **b** corresponding CL wavelength image with the characteristic recombination channels marked in the color-coded wavelength bar [88]

of 320 meV. In addition, luminescence from GaN donor acceptor pairs (DAP) at 378 nm, AlGaN layers at 340 nm and 343 nm, respectively, as well as near band edge (NBE) from GaN and Yellow luminescence (YL) from GaN can be obtained. The NBE GaN emission at 357 nm originates from the Si-doped GaN shell whereas the YL at 550 nm yields from the inner highly doped GaN core (proven by the CL wavelength image in Fig. 5.42) with orders of magnitude less intensity.

The local assignment of the different recombination channels with the real structure of the micro-column LED is accomplished using a STEM-CL map taken at $T = 14$ K. From this mapping, the panchromatic CL intensity on logarithmic scale (Fig. 5.43a) and the local distribution of peak wavelengths from 350 to 525 nm is depicted (Fig. 5.43b). The Si-doped inner GaN core can be obtained in the center, followed by the sequence of shell layers: AlGaN nucleation shell, n-GaN shell, InGaN QW, AlGaN EBL and outer p-GaN shell. Each shell appears with is characteristic emission which is marked in the image. The InGaN QW emits the highest CL intensity. In contrast, the highly doped inner GaN core shows two orders of magnitude less CL intensity due to a high density of point defects.

NBE GaN emission in the triangular-shaped region of the nucleation pyramid at the base of the micro-column is shifted to shorter wavelengths (354 nm) in comparison to the undoped GaN shell (357 nm) due to the high free carrier concentration in this area (high impurity incorporation). A complex domain structure is visible at the very top of the micro-column. CL intensity as well as the peak wavelength changes abruptly here. Self-limiting growth processes due to a change of surface facets ({1$\bar{1}$02} changes to {1$\bar{1}$01}) at the apex leading to different point defect concentration in certain regions and different In-incorporation in the QW.

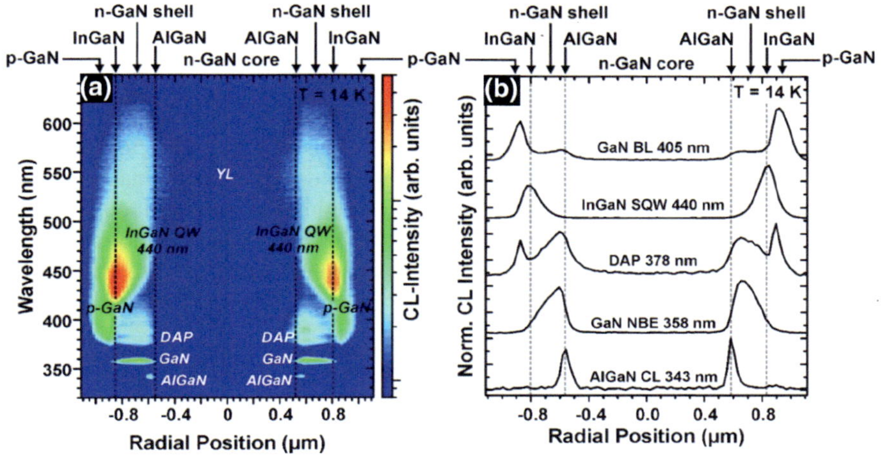

Fig. 5.44 **a** STEM-CL linescan across the micro-column LED taken at $T = 14$ K together with **b** monochromatic CL profiles of different recombination channels [88]

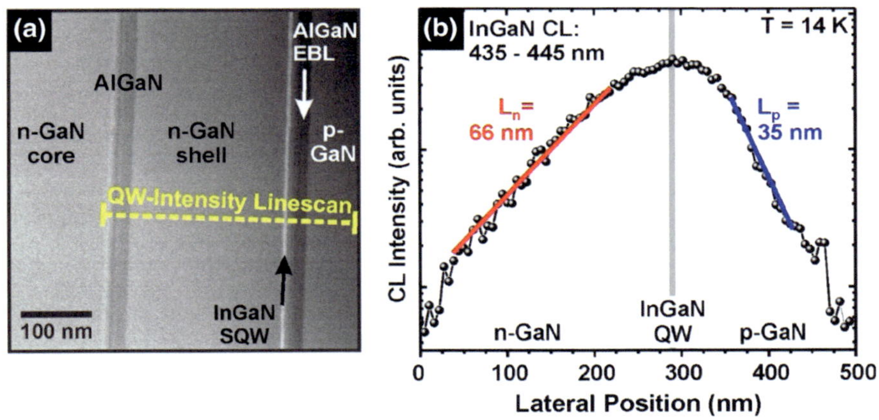

Fig. 5.45 **a** STEM-HAADF image of the shell layers with the position of the CL intensity profile marked with yellow dashed line and **b** monochromatic CL intensity of the InGaN emission at 435 nm versus position. The diffusion lengths are determined by exponential fits in the *n*- and *p*-barrier [88]

During the CL experiment, the ambipolar carrier transport and luminescence is determined by the diffusion and drift of excess carriers as well as competition of recombination channels and local concentration of non-radiative recombination centers. Figure 5.44 visualizes the evolution of luminescence across the micro-column. In the region of the *n*-core ($-0.56\ \mu\text{m} < x < 0.54\ \mu\text{m}$), only weak defect emission YL at $\lambda = 550$ nm is emitted (no GaN NBE emission). The following AlGaN nucleation layer shows emission at $\lambda = 343$ nm. Starting with the AlGaN shell at $x < -0.57\ \mu\text{m}$ and $x > 0.54\ \mu\text{m}$, respectively, NBE GaN emission at $\lambda = 357$ nm and GaN DAP emission at $\lambda = 378$ nm increases in intensity. The maximum of NBE und DAP intensity is reached at a position of $X = -0.62$ μm and $x = 0.64$ μm. Subsequently, both intensities monotonously decrease over a distance of $\Delta x = 200$ nm, whereas the InGaN QW emission at $\lambda = 435$ nm behaves complementary. The peak of the InGaN luminescence is reached at $x = -0.82$ μm as well as at $x = 0.84$ μm. The lateral diffusion of carriers into the quantum well competes with the recombination of GaN (NBE and DAP) within the barrier. The p-barrier emission starts at the position of the InGaN QW with blue luminescence at $\lambda = 400$ nm, increases monotonously up to a position at $x = -0.89$ μm and $x = 0.92$ μm, and extents up to the side facets of the micro-column. Simultaneously, the InGaN intensity decreases (Fig. 5.45).

TERS on InGaN/GaN LED Micro-column Structures

Tip enhanced Raman scattering (TERS) mapping of the previously discussed GaN micro-column based quantum well structures with a spatial resolution of about 30 nm was applied to obtain extensive information about local strain, chemical composition, charge accumulation, polymorphism, and In-clustering within the heterostructure [55]. A single vertically aligned nanorod was studied by approaching its apex with the STM tip in a TERS experiment. The surface sensitivity of TERS allows to study how the In-accumulation is distributed inside the quantum well at the apex of the nanorod. The far-field Raman spectrum, averaging over several nanorods, is dominated by the Raman signal of GaN. When approaching the tip, new peaks in the TERS spectra arise, which are attributed to the Raman signal from an InN cluster and its surrounding InGaN area. Those peaks are not present in the far-field spectra and their presence, in this particular experiment, is probably due to the matching resonance condition between tip plasmons and InN plasma frequencies.

Figure 5.46a, d show the Raman shift as a function of the spatial coordinates of the A_1(TO) phonon of the InGaN matrix around the InN cluster. The evidence of the InN cluster is given by the most prominent peak in TERS spectra in the region of 450–480 cm^{-1}. The Raman frequency of the TO mode of InN is shown as function of spatial coordinates in Fig. 5.46b. The intensity and the spectral position of this Raman mode clearly defines the localization and the composition of the InN cluster which is composed of two polytypes, h-InN and c-InN (Fig. 5.46c). From the spectral position of the A_1(TO) mode of the InGaN surrounding matrix, it is possible to estimate the local In-concentration around the InN cluster and the amount of strain close to the interface between the two different polytypes [55].

It was suggested that an InN cluster nucleates on top of a linear dislocation on the GaN core, which promoted the growth of the cluster into two different InN polytypes—an interpretation which was independently confirmed by HR-TEM images displayed in Fig. 5.47. Furthermore, the Raman map of the surface optical mode SO(A_1) frequency of the GaN cap layer provides additional information on the charge distribution due to the sensitivity of this Raman mode to charge accumulations. The SO(A_1) phonon frequency shifts to higher frequencies at spatial coordinates close to the hexagonal part of the InN cluster, while it shifts to lower frequency close to the cubic part (Fig. 5.46e). This behavior can be attributed to charge depletion and charge accumulation at the interface between h-InN and c-InN. This example

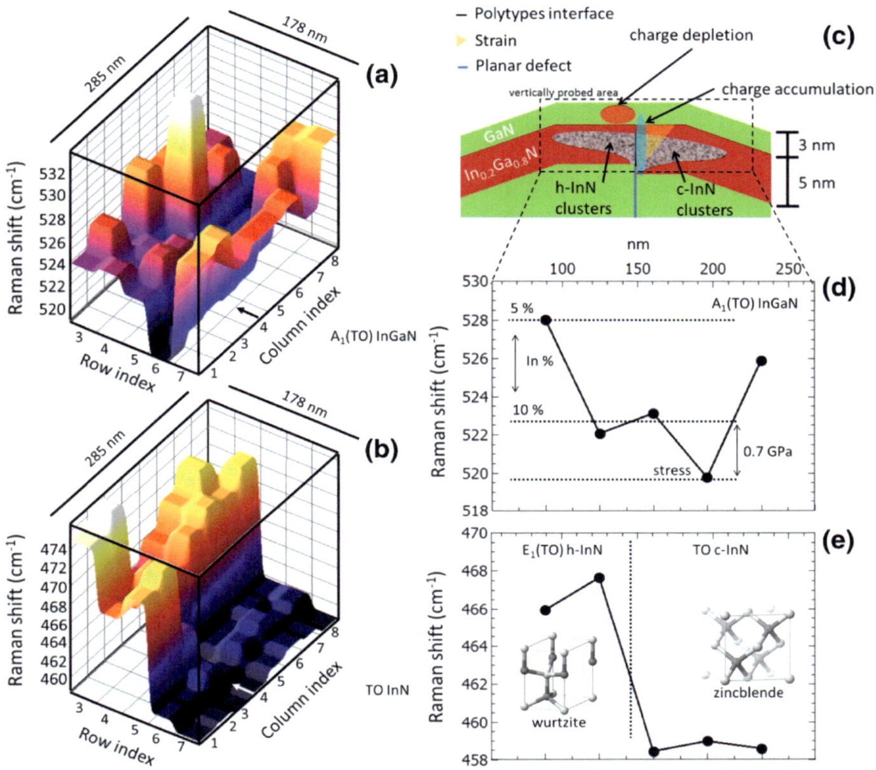

Fig. 5.46 **a** and **b** Three-dimensional visualization of the Raman frequencies as function of spatial coordinates of the A_1(TO) mode of InGaN and the TO modes of InN, respectively. **c** Simplified schematic illustration of the nanorod top region in cross sectional view. **d** and **e** Line scan profiles of the two Raman modes at the fourth column, indicated by the arrow in the three dimensional plot. Reprinted with permission from [55]. Copyright (2013) American Chemical Society

demonstrates the additional gain of information regarding composition, strain, clustering, and polymorphism, which becomes available due to the joint investigation of nanostructures by TERS and HR-TEM imaging.

5.4.4 *Shielding Electric Fields in Nanowire Based Quantum-Heterostructures*

The nanoscale characterization of nitride quantum well hetero-structures on the non-polar and semi-polar facets of micro-columns in the previous sections have demonstrated the beneficial reduction of piezo- and pyroelectric polarization fields in these directions which results in an absent or reduced quantum confined Stark

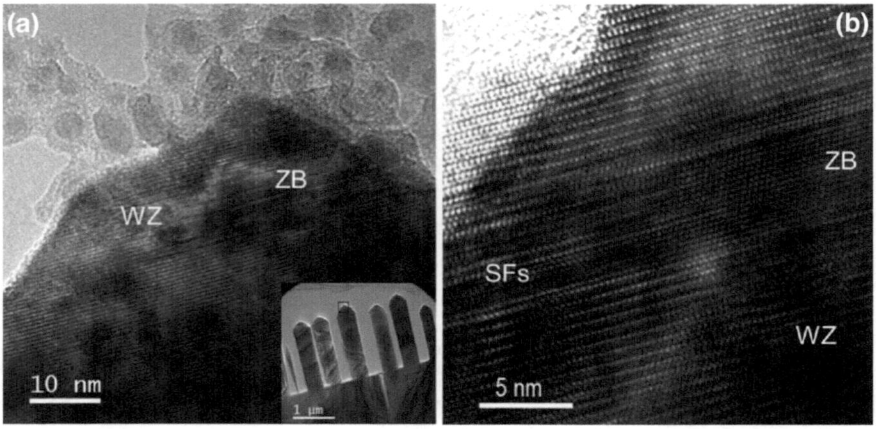

Fig. 5.47 **a** Cross-section HRTEM image of the top of a single vertically aligned GaN/InGaN nanorod. The inset shows a low resolution TEM image; the area of the HRTEM image is indicated by the square. **b** Cubic inclusions (ZB) and stacking faults (SFs) in the regular wurtzite (WZ) crystal structure are visible beneath the surface of the nanorods. Reprinted with permission from [55]. Copyright (2013) American Chemical Society

effect (QCSE) [150–152]. However, we have also shown in previous sections of this chapter how compositional fluctuations and rough interfaces of nonpolar quantum wells can result in spatial potential fluctuations with strong carrier localization on the nanometer scale. Obviously, it would be desirable to combine the benefits of *c*-plane growth with a simultaneous suppression of the intrinsic QCSE in this growth direction. To achieve this goal, Hönig et al. [153] have developed a new approach called IFGARD (internal field guarded active region design) that displaces the potential gradient in *c*-plane direction out of the active region, thus effectively suppressing the QCSE. Consequently, a larger overlap of the electron- and hole wave functions is achieved which results in an increased exciton recombination rate.

The first implementation of the IFGARD concept was realized in GaN nanowires with 40 times stacks of GaN nanodiscs (ND) embedded in AlN barriers. The nanowires were grown along the [000-1] direction. The shielding of the electric field gradient is achieved by adding GaN guard layers, which enclose the entire stack of GaN discs and AlN barriers. This rather counterintuitive approach, which adds an additional absorption layer at the top of the structures, proofs beneficial as the elimination of the electric field gradient within the active region overcompensates reabsorption losses in the guard layer [154]. Figure 5.48 shows scanning transmission electron microscopy images of NWs with 1 and 6 nm thick AlN barriers together with a schematic illustration of the structure. The lateral growth results in a tapered shape of the nanowire where the taper angle increases with increasing AlN barrier thickness. Consequently, the net interface charge at each side of the AlN barrier (indicated by red plus and blue minus symbols in Fig. 5.48) is not equal due to the different interface areas with increasing discrepancy for larger barrier thicknesses.

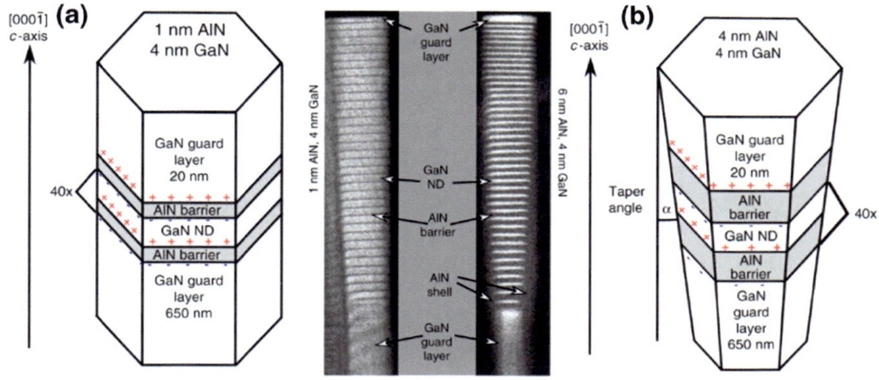

Fig. 5.48 Schematic illustration of nanowires (NWs) with embedded IFGARD nanodiscs with AlN barriers of 1 nm (**a**) and 4 nm (**b**) thickness. NWs with an increased barrier thickness are accompanied by an increased taper angle α due to lateral growth of the barrier material. Red "+" and blue "−" signs mark the positions of interface charges induced by the wurtzite crystal polarization. (center): HAADF-STEM images of the stacked GaN/AlN nanodiscs with 1 nm (left) and 6 nm barriers (right). Reprinted from [154]. Copyright (2018) licensed under CC BY 4.0

Since the electric field is completely shielded only for equal interface charges, the barrier thickness becomes a tuning parameter for the remaining electric field in the active regions of these nanowires.

The μ-PL spectra of two different series of nanowires with (i) varying GaN ND thickness and constant AlN barrier thickness and (ii) constant ND thickness and varying barrier thickness are displayed in Fig. 5.49. The left side of the figure (labeled QCSE regime) displays the PL spectra for four different nanowires with fixed GaN ND thickness of 4 nm and variable AlN barrier thickness between 1 and 6 nm. For the largest barrier thickness, the discrepancy between the interface charges on either side of the barrier is considerable, resulting in a weak shielding of the electric field in the active region and thus a strong shift of the luminescence peak to lower energies caused by the QCSE. With decreasing AlN barrier thickness, the electric field shielding increases due to approaching quantities of interface charges (interface areas) which results in a stronger suppression of the QCSE in the nanodiscs. This can be directly observed by the shift of the emission maximum from 2.76 to 3.34 eV for samples with constant ND thickness and decreasing barrier thickness between 6 and 1 nm, respectively.

A further increase of the emission energy is achieved by reducing the ND thickness from 4 to 1 nm, while keeping the barrier thickness constant at 1 nm. The corresponding PL emission shifts beyond the GaN guard layer emission at 3.47 eV (blue line in Fig. 5.49) due to the confinement of excitons in the thin nanodiscs. For an ND thickness of 1 nm, the emission maximum is observed at 4.04 eV. Figure 5.50

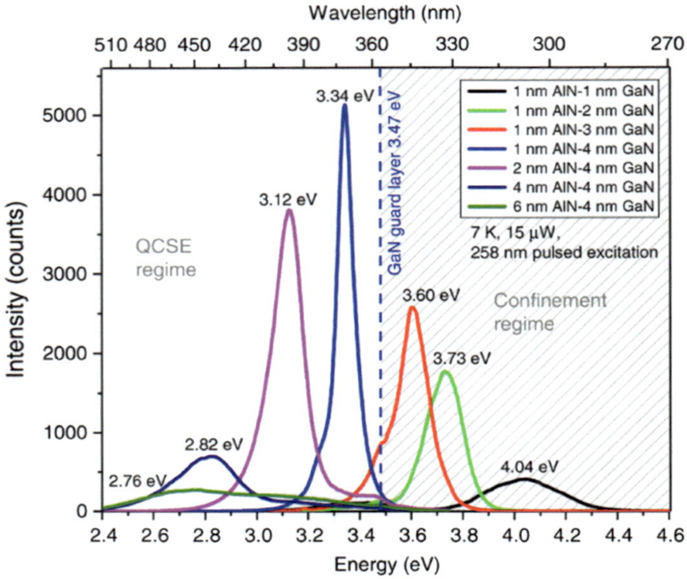

Fig. 5.49 Overview μ-PL spectra for nanowires with varying thickness of AlN barriers and GaN nanodiscs recorded at a sample temperature of 7 K. Reprinted from [154]. Copyright (2018) licensed under CC BY 4.0

correlates the emission energy of the two sample series together with their exciton confinement energies, electric field strength in the active region and measured recombination times.

Two main results are noteworthy: First, the electric field shielding with decreasing barrier thickness can be quantitatively expressed by a reduction of the remaining electric field from 2.61 MV/cm for the 6 nm thick barrier down to 0.38 MW/cm for the 1 nm thick barrier. This reduction of the electric field correlates with an increase of the PL recombination rate (reduction of decay time) by more than 2 orders of magnitude. Second, for a barrier thickness of 1 nm, the PL shift to larger energies with decreasing nanodisc thickness from 3 to 1 nm is exclusively governed by exciton confinement. The remaining electric field in the active region vanishes and the recombination time of 700 ps is independent of the nanodisc thickness. The discussed results show, that the IFGARD approach is capable of reducing the electric field inside the active region of polar nitride heterostructure and thus can be applied to increase emission intensity and recombination rates without sacrificing crystals quality or growth rate in non-polar or semi-polar growth directions [154].

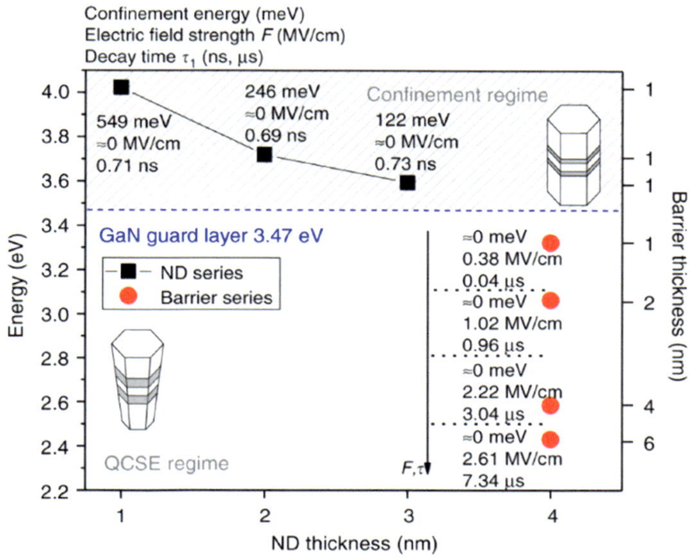

Fig. 5.50 Correlation between barrier and nanodisc (ND) thicknesses and ND emission energies (black squares and red dots indicate the confinement regime and QCSE regime, respectively). Quantitative values next to data points are the exciton confinement energies in meV, the unscreened electrical field strengths in MV/cm and the dominant decay times in ns/μs. Additionally, the geometric nanowire structure in the respective regimes is schematically illustrated. Reprinted from [154]. Copyright (2018) licensed under CC BY 4.0

5.4.5 Optical Properties and Charge Carrier Dynamics in 1D Quantum Wires

The previously discussed nanodisc heterostructures, despite being embedded in elongated microwires, still exhibit a 2D density of states due to the lateral extension of the microwires. In contrast, quantum wires (QWRs) stand out from the range of low dimensional structures as they combine superior carrier transport properties with a highly localized electronic density of states (1D) which even outperform 2D heterostructures [1, 2]. The combination of such 1D characteristics with the optical properties of group III-nitrides, e.g. an UV to infrared tunable direct band-gap and high exciton binding energies, lead to a strong perspective of nitride QWRs for the application in novel photonic devices. In order to reach the 1D-confinement regime, the diameter of the wire must be less than twice the bulk exciton-Bohr radius [155], which usually is in the order of a few nm.

So far, a variety of techniques exists for the realization of wires with diameters below 10 nm: (i) chemical etching [156], (ii) vapor-liquid-solid (VLS) growth [157–159], (iii) catalytic growth assisted by transition metals [160], (iv) template induced growth by carbon nanotube replacement [161] or (v) top-down selective area thermal sublimation [162]. However, none of them are able to meet the requirements like position control in combination with predefined orientation and the possibility of electrical contacting which all are fundamental prerequisites of a deterministic device production.

To overcome these obstacles, QWRs can be realized in an entirely self-assembled growth process by molecular beam epitaxy (MBE) on the side facets of GaN/AlN nanowire (NW) templates. Figure 5.51 shows the principle structure (a, b) and a scanning transmission electron microscopy (STEM) image (c, d) of the heterostructure in high angle annular dark-field (HAADF) contrast with GaN appearing in brighter

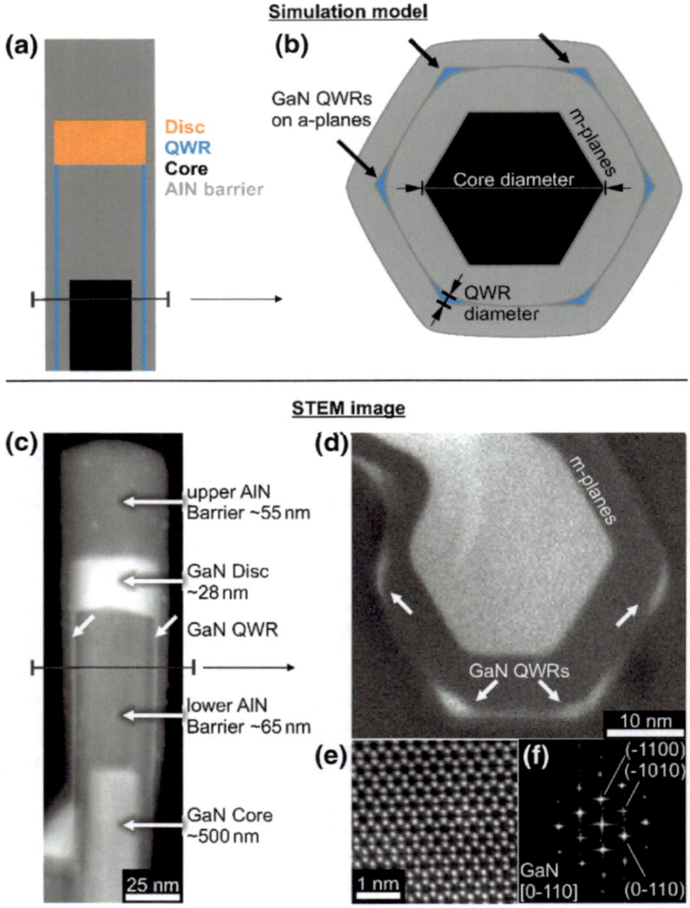

Fig. 5.51 **a** Schematic of the heterostructure for the numerical model and cross section (**b**) with embedded QWRs in the *a*-planes of the AlN inter-section. **c** STEM (HAADF) micrograph of the GaN/AlN NW hetero-structure on top of a single GaN NW plus cross section (**d**). The GaN QWRs (bright contrast) are visible laterally on the lower AlN barrier. **e**, **f** Atomic resolution STEM micrograph of a GaN core [0001] oriented similarly to the one shown in (**d**). Reprinted with permission from [163]. Copyright (2018) Royal Society of Chemistry

and AlN in darker contrast. Primarily, GaN NWs (core) with a length of 500 nm and a diameter of 25 nm serve as templates for an AlN/GaN/AlN heterostructure overgrowth. An AlN interlayer tends to soften the edges of the *m*-planes and lead to deviations from the strict hexagonal facetted shape [164], forming (11-20) facets (*a*-plane) at the six edges of the AlN barrier. In a subsequent growth step, GaN is primarily deposited on this *a*-planes, causing a self-limited QWR growth mechanism at the edges. In contrast to the GaN/AlN NWs, the QWRs posses a diameter, which is approximately one order of magnitude smaller with typical values between 2 and 3 nm (see HRTEM and schematic illustration in Fig. 5.51). Additionally to the lateral QWRs, an axial disc with a height of 28 nm is formed as shown in Fig. 5.51a, c.

Consequently, the different GaN sections provide several active regions leading to a complex emission spectrum. Time-integrated micro photoluminescence (μPL) and time-resolved micro photoluminescence (T-μPL) data in combination with numerical calculations allow the identification and assignment of the manifold emission features to the different spatial recombination centers within the NWs on a sub 30 nm scale [163]. It is shown that the high intense QWR luminescence is a result of an efficient charge carrier transfer between the different optically active regions, feeding the QWRs states with photo-excited charge carriers, which are generated in the larger GaN parts of the heterostructure.

Simulations predict extremely different internal electric fields in the different areas. While the field in the disc reaches the order of several MV/cm it nearly disappears within the QWRs and the main part of the core. Thus, the effective band gap in the disc region is lowered via the quantum confinement Stark-effect (QCSE). As the internal field is increasingly screened with an increased number of photo-excited charge carriers, the spectral position of the disc exhibits a pronounced dependence on the excitation power.

In Fig. 5.52 time-integrated and power dependent μPL emission spectra of two single NWs are shown. Each of the NWs contains six QWRs due to QWR formation at the smoothed edges of the GaN/AlN template. In addition to the up to six QWR related signals, both samples show two broadband emissions with different power dependence. The assignment of the signal origin to core, disc and QWR in Fig. 5.52 is carried out in anticipation of the results, which are presented below.

Both samples (Fig. 5.52a, b) exhibit emission lines at 3.47 eV, close to the GaN bulk bandedge. As the emission energy matches the signal of pure GaN NWs [164, 165] the features are attributed to the lower part of the GaN core which is not covered by an AlN shell during the growth process and consequently remains free of strain. Thereby, the absence of a power dependent emission energy, i.e. no QCSE, points to the lack of internal fields [163].

In addition to the GaN bandedge emission, another peak can be found in the range of 3.47–3.56 and 3.29–3.35 eV for sample A and B respectively, which shows a pronounced power dependence, shifting to higher energies for higher excitation powers. Here, charge carriers excited by photon absorption move to the interface charges of the opposite charge, thus screening the intrinsic internal fields and weakening the QCSE [166, 167]. As these internal fields are connected to the existence of interface charges on polar planes with moderate distances, the power dependent emission

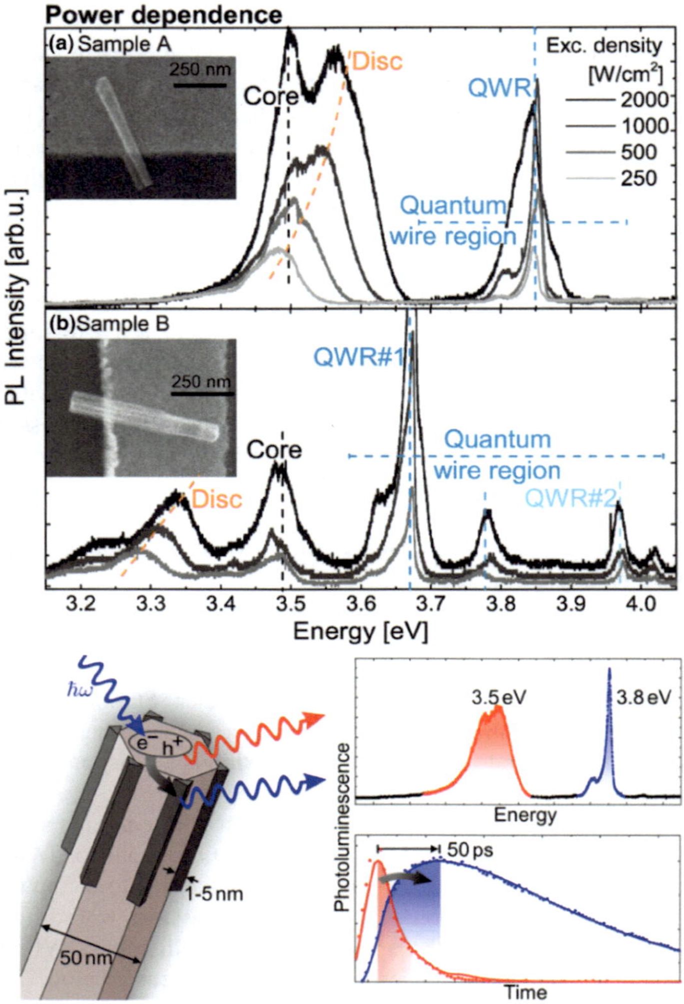

Fig. 5.52 Top: **a**, **b** Power-dependent single NW μPL spectra of sample A and B. The assignments of the spectral features to the core, the disc and the QWR are indicated. Insets show SEM images of the respective NWs. Bottom: Schematic illustration of the differences between spectral emission energy and decay dynamics for exciton recombination in the GaN nanowire (red) and quantum wire (blue). Reprinted with permission from [163]. Copyright (2018) Royal Society of Chemistry

peaks are determined by the geometric specifications of the disc. Besides the partially screened QCSE, the emission energy is determined via the interplay of confinement along the *c*-axis and compressive strain due to adjacent AlN layers. While the QCSE causes a redshift of the recombination energy with respect to the fundamental gap, confinement as well as strain lead to a counteracting blue-shift. Consequently, the actual recombination energy of the disc signal depends on properties like height and width of the disc as well as the thickness of the AlN barriers and thus, can be either blue-shifted (c.f. sample A) or redshifted (c.f. sample B) with respect to the bulk-like core emission.

Finally, the signals of QWRs can be found in the high energy range above 3.6 eV. Due to the 1D-confinement the recombination energy is strongly blue-shifted up to 4.6 eV, depending on the QWR height. Thereby, the recombination energy proofs to be sensitive to changes of the QWR height in the monolayer regime, for which reason the spectrum of even a single nanowire, which still contains 6 QWRS, may show several QWR emission features (Fig. 5.52b and [142]). In addition to height variations, also different localization centers can lead to the existence of a manifold QWR spectrum in accordance to observations in nanodiscs [168].

Another striking property of the QWRs is the unexpected high emission intensity, which in most cases is in the same order of magnitude as the disc and core signals. Considering the different volumes of the mentioned structures, the rate of absorbed photons in the QWR region should be significantly lower than in the much larger disc and core parts. However, a similar intensity indicates that photo-excited carriers, created in the disc or core, are transferred into the QWRs [163].

To find experimental evidence, the population dynamics of the different regions are investigated using time resolved μPL (TR-μPL). The data of both samples are presented in Fig. 5.53a, b. While the core and disc time evolution can be fitted using a biexponential decay, an additional rise time is observed in the case of the QWR signal. This rise time corresponds to the fast decay times of disc and core, substantiating the assumption of a transfer of charge carriers into the initially empty QWR states. Thus, the rather fast decay of the core and disc population within the first nanosecond, points to the loss of charge carriers due to transfer processes, dominating the optical decay. After this fast decay when the QWR states are occupied, both reservoirs are emptied much slower, representing the domination of optical losses. After the transfer losses also the screening of the internal fields is reduced and thus the remaining charge carriers are spatially separated, leading to only small overlaps of the electron and hole wavefunctions. This picture is supported by the findings of power dependent decay times for the disc emission which decrease with increasing excitation power (Fig. 5.53c).

In summary, data of μPL and TR-μPL measurements complemented by simulations of electronic properties, reveal that the optical processes are dominated by a complex carrier transfer process with striking efficiency. Thereby, the QWRs are fed via the large scaled reservoir of the GaN-core and -disc. This feeding process represents the key feature of the structure as it enables access to wires with diameters of just a few mono-layers via a macroscopic NW. Consequently, all well-established tools of processing like electrical contacting and position control can be used to integrate QWRs with their unique properties into electrical or photonic devices of a new generation.

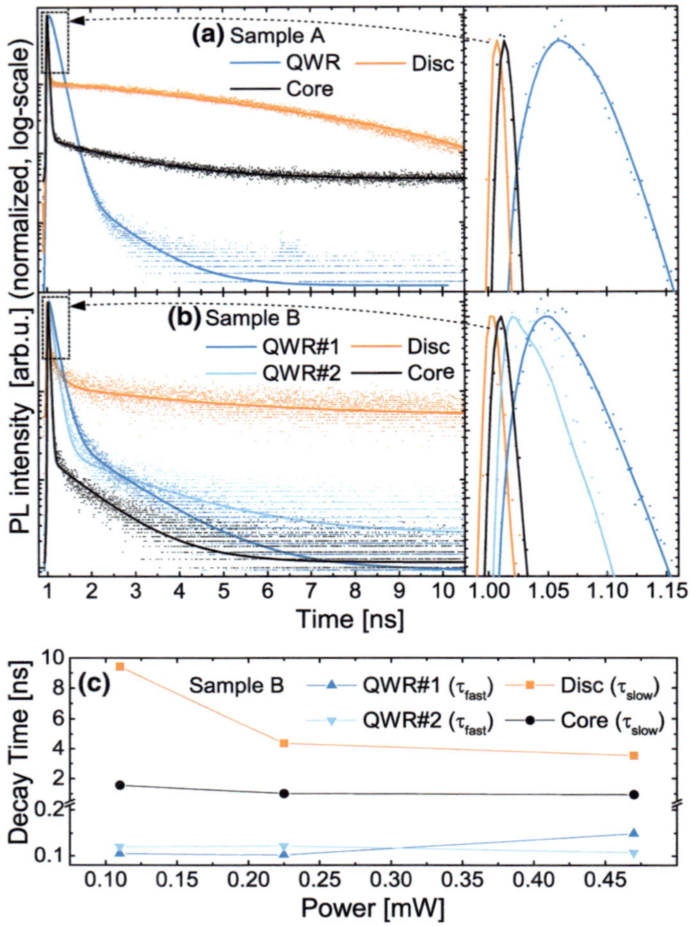

Fig. 5.53 **a** and **b** Time resolved μPL decay curves of the different emission features of the same single NWs as in Fig. 5.52a, b. Right: Magnified time scale to illustrate the initial QWR rise time within the first 150 ps. All transients were acquired at 2000 W cm^{-2} excitation power. **c** Decay times of the different emission features of sample B for different excitation powers. Reprinted with permission from [163]. Copyright (2018) Royal Society of Chemistry

5.5 Conclusion and Perspectives

In the first part of this chapter, we have introduced various state-of-the-art nanoscale characterization techniques including (scanning) transmission electron microscopy—cathodoluminescence down to liquid helium temperatures, tip enhanced Raman spectroscopy, UV enhanced continuous wave and pulsed micro-photoluminescence spectroscopy, X-ray diffraction and scanning tunneling microscopy and spectroscopy (Sect. 5.2). In combination with each other, these techniques provide unique insights into a multitude of nanostructure properties such as charge carrier excitation, relaxation, diffusion, and recombination dynamics, vibrational and structural information including strain, segregation, and clustering, and surface and interface morphology.

We have demonstrated the capabilities of these techniques on the examples of several nitride based nanostructures. In particular, we have focused on nitride based micro-and nanocolumns with and without embedded quantum wells, nanorods, and quantum wires, whereas nitride quantum dots and their quantum optical characterization are discussed in Chap. 12.

In the second part, we have discussed several approaches how to overcome well known limitations in the emission efficiency and spectral range of the emission energy in nitride nanostructures such as the formation of stacking faults in heteroepitaxial growth and internal electric fields in the polar growth conditions of nitride nanostructures. These include quantum wells grown along the nonpolar facets of GaN microcolumns (Sects. 5.4.1 and 5.4.3), the shielding of electrostatic fields by guard layers in the polar growth direction (Sect. 5.4.4), and the growth of quantum wires on the non-polar facts of nitride nanowires (Sect. 5.4.5). The presented results demonstrate the capabilities of modern nanoscale spectroscopy and microscopy techniques for the fundamental understanding of structural, electronic, and optical properties of nanostructures and can contribute to overcome current limitations in group III-nitride nanophotonic devices.

Acknowledgements We gratefully acknowledge the German Research Foundation (DFG) for financial support within the Research Instrumentation Program INST 272/148-1, the Collaborative Research Center SFB 787 "Semiconductor Nanophotonics: Materials, Models, Devices".

Many thanks to Silke Petzold (University of Magdeburg) for her work regarding the sample preparation.

References

1. Y. Arakawa, H. Sakaki, Appl. Phys. Lett. **40**, 939 (1982)
2. H. Sakaki, Jpn. J. Appl. Phys. **19**, L735 (1980)
3. Y. Arakawa, A. Yariv, IEEE J. Quantum Electron. **QE-22**, 1887 (1986)
4. A. Yariv, Appl. Phys. Lett. **53**, 1033 (1988)
5. S.J. Pennycook, A. Howie, Philos. Mag. A Phys. Condens. Matter Struct. Defects Mech. Prop. **41**(6), 809 (1980)
6. P.M. Petroff, R.A. Logan, A. Savage, J. Microsc. **118**(3), 255 (1980)
7. S.H. Roberts, J.W. Steeds, J. Cryst. Growth **59**(1–2), 312 (1982)
8. N. Yamamoto, J.C.H. Spence, D. Fathy, Philos. Mag. B Phys. Condens. Matter Stat. Mech. Electron. Opt. Magn. Prop. **49**(6), 609–629 (1984)
9. T. Mitsui, N. Yamamoto, K. Takemoto, O. Nittono, Jpn. J. Appl. Phys. **33**(3), L342 (1994)
10. X.H. Wu, C.R. Elsass, A. Abare, M. MacK, S. Keller, P.M. Petroff, S.P. Denbaars, J.S. Speck, S.J. Rosner, Appl. Phys. Lett. **72**(6), 692 (1998)
11. R. Gómez-Medina, N. Yamamoto, M. Nakano, F.J.G. De Abajo, New J. Phys. **10**, 105009 (2008)

12. S.K. Lim, M. Brewster, F. Qian, Y. Li, C.M. Lieber, S. Gradečak, Nano Lett. **9**(11), 3940 (2009)
13. L.F. Zagonel, S. Mazzucco, M. Tence, K. March, R. Bernard, B. Laslier, G. Jacopin, M. Tchernycheva, L. Rigutti, F.H. Julien, R. Songmuang, M. Kociak, Nano Lett. **11**(2), 568 (2011)
14. J.T. Griffiths, S. Zhang, B. Rouet-Leduc, W.Y. Fu, A. Bao, D. Zhu, D.J. Wallis, A. Howkins, I. Boyd, D. Stowe, M.J. Kappers, C.J. Humphreys, R.A. Oliver, Nano Lett. **15**(11), 7639 (2015)
15. M. Kociak, L.F. Zagonel, Ultramicroscopy **176**, 112 (2017)
16. H.P. Strunk, M. Albrecht, H. Scheel, J. Microsc. **224**(1), 79 (2006)
17. J.I. Deitz, A.T.M.G. Sarwar, S.D. Carnevale, T.J. Grassman, R.C. Myers, D.W. McComb, Microsc. Microanal. **24**(2), 93 (2018)
18. M. von Ardenne, Das Elektronen-Rastermikroskop. Theoretische Grundlagen. Z. Phys. **109**(9–10), 553–572 (1938)
19. M. von Ardenne, Das Elektronen-Rastermikroskop. Praktische Ausführung. Z. Tech. Phys. **19**, 407–416 (1938)
20. G. Schmidt, Optische Nanocharakterisierung GaN-basierter Quantenstrukturen für Mikrokavitäten. Doctoral thesis, Otto-von-Guericke-University Magdeburg, Magdeburg, 2017
21. J.I. Goldstein, J.L. Costley, G.W. Lorimer, R.J.B. Reed, Scan. Electron Microsc. **1**, 315 (1977)
22. L. Reimer, H. Kohl, *Transmission Electron Microscopy: Physics of Image Formation*. 36 of Springer Series in Optical Sciences, 5th edn. (Springer, New York, 2008)
23. D.B. Williams, C.B. Carter, J.C.H. Spence, *Transmission Electron Microscopy: A Textbook for Materials Science*, 2nd edn. (Springer, New York, 2009)
24. S.J. Pennycook, P.D. Nellist, *Scanning Transmission Electron Microscopy: Imaging and Analysis* (Springer, New York, 2011)
25. J. Cowley, Y. Huang, Ultramicroscopy **40**(2), 171 (1992)
26. M. Noltemeyer, F. Bertram, T. Hempel, B. Bastek, J. Christen, M. Brandt, M. Lorenz, M. Grundmann, F.H. Teherani, D.C. Look, D.J. Rogers, in *SPIE OPTO*, *SPIE Proceedings* (SPIE, 2012), p. 82630X
27. A.N. Polyakov, M. Noltemeyer, T. Hempel, J. Christen, M.A. Stepovich, Bull. Russ. Acad. Sci. Phys. **76**(9), 970 (2012)
28. A.N. Polyakov, M. Noltemeyer, T. Hempel, J. Christen, M.A. Stepovich, J. Surf. Invest. X-ray Synchrotron Neutron Tech. **6**(6), 901 (2012)
29. M. Noltemeyer, Dissertation, Otto-von-Guericke-Universität Magdeburg, Magdeburg, 2016
30. T. Malis, S.C. Cheng, R.F. Egerton, J. Electron Microsc. Tech. **8**(2), 193–200 (1988)
31. T. Kobayashi, T. Sugita, M. Koyama, S.-I. Takayanagi, IEEE Trans. Nucl. Sci. **19**(3), 324–333 (1972)
32. R. Egerton, *Electron Energy-Loss Spectroscopy in the Electron Microscope* (Springer, US, Boston and MA, 2011)
33. P. Perlin et al., Investigation of longitudinal-optical phonon-plasmon coupled modes in highly conducting bulk GaN. Appl. Phys. Lett. **67**(17), 2524–2526 (1995)
34. V.Y. Davydov et al., Phonon dispersion and Raman scattering in hexagonal GaN and AlN. Phys. Rev. B **58**(19), 12899–12907 (1998)
35. H. Harima, Properties of GaN and related compounds studied by means of Raman scattering. J. Phys. Condens. Matter **14**(38), R967–R993 (2002)
36. R. Kirste, S. Mohn, M.R. Wagner, J.S. Reparaz, A. Hoffmann, Phonon plasmon interaction in ternary group-III-nitrides. Appl. Phys. Lett. **101**(4), 041909 (2012)
37. R. Kirste et al., Compensation effects in GaN: Mg probed by Raman spectroscopy and photoluminescence measurements. J. Appl. Phys. **113**(10), 103504 (2013)
38. J.S. Reparaz et al., A novel contactless technique for thermal field mapping and thermal conductivity determination: two-laser Raman thermometry. Rev. Sci. Instrum. **85**(3), 034901 (2014)
39. Z. Zhang, S. Sheng, R. Wang, M. Sun, Tip-enhanced Raman spectroscopy. Anal. Chem. **88**(19), 9328–9346 (2016)

40. E. Poliani et al., Breakdown of far-field raman selection rules by light-plasmon coupling demonstrated by tip-enhanced raman scattering. J. Phys. Chem. Lett. **8**(22), 5462–5471 (2017)
41. R.V. Maximiano, R. Beams, L. Novotny, A. Jorio, L.G. Cançado, Mechanism of near-field Raman enhancement in two-dimensional systems. Phys. Rev. B Condens. Matter Mater. Phys. **85**(23), 235434 (2012)
42. A. Hartschuh, Tip-enhanced near-field optical microscopy, in *Handbook of Spectroscopy*, 2nd enlarged edn., vol 4, no 43 (2014), pp. 1585–1610
43. M.K. Schmidt, R. Esteban, A. González-Tudela, G. Giedke, J. Aizpurua, Quantum mechanical description of Raman scattering from molecules in plasmonic cavities. ACS Nano **10**(6), 6291–6298 (2016)
44. N.S. Mueller, S. Heeg, S. Reich, Surface-enhanced Raman scattering as a higher-order Raman process. Phys. Rev. A **94**(2), 023813 (2016)
45. Y. Saito, M. Motohashi, N. Hayazawa, S. Kawata, Stress imagining of semiconductor surface by tip-enhanced Raman spectroscopy. J. Microsc. **229**(2), 217–222 (2008)
46. N. Lee et al., High contrast scanning nano-Raman spectroscopy of silicon. J. Raman Spectrosc. **38**(6), 789–796 (2007)
47. R. Matsui, P. Verma, T. Ichimura, Y. Inouye, S. Kawata, Nanoanalysis of crystalline properties of GaN thin film using tip-enhanced Raman spectroscopy. Appl. Phys. Lett. **90**(6), 061906 (2007)
48. S. Berweger, C.C. Neacsu, Y. Mao, H. Zhou, S.S. Wong, M.B. Raschke, Optical nanocrystallography with tip-enhanced phonon Raman spectroscopy. Nat. Nanotechnol. **4**(8), 496–499 (2009)
49. J. Chen et al., Probing strain in bent semiconductor nanowires with raman spectroscopy. Nano Lett. **10**(4), 1280–1286 (2010)
50. N. Marquestaut, D. Talaga, L. Servant, P. Yang, P. Pauzauskie, F. Lagugné-Labarthet, Imaging of single GaN nanowires by tip-enhanced Raman spectroscopy. J. Raman Spectrosc. **40**(10), 1441–1445 (2009)
51. P.G. Gucciardi, J.C. Valmalette, Different longitudinal optical-transverse optical mode amplification in tip enhanced Raman spectroscopy of GaAs(001). Appl. Phys. Lett. **97**(26), 263104 (2010)
52. Y. Ogawa, Y. Yuasa, F. Minami, S. Oda, Tip-enhanced Raman mapping of a single Ge nanowire. Appl. Phys. Lett. **99**(5), 2–5 (2011)
53. Y. Ogawa, T. Toizumi, F. Minami, A.V. Baranov, Nanometer-scale mapping of the strain and Ge content of Ge/Si quantum dots using enhanced Raman scattering by the tip of an atomic force microscope. Phys. Rev. B Condens. Matter Mater. Phys. **83**(8), 081302 (2011)
54. J.S. Reparaz et al., Probing local strain and composition in Ge nanowires by means of tip-enhanced Raman scattering. Nanotechnology **24**(18), 185704 (2013)
55. E. Poliani et al., Nanoscale imaging of InN segregation and polymorphism in single vertically aligned InGaN/GaN multi quantum well nanorods by tip-enhanced Raman scattering. Nano Lett. **13**(7), 3205–3212 (2013)
56. H. Rohrer, G. Binnig, Helv. Phys. Acta **55**(6), 726 (1982)
57. G. Binnig, H. Rohrer, Ch. Gerber, E. Weibel, Phys. Rev. Lett. **49**, 57 (1982)
58. A.R. Smith, R.M. Feenstra, D.W. Greve, M.-S. Shin, M. Skowronski, J. Neugebauer, J.E. Northrup, J. Vac. Sci. Technol. B **16**(4), 2242 (1998)
59. R. Held, G. Nowak, B.E. Ishaug, S.M. Seutter, A. Parkhomovsky, A.M. Dabrian, P.I. Cohen, I. Grzegory, S. Porowski, J. Appl. Phys. **85**, 7697 (1999)
60. R.M. Feenstra, P. Mårtensson, Phys. Rev. Lett. **61**(4), 447 (1988)
61. R.M. Feenstra, Semicond. Sci. Technol. **9**(12), 2157 (1994)
62. H. Eisele, O. Flebbe, T. Kalka, C. Preinesberger, F. Heinrichsdorff, A. Krost, D. Bimberg, M. Dähne-Prietsch, Appl. Phys. Lett. **75**(1), 106 (1999)
63. Ch. Schulz, Th. Schmidt, J.I. Flege, N. Berner, Ch. Tessarek, D. Hommel, J. Falta, Phys. Stat. Sol. C **6**(2), 305 (2009)
64. Ch. Schulz, S. Kuhr, H. Geffers, Th. Schmidt, J.I. Flege, T. Aschenbrenner, D. Hommel, J. Falta, J. Vac. Sci. Technol. A **29**(1), 11013 (2011)

65. L.F.J. Piper, T.D. Veal, M. Walker, I. Mahboob, C.F. McConville, H. Lu, W.J. Schaff, J. Vac. Sci. Technol. A **23**(4), 617 (2005)
66. T. Ohashi, Y. Saito, T. Maruyama, Y. Nanishi, J. Cryst. Growth **237–239**, 1022 (2002)
67. H. Eisele, Ph. Ebert, Phys. Stat. Sol. RRL **6**(9–10), 359 (2012)
68. S. Zhao, S. Fathololoumi, K.H. Bevan, D.P. Liu, M.G. Kibria, Q. Li, G.T. Wang, H. Guo, Z. Mi, Nano Lett. **12**(6), 2877 (2012)
69. D. Krüger, S. Kuhr, T. Schmidt, D. Hommel, J. Falta, Phys. Stat. Sol. RRL **3**(4), 91 (2009)
70. L. Ivanova, S. Borisova, H. Eisele, M. Dähne, A. Laubsch, Ph. Ebert, Appl. Phys. Lett. **93**(19), 192110 (2008)
71. Ph. Ebert, L. Ivanova, S. Borisova, H. Eisele, A. Laubsch, M. Dähne, Appl. Phys. Lett. **94**, 062104 (2009)
72. H. Eisele, J. Schuppang, M. Schnedler, M. Duchamp, C. Nenstiel, V. Portz, T. Kure, M. Bügler, A. Lenz, M. Dähne, A. Hoffmann, S. Gwo, S. Choi, J.S. Speck, R.E. Dunin-Borkowski, Ph. Ebert, Phys. Rev. B **94**(24), 245201 (2016)
73. A. Dadgar, A. Strittmatter, J. Bläsing, M. Poschenrieder, O. Contreras, P. Veit, T. Riemann, F. Bertram, A. Reiher, A. Krtschil, A. Diez, T. Hempel, T. Finger, A. Kasic, M. Schubert, D. Bimberg, F.A. Ponce, J. Christen, A. Krost, Phys. Stat. Sol. C **0**(6), 1583 (2003)
74. Ph. Ebert, S. Schaafhausen, A. Lenz, A. Sabitova, L. Ivanova, M. Dähne, Y.-L. Hong, S. Gwo, H. Eisele, Appl. Phys. Lett. **98**(6), 062103 (2011)
75. L. Lymperakis, P.H. Weidlich, H. Eisele, M. Schnedler, J.-P. Nys, B. Grandidier, D. Stiévenard, R.E. Dunin-Borkowski, J. Neugebauer, Ph. Ebert, Appl. Phys. Lett. **103**(15), 152101 (2013)
76. M. Franz, S. Appelfeller, H. Eisele, Ph. Ebert, M. Dähne, Phys. Rev. B **99**, 195306 (2019)
77. M. Schnedler, V. Portz, H. Eisele, R.E. Dunin-Borkowski, Ph. Ebert, Phys. Rev. B **91**(20), 205309 (2015)
78. M. Yoshizawa, A. Kikuchi, M. Mori, N. Fujita, K. Kishino, Jpn. J. Appl. Phys. **36**(4B), L459 (1997)
79. M. Sanchez-Garcia, E. Calleja, E. Monroy, F. Sanchez, F. Calle, E. Munoz, R. Beresford, J. Cryst. Growth **183**(1), 23 (1998)
80. D. Zubia, S.D. Hersee, J. Appl. Phys. **85**(9), 6492 (1999)
81. E. Calleja, M.A. Sanchez-Garca, F.J. Sanchez, F. Calle, F.B. Naranjo, E. Munoz, U. Jahn, K. Ploog, Phys. Rev. B **62**, 16826 (2000)
82. S.D. Hersee, X. Sun, X. Wang, Nano Lett. **6**(8), 1808 (2006)
83. C. Nenstiel, M. Bügler, G. Callsen, F. Nippert, T. Kure, S. Fritze, A. Dadgar, H. Witte, J. Bläsing, A. Krost, A. Hoffmann, Phys. Stat. Sol. RRL **9**(12), 716 (2015)
84. R.S. Wagner, W.C. Ellis, Appl. Phys. Lett. **4**(5), 89 (1964)
85. T. Kuykendall, P. Pauzauskie, S. Lee, Y. Zhang, J. Goldberger, P. Yang, Nano Lett. **3**(8), 1063 (2003)
86. Q. Li, G.T. Wang, Appl. Phys. Lett. **93**(4), 043119 (2008)
87. B. Liu, Y. Bando, C. Tang, F. Xu, D. Golberg, Appl. Phys. Lett. **87**(7), 073106 (2005)
88. M. Müller, Ph.D. thesis, Otto-von-Guericke-University Magdeburg, Germany, 2018
89. R. Koester, J.S. Hwang, C. Durand, D.L.S. Dang, J. Eymery, Nanotechnology **21**(1), 015602 (2010)
90. K. Kishino, A. Kikuchi, H. Sekiguchi, S. Ishizawa, in *SPIE Conference Proceedings*, vol 6473 (2007), p. 64730T
91. A.-L. Bavencove, G. Tourbot, J. Garcia, Y. Desieres, P. Gilet, F. Levy, B. Andre, B. Gayral, B. Daudin, L.S. Dang, Nanotechnology **22**(34), 345705 (2011)
92. S. Krylyuk, D. Paramanik, M. King, A. Motayed, J.-Y. Ha, J.E. Bonevich, A. Talin, A.V. Davydov, Appl. Phys. Lett. **101**(24), 241119 (2012)
93. P. Shields, M. Hugues, J. Zuniga-Perez, M. Cooke, M. Dineen, W. Wang, F. Causa, D. Allsopp, Phys. Stat. Sol. (c) **9**(3–4), 631 (2012)
94. C.-Y. Wang, L.-Y. Chen, C.-P. Chen, Y.-W. Cheng, M.-Y. Ke, M.-Y. Hsieh, H.-M. Wu, L.-H. Peng, J. Huang, Opt. Express **16**(14), 10549 (2008)
95. T. Schimpke, M. Mandl, I. Stoll, B. Pohl-Klein, D. Bichler, F. Zwaschka, J. Strube-Knyrim, B. Huckenbeck, B. Max, M. Müller, P. Veit, F. Bertram, J. Christen, J. Hartmann, A. Waag, H.-J. Lugauer, M. Strassburg, Phys. Stat. Sol. (a) **213**(6), 1577 (2016)

96. K. Kishino, H. Sekiguchi, A. Kikuchi, J. Cryst. Growth **311**(7), 2063 (2009)
97. A. Urban, J. Malindretos, J.-H. Klein-Wiele, P. Simon, A. Rizzi, New J. Phys. **15**(5), 053045 (2013)
98. T. Eriksson, K.-D. Lee, B. Heidari, P. Rode, W. Bergbauer, M. Mandl, C. Kolper, M. Strassburg, in *SPIE Conference Proceedings*, vol 7970 (2011), p. 797015
99. C.-H. Liao, W.-M. Chang, H.-S. Chen, C.-Y. Chen, Y.-F. Yao, H.-T. Chen, C.-Y. Su, S.-Y. Ting, Y.-W. Kiang, C.C. Yang, Opt. Express **20**(14), 15859 (2012)
100. B. Cord, J. Yang, H. Duan, D.C. Joy, J. Klingfus, K.K. Berggren, J. Vac. Sci. Technol. B **27**(6), 2616 (2009)
101. A.E. Grigorescu, C.W. Hagen, Nanotechnology **20**(29), 292001 (2009)
102. T. Eriksson, S. Yamada, P.V. Krishnan, S. Ramasamy, B. Heidari, Microelectron. Eng. **88**, 293 (2011)
103. K. Kishino, S. Sekiguchi, A. Kikuchi, J. Cryst. Growth **311**, 2063 (2009)
104. H. Sekiguchi, K. Kishino, A. Kikuchi, Appl. Phys. Express **1**, 124002 (2008)
105. S. Li, A. Waag, J. Appl. Phys. **111**(7), 071101 (2012)
106. M. Mandl, X. Wang, T. Schimpke, C. Kölper, M. Binder, J. Ledig, A. Waag, X. Kong, A. Trampert, F. Bertram, J. Christen, F. Barbagini, E. Calleja, M. Strassburg, Phys. Stat. Sol. Rapid Res. Lett. **7**(10), 800 (2013)
107. S. Albert, A.M. Bengoechea-Encabo, F. Barbagini, D. Lopez-Rormero, M.A. Sanchez-Garcia, E. Calleja, P. Lefebvre, X. Kong, U. Jahn, A. Trampert, M. Müller, F. Bertram, G. Schmidt, P. Veit, S. Petzold, J. Christen, P. De Mierry, J. Zuniga-Perez, Int. J. High Speed Electron. Syst. **23**(03), 1450020 (2014)
108. A. Laubsch, M. Sabathil, W. Bergbauer, M. Strassburg, H. Lugauer, M. Peter, S. Lutgen, N. Linder, K. Streubel, J. Hader, J.V. Moloney, B. Pasenow, S.W. Koch, Phys. Stat. Sol. (c) **6**(S2), 913 (2009)
109. E. Kioupakis, P. Rinke, K.T. Delaney, C.G.V. deWalle, Appl. Phys. Lett. **98**(16), 161107 (2011)
110. M. Binder, A. Nirschl, R. Zeisel, T. Hager, H.-J. Lugauer, M. Sabathil, D. Bougeard, J. Wagner, B. Galler, Appl. Phys. Lett. **103**(7), 071108 (2013)
111. J. Iveland, L. Martinelli, J. Peretti, J.S. Speck, C. Weisbuch, Phys. Rev. Lett. **110**, 177406 (2013)
112. M.S. Mohajerani, S. Khachadorian, T. Schimpke, C. Nenstiel, J. Hartmann, J. Ledig, A. Avramescu, M. Strassburg, A. Homann, A. Waag, Appl. Phys. Lett. **108**(9), 091112 (2016)
113. M. Müller, P. Veit, F.F. Krause, T. Schimpke, S. Metzner, F. Bertram, T. Mehrtens, K. Muller-Caspary, A. Avramescu, M. Strassburg, A. Rosenauer, J. Christen, Nano Lett. **16**(9), 5340 (2016)
114. S. Keller, B.P. Keller, D. Kapolnek, A.C. Abare, H. Masui, L.A. Coldren, U.K. Mishra, S.P.D. Baars, Appl. Phys. Lett. **68**(22), 3147 (1996)
115. T. Schimpke, A. Avramescu, A. Koller, A. Fernando-Saavedra, J. Hartmann, J. Ledig, A. Waag, M. Strassburg, H.-J. Lugauer, J. Cryst. Growth **465**, 34 (2017)
116. H. Fang, Z.J. Yang, Y. Wang, T. Dai, L.W. Sang, L.B. Zhao, T.J. Yu, G.Y. Zhang, J. Appl. Phys. **103**(1), 014908 (2008)
117. M. Gibbon, J.P. Stagg, C.G. Cureton, E.J. Thrush, C.J. Jones, R.E. Mallard, R.E. Pritchard, N. Collis, A. Chew, Semicond. Sci. Technol. **8**(6), 998 (1993)
118. T. Wunderer, M. Feneberg, F. Lipski, J. Wang, R.A.R. Leute, S. Schwaiger, K. Thonke, A. Chuvilin, U. Kaiser, S. Metzner, F. Bertram, J. Christen, G.J. Beirne, M. Jetter, P. Michler, L. Schade, C. Vierheilig, U.T. Schwarz, A.D. Drager, A. Hangleiter, F. Scholz, Phys. Stat. Sol. (b) **248**(3), 549 (2011)
119. C. Mounir, T. Schimpke, G. Rossbach, A. Avramescu, M. Strassburg, U.T. Schwarz, J. Appl. Phys. **120**(15), 155702 (2016)
120. T. Nobis, M. Grundmann, Phys. Rev. A **72**, 063806 (2005)
121. S. Chichibu, K. Wada, S. Nakamura, Appl. Phys. Lett. **71**(16), 2346 (1997)
122. S. Chichibu, T. Azuhata, T. Sota, S. Nakamura, Appl. Phys. Lett. **69**(27), 4188 (1996)

123. D.M. Graham, A. Soltani-Vala, P. Dawson, M.J. Godfrey, T.M. Smeeton, J.S. Barnard, M.J. Kappers, C.J. Humphreys, E.J. Thrush, J. Appl. Phys. **97**(10), 103508 (2005)
124. M.J. Galtrey, R.A. Oliver, M.J. Kappers, C.J. Humphreys, P.H. Clifton, D. Larson, D.W. Saxey, A. Cerezo, J. Appl. Phys. **104**(1), 013524 (2008)
125. D. Watson-Parris, M.J. Godfrey, P. Dawson, R.A. Oliver, M.J. Galtrey, M.J. Kappers, C.J. Humphreys, Phys. Rev. B **83**, 115321 (2011)
126. S. Schulz, D.P. Tanner, E.P. O'Reilly, M.A. Caro, T.L. Martin, P.A.J. Bagot, M.P. Moody, F. Tang, J.T. Griths, F. Oehler, M.J. Kappers, R.A. Oliver, C.J. Humphreys, D. Sutherland, M.J. Davies, P. Dawson, Phys. Rev. B **92**, 235419 (2015)
127. L. Bellaiche, T. Mattila, L.-W. Wang, S.-H. Wei, A. Zunger, Appl. Phys. Lett. **74**(13), 1842 (1999)
128. L.-W. Wang, Phys. Rev. **63**, 245107 (2001)
129. S.F. Chichibu, A. Uedono, T. Onuma, B.A. Haskell, A. Chakraborty, T. Koyama, P.T. Fini, S. Keller, S.P. DenBaars, J.S. Speck, U.K. Mishra, S. Nakamura, S. Yamaguchi, S. Kamiyama, H. Amano, I. Akasaki, J. Han, T. Sota, Nat. Mater. **5**(10), 810 (2006)
130. Y.-R. Wu, R. Shivaraman, K.-C. Wang, J.S. Speck, Appl. Phys. Lett. **101**(8), 083505 (2012)
131. C. Humphreys, J. Griths, F. Tang, F. Oehler, S. Findlay, C. Zheng, J. Etheridge, T. Martin, P. Bagot, M. Moody, D. Sutherland, P. Dawson, S. Schulz, S. Zhang, W. Fu, T. Zhu, M. Kappers, R. Oliver, Ultramicroscopy **176**, 93 (2017)
132. T.M. Smeeton, M.J. Kappers, J.S. Barnard, M.E. Vickers, C.J. Humphreys, Appl. Phys. Lett. **83**(26), 5419 (2003)
133. T. Li, E. Hahn, D. Gerthsen, A. Rosenauer, A. Strittmatter, L. Reißmann, D. Bimberg, Appl. Phys. Lett. **86**(24), 241911 (2005)
134. A. Rosenauer, T. Mehrtens, K. Müller, K. Gries, M. Schowalter, P.V. Satyam, S. Bley, C. Tessarek, D. Hommel, K. Sebald, M. Seyfried, J. Gutowski, A. Avramescu, K. Engl, S. Lutgen, Ultramicroscopy **111**(8), 1316 (2011)
135. F.F. Krause, J.-P. Ahl, D. Tytko, P.-P. Choi, R. Egoavil, M. Schowalter, T. Mehrtens, K. Müller-Caspary, J. Verbeeck, D. Raabe, J. Hertkorn, K. Engl, A. Rosenauer, Ultramicroscopy **156**, 29 (2015)
136. T. Bartel, C. Kisielowski, Ultramicroscopy **108**(11), 1420 (2008)
137. H. Schömig, S. Halm, A. Forchel, G. Bacher, J. Off, F. Scholz, Phys. Rev. B **92**, 106802 (2004)
138. A. Bell, J. Christen, F. Bertram, F.A. Ponce, H. Marui, S. Tanaka, Appl. Phys. Lett. **84**(1) (2004)
139. S. Marcinkevicius, K.M. Kelchner, S. Nakamura, S.P. DenBaars, J.S. Speck, Phys. Stat. Sol. (c) **11**(3–4), 690 (2014)
140. Y. Narukawa, Y. Kawakami, M. Funato, S. Fujita, S. Fujita, S. Nakamura, Appl. Phys. Lett. **70**(8), 981 (1997)
141. G. Schmidt, M. Müller, P. Veit, S. Metzner, F. Bertram, J. Hartmann, H. Zhou, H.-H. Wehmann, A. Waag, J. Christen, Sci. Rep. **8**, 16026 (2018)
142. J. Arbiol et al., Nanoscale **4**, 7517 (2012)
143. J.T. Griffiths et al., Appl. Phys. Lett. **110**, 172105 (2017)
144. I. Griffths, D. Cherns, X. Wang, H.-H. Wehman, M. Mandl, M. Strassburg, A. Waag, Phys. Stat. Sol. (c) **11**(3–4), 425 (2014)
145. S.-Y. Bae, K. Lekhal, H.-J. Lee, J.-W. Min, D.-S. Lee, Y. Honda, H. Amano, Phys. Stat. Sol. (b), 1600722 (2017)
146. C. Tessarek, M. Heilmann, E. Butzen, A. Haab, H. Hardtdegen, C. Dieker, E. Spiecker, S. Christiansen, Cryst. Growth Des. **14**(3), 1486 (2014)
147. J. Hartmann, X. Wang, H. Schuhmann, W. Dziony, L. Caccamo, J. Ledig, M.S. Mohajerani, T. Schimpke, M. Bähr, G. Lilienkamp, W. Daum, M. Seibt, M. Straburg, H.-H. Wehmann, A. Waag, Phys. Stat. Sol. (a) **212**(12), 2830 (2015)
148. A.M. Fischer, Z. Wu, K. Sun, Q. Wei, Y. Huang, R. Senda, D. Iida, M. Iwaya, H. Amano, F.A. Ponce, Appl. Phys. Express **2**(4), 041002 (2009)

149. F. Wu, Y.-D. Lin, A. Chakraborty, H. Ohta, S.P. DenBaars, S. Nakamura, J.S. Speck, Appl. Phys. Lett. **96**(23), 231912 (2010)
150. V. Fiorentini, F. Bernardini, F. Della Sala, A. Di Carlo, P. Lugli, Effects of macroscopic polarization in III-V nitride multiple quantum wells. Phys. Rev. B **60**(12), 8849–8858 (1999)
151. N. Grandjean, B. Damilano, S. Dalmasso, M. Leroux, M. Laügt, J. Massies, Built-in electric-field effects in wurtzite AlGaN/GaN quantum wells. J. Appl. Phys. **86**(7), 3714 (1999)
152. M. Leroux et al., Quantum confined Stark effect due to built-in internal polarization fields in (Al, Ga)N/GaN quantum wells. Phys. Rev. B **58**(20), R13371–R13374 (1998)
153. G.M.O. Hönig, S. Westerkamp, A. Hoffmann, G. Callsen, Shielding electrostatic fields in polar semiconductor nanostructures. Phys. Rev. Appl. **7**(2), 024004 (2017)
154. S. Schlichting et al., Suppression of the quantum-confined Stark effect in polar nitride heterostructures. Commun. Phys. **1**(1), 48 (2018)
155. Y. Kayanuma, Phys. Rev. B **38**, 9797–9805 (1988)
156. L.T. Canham, Appl. Phys. Lett. **57**, 1046 (1990)
157. M. Yazawa, M. Koguchi, A. Muto, M. Ozawa, K. Hiruma, App. Phys. Lett. **61**, 2051 (1992)
158. J. Hu, T.W. Odom, C.M. Lieber, Acc. Chem. Res. **32**, 435 (1999)
159. X.F. Duan, C.M. Lieber, Adv. Mater. **12**, 298 (2000)
160. M.H. Huang, S. Mao, H. Feick, H. Yan, Y. Wu, H. Kind, E. Weber, R. Russo, P. Yang, Science **292**, 1897 (2001)
161. W. Han, S. Fan, L. Qunqing, Y. Hu, Science **277**, 1287 (1997)
162. B. Damilano, J. Brault, B. Alloing, J. Massies, Nano Lett. **16**, 1863 (2016)
163. J. Müßener, LATh. Greif, S. Kalinowski, G. Callsen, P. Hille, J. Schörmann, M.R. Wagner, A. Schliwa, S. Martí-Sánchez, J. Arbiol, A. Hoffmann, M. Eickhoff, RSC Nanoscale **10**, 5591 (2018)
164. F. Furtmayr, M. Vielemeyer, M. Stutzmann, J. Arbiol, S. Estrad, F. Peir, J.R. Morante, M. Eickhoff, J. Appl. Phys. **104**, 034309 (2008)
165. J. Schörmann, P. Hille, M. Schäfer, J. Müßener, P. Becker, P.J. Klar, M. Kleine-Boymann, M. Rohnke, M. De La Mata, J. Arbiol, D.M. Hofmann, J. Teubert, M. Eickhoff, J. Appl. Phys. **114**, 103505 (2013)
166. D.A.B. Miller, D.S. Chemla, T.C. Damen, A.C. Gossard, W. Wiegmann, T.H. Wood, C.A. Burrus, Phys. Rev. Lett. **53**, 2173 (1984)
167. J. Renard, R. Songmuang, G. Tourbot, C. Bougerol, B. Daudin, B. Gayral, Phys. Rev. B **80**, 121305 (2009)
168. L. Rigutti, J. Teubert, G. Jacopin, F. Fortuna, M. Tchernycheva, A. De Luna Bugallo, F.H. Julien, F. Furtmayr, M. Stutzmann, M. Eickhoff, Phys. Rev. B **82**, 235308 (2010)

Chapter 6
Theory of Spectroscopy and Light Emission of Semiconductors Nanostructures

Sandra C. Kuhn, Alexander Carmele, Andreas Knorr and Marten Richter

Abstract Due to their tunable optoelectronic properties, quantum confined electronic excitations in semiconductor quantum dots offer a versatile platform to design nanophotonic device applications. To address and control individual electronic excitations such as excitons, the fundamental Coulomb interaction between the electronic states as well as their coupling to other quasiparticles such as phonons and photons is of interest. In this chapter, we develop a theory of quantum dot spectroscopy and study coupled quantum dot-cavity structures with respect to their correlated photon emission statistics. To account for the surrounding material of the quantum emitter, we include electron-phonon interaction as well as analyze transitions between localized bound quantum dot states and delocalized states of the host medium. Coherent couplings between different quantum dots and the underlying microscopic coupling mechanisms are investigated using two-dimensional spectroscopy.

6.1 Introduction

The goal of the quantum description of solid state based quantum dot materials and devices for applications in nanophotonics is to investigate the advantages and drawbacks of semiconductor light emitters in comparison to atomic systems. As an introduction to the field we first review fundamental theoretical concepts.

S. C. Kuhn · A. Carmele · A. Knorr · M. Richter (✉)
Institut für Theoretische Physik, Nichtlineare Optik und Quantenelektronik,
Technische Universität Berlin, Hardenbergstrasse 36, 10623 Berlin, Germany
e-mail: marten.richter@tu-berlin.de

S. C. Kuhn
e-mail: kuhn@itp.tu-berlin.de

A. Carmele
e-mail: alex@itp.tu-berlin.de

A. Knorr
e-mail: andreas.knorr@physik.tu-berlin.de

M. Kneissl et al. (eds.), *Semiconductor Nanophotonics*, Springer Series
in Solid-State Sciences 194, https://doi.org/10.1007/978-3-030-35656-9_6

In Sect. 6.2, light emission spectra of semiconductor quantum dots in their solid state environment are discussed. This involves the description of phonon assisted light emission, in particular the occurency of phonon sidebands caused by induced and spontaneous phonon emission. Light emission and photon statistics can be tuned by embedding single quantum dots in cavities, cf. Sect. 6.3. As an example, we focus on correlated few photon emission such as entangled photon pairs. First, we discuss a joint theory-experiment study on the two-photon dressing of a three-level system showing that the time ordering of the cascaded photons can be removed or conserved, demonstrating sources of orthogonally polarized, bunched or antibunched photon pairs. Second, the investigation of the photon emission into different spatial directions of a micropillar cavity shows a temporally ordered generation of two photons constituting a heralded single photon source. Third, to attack the transition of the single (few) emitter to many emitter limit in quantum optical devices we present an exact numerical approach, capable of solving the full density matrix up to hundred identical quantum dots coupled to an optical cavity mode.

In Sect. 6.4, using the spectral signatures of intraband transitions between bound quantum dot states and reservoir states, an analytical formula for reconstructing the bound quantum dot ground state wave function from experimental data is derived. Also, Coulomb effects occurring for intraband transitions open the opportunity to study new interband excitonic states. To treat many particle system-reservoir correlations, we present in Sect. 6.5 a novel theoretical factorization scheme of solid state continuum reservoir states and localized quantum dot states. The procedure utilizes an exact diagonalization scheme for the localizied states inside the quantum dot.

In Sect. 6.6, we develop a theoretical technique to describe nonlinear two-dimensional spectroscopy: In particular, we study a method to distinguish electrostatic couplings and excitation transfer processes (Förster and Dexter interaction). For a disordered GaAs quantum well, i.e. involving confinement of excitons, we analyzed the contribution of exciton-phonon interactions to spectral diffusion.

6.2 State of the Art of Microscopic Description of Quantum Dots

This chapter is focused on the theoretical formulations used for the microscopic description of optical processes in quantum dots (QDs). A QD model system is introduced in Sect. 6.2.1 and the corresponding interaction Hamiltonians are introduced in Secs. 6.2.2, 6.2.3 and 6.2.4.

6.2.1 Quantum Dot Model

The many particle dynamics in optically excited semiconductor nanostructures is governed by the dynamics of carriers (electrons, holes), lattice and optical (quantum) fields and their quasi-particles (excitons, phonons, photons), cf. Fig. 6.1a. The electronic system is described by the single particle states $\phi_i(\mathbf{r})$ with energy ε_i in the conduction band c and in the valence band v. Due to the expitaxial growth process, QDs are typically embedded in a wetting layer or bulk material. Because of the quantum confinement of the carriers discrete, bound quantum states are formed inside the QD [1] as well as unbound continuous scattering states delocalized in the barrier material. Depending on the dimensionality of the system, the electronic index $i = \lambda$ describes the discrete QD states or continuous bands in a quantum well or bulk material $i = (\lambda, \mathbf{k})$, with $\lambda = c, v$ denoting the band index and $\mathbf{k}$ the momentum. ΔE_{cv} denotes the QD gap energy between the lowest conduction and highest valence band state of the QD, cf. Fig. 6.1a: In this introduction, for simplicity, the confined states in the QD are restricted to one level in the valence and conduction band, respectively, but more involved models are available [2, 3]. The energy difference between the lowest (highest) QD state and the lowest (highest) energy bulk state in the conduction (valence) band is ΔE_c (ΔE_v).

To describe the corresponding Hamiltonians, $a_i^{(\dagger)}$ is introduced as the fermionic annihilation (creation) operator of an electron in the state i with energy ε_i. The electrons interact with phonons, which are described by the bosonic annihilation (creation) operators $b_{\mathbf{q}}^{(\dagger)}$ with energy dispersion $\omega_{\mathbf{q}}$ and $\mathbf{q}$ including the phonon mode and momentum. Photons of mode μ and frequency ω_μ are described by $c_\mu^{(\dagger)}$. The interaction-free Hamiltonian reads $H_0 = \sum_i \varepsilon_i a_i^\dagger a_i + \sum_{\mathbf{q}} \hbar\omega_{\mathbf{q}} b_{\mathbf{q}}^\dagger b_{\mathbf{q}} + \sum_\mu \hbar\omega_\mu c_\mu^\dagger c_\mu$. The relevant interactions are briefly discussed in the following sections.

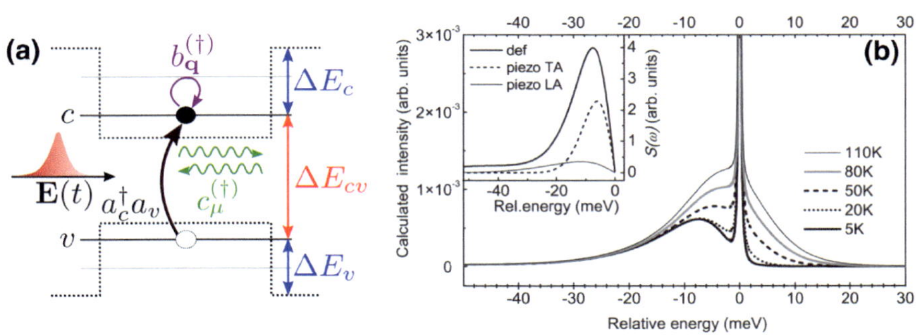

Fig. 6.1 **a** QD level scheme in electron picture. **b** Calculated temperature dependent fluorescence spectrum of a single emission line (zero-phonon-line peak intensity for the 5 K curve is normalized to 1). The inset shows spectral densities $S(\omega)$ for a model QD with a height of 1.4 nm, base length of 7 nm, and emission energy of 4.03 eV for deformation potential coupling (def), and piezo electric coupling for TA and LA phonons. ©2012 American Physical Society, reprinted from [4]

6.2.2 Electron-Light Interaction

The interaction of electrons with light constitutes the fundamental coupling process in opto-electronic devices. Depending on the application, either a semi-classical or a quantized description of the electron-light interaction is required. The interaction of the quantized electronic system with a classical external optical field $\mathbf{E}(t)$ in dipole approximation [5] (used in Sects. 6.4 and 6.6) is described by:

$$H_{\mathrm{el-light}} = -\sum_{i,j} \mathbf{d}_{ij} \cdot \mathbf{E}(t) a_i^{\dagger} a_j + \mathrm{h.a.}, \tag{6.1}$$

in which $\mathbf{d}_{ij} = q \int \mathrm{d}\mathbf{r} \phi_i^*(\mathbf{r}) \mathbf{r} \phi_j(\mathbf{r})$ is the dipole moment, where q denotes the electron charge and $\phi_i(\mathbf{r})$ the electron wave function in the quantum state i and energy ε_i.

The interaction with a quantized electromagnetic field reads [6]:

$$H_{\mathrm{el-pt}} = -\sum_{\mu} \sum_{i,j} M_{ij}^{\mu} a_i^{\dagger} a_j c_{\mu}^{\dagger} + \mathrm{h.a.}, \tag{6.2}$$

where $M_{ij}^{\mu} = i\sqrt{\frac{\hbar\omega_{\mu}}{2\varepsilon_0}} \mathbf{f}_{\mu}(\mathbf{r}_{\mathrm{QD}}) \cdot \mathbf{d}_{ij}$ is the electron-photon coupling element with $\mathbf{f}_{\mu}(\mathbf{r}_{\mathrm{QD}})$ being the mode eigenfunction, r_{QD} the position of the QD and ω_{μ} the photon frequency [7]. A quantized description is used to analyze coupled QD-cavity structures in Sect. 6.3.

6.2.3 Electron-Phonon Interaction

A quantized description of the interaction of electrons with lattice vibrations (phonons) is required for the modeling of low dimensional structures, since it represents a fundamental source of decoherence [8] and thermalization [9] in these structures. The electron-phonon coupling Hamiltonian reads [10, 11]:

$$H_{\mathrm{el-ph}} = \sum_{i,j,\mathbf{q}} g_{ij}^{\mathbf{q}} a_i^{\dagger} a_j (b_{\mathbf{q}} + b_{-\mathbf{q}}^{\dagger}), \tag{6.3}$$

where $g_{ij}^{\mathbf{q}}$ denotes the electron-phonon coupling element. In general, different coupling mechanisms such as deformation potential coupling and piezoelectric coupling for longitudinal-acoustic (LA) phonons and deformation potential coupling and Fröhlich coupling for longitudinal-optical (LO) phonons contribute to the electron-phonon coupling element [4, 10–13].

The independent-boson model acts as an example for an exact solution of electron-phonon interaction in linear optics [14]. Here, a two-level QD (cf. Fig. 6.1a) interacts with a phonon bath diagonally in the electronic states after δ-pulse excitation:

$\mathbf{d}_{ij} \cdot \mathbf{E}(t) \propto \delta(t)$ preparing a polarization p_0 as initial condition. The corresponding dynamics can be solved analytically for the microscopic interband polarization $p(t) = \langle a_v^\dagger a_c \rangle(t)$ [15]:

$$p(t)|_{t\geq 0} = p_0 \exp\left\{ \left(-i\omega_{\mathrm{QD}} + i\sum_{\mathbf{q}} \frac{|\tilde{g}_{vc}^{\mathbf{q}}|^2}{\hbar^2\omega_{\mathbf{q}}} \right) t \right.$$
$$\left. - \sum_{\mathbf{q}} \frac{|\tilde{g}_{vc}^{\mathbf{q}}|^2}{\hbar^2\omega_{\mathbf{q}}^2} \left[n_{\mathbf{q}} \mathrm{e}^{i\omega_{\mathbf{q}}t} + (n_{\mathbf{q}}+1)\mathrm{e}^{-i\omega_{\mathbf{q}}t} \right] \right\}, \tag{6.4}$$

with the interband gap energy $\Delta E_{cv} = \hbar\omega_{\mathrm{QD}} = \varepsilon_c - \varepsilon_v$, the Bose-Einstein-distribution $n_{\mathbf{q}} = [\exp(\hbar\omega_{\mathbf{q}}/k_{\mathrm{B}}T) - 1]^{-1}$ and $\tilde{g}_{vc}^{\mathbf{q}} = g_{vv}^{\mathbf{q}} - g_{cc}^{\mathbf{q}}$. For the two-level system $p(t)$ is related to linear absorption $\alpha(\omega) = \frac{\omega}{nc\varepsilon_0}\Im\left[\frac{P(\omega)}{E(\omega)}\right]$ via the macroscopic polarisation $P(t) = \sum_{ij} d_{ij} a_i^\dagger a_j$ (one polarization direction), but $p(t)$ can also be connected to fluorescence for the two level system [16]. Equation (6.4) contains a polaron energy shift as well as time- and temperature-dependent line shape contributions: They lead to an asymmetric deviation from standard, radiatively broadened Lorentzian fluorescence line shape in terms of phonon sidebands for acoustic and phonon satellites for optical phonons through phonon emission ($n_{\mathbf{q}} + 1$-term) and absorption ($n_{\mathbf{q}}$-term).

Figure 6.1b shows numerical simulations of GaN/AlN QD fluorescence for different types of electron-phonon interaction and different temperatures using the independent Boson model with realistic wave functions based on 8-band $\mathbf{k} \cdot \mathbf{p}$ theory [17]. The results for the acoustic sidebands show good agreement of experiment and theory for varying temperatures, using only independent calculated parameters for the phonon coupling element, cf. Sect. 12.4.3 in Chap. 12.

6.2.4 Coulomb Interaction

The Coulomb interaction between electrons and holes is fundamental for analyzing the formation of electron-hole quasiparticles like excitons, biexitons or trions [18, 19] as well as for higher densities scattering of charge carriers [20, 21]. The Coulomb coupling in second quantization reads:

$$H_{\mathrm{C}} = \frac{1}{2}\sum_{1234} V_{1234} a_1^\dagger a_2^\dagger a_3 a_4, \tag{6.5}$$

where 1, 2, 3 and 4 are multi-indices including all quantum number states such as the band index or the spin configuration. The calculation of a Coulomb matrix element:

$$V_{1234} = q^2 \int d^3r \int d^3r' \phi_1^*(\mathbf{r})\phi_2^*(\mathbf{r}')G(\mathbf{r},\mathbf{r}')\phi_3^*(\mathbf{r}')\phi_4^*(\mathbf{r}), \tag{6.6}$$

with the Green's function $G(\mathbf{r}, \mathbf{r}')$ being the generalized Coulomb interaction potential with respect to the dielectric environment, cf. Sect. 6.4. Equation (6.6) requires—in general—the solution of a six-dimensional spatial integral, which is often performed numerically. Depending on the size and complexity of the investigated system, the computation of the Coulomb coupling elements can significantly limit the quantum mechanical calculations. In Sect. 6.4 we use a Poisson Green's function method to efficiently calculate Coulomb coupling elements numerically [22], which is based on the solution of a generalized Poisson equation via a Green's function formulation.

The Coulomb interaction between the charge carriers within a single QD and between different quantum emitters leads to coupled states and a renormalization of the energies. Therefore, in Sects. 6.3, 6.5 and 6.6 the localized electronic states are transformed into an excitonic basis resulting in many-particle eigenstates: The ground state G, exciton states X and biexciton states B, cf. Fig. 6.2a in Sect. 6.3.

6.3 Coupled Quantum Dot-Cavity Structures

QD-cavity structures as shown in Fig. 6.2b, consisting of QD material in a cavity which supports confined modes of electromagnetic field, are fundamental for the study of cavity quantum electrodynamics in a solid state environment and used to enhance the performance of QD based optoelectronic devices. In particular, they represent promising candidates for devices used in quantum information such as high-quality single photon sources or sources of entangled photon pairs.

6.3.1 Correlation Function and Master Equation

To discriminate between different photon statistics, the observable of interest is the two-time second order correlation function $g^{(2)}(t, \tau)$ measured in a Hanbury-Brown-Twiss-experiment [23]:

$$g^{(2)}(t, \tau) = \frac{\langle : I(t)I(t+\tau) : \rangle}{\langle I(t)\rangle\langle I(t+\tau)\rangle}, \tag{6.7}$$

where $I(t) = \mathbf{E}^-\mathbf{E}^+$ is the photon intensity operator: $\mathbf{E}^{+(-)} = (-)\sum_\mu \sqrt{\frac{\hbar\omega_\mu}{2\varepsilon_0}}\mathbf{f}_\mu^{(*)}c_\mu^{(\dagger)}$ with the mode eigenfunction $\mathbf{f}_\mu^{(*)}$ [7]. Note, the product of operators in (6.7) is normal ordered as defined by $: b^\dagger b :=: bb^\dagger := b^\dagger b$ for bosonic operators.

Here, second order correlation functions describing the *steady state* (reached for large times t) are considered. In this case, the result of the second order correlation function $g^{(2)}(t, \tau)$ does not depend on t but on the time delay τ: $g^{(2)}(\tau)$. The second order correlation function $g^{(2)}(\tau)$ gives the probability of detecting one photon and a second photon with the delay time τ. An emission of bunched photons is reflected by $g^{(2)}(0) > 1$ indicating an emission of two or more photons simultaneously. In contrast, $g^{(2)}(0) < 1$ indicates anti-bunched single photon emission, while $g^{(2)}(0) = 1$ represents the coherent limit [24].

The QD-cavity system represents an open quantum system, whose quantum dynamics is often described by the Liouville-von-Neumann equation (master equation) [25]:

$$\partial_t \rho = -\frac{\iota}{\hbar}[H, \rho] + \sum_i \mathscr{D}[J_i]\rho, \tag{6.8}$$

with the density matrix ρ and the Lindblad dissipators $\mathscr{D}[J_i]\rho = 2J_i \rho J_i^\dagger - \{J_i^\dagger J_i, \rho\}$ representing Markovian interaction processes with the environment such as electrical pumping, cavity losses or radiative decay. The Lindblad approach (6.8) is well studied in a variety of applications [26, 27] and is used in Sect. 6.3.2 to investigate polarization-entanglement and path-controlled time reordering of paired photons an a biexciton cascade. Spatial cross correlations of weakly and strongly coupled modes are discussed in Sect. 6.3.3 with respect to single, bunched and heralded QD photon sources. Section 6.3.4 focusses on the effective description of the few and many emitter limit and application to many emitter nanolasing using the Lindblad approach.

6.3.2 *Polarization-Entanglement*

In our investigation, we focus on the emission statistics of entangled photon pairs, i.e. two photon states which cannot be simply factorized into their individual photon contributions. Photon pairs generated in a biexcitonic decay, i.e. a two stage cascade process are entangled if the which-path information is lost during the emission process [28–30], cf. Fig. 6.2a. A typical example is the Bell state: $1/\sqrt{2}(|X_H\rangle|X_V\rangle \pm |X_V\rangle|X_H\rangle)$ [31]. Despite the proof-of-principle demonstration of photon entanglement in atom-molecular quantum optics, it is a major goal to implement the generation of polarization-entangled photon pairs in semiconductor nanostructures due to their scalability [32] and the possibility to pump the device electrically via the surrounding wetting layer [33]. However, the semiconductor specific environment also induces detrimental effects in electrically-driven entangled photon sources, which establish a which-path information, namely non-radiative dissipation via phonon emission and exciton fine-structure splitting due to geometrical QD asymmetries.

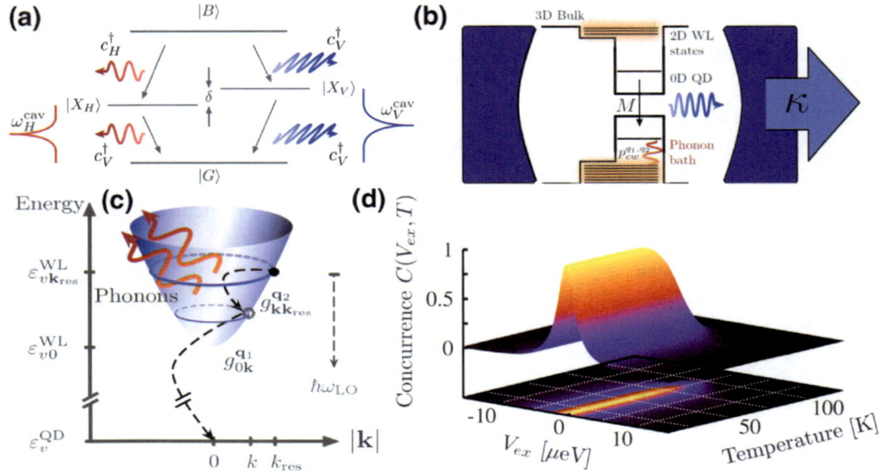

Fig. 6.2 Generation of polarization-entangled photon pairs: **a** cascade with a fine-structure splitting $\delta \neq 0$ (without biexcitonic shift). There are two possible paths: either two photons with vertical V or horizontal H polarization are emitted into a cavity mode $\omega^{\text{cav}}_{V/H}$. **b** The setup consists of a two-mode cavity for horizontal and vertical photon modes, and a QD with excitonic and biexcitonic states which are coupled to the surrounding wetting layer material. **c** The coupling between the QD carrier and the carriers within the wetting layer is mediated via two-phonon emission processes in a higher-order Markovian processes, i.e. the full process is energy conserving whereas the steps in between are not. **d** The two-photon correlation function as a measure for concurrence shows that even for vanishing fine-structure splitting $V_{\text{ex}} = \hbar\omega_H - \hbar\omega_V$ the carriers decay non-radiatively with increasing temperature and deplenish the photon generation from the biexciton state.

First, cf. Sect. 6.3.2.1, we show that phonon-induced electronic transitions of excitonic states to the surrounding wetting layer material leads to non-radiative decay of entanglement with decreasing confinement and increasing temperature. Second, cf. Sect. 6.3.2.2, we show that even in the presence of a fine-structure splitting the which-path information can be erased via quantum chronology engineering (photon post-selection) for a two-photon cascade process of a driven-biexciton state. Note, in Sect. 12.4.5 in Chap. 12 further results of the photon emission from the biexciton cascade are presented.

6.3.2.1 Influence of Exciton-Phonon Interaction

To discuss the influence of the electron-phonon interaction for entanglement generation, we study the generation of photon pairs emitted by a biexciton cascade in the presence of phonon-scattering imposed by longitudinal-optical phonons [34]. Due to the strong electronic confinement in QDs, typically, a single phonon emission is not sufficient to couple the confined electronic states to the QD-emitting wetting layer.

Therefore, the next order leading process is a higher-order Markovian two-phonon emission, cf. Fig. 6.2c. A temperature and phonon-coupling strength dependent is given via the effective two-phonon coupling element [34]:

$$M(\mathbf{q}_1, \mathbf{q}_2) = \sum_{\mathbf{k}, \mathbf{k} \neq \mathbf{k}_{res}} \frac{g^{\mathbf{q}_1}_{\mathbf{k},0} g^{\mathbf{q}_2}_{\mathbf{k},\mathbf{k}_{\text{res}}}}{\omega^{\mathbf{k}}_{\text{WL}} - \omega_{\text{QD}} - \omega_{\text{LO}}}, \tag{6.9}$$

where the single-phonon coupling element is denoted by $g^{\mathbf{q}_1}_{\mathbf{k},\mathbf{k}'}$, cf. Fig. 6.2c. This coupling matrix represents the strength of emitting two photons on the expense of scattering an electron from the wetting layer to a quantum confined state. Within this model, an effective Lindblad equation can be derived [25]:

$$\dot{\rho} = -i\,[H_{\text{CC}}/\hbar, \rho] + \mathscr{D}[\sqrt{\gamma_{\text{LO}}}\sigma_{GB}]\rho + \sum_{i=H,V} \mathscr{D}[\sqrt{\kappa}a_i]\rho. \tag{6.10}$$

Here, the light-matter interaction Hamiltonian is given via

$$H_{\text{CC}} = M \sum_{i=H,V} \left(\sigma_{GX_i} c_i^\dagger + \sigma_{X_i B} c_i^\dagger + \text{h.a.}\right), \tag{6.11}$$

with the biexciton state B, the intermediate exciton states in the horizontal and vertical polarization-basis X_H/X_V and ground state G, denoted with the flip-operators $\sigma_{ij} = |i\rangle\langle j|$. The cavity modes c_H, c_V for the generated photons are damped equally strong for both polarizations with cavity loss κ, and the decay of biexcitons into unbound electron hole excitations is [34, 35]

$$\gamma_{\text{LO}} = \int \text{d}\mathbf{q}_1 \int \text{d}\mathbf{q}_2 M(\mathbf{q}_1, \mathbf{q}_2) \left(M^*(\mathbf{q}_1, \mathbf{q}_2) + M^*(\mathbf{q}_2, \mathbf{q}_1)\right). \tag{6.12}$$

In Fig. 6.2d, the concurrence, which is proportianal to the averaged two-photon detection probability [36, 37],

$$C = 2\frac{1}{T}\left|\int_0^T \text{Tr}\left[c_V^\dagger c_V^\dagger c_H c_H \rho(t)\right] \text{d}t\right|, \tag{6.13}$$

vanishes for a finite fine-structure splitting and large temperature. These results show that the confinement potential of the biexciton state is of great importance to steer the entangled pair emission process at elevated temperature through the fine-structure splitting [38, 39].

6.3.2.2 Path-Controlled Time Reordering of Paired Photons in a Biexciton Cascade

In addition to phonon-induced non-radiative losses, typical InGaAs QDs have also imperfect cylindrical symmetry which lifts the degeneracy of the exciton states via the occurrence of fine-structure splitting [40, 41]. This fine-structure has an additional detrimental effect on the degree of entanglement, resulting in vanishing off-diagonal density matrix elements due to a time-integrated detection process [30]. Various methods have been proposed to circumvent this difficulty either in recovering the degeneracy of the intermediate exciton state via applied strain [42], quantum feedback [36, 43], external fields, deterministic nanofabrication [44] or in replacing time-integrated quantum state tomography with a time-resolved detection process allowing for post-selection [40, 41]. Another feasible proposal relies on the manipulation of quantum chronology [45], which correlates two photons from subsequent biexciton cascades to overcome the which-path information not within but across detection events. Here, we discuss how this kind of time-reordering can be achieved without bulky linear optics element [46]. The idea takes advantage of the observation that photons emitted from a laser-driven source show spectrally-addressable side peaks in the spectrum [47]. Via measurement-based spectral selection of the emitted photons, the photon-photon correlation functions shows time-delay symmetric photon bunching: $g^{(2)}(\tau) = g^{(2)}(-\tau)$ for $\tau \to 0$, which is a signature of a time-reordered emission process. Typically, a biexciton photon is followed by an exciton photon (bunching) but an exciton photon cannot be followed immediately by a biexciton photon due to necessary repopulation dynamics. In the strong-driving limit, this condition is lifted when side peak photons become addressable via spectral selection, i.e. when the Mollow side peaks are not hidden beneath the natural linewidth of the biexciton-exciton and exciton-ground state transitions. In the following, we describe this phenomenon in a four-level system with a biexciton driven in two-photon resonance [46, 48, 49].

The QD is modelled as a four-level system consisting of a ground state $|G\rangle$, two excitonic states $|H\rangle$, $|V\rangle$ and a biexcitonic state $|B\rangle$, cf. Fig. 6.3a. The ground state energy $\omega_G = 0$ is set to zero. We consider the case in which an external laser field drives only the horizontal polarization with frequency ω_L and amplitude Ω_H. The full Hamiltonian reads in the two-photon resonance rotating frame and rotating-wave limit:

$$H_R/\hbar = \Delta\left(\sigma_{HH} + \sigma_{VV}\right) + \sqrt{\Delta\Omega/2}\left(\sigma_{HG} + \sigma_{BH} + \text{h.a.}\right), \tag{6.14}$$

with $\Delta = -\Delta_B$ binding energy of the biexciton and $\sigma_{ij} = |i\rangle\langle j|$, and a renormalized Rabi frequency: $\Omega_L =: \sqrt{\Delta\Omega/2}$. We assume a radiative decay Γ of the electronic system via photon emission into a Markovian continuum described by the master equation:

$$\dot{\rho} = -i\left[H_R/\hbar, \rho\right] + \frac{\Gamma}{2}\sum_{i=H,V}\left(\mathscr{D}[\sigma_{Gi}] + \mathscr{D}[\sigma_{iB}]\right)\rho, \tag{6.15}$$

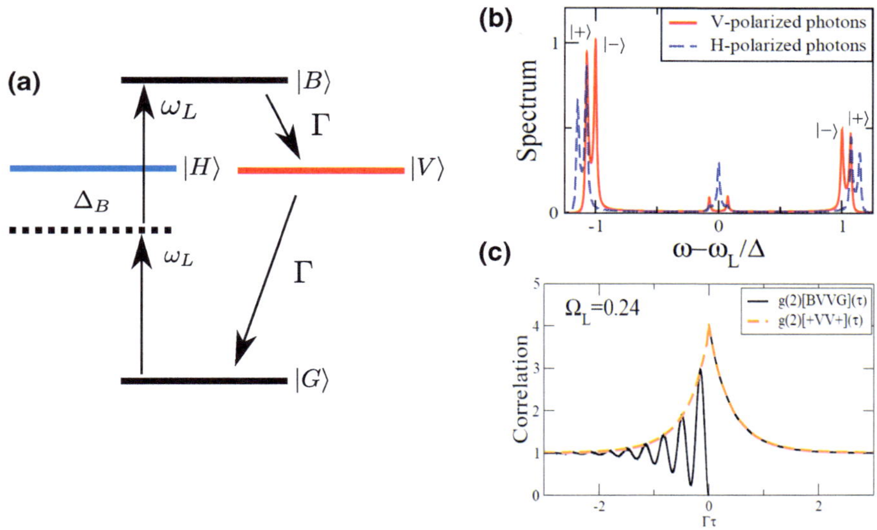

Fig. 6.3 Time-reordering in biexciton cascade: **a** the setup consists of a two-photon excitation process in resonance with the biexciton energy with horizontal polarization. **b** The power spectrum of the four-level system. In the strong driving limit, the biexciton and exciton peaks split and can be addressed separately via frequency filtering. **c** The two-photon correlation function in the strong driving limit. When biexciton- and exciton-photons are correlated (black, solid line), the correlation function shows bunching in the biexciton-exciton direction $\tau > 0$, and anti-bunching in the exciton-biexciton direction $\tau \leq 0$. If the biexciton- and exciton-peaks are frequency-filtered (orange, dashed line), dressed-state selective vanishing time-reordering is observed. Bunching in both directions

assuming the biexciton decay to be double as fast as the exciton decay and using the standard Lindblad form $\mathscr{D}[J]\rho = 2J\rho J^\dagger - \{J^\dagger J, \rho\}$.

In Fig. 6.3b, the power spectrum $S_i(\omega) = \mathrm{Re}\Big[\lim_{t\to\infty} \int_0^\infty \langle c^\dagger(t)c(t+\tau)\rangle \exp[-i\omega\tau]d\tau\Big]$ with $c(t) := \sigma_{iB}(t) + \sigma_{Gi}(t)$ $(i = H, V)$ is calculated via the quantum regression theorem and the results are depicted in the strong driving limit [46]. The biexciton and exciton peaks for both polarizations exhibit a splitting due to the strong external driving. After a two-fold adiabatic elimination of coherences, the two-time correlation function and vice versa the spectrum can be computed analytically:

$$
\begin{aligned}
S_V(\omega)/\alpha = &\frac{\Omega_R\Gamma+\frac{\Gamma}{2}(\omega-\omega_L-\mathrm{d}-\frac{\Omega}{2}+\Omega_R)}{\Gamma^2+[\omega-\omega_L-(\Delta+\frac{\Omega}{2}-\Omega_R)]^2} + \frac{\Omega_R\Gamma-\frac{\Gamma}{2}(\omega-\omega_L-\Delta-\frac{\Omega}{2}-\Omega_R)}{\Gamma^2+[\omega-\omega_L-(\Delta+\frac{\Omega}{2}+\Omega_R)]^2} \\
&+ \frac{\Omega_R\Gamma+\frac{\Gamma}{2}(\omega-\omega_L+\Delta+\frac{\Omega}{2}+\Omega_R)}{\Gamma^2+[\omega-\omega_L-(-\Delta-\frac{\Omega}{2}-\Omega_R)]^2} + \frac{\Omega_R\Gamma-\frac{\Gamma}{2}(\omega-\omega_L+\Delta+\frac{\Omega}{2}-\Omega_R)}{\Gamma^2+[\omega-\omega_L-(-\Delta-\frac{\Omega}{2}+\Omega_R)]^2},
\end{aligned}
\tag{6.16}
$$

with the effective two-photon Rabi frequency $\Omega_R^2 = (\Gamma^2 - \Omega^2)/4$ and $\alpha = \Omega^2/[8\Omega_R (\Omega^2 + \Gamma^2)]$. In this formula, emission peaks around $\omega - \omega_L = 0$ are not included due to the adiabatic elimination.

After characterization of the quantum light source via the power spectral density $S(\omega)$, several photon-photon correlation combination can be discussed by choosing a corresponding eigenbasis (bare state $|G\rangle$, $|B\rangle$, $|H\rangle$, $|V\rangle$ or dressed state basis $|+\rangle$, $|-\rangle$, $|0\rangle$, $|V\rangle$ obtained via diagonalization of the Hamiltonian). For example, if the experiment does not filter spectrally, the bare state basis can be chosen to conveniently describe the experiment, and the four-level system will exhibit symmetric anti-bunching around $\tau = 0$ as expected from an effective two-level system, even in the strong-driving limit. The biexciton-exciton correlation reads:

$$g^{(2)}_{BVVG}(\tau) = 2e^{-\Gamma\tau}\left(1 + \cosh(\Gamma\tau) + \frac{\Gamma^2}{\Omega^2}[1 + \cos(\Omega\tau)]\right) \tag{6.17}$$

with $g^{(2)}_{BVVG}(\tau) = \frac{\langle\sigma_{BV}(t)\sigma_{VG}(t+\tau)\sigma_{GV}(t+\tau)\sigma_{VB}(t)\rangle}{\langle\sigma_{BV}(t)\sigma_{VB}(t)\rangle\langle\sigma_{VG}(t)\sigma_{GV}(t)\rangle}$. The correlation functions unravels the intuitive result that after measuring a biexciton photon, the probability to measure an exciton photon without delay is strongly increased: $g^{(2)}_{BVVG}(0) = 4\left(1 + \frac{\Gamma^2}{\Omega^2}\right)$. However, the reverse process, i.e. first an exciton photon is measured, afterwards a biexciton photon, is less likely and vanishes even for $\tau \to 0$:

$$g^{(2)}_{VGBV}(\tau) = 1 + e^{-\Gamma\tau}\left(e^{-\Gamma\tau} - 2\cos(\Omega\tau)\right) \tag{6.18}$$

with $g^{(2)}_{VGBV}(0) = 0$ for all driving strengths and finite decays. In Fig. 6.3c, both correlations are plotted (solid, black line) for a driving strength of $\Omega_L = 0.24\,\mathrm{ps}^{-1}$. Clearly, the bunching is seen for $\tau > 0$ when the measurement is conditioned on the detection of a biexciton photon first but anti-bunching for $\tau < 0$ with visible, due to repumping enforced, Rabi oscillations.

In addition to the bare state basis, we can also choose the diagonal Eigenbasis of the Hamiltonian, i.e. the dressed state basis, to conveniently model spectrally-filtered photon-photon signals. For example, if only V-polarized photons are detected, i.e. $|H\rangle$ contributions are omitted, the corresponding dressed state reads: $|\pm\rangle = (|G\rangle \pm |B\rangle)/\sqrt{2}$, cf. Fig. 6.3b. So, instead of discriminating only between biexciton and exciton photons, the experiment selects spectrally in addition between the biexcitonic and excitonic Eigenstates (left or right peak of either exciton or biexciton peak), cf. Interestingly, the dressed-state photons are also correlated, and we yield for the direction biexciton, then exciton emissions:

$$g^{(2)}_{+VV+}(\tau) = 2e^{-\Gamma\tau}\left(1 + \cosh(\Gamma\tau) + \alpha[1 + \cos(\Omega\tau)]\right), \tag{6.19}$$

and for the exciton, then biexciton emission:

$$g^{(2)}_{V++V}(\tau) = 1 + 2\frac{1+\alpha}{1+2\alpha}\left[e^{-2\Gamma\tau}\left(1 - \frac{1}{2}\frac{1}{1+\alpha}\right) + e^{-\Gamma\tau}\left(1 - \frac{\alpha\cos(\Omega\tau)}{1+\alpha}\right)\right], \tag{6.20}$$

with $\alpha := \Gamma^2/\Omega^2$. A remarkable phenomenon takes place, if we drive the four-level system so strongly that we can safely assume $\alpha \to 0$. In this regime,

$$g^{(2)}_{+VV+}(\tau) \to 1 + 2e^{-2\Gamma\tau} + e^{-\Gamma\tau} \leftarrow g^{(2)}_{V++V}(\tau), \tag{6.21}$$

and the correlation function looses the information which photon from which frequency has been emitted first.

In conclusion, we have demonstrated that the strong-driving limit provides a feasible approach to erase the time-ordering of two-photon emissions in a four-level system with practical application in quantum information science. Thereby, photons from separate emission processes can be cross-correlated and used to establish polarization-entanglement even in the presence of fine-structure splitting. This allows to circumvent practical limitations in generating polarization- entangled photon pairs in semiconductor devices, operating in the low-temperature limit.

6.3.3 *Spatial Cross Correlation of Weakly and Strongly Coupled Modes: Single, Bunched and Heralded QD Photon Sources*

QD-micropillar structures constitute of a single QD embedded in a cavity. In this section a QD-micropillar structure is used as model system allowing the out-coupling of photons into *two* spatial directions. Based on this system we propose a time-bin heralded single photon source. For this purpose, we theoretically analyze the photon-photon correlation function (6.23) characterizing the emission into two different spatial directions of a QD in a micropillar cavity, cf. Fig. 6.4a: axial bound cavity modes (strong light confinement in cavity axis) and lateral propagating modes (continuum of modes emitting perpendicular to the cavity axis). The QD is described by an excitonic four-level scheme as in the last section consisting of a ground state $|G\rangle$, two spin-degenerated exciton states $|X_{\sigma=\uparrow,\downarrow}\rangle$ and a biexcitonic state $|B\rangle$. The related second order correlation function for all combinations of both spatial emission directions $x, x' = ax, lat$ reads:

$$g^{(2)}_{x-x'}(\tau) = \frac{\langle \mu_x(t)\mu_{x'}(t+\tau)\mu^\dagger_{x'}(t+\tau)\mu^\dagger_x(t)\rangle}{\langle \mu_x(t)\mu^\dagger_x(t)\rangle\langle \mu_{x'}(t)\mu^\dagger_{x'}(t)\rangle}, \tag{6.22}$$

where μ_x and $\mu_{x'}$ are the photon source field expressions, which are directly proportional to the outside field. Depending on the emission direction, μ_x are either the excitonic QD dipoles $|X_\sigma\rangle\langle G|$ and $|B\rangle\langle X_\sigma|$ directly: $\mu_{\text{lat}} \equiv \sum_\sigma (|X_\sigma\rangle\langle G| + |B\rangle\langle X_\sigma|)$ (lateral direction) or the axial cavity mode photons $b^\dagger_m$: $\mu_{\text{ax}} \equiv \sum_m b^\dagger_m$ (axial direction), cf. appendix of [50]. Introducing a time delay τ dependent reduced system operator $\xi_x(t+\tau) \equiv \mathscr{U}(t+\tau, t)\left(\mu^\dagger_x \rho_0(t)\mu_x\right)$ for the initial system density matrix $\rho_0(t) = \rho_0$ via a Liouville-space propagator $\mathscr{U}(t+\tau, t)$, the two-time correlation function reads:

$$g^{(2)}_{x-x'}(\tau) = \frac{tr_S(\mu^\dagger_{x'}\rho_x(t+\tau)\mu_{x'})}{tr_S(\mu^\dagger_x\rho_0\mu_x)tr_S(\mu^\dagger_{x'}\rho_0\mu_{x'})}. \tag{6.23}$$

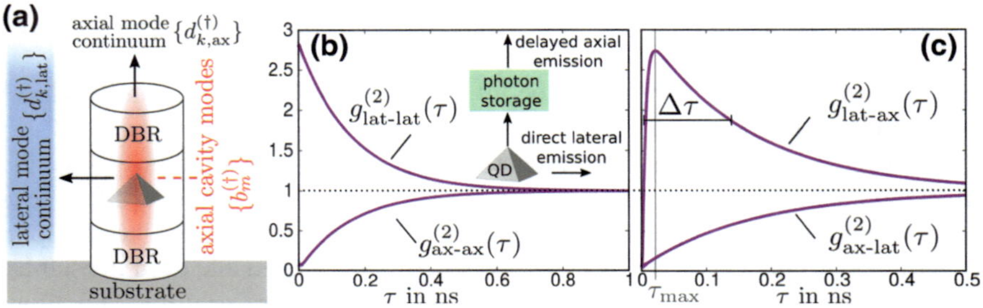

Fig. 6.4 **a** QD-micropillar cavity: a single QD is embedded between two DBRs requiring an emission in axial and lateral direction. The QD directly couples to the lateral mode continuum $\{d^{(\dagger)}_{k,lat}\}$. The axial cavity modes $\{b^{(\dagger)}_m\}$ couple to an axial mode continuum $\{d^{(\dagger)}_{k,ax}\}$. **b**, **c** Second order correlation function: **a** $g^{(2)}_{ax-ax}(\tau)$ shows anti-correlation and $g^{(2)}_{lat-lat}(\tau)$ correlation. **b** $g^{(2)}_{lat-ax}(\tau)$ reaches a maximum at τ_{max} for the emission in axial direction. $g^{(2)}_{ax-lat}(\tau)$ shows anti-correlation. Reprinted with permission from [50], (The Optical Society)

Based on the calculations of second order correlation function (6.23) two types of photon emission statistics from a single quantum light device are proposed:

(i) The unidirectional correlation function $g^{(2)}_{x-x}(\tau)$ shows a single photon emission into the axial direction ($g^{(2)}_{\mathrm{ax-ax}}(\tau)$, Fig. 6.4b and a two-photon emission into the lateral (weak QD-light coupling) direction ($g^{(2)}_{\mathrm{lat-lat}}(\tau)$, Fig. 6.4b.
(ii) The photon pair emission described by the lateral and axial cross correlation $g^{(2)}_{\mathrm{lat-ax}}(\tau)$ indicates a temporally ordered generation of two photons (first lateral and second axial) within a defined time bin, Fig. 6.4c. In particular, this allows to use a light source for the preparation of time-bin/heralded photon pairs (heralded single photon source). To determine, whether the heralded axial photons are emitted as bunched or as the desired single photons, we calculated the third order correlation function $g^{(3)}_{\mathrm{lat-ax-ax}}(\tau')$. The results show anti-bunching and thus confirm the emission of heralded single photons [50].

Our results of the photon-photon correlations offer interesting perspectives to use one and the same solid state based quantum light device as sources of single-, two- and heralded photon sources by using two spatially distinct emission channels, cp. Fig. 6.4a.

6.3.4 *Effective Description of the Few and Many Emitter Limit and Application to Many Emitter Nanolasing*

So far, we have focused on the dynamics of a single QD emitter coupled to bound photon mode(s). In many photonic devices such as QD cavities or nanolasers, few or many quantum emitter couple to the same bound photon modes. However, in particular, for intermediate numbers of quantum emitter and modes, the states of the quantum emitter and the photon number states of the modes are highly correlated [51]. In this case, the Liouville-von-Neumann equation with Lindblad dissipators for the product space of quantum emitters and photon number states is a common approach, the closed electron-photon system but a numerical solution the number of emitters N is limited by memory constraints: Consider N emitters described by two level systems, their Hilbert space scales exponentially as 2^N and the Liouville space scales as 2^{2N}. Together with the photonic part of the Liouville space already very small numbers of emitters N prevent a numerical simulation.

In many photonic nanostructures the quantum emitters are similar and often couple with comparable strength to the photon modes, so that, it is natural to assume identical emitter Liouvillians and one obtains a permutational symmetric system, where the individual emitter is not distinguishable from other quantum emitters. [54] is one example from recent independent reinventions and applications [52, 53, 55–67] of permutation symmetric methods [68–70]. In the permutation symmetric case, it does not matter, which quantum emitters are in a particular Liouville space state, but only how many quantum emitters are in this state. In [54], it was shown for two level emitters, that the number of different density matrix elements scales with a third order polynomial $\propto N^3$ in the emitter number N. If we ignore the photon part for a moment, the density matrix for the two level quantum emitters can be described by the elements $\mathscr{P}[n_{11}, n_{10}, n_{01}]$ with numbers of emitters in density matrix states: excited state population n_{11}, in a coherence between excited and ground state n_{10} or between ground state and excited state n_{01} resulting in the overall N^3 scaling of elements $\mathscr{P}[n_{11}, n_{10}, n_{01}]$. It is straight forward to derive equations of motion for $\mathscr{P}[n_{11}, n_{10}, n_{01}]$ [52, 54]. Exemplary, Fig. 6.5a illustrates, how radiative decay coming from a Lindblad operator enters into the equations of motion. In [52], a diagrammatic language was developed, that generalizes the permutation symmetric methods to multi level systems. In Fig. 6.5a the bubbles represent the numbers of quantum emitters in a particular density matrix state. Arrows pointing between two bubbles represent processes, that convert one density matrix state into another. Arrows pointing outward represent dephasing processes. The diagram representing a two level system with radiative decay and electron-photon coupling to a single photon mode is shown in Fig. 6.5b. The extension to a three level system in 6.5c is straight forward. The number of bubbles connected to the ground state bubble n_{00} determines the order of the polynomial scaling in the emitter number N. The PsiQuaSP library [53] allows to translate the diagrams into equations of motion and to solve them in steady state or dynamically. Applications of this framework include quantum phase transitions [63] and also nanolaser dynamics [54].

As an example, Fig. 6.6 shows the $g^{(2)}$ function of an incoherently pumped nanolaser setup calculated with the permutational symmetric method [54]. By increasing the incoherent pump rate P the statistics of both the exciton and plasmon system go from initially thermal light $g^{(2)} \approx 2$ to coherent lasing $g^{(2)} \approx 1$ back

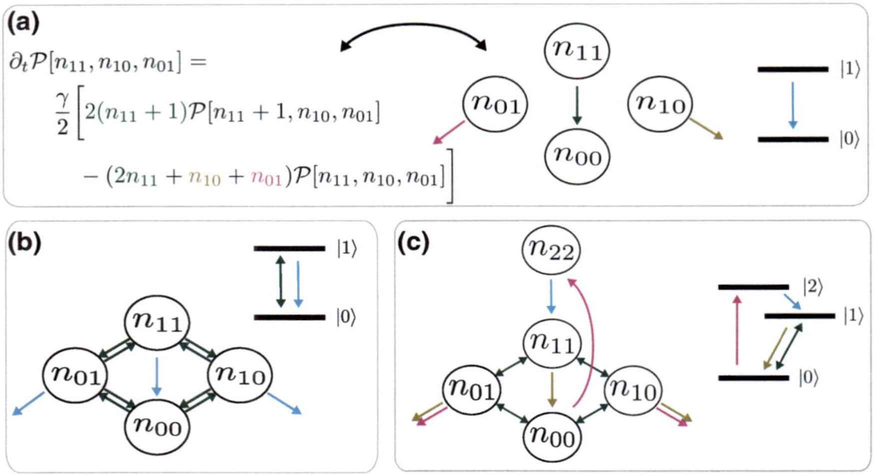

Fig. 6.5 Diagrammatic representation **a** of radiative decay for a two level system translated into an equation of motion [52], **b** two level system with electron-photon interaction Hamiltonian and radiative decay Lindblad, **c** three level laser with incoherent driving and relaxation (pink, blue), laser mode (green) and radiative decay (yellow). Reprinted, rearranged [53], license http://creativecommons.org/licenses/by/4.0/

Fig. 6.6 Excitonic and plasmonic $g^{(2)}$ of a set of quantum emitters coupled to plasmon mode for 10 quantum emitters and 30 quantum emitters. ©2015 American Physical Society, reprinted from [54]

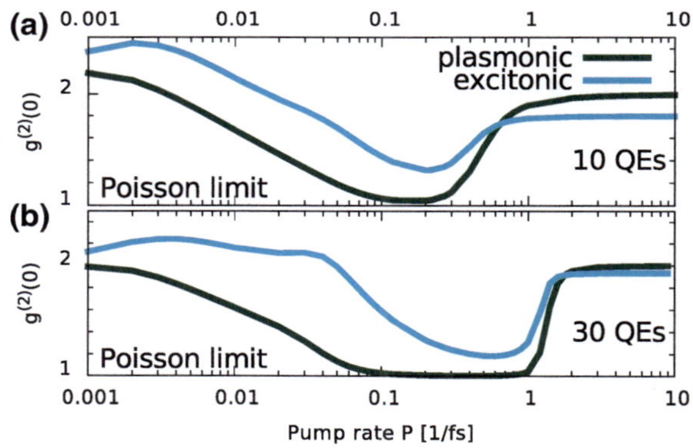

to thermal light. By increasing the emitter number the pump rate range for achieving coherent light is much broader and thus the coherent state is more stable. On the other hand the excitonic correlation function does not reach a value indicating full coherence ($g^{(2)} = 1$) due to its non bosonic nature.

6.4 Intraband Transitions Between Bound Quantum Dot States and States of the Host Medium

This section is devoted to bound-continuum intraband optical transitions of QDs embedded in a bulk material to study the character of elementary excitations consisting of delocalized and localized electronic states. First, the QD-continuum model system is introduced as well as an experimental setup (pump-probe experiment) to address the intraband transitions. Subsequently, we present an all-optical formula to reconstruct the QD wave function from bound-continuum intraband absorption spectra. The influence of Coulomb coupling on the intraband transitions is discussed in the last part of this section.

6.4.1 Quantum Dot-Continuum Model System and Pump Probe Setup

Bound to continuum intraband transitions in QD emitters occur between quantum confined QD states and the spatially delocalized continuum states of the QD host medium, cf. blue dashed arrows in Fig. 6.7. As a model system we use a self-organized InAs QD (Volmer-Weber growth mode) with discrete QD states i, which is embedded in a GaAs bulk material with quasi-continuous bulk states $\mathbf{k}$ in the conduction c and in the valence band v, respectively. Depending on the carrier system (electron, hole) the intraband absorption spectrum exhibits a range of different spectral contributions, cf. Fig. 6.7. However, the spectral range of the electronic conduction band transitions in the intraband absorption spectrum is usually well separated from the hole continuum band transitions and thus separately addressable in experiments. Therefore, we focus on the electron contributions only.

Intraband transitions can be experimentally addressed e.g., via pulsed pump-probe (PP) experiments, which are nonlinear two step processes [72–75]: First, a pump pulse (external ligth field) with high intensity excites the system within a defined energy range determined by the pulse frequency. The pump pulse creates electron and hole populations in the resonant system states, e.g. in the confined ground states of a QD. After a delay time, in a second step a weak probe pulse tests the pump induced carrier populations. This way, for example, intraband transitions inside a QD or between QD and surrounding bulk material are accessible, cf. Fig. 6.7. To theoretically describe the system interaction with the classical pump and probe field $\mathbf{E}(t)$, a semi-classical electron light interaction is included for each pulse via (6.1). Here, for the external field $\mathbf{E}(t) = \tilde{\mathbf{E}}(t)\exp(-i\omega t)$ a modulated Gaussian with the slowly varying time dependent amplitude $\tilde{\mathbf{E}}(t) = \mathbf{E}_0 \exp[-(t - t_0)^2/\tau^2)]$ is used, where $\mathbf{E}_0$ denotes the amplitude, t_0 the temporal offset, τ the length and ω the frequency.

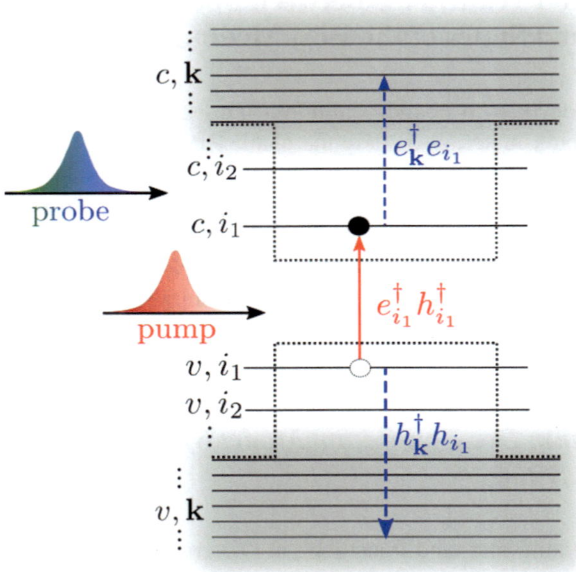

Fig. 6.7 QD-continuum model system: discrete QD states i and quasi-continuous bulk states $\mathbf{k}$ in the conduction c and in the valence band v are used as disjoint index sets for the electrons and holes, respectively. The blue dashed arrows mark the bound to continuum intraband transitions, which are tested by the probe pulse. ©2014 American Physical Society, slightly modified from [71]

6.4.2 All-Optical Reconstruction of Quantum Dot Wave Functions

To construct high quality QD devices precise information about the structural properties of the QD is required. Here, we show that bound to continuum intraband transitions can be used to obtain information about the spatial extension of the QD wave function. We present an all-optical, measurable reconstruction formula to explicitly calculate the bound QD ground state wave function from a bound to continuum intraband absorption spectrum [71]. Since electron and hole intraband transitions can be spectrally well separated, the electron and hole wave functions can be separately extracted. The proposed method relies on the fact that, using a free particle (plane wave) approximation for the continuum states, the intraband dipole moment $d_{i\mathbf{k}}^{\mu\mu}$ with $\mu = e, h$ is directly connected to the derivative of the QD wave function $\widetilde{\xi}^*_{\mu,i}(\mathbf{k})$ in the wave number domain $\mathbf{k}$ [71]):

$$d_{i\mathbf{k}}^{\mu\mu} = \frac{q}{\sqrt{V_b}} \int \xi^*_{\mu,i}(\mathbf{r})\, r_x\, \mathrm{e}^{i\mathbf{k}\cdot\mathbf{r}} \mathrm{d}^3\mathbf{r} = -\frac{iq}{\sqrt{V_b}} \frac{\partial}{\partial_{k_x}} \widetilde{\xi}^*_{\mu,i}(\mathbf{k}), \tag{6.24}$$

where, $\xi^*_{\mu,i}(\mathbf{r})$ ($\widetilde{\xi}^*_{\mu,i}(\mathbf{k})$) is the QD wave function in real (Fourier) space, $r_x = \mathbf{e}_x \cdot \mathbf{r}$, $\frac{\partial}{\partial_{k_x}} = \mathbf{e}_x \cdot \nabla_{\mathbf{k}}$, V_b is the bulk volume and q is the electron charge. Based on (6.24) we derived the reconstruction formula [71]:

$$\pm \frac{(2m_e)^{\frac{1}{4}}}{\sqrt{B\hbar}} \int\limits_0^{k_0} \left(\frac{\tilde{\alpha}^i_\mu(k'_0)}{k'_0} \right)^{\frac{1}{2}} dk'_0 = \left(\widetilde{\xi}_{\mu,i}(k_0) - \widetilde{\xi}_{\mu,i}(0) \right). \tag{6.25}$$

For a compact notation, we define $\tilde{\alpha}^i_\mu(\omega) = \frac{\alpha^i_\mu(\omega)}{\omega}$ and $k_0(\omega) = \sqrt{\frac{2m_e(\hbar\omega - \Delta E^i_\mu)}{\hbar^2}}$ with the effective electron mass m_e and the energy difference ΔE^i_μ between the QD state i and the lowest (highest) continuum state of the conduction (valence) band. In $B = \frac{q^2 m_\mu^{\frac{3}{2}}}{3\sqrt{2}\pi \hbar^3 n c \varepsilon_0}$ the refractive index n, the speed of light c, the vacuum permittivity ε_0 and the effective electron/hole masses enter. Equation (6.25) offers the possibility to reconstruct the Fourier transform of the QD ground state wave function $\widetilde{\xi}_{\mu,i}(k_0)$ from a measured intraband spectrum $\tilde{\alpha}^i_\mu(k_0(\omega))$. Since (6.24) is derived using free particles and the validity of (6.25) needs to be investigated for correct unbound states resembling their character as scattering states of the QD potential: For this purpose, first, the radial Schrödinger equation is solved for a QD confinement potential $V(r)$. Second, the bound to continuum intraband spectrum $\alpha^i_e(\omega)$ is calculated using the resulting wave functions and used as in input for the reconstruction formula (6.25). Finally, a comparison of the original (Fig. 6.8a) and the reconstructed QD ground state wave function (Fig. 6.8b) demonstrates the validity of the reconstruction formula beyond plane waves. Apart from an offset, the QD wave functions can be extracted with good quality from the intraband absorption spectra. The developed scheme, here illustrated for a QD embedded in bulk material, can be transferred with small modifications to other confined nanostructures, for instance to QDs embedded in two dimensional host materials, to defects in quantum wells and similar systems.

6.4.3 *Influence of Coulomb Coupling on Bound-Continuum Intraband Transitions*

After the probe pulse tests the intraband transition of an electron from a bound QD state to a state of the conduction band continuum, cf. Fig. 6.7, the hole is still inside the QD interacting via Coulomb with the delocalized electron. This raises the question, whether these Coulomb interactions between bound QD and unbound continuum states of the host material lead to the formation of new excitonic states. To answer this question, different bound to continuum quasi particles such as excitons, biexciton or trions are investigated to determine their characteristic spectral signatures such as energy shifts, variations of the line width and splittings [77]. Especially, we predict the existence of bound excitons consisting of a localized carrier inside the QD and a delocalized carrier of the continuum.

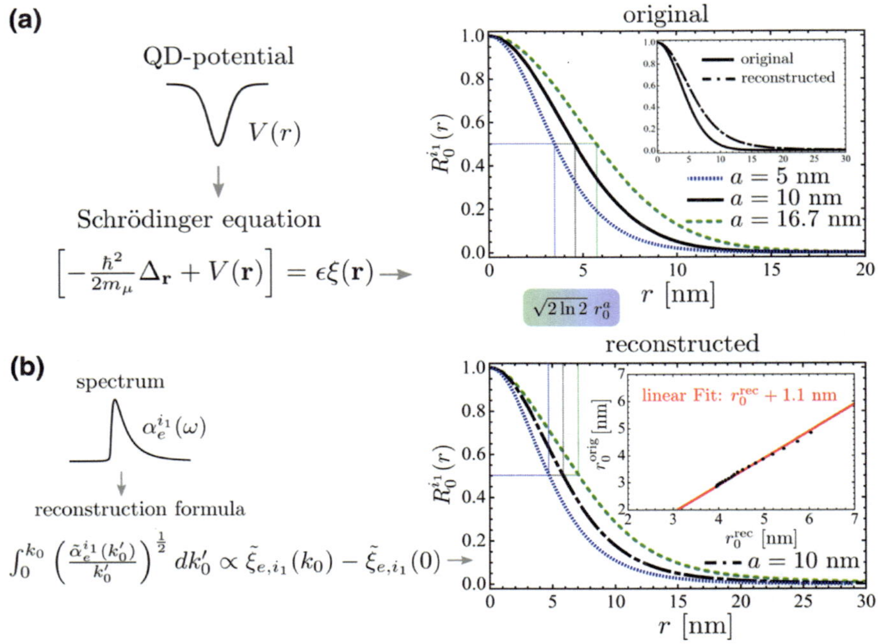

Fig. 6.8 Comparison of the original (**a**) and the reconstructed (**b**) QD electron ground state wave function for different spatial extensions a of the radial symmetric QD confinement potential $V(r)$: **a** wave functions are the solutions of the Schrödinger equation. **b** Wave functions are extracted from intraband absorption spectra calculated beyond the plane wave approximation. The full width at half maximum of $R_0^{i_1}(r)$ gives the localization lengths r_0^{rec} and r_0^{orig}, which are plotted against each other in the inset of (**b**). [76] reprinted with permission

In addition to intraband absorption spectra (Fig. 6.9b) showing different quasi particle excitations, we explore our results by the bound to continuum exciton wave function (Fig. 6.9a, c) to get information about the localization of the unconventional bound continuum exciton. For this, spectrally narrow pulses excite selectively different exciton states (arrows in Fig. 6.9). The results show signatures of a Coulomb induced formation of a bound excitonic state, which is spatially localized in the vicinity of the QD. Especially for the energetically higher states, the spatial extension of the bound excitons can increase up to 100 nm, which is much larger than the QD extension Δ_{QD} of 10 nm. This large extension of the exciton wave function opens a way to study new coupling between excitons in different distant QDs.

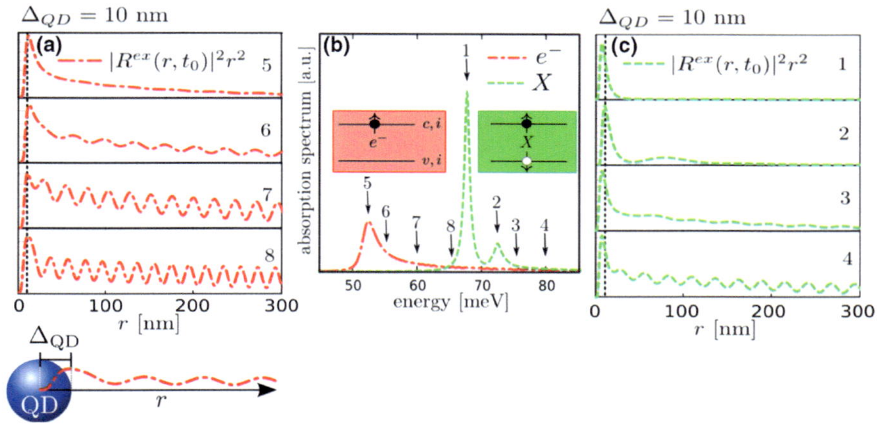

Fig. 6.9 **b** Bound to continuum intraband absorption spectra combared for single electron e^- (Coulomb free) and exciton X (Coulomb coupled) occupation of the QD. The exciton spectrum is shifted and shows signatures (splitting) of a bound to continuum exciton due to Coulomb interaction. **a** Coulomb free and **c** Coulomb coupled exciton probability for several optical excitation energies. ©2017 Society of Photo-Optical Instrumentation Engineers (SPIE), reprinted from [78]

6.5 Hybrid Density Matrix Approach as a Factorization Scheme for Many-Body Systems

Nanoscale semiconductor quantum emitters couple to an embedding bulk, quantum well or wetting layer carrier reservoir that play a significant role for stability, electrical pumping, dephasing and scalability. The total structure made of quantum emitter and environment represents an important example of a hybrid system, which consist of a subsystem with localized, discrete states and a subsystem with quasi continuous states. However, both subsystems (QD and carrier reservoir) constitute many-body systems with different properties: Localized few particle configurations in the quantum emitter, such as (bi-)excitons or trions exhibit strong fluctuations with respect to the mean occupation number. In contrast, the reservoir has a high density of occupied states and fluctuations compared to the average occupation are of minor importance. Therefore, the subsystems require a description within different approximations.

From this point of view, we developed a factorization scheme for many body correlations using a projection operator technique, which combines the advantages of conventional correlation expansions and an exact diagonalization scheme [79]. In particular, the correlations of the reservoir are treated in a higher order mean field approach using single particle occupations while the correlation for the localized quantum emitter state are described non-perturbatively via an exact diagonalization scheme.

The complete system state is expanded in fermionic many-particle Fock states $|\mathbf{n}_{I_eI_h}, \nu\rangle$ consisting of a set of quantum emitterstates $\mathbf{n}_{I_eI_h}$ with I_e electrons and

I_h holes and continuum states ν. The projection operator $P^{I_eI_h} = \sum_{\mathbf{n}_{I_eI_h},\nu} |\mathbf{n}_{I_eI_h}, \nu\rangle \langle \mathbf{n}_{I_eI_h}, \nu|$ projects into the localized electron and hole many-particle states. With respect to the internal Coulomb interaction of the quantum emitter, the local states inside the quantum emitter are transformed into new excitonic eigenstates

$$X_{\mathbf{j}}^{I_eI_h\dagger} = \sum_{\mathbf{j}} c_{\mathbf{ij}}^{I_eI_h*} P^{I_eI_h} e_{x_1}^{\dagger} \ldots e_{x_{I_e}}^{\dagger} h_{y_1}^{\dagger} \ldots h_{y_{I_h}}^{\dagger} P^{00}, \tag{6.26}$$

with the expansion coefficients $c_{\mathbf{ij}}^{I_eI_h*}$.

On this basis, the developed hybrid density matrix approach resembles a combination of a typical Boltzmann like treatment of electron ($\mu = e$) and hole ($\mu = h$) occupations $f_\nu^\mu = \langle \mu_\nu^\dagger \mu_\nu \rangle$ of the reservoir state μ, including their Pauli-Blocking and density matrix contributions for the reduced density matrix of the quantum emitter system $\rho_{\mathbf{ij}}^{I_eI_h,I_e'I_h'}$ [79]:

$$\rho_{\mathbf{ij}}^{I_eI_h,I_e'I_h'} = \langle P^{I_eI_h} X_{\mathbf{i}}^{I_eI_h\dagger} X_{\mathbf{j}}^{I_e'I_h'} P^{I_e'I_h'} \rangle, \tag{6.27}$$

where $\mathbf{i}$ and $\mathbf{j}$ denote the quantum emitter configurations. To derive equations of motion for the system observables f_ν^μ and $\rho_{\mathbf{ij}}^{I_eI_h,I_e'I_h'}$, the reservoir correlations are factorized within a Hartree-Fock-approach and reservoir assisted system correlations (continuum contributions $A^\dagger A$ and quantum emitter contributions $X^\dagger X$) using the factorization rule $\langle A^\dagger A X^\dagger X \rangle = \langle A^\dagger A \rangle \langle X^\dagger X \rangle + \langle A^\dagger A X^\dagger X \rangle^c$.

As an example, a scheme of the hybrid density matrix approach is shown in Fig. 6.10 for scattering contributions resulting from Auger processes described by the Coulomb Hamiltonian $H_{\text{Auger},1} = \sum_{n_1,n_2,n_3,\nu_1} V_{e\ e}^{n_1n_2n_3\nu_1} e_{n_1}^\dagger e_{n_2}^\dagger e_{n_3} e_{\nu_1}$ for the capture of one carrier from the reservoir state ν_1 into a quantum emitter state n_1 and a carrier relaxation within the quantum emitter, cf. Fig. 6.10a. Figure 6.10b demonstrates the hybrid density matrix approach for this scattering process: The temporal evolution of the system observables $\rho_{\mathbf{ii}}^{I_eI_h}$ and f_ν^e is determined by reservoir assisted system correlations such as $\langle X_{\mathbf{i}}^{I_eI_h\dagger} e_{\nu_1}^\dagger X_{\mathbf{j}}^{I_e+1I_h} \rangle$, cf. Fig. 6.10b. To solve the hierarchy problem, the assisted correlations in the second order of equation of motion (green boxes in Fig. 6.10b) can be factorized. In contrast to pure Boltzmann with Pauli blocking terms and pure density matrix equations, in the approach developed here a hybrid form enters like $(1 - f_{\nu_1}^{\mu_1})\rho_{\mathbf{jj}}^{I_e+1I_h}$ (cf. Fig. 6.10b green box outlined in red) allowing a dynamical treatment of the reservoir which is required for the description of time modulated reservoir dynamics in electrically pumped QD devices [79]. Furthermore, however, the approach is capable of including Non-Markovian effects and thus it goes beyond the Markovian approximation typically used in Lindblad formalism to describe scattering processes. The method is illustrated on the example of Coulomb scatterings between quantum emitter and surrounding bulk material (reservoir). However, the approach is also applicable to other systems and interactions mechanisms such as electron-phonon interactions.

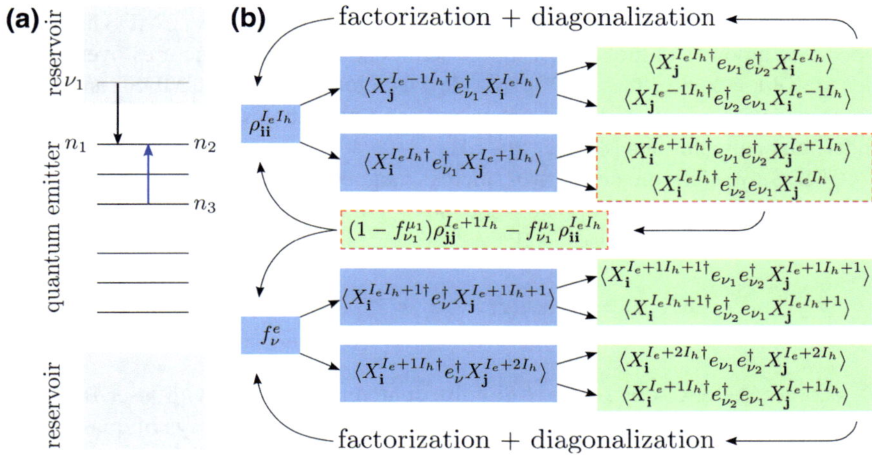

Fig. 6.10 **a** Coulomb induced electron capture via Auger-process $H_{\mathrm{Auger},1}$ of one electron from the reservoir state ν_1 into a quantum emitter state n_1 and a carrier relaxation within the quantum emitter. **b** Scheme of the hybrid density matrix approach for the process described in (**a**). ©2015 American Physical Society, reprinted from [79]

6.6 Two-Dimensional Spectroscopy in Semiconductor Nanostructures

In four-wave mixing spectroscopy, ultrafast sequences of short pulses with precisely tuned durations, time intervals, frequencies, phases and polarizations are applied to the sample. By spreading the optical response across two dimensions, multidimensional coherent spectroscopy allows a detailed insight into the dynamical and structural properties of nanoemitters in third order of the electric field [80–82].

6.6.1 Theory of Four-Wave Mixing Spectroscopy

Higher-order excitonic states in semiconductors are relevant for many quantum information applications such as quantum storage, quantum memory, and quantum entanglement generation via biexciton cascades. However, most nanostructures exhibit highly packed levels and couplings that often remain unresolved in conventional 1D spectroscopy techniques such as photoluminescence and pump-probe. Nonlinear coherent multidimensional spectroscopy (CMDS) offers additional degrees of freedom for accessing correlations homogen/inhomogen broadening and further quasiparticle properties [83–90]: Specific higher exciton correlations can be addressed by controlling the pulse sequence and spreading the opticalresponse across two frequency axes. Starting as a probe mainly for chemical properties, CMDS has developed into a powerful tool to investigate semiconductor nanostructures over the last decade [80, 82, 86, 88, 91–95]. Beside far-field techniques, CMDS can also be combined with micro- and nano-optics and nano-plasmonics in order to characterize individual nanoemitters [81, 92, 96–101].

CMDS transfers the concept of nuclear magnetic resonance spectroscopy from the radio frequency to the optical frequency regime [102]. The sample is excited by a sequence of ultrafast pulses [82]:

$$\mathbf{E}(\mathbf{r}, t) = \sum_j \mathscr{E}_j(\mathbf{r}, t - \tau_j) e^{-i\omega_j(t-\tau_j)+i\varphi_j} + c.c. \tag{6.28}$$

$\mathscr{E}_j$ denotes the envelope of the j-th pulse centered at time τ_j with laser frequency ω_j. The phases φ_j are used to separate different quantum pathways of quasiparticle excitations that can be extracted, e.g., via phase cycling [102]. different $\varphi_i = \mathbf{k}_i \cdot \mathbf{r}$, by repeated measurements e.g. choosing a combination cycling phases φ_i or by a temporal modulation $\varphi_i = \Omega_i T$.

In a four-wave mixing (FWM) experiment, the optical response in third order of the electric field ($\chi^{(3)}$ limit) is probed, yielding the common CMDS signals such as photon echo or double quantum coherence, cf. Fig. 6.11. The α-component of the third-order contribution to the optical polarization reads:

$$\begin{aligned} P_\alpha^{(3)}(\mathbf{r}, t) = \int_0^\infty \mathrm{d}t_3 \int_0^\infty \mathrm{d}t_2 \int_0^\infty \mathrm{d}t_1 \sum_{\beta\gamma\delta=1}^{3} R^{(3)}_{\alpha\beta\gamma\delta}(t_3, t_2, t_1) \\ \times E_\beta(\mathbf{r}, t - t_3) E_\gamma(\mathbf{r}, t - t_3 - t_2) E_\delta(\mathbf{r}, t - t_3 - t_2 - t_1), \end{aligned} \tag{6.29}$$

with the time intervals t_i between the optical field interactions and the third order response function $R^{(3)}_{\alpha\beta\gamma\delta}(t_3, t_2, t_1)$ [82]. $P_\alpha^{(3)}(\mathbf{r}, t)$ includes different signals associated with specific phase combinations [82]: photon echo $\phi_I = -\varphi_1 + \varphi_2 + \varphi_3$, non-rephasing photon echo $\phi_{II} = \varphi_1 - \varphi_2 + \varphi_3$, and double quantum coherence $\phi_{III} = \varphi_1 + \varphi_2 - \varphi_3$. The different excitation pathways representing the evolution of the density matrix and contributing to the respective signal are called Liouville space pathways [82]. They are represented by double-sided Feynman diagrams (cf. Fig. 6.11) that are read from bottom to top. Starting with the system in the ground state ρ_{gg}, each system-light interaction denoted by a wavy arrow changes the state of the density matrix basis elements [16]. In the case of double quantum coherence (bottom left panel of Fig. 6.11), the density matrix in both ESA pathways is in a coherence between the ground state g and a single exciton state e after the arrival of the first pulse. The second pulse creates a coherence between the ground state and a two-exciton state f. The third laser pulse acts differently in the two pathways: It generates a $f \leftrightarrow e'$ coherence in pathway 1 and a $e' \leftrightarrow g$ coherence in pathway 2. The last, outgoing pulse represents the signal field. The third-order CMDS signal for a signal phase ϕ is often detected using heterodyning with a fourth local oscillator

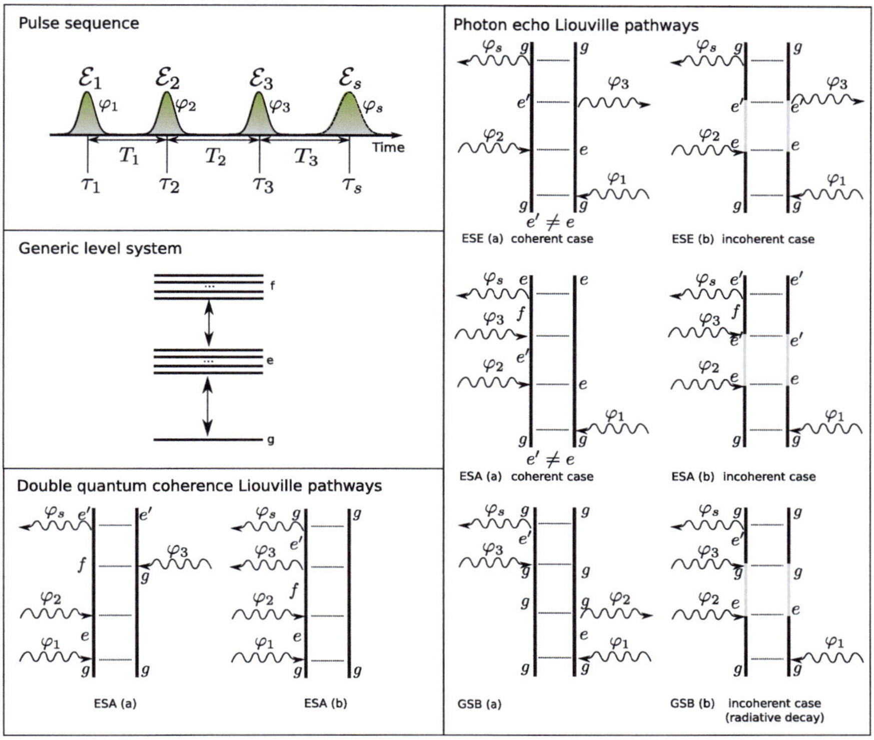

Fig. 6.11 Typical pulse sequence applied in a heterodyne-detected FWM experiment for CMDS (top left panel), probing the three-band model composed of system ground state g and the singly and doubly excited state manifolds e and f (middle left panel). Feynman diagrams represent the Liouville space pathways denoted as excited state absorption (ESA), excited state emission (ESE), and ground state bleaching (GSB) that contribute to the double quantum coherence (bottom left panel) and photon echo signals (right panel)

pulse $\mathcal{E}_s$ centered at time τ_s with frequency ω_s. The signal depends on the delay times T_1, T_2, and T_3 between the four pulses (with $T_i \equiv \tau_{i+1} - \tau_i$):

$$S_\phi^{(3)}(T_3, T_2, T_1) = \int_{-\infty}^{\infty} \mathrm{d}t \mathbf{P}_\phi^{(3)}(t) \cdot \mathcal{E}_s^*(t - \tau_s) e^{\imath\omega_s(t-\tau_s)}. \tag{6.30}$$

Specific two-dimensional CMDS signals are obtained by a Fourier transform with respect to two positive delay times T_i and plotting over the corresponding frequencies Ω_i. For photon echo ϕ_I, usually T_1 and T_3 are Fourier transformed into Ω_1 and Ω_3, which can be interpreted as excitation and emission frequency. The double quantum coherence spectrum is depicted either as a function of Ω_1 and Ω_2 [103] or of Ω_2 and Ω_3 [80], where in the former case the interpretation is more straightforward (since the density matrix evolves identically during T_1 and T_2 for both Liouville pathways contributing to the double quantum coherence), but in the latter case the measurement is more convenient (since the Ω_3 frequency can be directly recorded by a spectrometer).

6.6.2 Mechanisms of Coulomb Interaction in Quantum Dots

To illustrate how CMDS as introduced in Sect. 6.6.1 can detect different coupling mechanisms in nanostructures, Sects. 6.6.3 and 6.6.4 treat the example of two interacting individual QDs, cf. Sect. 6.2.1. In principle, three main Coulomb-induced interaction types between QDs can be distinguished, as shown schematically in Fig. 6.12 with the corresponding Hamiltonians: (a) The electrostatic monopole-monopole interaction only lead to an energy renormalization of the single- and two-exciton states and do not change the state configurations themselves. (b) Förster transfer processes originate from the dipole-dipole coupling in the Taylor expansion of the Coulomb interaction on the unit celllength scale [103]. They transfer the

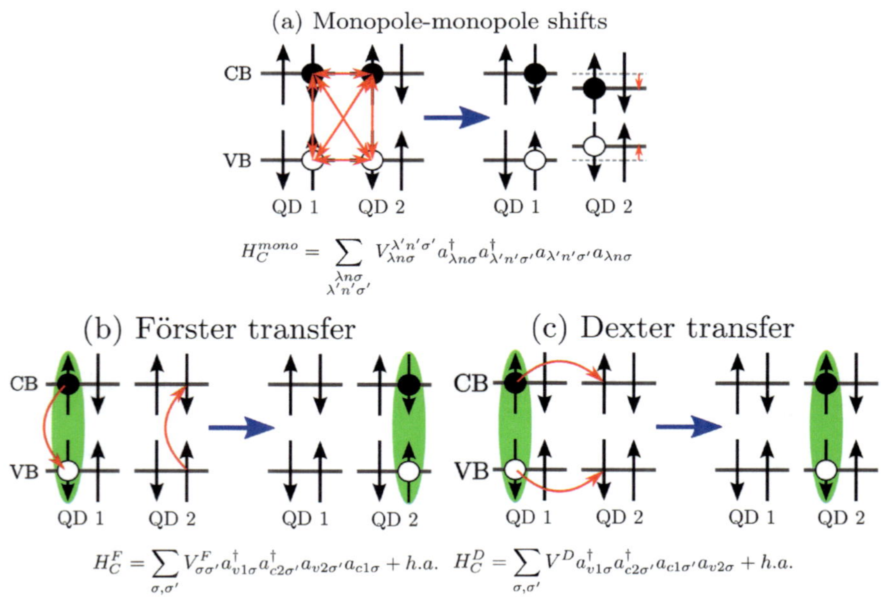

Fig. 6.12 Coulomb coupling between two QDs modeled as spin-degenerate two-level systems: **a** monopole-monopole shifts, **b** Förster-type energy transfer, **c** Dexter-type energy transfer. Excitons are marked as green-shaded areas. $a^{(\dagger)}_{\lambda n\sigma}$ denotes the annihilation (creation) operator of an electron with spin $\sigma = \uparrow, \downarrow$ in band $\lambda = c, v$ of the n-th QD as introduced in Sect. 6.2.1 [81, 103, 104]

excitation energy from one QD (donor) to the other (acceptor) via dipole coupling. (c) Dexter interaction causes a direct transfer of the electron and hole that form the exciton from one QD to the other and therefore requires an overlap between the source and destination electron and hole wave functions.

6.6.3 Phase-Referenced 2D Spectroscopy of Coherently Coupled Individual QDs

In this chapter, photon echo (ϕ_I) and double quantum coherence (ϕ_{III}) spectra of a QD molecule are studied, cf. Fig. 6.13. (Note that ω_i in the plot corresponds to the Fourier transformed i-th pulse delay time, introduced as $\hbar\Omega_i$ in Sect. 6.6.1.) The two QDs in this nanostructure are probably relatively far apart, so that Förster and Dexter processes are of minor importance compared to the monopole-monopole terms illustrated in Fig. 6.12a, (cf. [80]). They cause so-called biexciton shifts, i.e., the energy of a biexciton is altered with respect to the sum of the two single-exciton transitions it is composed of. Two types of biexciton shifts are distinguished: The intradot biexciton binding energies Δ_1 and Δ_2 for biexcitons formed of two excitons within the same QD 1 or 2, respectively, and the interdot biexciton binding energy Δ_{12} for two-exciton states composed of one exciton in QD 1 and the other in QD 2.

In the considered experiment, the second delay time T_2 is fixed at zero delay. Only the coherent ESE, ESA, and GSB pathways contribute to the photon echo signal, i.e., the coherent case (a) of the photon echo Feynman diagrams in Fig. 6.11. All three pathways exhibit a coherence between the ground state g and a single exciton state e between the first and second pulse, such that only $g \to e$ transition energies are identified along the ω_1 axis in the photon echo signals of Fig. 6.13a, b. The ω_3 axis displays the FWM frequency recorded by a spectrometer. Along this axis, ground state to single exciton ($g \to e$ from the ESE and GSB pathways) and single exciton to two exciton ($e \to f$ from the ESA pathway) transitions appear, such that the photon echo of all single exciton resonances shows up along the diagonal $\omega_1 = \omega_3$. For the QD system, this leads to signatures GX_1 and GX_2 along the diagonal in Fig. 6.13a, b at the single-exciton resonance energies $E_1 = 1359.7$ meV of QD 1 and $E_2 = 1358.95$ meV of QD 2. Each QD exhibits two circularly polarized transitions (spin up and spin down, cf. Fig. 6.12) at almost the same energy that are only slightly detuned by spin-orbit coupling, leading to split-up of each exciton resonance GX_1 and GX_2 along the diagonal into clusters of four peaks.

Off-diagonal contributions to the photon echo signal arise if the lower $g \leftrightarrow e$ and upper $e \leftrightarrow f$ transitions are different, e.g., through biexciton shifts. If biexciton shifts were negligible compared to the linewidth, the ESE and GSB pathways would interfere destructively with the ESA pathway causing a cancellation of the off-diagonal peaks. The biexciton shifts in the considered QD molecule are large enough to let off-diagonal contributions appear: The single-exciton to intradot-biexciton coherence peaks X_1B_1 and X_2B_2 in Fig. 6.13a, b are redshifted along the ω_3 axis by the intradot biexciton binding energies $\Delta_1 = -3.3$ meV and $\Delta_2 = -3.6$ meV,

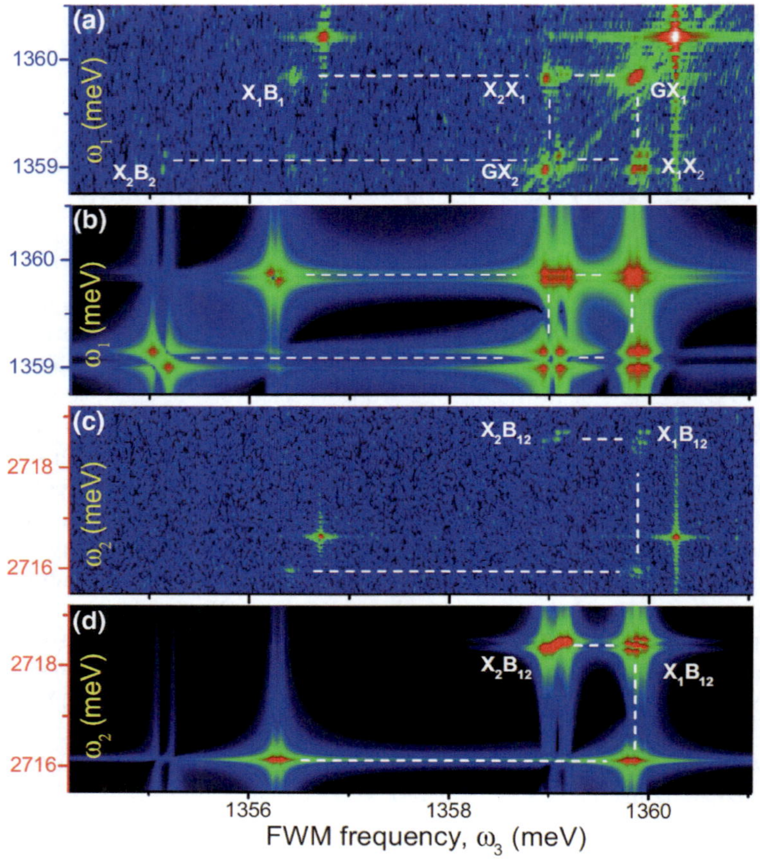

Fig. 6.13 **a** Measured and **b** simulated two-dimensional photon echo spectra of a QD molecule. **c** Measured and **d** simulated two-dimensional double quantum coherence spectra of the QD molecule. ©2017 American Physical Society, reprinted from [80]

respectively. Furthermore, two off-diagonal cross-peaks X_2X_1 and X_1X_2 appear at the spectral positions ($\omega_3 = E_2$; $\omega_1 = E_1$) and ($\omega_3 = E_1$; $\omega_1 = E_2$), respectively indicating an electrostatic interaction between an exciton in QD 1 and an exciton in QD 2. This proves coherent interdot coupling between the two QDs with a small biexciton binding energy of the order $\Delta_{12} = 90\ \mu$eV [80].

Only two diagrams (ESA (a) and ESA (b) depicted in Fig. 6.11) contribute to the double quantum coherence spectrum of Fig. 6.13c, d. Both pathways are in a ground state g to biexciton f coherence between the second and third pulse, so that the position of a peak on the corresponding $\hbar\Omega_2 = \omega_2$ axis indicates the energy of the involved biexciton state. The interdot coupling between the two QDs is indicated by the peak pair X_1B_{12} and X_2B_{12} at the interaction-shifted two-exciton transition energy $\omega_2 = E_1 + E_2 + \Delta_{12} = 2718.74$ meV with FWM frequencies $\omega_3 = E_1$ and $\omega_3 = E_2$. e' coherence after the third pulse ($\hbar\Omega_3 = \omega_3$ axis), so that the associated peaks in Fig. 6.13c, d indicate the single exciton energy reduced by the biexciton binding energy of the excitons forming the biexciton state. The ESA (b) quantum pathway is in a ground state g to single exciton e coherence after the third pulse, therefore the related peaks indicate directly the single exciton energy of the single exciton states from which the biexciton is formed. The photon echo and the double quantum coherence in Fig. 6.13 are aligned, so that the diagonal contribution on the photon echo indicate, if the transition is a ground state to single exciton contribution.

6.6.4 *Förster and Dexter Transfer Processes in Coupled Nanostructures*

If only the electrostatic monopole-monopole interaction is present cf. Sect. 6.6.3, the single exciton and two-exciton states depicted in Fig. 6.14a could be treated by only considering the energy shifts as for the QD molecule in the last subsection. However, excitation energy and charge transfer processes such as Förster and Dexter interaction (cf. Fig. 6.12b, c) transfer the excitation from one QD to the other QD. Consequently, the energy eigenstates of the system are superpositions of the uncoupled states, depicted in Fig. 6.14a.

The Förster coupling element $V^F_{\sigma\sigma'}$ [81, 103] (cf. Fig. 6.12b) depends on the spins σ and σ' of the source and final electron/hole states, since the microscopic dipole elements of the respective interband transitions enter the dipole-dipole interaction. Therefore, dark exciton configurations (with zero interband dipole moment) do not participate in Förster transfer processes. A bright exciton (i.e., an electron-hole pair with opposite spins of the electron and the hole) recombines in one QD, while the excitation energy is transferred to the other QD where a new bright exciton is created. Depending on the relative orientation of the crystal axis of the two QDs, the Förster transfer process is spin preserving or spin flipping [103, 105] or something in between: As determined by $V^F_{\sigma\sigma'}$, the spin configuration of the exciton transferred to the acceptor QD can be preserved ($\sigma = \sigma'$) or changed ($\sigma \neq \sigma'$) compared to the initial exciton spin configuration in the donor QD.

Dexter excitation transfer requires electronic wave function overlap between the two QDs and it dominates over Förster transfer for shorter distances between the two QDs. The transferred electron and hole do not change their spins, which is reflected in the Dexter Hamiltonian, cf. (Fig. 6.12c). Beside the usual bright excitons, Dexter processes can transfer dark excitons (with identical spins of the electron and hole) from one QD to the other, what Förster can not achieve due to the involved interband dipole moments, necessarily involving only bright excitons.

Thus, spin selection rules of Förster and Dexter processes are very different and should allow to distinguish the two processes spectroscopically by their peak pattern. It is possible to distinguish the signatures of spin-preserving and spin-flipping

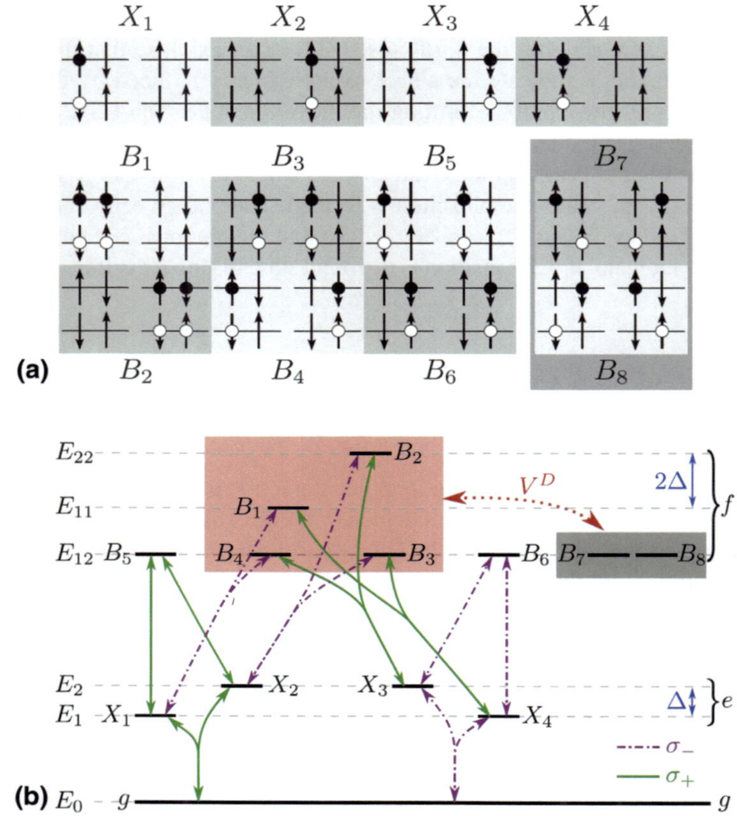

Fig. 6.14 **a** Exciton X_i and biexciton B_i electron-hole configurations for two coupled QDs (only S–S states) including four bright single excitons X_1–X_4, six bright two excitons B_1–B_6, and two dark two exciton states B_7 and B_8. **b** Level scheme for possible excitation pathways of the uncoupled system. E_0 is the ground state energy, E_n are single exciton energies and E_{nm} are the two exciton states. Δ is the band gap detuning between the QDs. Green and purple show the possible excitation for right-hand and left-hand circularly polarized light, respectively. ©2015 American Physical Society, reprinted from [103]

Förster processes in linear absorption spectra (Fig. 6.15(i)a, d): If the coupling is spin-preserving, only the spin-up excitons X_1 and X_2 (cf. Fig. 6.14a) are optically accessible using σ_+ polarization, as illustrated in the excitation scheme of Fig. 6.14b. In contrast, spin-flipping Förster coupling involves all four bright excitons X_1–X_4, leading to four absorption peaks. However, in linear absorption it is not possible to distinguish spin-preserving Förster and Dexter transfer (Fig. 6.15(i)a, (ii)a) since the overall peak pattern looks alike. The main reason for the failure of linear optics to separate Dexter and Förster dominating transfer processes is that only bright excitons and not two-exciton states are probed in linear optics.

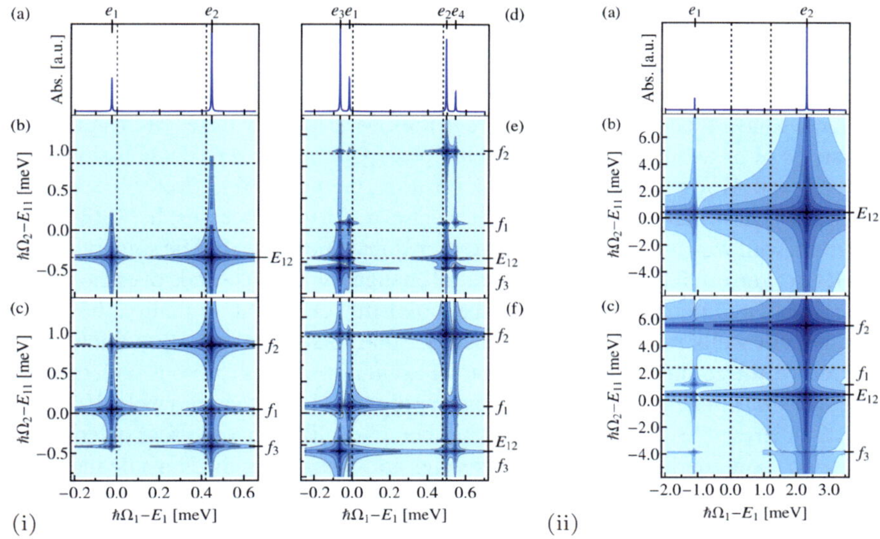

Fig. 6.15 **i**, **ii** Show a set of spectra for (i) the Förster dominated case (spin-preserving Förster coupling (**a**)–(**c**) and spin-flipping (**d**)–(**f**)) and for **ii** the Dexter dominated case. **a**, **d** show the linear absorption spectrum, **b**, **e** double quantum coherence spectra with co-circular excitation $\sigma_+\sigma_+\sigma_+\sigma_+$ and **c**, **f** double quantum coherence spectra with cross-circular excitation $\sigma_+\sigma_-\sigma_+\sigma_-$. ©2015 American Physical Society, reprinted from [103]

Double quantum coherence CMDS (cf. Fig. 6.11 in Sect. 6.6.1) is specifically designed to probe biexcitons and the connected exciton states and should be the ideal spectroscopy to determine the transfer process. A co-circular excitation $\sigma_+\sigma_+\sigma_+\sigma_+$ leads also for the double quantum coherence to the same type of peak pattern for spin-preserving Förster and Dexter processes (cf. Fig. 6.15(i)b, (ii)b). The co-circular pulse configuration excites only electrons of the same spin, thus excluding biexcitons within one QD that would allow a Dexter transfer to and thus hybridization with dark two exciton configurations (c.f. dark states B_7 and B_8 in Fig. 6.14a). This prevents signatures that differentiate Förster and Dexter processes. However, for cross-circular $\sigma_+\sigma_-\sigma_+\sigma_-$ polarization, a differentiation of Förster and Dexter transfer is indeed possible: The key signatures are at the naked two exciton energy E_{12}, where the Förster dominated spectra show no resonance for cross-circular excitation but the Dexter dominated spectra show a signature. Here, Dexter coupling leads to a hybridization between the bright biexciton configurations B_1 to B_4 and two dark biexciton states B_7 and B_8, cf. Fig. 6.14. The Dexter shifts associated with the hybridization cancel for some of the two-exciton states leading to a peak at the uncoupled resonance energy E_{12}. Furthermore, the state at this energy gets an enhanced oscillator strength compared to the shifted hybridized states. The overall structure of the peak patterns for cross-circular pulse polarizations thus allow to uniquely determine dominances of spin-preserving and spin-flip Förster dominated transfer or Dexter dominated transfer.

6.6.5 Localization Dynamics of Excitons in Disordered Semiconductor Quantum Wells

So far, we have focused on the CMDS capabilities to identify the exciton and biexciton states of nanostructures using specific peak patterns for zero delay waiting time $T = T_2 = 0$. However, beside the peak position, the peak shape over the delay time T_2 contains information about so called spectral diffusion processes: In [106] a GaAs quantum well was studied using a two-dimensional photon echo. The signature of the 1s HH exciton (cf. Fig. 6.16) shows spectral changes of its peak shape over the waiting time $T = T_2$. One reason for such a spectral diffusion is that, in reality, the growth of quantum wells results always in disorder due to surface and alloy fluctuations. Therefore, the common picture that optical excitation creates excitons with center of mass (COM) momentum $\mathbf{Q} = 0$ does only apply to an ideal quantum well [106]. Typically, an ensemble of exciton COM states being localized in a disorder potential can be found. Low energy exciton disorder states are stronger localized inside the potential and have higher oscillator strength. The number of exciton states increases for higher energies, especially above the mobility edge [106], but their oscillator strength is decreasing. The distribution of oscillator strength can be seen along the diagonal in Fig. 6.16(i)a, (ii)a for $T_2 = 0$ ps, where above 1545 meV almost no contribution of higher energy COM exciton states is visible. In contrast, in the photon echo the transition from high to low oscillator strength is more pronounced, since the dipole enters in fourth order. Inspecting Fig. 6.16 for longer delay times the shape of the peak changes considerably. For example, an additional feature appears for high excitation energy and low emission energy, where no contribution was visible along the diagonal (marked in Fig. 6.16(i)d with a circle). This spectral redistribution is caused by incoherent exciton-phonon relaxation inside the disorder potential due to acoustical phonons [106]. The contributing diagrams are the incoherent ESE (b), ESA (b) and GSB (b) processes in Fig. 6.11, where an exciton initially excited at one spectral position relaxes during T_2 to another COM exciton 1s state inside the disorder potential due to acoustic phonon interaction. For the excitation energy of the additional off-diagonal feature (see circle in Fig. 6.16 (i)d) the oscillator strength of the exciton states is so small, that no contribution is visible for $T_2 = 0$ ps since the dipole moment enters in fourth order to the signal. On the other hand, for $T_2 = 30$ ps, the excitons initially created by the first two pulses on the higher energy side have mainly relaxed toward lower energy exciton COM states. These states are more localized and have a high oscillator strength. Thus, for off-diagonal features the initial dark exciton states enter the optical signal in second order of the dipole moment and the lower energy exciton states also enter the optical signal in second order of the dipole moment. This results in a higher signal at the off-diagonal feature than at the diagonal, where the dipole moment of the low oscillator strength exciton state enters in fourth order. Overall, the shape of the 1s HH exciton signature over the waiting time T_2 is determined by the exciton relaxation processes, which can be formulated using a relaxation Green's function. In [107], a method was proposed to reconstruct

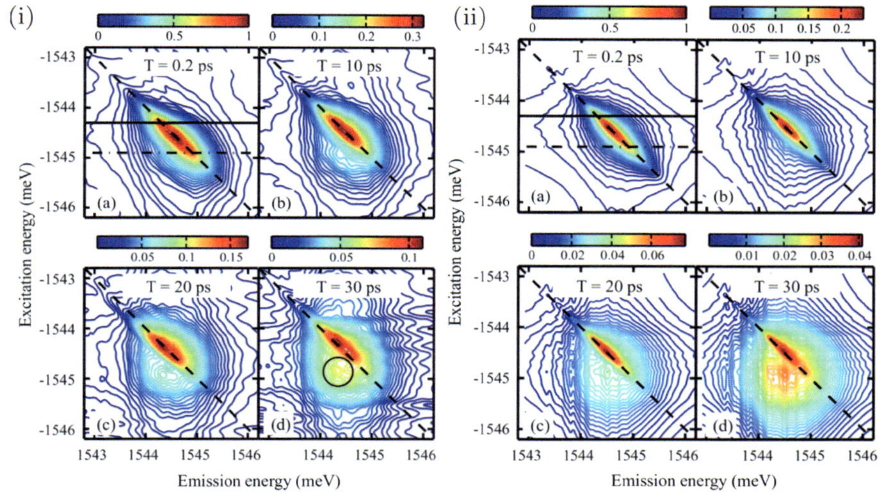

Fig. 6.16 **i** Measured and **ii** calculated two-dimensional photon echo spectra of a disordered GaAs quantum well for different delay times $T = T_2$ between the second and third center on the 1s HH transition [106]. ©2017 American Physical Society, reprinted from [106]

the Green's function from two-dimensional photon echo data [107] and to obtain enhanced insight into the relaxation process, which is here considerably masked by the homogeneous broadening.

6.7 Conclusion

In conclusion, light emission and the photon statistics of semiconductor quantum dots are dominated by intrinsic (electron-phonon) and, for coupled nanostructures, extrinsic (quantum dot-quantum dot) coupling. This is seen in emission and absorption, where dominant sidebands caused by induced and spontaneous emission of phonons and Coulomb interaction induced dot-dot coupling occur.

Several aspects of the photon statistics of entangled and heralded quantum dot photon sources were discussed:

First, the limitations of photon pair entanglement using biexciton cascade emission was studied, specifically addressing a finite fine-structure splitting and exciton-phonon coupling as a dissipative process. In particular, a joint theory-experiment study on the two-photon dressing of a three-level system has shown that the time ordering of the cascaded photons can be removed or conserved, demonstrating sources of orthogonally polarized, bunched or antibunched photon pairs.

Second, the investigation of the photon emission into different spatial directions of a micropillar cavity has indicated a temporally ordered generation of two photons constituting a heralded single photon source.

Third, to attack the transition of the single (few) emitter to many emitter limit we presented an exact numerical approach, capable of solving the full density matrix up to hundred identical quantum dots coupled to an optical cavity mode. In semi-classical spectroscopy, we studied intraband transitions and signatures of interdot coupling:

First, the spectral signatures of intraband transitions between bound quantum dot states and reservoir states were used to derive an analytical formula for reconstructing the bound quantum dot ground state wave function from experimental data. Additionally, Coulomb effects on intraband transitions were introduced to discuss new interband excitonic states.

Second, to treat many particle system-reservoir correlations, we presented a novel theoretical factorization scheme of solid state continuum reservoir states and localized quantum dot states, utilizing an exact diagonalization scheme for the localizied states inside the quantum dot.

Third, for quantum dots, we developed a method to distinguish electrostatic couplings as well as excitation transfer processes such as Förster and Dexter interaction using nonlinear two-dimensional spectroscopy. In this context, also, we analyzed the contribution of exciton-phonon interactions to spectral diffusion for a disordered GaAs quantum well as seen in corresponding experiments.

Acknowledgements We gratefully acknowledge financial support from the Deutsche Forschungsgemeinschaft (DFG) through Sonderforschungsbereich 787 Projekt B1 (43659573).

References

1. D. Bimberg, M. Grundmann, N. Ledentsov, *Quantum Dot Heterostructures* (Wiley, 1999)
2. B. Lingnau, K. Lüdge, B. Herzog, M. Kolarczik, Y. Kaptan, U. Woggon, N. Owschimikow, Phys. Rev. B **94**, 014305 (2016). https://doi.org/10.1103/PhysRevB.94.014305
3. J. Seebeck, T.R. Nielsen, P. Gartner, F. Jahnke, Phys. Rev. B **71**, 125327 (2005). https://doi.org/10.1103/PhysRevB.71.125327
4. I.A. Ostapenko, G. Hönig, S. Rodt, A. Schliwa, A. Hoffmann, D. Bimberg, M.R. Dachner, M. Richter, A. Knorr, S. Kako, Y. Arakawa, Phys. Rev. B **85**, 081303 (2012). https://doi.org/10.1103/PhysRevB.85.081303
5. H. Haug, S. Koch, *Quantum Theory of the Optical and Electronic Properties of Semiconductors*, 5th edn. (2009). https://doi.org/10.1088/1742-6596/248/1/012018
6. G. Kießlich, A. Wacker, E. Schöll, S.A. Vitusevich, A.E. Belyaev, S.V. Danylyuk, A. Förster, N. Klein, M. Henini, Phys. Rev. B **68**, 125331 (2003). https://doi.org/10.1103/PhysRevB.68.125331
7. S. Franke, S. Hughes, M.K. Dezfouli, P.T. Kristensen, K. Busch, A. Knorr, M. Richter, arXiv:1808.06392(2018)
8. P. Tighineanu, C.L. Dreeßen, C. Flindt, P. Lodahl, A.S. Sørensen, Phys. Rev. Lett. **120**, 257401 (2018). https://doi.org/10.1103/PhysRevLett.120.257401
9. S. Bounouar, M. Müller, A.M. Barth, M. Glässl, V.M. Axt, P. Michler, Phys. Rev. B **91**, 161302 (2015). https://doi.org/10.1103/PhysRevB.91.161302

10. G.D. Mahan, *Many-Particle Physics* (Plenum Press, New York, 1990)
11. G. Czycholl, *Theoretische Festkörperphysik* (Springer, 2008)
12. K. Kaasbjerg, K.S. Thygesen, K.W. Jacobsen, Phys. Rev. B **85**, 115317 (2012). https://doi.org/10.1103/PhysRevB.85.115317
13. K. Kaasbjerg, K.S. Thygesen, A.P. Jauho, Phys. Rev. B **87**, 235312 (2013)
14. B. Krummheuer, V.M. Axt, T. Kuhn, Phys. Rev. B **65**, 195313 (2002). https://doi.org/10.1103/PhysRevB.65.195313
15. J. Förstner, C. Weber, J. Danckwerts, A. Knorr, Physica Status Solidi (b), **238**(3), 419. https://doi.org/10.1002/pssb.200303155
16. S. Mukamel, *Principles of Nonlinear Optical Spectroscopy* (Oxford University Press, New York, 1995)
17. A. Schliwa, M. Winkelnkemper, D. Bimberg, Phys. Rev. B **76**, 205324 (2007). https://doi.org/10.1103/PhysRevB.76.205324
18. D. Reuter, P. Kailuweit, A.D. Wieck, U. Zeitler, O. Wibbelhoff, C. Meier, A. Lorke, J.C. Maan, Phys. Rev. Lett. **94**, 026808 (2005). https://doi.org/10.1103/PhysRevLett.94.026808
19. V.A. Fonoberov, E.P. Pokatilov, A.A. Balandin, Phys. Rev. B **66**, 085310 (2002). https://doi.org/10.1103/PhysRevB.66.085310
20. D. Nikonov, A. Imamoğlu, L. Butov, H. Schmidt, Phys. Rev. Lett. **79**, 4633 (1997). https://doi.org/10.1103/PhysRevLett.79.4633
21. T.R. Nielsen, P. Gartner, M. Lorke, J. Seebeck, F. Jahnke, Phys. Rev. B **72**, 235311 (2005). https://doi.org/10.1103/PhysRevB.72.235311
22. A. Zimmermann, S. Kuhn, M. Richter, Phys. Rev. B **93**, 035308 (2016). https://doi.org/10.1103/PhysRevB.93.035308
23. A. Carmele, A. Knorr, M. Richter, Phys. Rev. B **79**, 035316 (2009). https://doi.org/10.1103/PhysRevB.79.035316
24. M. Fox, in *Quantum Optics: An Introduction*. Oxford Master Series in Physics (OUP Oxford, 2006)
25. H. Breuer, F. Petruccione, *The Theory of Open Quantum Systems* (Oxford University Press, 2002)
26. M. Berman, R. Kosloff, H. Tal-Ezer, J. Phys. A Math. General **25**(5), 1283 (1992). http://stacks.iop.org/0305-4470/25/i=5/a=031
27. C. Gies, F. Jahnke, W.W. Chow, Phys. Rev. A **91**, 061804 (2015). https://doi.org/10.1103/PhysRevA.91.061804
28. J. Clauser, A. Shimony, Rep. Progress Phys. **41**(12), 1881 (1978)
29. A. Aspect, P. Grangier, G. Roger, Phys. Rev. Lett. **47**(7), 460 (1981)
30. A. Carmele, A. Knorr, Phys. Rev. B **84**(7), 075328 (2011)
31. O. Benson, C. Santori, M. Pelton, Y. Yamamoto, Phys. Rev. Lett. **84**, 2513 (2000). https://doi.org/10.1103/PhysRevLett.84.2513
32. O. Benson, C. Santori, M. Pelton, Y. Yamamoto, Phys. Rev. Lett. **84**(11), 2513 (2000)
33. C. Salter, R. Stevenson, I. Farrer, C. Nicoll, D. Ritchie, A. Shields, Nature **465**(7298), 594 (2010)
34. M.R. Dachner, E. Malic, M. Richter, A. Carmele, J. Kabuss, A. Wilms, J.E. Kim, G. Hartmann, J. Wolters, U. Bandelow, et al., Physica Status Solidi (b) **247**(4), 809 (2010)
35. A. Carmele, F. Milde, M.R. Dachner, M.B. Harouni, R. Roknizadeh, M. Richter, A. Knorr, Phys. Rev. B **81**(19), 195319 (2010)
36. S.M. Hein, F. Schulze, A. Carmele, A. Knorr, Phys. Rev. Lett. **113**(2), 027401 (2014)
37. N. Akopian, N. Lindner, E. Poem, Y. Berlatzky, J. Avron, D. Gershoni, B. Gerardot, P. Petroff, Phys. Rev. Lett. **96**(13), 130501 (2006)
38. G. Callsen, A. Carmele, G. Hönig, C. Kindel, J. Brunnmeier, M. Wagner, E. Stock, J. Reparaz, A. Schliwa, S. Reitzenstein et al., Phys. Rev. B **87**(24), 245314 (2013)
39. T. Heindel, A. Thoma, M. von Helversen, M. Schmidt, A. Schlehahn, M. Gschrey, P. Schnauber, J.H. Schulze, A. Strittmatter, J. Beyer et al., Nat. Commun. **8**, 14870 (2017)
40. R. Winik, D. Cogan, Y. Don, I. Schwartz, L. Gantz, E.R. Schmidgall, N. Livneh, R. Rapaport, E. Buks, D. Gershoni, Phys. Rev. B **95**, 235435 (2017). https://doi.org/10.1103/PhysRevB.95.235435

41. S. Bounouar, C. de la Haye, M. Strauß, P. Schnauber, A. Thoma, M. Gschrey, J.H. Schulze, A. Strittmatter, S. Rodt, S. Reitzenstein, Appl. Phys. Lett. **112**, 153107 (2018). https://doi.org/10.1063/1.5020242
42. J. Zhang, J.S. Wildmann, F. Ding, R. Trotta, Y. Huo, E. Zallo, D. Huber, A. Rastelli, O.G. Schmidt, Nat. Commun. **6**, 10067 (2015)
43. Y. Lu, N.L. Naumann, J. Cerrillo, Q. Zhao, A. Knorr, A. Carmele, Phys. Rev. A **95**(6), 063840 (2017)
44. R.J. Young, R.M. Stevenson, P. Atkinson, K. Cooper, D.A. Ritchie, A.J. Shields, New J. Phys. **8**(2), 29 (2006)
45. J.E. Avron, G. Bisker, D. Gershoni, N.H. Lindner, E.A. Meirom, R.J. Warburton, Phys. Rev. Lett. **100**(12), 120501 (2008)
46. S. Bounouar, M. Strauß, A. Carmele, P. Schnauber, A. Thoma, M. Gschrey, J. Schulze, A. Strittmatter, S. Rodt, A. Knorr, S. Reitzenstein, Phys. Rev. Lett. **118**(23), 233601 (2017)
47. C. Schrama, G. Nienhuis, H. Dijkerman, C. Steijsiger, H. Heideman, Phys. Rev. A **45**(11), 8045 (1992)
48. F. Hargart, M. Müller, K. Roy-Choudhury, S. Portalupi, C. Schneider, S. Höfling, M. Kamp, S. Hughes, P. Michler, Phys. Rev. B **93**(11), 115308 (2016)
49. P. Ardelt, M. Koller, T. Simmet, L. Hanschke, A. Bechtold, A. Regler, J. Wierzbowski, H. Riedl, J. Finley, K. Müller, Phys. Rev. B **93**(16), 165305 (2016)
50. S.C. Kuhn, A. Knorr, S. Reitzenstein, M. Richter, Opt. Express **24**(22), 25446 (2016). https://doi.org/10.1364/OE.24.025446, http://www.opticsexpress.org/abstract.cfm?URI=oe-24-22-25446
51. M. Richter, A. Carmele, A. Sitek, A. Knorr, Phys. Rev. Lett. **103**, 087407 (2009). https://doi.org/10.1103/PhysRevLett.103.087407
52. M. Gegg, M. Richter, New J. Phys. **18**(4), 043037 (2016). http://stacks.iop.org/1367-2630/18/i=4/a=043037
53. M. Gegg, M. Richter, Sci. Rep. **7**(1), 16304 (2017). https://doi.org/10.1038/s41598-017-16178-8
54. M. Richter, M. Gegg, T.S. Theuerholz, A. Knorr, Phys. Rev. B **91**, 035306 (2015). https://doi.org/10.1103/PhysRevB.91.035306
55. B. Chase, J. Geremia, Phys. Rev. A **78**, 052101 (2008)
56. B. Baragiola, B. Chase, J. Geremia, Phys. Rev. A **81**, 032104 (2010)
57. S. Hartmann, Quantum Inf. Comput. **16**, 1333 (2016)
58. M. Xu, D. Tieri, M. Holland, Phys. Rev. A **87**, 062101 (2013)
59. L. Novo, T. Moroder, O. Gühne, Phys. Rev. A **88**, 012305 (2013)
60. F. Damanet, D. Braun, J. Martin, Phys. Rev. A **94**, 033838 (2016)
61. Z.X. Gong, M. Xu, M. Foss-Feig, J. Thompson, A. Rey, M. Holland, A. Gorshkov, arXiv:1611.00797 (2016)
62. P. Kirton, J. Keeling, Phys. Rev. Lett. **118**, 123602 (2017)
63. M. Gegg, A. Carmele, A. Knorr, M. Richter, New J. Phys. **20**(1), 013006 (2018). http://stacks.iop.org/1367-2630/20/i=1/a=013006
64. N. Shammah, N. Lambert, F. Nori, S. De Liberato, Phys. Rev. A **96**, 023863 (2017)
65. P. Kirton, J. Keeling, New J. Phys. **20**(1), 015009 (2018). http://stacks.iop.org/1367-2630/20/i=1/a=015009
66. N. Shammah, S. Ahmed, N. Lambert, S. De Liberato, F. Nori, arXiv:1805.05129 (2018)
67. T. Warnakula, M.I. Stockman, M. Premaratne, JOSA B **35**(6), 1397 (2018)
68. S. Sarkar, J. Satchell, J. Phys. A: Math. Gen. **20**, 2147 (1987)
69. S. Sarkar, J. Satchell, Europhys. Lett. **3**, 797 (1987)
70. H. Carmichael, *Statistical Methods in Quantum Optics I: Master Equations and Fokker-Planck Equations* (Springer, 2002)
71. S.C. Kuhn, A. Knorr, M. Richter, N. Owschimikow, M. Kolarczik, Y.I. Kaptan, U. Woggon, Phys. Rev. B **89**, 201414 (2014). https://doi.org/10.1103/PhysRevB.89.201414
72. F. Quochi, M. Dinu, L.N. Pfeiffer, K.W. West, C. Kerbage, R.S. Windeler, B.J. Eggleton, Phys. Rev. B **67**, 235323 (2003). https://doi.org/10.1103/PhysRevB.67.235323

73. G. Dasbach, T. Baars, M. Bayer, A. Larionov, A. Forchel, Phys. Rev. B **62**, 13076 (2000). https://doi.org/10.1103/PhysRevB.62.13076
74. S. Dommers, V.V. Temnov, U. Woggon, J. Gomis, J. Martinez-Pastor, M. Laemmlin, D. Bimberg, Appl. Phys. Lett. **90**, 033508 (2007)
75. J. Gomis-Bresco, S. Dommers, V.V. Temnov, U. Woggon, M. Laemmlin, D. Bimberg, E. Malic, M. Richter, E. Schöll, A. Knorr, Phys. Rev. Lett. **101**, 256803 (2008). https://doi.org/10.1103/PhysRevLett.101.256803
76. S.C. Kuhn, Theory of optical and dissipative processes in quantum dots. Ph.D. thesis, Technische Universität Berlin (2016)
77. S.C. Kuhn, M. Richter, Phys. Rev. B **90**, 125308 (2014). https://doi.org/10.1103/PhysRevB.90.125308
78. S.C. Kuhn, M. Richter, Proc.SPIE **9746**, 9746 (2016). https://doi.org/10.1117/12.2207635
79. S.C. Kuhn, M. Richter, Phys. Rev. B **91**, 155309 (2015). https://doi.org/10.1103/PhysRevB.91.155309
80. V. Delmonte, J.F. Specht, T. Jakubczyk, S. Höfling, M. Kamp, C. Schneider, W. Langbein, G. Nogues, M. Richter, J. Kasprzak, Phys. Rev. B **96**, 041124 (2017). https://doi.org/10.1103/PhysRevB.96.041124
81. J.F. Specht, M. Richter, Appl. Phys. B **122**(4), 97 (2016). https://doi.org/10.1007/s00340-016-6368-1
82. D. Abramavicius, B. Palmieri, D.V. Voronine, F.Šanda, S. Mukamel, Chemical Reviews **109**(6), 2350 (2009). https://doi.org/10.1021/cr800268n
83. V. Chernyak, W.M. Zhang, S. Mukamel, J. Chem. Phys. **109**(21), 9587 (1998). http://scitation.aip.org/content/aip/journal/jcp/109/21/10.1063/1.477621
84. T. Brixner, J. Stenger, H.M. Vaswani, M. Cho, R.E. Blankenship, G.R. Fleming, Nature **434**(7033), 625 (2005)
85. G.S. Engel, T.R. Calhoun, E.L. Read, T.K. Ahn, T. Mancal, Y.C. Cheng, R.E. Blankenship, G.R. Fleming, Nature **446**, 782 (2007)
86. T. Zhang, I. Kuznetsova, T. Meier, X. Li, R.P. Mirin, P. Thomas, S.T. Cundiff, PNAS **104**(36), 14227 (2007)
87. A. Nemeth, F. Milota, T. Mancal, T. Pullerits, J. Sperling, J. Hauer, H.F. Kauffmann, N. Christensson, J. Chem. Phys. **133**(9) (2010). https://doi.org/10.1063/1.3474995
88. J. Tollerud, J.A. Davis, JOSA B **33**(7), C108 (2016). http://josab.osa.org/abstract.cfm?URI=josab-33-7-C108
89. R.D. Mehlenbacher, T.J. McDonough, M. Grechko, M.Y. Wu, M.S. Arnold, M.T. Zanni, Nature Comm. **6** (2015)
90. E. Cassette, J.C. Dean, G.D. Scholes, Small **12**(16), 2234 (2016)
91. G. Moody, M.E. Siemens, A.D. Bristow, X. Dai, A.S. Bracker, D. Gammon, S.T. Cundiff, Phys. Rev. B **83**, 245316 (2011). https://doi.org/10.1103/PhysRevB.83.245316
92. J. Kasprzak, B. Patton, V. Savona, W. Langbein, Nat. Photonics **5**(1), 57 (2010). https://doi.org/10.1038/nphoton.2010.28410.1038/nphoton.2010.284, https://www.nature.com/articles/nphoton.2010.284#supplementary-information
93. E. Harel, S.M. Rupich, R.D. Schaller, D.V. Talapin, G.S. Engel, Phys. Rev. B **86**, 075412 (2012). https://doi.org/10.1103/PhysRevB.86.075412
94. G. Moody, R. Singh, H. Li, I.A. Akimov, M. Bayer, D. Reuter, A.D. Wieck, A.S. Bracker, D. Gammon, S.T. Cundiff, Phys. Rev. B **87**, 041304 (2013). https://doi.org/10.1103/PhysRevB.87.041304
95. G. Moody, S.T. Cundiff, Adv. Phys. X **2**(3), 641 (2017). https://doi.org/10.1080/23746149.2017.1346482. PMID: 28894306
96. M. Aeschlimann, T. Brixner, A. Fischer, C. Kramer, P. Melchior, W. Pfeiffer, C. Schneider, C. Strüber, P. Tuchscherer, D.V. Voronine, Science 1209206 (2011)
97. M. Richter, F. Schlosser, M. Schoth, S. Burger, F. Schmidt, A. Knorr, S. Mukamel, Phys. Rev. B **86**, 085308 (2012). https://doi.org/10.1103/PhysRevB.86.085308
98. F. Schlosser, A. Knorr, S. Mukamel, M. Richter, New J. Phys. **15**(2), 025004 (2013). http://stacks.iop.org/1367-2630/15/i=2/a=025004

99. M. Krecik, S.M. Hein, M. Schoth, M. Richter, Phys. Rev. A **92**, 052113 (2015). https://doi.org/10.1103/PhysRevA.92.052113
100. E.W. Martin, S.T. Cundiff, Phys. Rev. B **97**, 081301 (2018). https://doi.org/10.1103/PhysRevB.97.081301
101. S. Goetz, D. Li, V. Kolb, J. Pflaum, T. Brixner, Opt. Express **26**(4), 3915 (2018). https://doi.org/10.1364/OE.26.003915, http://www.opticsexpress.org/abstract.cfm?URI=oe-26-4-3915
102. P. Tian, D. Keusters, Y. Suzaki, W.S. Warren, Science **300**(5625), 1553 (2003). https://doi.org/10.1126/science.1083433, http://science.sciencemag.org/content/300/5625/1553
103. J.F. Specht, A. Knorr, M. Richter, Phys. Rev. B **91**, 155313 (2015). https://doi.org/10.1103/PhysRevB.91.155313
104. M. Richter, K.J. Ahn, A. Knorr, A. Schliwa, D. Bimberg, M.E.A. Madjet, T. Renger, Physica Status Solidi (b) **243**(10), 2302. https://doi.org/10.1002/pssb.200668053
105. G.D. Scholes, D.L. Andrews, Phys. Rev. B **72**, 125331 (2005). https://doi.org/10.1103/PhysRevB.72.125331
106. R. Singh, M. Richter, G. Moody, M.E. Siemens, H. Li, S.T. Cundiff, Phys. Rev. B **95**, 235307 (2017). https://doi.org/10.1103/PhysRevB.95.235307
107. M. Richter, R. Singh, M. Siemens, S.T. Cundiff, Sci. Adv. **4**, 6 (2018). https://doi.org/10.1126/sciadv.aar7697, http://advances.sciencemag.org/content/4/6/eaar7697

Chapter 7
Multi-dimensional Modeling and Simulation of Semiconductor Nanophotonic Devices

Markus Kantner, Theresa Höhne, Thomas Koprucki, Sven Burger, Hans-Jürgen Wünsche, Frank Schmidt, Alexander Mielke and Uwe Bandelow

Abstract Self-consistent modeling and multi-dimensional simulation of semiconductor nanophotonic devices is an important tool in the development of future integrated light sources and quantum devices. Simulations can guide important technological decisions by revealing performance bottlenecks in new device concepts, contribute to their understanding and help to theoretically explore their optimization potential. The efficient implementation of multi-dimensional numerical simulations for computer-aided design tasks requires sophisticated numerical methods and modeling techniques. We review recent advances in device-scale modeling of quantum dot based single-photon sources and laser diodes by self-consistently coupling the optical Maxwell equations with semi-classical carrier transport models using semi-classical and fully quantum mechanical descriptions of the optically active region, respectively. For the simulation of realistic devices with complex, multi-dimensional geometries, we have developed a novel hp-adaptive finite element approach for the optical Maxwell equations, using mixed meshes adapted to the multi-scale properties of the photonic structures. For electrically driven devices, we introduced novel discretization and parameter-embedding techniques to solve the drift-diffusion system for strongly degenerate semiconductors at cryogenic temperatures. Our methodical advances are demonstrated on various applications, including vertical-cavity surface-emitting lasers, grating couplers and single-photon sources.

M. Kantner (✉) · T. Koprucki · H.-J. Wünsche · A. Mielke · U. Bandelow
Weierstraß-Institut für Angewandte Analysis und Stochastik,
Mohrenstraße 39, 10117 Berlin, Germany
e-mail: kantner@wias-berlin.de

T. Höhne · S. Burger · F. Schmidt
Zuse-Institut Berlin, Takustraße 7, 14195 Berlin, Germany

H.-J. Wünsche
Ferdinand-Braun-Institut, Leibniz-Institut für Höchstfrequenztechnik,
Gustav-Kirchhoff-Str. 4, 12489 Berlin, Germany

A. Mielke
Institut für Mathematik, Humboldt-Universität zu Berlin,
Rudower Chaussee 25, 12489 Berlin, Germany

M. Kneissl et al. (eds.), *Semiconductor Nanophotonics*, Springer Series in Solid-State Sciences 194, https://doi.org/10.1007/978-3-030-35656-9_7

7.1 Introduction

Nanophotonic devices based on semiconductor nanostructures embedded in optical micro-resonators enable a broad range of applications, including optical telecommunication and signal processing, photovoltaics, optical interconnects, medical imaging and sensing [1–5]. The optimal design of novel devices tailored to specific applications is a considerable challenge, that can be assisted by mathematical modeling and multi-dimensional numerical simulations. This requires efficient multi-physics TCAD (technology computer-aided design) simulation tools, which allow for the calculation of electromagnetic fields in photonic resonators, the electronic structure of semiconductor materials, the carrier transport dynamics in complex device geometries and the interaction of light with electron-hole excitations in the optically active nanostructures. TCAD simulation tools enable detailed studies of the complex interplay of various processes in opto-electronic devices on different time and length scales. Numerical simulations can provide information that are difficult to access experimentally, help to reveal performance bottlenecks, reduce the development costs of new prototypes, and allow to assess the optimization potential of particular design adaptations.

This chapter provides a survey on modeling fundamentals, recent advances in numerical methods and several example applications. Basic concepts of semi-classical carrier transport theory and optical field calculations are reviewed in Sect. 7.2. In Sect. 7.3, the focus lies on coupled opto-electronic modeling of specific nanophotonic devices with quantum dot (QD) active regions. Here we outline a rate equation based semi-classical QD laser model (Sect. 7.3.1) and a hybrid quantum-classical modeling approach developed for QD-based single-photon emitting diodes (Sect. 7.3.2). Section 7.4 provides a review of several recent advances in the development of numerical methods for the drift-diffusion system and Maxwell's equations. In particular, we highlight new discretization and path following schemes for strongly degenerate semiconductors at cryogenic operation temperatures and mixed finite element methods for the vectorial Maxwell equations with controlled high accuracy beyond the limits of standard tools. The newly developed numerical methods and modeling approaches are demonstrated with selected example applications in Sect. 7.5, including QD-based single-photon sources (Sects. 7.5.1, 7.5.4), vertical-cavity surface-emitting lasers (Sect. 7.5.2) and grating couplers (Sect. 7.5.3).

7.2 Basic Concepts

The mathematical modeling of semiconductor opto-electronic devices is a multi-physics problem, that requires the self-consistent solution of several different physical models, see Fig. 7.1. The key problems are the description of the charge transport in semiconductor devices and the calculation of optical fields in photonic resonators, which will be reviewed in Sects. 7.2.1 and 7.2.2, respectively. These two "building

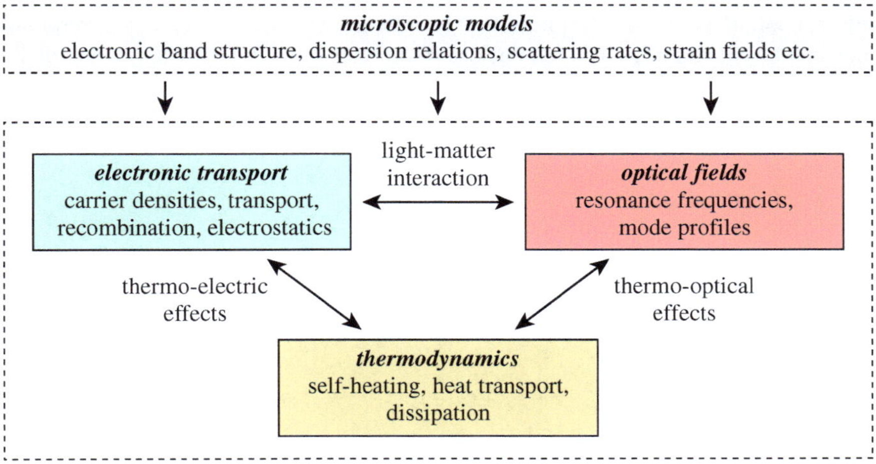

Fig. 7.1 Modeling of semiconductor opto-electronic devices is a multi-physics problem that requires the self-consistent coupling of physical models for the transport of charge carriers, optical fields and heating effects

blocks" are coupled by appropriate models for the light-matter interaction in the optically active region. As the spatial and temporal scales involved in the problem vary drastically between different devices (e.g., large-scale broad area lasers, miniaturized nanolasers or single-photon emitting diodes), there is no general model for the description of the active region of opto-electronic devices. Hence, one has to resort to approximations that are tailored to the characteristics of the particular device at hand. Moreover, in the field of modeling of nanophotonic devices, one distinguishes between a semi-classical theory (quantization of matter, classical optical fields) for conventional semiconductor lasers and a fully quantum mechanical theory in quantum optical devices. In Sect. 7.3, the opto-electronic coupling is presented in detail for vertical-emitting QD lasers (Sect. 7.3.1) and a QD-based single-photon emitting diode (Sect. 7.3.2). As self-heating and other thermal effects often play a significant role in opto-electronic devices [6–9], the generation and transport of heat is considered as a further building block. The fully coupled thermo-opto-electronic problem allows for the description of various thermo-electric and thermo-optical coupling effects. The consistent coupling of the physical models as sketched in Fig. 7.1 is a non-trivial task and requires particular attention to guarantee important conservation laws and consistency with fundamental thermodynamic principles.

The macroscopic models for charge and heat transport as well as the models for the optical fields involve several material parameters and phenomenological relations, that need to be provided by either experimental data or pre-computed microscopic calculations. This comprises, e.g., the electronic band structure, carrier scattering rates, strain fields and optical dispersion relations. As above, there is no general approach and the particularities depend on the device of interest. If certain microscopic degrees of freedom strongly interact with the macroscopic variables, they can not be regarded as fixed background parameters and need to be included as dynamically changing variables in the self-consistently coupled model.

7.2.1 *Electronic Transport*

The transport of charge carriers in semiconductor devices is described by the van Roosbroeck system [10, 11], which is a system of three nonlinearly coupled partial differential equations on the computational domain $\Omega \in \mathbb{R}^d$, $d \in \{1, 2, 3\}$ that reads

$$-\nabla \cdot \varepsilon \nabla \phi = q\left(C + p - n\right), \tag{7.1a}$$

$$\partial_t n - \frac{1}{q}\nabla \cdot \mathbf{j}_n = -R, \tag{7.1b}$$

$$\partial_t p + \frac{1}{q}\nabla \cdot \mathbf{j}_p = -R. \tag{7.1c}$$

The electrostatic interaction between the carriers is described by Poisson's equation (7.1a), that determines the electrostatic potential ϕ generated by the (net-)charge density $q\left(C + p - n\right)$. Here, q denotes the elementary charge, n is the density of electrons, p is the density of holes and $C = N_D^+ - N_A^-$ is the built-in doping profile. The dielectric constant $\varepsilon = \varepsilon_0 \varepsilon_s$ is given by the vacuum permittivity ε_0 and the relative permittivity of the material ε_s in the static limit. The continuity equations (7.1b)–(7.1c) describe the transport and recombination dynamics of the electrons and holes, respectively, where $\mathbf{j}_{n/p}$ are the charge current densities and R is the (net-)recombination rate.

7.2.1.1 Carrier Densities, Current Densities and Degeneration Effects

Due to fast intraband scattering processes, the carrier densities in each band thermalize on a very short time scale such that they can be described by *quasi-equilibrium* distribution functions. The carrier densities are given via the state equations

$$n = N_c \mathscr{F}\left(\frac{\mu_c + q\phi - E_c}{k_B T}\right), \qquad p = N_v \mathscr{F}\left(\frac{E_v - q\phi - \mu_v}{k_B T}\right), \tag{7.2}$$

where the quasi-Fermi energies μ_c and μ_v describe the filling of the respective bands. Here, E_c and E_v denote the conduction and the valence band edge energy, T is the absolute temperature and k_B is Boltzmann's constant. The conduction band and valence band effective density of states N_c and N_v as well as the function $\mathscr{F}$ depend on the electronic density of states and the statistical distribution function. Assuming Fermi–Dirac statistics and parabolic energy bands (effective mass approximation), the function $\mathscr{F}$ is given by the Fermi–Dirac integral

$$\mathscr{F}(\eta) = F_\nu(\eta) = \frac{1}{\Gamma(\nu+1)} \int_0^\infty \mathrm{d}\zeta \, \frac{\zeta^\nu}{\exp(\zeta-\eta)+1}, \tag{7.3}$$

where, e.g., $\nu = 1/2$ in 3D bulk materials or $\nu = 0$ in 2D systems (quantum wells). In the low density limit (i.e., $\eta < -1$), degeneration effects due to Fermi–Dirac statistics are negligible such that $\mathscr{F}(\eta) \approx \exp(\eta)$, see Fig. 7.2. This corresponds to the classical Maxwell–Boltzmann distribution in kinetic gas theory.

The current densities are driven by the gradients of the quasi-Fermi energies

$$\mathbf{j}_n = M_n n \nabla \mu_c, \qquad \mathbf{j}_p = M_p p \nabla \mu_v, \tag{7.4a}$$

where M_n and M_p are the respective carrier mobilities. In the thermodynamic equilibrium, the quasi-Fermi energies become a common global constant $\mu_{\mathrm{eq}} = \text{const.}$, such that the (net-)current flux is zero. Using the carrier density relations (7.2), the current densities can be cast in the drift-diffusion form

$$\mathbf{j}_n = -qM_n n\nabla\phi + qD_n(n)\nabla n, \qquad \mathbf{j}_p = -qM_p p\nabla\phi - qD_p(p)\nabla p, \tag{7.4b}$$

where the diffusion coefficients $D_{n/p}$ are connected with the carrier mobilities via the *generalized* Einstein relations

$$D_n(n) = \frac{k_B T M_n}{q} g\left(\frac{n}{N_c}\right), \qquad D_p(p) = \frac{k_B T M_p}{q} g\left(\frac{p}{N_v}\right). \tag{7.5}$$

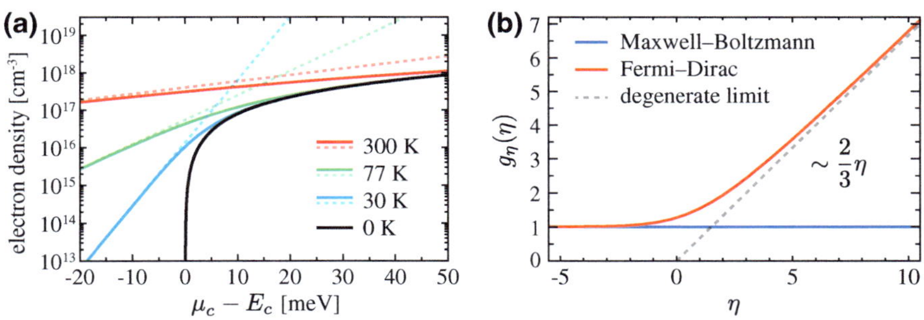

Fig. 7.2 **a** Electron density in GaAs as a function of the quasi-Fermi energy μ_c for different temperatures. The carrier density calculated using the Fermi–Dirac integral (solid lines) is compared with the Maxwell–Boltzmann approximation (dashed lines). **b** Plot of the degeneracy factor $g_\eta(\eta) = g(\mathscr{F}(\eta))$ according to (7.6), which leads to an enhancement of the diffusion current. Reprinted, with permission, from [12]. © 2016 Springer Science+Business Media

The expressions (7.5) feature the carrier density-dependent *degeneracy factor*

$$g(x) = x(\mathscr{F}^{-1})'(x), \tag{7.6}$$

which describes a nonlinear enhancement (because of $g(x) \geq 1$) of the diffusion current in the case of degenerate carrier statistics. For $\mathscr{F}(\eta) = \exp(\eta)$, one recovers the classical result $g \equiv 1$ corresponding to linear diffusion.

7.2.1.2 Recombination Rate Models

The (net-)recombination rate R in (7.1b)–(7.1c) describes various radiative and non-radiative recombination and generation processes that lead to the annihilation or creation of electron-hole pairs. The dominant processes in opto-electronic semiconductor devices are illustrated in Fig. 7.3. For Shockley–Read–Hall recombination, spontaneous radiative recombination and Auger recombination (see Fig. 7.3a–c), the recombination rate takes the form [13]

$$R = \left(1 - \exp\left(-\frac{\mu_c - \mu_v}{k_B T}\right)\right) \sum_{\alpha} r_{\alpha}(\mu_c, \mu_v, \phi), \tag{7.7}$$

where α labels the various processes and r_α are process specific (non-negative) recombination rates, that are derived from quantum mechanical perturbation theory (i.e., Fermi's golden rule) [14, 15]. The first factor in (7.7) controls the ratio between carrier generation and recombination and vanishes under thermodynamic equilibrium conditions $\mu_c \equiv \mu_v \equiv \mu_{\text{eq}}$. This reflects the detailed balance condition due to the microscopic reversibility of the underlying kinetic equations. For details on the recombination rate models r_α we refer to [11, 16, 17]. Note that the stimulated recombination rate, see Fig. 7.3d, is not of type (7.7), as it depends also on the photon number in the laser mode. We will come back to this in Sect. 7.3.1.

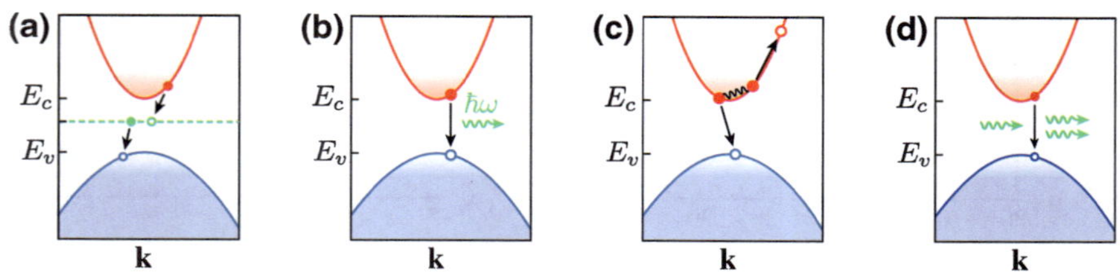

Fig. 7.3 Recombination processes in opto-electronic semiconductor devices: **a** Shockley–Read–Hall recombination of an electron-hole pair via relaxation to an intermediate trap state within the band gap, **b** spontaneous radiative recombination, **c** Auger recombination via an additional intraband scattering process and **d** stimulated recombination. The picture is taken from [13]

7.2.1.3 Boundary Conditions

The van Roosbroeck system (7.1) must be supplemented with boundary conditions modeling electrical contacts (Ohmic contacts, Schottky contacts, gate contacts), semiconductor-insulator interfaces or artificial boundaries of the computational domain $\partial\Omega$. These are typically mixed conditions on disjoint parts of the boundary given by either Neumann, Dirichlet or Robin boundary conditions. A generic approach that is very useful for the implementation in numerical software tools are Robin boundary conditions (boundary conditions of third kind)

$$\mathbf{n} \cdot \nabla u + \alpha_u \left(u - u_0 \right) = 0, \tag{7.8}$$

where $u \in (\phi, \mu_c, \mu_v)$ is any of the basic variables, and $\mathbf{n}$ is the outward-oriented normal vector on $\partial\Omega$. The condition (7.8) is a weighted combination of homogeneous Neumann boundary conditions ($\alpha_u \to 0$: $\mathbf{n} \cdot \nabla u = 0$) and Dirichlet boundary conditions ($|\alpha_u| \to \infty$: $u = u_0$). For details on specific boundary condition models we refer to [11, 17, 18].

7.2.2 *Optical Fields*

Optical processes in photonic devices are mathematically described by Maxwell's equations in terms of electric fields $\mathbf{E}(\mathbf{r}, t)$ and magnetic fields $\mathbf{H}(\mathbf{r}, t)$ in space $\mathbf{r}$ and time t. We briefly sketch the models used in photonics, more detailed information can be found, e.g., in [19].

Here, we are interested in solutions for specific frequencies ω, therefore the time dependency of the real electric field $\mathbf{E}(\mathbf{r}, t)$ is expressed by means of the complex electric field $\underline{\mathbf{E}}$ with $\mathbf{E}(\mathbf{r}, t) = \mathrm{Re}\left(\underline{\mathbf{E}} \exp\left(-i\omega t\right)\right)$. Since all field quantities are expressed in this way, the underline character is dropped and all fields and current densities are complex vectors. In its simplest form, i.e., when the entire electromagnetic field is generated by an impressed current density $\mathbf{J}_{\mathrm{source}}$ of frequency ω, Maxwell's equations yield

$$\nabla \times \mu(\mathbf{r})^{-1} \nabla \times \mathbf{E}(\mathbf{r}) - \omega^2 \varepsilon(\mathbf{r}) \mathbf{E}(\mathbf{r}) = i\omega \mathbf{J}_{\mathrm{source}}(\mathbf{r}). \tag{7.9}$$

In (7.9) we identify the material properties as permittivity $\varepsilon = \varepsilon_r \varepsilon_0$ and permeability $\mu = \mu_r \mu_0$, where $\varepsilon_0(\mu_0)$ and $\varepsilon_r(\mu_r)$ are the free space and relative permittivities (permeabilities). For all relevant optical materials, the relative permeability is $\mu_r = 1$, and the refractive index n is given by $n = \sqrt{\varepsilon_r}$. For the applications discussed in this chapter, (7.9) is enhanced to cover three different settings: (i) light scattering simulations, where external source fields $\mathbf{E}_{\mathrm{source}}$ are present, (ii) resonance mode computations, for investigation of the modal structure of the setup, and (iii) propagation mode computation for investigation of waveguide structures. To cope with these situations, Maxwell's equations are used in properly adapted forms.

In *scattering problems on unbounded domains* a source field $\mathbf{E}_{\text{source}}$ is given which travels through (unbounded) space, being itself a solution to Maxwell's equations. Without a scattering object this would be the only field, i.e., the total field $\mathbf{E} = \mathbf{E}_{\text{source}}$. But if it hits an object, a scattered field $\mathbf{E}_{\text{scatt}}$ is generated. We compute the field $\mathbf{E} = \mathbf{E}_{\text{source}} + \mathbf{E}_{\text{scatt}}$ inside a computational domain Ω which we also call the interior domain, in contrast to the exterior domain Ω_{ext} surrounding Ω. Further we denote the boundary of the computational domain with $\partial\Omega$.

Let the source field $\mathbf{E}_{\text{source}}(\mathbf{r})$, with $\mathbf{r}$ in the exterior domain, $\mathbf{r} \in \Omega_{\text{ext}} = \mathbb{R}^3 \setminus \bar{\Omega}$, be given. Then the general scattering problem in 3D is defined by the following conditions:

1. Interior problem. The field on Ω obeys Maxwell's equations, (7.9).
2. Exterior problem. The field in the exterior domain is a superposition of the source and the scattered fields $\mathbf{E}_{\text{source}}$ and $\mathbf{E}_{\text{scatt}}$. Both obey Maxwell's equations.
3. Continuity of the tangential data and their normal derivative along the boundary of the computational domain is fulfilled.
4. In the case of time-harmonic fields, the Silver–Müller radiation condition for the scattered field holds true uniformly in all directions $\mathbf{r}$,

$$\lim_{|\mathbf{r}|\to\infty} |\mathbf{r}| \left[(\nabla \times \mathbf{E}_{\text{scatt}}(\mathbf{r})) \times \frac{\mathbf{r}}{|\mathbf{r}|} - i\frac{\omega}{c}\mathbf{E}_{\text{scatt}}(\mathbf{r}) \right] = 0, \tag{7.10}$$

where c is the speed of light in the exterior domain. A typical application of scattering problems is the computation of light fields in single-photon source devices, see Sects. 7.3.2 and 7.5.1 of this chapter and Sect. 8.2.

In *resonance problems* we consider (7.9) without sources (i.e., $\mathbf{J}_{\text{source}} = \mathbf{0}$) as an eigenvalue problem with the unknown eigenvalue ω and the unknown resonance mode, the eigenvector $\mathbf{E}$. Note that in the case of active, dispersive media with complex-valued dielectric function $\varepsilon(\mathbf{r}, \omega)$ or leaky optical cavities (open boundary conditions, e.g., Silver–Müller radiation condition), (7.9) is in general a non-Hermitian eigenvalue problem featuring complex resonance frequencies $\omega \in \mathbb{C}$. Here, $\text{Re}\,(\omega)$ is the radial frequency of the optical field and $\text{Im}\,(\omega)$ corresponds to the width of the resonance (i.e., the decay rate of the leaky optical mode) [20]. A special challenge is the treatment of the exterior domain. Either the boundary conditions on the boundary of the computational domain are explicitly known, e.g., as a magnetic or electric wall, or methods from scattering problems are applied. A typical application of resonance problems is the computation of cavity modes in semiconductor lasers. In Sect. 7.3.1 we outline the self-consistent coupling of the optical resonance problem with the electronic transport problem for the simulation of QD lasers. Numerical results on cavity modes in vertical-cavity surface-emitting lasers (VCSELs) are given in Sect. 7.5.2.

In *mode propagation problems*, again we consider (7.9) without sources. This time we specialize the problem further to z-invariant problems: $\varepsilon = \varepsilon(x, y)$. We take a fixed frequency ω and look for solutions of the type $\mathbf{E}(x, y, z) = \mathbf{E}(x, y)e^{ik_z z}$ with a (unknown) phase velocity k_z. We obtain an eigenvalue problem for the waveguide mode $\mathbf{E}(x, y)$ together with its corresponding eigenvalue k_z. A typical application of mode propagation problems is the computation of propagation modes in active devices and in integrated optical setups, see Sects. 7.3.1 and 7.5.3.

7.2.3 Thermodynamics

Thermal effects can play an important role in opto-electronic semiconductor devices. This includes, e.g., thermal lensing and temperature-dependence of the optical gain in semiconductor lasers or self-heating in light-emitting diodes [21]. Next to an accurate modeling of thermal effects, the thermodynamic consistency of the coupled system, i.e., the consistency of the model equations with fundamental laws of (non-)equilibrium thermodynamics, is a major modeling issue.

In the framework of macroscopic device simulation, it is usually assumed that the transport and generation of heat is governed by the classical heat transport equation. In combination with the van Roosbroeck system (7.1), this enables the self-consistent description of several thermo-electric cross effects including Joule heating, recombination heating and the Thomson–Peltier effect [7, 22–24]. Moreover, the fully consistent thermo-electric model requires modifications in the electronic transport equations presented in Sect. 7.2.1. In particular, the current densities (7.4a) must be supplemented with an additional driving force $\sim \nabla T$ to account for the Seebeck effect. We refer to [22, 23, 25] for details.

The full thermo-opto-electronic model (see Fig. 7.1) is required to be consistent with fundamental principles of (non-)equilibrium thermodynamics. This comprises the consistency with the thermodynamic equilibrium (detailed balance, microscopic reversibility), the consistency with the second law of thermodynamics (non-negative entropy production rate), Onsager's reciprocal relations [26] and the preservation of conserved quantities (e.g., total charge, energy). The GENERIC (*General Equation for Non-Equilibrium Reversible-Irreversible Coupling*) formalism [27, 28] of non-equilibrium thermodynamics provides a universal approach to derive coupled and thermodynamically consistent multi-physics models for, e.g., nanophotonic semiconductor devices, where fundamental physical properties are encoded in the mathematical structure and symmetry relations of the operators and functionals driving the system's evolution [28, 29].

7.3 Quantum Dot Based Light-Emitting Devices

In this section, two different types of nanophotonic light-emitting diodes are considered, which are both based on semiconductor quantum dots embedded in dielectric microcavities. First, in Sect. 7.3.1, we describe some essential ideas on a comprehensive simulation approach to QD-based micro-cavity lasers such as VCSELs [30]. The model is based on a multi-species [31] description of the transport, scattering [32] and recombination dynamics of charge carriers in extended (bulk) or localized states (wetting layer (WL), QDs) coupled with the vectorial Maxwell equations (7.9). In contrast to these laser devices, which are based on macroscopic QD ensembles, the single-photon sources considered in Sect. 7.3.2 contain only a single QD. For this case, we have developed a new hybrid quantum-classical modeling approach [33–35], which combines the building blocks of Sect. 7.2 with a Markovian quantum master equation [36] describing the evolution of the coupled QD-photon system.

7.3.1 Quantum Dot Lasers

Semiconductor QD lasers have been extensively studied for applications in telecommunication systems and were shown to have a variety of excellent properties that are essentially linked to the discrete energy spectrum associated with the three-dimensional carrier confinement provided by the QDs. Among others, this includes enhanced opto-electronic efficiency and differential gain, increased temperature stability, reduced threshold current and broader modulation bandwidth [3, 37].

In this section, we consider a semiconductor laser based on an ensemble of QDs embedded in an electrically pumped micro-cavity. Most prominent examples are VCSELs, but also micro-disk and micro-pillar resonators [38] belong to this type of lasers. These devices have a rather complicated geometrical structure (in comparison with conventional Fabry–Pérot lasers) and their accurate mathematical description requires a multi-dimensional simulation approach. A semi-classical description of such devices is achieved by combining the van Roosbroeck system and the vectorial (optical) Maxwell equations introduced in Sects. 7.2.1 and 7.2.2 with microscopic models for the nanostructured gain medium.

The optical field in the laser cavity is subject to the wave equation [39]

$$\nabla \times \nabla \times \mathbf{E}(\mathbf{r}, t) + \frac{1}{c_0^2} \frac{\partial^2}{\partial t^2} \int_0^\infty \mathrm{d}\tau \, \varepsilon_r(\mathbf{r}, t, \tau) \, \mathbf{E}(\mathbf{r}, t - \tau) = -\mu_0 \frac{\partial}{\partial t} \mathbf{j}_{\mathrm{sp}}(\mathbf{r}, t) . \tag{7.11}$$

The response function $\varepsilon_r(\mathbf{r}, t, \tau)$ of the semiconductor depends parametrically on the slowly varying carrier densities in the active region (time scale t) and fast dephasing processes, which lead to polarization decay (time scale τ). Here, the response function describes both, the optical gain as well as cavity losses. The fluctuating current density $\mathbf{j}_{\mathrm{sp}}(\mathbf{r}, t)$ induces a spontaneous polarization that is the source of spontaneous emission. The electric field is expanded in quasi-normal modes (see Sect. 7.2.2). For the sake of simplicity, we focus on the single-mode case $\mathbf{E}(\mathbf{r}, t) = \frac{1}{2} e^{-i\omega t} A(t) \, \mathbf{E}(\mathbf{r}) + \text{c.c.}$ throughout this section, where the modal field $\mathbf{E}(\mathbf{r})$ and its complex resonance frequency ω solve the source-free Helmholtz equation (7.9)

$$\nabla \times \nabla \times \mathbf{E}(\mathbf{r}) - \frac{\omega^2}{c_0^2} \tilde{\varepsilon}_r(\mathbf{r}, t, \omega) \, \mathbf{E}(\mathbf{r}) = 0 \tag{7.12}$$

with $\tilde{\varepsilon}_r(\mathbf{r}, t, \omega) = \int_0^\infty \mathrm{d}\tau\, e^{i\omega\tau} \varepsilon_r(\mathbf{r}, t, \tau)$ and outgoing wave conditions (7.10) on $\partial\Omega$. The evolution equation for the slowly varying mode amplitude $A(t)$ is derived under rather general conditions following [39, 40]. With an appropriate normalization of the modal field, this can be cast into the photon number equation

$$\dot{N}_{\mathrm{ph}} = 2\ \mathrm{Im}(\omega)\, N_{\mathrm{ph}} + r_{\mathrm{sp}}. \tag{7.13}$$

In the single mode approach outlined here, only a small additive term describing the coupling of modes under non-adiabatic conditions is omitted, which can be safely neglected in nanoresonators [39]. Expressions for the spontaneous emission r_{sp} into the given mode, which is a very small quantity in the lasers under consideration, can be found in [39–41]. Consequently, stationary lasing with a large N_{ph} requires an extremely small $\mathrm{Im}(\omega)$ that thus can be disregarded everywhere except in the photon number equation (7.13). Note that $\mathrm{Im}(\omega)$ contains both, positive contributions from the amplification by the QDs as well as negative contributions from optical cavity losses. In this sense, the classical threshold condition corresponds to $\mathrm{Im}(\omega) = 0$.

The coupling of the electronic system to the optical field equations is mainly given by the stimulated recombination rate

$$R_{\mathrm{stim}} = \frac{\varepsilon_0}{\hbar} \int_\Omega \mathrm{d}^3r\ \mathrm{Im}\left(\tilde{\varepsilon}_{r,\mathrm{QD}}(\mathbf{r}, t, \omega)\right) |\mathbf{E}(\mathbf{r})|^2\, N_{\mathrm{ph}}(t), \tag{7.14}$$

where $\tilde{\varepsilon}_{r,\mathrm{QD}}(\mathbf{r}, t, \omega)$ is the contribution of the QDs to the permittivity function in (7.12). In order to provide gain, i.e., $\mathrm{Im}\left(\tilde{\varepsilon}_{r,\mathrm{QD}}(\mathbf{r}, t, \omega)\right) > 0$, the QDs must be sufficiently occupied with electron-hole pairs. Here, the carrier transport from the contacts of the diode to the QDs comes into play. Current flow towards the active layer is governed by the van Roosbroeck system (7.1) introduced in Sect. 7.2.1. Within the active layer, the dynamics of the electronic system is more complex. The active region has a complicated electronic density of states that features bulk scattering states, wetting layer scattering states confined in one dimension but extended in the other two dimensions, and QD states that are completely confined to individual QDs. The basic idea of a multi-species model (see Fig. 7.4) is to regard carriers in different types of states as different distinguishable carrier species [31]. Just as in the remaining parts of the device, the flow of bulk carriers within the active layer is also described by the van Roosbroeck system. The coupling of bulk carriers to WL and QD carriers is achieved by additional capture/escape rates $C_{n/p}^{\mathrm{cap}}$ added on the usual recombination rate R on the right hand side of the continuity equations (7.1b)–(7.1c). In the most simple case, quasi-equilibrium distributions can be assumed, such that these rates are effectively modeled as [6, 42, 43]

$$C_n^{\mathrm{cap}} = \frac{n}{\tau_n}\left(1 - \frac{w_e}{N_{e,\mathrm{WL}}^{\mathrm{sat}}}\right)\left(1 - \exp\left(-\frac{\mu_c - \mu_c^{\mathrm{WL}}}{k_B T}\right)\right)\Theta(\mathbf{r}),$$

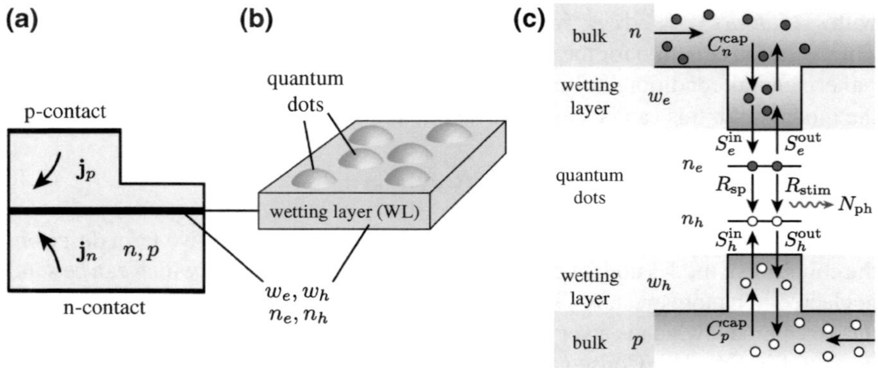

Fig. 7.4 **a** Sketch of a pin-diode laser structure with an embedded QD layer as optically active gain medium. **b** The InAs-QDs are grown on a WL acting as an additional 2D-like carrier reservoir. **c** Schematic illustration of scattering and recombination processes in the active region of the multi-species model. Carriers are injected from the bulk into the WL states from where they enter the QDs by Coulomb scattering. Only the QD carriers contribute to the stimulated recombination. Reprinted, with permission, from [31]. © 2016 Springer Science+Business Media

where τ_n is the characteristic scattering time, μ_c^{WL} is the chemical potential of the WL electrons and $\Theta(\mathbf{r})$ is an indicator function that is one within the active region and zero outside. The second factor describes a nonlinear reduction of the scattering rate due to state filling, where w_e is the sheet electron density in the WL and $N_{e,\text{WL}}^{\text{sat}}$ is a saturation density. An analogous expression holds for holes. In order to account for electrostatic interaction between bound and continuum carriers, the right hand side of Poisson's equation (7.1a) must be supplemented with the WL and QD carrier densities. Carrier transport within the WL is described by 2D versions of the continuity equations (7.1b), (7.1c) for the WL sheet carrier densities w_e and w_h supplemented accordingly by capture/escape terms describing the scattering to both bulk and QD states, see Fig. 7.4c. We refer to [31] for further details on the multi-species approach.

Finally, kinetic equations describing the occupation dynamics of the QD states have to be established. For conventional laser models, this implies a drastic increase of complexity, because the QDs can be occupied in several different configurations whose energies exhibit a rather wide inhomogeneous broadening. Different such scenarios have been treated in the literature [44–46], see also Chaps. 2 and 4 of this book. Here we restrict ourselves to a rather simple approach based on conventional rate equations describing the dynamics of the (average) QD electron and hole occupation probabilities n_e and n_h

$$\partial_t n_e = S_e(n_e, n_h, w_e, w_h) - R_{\text{stim}}\left(n_e, n_h, w_e, w_h, N_{\text{ph}}\right) - R_{\text{sp}}(n_e, n_h), \quad (7.15a)$$
$$\partial_t n_h = S_h(n_e, n_h, w_e, w_h) - R_{\text{stim}}\left(n_e, n_h, w_e, w_h, N_{\text{ph}}\right) - R_{\text{sp}}(n_e, n_h), \quad (7.15b)$$

where $S_\nu = S_\nu^{\text{in}}(1 - n_\nu) - S_\nu^{\text{out}} n_\nu$ for $\nu \in \{e, h\}$ are the net-capture rates into the QDs and R_{stim} and R_{sp} are the total stimulated and spontaneous recombination rates.

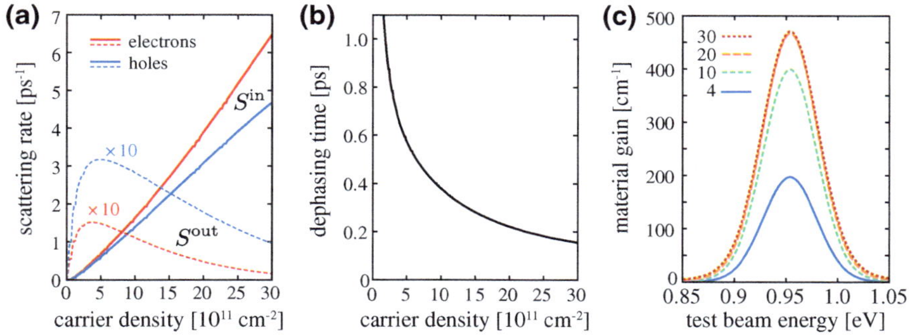

Fig. 7.5 **a** Calculated in (solid) and out-scattering (dashes) rates of electrons (*solid*) and holes (*dotted*) from the WL to the QD states and **b** dephasing time $T_2 = \Gamma_2^{-1}$ as a function of the WL sheet density ($w_e = w_h$). **c** Material gain spectra for WL densities $\{4, 10, 20, 30\} \times 10^{11}\,\mathrm{cm}^{-2}$. Reprinted, with permission, from [31]. © 2016 Springer Science+Business Media

The scattering rates $S_\nu^{\rm in}$ from the 2D (WL) and 3D (bulk) reservoirs due to Coulomb interaction can be computed microscopically using *Fermi's Golden Rule* in second order screened Born–Markov approximation, where the screening can be treated in the stationary limit using the *Lindhard formula* [32, 47, 48]. The computations involve high dimensional integrals, that can be evaluated efficiently using the quasi-Monte Carlo method [32, 49]. Assuming thermalized reservoirs (quasi-equilibrium), the in- and out-scattering rates are related by the detailed balance principle involving the respective reservoir's quasi-Fermi energy, e.g., $S_e^{\rm out} \propto S_e^{\rm in} \exp\left(-\mu_c^{\rm WL}/k_B T\right)$ [45]. The characteristic dependency of the microscopically calculated Auger scattering rates (into a bound InGaAs-QD state) on the reservoir carrier density is shown in Fig. 7.5a, see [31, 32] for details. The permittivity function describing the interband polarization of the QD layer can be approximated as

$$\tilde{\varepsilon}_{r,\rm QD}(\mathbf{r}, t, \omega) = \frac{|d_{c,v}|^2}{\varepsilon_0} N_{\rm QD}\left(n_e + n_h - 1\right) \int_{\mathbb{R}} \mathrm{d}E\, \frac{G(E)}{i\left(E - \hbar\omega\right) + \hbar\Gamma_2\left(w_e, w_h, E\right)}, \tag{7.16}$$

where $d_{c,v}$ is the interband dipole moment, $N_{\rm QD}$ is the QD (volume) density, $G(E)$ is a Gaussian distribution function describing the inhomogeneous broadening of the QD ensemble and $\Gamma_2 = T_2^{-1}$ is the microscopically computed carrier-dependent dephasing rate [31, 50], see Fig. 7.5b. The imaginary part of (7.16) is the material gain (with Lorentzian line shape) that enters the stimulated recombination rate (7.14). Gain spectra for different reservoir carrier densities are shown in Fig. 7.5c.

7.3.2 Single-Photon Sources

Emitters of non-classical light, such as single-photon sources and sources of entangled photon pairs, are key components for many applications in the field of optical quantum communication, information processing, computing and metrology [51, 52]. Semiconductor device simulation can assist the development of engineered devices with optimized photon extraction efficiency and efficient current injection schemes. In this section, we discuss a comprehensive modeling approach for electrically driven QD-based single-photon emitting diodes. Our model is again based on the van Roosbroeck system (7.1) and Maxwell's wave equation (7.9), however, here the electro-optical coupling is described by a fully quantum mechanical model for the coupled QD-photon system. The approach is demonstrated by applications presented in Sect. 7.5.1 and Sect. 8.2.

7.3.2.1 Photon Extraction Efficiency and Purcell Factor

The single-photon emission of a QD is modeled as radiation of a harmonically oscillating electric dipole polarized in the plane perpendicular to its growth axis [53]. The frequency domain electric current density $\mathbf{J}_{\text{source}}$ of such a dipole with strength vector $\mathbf{j}$ at a position $\mathbf{r}_0$ is given by

$$\mathbf{J}_{\text{source}}(\mathbf{r}) = \mathbf{j}\,\delta(\mathbf{r} - \mathbf{r}_0), \tag{7.17}$$

which enters the time-harmonic wave equation (7.9) as a source term. The spontaneous emission rate of the QD depends on the local density of photonic states of its specific surrounding. This is denoted as Purcell effect [54] and quantified by the dimensionless Purcell factor F_P. That is the total emitted power P_{source} of the dipole source in a structured area relative to its emission power P_{bulk} in a homogeneous environment with the same background material (with refractive inde n_b), and is calculated as follows [55]:

$$F_P = \frac{P_{\text{source}}}{P_{\text{bulk}}} = P_{\text{source}} \left(\frac{n_b \mu_0 \omega^2}{12\pi c_0} |\mathbf{j}|^2 \right)^{-1}. \tag{7.18}$$

The Purcell factor is closely related to the *quality factor* $Q \propto F_P$ which characterizes the leakiness of a device. A typical way to determine the Q-factor is to calculate the corresponding resonance frequency $\omega = \text{Re}\,(\omega) + i\,\text{Im}\,(\omega)$ which is an eigenvalue solution to a *resonance problem*, see Sect. 7.2.2. With this, the Q-factor is obtained from

$$Q = -\frac{\text{Re}\,(\omega)}{2\,\text{Im}\,(\omega)} = \text{Re}\,(\omega)\,\tau, \tag{7.19}$$

where τ is identified as the resonance mode lifetime.

Furthermore, one is often interested in the extraction efficiency η, i.e., in the amount of emitted light of power P_{out} that is coupled to a specific direction and numerical aperture or into a specific mode. Thus, η can be calculated as

$$\eta = \frac{P_{\text{out}}}{P_{\text{source}}}. \tag{7.20}$$

A numerical study on the optimization of geometry parameters for the maximization of the light extraction efficiency from a QD-based single-photon source is reviewed in Sect. 8.2.

7.3.2.2 Hybrid Quantum-Classical Modeling of Quantum Light-Emitting Diodes

This section describes a comprehensive modeling approach for the device-scale simulation of QD-based electrically driven single-photon sources. In contrast to the semi-classical approach followed in the laser model described in Sect. 7.3.1, the description of the quantum optical properties of the radiation generated by non-classical light sources requires a fully quantum mechanical modeling approach in order to access, e.g., the second-order auto-correlation function $g^{(2)}(\tau)$. This can be achieved using the second quantization formalism to describe the dynamics of the bound QD carriers coupled to the cavity photons of the optical modes supported by the resonator [37, 47]. The QD-photon system is an open quantum system that is coupled to its macroscopic environment (continuum carriers in the surrounding semiconductor, background radiation, phonons). This leads to dissipative dynamics that are conveniently described by a Markovian quantum master equation in Lindblad form for the (reduced) density matrix ρ of the open system [36].

The *hybrid quantum-classical* modeling approach developed in [33, 56] is based on the self-consistent coupling of the Lindblad master equation with the van Roosbroeck system. This allows for a multi-scale simulation of electrically driven quantum light sources, where the quantum optical properties of the QD-photon system and the spatially resolved injection current in realistic semiconductor device geometries can be calculated out of one box. The model equations read [33]

$$-\nabla \cdot \varepsilon \nabla \phi = q\left(C + p - n + Q(\rho)\right), \tag{7.21a}$$

$$\partial_t n - \frac{1}{q}\nabla \cdot \mathbf{j}_n = -R - S_n\left(\mu_c, \mu_v, \phi; \rho\right), \tag{7.21b}$$

$$\partial_t p + \frac{1}{q}\nabla \cdot \mathbf{j}_p = -R - S_p\left(\mu_c, \mu_v, \phi; \rho\right), \tag{7.21c}$$

$$\partial_t \rho = -\frac{\mathrm{i}}{\hbar}\left[\mathcal{H}, \rho\right] + \mathcal{D}\left(\mu_c, \mu_v, \phi\right)\rho. \tag{7.21d}$$

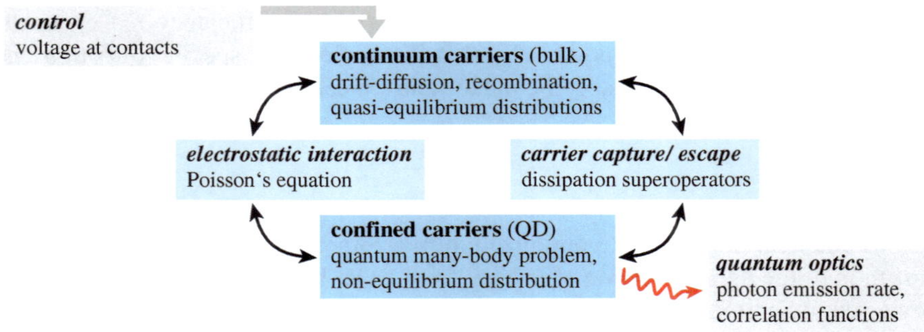

Fig. 7.6 Schematic representation of the hybrid quantum-classical model system (7.21). A quantum system described by a Lindblad-type master equation is self-consistently coupled to the semi-classical transport equations for the freely roaming continuum carriers. Both subsystems exchange charge by capture and escape of carriers and interact via their self-consistently generated electric field. Adapted, with permission, from [33]. © 2017 American Physical Society

The model is based on a separation of freely roaming continuum carriers and bound carrier confined to the QD (*Born approximation*). The transport and recombination of continuum carriers is described by the van Roosbroeck system (7.21a)–(7.21c), which is extended by additional terms that constitute the coupling to the open quantum system. The coupling terms describe the bound charge density Q and the (net-)scattering rates S_n and S_p between the continuum and bound QD states. The state of the quantum system is described by the (reduced) density matrix ρ, which evolves according to the Lindblad master equation (7.21d). A schematic illustration of the model system (7.21) is given in Fig. 7.6.

The Hamiltonian $\mathcal{H}$ of the open quantum system must be chosen appropriately to the particular problem at hand. In quantum optical problems, a typical Hamiltonian takes the form

$$\mathcal{H} = \mathcal{H}_0 + \mathcal{H}_{\mathrm{Coul}} + \mathcal{H}_{\mathrm{LM}}. \tag{7.22a}$$

The single-particle energies of the electrons, holes and photons are given by

$$\mathcal{H}_0 = \sum_i \varepsilon_{e,i} e_i^\dagger e_i + \sum_i \varepsilon_{h,i} h_i^\dagger h_i + \sum_i \hbar\omega_i a_i^\dagger a_i, \tag{7.22b}$$

where $\varepsilon_{\lambda,i}$ ($\lambda \in \{e, h\}$) denote bound QD energy levels and $\hbar\omega_i$ are the energies of the cavity photons. The fermionic creation and annihilation operators e_i and $e_i^\dagger$ (or h_i and $h_i^\dagger$) obey the usual anti-commutator relations, whereas the canonical commutator relations hold for the bosonic photon operators a_i and $a_i^\dagger$. The Coulomb interaction between the bound carriers is described by [47, 57]

$$\mathcal{H}_{\mathrm{Coul}} = \frac{1}{2} \sum_{i,j,k,l} \left(V_{i,j,k,l}^{e,e} e_i^\dagger e_j^\dagger e_k e_l + V_{i,j,k,l}^{h,h} h_i^\dagger h_j^\dagger h_k h_l - 2V_{i,j,k,l}^{e,h} e_i^\dagger h_j^\dagger h_k e_l \right), \tag{7.22c}$$

where $V_{i,j,k,l}^{\lambda,\lambda'}$ are the Coulomb matrix elements. The light-matter interaction Hamiltonian (in the dipole- and rotating wave approximation) reads [37]

$$\mathcal{H}_{\mathrm{LM}} = \sum_{i,j,k} \left(g_{i,j,k} e_i h_j a_k^\dagger + g_{i,j,k}^* a_k h_j^\dagger e_i^\dagger \right), \tag{7.22d}$$

where $g_{i,j,k}$ are the light-matter coupling coefficients (including also the optical selection rules). A central feature of the Hamiltonian (7.22) is

$$[\mathcal{H}, N] = 0, \tag{7.23}$$

where $N = \sum_i e_i^\dagger e_i - \sum_i h_i^\dagger h_i$ is the charge number operator. Hence, the Hamiltonian part of (7.21d) conserves the charge of the quantum system.

Irreversible processes (e.g., capture/escape of carriers from continuum states to the QD, intraband carrier relaxation, spontaneous emission, pure dephasing and out-coupling of cavity photons) that arise due to the coupling of the open quantum system to its macroscopic environment are described by the dissipation superoperator

$$\mathcal{D}(\mu_c, \mu_v, \phi)\,\rho = \sum_{\alpha\in\mathcal{I}} \sum_{\nu,\ell} \gamma_\alpha(\nu, \ell; \mu_c, \mu_v, \phi)\, L_{A_\alpha(\nu,\ell)}\rho, \tag{7.24}$$

where γ_α is a microscopic transition rate that depends on the state (μ_c, μ_v, ϕ) of the macroscopic system, $L_A\rho = A\rho A^\dagger - \frac{1}{2}\{A^\dagger A, \rho\}$ is the Lindblad superoperator, $A_\alpha(\nu, \ell)$ is a quantum jump operator and α labels the respective dissipative processes. The jump operators are required to obey the eigenoperator relations

$$[\mathcal{H}, A_\alpha(\nu, \ell)] = -h\nu A_\alpha(\nu, \ell), \qquad [N, A_\alpha(\nu, \ell)] = -\ell A_\alpha(\nu, \ell), \tag{7.25}$$

where $h\nu$ is the transition energy (with Planck's constant h) and $\ell \in \mathbb{Z}$ quantifies the amount of charge exchanged by the interaction. For each *forward* process $A_\alpha(\nu, \ell)$, the dissipation superoperator (7.24) involves also the corresponding *backward* process $A_\alpha^\dagger(\nu, \ell) = A(-\nu, -\ell)$. The respective forward and backward transition rates are related by a (generalized) Kubo–Martin–Schwinger condition [13] and satisfy the quantum detailed balance relation in the thermodynamic equilibrium. Due to its additive structure, the dissipation superoperator (7.24) can be separated as

$$\mathcal{D}\rho = \mathcal{D}_0\rho + \mathcal{D}_e\rho + \mathcal{D}_h\rho, \tag{7.26}$$

where the index set $\mathcal{I}$ was decomposed into disjoint subsets $\mathcal{I} = \mathcal{I}_0 \cup \mathcal{I}_e \cup \mathcal{I}_h$. This corresponds to a classification of the admitted dissipative processes with respect to their action on the charge of the quantum system. Here, $\mathcal{D}_0$ leaves the charge of the quantum system invariant (i.e., $\ell = 0$), whereas $\mathcal{D}_e$ and $\mathcal{D}_h$ describe the exchange of electrons and holes (with $\ell = \pm 1$ for one-particle exchange processes), respectively.

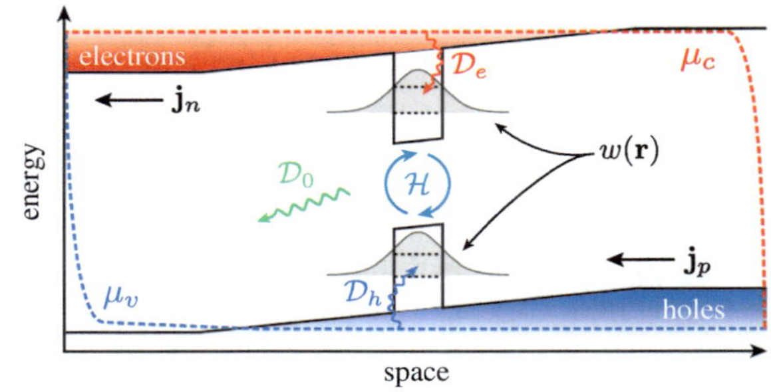

Fig. 7.7 Illustration of the coupling scheme and the spatial profile function $w(\mathbf{r})$ for a single QD embedded in a pin-diode structure. Adapted, with permission, from [33]. © 2017 American Physical Society

Finally, the quantum system couples back to the drift-diffusion system via the QD charge density Q and the scattering rates $S_{n,p}$. The charge density of a single QD can be described by the expectation value of the charge number operator

$$Q(\rho) = -w(\mathbf{r})\,\mathrm{tr}(N\rho), \tag{7.27}$$

where $w(\mathbf{r})$ is a (normalized) spatial profile function that describes the localization of the QD carriers and phenomenologically replaces the absolute square of the many-body QD carriers wave function (see Fig. 7.7). The macroscopic (net-)scattering rates in (7.21b)–(7.21c) read

$$S_n(u_w, \rho) = +w(\mathbf{r})\,\mathrm{tr}(N\mathcal{D}_e(u_w)\,\rho), \tag{7.28a}$$

$$S_p(u_w, \rho) = -w(\mathbf{r})\,\mathrm{tr}(N\mathcal{D}_h(u_w)\,\rho), \tag{7.28b}$$

where $u_w = (\langle\mu_c\rangle_w, \langle\mu_v\rangle_w, \langle\phi\rangle_w)$ (with $\langle\phi\rangle_w = \int \mathrm{d}^3r\, w(\mathbf{r})\,\phi(\mathbf{r}, t)$ etc.) denotes the spatially averaged state of the macroscopic system in the vicinity of the QD. The scattering rates (7.28) summarize all microscopic charge transfer processes between the macroscopic and the quantum mechanical subsystem and are—just like the bound charge density (7.27)—localized according to the spatial profile function w. The microscopic scattering rates for the capture and escape of electrons $\gamma_{\alpha\in\mathcal{I}_e}$ and holes $\gamma_{\alpha\in\mathcal{I}_h}$ are obtained by microscopic calculations [32, 48, 58–60] and need to be parametrized in terms of the macroscopic potentials u_w. With the choice of (7.27)–(7.28), the hybrid system (7.21) inherently preserves the total charge [33].

7.3.2.3 Extension by Schrödinger–Poisson system

The hybrid system (7.21) can be extended by a self-consistent Schrödinger–Poisson system for the QD carriers to account for the quantum confined Stark effect [34, 35]. This extension is necessary to describe (bias-dependent) energy shifts of the QD excitons and the detuning of the light-matter interaction. To this end, the system (7.21) must be coupled to a (stationary) Schrödinger equation for each band

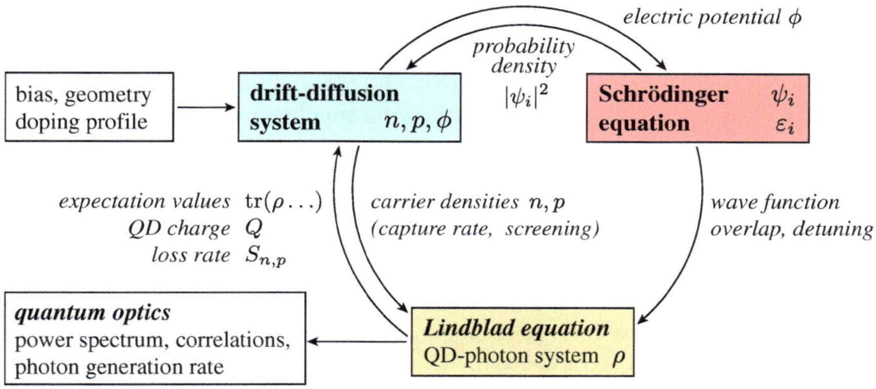

Fig. 7.8 Schematic representation of the extended hybrid quantum-classical modeling approach featuring a self-consistent Schrödinger–Poisson system, see Sect. 7.3.2.3 for details. Reprinted, with permission, from [34]. © 2019 Society of Photo-Optical Instrumentation Engineers (SPIE)

$$H_e^0(\phi)\,\psi_{e,i} = \varepsilon_{e,i}\psi_{e,i}, \qquad\qquad H_h^0(\phi)\,\psi_{h,i} = \varepsilon_{h,i}\psi_{h,i}, \tag{7.29}$$

which describes the (envelope) wave functions $\psi_{\lambda,i}$ and one-particle energies $\varepsilon_{\lambda,i}$, $\lambda \in \{e, h\}$, of the QD electrons and holes in a suitably chosen confinement potential. The Schrödinger equations (7.29) are posed on a subdomain $\Omega_0 \subseteq \Omega$ of the full computational domain and are supplemented with outgoing wave conditions on $\partial\Omega_0$. The one-particle Hamiltonians $H_{e,h}^0(\phi)$ depend on the device's internal electric potential ϕ in the vicinity of the QD, which is subject to Poisson's equation (7.21a). As a result, the (previously static) one-particle energies in the Hamiltonian (7.22b) are replaced by the eigenvalues of (7.29), which depend dynamically on the state of the QD's macroscopic environment. Due to the dependency on the electric potential, this is in general a non-Hermitian eigenvalue problem [61], whose solutions are quasi-bound resonance modes with complex energy eigenvalues. The imaginary part of the eigenvalues describes the resonance's decay rate. Close to flat band conditions, however, the decay rate is very small and can be safely neglected such that the wave functions are normalized as $\int_{\Omega_0} \mathrm{d}^3r\,\left|\psi_{\lambda,i}(\mathbf{r})\right|^2 = 1$ for $\lambda \in \{e, h\}$. In this extended version of the hybrid quantum-classical model, the spatial profile function $w(\mathbf{r})$ localizing the coupling terms (7.27)–(7.28) is replaced by the absolute squares of the (numerically calculated) QD wave functions $\left|\psi_{\lambda,i}(\mathbf{r})\right|^2$ [34]. With the correspondingly modified QD charge density $Q(\rho)$ entering Poisson's equation (7.21a), this yields a self-consistently coupled Schrödinger–Poisson problem. A schematic representation of the coupling scheme of the extended system is shown in Fig. 7.8.

The extended system (7.21) and (7.29) allows for several additional modeling refinements: Just like the single-particle energies $\varepsilon_{\lambda,i}$, also the Coulomb matrix elements $V_{i,j,k,l}^{\lambda,\lambda'}$ in (7.22c) depend on the state of the QD's macroscopic environment, because of (i) the variation of the wave function envelopes in the self-consistently calculated electric potential and (ii) because of screening by the continuum charge carriers in the vicinity of the QD. In [34], the plasma screening has been taken into account using the static, long wavelength limit of Lindhard's formula [47]. Finally, the light-matter coupling constants $g_{i,j,k}$ (in (7.22d)) and the spontaneous emission rates (corresponding $\gamma_{\alpha\in\mathcal{I}_0}(\nu,\ell)$ in $\mathcal{D}_0$) can be supplemented with the respective wave function overlap integrals.

7.3.2.4 Electro-Optical Coupling

The key process in QD-based quantum light sources is the spontaneous emission of single photons due to the recombination of bound electron-hole pairs. This is a dissipative process, that is contained in the (charge-conserving) dissipation superoperator $\mathcal{D}_0$. In the limit of very leaky, low Q resonators, the corresponding microscopic transition rate is given by the Weisskopf–Wigner decay rate [36]

$$\gamma_{\alpha\in\mathcal{I}_0}(\nu,\ell=0)=F_P\,\frac{(2\pi)^3 n_b|d_{c,v}|^2\nu^3}{3\pi\hbar\varepsilon_0 c_0^3}\left(1+n_{\mathrm{ph}}(h\nu)\right), \tag{7.30}$$

where $d_{c,v}$ is the interband dipole moment, n_b is the refractive index of the background material and $n_{\mathrm{ph}}(h\nu)=(\exp(\beta h\nu)-1)^{-1}$ is the thermal photon number at the respective electronic transition energy $h\nu$. The Purcell factor F_P defined in (7.18) quantifies the modification of the free space decay rate due a modified local density of photonic states at the position of the QD.

7.3.2.5 Consistency with Thermodynamic Principles

The hybrid quantum-classical model system (7.21) has several appealing thermodynamic properties, which served as a major principle in the construction of the coupled system [13, 33]. Most strikingly, the system (7.21) satisfies the second law of thermodynamics, as it was shown to have a non-negative entropy production rate [33]. Moreover, it is consistent with the thermodynamic equilibrium, where the solution obeys the (quantum) detailed balance relation (microscopic reversibility) and minimizes the equilibrium grand potential

$$\Phi_{\mathrm{eq}}(n,p,\rho)=\mathcal{F}(n,p,\rho)-\mu_{\mathrm{eq}}\int_{\Omega}\mathrm{d}^3r\,\left(n-p-Q(\rho)\right).$$

Here, $\mathcal{F}(n,p,\rho)$ is the hybrid free energy functional that comprises both classical and quantum mechanical energy contributions [33].

The hybrid system is consistent with the GENERIC formalism (see Sect. 7.2.3) of non-equilibrium thermodynamics [62]. The construction is based on gradient structures for both the van Roosbroeck system (see [62, 63]) and the dissipation superoperator (7.24). The latter was first described in [64], where the central object is the *Onsager operator*

$$\mathbb{K}(\rho)\,\square = \frac{1}{2}\beta\tilde{\gamma}\left(\left[A^{\dagger}, \mathcal{C}_{\rho}^{-\beta h\nu}[A, \square]\right] + \left[A, \mathcal{C}_{\rho}^{\beta h\nu}[A^{\dagger}, \square]\right]\right), \tag{7.31}$$

which involves the *tilted* Kubo–Mori operator

$$\mathcal{C}_{\rho}^{\alpha} A = e^{-\frac{1}{2}\alpha} \int_{0}^{1} \mathrm{d}s\, e^{s\alpha} \rho^{s} A \rho^{1-s}. \tag{7.32}$$

The transition rate $\tilde{\gamma}$ in (7.31) is connected with the scattering rates in (7.24) via $\gamma = \tilde{\gamma}\, \exp\left(\pm\frac{1}{2}\beta h\nu\right)$. For a quantum jump operator A obeying $[\mathcal{H}, A] = -h\nu A$, the Onsager operator (7.31) maps the derivative of the quantum mechanical free energy contribution $\mathrm{D}_{\rho}\mathcal{F} = \mathcal{H} + k_B T\,(\log\rho + I)$ (where I is the identity operator) to the dissipation superoperator

$$\partial_t \rho|_{\mathrm{diss}} = -\mathbb{K}(\rho)\,\mathrm{D}_{\rho}\mathcal{F} = \tilde{\gamma}\left(e^{\frac{1}{2}\beta h\nu} L_A \rho + e^{-\frac{1}{2}\beta h\nu} L_{A^{\dagger}} \rho\right).$$

Within the GENERIC approach, the non-negativity of the corresponding dissipation rate is guaranteed by construction and follows immediately from the non-negativity of the Onsager operator $\mathbb{K}(\rho) \geq 0$ [64]

$$\begin{aligned} -\frac{1}{T}\frac{\mathrm{d}\mathcal{F}}{\mathrm{d}t} &= -\frac{1}{T}\mathrm{tr}\left(\mathrm{D}_{\rho}\mathcal{F}\,\partial_t \rho\right) = -\frac{1}{T}\mathrm{tr}\left(\mathrm{D}_{\rho}\mathcal{F}\left(-\frac{i}{\hbar}[\mathcal{H}, \rho] - \mathbb{K}(\rho)\,\mathrm{D}_{\rho}\mathcal{F}\right)\right) \\ &= +\frac{1}{T}\mathrm{tr}\left(\mathrm{D}_{\rho}\mathcal{F}\,\mathbb{K}(\rho)\;\mathrm{D}_{\rho}\mathcal{F}\right) \geq 0. \end{aligned}$$

Onsager's reciprocal relations are reflected by the self-adjointness $\mathbb{K}(\rho) = \mathbb{K}^{\dagger}(\rho)$. Without the gradient structure, the consistency with the second law of thermodynamics relies on *Spohn's inequality* [65].

7.4 Numerical Methods

7.4.1 Numerical Methods for the Drift-Diffusion Equations

The numerical solution of the van Roosbroeck system requires a highly specialized discretization scheme, in order to preserve essential structural properties of the continuous system on the discrete level. Due to sudden jumps of the doping concentration (e.g., pn-junctions) or the material composition (heterostructures), the equations are typically stiff, such that it is highly desirable to obtain stable solutions also on rather coarse grids without additional smallness conditions. These problems have been, to a large extent, overcome by the finite volume discretization method together with the flux discretization scheme developed by Scharfetter and Gummel [66]. In this

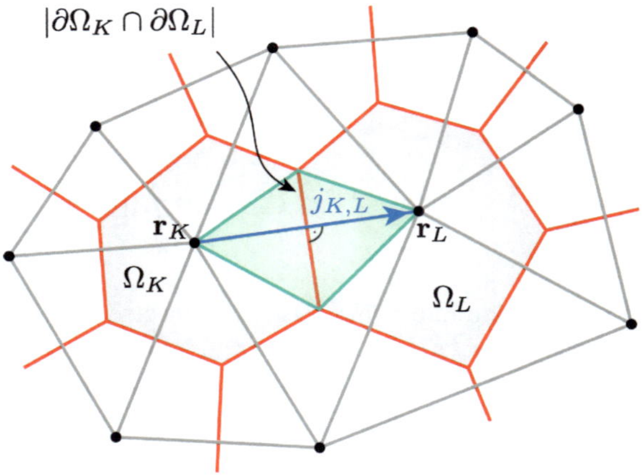

Fig. 7.9 Delaunay triangulation (*grey lines*) of a given point set (*black dots*) and corresponding Voronoï cells (*red lines*). The discrete flux (*blue arrow*) between the two control volumes K and L is given by the projection $j_{K,L}$ along the edge $\overline{KL}$ (see text). The flux is weighted by the edge factor $s_{K,L}$. The picture is taken from [13]

section, we outline the finite volume Scharfetter–Gummel method for the numerical solution of the van Roosbroeck system (7.1). Moreover, we discuss generalizations of the approach for strongly degenerate semiconductors and its application at cryogenic operation temperatures.

7.4.1.1 Finite Volume Method

In this section, we briefly sketch the finite volume method on unstructured grids using *restricted* Voronoï cells as control volumes. For a detailed introduction we refer to [67]. We assume a partition of the computational domain $\Omega = \bigcup_K \Omega_K$ into control volumes Ω_K. These control volumes are typically constructed from the dual mesh of a boundary conforming Delaunay triangulation [68], whose elements are called restricted Voronoï cells. The discretization of the continuous equations is achieved via integration over the control volumes and application of the divergence theorem. In the following, we use the electrostatic potential ϕ and the quasi-Fermi potentials $\phi_n = -\mu_c/q, \phi_p = -\mu_v/q$ as basic variables. In the most simple case (homogeneous material, homogeneous Neumann boundary conditions), we obtain the discrete van Roosbroeck system as [67]

$$F_{\phi_K^{(i)}} = \sum_{L \in \mathcal{N}(K)} s_{K,L} D_{K,L}^{(i)} - q\,|\Omega_K| \left(C_K + p_K^{(i)} - n_K^{(i)} \right), \tag{7.33a}$$

$$F_{\phi_{n,K}^{(i)}} = q\,|\Omega_K| \left(n_K^{(i)} - n_K^{(i-1)} \right) + \Delta t_i \Big(q\,|\Omega_K|\, R_K^{(i)} - \sum_{L \in \mathcal{N}(K)} s_{K,L} j_{n,K,L}^{(i)} \Big), \tag{7.33b}$$

$$F_{\phi_{p,K}^{(i)}} = q\,|\Omega_K| \left(p_K^{(i)} - p_K^{(i-1)} \right) + \Delta t_i \Big(q\,|\Omega_K|\, R_K^{(i)} + \sum_{L \in \mathcal{N}(K)} s_{K,L} j_{p,K,L}^{(i)} \Big). \tag{7.33c}$$

The discrete carrier densities are

$$n_K^{(i)} = n(\phi_K^{(i)}, \phi_{n,K}^{(i)}) = N_c \mathscr{F}(\eta_{n,K}^{(i)}), \qquad \eta_{n,K}^{(i)} = -\frac{E_c - q\phi_K^{(i)} + q\phi_{n,K}^{(i)}}{k_B T},$$

$$p_K^{(i)} = p(\phi_K^{(i)}, \phi_{p,K}^{(i)}) = N_v \mathscr{F}(\eta_{p,K}^{(i)}), \qquad \eta_{p,K}^{(i)} = +\frac{E_v - q\phi_K^{(i)} + q\phi_{p,K}^{(i)}}{k_B T},$$

with nodal values $u_K^{(i)} = u\,(\mathbf{r}_K, t_i)$, $u \in \{\phi, \phi_n, \phi_p\}$, evaluated at the K-th node $\mathbf{r}_K$ and time t_i (see Fig. 7.9). The discrete recombination rate reads $R_K^{(i)} = R(\phi_K^{(i)}, \phi_{n,K}^{(i)}, \phi_{p,K}^{(i)})$. The geometrical factors are the volume $|\Omega_K|$ of the K-th cell and the edge factor $s_{K,L} = |\partial\Omega_K \cap \partial\Omega_L| \,/\, \|\mathbf{r}_K - \mathbf{r}_L\|$, see Fig. 7.9. For time-discretization, we employ an implicit Euler method with time step $\Delta t_i = t_i - t_{i-1}$. The summation in the flux terms extends over all neighboring cells $L \in \mathcal{N}(K)$ adjacent to the K-th cell and involves the discrete flux projections

$$\begin{aligned} D_{K,L}^{(i)} &= (\mathbf{r}_L - \mathbf{r}_K) \cdot \mathbf{D}^{(i)} = -\varepsilon(\phi_L^{(i)} - \phi_K^{(i)}), \\ j_{n,K,L}^{(i)} &= (\mathbf{r}_L - \mathbf{r}_K) \cdot \mathbf{j}_n^{(i)}, \\ j_{p,K,L}^{(i)} &= (\mathbf{r}_L - \mathbf{r}_K) \cdot \mathbf{j}_p^{(i)}. \end{aligned}$$

The discretization of the carrier fluxes $j_{n/p,K,L}^{(i)}$ requires special attention and will be carried out using the Scharfetter–Gummel method outlined in Sect. 7.4.1.2. The linear system

$$\mathbf{F}\left(\phi_1 \dots \phi_N, \phi_{n,1} \dots \phi_{n,N}, \phi_{p,1} \dots \phi_{p,N}\right) = \mathbf{0} \tag{7.34}$$

is solved using a Newton iteration method. Alternatively, a Gummel iteration [69] or combined Gummel–Newton iteration [70] can be used, which rely on decoupling, linearization and block-wise solution of the full problem.

7.4.1.2 Generalized Scharfetter–Gummel Scheme for Fermi–Dirac Statistics

The typically exponentially varying doping profiles (and therefore also carrier densities) in semiconductor devices lead to numerical instabilities when using the standard finite difference discretization. In particular, the naive discretization approach results in spurious oscillations and may cause unphysical results such as negative carrier densities [67, 71]. A robust discretization scheme for the drift-diffusion current was introduced by Scharfetter and Gummel [66], who explicitly solved the current density expressions (7.4b) as a separate differential equation along the edge $\overline{KL}$ between two adjacent nodes of the mesh (we drop the time step index)

$$\frac{q\left(\phi_L-\phi_K\right)}{k_B T}=\int_{n_K}^{n_L} \mathrm{d}n \, \frac{g\left(n/N_c\right)}{\frac{j_{n,K,L}}{q M_n\left(\phi_L-\phi_K\right)}+n}. \tag{7.35}$$

Assuming a constant discrete flux $j_{n,K,L}$, the integral (7.35) can be carried out analytically in the case of Maxwell–Boltzmann statistics ($g \equiv 1$), leading to a discrete current formula

$$j_{n,K,L}=M_n k_B T\left(n_L B\left(\frac{q\left(\phi_L-\phi_K\right)}{k_B T}\right)-n_K B\left(-\frac{q\left(\phi_L-\phi_K\right)}{k_B T}\right)\right). \tag{7.36}$$

The exponential terms in the Bernoulli function $B\left(x\right)=x/\left(e^x-1\right)$ reflect the characteristics of the doping profile and allow for numerically stable calculations. The so-called *Scharfetter–Gummel scheme* (7.36) smoothly interpolates between the central difference scheme for diffusive currents and the upwind scheme for convective flows. With the advent of the Scharfetter–Gummel discretization it became possible to accurately deal with the stiff solutions of the van Roosbroeck system on coarse grids such that practically relevant devices could be studied with a reasonable computational effort [72]. With some modifications, the method is used in many TCAD simulation tools [73, 74].

In the case of degenerate carrier statistics, the classical Scharfetter–Gummel scheme (7.36) breaks down and generalizations of the scheme are required. This is important, e.g., in semiconductor lasers operating at high carrier densities, organic semiconductors with Gaussian quasi-bands [75] or quantum light emitting diodes operating at extremely low temperatures [76, 77]. For weakly degenerate semiconductors following the Blakemore approximation [78], an exact generalization of the Scharfetter–Gummel scheme was obtained in [79], however, in the strongly degenerate case no closed solution of the problem (7.35) exists, such that one has to resort

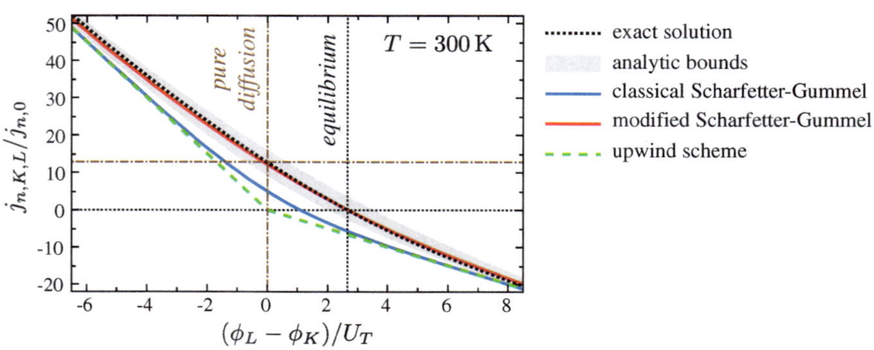

Fig. 7.10 Comparison of the classical and the modified Scharfetter–Gummel schemes (7.36) and (7.37) with a numerically exact solution of (7.35). The nodal carrier densities are $n_K = 1\times10^{18}\,\mathrm{cm}^{-3}$ and $n_L = 3\times10^{18}\,\mathrm{cm}^{-3}$. GaAs material parameters are taken from [17]. The current density projection is scaled with $j_{n,0} = M_n k_B T N_c$. The picture is taken from [13]

to approximations. We highlight the approach developed in [80, 81], which relies on averaging of the degeneracy factor $g \to g_{n,K,L}$ along the edge. This leads to

$$j_{n,K,L} = M_n k_B T g_{n,K,L} \left(n_L B \left(\frac{q(\phi_L - \phi_K)}{k_B T g_{n,K,L}} \right) - n_K B \left(- \frac{q(\phi_L - \phi_K)}{k_B T g_{n,K,L}} \right) \right) \tag{7.37a}$$

with

$$g_{n,K,L} = \frac{\eta_{n,L} - \eta_{n,K}}{\log(\mathscr{F}(\eta_{n,L})) - \log(\mathscr{F}(\eta_{n,K}))}. \tag{7.37b}$$

The modified Scharfetter–Gummel scheme (7.37) preserves many structural properties such as the consistency with the thermodynamic equilibrium and the drift- or diffusion dominated limits, while being highly accurate independent of the actual carrier statistics or the density of states function [81–83]. Figure 7.10 shows a comparison of the classical and the modified Scharfetter–Gummel schemes (7.36) and (7.37) with a numerical solution of the integral equation (7.35) obtained using Gauss–Kronrod quadrature [82, 84] in the strongly degenerate limit. We refer to [85–88] for other generalizations of the Scharfetter–Gummel scheme. The scheme (7.37) has been used for the simulation of the current flow in single-photon emitting diodes operating at cryogenic temperatures, in [33, 34] and Sects. 7.5.1, 7.5.4 and 8.2.

7.4.1.3 Electronic Transport Simulation at Cryogenic Temperatures

At cryogenic temperatures the numerical solution of the system (7.33) is a challenging task and subject to serious convergence issues. Next to strong degeneration effects and sharp internal layers at pn-junctions and heterointerfaces, the major obstacle in achieving convergent solutions is the extreme depletion of minority carrier densities in the cryogenic limit. The temperature dependence of the carrier densities is given in (7.2) and plotted in Fig. 7.2a. As a consequence of the very low thermally activated

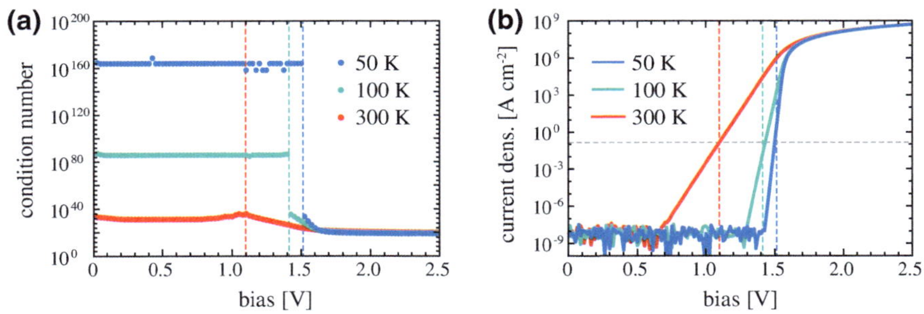

Fig. 7.11 **a** Condition number estimate of the Jacobian for a pin-diode problem at different temperatures and corresponding **b** current–voltage curves. The sudden drop of the condition number appears at the diode's threshold voltage. Adapted, with permission, from [12]. © 2016 Springer Science+Business Media

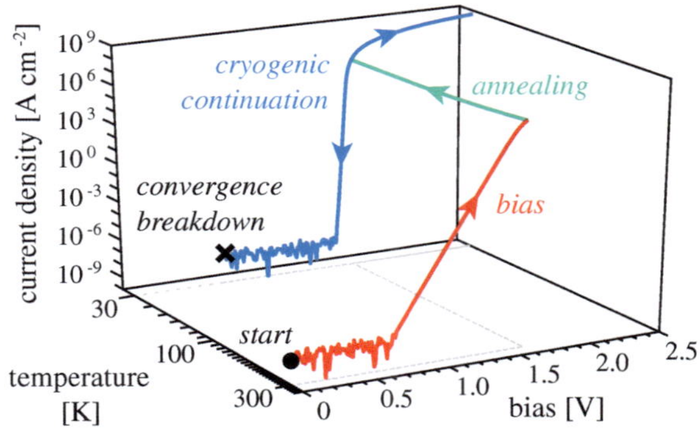

Fig. 7.12 Continuation scheme for transport simulation at cryogenic temperatures. After reaching flat band conditions at an elevated temperature, a solution at the cryogenic working point is obtained by successively lowering the device temperature. Reprinted, with permission, from [12]. © 2016 Springer Science+Business Media

carrier densities, the Jacobian matrix of the discrete van Roosbroeck system (7.33) is singular in finite precision arithmetics [12, 89, 90], such that the numerical solution of (7.34) is extremely sensitive to round-off errors and unfeasible using standard routines.

Indeed, the condition number estimate of the Jacobian matrix for a simple 1D-pin diode problem shown in Fig. 7.11a reveals an exponential increase of the condition number for decreasing temperatures, as long as the diode is biased below the threshold voltage. Beyond threshold, the condition number of the cryogenic problems suddenly drops to the rather moderate value of the room temperature problem (300 K), which can be solved with standard methods. The drop of the condition number correlates with the onset of current flow as shown in Fig. 7.11b. This observation motivates the two-step temperature embedding scheme introduced in [12]. Starting from the thermodynamic equilibrium solution at an elevated temperature (e.g., room temperature), the bias is raised beyond the device's threshold voltage until flat band conditions are attained. In a second step, the temperature parameter is successively decreased until the cryogenic working point is reached. Finally, the actual computations of interest can be carried out in the vicinity of flat band conditions. The approach is illustrated in Fig. 7.12 and circumvents the numerically intractable regime, by avoiding configurations with extreme carrier depletion.

We refer to [12] for a discussion of the application of the temperature embedding scheme in transport simulations in wide gap semiconductors (e.g., III-nitrides, SiC), where similar numerical problems exist.

7.4.2 Finite-Element Approach to Maxwell's Equations

In this section we sketch the finite-element approach to Maxwell's equations by stating (7.9) in its weak form in Sect. 7.4.2.1 and by summarizing in Sect. 7.4.2.2 the finite-element expansion and specific implementations for applications discussed in Sect. 7.5.

7.4.2.1 Variational Formulation

We start from the **curl**-**curl**-equation (7.9) for the electric field without sources, multiply with the complex conjugate of a vectorial test function **v**, and integrate

$$\int_{\Omega} \mathrm{d}V\, \mathbf{v}^* \cdot \left(\nabla \times \mu^{-1} \nabla \times \mathbf{E}\right) - \omega^2 \int_{\Omega} \mathrm{d}V\, \mathbf{v}^* \cdot \varepsilon \mathbf{E} = 0. \tag{7.38}$$

Integration by parts yields

$$\int_{\Omega} \mathrm{d}V\, \left(\nabla \times \mathbf{v}^*\right) \cdot \mu^{-1} \left(\nabla \times \mathbf{E}\right) - \int_{\partial\Omega} \mathrm{d}\mathbf{A} \cdot \left(\mathbf{v}^* \times \left[\mu^{-1} \nabla \times \mathbf{E}\right]\right) - \omega^2 \int_{\Omega} \mathrm{d}V\, \mathbf{v}^* \cdot \varepsilon \mathbf{E} = 0. \tag{7.39}$$

Expressions of this type are the building blocks for the variational formulation of the Maxwell scattering problem. Let the source field data $\mathbf{E}_{\text{source}}$ and $\nabla \times \mathbf{E}_{\text{source}}$ at the boundary $\partial\Omega$ of the computational domain Ω be given. A typical variational form of the scattering problem reads, see e.g., [19]:

Find $\mathbf{E} \in U$:

$$a(\mathbf{v}, \mathbf{E}) = \int_{\partial\Omega} \mathrm{d}\mathbf{A} \cdot \left(\mathbf{v}^* \times \left[\left(\mu^c\right)^{-1} \nabla \times \mathbf{E}_{\text{source}}\right]\right) + a_{\text{ext}}\left(\mathbf{v}, \mathbf{g}\right) \tag{7.40}$$

$\forall \mathbf{v} \in U$ with

$$\begin{aligned}
a(\cdot,\cdot) &= a_{\text{int}}(\cdot,\cdot) + a_{\text{ext}}(\cdot,\cdot), \\
a_{\text{int}}(\mathbf{v}, \mathbf{E}) &= \int_{\Omega} \mathrm{d}V\, \left(\nabla \times \mathbf{v}^*\right) \cdot \mu^{-1} \left(\nabla \times \mathbf{E}\right) - \omega^2 \int_{\Omega} \mathrm{d}V\, \varepsilon \mathbf{v}^* \cdot \mathbf{E}, \\
a_{\text{ext}}(\mathbf{v}, \mathbf{g}) &= \int_{\Omega_{\text{ext}}} \mathrm{d}V\, \left(\nabla \times \mathbf{v}^*\right) \cdot \left(\mu^c\right)^{-1} \left(\nabla \times \mathbf{E}\right) - \omega^2 \int_{\Omega_{\text{ext}}} \mathrm{d}V\, \varepsilon^c \mathbf{v}^* \cdot \mathbf{E}
\end{aligned}$$

and **g** is an arbitrary tangentially continuous function with

$$\mathbf{g}(\partial\Omega) = \mathbf{E}_{\text{source}}(\partial\Omega), \qquad \mathbf{g}(\infty) = 0,$$

and

$$\left(\mu^c\right)^{-1} = \frac{1}{|J|} J^T \mu^{-1} J, \qquad \varepsilon^c = |J|\, J^{-1} \varepsilon J^{-T}.$$

The function space U ensures the tangential continuity of the vectorial fields and the existence of the first weak derivative. The exterior domain Ω_{ext} undergoes a special treatment in terms of complex continuation of the spatial coordinates to ensure an exponential decay of the scattered field. Practically this is achieved by a transformation of the material properties μ and ε, where J is the Jacobian of this transformation.

7.4.2.2 Patch-Wise Tangential Continuous, Polynomial Approximation on Unstructured Grids

The finite element method (FEM) is based on a *variational formulation* of Maxwell's equations constructed from integral expressions on the computational domain, such as (7.40). The FEM does not use any approximation to Maxwell's equations itself, except the discretization of the geometry. However, it approximates the solution space in which one seeks a reasonable approximation to the exact solution.

The solution space of FEM is constructed by subdividing the computational domain into small, easy-to-handle patches and by providing a number of tailored polynomials on each patch for the approximation of the solution. The patches together with these local polynomials are called *finite elements*. Common examples of finite elements are triangles and rectangles in 2D and tetrahedrons and cubes in 3D. The polynomials are mostly constant, linear, quadratic and cubic expressions. These locally defined polynomial spaces are pieced together in a way that tangential continuity of the electric and magnetic field across the boundaries of neighboring patches are ensured. Once these local approximations with proper continuity conditions have been realized, these are inserted into the variational equation. Finally, this yields a linear system whose solution is a piecewise polynomial approximation of the exact solution. In summary, the two basic steps in each finite element method for Maxwell's equations are:

1. Find a variational formulation of Maxwell's equations.
2. Construct a suitable set of finite elements based on polynomials defined on local geometric patches. This transforms the variational formulation into a discrete, algebraic problem.

This concept has a number of remarkable properties:

- Complex geometrical shapes can be accurately treated, e.g., roundings can be approximated well and easily.
- The finite element mesh can easily be refined and adapted to properties of the solution, e.g., for resolving singularities at corners.
- High order approximations are available and ensure fast convergence.

In order to achieve improved solution methods for Maxwell's equations in 3D we addressed a series of numerical problems:

Hierarchic hp-adaptive finite elements. Due to the multi-scale properties of typical nanophotonic structures, conventional finite element discretizations based on tetrahedrons or hexahedrons are ineffective. We found it necessary to use mixed meshes composed of tetrahedrons, hexahedrons, prisms and pyramids. Following [91], we

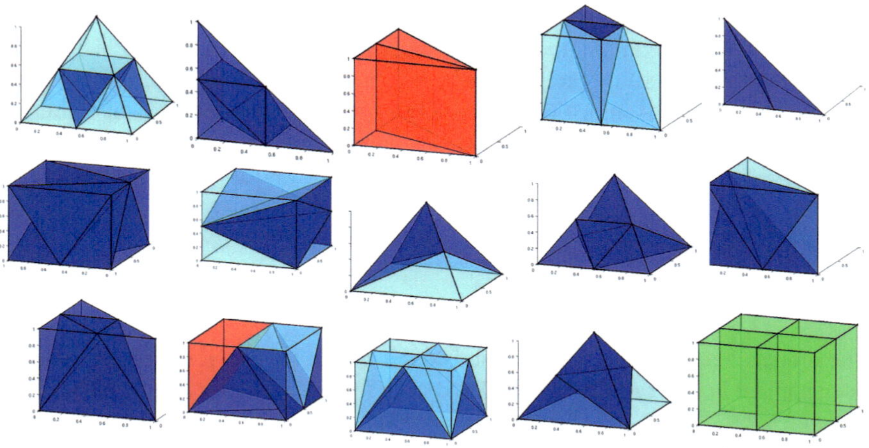

Fig. 7.13 Some examples of decompositions of pyramids, prisms, tetrahedrons, and hexagons. An efficient modeling of nontrivial geometries requires a careful bookkeeping of the elements and their neighborhood relations

used an advanced setting based on hierarchical finite elements up to order 9, where the finite elements are associated to the geometric entities edge, face and cell. E.g., for prism elements, this gives the possibility to use high-order finite elements in lateral direction, where the lateral size of the element can be large, and low-order finite elements in vertical direction, where the height of the elements must be small due to the thin layers. See Fig. 7.15 for a typical model of a VCSEL which contains multiple length scales. We also use adaptive schemes which allow to choose between local refinement of the mesh element size h or local refinement of the polynomial order p of the finite elements, or combinations thereof [92, 93]. This results in discretizations on unstructured, locally refined meshes with polynomial order varying over the different mesh elements, so-called hp-FEM discretizations [94].

Modeling the geometry. For the optimization of the entire structure it is necessary to define a parameterized model of the geometry. Whereas for standard problems advanced computer-aided design (CAD) tools are available, the analysis of VCSELs and other optical structures requires a number of additional features: These are the use of mixed meshes, the modeling of isolated structures embedded into heterogeneous and unbounded exterior domains, and the treatment of exact periodic boundaries. Based on the open source kernels OpenCascade[1] and PythonOCC[2] we developed a CAD-tool which meets these needs plus the standard operations like set operations, extrusions, sweeping, revolving. Examples of mesh elements and their refinement are displayed in Fig. 7.13.

[1]www.opencascade.com.

[2]www.pythonocc.org.

Specific implementations for rotationally symmetric geometries and for highly singular fields. We have implemented FEM based methods in order to treat the singular fields in the vicinity of dipole sources by a subtraction-field approach. Further, this treatment has been extended to rotationally symmetric geometries, where the field radiated by the dipole can be expanded in a series of rotationally symmetric functions [53]. This approach has been applied in order to model single-photon sources [95].

7.5 Applications

The models and computational methods described in the previous subsections have been applied in order to simulate and to design various nanoelectronic and nanophotonic devices. This includes active devices like VCSELs [96–99], edge-emitters [8, 9, 31, 100–102], QD-based single-photon devices [33, 53, 76, 95, 103–108] and organic light emitting diodes [21, 109]. Also passive devices have been investigated, including metallic and dielectric nanoantennas [110–114] and -resonators [115–121], dielectric, nanostructured waveguides [92, 122, 123], plasmonic and dielectric metamaterials [124–127], and couplers for integrated optics applications [112, 128, 129]. In this section we shortly outline some of these applications.

7.5.1 Quantum Dot Single-Photon Sources

Single-photon sources are fundamental building blocks of future quantum communication networks [5, 130]. Self-assembled semiconductor QDs are excellent candidates for the realization of single-photon sources in the solid state as they can be easily integrated in dielectric micro-resonators and provide outstanding optical properties in terms of the suppression of multi-photon emission events and a high degree of photon indistinguishability [52, 131, 132]. The emission properties of the QD, in particular the rate of spontaneous emission [54], can be enhanced by tailoring the geometry of its surrounding structure. Moreover, the efficiencies with which the generated photons are extracted into a specific direction or coupled into an optical fiber depend strongly on the geometry of the photonic resonator [53, 103, 133].

By numerical simulations of the scattering problem described in Sect. 7.2.2 for a harmonically oscillating electric dipole (see Sect. 7.3.2.1), we have optimized the geometry of the QD-based single-photon source featuring a microlens structure at the surface. The optimized geometry is shown along with the energy density of the scattered electromagnetic field in Fig. 7.14a. The geometry optimization in the high-dimensional parameter space (lens diameter, lens height, QD position, fiber distance and fiber core diameter) was carried out using a Bayesian optimization approach [53, 134]. For a detailed description of the numerical method and optimization results, we refer to Sect. 8.2.

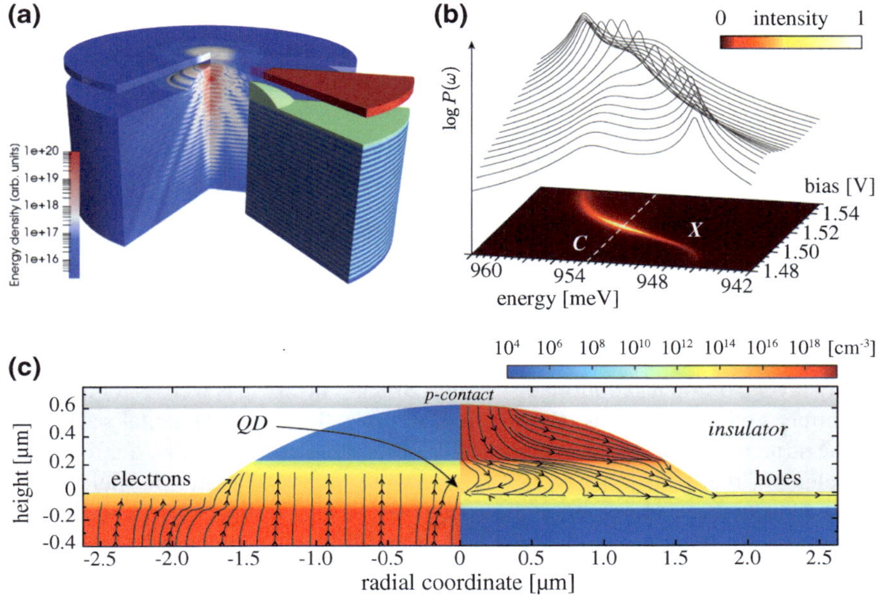

Fig. 7.14 Simulation of a QD-based single-photon source featuring a monolithically fabricated microlens structure at the surface. **a** Numerically optimized geometry (lens diameter, lens height, QD position, fiber distance) for optimal photon extraction into a single mode fiber. The energy density of the scattered electromagnetic field is color-coded. The picture is adapted with permission from [53]. © 2018 Optical Society of America (OSA). **b** Calculated bias dependence of the power spectrum $P(\omega)$ in a QD-based single-photon emitting diode using the hybrid modeling approach described in Sects. 7.3.2.2–7.3.2.3. The exciton line (X) is blue shifted due to a combination of the quantum confined Stark effect and plasma screening. The resonance with the cavity mode (C) appears at $U \approx 1.516\,\mathrm{V}$ [34]. **c** 2D cross section of the spatially resolved current injection at $I = 2.75\,\mathrm{nA}$ ($U = 1.49\,\mathrm{V}$ DC bias) and $T = 30\,\mathrm{K}$. The geometry of the p-i-n diode structure is the same as in Fig. 7.14a. Left: electron density n (color coded) and current density $\mathbf{j}_n$ (arrows point in the direction of electron flux). Right: hole density p (color coded) and current density $\mathbf{j}_p$. In the low-injection regime, the scattering of continuum carriers to the QD is the dominant loss term in the transport equations, such that the QD notably contributes to current guiding. Adapted with permission from [34]. © 2019 Society of Photo-Optical Instrumentation Engineers (SPIE)

In the interest of compactness and scalability, the realization of electrically driven single-photon sources is highly attractive for practical applications. We employed the hybrid quantum-classical modeling approach [33, 34, 56] described in Sect. 7.3.2.2 to simulate the electro-optical performance of the electrically contacted single-photon source shown in Fig. 7.14c (same geometry as in Fig. 7.14a). Our device-scale simulation approach allows for a comprehensive characterization of the single-photon emitting diode, as it describes the spatially resolved carrier injection into the QD, see Fig. 7.14c, along with the quantum optical features (e. g., second-order correlation function, photon generation rate, line broadening due to excitation-induced dephasing etc.) of the coupled QD-photon system. With the inclusion of the Schrödinger–

Poisson system (see Sect. 7.3.2.3), the approach allows to describe line shifts of the exciton due to a combination of the quantum confined Stark effect and plasma screening [34], see Fig. 7.14b. A detailed description of the numerical results is given Sect. 8.2.

7.5.2 *Vertical-Cavity Surface-Emitting Lasers*

VCSELs are highly miniaturized light sources with unique properties and with important applications, e.g., in optical communications [136]. The optical design of a VCSEL cavity includes finding optimum parameters for dimensions of the various cavity layers, of top and bottom distributed Bragg reflectors, as well as for apertures or further optical structures. For the design, typically the fundamental resonances of the structure can be used, as physically observable quantities like the resonance wavelength, the quality-factor, and the properties of the radiated far-field can be deduced from these. However, solving the Maxwell eigenvalue problem corresponding to (7.9) in this case is involved due to the different geometrical scales of the problem: thin layers of few nanometers thickness need to be resolved at a wavelength around one micron and at a total size of the device of about 10 μm in three dimensions [137], see Fig. 7.15a. In some cases, cylindrical symmetry can be used

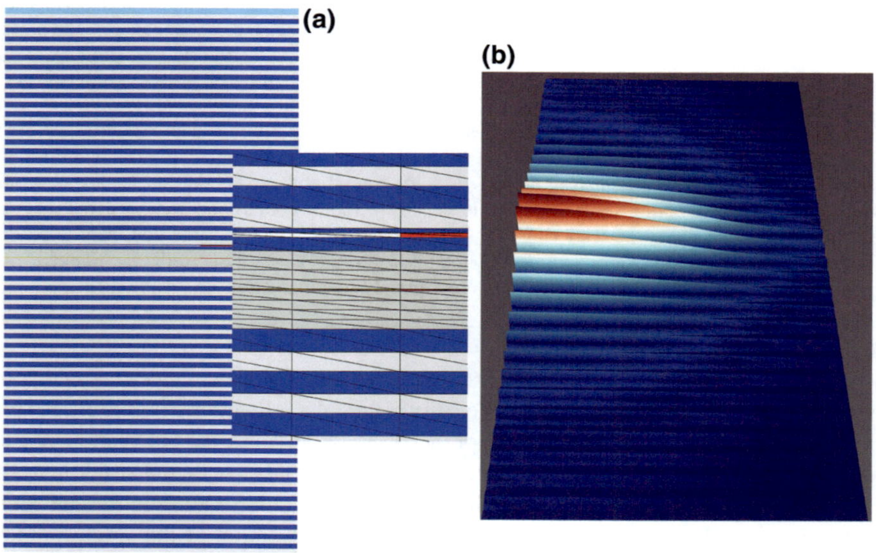

Fig. 7.15 **a** 2D cross section of a cylindrically symmetric VCSEL (setup as in [135]), the inset shows a detail of the finite element triangulation, including part of the 5 nm thin and 6 μm wide gain region (*orange color*). **b** Visualization of the computed electric field of the fundamental mode with a resonance wavelength of $\lambda = 980\,\text{nm}$

to reduce the numerical complexity, other cases do not involve such symmetries. We have demonstrated that using higher-order, adaptive finite-element methods can yield relative accuracies of the resonance wavelength of better than 10^{-8} and of the quality-factor of better than 10^{-5} [96] for a standard, rotationally symmetric VCSEL setup, following a design from the literature [137]. Further, convergence to a level of better than 10^{-5} for the resonance wavelength and 10^{-2} for the quality-factor has been demonstrated for a fully 3D setup, using elements of fourth polynomial order on a tetrahedral discretization [96]. In the same context, also coupled thermo-optical simulations have been demonstrated. These allow to quantitatively determine the impact of thermal lensing in VCSELs.

The developed methods have been applied to design single-mode VCSELs by engineering of the oxide-aperture geometry and to interpret experimental results from realized structures [97, 98]. A numerically computed electric field distribution is plotted in Fig. 7.15b. For further details on the engineering of the opto-electronic design of VCSELs we refer to Chaps. 10 and 11 of this book.

7.5.3 Grating Couplers

By using nanostructured high refractive-index materials photonic integrated circuits can be realized with a small footprint. However, it is challenging to couple light between optical fibers and such integrated circuits. For this purpose grating couplers have been proposed to be a viable, efficient and broadband solution [138]. Optical design of grating couplers typically aims at reducing losses and thereby maximizing the coupling efficiency, which is typically defined as the optical power flux in an output waveguide directing the light from the coupler to the photonic integrated circuit divided by the optical power flux in the input optical fiber. The coupling efficiency is derived from the electromagnetic near field distribution which is obtained from simulating (7.9) in the form of a scattering problem. Therein the source field typically is the propagation mode in the optical waveguide, and the power flux in the output waveguide is obtained from the overlap integral between the near field distribution with the guided propagation modes of the output waveguide.

When a 2-dimensional (2D) model is sufficient, triangular meshes and 2D finite elements can be used, such that typical computation times are in the range of few seconds, which enables fast optimizations and parameter studies. However, for accurate determination of coupling efficiencies often a full 3D model is required instead. As the computational domain in this case is relatively large (several tens of micrometers in each dimension) the problem is computationally demanding. For numerically solving this problem in 3D we use higher-order finite elements on prismatoidal and tetrahedral meshes. We have shown that this allows for obtaining results with relative numerical errors of the coupling efficiency as small as about 10^{-3} for typical setups [128]. Figure 7.16 shows a prismatoidal mesh and visualizations of computed field distributions for a 3D grating coupler model. Also, results from a related convergence study are shown. Further, the developed methods have been used to design grating

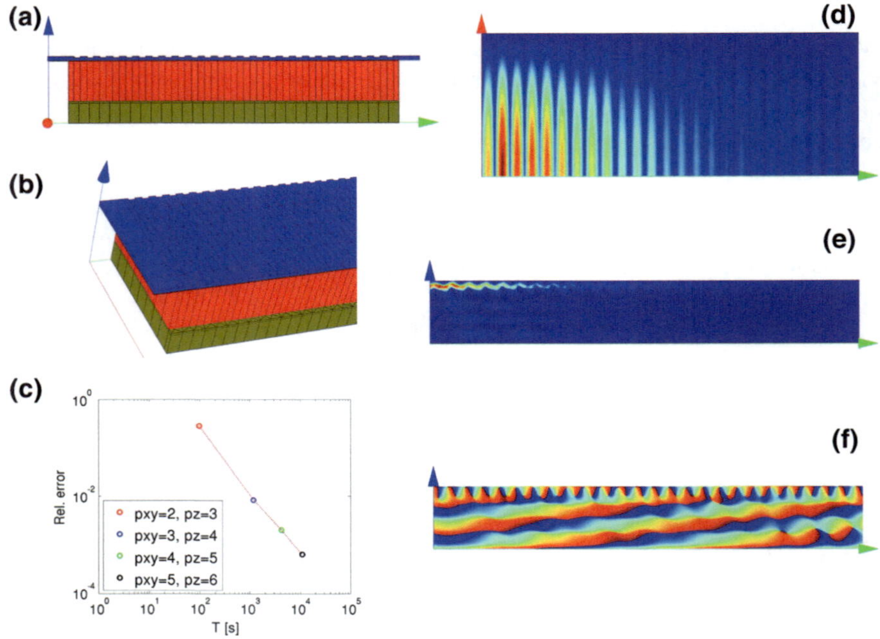

Fig. 7.16 **a** Visualization of a part of the mesh discretizing a 3D grating coupler model (*blue:* Si waveguide, *red:* SiO_2 buffer layer, *green:* Si substrate, coordinate axes: *red/green/blue: x/y/z*). **b** Detail of the mesh at oblique viewing angle. **c** Convergence: relative error of the field energy in simulations with varied finite element degree (p_{xy} (p_z): polynomial order of the finite elements defined on prismatoidal mesh elements in horizontal (vertical) direction) as a function of CPU time T (specific problem setup as defined in [128]). **d** Visualization of the computed field intensity distribution in a cross-section through the waveguide layer. **e** Visualization of the computed field intensity distribution in a vertical cross section through the center of the device. **f** Visualization of the corresponding phase distribution of the electromagnetic field

couplers capable of coupling both, fundamental and higher order modes from and to multimode fibers [129, 139]. In a related context, the methods have also been used to design and optimize coupling between integrated optical waveguides and plasmonic waveguides [112].

7.5.4 *Efficient Current Injection into Oxide-Confined Pn-Diodes*

Current injection into single QDs embedded in vertical pn-diodes featuring oxide apertures is essential to realize small-scale, integrable single-photon emitting diodes based on the buried stressor growth technique [140–144]. This requires efficient electrical pumping of sub-micron sized regions under pulsed excitation to achieve

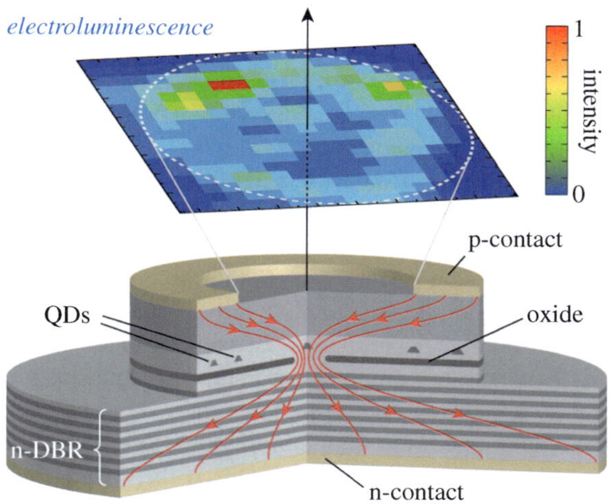

Fig. 7.17 Illustration of a single-photon emitting diode based on site-controlled nucleated QDs using the buried stressor approach. The electroluminescence map reveals optical activity of parasitic QDs in the outer region of the mesa, which indicates insufficient current confinement by the oxidized aperture. The naively expected current path is indicated by red arrows. The electroluminescence map is reprinted from [76]

control over the carrier population in the central QDs. In [76] we reported on experimental and theoretical evidence for a rapid lateral spreading of the injection current after passing the oxide aperture in conventional pin-diode designs. This observation clearly contradicts the naively expected current flow indicated by the red arrows in Fig. 7.17 and indicates a breakdown of the current confining properties of oxidized apertures at low injection currents and cryogenic temperatures (which are the typical operation conditions of single photon emitting diodes). A theoretical analysis of the phenomenon by means of numerical simulations of the van Roosbroeck system (7.1) revealed the absence of carrier recombination in the intrinsic region of the diode as the primary reason for the lateral current spreading. Due to their long life time, the electrons have an enhanced diffusion length, that enables the excitation of parasitic QDs in the outer regions of the mesa, see Fig. 7.17. By an alternative design employing p-doping down to the oxide layer, the current spreading can be suppressed resulting in an enhanced current confinement and increased injection efficiencies [76]. For a detailed discussion of this application we refer to Sect. 3.6.

7.6 Conclusion and Outlook

The computer-aided design and optimization of nanophotonic devices requires efficient numerical methods and modeling techniques. We have developed novel modeling approaches for the multi-dimensional, device-scale simulation of single-photon sources and laser diodes based on semiconductor QDs. To this end, we have coupled the optical Maxwell equations with semi-classical transport theory (van Roosbroeck system) by appropriate models for the optically active region. This includes a multi-species model for semiconductor lasers with QD active region featuring a semi-classical description of the optical response and microscopically calculated Coulomb effects (scattering, gain, dephasing). For QD-based single-photon emitting diodes, we developed a hybrid quantum-classical modeling approach, which combines the spatially resolved description of the carrier transport and optical fields with a fully quantum mechanical model for the light-matter interaction based on quantum master equations (featuring a dissipative Jaynes–Cummings model).

The application of our models to the numerical simulation of realistic devices with complex, multi-dimensional geometries and heterostructure designs asks for efficient discretization methods. For the solution of Maxwell's equations we introduced a novel hierarchic, hp-adaptive finite element approach using mixed meshes adapted to the multi-scale properties of the photonic structures. This allows for improved performance in comparison to standard finite element methods, and for numerical solutions with controlled accuracy. For the electronic transport problem, the cryogenic operation temperatures of QD-based single-photon sources impose several numerical challenges due to (i) the strong degeneration of the electron-hole plasma and (ii) the extreme depletion of the minority carrier densities. We presented a generalized finite volume Scharfetter–Gummel scheme, to accurately discretize the drift-diffusion equations for strongly degenerate semiconductors, taking the Pauli-blocking fully into account (Fermi–Dirac statistics). In addition, we developed a temperature-embedding method to overcome the stability issues of the discrete carrier transport problem at very low temperatures. The modeling approaches and numerical techniques are demonstrated on various applications, including VCSELs, grating couplers and single-photon sources.

The demonstrated advances in self-consistent modeling and multi-dimensional numerical simulation of nanophotonic devices can reveal the performance bottlenecks in new device concepts, contribute to their understanding and help to theoretically explore their optimization potential. This makes numerical simulation an important tool, guiding important technological decisions in the development of future nanophotonic and quantum devices, such as on-chip single-photon sources and nanolasers for integrated quantum photonic circuits.

Acknowledgements This work has been supported by the German Research Foundation (DFG) within the collaborative research center SFB 787 *Semiconductor Nanophotonics* under grant B4. The authors would like to thank Patricio Farrell, Jürgen Fuhrmann, Philipp Gutsche, Jan Pomplun, Nella Rotundo, Alexander Wilms, Benjamin Wohlfeil and Lin Zschiedrich for excellent collaboration and valuable discussions.

References

1. P. Bhattacharya, Z. Mi, Proc. IEEE **95**(9), 1723 (2007). https://doi.org/10.1109/JPROC.2007.900897
2. P. Michler (ed.), in *Single Semiconductor Quantum Dots*. NanoScience and Technology (Springer, Berlin, Heidelberg, 2009). https://doi.org/10.1007/978-3-540-87446-1
3. D. Bimberg, U.W. Pohl, Mater. Today **14**(9), 388 (2011). https://doi.org/10.1016/S1369-7021(11)70183-3

4. P. Lodahl, S. Mahmoodian, S. Stobbe, Rev. Mod. Phys. **87**(2), 347 (2015). https://doi.org/10.1103/revmodphys.87.347
5. P. Michler (ed.), in *Quantum Dots for Quantum Information Technologies*. Springer Series in Nano-Optics and Nanophotonics (Springer, Cham, 2017). https://doi.org/10.1007/978-3-319-56378-7
6. M. Streiff, A. Witzig, M. Pfeiffer, P. Royo, W. Fichtner, IEEE J. Sel. Top. Quantum Electron. **9**, 879 (2003). https://doi.org/10.1109/JSTQE.2003.818858
7. U. Bandelow, H. Gajewski, R. Hünlich, in *Optoelectronic Devices*, ed. by J. Piprek (Springer, New York, 2005), Chap. 3, pp. 63–85. https://doi.org/10.1007/0-387-27256-9_3
8. H. Wenzel, P. Crump, H. Ekhteraei, C. Schultz, J. Pomplun, S. Burger, L. Zschiedrich, F. Schmidt, G. Erbert, in *2011 11th International Conference on Numerical Simulation of Optoelectronic Devices (NUSOD)* (IEEE, 2011), p. 143. https://doi.org/10.1109/nusod.2011.6041183
9. J. Pomplun, H. Wenzel, S. Burger, L. Zschiedrich, M. Rozova, F. Schmidt, P. Crump, H. Ekhteraei, C.M. Schultz, G. Erbert, Proc. SPIE **8255**, 825510 (2012). https://doi.org/10.1117/12.909330
10. W.W. van Roosbroeck, Bell Syst. Tech. J. **29**(4), 560 (1950). https://doi.org/10.1002/j.1538-7305.1950.tb03653.x
11. S. Selberherr, *Analysis and Simulation of Semiconductor Devices* (Springer, Vienna, 1984). https://doi.org/10.1007/978-3-7091-8752-4
12. M. Kantner, T. Koprucki, Opt. Quantum. Electron. **48**(12), 543 (2016). https://doi.org/10.1007/s11082-016-0817-2
13. M. Kantner, Modeling and simulation of electrically driven quantum dot based single-photon sources: From classical device physics to open quantum systems. Ph.D. thesis, Technical University Berlin, Berlin (2018). https://doi.org/10.14279/depositonce-7516
14. K. Hess, *Advanced Theory of Semiconductor Devices*, 2nd edn. (Wiley-IEEE Press, New York, 2000). https://doi.org/10.1109/9780470544105
15. C. Jacoboni, *Theory of Electron Transport in Semiconductors* (Springer, Berlin, Heidelberg, 2010). https://doi.org/10.1007/978-3-642-10586-9
16. S.M. Sze, *Physics of Semiconductor Devices*, 2nd edn. (Wiley, New York, 1981). https://doi.org/10.1002/0470068329
17. V. Palankovski, R. Quay, *Analysis and Simulation of Heterostructure Devices*. Series in Computational Microelectronics (Springer, Vienna, 2004). https://doi.org/10.1007/978-3-7091-0560-3
18. D. Schröder, *Modelling of Interface Carrier Transport for Device Simulation*. Series in Computational Microelectronics (Springer, Vienna, 1994). https://doi.org/10.1007/978-3-7091-6644-4
19. F. Schmidt, in *Handbook of Optoelectronic Device Modeling and Simulation: Fundamentals, Materials, Nanostructures, LEDs, and Amplifiers*, vol. 2, ed. by J. Piprek (CRC Press, Taylor & Francis Group, Boca Raton, 2017), Chap. Photonics, pp. 807–852. https://doi.org/10.4324/9781315152318-27
20. E.S.C. Ching, P.T. Leung, A.M. van den Brink, W.M. Suen, S.S. Tong, K. Young, Rev. Mod. Phys. **70**(4), 1545 (1998). https://doi.org/10.1103/revmodphys.70.1545
21. A. Fischer, T. Koprucki, K. Gärtner, M.L. Tietze, J. Brückner, B. Lüssem, K. Leo, A. Glitzky, R. Scholz, Adv. Funct. Mater. **24**(22), 3367 (2014). https://doi.org/10.1002/adfm.201303066
22. G.K. Wachutka, IEEE Trans. Comput. Aided Design Integr. Circuits Syst. **9**(11), 1141 (1990). https://doi.org/10.1109/43.62751
23. U. Lindefelt, J. Appl. Phys. **75**(2), 942 (1994). https://doi.org/10.1063/1.356450
24. G. Albinus, H. Gajewski, R. Hünlich, Nonlinearity **15**(2), 367 (2002). https://doi.org/10.1088/0951-7715/15/2/307

25. M. Kantner, J. Comput. Phys. **402**, 109091 (2020). https://doi.org/10.1016/j.jcp.2019.109091
26. L. Onsager, Phys. Rev. **37**(4), 405 (1931). https://doi.org/10.1103/PhysRev.37.405
27. M. Grmela, H.C. Öttinger, Phys. Rev. E **56**, 6620 (1997). https://doi.org/10.1103/PhysRevE.56.6620
28. A. Mielke, in *Recent Trends in Dynamical Systems*, ed. by A. Johann, H.P. Kruse, F. Rupp, S. Schmitz, no. 35 in Springer Proceedings in Mathematics & Statistics (Springer, Basel, 2013), Chap. 21, pp. 555–585. https://doi.org/10.1007/978-3-0348-0451-6_21
29. A. Mielke, in *Mathematical Results in Quantum Mechanics*, ed. by P. Exner, W. König, H. Neidhardt (World Scientific, Singapore, 2015), pp. 331–348. https://doi.org/10.1142/9789814618144_0029
30. R. Michalzik (ed.), *VCSELs–Fundamentals, Technology and Applications of Vertical-Cavity Surface-Emitting Lasers*, Springer Series in Optical Sciences, vol. 166 (Springer, Berlin, Heidelberg, 2013). https://doi.org/10.1007/978-3-642-24986-0
31. T. Koprucki, A. Wilms, A. Knorr, U. Bandelow, Opt. Quantum. Electron. **42**(11), 777 (2011). https://doi.org/10.1007/s11082-011-9479-2
32. A. Wilms, P. Mathé, F. Schulze, T. Koprucki, A. Knorr, U. Bandelow, Phys. Rev. B **88**, 235421 (2013). https://doi.org/10.1103/PhysRevB.88.235421
33. M. Kantner, M. Mittnenzweig, T. Koprucki, Phys. Rev. B **96**(20), 205301 (2017). https://doi.org/10.1103/PhysRevB.96.205301
34. M. Kantner, Proc. SPIE **10912**, 109120U (2019). https://doi.org/10.1117/12.2515209
35. M. Kantner, T. Koprucki, H.-J. Wünsche, U. Bandelow, in Proceedings of the 24th International Conference on Simulation of Semiconductor Processes and Devices (SISPAD 2019), pp. 355–358 (2019). https://doi.org/10.1109/SISPAD.2019.8870459
36. H.P. Breuer, F. Petruccione, *The Theory of Open Quantum Systems* (Oxford University Press, Oxford, 2002). https://doi.org/10.1093/acprof:oso/9780199213900.001.0001
37. W.W. Chow, F. Jahnke, Prog. Quantum Electron. **37**(3), 109 (2013). https://doi.org/10.1016/j.pquantelec.2013.04.001
38. S. Reitzenstein, A. Forchel, J. Phys. D: Appl. Phys. **43**(3), 033001 (2010). https://doi.org/10.1088/0022-3727/43/3/033001
39. G.A. Baraff, R.K. Smith, Phys. Rev. A **61**(4), 043808 (2000). https://doi.org/10.1103/PhysRevA.61.043808
40. H. Wenzel, H.J. Wünsche, IEEE J. Quantum Electron. **30**(9), 2073 (1994). https://doi.org/10.1109/3.309867
41. A. Witzig, Modeling the optical processes in semiconductor lasers. Ph.D. thesis, ETH Zürich, Zürich (2002). https://doi.org/10.3929/ethz-a-004407405
42. S. Steiger, R.G. Veprek, B. Witzigmann, J. Comput. Electron. **7**(4), 509 (2008). https://doi.org/10.1007/s10825-008-0261-z
43. M. Grupen, K. Hess, IEEE J. Quantum Electron. **34**(1), 120 (1998). https://doi.org/10.1109/3.655016
44. W.W. Chow, S.W. Koch, IEEE J. Quantum Electron. **41**, 495 (2005). https://doi.org/10.1109/JQE.2005.843948
45. K. Lüdge, E. Schöll, IEEE J. Quantum Electron. **45**(11), 1396 (2009). https://doi.org/10.1109/jqe.2009.2028159
46. E. Malić, K.J. Ahn, M.J.P. Bormann, P. Hövel, E. Schöll, A. Knorr, M. Kuntz, D. Bimberg, Appl. Phys. Lett. **89**(10), 101107 (2006). https://doi.org/10.1063/1.2346224
47. H. Haug, S.W. Koch, *Quantum Theory of the Optical and Electronic Properties of Semiconductors*, 4th edn. (World Scientific, Singapore, 2004). https://doi.org/10.1142/5394
48. T.R. Nielsen, P. Gartner, F. Jahnke, Phys. Rev. B **69**, 235314 (2004). https://doi.org/10.1103/PhysRevB.69.235314

49. A. Wilms, D. Breddermann, P. Mathé, Phys. Status Solidi C **9**(5), 1278 (2012). https://doi.org/10.1002/pssc.201100101
50. A. Wilms, Coulomb induced interplay of localized and reservoir carriers in semiconductor quantum dots. Ph.D. thesis, Technical University Berlin (2013). https://doi.org/10.14279/depositonce-3530
51. C. Santori, D. Fattal, Y. Yamamoto, *Single-photon Devices and Applications* (Wiley, Weinheim, 2010)
52. S. Buckley, K. Rivoire, J. Vučković, Rep. Prog. Phys. **75**(12), 126503 (2012). https://doi.org/10.1088/0034-4885/75/12/126503
53. P.I. Schneider, N. Srocka, S. Rodt, L. Zschiedrich, S. Reitzenstein, S. Burger, Opt. Express **26**, 8479 (2018). https://doi.org/10.1364/oe.26.008479
54. E.M. Purcell, Phys. Rev. **69**, 681 (1946). https://doi.org/10.1007/978-1-4615-1963-8_40
55. D.J. Griffiths, *Introduction to Electrodynamics* (Cambridge University Press, Cambridge, 2017). https://doi.org/10.1017/9781108333511
56. M. Kantner, M. Mittnenzweig, T. Koprucki, Proc. SPIE **10526**, 1052603 (2018). https://doi.org/10.1117/12.2289185
57. N. Baer, P. Gartner, F. Jahnke, Eur. Phys. J. B **42**(2), 231 (2004). https://doi.org/10.1140/epjb/e2004-00375-6
58. E. Malić, M.J.P. Bormann, P. Hövel, M. Kuntz, D. Bimberg, A. Knorr, E. Schöll, IEEE, J. Sel. Top. Quantum Electron. **13**(5), 1242 (2007). https://doi.org/10.1109/ISLC.2006.1708081
59. I. Magnúsdóttir, A.V. Uskov, S. Bischoff, B. Tromborg, J. Mørk, J. Appl. Phys. **92**(10), 5982 (2002). https://doi.org/10.1063/1.1512694
60. R. Ferreira, G. Bastard, *Capture and Relaxation in Self-Assembled Semiconductor Quantum Dots*. 2053–2571 (Morgan & Claypool Publishers, San Rafael, CA, 2015). https://doi.org/10.1088/978-1-6817-4089-8
61. N. Moiseyev, *Non-Hermitian Quantum Mechanics* (Cambridge University Press, Cambridge, 2011). https://doi.org/10.1017/CBO9780511976186
62. M. Kantner, M. Mittnenzweig, A. Mielke, N. Rotundo, in *Topics in Applied Analysis and Optimisation*, ed. by M. Hintermüller and J. Rodrigues (Springer, Cham, 2019), pp. 269–293. https://doi.org/10.1007/978-3-030-33116-0_11
63. A. Mielke, Nonlinearity **24**, 1329 (2011). https://doi.org/10.1088/0951-7715/24/4/016
64. M. Mittnenzweig, A. Mielke, J. Stat. Phys. **167**(2), 205 (2017). https://doi.org/10.1007/s10955-017-1756-4
65. H. Spohn, J. Math. Phys. **19**(5), 1227 (1978). https://doi.org/10.1063/1.523789
66. D.L. Scharfetter, H.K. Gummel, IEEE Trans, Electron Dev. **16**(1), 64 (1969). https://doi.org/10.1109/t-ed.1969.16566
67. P. Farrell, N. Rotundo, D.H. Doan, M. Kantner, J. Fuhrmann, T. Koprucki, in *Handbook of Optoelectronic Device Modeling and Simulation: Lasers, Modulators, Photodetectors, Solar Cells, and Numerical Methods*, vol. 2, ed. by J. Piprek (CRC Press, Taylor & Francis Group, Boca Raton, 2017), Chap. 50, pp. 733–771. https://doi.org/10.4324/9781315152318-25
68. H. Si, K. Gärtner, J. Fuhrmann, Comput. Math. Math. Phys. **50**(1), 38 (2010). https://doi.org/10.1134/S0965542510010069
69. H.K. Gummel, IEEE Trans. Electron Dev. **11**(10), 455 (1964). https://doi.org/10.1109/T-ED.1964.15364
70. H. Gajewski, K. Gärtner, J. Appl. Math. Mech. **72**(1), 19 (1992). https://doi.org/10.1002/zamm.19920720103
71. F. Brezzi, L.D. Marini, P. Pietra, Comput. Methods Appl. Mech. Eng. **75**(1–3), 493 (1989). https://doi.org/10.1016/0045-7825(89)90044-3
72. P.A. Markowich, in *The Stationary Semiconductor Device Equations*. Series in Computational Microelectronics (Springer, Vienna, 1986). https://doi.org/10.1007/978-3-7091-3678-2

73. Silvaco International, *Atlas User's Manual* (Santa Clara, CA, 2016)
74. Synopsys Inc, *Sentaurus Device UserGuide* (Mountain View, CA, 2010)
75. S.L.M. van Mensfoort, R. Coehoorn, Phys. Rev. B **78**(8), 085207 (2008). https://doi.org/10.1103/PhysRevB.78.085207
76. M. Kantner, U. Bandelow, T. Koprucki, J.H. Schulze, A. Strittmatter, H.J. Wünsche, IEEE Trans. Electron Dev. **63**(5), 2036 (2016). https://doi.org/10.1109/ted.2016.2538561
77. J.D. Cressler, H.A. Mantooth (eds.), *Extreme Environment Electronics* (CRC Press, Taylor & Francis Group, Boca Raton, 2012). https://doi.org/10.1201/b13001
78. J.S. Blakemore, Solid-State Electron. **25**(11), 1067 (1982). https://doi.org/10.1016/0038-1101(82)90143-5
79. T. Koprucki, K. Gärtner, Opt. Quantum. Electron. **45**(7), 791 (2013). https://doi.org/10.1007/s11082-013-9673-5
80. M. Bessemoulin-Chatard, Numer. Math. **121**(4), 637 (2012). https://doi.org/10.1007/s00211-012-0448-x
81. T. Koprucki, N. Rotundo, P. Farrell, D.H. Doan, J. Fuhrmann, Opt. Quantum. Electron. **47**(6), 1327 (2015). https://doi.org/10.1007/s11082-014-0050-9
82. P. Farrell, M. Patriarca, J. Fuhrmann, T. Koprucki, Opt. Quant. Electron. **50**, 101 (2018). https://doi.org/10.1007/s11082-018-1349-8
83. P. Farrell, T. Koprucki, J. Fuhrmann, J. Comput. Phys. **346**, 497 (2017). https://doi.org/10.1016/j.jcp.2017.06.023
84. M. Patriarca, P. Farrell, J. Fuhrmann, T. Koprucki, Comput. Phys. Commun. **235**, 40 (2019). https://doi.org/10.1016/j.cpc.2018.10.004
85. J. Fuhrmann, Comput. Phys. Commun. **196**, 166 (2015). https://doi.org/10.1016/j.cpc.2015.06.004
86. Z. Yu, D. Chen, L. So, R.W. Dutton, *PISCES-2ET 2D Device Simulator* (Integrated Circuits Laboratory, Stanford University, Stanford, Tech. rep., 1994)
87. H. Gajewski, Mitt. Ges. Angew. Math. Mech. **16**(1), 35 (1993)
88. H. Gajewski, K. Gärtner, J. Appl. Math. Mech. **76**(5), 247 (1996). https://doi.org/10.1002/zamm.19960760502
89. D.M. Richey, J.D. Cressler, R.C. Jaeger, J. Phys. IV France **04**(C6), C6 (1994). https://doi.org/10.1051/jp4:1994620
90. S. Selberherr, IEEE Trans. Electron Dev. **36**(8), 1464 (1989). https://doi.org/10.1109/16.30960
91. M. Bergot, M. Duruflé, J. Comput. Phys. **232**(1), 189 (2013). https://doi.org/10.1016/j.jcp.2012.08.005
92. J. Pomplun, S. Burger, L. Zschiedrich, F. Schmidt, Phys. Status Solidi B **244**, 3419 (2007). https://doi.org/10.1002/pssb.200743192
93. S. Burger, L. Zschiedrich, J. Pomplun, S. Herrmann, F. Schmidt, Proc. SPIE **9424**, 94240Z (2015). https://doi.org/10.1117/12.2085795
94. I. Babuška, M.R. Dorr, Numer. Math. **37**(2), 257 (1981). https://doi.org/10.1007/BF01398256
95. N. Srocka, A. Musiał, P.I. Schneider, P. Mrowiński, P. Holewa, S. Burger, D. Quandt, A. Strittmatter, S. Rodt, S. Reitzenstein, G. Sek, AIP Adv. **8**, 085205 (2018). https://doi.org/10.1063/1.5038137
96. M. Rozova, J. Pomplun, L. Zschiedrich, F. Schmidt, S. Burger, Proc. SPIE **8255**, 82550K (2012). https://doi.org/10.1117/12.906372
97. V. Shchukin, N. Ledentsov Jr., J. Kropp, G. Steinle, N. Ledentsov, S. Burger, F. Schmidt, IEEE J. Quantum Electron. **50**, 990 (2014). https://doi.org/10.1109/jqe.2014.2364544
98. V.A. Shchukin, N.N. Ledentsov, J.R. Kropp, G. Steinle, N.N. Ledentsov Jr., K.D. Choquette, S. Burger, F. Schmidt, Proc. SPIE **9381**, 93810V (2015). https://doi.org/10.1117/12.2077012
99. T. Höhne, L. Zschiedrich, N. Haghighi, J.A. Lott, S. Burger, Proc. SPIE **106821**, 106821U (2018). https://doi.org/10.1117/12.2307200
100. J. Pomplun, S. Burger, F. Schmidt, A. Schliwa, D. Bimberg, A. Pietrzak, H. Wenzel, G. Erbert, Phys. Status Solidi B **247**, 846 (2010). https://doi.org/10.1002/pssb.200945451

101. D. Peschka, M. Thomas, A. Glitzky, R. Nürnberg, K. Gärtner, M. Virgilio, S. Guha, T. Schroeder, G. Capellini, T. Koprucki, I.E.E.E. Photon, J. **7**(3), 1 (2015). https://doi.org/10.1109/jphot.2015.2427093
102. D. Peschka, M. Thomas, A. Glitzky, R. Nürnberg, M. Virgilio, S. Guha, T. Schroeder, G. Capellini, T. Koprucki, Opt. Quant. Electron. **48**(2), 156 (2016). https://doi.org/10.1007/s11082-016-0394-4
103. M. Gschrey, A. Thoma, P. Schnauber, M. Seifried, R. Schmidt, B. Wohlfeil, L. Krüger, J.H. Schulze, T. Heindel, S. Burger, F. Schmidt, A. Strittmatter, S. Rodt, S. Reitzenstein, Nat. Commun. **6**, 7662 (2015). https://doi.org/10.1038/ncomms8662
104. P. Schnauber, A. Thoma, C.V. Heine, A. Schlehahn, L. Gantz, M. Gschrey, R. Schmidt, C. Hopfmann, B. Wohlfeil, J.H. Schulze, A. Strittmatter, T. Heindel, S. Rodt, U. Woggon, D. Gershoni, S. Reitzenstein, Technologies **4**(1), 1 (2016). https://doi.org/10.3390/technologies4010001
105. P. Schnauber, J. Schall, S. Bounouar, T. Höhne, S.I. Park, G.H. Ryu, T. Heindel, S. Burger, J.D. Song, S. Rodt, S. Reitzenstein, Nano Lett. **18**, 2336 (2018). https://doi.org/10.1021/acs.nanolett.7b05218
106. T. Höhne, P. Schnauber, S. Rodt, S. Reitzenstein, S. Burger, Phys. Status Solidi B **256**, 1800437 (2019). https://doi.org/10.1002/pssb.201800437
107. K. Żołnacz, A. Musiał, N. Srocka, J. Große, M.J. Schlösinger, P.-I. Schneider, O. Kravets, M. Mikulicz, J. Olszewski, K. Poturaj, G. Wójcik, P. Mergo, K. Dybka, M. Dyrkacz, M. Dłubek, S. Rodt, S. Burger, L. Zschiedrich, G. Sęk, S. Reitzenstein, W. Urbańczyk, Method for direct coupling of a semiconductor quantum dot to an optical fiber for single-photon source applications. Opt. Express **27**(19), 26772–26785 (2019). https://doi.org/10.1364/OE.27.026772
108. P. Mrowiński, P. Schnauber, P. Gutsche, A. Kaganskiy, J. Schall, S. Burger, S. Rodt, S. Reitzenstein, Directional emission of a deterministically fabricated quantum Dot–Bragg reflection multimode waveguide system. ACS Photonics **6**(9), 2231–2237 (2019). https://doi.org/10.1021/acsphotonics.9b00369
109. A. Fischer, P. Pahner, B. Lüssem, K. Leo, R. Scholz, T. Koprucki, J. Fuhrmann, K. Gärtner, A. Glitzky, Org. Electron. **13**(11), 2461 (2012). https://doi.org/10.1016/j.orgel.2012.06.046
110. M. Richter, F. Schlosser, M. Schoth, S. Burger, F. Schmidt, A. Knorr, S. Mukamel, Phys. Rev. B **86**, 085308 (2012). https://doi.org/10.1103/physrevb.86.085308
111. V.E. Babicheva, S.S. Vergeles, P.E. Vorobev, S. Burger, J. Opt. Soc. Am. B **29**, 1263 (2012). https://doi.org/10.1364/josab.29.001263
112. G. Kewes, A.W. Schell, R. Henze, R.S. Schonfeld, S. Burger, K. Busch, O. Benson, Appl. Phys. Lett. **102**, 051104 (2013). https://doi.org/10.1063/1.4790824
113. A. Abass, P. Gutsche, B. Maes, C. Rockstuhl, E.R. Martins, Opt. Express **24**(17), 19638 (2016). https://doi.org/10.1364/oe.24.019638
114. C. Becker, S. Burger, C. Barth, P. Manley, K. Jäger, D. Eisenhauer, G. Köppel, P. Chabera, J. Chen, K. Zheng, T. Pullerits, ACS Photonics **5**, 4668 (2018). https://doi.org/10.1021/acsphotonics.8b01199
115. M. Karl, B. Kettner, S. Burger, F. Schmidt, H. Kalt, M. Hetterich, Opt. Express **17**, 1144 (2009). https://doi.org/10.1364/oe.17.001144
116. B. Maes, J. Petráček, S. Burger, P. Kwiecien, J. Luksch, I. Richter, Opt. Express **21**, 6794 (2013). https://doi.org/10.1364/oe.21.006794
117. J.R. de Lasson, L.H. Frandsen, P. Gutsche, S. Burger, O.S. Kim, O. Breinbjerg, A. Ivanskaya, F. Wang, O. Sigmund, T. Häyrynen, A.V. Lavrinenko, J. Mork, N. Gregersen, Opt. Express **26**, 11366 (2018). https://doi.org/10.1364/oe.26.011366
118. G. Kewes, F. Binkowski, S. Burger, L. Zschiedrich, O. Benson, ACS Photonics **5**, 4089 (2018). https://doi.org/10.1021/acsphotonics.8b00766
119. L. Zschiedrich, F. Binkowski, N. Nikolay, O. Benson, G. Kewes, S. Burger, Phys. Rev. A **98**, 043806 (2018). https://doi.org/10.1103/PhysRevA.98.043806
120. F. Binkowski, L. Zschiedrich, M. Hammerschmidt, S. Burger, Modal analysis for nanoplasmonics with nonlocal material properties. Phys. Rev. B. **100**(15), 155406 (2019). https://doi.org/10.1103/PhysRevB.100.155406

121. P. Lalanne, W. Yan, A. Gras, C. Sauvan, J.-P. Hugonin, M. Besbes, G. Demésy, M.D. Truong, B. Gralak, F. Zolla, A. Nicolet, F. Binkowski, L. Zschiedrich, S. Burger, J. Zimmerling, R. Remis, P. Urbach, H.T. Liu, T. Weiss, Quasinormal mode solvers for resonators with dispersive materials. J. Opt. Soc. Am. A **36**(4), 686–704 (2019). https://doi.org/10.1364/JOSAA.36.000686
122. R. Holzlöhner, S. Burger, P.J. Roberts, J. Pomplun, J. Europ. Opt. Soc: Rap. Comm. **1**, 06011 (2006). https://doi.org/10.2971/jeos.2006.06011
123. J. Bethge, G. Steinmeyer, S. Burger, F. Lederer, R. Iliew, J. Light. Technol. **27**, 1698 (2009). https://doi.org/10.1109/jlt.2009.2021583
124. P. Gutsche, R. Mäusle, S. Burger, Photonics **3**, 60 (2016). https://doi.org/10.3390/photonics3040060
125. P. Gutsche, L.V. Poulikakos, M. Hammerschmidt, S. Burger, F. Schmidt, Proc. SPIE **9756**, 97560X (2016). https://doi.org/10.1117/12.2209551
126. D. Werdehausen, I. Staude, S. Burger, J. Petschulat, T. Scharf, T. Pertsch, M. Decker, Opt. Mater. Express **8**, 3456 (2018). https://doi.org/10.1364/OME.8.003456
127. D. Werdehausen, S. Burger, I. Staude, T. Pertsch, M. Decker, Dispersion-engineered nanocomposites enable achromatic diffractive optical elements. Optica. **6**(8),1031–1038 (2019). https://doi.org/10.1364/OPTICA.6.001031
128. B. Wohlfeil, S. Burger, C. Stamatiadis, J. Pomplun, F. Schmidt, L. Zimmermann, K. Petermann, Proc. SPIE **8988**, 89880K (2014). https://doi.org/10.1117/12.2044461
129. B. Wohlfeil, G. Rademacher, C. Stamatiadis, K. Voigt, L. Zimmermann, K. Petermann, I.E.E.E. Photon, Technol. Lett. **28**, 1241 (2016). https://doi.org/10.1109/lpt.2016.2514712
130. H.J. Kimble, Nature **453**(7198), 1023 (2008). https://doi.org/10.1038/nature07127
131. I. Aharonovich, D. Englund, M. Toth, Nat. Photonics **10**(10), 631 (2016). https://doi.org/10.1038/nphoton.2016.186
132. N. Somaschi, V. Giesz, L. de Santis, J.C. Loredo, M.P. Almeida, G. Hornecker, S.L. Portalupi, T. Grange, C. Antón, J. Demory, C. Gómez, I. Sagnes, N.D. Lanzillotti-Kimura, A. Lemaítre, A. Auffeves, A.G. White, L. Lanco, P. Senellart, Nat. Photonics **10**, 340 (2016). https://doi.org/10.1038/nphoton.2016.23
133. W.L. Barnes, G. Björk, J.M. Gérard, P. Jonsson, J.A.E. Wasey, P.T. Worthing, V. Zwiller, Eur. Phys. J. D **18**(2), 197 (2002). https://doi.org/10.1140/epjd/e20020024
134. P.I. Schneider, X. Garcia Santiago, V. Soltwisch, M. Hammerschmidt, S. Burger, C. Rockstuhl, ACS Photonics **6**(11), 2726–2733 (2019). https://doi.org/10.1021/acsphotonics.9b00706
135. P. Bienstman, R. Baets, J. Vukusic, A. Larsson, M.J. Noble, M. Brunner, K. Gulden, P. Debernardi, L. Fratta, G.P. Bava, H. Wenzel, B. Klein, O. Conradi, R. Pregla, S.A. Riyopoulos, J.F.P. Seurin, S.L. Chuang, IEEE J. Quantum Elect. **37**, 1618 (2001). https://doi.org/10.1109/3.970909
136. K. Iga, IEEE, J. Sel. Top. Quantum Electron. **6**, 1201 (2000). https://doi.org/10.1109/2944.902168
137. M. Dems, I.S. Chung, P. Nyakas, S. Bischoff, K. Panajotov, Opt. Express **18**, 16042 (2010). https://doi.org/10.1364/oe.18.016042
138. D. Taillaert, P. Bienstman, R. Baets, Opt. Lett. **29**, 2749 (2004). https://doi.org/10.1364/ol.29.002749
139. B. Wohlfeil, C. Stamatiadis, M. Jäger, L. Zimmermann, S. Burger, K. Petermann, in *2014 The European Conference on Optical Communication (ECOC)* (2014), pp. 1–3. https://doi.org/10.1109/ecoc.2014.6963980
140. W. Unrau, D. Quandt, J.H. Schulze, T. Heindel, T.D. Germann, O. Hitzemann, A. Strittmatter, S. Reitzenstein, U.W. Pohl, D. Bimberg, Appl. Phys. Lett. **101**(21), 211119 (2012). https://doi.org/10.1063/1.4767525
141. A. Strittmatter, A. Holzbecher, A. Schliwa, J.H. Schulze, D. Quandt, T.D. Germann, A. Dreismann, O. Hitzemann, E. Stock, I.A. Ostapenko, S. Rodt, W. Unrau, U.W. Pohl, A. Hoffmann, D. Bimberg, V.A. Haisler, Phys. Status Solidi A **209**(12), 2411 (2012). https://doi.org/10.1002/pssa.201228407

142. A. Strittmatter, A. Schliwa, J.H. Schulze, T.D. Germann, A. Dreismann, O. Hitzemann, E. Stock, I.A. Ostapenko, S. Rodt, W. Unrau, U.W. Pohl, A. Hoffmann, D. Bimberg, V.A. Haisler, Appl. Phys. Lett. **100**(9), 093111 (2012). https://doi.org/10.1063/1.3691251
143. F. Kießling, T. Niermann, M. Lehmann, J.H. Schulze, A. Strittmatter, A. Schliwa, U.W. Pohl, Phys. Rev. B **91**(7), 075306 (2015). https://doi.org/10.1103/physrevb.91.075306
144. M. Strauß, A. Kaganskiy, R. Voigt, P. Schnauber, J.H. Schulze, S. Rodt, A. Strittmatter, S. Reitzenstein, Appl. Phys. Lett. **110**(11), 111101 (2017). https://doi.org/10.1063/1.4978428

Chapter 8
Deterministic Quantum Devices for Optical Quantum Communication

Sven Rodt, Philipp-Immanuel Schneider, Lin Zschiedrich, Tobias Heindel, Samir Bounouar, Markus Kantner, Thomas Koprucki, Uwe Bandelow, Sven Burger and Stephan Reitzenstein

Abstract Photonic quantum technologies are based on the exchange of information via single photons. The information is typically encoded in the polarization of the photons and security is ensured intrinsically via principles of quantum mechanics such as the no-cloning theorem. Thus, all optical quantum communication networks rely crucially on the availability of suitable quantum-light sources. Such light sources with close to ideal optical and quantum optical properties can be realized by self-assembled semiconductor quantum dots. These high-quality nanocrystals are predestined single-photon emitters due to their quasi zero-dimensional carrier confinement. Still, the development of practical quantum-dot-based sources of single photons and entangled-photon pairs for applications in photonic quantum technology and especially for the quantum-repeater scheme is very demanding and requires highly advanced device concepts and deterministic fabrication technologies. This is mainly explained by their random position and emission energy as well as by the low photon-extraction efficiency in simple planar device configurations.

8.1 Introduction

Quantum-light sources are key building blocks for applications in photonic quantum technology. These quantum devices can provide single-photons "on-demand" which act as information carriers in quantum communication networks [1–3] (see also

S. Rodt (✉) · T. Heindel · S. Bounouar · S. Reitzenstein
Institute of Solid State Physics, Technische Universität Berlin, Hardenbergstrasse 36, 10623 Berlin, Germany
e-mail: srodt@physik.tu-berlin.de

P.-I. Schneider · L. Zschiedrich
JCMwave GmbH, Bolivarallee 22, 14050 Berlin, Germany

M. Kantner · T. Koprucki · U. Bandelow
Weierstraß-Institut für Angewandte Analysis und Stochastik, Mohrenstraße 39, 10117 Berlin, Germany

S. Burger
Zuse-Institut Berlin, Takustraße 7, 14195 Berlin, Germany

M. Kneissl et al. (eds.), *Semiconductor Nanophotonics*, Springer Series in Solid-State Sciences 194, https://doi.org/10.1007/978-3-030-35656-9_8

Chap. 9). The information is usually encoded in the polarization, phase, frequency, or time bin of the individual photons [4] and the no-cloning theorem of quantum mechanics [5] ensures, in principle, secure communication even in the presence of eavesdropping. In case of point-to-point quantum key distribution (QKD) protocols such as the BB84 protocol [6] (see Sect. 9.5.1 for details) it is sufficient to generate single photons with high purity and high emission rate to implement simple communication networks [7] as well as larger networks based on trusted nodes [8]. Beyond that, advanced quantum communication schemes like measurement device independent (MDI) QKD [9] ensure ultimate security in realistic application scenarios with non-ideal detectors through entanglement distribution via Bell-state measurements which require quantum-light sources emitting indistinguishable photons. The quantum repeater is even more complex but it is the key technology for long-distance quantum communication [2, 10]. Similar to MDI-QKD, this concept is based on entanglement distribution and typically requires sources of entangled photon pairs as well as quantum memories for upscaling to an extended network of multiple quantum nodes interconnected via Bell-state measurements. As a consequence, for the MDI-QKD and the quantum repeater concepts the entanglement distribution via Bell-state measurements sets stringent requirements on the quantum-light sources. Besides delivering single photons with high purity and high emission rate on demand, they have to emit highly indistinguishable photons with matching wavelength on a scale of the homogeneous linewidth to ensure successful Bell-state measurements between remote sources in a large-scale quantum network.

In this context, self-assembled semiconductor quantum dots (QDs) in the GaAs material system (cf. Chap. 3) are excellent candidates to realize quantum-light sources for quantum communication networks [11, 12]. These nanoscale objects provide quasi 3D quantum confinement of electrons and holes which recombine with a typical radiative lifetime of 1 ns and high quantum efficiency of >90% under emission of single photons [13]. However, the efficient extraction of the photons from the semiconductor host material and their collection by external optical elements constitutes a great challenge. This is partly due to the effects of total internal reflection at the semiconductor-air interface which limits the photon-extraction efficiency to values on the order of 1% in simple planar geometries [14, 15]. To overcome this drawback different photon extraction strategies have been proposed and implemented. They comprise microcavity based approaches [16, 17], waveguiding concepts [18, 19], geometrical concepts [20–22] and combinations thereof [23]. Of special interest are solutions providing broadband enhancement of emission which is in general user-friendly because it does not require precise spectral matching of the QD and a cavity mode and leaves more flexibility when it comes to Bell-state measurements between remote sources. Moreover, broadband enhancement is crucial for realizing bright sources of polarization-entangled photon pairs based on the radiative biexciton-exciton decay in QDs [24]. Beyond that, close to ideal quantum properties in terms of photon purity, indistinguishability and entanglement fidelity can only be ensured by highly optimized device designs, extremely careful sample fabrication and advanced excitation schemes. For instance, the contribution of off-resonant emitters in cavity-based single-photon sources can lead to non-ideal purity at high excitation rates [25], and spectral diffusion caused by charged surface states and access carriers in the environment of the QD can reduce the degree of indistinguishability [26]. Apart from these technological aspects, spectroscopic methods have evolved significantly in recent years. Especially resonant excitation schemes such as resonance fluorescence spectroscopy [27] and resonant two-photon absorption [28] have led to the generation of single photons with close to ideal quantum properties. The enormous progress in the development of QD-based quantum-light sources has already enabled first applications in point-to-point quantum communication [29] but further progress is necessary to implement fully functional and multipartite quantum networks.

This chapter presents recent advances in the design, development and optical characterization of deterministically fabricated quantum-light sources based on single InGaAs QDs (cf. Chap. 3) embedded into monolithic microlenses and planar waveguides. The design of these quantum devices is optimized via modeling in the framework of a finite-elements method (cf. Chap. 7) to maximize their broadband photon-extraction efficiency and waveguide coupling strength. In case of microlenses, the emission efficiency is maximized by using highly reflective back-side mirrors in combination with the curved upper surface of the microlens to direct photons towards the collecting optics normal to the sample surface. To fully profit from this concept, the emitter needs to be located in the center of the microlens with a lateral positioning accuracy better than 50 nm. In light of the random growth positions of self-assembled QDs this tight design rule cannot be met using standard electron-beam lithography (EBL) nanofabrication which would lead to process yields below 1%. We tackle this issue by an unique nanofabrication platform named in-situ *EBL* in which we first use low-temperature cathodoluminescence spectroscopy (CL) to select suitable QDs based on their emission wavelength and emission intensity and subsequently integrate them into photonic nanostructures by low-temperature EBL in the same electron microscope. We present details on the development of this deterministic fabrication technology and how it is applied to realize quantum-light sources with excellent optical and quantum-optical properties in terms of photon-extraction efficiency, photon purity, and photon indistinguishability. Such QD-microlenses are also very attractive candidates for photon-pair emission via the biexciton-exciton radiative cascade. In this context we discuss twin-photon emission and entangled photon pair generation of QD-microlenses with a focus on coherent triggering the emission via two-photon absorption. With respect to future applications in quantum communication networks we present strain-induced spectral fine-tuning, emission at telecom-wavelengths as well as the stand-alone and user-friendly operation of a QD-microlens-based quantum-light source providing single-photons via direct fiber coupling. Another interesting aspect of quantum nanophotonics are QD-waveguide circuits which we realize deterministically by means of in-situ EBL with high potential to realize complex photonic circuits with high quantum functionality in the future. The experiments are accompanied by numerical modeling to reveal the underlying physical processes (cf. Chaps. 4, 6, and 7). The chapter concludes with an outlook on further developments and perspectives of our approach for QD-based quantum-light sources and quantum communication networks.

8.2 Numerical Modeling and Optimization of Quantum Devices for the Generation and Distribution of Single Photons

In the following we present methods to simulate and optimize the design of quantum devices for the generation and distribution of single photons. We review the application of a numerical method to optimize the broadband photon-extraction efficiency of a QD-based single-photon source [30–32] in Sects. 8.2.1–8.2.3. The numerical method and simulation results for an electrically driven single-photon source are presented in Sect. 8.2.4.

8.2.1 A Setup for a QD-Based Fiber-Coupled Single-Photon Source

As an application example and in order to tackle a relevant wavelength regime for fiber-based data transmission, we present the optimization of a single-photon source based on a QD emitting at a vacuum wavelength of $\lambda = 1300$ nm within the telecom O-band. The QD is embedded into a mesa structure consisting of GaAs (refractive index $n_{\mathrm{GaAs}} = 3.4$). A back-side distributed Bragg reflector (DBR) consisting of alternating layers of GaAs and AlGaAs ($n_{\mathrm{AlGaAs}} = 3.0$) reflects the light emitted by the QD back into the upper hemisphere. The light is then coupled into an optical fiber with enlarged numerical aperture (NA) above the QD consisting of a homogeneous fiber core and a homogeneous fiber cladding ($n_{\mathrm{core}} = 1.5, n_{\mathrm{clad}} = 1.45$, $\mathrm{NA} \equiv \sqrt{n_{\mathrm{core}}^2 - n_{\mathrm{clad}}^2} = 0.38$). A schematic drawing of the setup is shown in Fig. 8.1.

8.2.2 Numerical Method for the Efficient Simulation of Optical Devices with Embedded QDs

The self-assembled QD under consideration consists of InGaAs and has a typical extension of 20 nm in horizontal direction and 5 nm in vertical direction. The QD is modeled in dipole approximation by an oscillating point-like current density

$$\mathbf{J}(\mathbf{r}, t) = \mathrm{Re}\{-i\omega\mathbf{p}\delta(\mathbf{r} - \mathbf{r}_{\mathrm{QD}})e^{-i\omega t}\}. \tag{8.1}$$

Here, $\mathbf{r}_{\mathrm{QD}}$ is the position of the QD and ω its emission frequency. The QD has a larger lateral than vertical extension. Hence, the electronic state is excited horizontally and the dipole moment $\mathbf{p}$ lies in the horizontal plane.

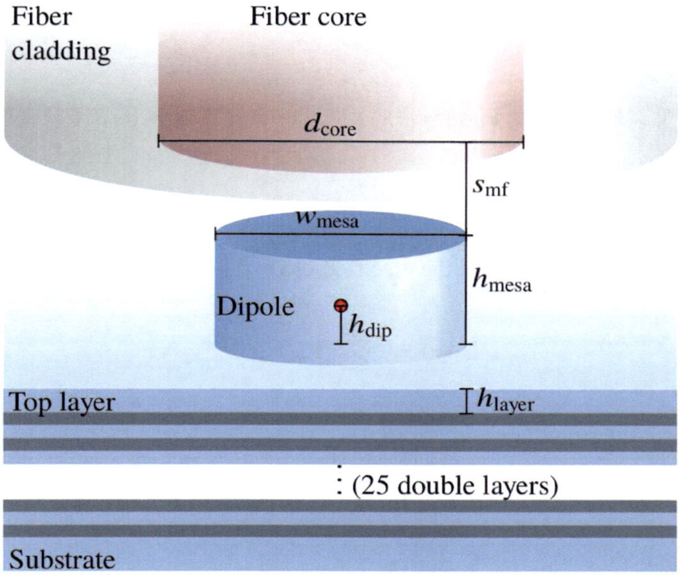

Fig. 8.1 The considered system consists of a QD (red point) embedded into a diffracting structure (GaAs, blue), a Bragg reflector (alternating layers of GaAs (blue) and $Al_{0.9}Ga_{0.1}As$ (dark gray)), and an optical fiber with homogeneous fiber core (orange) and fiber cladding (light gray). The Bragg reflector is grown on a substrate made of GaAs (blue) and has a GaAs top layer (blue). The system is parametrized by 6 geometrical parameters, the diameter of the fiber core (d_{core}), the width and height of the mesa (w_{mesa}, h_{mesa}), the elevation of the dipole (h_{dip}), the distance between mesa and fiber (s_{mf}), and the top-layer-thickness (h_{layer}). Adapted with permission from [31] (OSA)

Since the considered source current density $\mathbf{J}(\mathbf{r}, t)$ is time-harmonic (see (8.1)), the same holds for the electromagnetic fields $\mathbf{E}(\mathbf{r}, t)$ and $\mathbf{H}(\mathbf{r}, t)$. The time-harmonic Maxwell's equations can be cast into a second-order form for the electric field

$$\nabla \times \mu^{-1} \nabla \times \mathbf{E}(\mathbf{r}, \omega) - \epsilon \omega^2 \mathbf{E}(\mathbf{r}, \omega) = i\omega \mathbf{J}(\mathbf{r}, \omega), \tag{8.2}$$

with the permeability $\mu = \mu(\mathbf{r}, \omega)$ and the permittivity $\epsilon = \epsilon(\mathbf{r}, \omega)$. The imaginary part $i\sigma/\omega$ of the permittivity tensor is proportional to the conductivity $\sigma(\mathbf{r}, \omega)$ of the material. For simplicity we will henceforth implicitly assume that the fields and material tensors depend on ω.

The electric field produced by the dipole emitter defined in (8.1) diverges for $\mathbf{r} \to \mathbf{r}_{QD}$. Hence, a direct finite-element discretization of $\mathbf{E}$ suffers from a slow numerical convergence. To cure this, the electric field is expressed as a sum $\mathbf{E} = \mathbf{E}_s + \mathbf{E}_c$ of the singular dipole field $\mathbf{E}_s$ and a correction field $\mathbf{E}_c$ [33]. The singular field is chosen such that it is a solution of

$$\nabla \times \mu_d^{-1} \nabla \times \mathbf{E}_s(\mathbf{r}) - \epsilon_d \omega^2 \mathbf{E}_s(\mathbf{r}) = i\omega \mathbf{J}(\mathbf{r}), \tag{8.3}$$

where $\mu_d = \mu(\mathbf{r}_{\text{QD}})$ and $\epsilon_d = \epsilon(\mathbf{r}_{\text{QD}})$ are constant. Then, Maxwell's equation for the correction field reads [33]

$$\nabla \times \mu^{-1} \nabla \times \mathbf{E}_c(\mathbf{r}) - \epsilon\omega^2 \mathbf{E}_c(\mathbf{r}) = -\nabla \times (\mu^{-1} - \mu_d^{-1}) \nabla \times \mathbf{E}_s(\mathbf{r}) + (\epsilon - \epsilon_d)\omega^2 \mathbf{E}_s(\mathbf{r}). \tag{8.4}$$

For the considered setup, the material tensors are rotationally symmetric, i.e. in cylindrical coordinates it holds $\mu(r, z, \phi) = \mu(r, z)$ and $\epsilon(r, z, \phi) = \epsilon(r, z)$ with $r = |\mathbf{r}|$. Therefore, we expand the electric field in a series of Fourier modes

$$\mathbf{E}_c(\mathbf{r}) = \sum_{n=-\infty}^{\infty} \mathbf{E}_n(r, z) e^{in\phi}. \tag{8.5}$$

Likewise, the right hand side of (8.4),

$$\mathbf{f}(\mathbf{r}) = -\nabla \times (\mu^{-1} - \mu_d^{-1}) \nabla \times \mathbf{E}_s(\mathbf{r}) + (\epsilon - \epsilon_d)\omega^2 \mathbf{E}_s(\mathbf{r}), \tag{8.6}$$

can be expanded into a series

$$\mathbf{f}(\mathbf{r}) = \sum_{n=-\infty}^{\infty} \mathbf{f}_n(r, z) e^{in\phi} \quad \text{with} \quad \mathbf{f}_n(r, z) = \frac{1}{2\pi} \int_0^{2\pi} \mathrm{d}\phi \, \mathbf{f}(\mathbf{r}) \, e^{-in\phi}. \tag{8.7}$$

Inserting (8.5) and (8.6) into (8.4) and integrating over $\frac{1}{2\pi} \int \mathrm{d}\phi \, e^{-im\phi}$ yields independent 2-dimensional differential equations for every Fourier mode $m \in \mathbb{Z}$, i.e.

$$\tilde{\nabla} \times \mu^{-1} \tilde{\nabla} \times \mathbf{E}_m(r, z) - \epsilon\omega^2 \mathbf{E}_m(r, z) - \mathbf{f}_m(r, z) = 0, \tag{8.8}$$

where $\tilde{\nabla}$ arises from ∇ by the replacement $\partial_\phi \to im$. In numerical computations a finite number of Fourier modes is chosen automatically by an adaptive algorithm such that the fields converge within a chosen level of accuracy.

The geometry of the waveguide is invariant in z-direction. The eigenmodes $\mathbf{E}_n(\mathbf{r})$ for a given frequency ω are solutions of (8.2) with no source current, i.e.

$$\nabla \times \mu^{-1} \nabla \times \mathbf{E}_n(\mathbf{r}) - \epsilon\omega^2 \mathbf{E}_n(\mathbf{r}) = 0 \quad \text{with} \quad \mathbf{E}_n(\mathbf{r}) = \mathbf{E}_n(r, \phi) e^{ik_n z}. \tag{8.9}$$

For practical reasons, one is interested in the fraction of the light-field energy that is coupled into the fundamental mode of the waveguide. In order to define the coupling efficiency, we expand the electromagnetic field, which is scattered into the waveguide, into a series of waveguide eigenmodes, i.e.,

$$\mathbf{E}_{\text{scatt}} = \sum_n \langle \mathbf{E}_n, \mathbf{E}_{\text{scatt}} \rangle \, \mathbf{E}_n. \tag{8.10}$$

As scalar product we employ the following overlap integral:

$$\langle \mathbf{E}_1, \mathbf{E}_2 \rangle = \frac{1}{2i\omega} \int d\mathbf{n} \cdot (\mathbf{E}_1 \times \mu^{-1} \nabla \times \mathbf{E}_2) = \frac{1}{2} \int d\mathbf{n} \cdot (\mathbf{E}_1 \times \mathbf{H}_2), \quad (8.11)$$

where the integration is performed over a cross section of the waveguide. In the case of guided modes in a loss-free non-active waveguide the total power emitted into the waveguide is given as

$$P = \sum_n | \langle \mathbf{E}_n, \mathbf{E}_{\text{scatt}} \rangle |^2. \quad (8.12)$$

The coupling efficiency towards a specific mode n is defined as the power flux $P_n = | \langle \mathbf{E}_n, \mathbf{E}_{\text{scatt}} \rangle |^2$ to this mode divided by the total emitted power of the dipole P_{tot}. In the following the coupling efficiency

$$\eta = \frac{P_1 + P_2}{P_{\text{tot}}} \quad (8.13)$$

into the two degenerate fundamental eigenmodes $\mathbf{E}_1$ and $\mathbf{E}_2$ of the waveguide is considered.

8.2.3 *Numerical Optimization of the Light Extraction from a Single-Photon Source*

The optimization of the device design by numerical modeling is an important tool in the field of computational physics as well as in nano-optics. The increase in complexity and the reduction of feature sizes of structures based on todays nano-fabrication technologies make an ad-hoc design no longer feasible. The optimization objective function is in general non-convex and its computation is resource demanding such that the right choice for the optimization method is crucial to obtain the best result. We have benchmarked four global methods for the optimization of the coupling efficiency of the single-photon source: downhill simplex optimization [34], particle swarm optimization [35], differential evolution [36], and Bayesian optimization [37]. The results [32] show that Bayesian optimization, mainly known from machine learning applications, obtains significantly better results in a fraction of the run times of the other optimization methods.

8.2.3.1 Bayesian Optimization

Bayesian optimization is based on a stochastic model of the objective function. The method derives promising-parameter samples by means of Bayesian inference based on *all* previous function evaluations. This is in contrast to the approaches mentioned before, which only use few historic data points to determine new samples. This statistical inference can drastically reduce the number of required simulations.

As statistical model, we employ a Gaussian process (GP) that is updated with each new evaluation of the objective function through GP regression [37, 38]. The GP allows one to identify parameter values with the largest expected improvement $\mathrm{EI}(\mathbf{x})$. After exploiting a possible improvement within a local minimum, other parts of the parameter space are automatically scanned until a better minimum is found. The search process for the next sampling point

$$\mathbf{x}^* = \underset{\mathbf{x}\in\mathcal{X}}{\operatorname{argmax}}\,[\mathrm{EI}(\mathbf{x})]$$

is itself an optimization problem that can be computationally demanding. Usually, the evaluation of the objective function, which requires to solve Maxwell's equations, is much more time consuming than an evaluation of the GP. However, the evaluation time of the GP grows at least quadratically with the number of observations. It is, therefore, important to balance the effort to find $\mathbf{x}^*$ with the effort to compute the objective function. We do this by adapting the effort of finding a new sampling position to the computation time of the objective function.

8.2.3.2 Optimization Results

Here, we review results of the optimization of the single-photon source introduced in Sect. 8.2.1 as obtained in a benchmark study [32]. The optimization runs were performed for a fixed number of 2,500 iterations. In order to evaluate the average performance of the different optimization methods, each run was repeated 6 times with different initial conditions. The optimization was performed on a machine with 6-core Intel Xeon CPUs running at 3.2 GHz with 11 GB of RAM. It was configured to perform at most 4 parallel computations of the objective function.

Figure 8.2b, c compare the performance of Bayesian optimization, particle swarm optimization, differential evolution, and the downhill simplex method for the maximization of the coupling efficiency of the emitted light into the optical fiber. Bayesian optimization clearly outperforms all other methods. After only 500 iterations and a computation time of 3.5 h it finds a setup with an average coupling efficiency of 54% while the other optimization methods reach only average efficiencies below 52% after 2,500 iteration and a computation time of 10 h.

A comparison of Fig. 8.2b, c shows that Bayesian optimization has a significant computational overhead compared to the other optimization approaches, which can determine the next sampling point on a negligible time scale. For Bayesian optimization the calculation of the next sampling point took on average 21 s while the simulations itself took 58 s. Despite this significant overhead, the sampling strategy of Bayesian optimization leads to much smaller total computation times. Even down to simulation times of only 2.2 s, Bayesian optimization would still converge faster

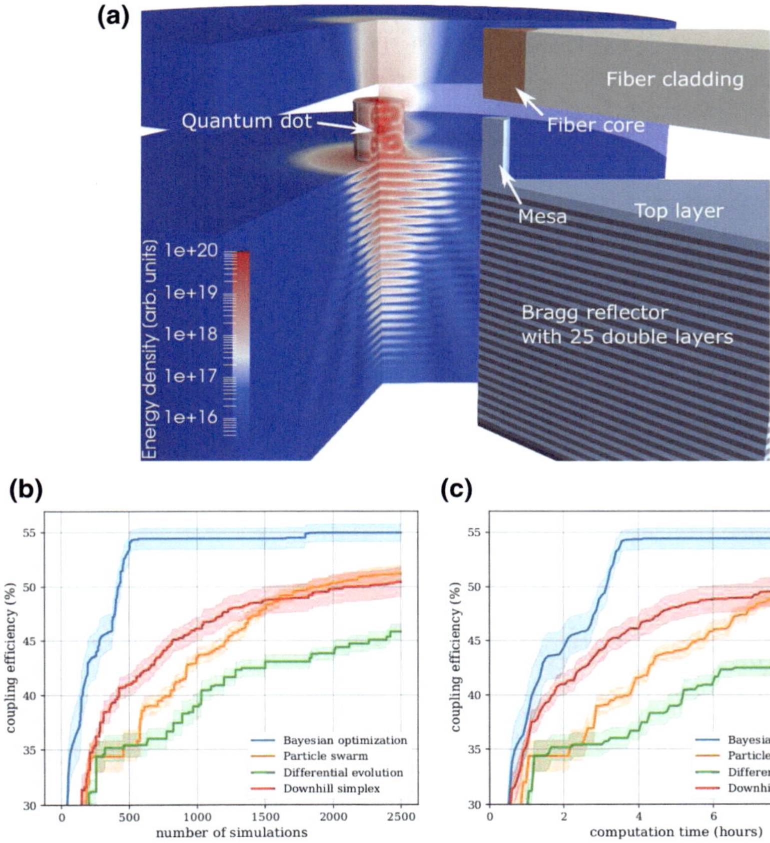

Fig. 8.2 **a** Visualization of the energy density of the light field of the optimized fiber-coupled single-photon source with a coupling efficiency of 60%. A cut through the geometry is shown in front of the energy-density plot. The single-photon source consists of a QD dipole source embedded into a mesa structure (GaAs, blue), a distributed Bragg reflector (alternating layers of GaAs (blue) and $Al_{0.9}Ga_{0.1}As$ (gray)), and an optical fiber with a homogeneous fiber core (brown) and fiber cladding (light gray). The Bragg reflector is grown on a substrate made of GaAs and has a GaAs top layer (blue). The six optimized geometrical parameters are the mesa height $h_{\text{mesa}} = 1217\,\text{nm}$, mesa width $w_{\text{mesa}} = 1044\,\text{nm}$, top-layer thickness $h_{\text{layer}} = 190\,\text{nm}$, dipole elevation within the mesa $h_{\text{dip}} = 613\,\text{nm}$, fiber-core diameter $d_{\text{core}} = 1937\,\text{nm}$, and mesa-fiber distance $s_{\text{mf}} = 352\,\text{nm}$. **b** Best simulated coupling efficiency for different optimization approaches as a function of the number of simulations averaged over six independent optimization runs. The shading indicates the uncertainty of the average. **c** Same as (**b**) but the best coupling efficiency is shown as a function of the total computation time. Bayesian optimization has a significant computational overhead for calculating the next sampling point. Nevertheless, it outperforms the other optimization approaches. Adapted with permission from [31] (OSA)

than the second-best method, particle swarm. As such, Bayesian optimization should be advantageous for most 2D and almost all 3D simulation problems.

The computed optimal configuration [32] is shown in Fig. 8.2a. The setup achieves a remarkably large coupling efficiency of 60% to the optical fiber. In a previous work [31], we tried to improve the coupling efficiency of the same system by performing numerical scans of all parameters of the geometry apart from the top-layer thickness h_{layer}, which was set to a fix value of 195 nm. A full parameter scan of all five parameters with only 10 values per parameter would have required 100,000 simulations. Therefore, we restricted the scan to a small fraction of the physically realizable parameter space. E.g., the parameter scan of the dipole elevation was restricted to values between 0 and 50 nm in steps of 10 nm, while for the current optimization we considered values between 0 and 1000 nm. Due to the previous restriction a maximal coupling efficiency of only 23% was obtained [31]. This demonstrates that an optimization within a large space of realizable system parameters is important to assess the technological potential of a nano-optical system. We have also used the developed simulation and optimization methods for the design and analysis of experimentally realized on-chip waveguides [30, 39].

8.2.4 Numerical Simulation of a QD-Based Single-Photon Emitting Diode—The Role of Electrical Carrier Injection

Semiconductor QDs can be easily integrated into electrically operated p-i-n diodes, which makes them excellent candidates for the realization of electrically-driven quantum-light sources [40–44]. While electrical excitation is highly attractive in the interest of practicability, compactness and scalability, the realization of high-quality electrically driven devices remains very challenging since the coupling of the QD to fluctuating continuum charge carriers and phonons typically degrades the device performance. Numerical device simulation can help to assist the development and optimization of current injection into devices by providing insight into the principles of operation, identifying performance bottlenecks and assessing the impact of design variations.

In Sect. 8.2.4.1 we describe a hybrid quantum-classical modeling approach for the simulation of electrically driven QD-based single-photon emitting diodes. Numerical results, including, e.g., the spatially resolved simulation of the injection current and the device's internal electric potential, the second-order correlation function and the bias dependency of the emission spectrum are presented in Sect. 8.2.4.2.

8.2.4.1 Hybrid Quantum-Classical Modeling Approach

The simulation of electrically-driven quantum-light sources on all involved length scales requires a combination of classical device physics (semi-classical carrier transport theory) and cavity quantum electrodynamics. The numerical simulations presented and described in the following are based on the *hybrid quantum-classical modeling approach* developed in [45–47], which self-consistently couples the drift-diffusion system to a Markovian quantum master equation (Lindblad master equation [48]) for the coupled QD-photon system. The model equations read:

$$-\nabla \cdot \varepsilon_0 \varepsilon_s \nabla \phi = q\left(C + p - n\right) + Q\left(\rho, \psi_e, \psi_h\right), \tag{8.14a}$$

$$\partial_t n - \frac{1}{q}\nabla \cdot \mathbf{j}_n = -R\left(n, p\right) - S_n\left(\rho, n, p, \psi_e, \psi_h\right), \tag{8.14b}$$

$$\partial_t p + \frac{1}{q}\nabla \cdot \mathbf{j}_p = -R\left(n, p\right) - S_p\left(\rho, n, p, \psi_e, \psi_h\right), \tag{8.14c}$$

$$H_\alpha^0\left(\phi\right)\psi_\alpha = \varepsilon_\alpha \psi_\alpha \quad \left(\alpha \in \{e, h\}\right), \tag{8.14d}$$

$$\partial_t \rho = -\frac{i}{\hbar}\left[\mathcal{H}\left(n, p, \psi_e, \psi_h\right), \rho\right] + \mathcal{D}\left(\rho, n, p, \psi_e, \psi_h\right). \tag{8.14e}$$

The system (8.14) consists of the drift-diffusion system (8.14a)–(8.14c), a one-particle Schrödinger equation (8.14d) for each of the electrons and holes bound in the QD, and a Lindblad master equation (8.14e) for the quantum statistical operator ρ. The latter describes the microscopic dynamics of the coupled QD-photon system. We refer to Sect. 7.3.2 for a detailed explanation of the model equations and its underlying theoretical concepts.

Poisson's equation (8.14a) describes the electrostatic interaction between the free and bound charges of the system, where ϕ is the electric potential, C is the built-in doping profile, n is the density of conduction band electrons, p is the density of valence band holes, Q is the charge density of QD-bound carriers, q is the elementary charge, ε_0 is the vacuum permittivity and ε_s is the dielectric constant of the semiconductor. The continuity equations (8.14b)–(8.14c) describe the transport and recombination dynamics of the bulk electrons and holes, where the electrical current densities $\mathbf{j}_{n/p}$ are modeled by the usual drift-diffusion expressions [49] and R is the recombination rate of continuum charge carriers (Shockley–Read–Hall recombination, spontaneous emission, Auger recombination). The loss terms S_n and S_p account for the capture of continuum electrons and holes to (bound) QD states.

The (envelope) wave functions and single-particle energies of the QD electrons and holes are modeled by the stationary Schrödinger equation (8.14d) using

$$H_\alpha^0 = -\frac{\hbar^2}{2}\nabla \cdot \frac{1}{m_\alpha^*}\nabla + U_\alpha\left(\mathbf{r}\right) + z_\alpha q \phi\left(\mathbf{r}\right), \tag{8.15}$$

for the respective bands $\alpha \in \{e, h\}$ ($z_e = -1, z_h = +1$). The confinement potential

$$U_\alpha(\mathbf{r}) = \frac{m^*_{\alpha,\mathrm{QD}}}{2}\omega_\alpha^2\left(x^2+y^2\right) - \begin{cases} U_\alpha^0 & |z-z_0| \le h_{\mathrm{QD}}/2 \\ 0 & \text{else} \end{cases}$$

for $\alpha \in \{e, h\}$ (with effective mass $m^*_\alpha(z) = m^*_{\alpha,\mathrm{QD}}$ for $|z-z_0| \le h_{\mathrm{QD}}/2$ and $m^*_\alpha(z) = m^*_{\alpha,\mathrm{barrier}}$ else), is a good approximation to the energy eigenvalues and wave functions of lens-shaped $In_xGa_{1-x}As$-QDs [50]. In order to account for the quantum confined Stark effect (QCSE), the Hamiltonian (8.15) couples to the device's internal electric potential ϕ, which itself is subject to Poisson's equation (8.14a). Here we choose $U_e^0 = 477\,\mathrm{meV}$, $U_h^0 = 318\,\mathrm{meV}$, $m^*_{e,\mathrm{QD}} = m^*_{e,\mathrm{barrier}} = 0.067\,m_0$ (m_0 is the free electron mass), $m^*_{h,\mathrm{QD}} = 0.15\,m_0$, $m^*_{h,\mathrm{barrier}} = 0.51\,m_0$, $h_{\mathrm{QD}} = 3.57\,\mathrm{nm}$ and $\left(\hbar/(m^*_{\alpha,\mathrm{QD}}\omega_\alpha)\right)^{1/2} = 12.5\,\mathrm{nm}$ (for both $\alpha \in \{e, h\}$).

The QD-photon system is modeled by a dissipative Jaynes–Cummings model, which is illustrated in Fig. 8.3. The Hamiltonian in (8.14e) is taken as

$$\mathcal{H} = \varepsilon_e e^\dagger e + \varepsilon_h h^\dagger h + \hbar\omega_0 a^\dagger a + \hbar g\left(e^\dagger h^\dagger a + a^\dagger h e\right) - V_{e,h} e^\dagger h^\dagger h e. \qquad (8.16)$$

The first two terms describe the single-particle energy contributions of QD carriers, where the electron and hole creation ($e^\dagger$, $h^\dagger$) and annihilation operators (e, h) obey the usual fermionic anti-commutator relations. The single-particle energies $\varepsilon_\alpha = \mathrm{Re}\left(\int \mathrm{d}^3r\, \psi^*_\alpha(\mathbf{r})\, H^0_\alpha \psi_\alpha(\mathbf{r})\right)$ for $\alpha \in \{e, h\}$ are obtained from solving (8.14d). The third term models the energy of the cavity photons, where the photon creation ($a^\dagger$) and annihilation operators (a) satisfy the canonical commutator relation and $\hbar\omega_0 = 0.9537\,\mathrm{eV}$ (vacuum wavelength $\lambda = 1.3\,\mu\mathrm{m}$). The fourth term in (8.16) describes the interaction of the QD carriers with the cavity photons (in dipole and rotating wave approximation), where the light-matter coupling coefficient is

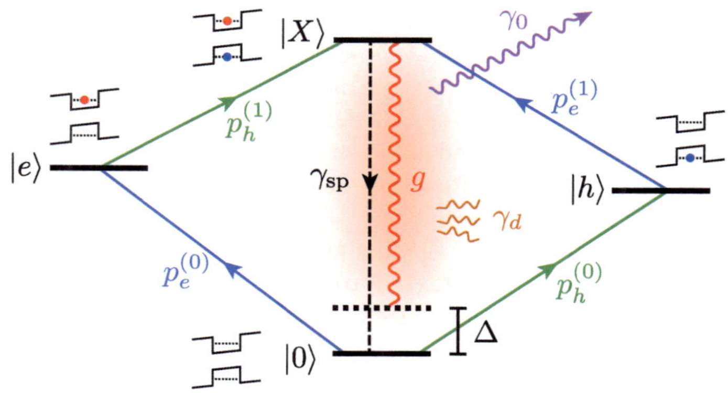

Fig. 8.3 Dissipative QD-photon system described by the Lindblad master equation (8.14e). The QD features 4 electronic states: empty QD $|0\rangle$, single electron $|e\rangle$, single hole $|h\rangle$ and (bright) exciton $|X\rangle$. The detuning Δ is calculated self-consistently using a device-scale modeling approach, see Sect. 8.2.4.1 for details. © 2019 The International Society for Optics and Photonics (SPIE). Reprinted, with permission, from [47]

$$g = d_{c,v}\sqrt{\frac{\omega_0}{\hbar\varepsilon_0 n_r^2 V_0}} \int \mathrm{d}^3r\, \psi_e^*(\mathbf{r})\, \psi_h(\mathbf{r}) .$$

Here, $d_{c,v} = 1.6 \times 10^{-19}$ C nm is the interband dipole moment, $n_r = 3.4$ is the refractive index (at the optical frequency ω_0) and the optical mode volume is taken as $V_0 = (4\,\mu\mathrm{m})^3$. Moreover, g involves the overlap integral of the QD wave functions. The last term in the Hamiltonian (8.16) describes the Coulomb interaction of the bound QD carriers, where the exciton binding energy is calculated from

$$V_{e,h}(\psi_e, \psi_h, n, p) = \int \mathrm{d}^3r \int \mathrm{d}^3r'\, \psi_e^*(\mathbf{r})\, \psi_h^*(\mathbf{r}')\, V\left(\|\mathbf{r}-\mathbf{r}'\|, n, p\right) \psi_h(\mathbf{r}')\, \psi_e(\mathbf{r})$$

with the screened Coulomb potential $V(r, n, p) = e^{-\kappa(n,p)r} q^2 / \left(4\pi\varepsilon_0\varepsilon_{\mathrm{bg}} r\right)$. We assume a background permittivity $\varepsilon_{\mathrm{bg}} = 13.9$. The screening wave number κ depends on the carrier densities in the vicinity of the QD and is modeled using the static, long wavelength limit of Lindhard's formula [51]

$$\kappa^2(n, p) = \frac{q^2}{\varepsilon_0\varepsilon_{\mathrm{bg}}} \left(\int \mathrm{d}^3r\, |\psi_e(\mathbf{r})|^2 \frac{\partial n(\mathbf{r})}{\partial \mu_c} + \int \mathrm{d}^3r\, |\psi_h(\mathbf{r})|^2 \frac{\partial p(\mathbf{r})}{\partial \mu_v} \right).$$

Hence, next to the QCSE, also the screening of the Coulomb interaction leads to a shift of the QD exciton energy (and the detuning $\Delta = \varepsilon_e + \varepsilon_h - V_{e,h} - \hbar\omega_0$), depending on the state of the QD's macroscopic environment.

The coupling of the quantum system to its classical environment leads to irreversible processes, which are described by the dissipation superoperator

$$\mathcal{D}(\rho) = \mathcal{D}_e(\rho) + \mathcal{D}_h(\rho) + \mathcal{D}_0(\rho) . \tag{8.17}$$

Following [45], the dissipation superoperator is decomposed into processes $\mathcal{D}_{e,h}(\rho)$ that change the charge of the quantum system (capture of electrons or holes) and charge-conserving processes $\mathcal{D}_0(\rho)$ (out-coupling of cavity photons, spontaneous decay, pure dephasing). The corresponding operators read

$$\mathcal{D}_e(\rho) = p_e^{(0)} L_{e^\dagger(1-n_h)}\rho + p_e^{(1)} L_{e^\dagger n_h}\rho, \tag{8.18a}$$

$$\mathcal{D}_h(\rho) = p_h^{(0)} L_{h^\dagger(1-n_e)}\rho + p_h^{(1)} L_{h^\dagger n_e}\rho, \tag{8.18b}$$

$$\mathcal{D}_0(\rho) = \gamma_c L_a\rho + \gamma_{\mathrm{sp}} L_{eh}\rho + \gamma_d L_{n_e n_h}\rho \tag{8.18c}$$

where $L_A\rho = A\rho A^\dagger + \frac{1}{2}\left\{A^\dagger A, \rho\right\}$ is the Lindblad superoperator [48]. The capture rates

$$p_e^{(j)} = \Gamma_e^{(j)} \int_{\Omega_0} \mathrm{d}^3r\, |\psi_e(\mathbf{r})|^2\, n(\mathbf{r}), \qquad p_h^{(j)} = \Gamma_h^{(j)} \int_{\Omega_0} \mathrm{d}^3r\, |\psi_h(\mathbf{r})|^2\, p(\mathbf{r}) \tag{8.19}$$

($j = 0, 1$) depend on the carrier densities n, p in the vicinity of the QD. Here, the wave functions are used as weight functions to compute integral averages of the carrier densities. The capture rate coefficients are set to $\Gamma^{(0)}_{e/h} = 10^{-12}\,\mathrm{m}^3\mathrm{s}^{-1}$ (from [52]) for the *primary* capture processes. In order to phenomenologically account for the Coulomb enhancement [53] of the *secondary* capture processes (i.e., scattering into the charged QD), we assume $\Gamma^{(1)}_{e/h} = 10 \times \Gamma^{(0)}_{e/h}$. The photon out-coupling rate is $\gamma_0 = 5 \times 10^{12}\,\mathrm{s}^{-1}$ and the pure dephasing rate is $\gamma_d = 10^{11}\,\mathrm{s}^{-1}$. Spontaneous emission of the exciton into waste modes is modeled using the Weisskopf–Wigner decay rate [48]

$$\gamma_{\mathrm{sp}} = \frac{n_r d^2_{c,v} \omega_0^3}{3\pi\varepsilon_0 \hbar c_0^3} \int_{\Omega_0} \mathrm{d}^3 r\, \psi_e^*(\mathbf{r})\, \psi_h(\mathbf{r})\,.$$

Finally, due to charging of the QD and scattering of continuum carriers to bound states, the quantum system couples back to the drift-diffusion system (8.14a)–(8.14c). Following [45, 47], the QD charge density Q and the loss rates S_n, S_p are taken as

$$Q = q\,|\psi_h(\mathbf{r})|^2 \langle h^\dagger h \rangle - q\,|\psi_e(\mathbf{r})|^2 \langle e^\dagger e \rangle, \tag{8.20a}$$

$$S_n = |\psi_e(\mathbf{r})|^2 \left(p_e^{(0)} \langle e e^\dagger h h^\dagger \rangle + p_e^{(1)} \langle e e^\dagger h^\dagger h \rangle \right), \tag{8.20b}$$

$$S_p = |\psi_h(\mathbf{r})|^2 \left(p_h^{(0)} \langle e e^\dagger h h^\dagger \rangle + p_h^{(1)} \langle e^\dagger e h h^\dagger \rangle \right), \tag{8.20c}$$

where $\langle \ldots \rangle = \mathrm{tr}\,(\rho \ldots)$ indicates the expectation value.

8.2.4.2 Simulation Results

We employ the hybrid model (8.14) to calculate the stationary characteristics (DC bias) of a single-photon emitting diode featuring a single QD and a top-mesa structure, see e.g. Fig. 8.1. The geometry parameters for the mesa are taken from [31]. The top-contact is assumed to consist of an optically transparent material (e. g., ITO), placed on an insulator layer with low relative permittivity $\varepsilon_s = 3.0$. The GaAs material parameters and mobility models are taken from [54, 55], the doping densities are set to $N_D^+ = 2 \times 10^{18}\,\mathrm{cm}^{-3}$ and $N_A^- = 1 \times 10^{19}\,\mathrm{cm}^{-3}$. The full system (8.14a)–(8.14e) is solved by repeatedly solving (8.14a)–(8.14c) → (8.14d) → (8.14e) → (8.14a)–(8.14c) → … until convergence is reached. The numerical method for the drift-diffusion system is described in Sect. 7.4.1.3.

Figure 8.4a, b shows 2D cross sections of the carrier and current densities at $U = 1.485$ V. In this low-injection regime, the QD notably contributes to current guiding, since the scattering of continuum carriers to the QD is the dominant loss term on the right hand side of the continuity equation (8.14b)–(8.14c), see Fig. 8.4c. The electric potential ϕ is shown in Fig. 8.4d. At the QD position, a weak electric field is observed in vertical direction, which causes energy shifts of the QD confined carriers via the QCSE. The occupation probability of the QD versus the injection current is shown in Fig. 8.5a. At low current ($I \lesssim 1$ nA), the QD is predominantly empty or populated

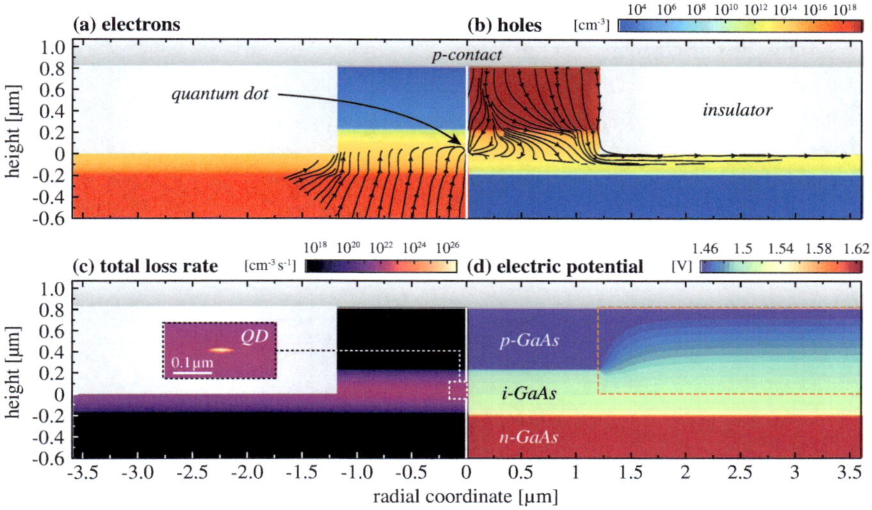

Fig. 8.4 2D cross sections at stationary injection $U = 1.485\,\text{V}$ ($I = 1.2\,\text{nA}$) and $T = 30\,\text{K}$. The QD is located on the symmetry axis at $z = 30\,\text{nm}$. **a** Electron density n (color coded) and current density $\mathbf{j}_n$ (arrows point in the direction of electron flux) and **b** hole density p (color coded) and current density $\mathbf{j}_p$. In the low-injection regime, the scattering of continuum carriers to the QD contributes notably to the current guiding. **c** Recombination rate and scattering losses $R + S_n + S_p$ of continuum carriers. **d** Electric potential ϕ in the diode and the insulator

by a single carrier; at higher currents ($I \gtrsim 1\,\mu\text{A}$) the exciton occupation probability approaches unity. Note that here only the electronic states up to the exciton (as shown in Fig. 8.3) are taken into account. We refer to [45] for a more comprehensive model, where also dark excitons, trions and the biexciton are considered. The increase of the exciton population drives the photon emission rate $\Gamma_0 = \eta_{\text{ext}}\,\gamma_0\langle a^\dagger a\rangle$, see Fig. 8.5c. Here, $\eta_{\text{ext}} = 0.232$ is the photon-extraction efficiency [31] (cf. Sect. 8.2.2). The power spectrum

$$P(\omega) = \frac{1}{2\pi}\int_{-\infty}^{\infty} \mathrm{d}\tau\, e^{-i\omega\tau}\langle a^\dagger(\tau)\, a(0)\rangle \tag{8.21}$$

(Fourier transform of the first-order autocorrelation function) is plotted as a function of the applied bias in Fig. 8.5b. With increasing bias, the exciton line is blue shifted by a few meV due to a combination of the QCSE and plasma screening. At $U \approx 1.509\,\text{V}$, the exciton line crosses the (broad) optical resonator mode, which leads to a peak in the photon emission rate of about 8.4 MHz (see Fig. 8.5c). In the high injection regime ($U \gtrsim 1.52\,\text{V}$), excitation-induced dephasing leads to a notable broadening of the exciton line as shown in Fig. 8.5b. The single-photon purity of the diode is given by the second-order autocorrelation function

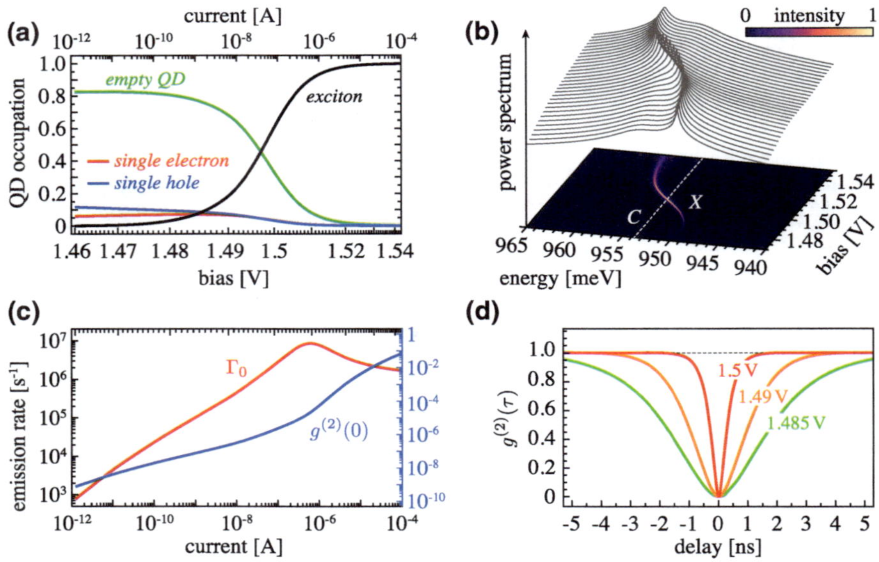

Fig. 8.5 **a** Occupation probability of QD states versus the injection current. **b** Power spectrum $P(\omega)$ versus applied bias. The exciton line (X) is blue-shifted with increasing bias due to the QCSE and plasma screening. The resonance with the optical cavity mode (C) appears at $U \approx 1.509\,\text{V}$. **c** Photon emission rate $\Gamma_0 = \eta_{\text{ext}}\,\gamma_0 \langle a^\dagger a \rangle$ and single-photon purity $g^{(2)}(0)$ versus the injection current. **d** Time-resolved second-order correlation function $g^{(2)}(\tau)$ at different applied voltages

$$g^{(2)}(\tau) = \frac{\langle a^\dagger(0)\, a^\dagger(\tau)\, a(\tau)\, a(0) \rangle}{\langle a^\dagger(\tau)\, a(0) \rangle^2}, \tag{8.22}$$

which is plotted in Fig. 8.5d for different bias values. While the equal-time expectation value $g^{(2)}(0)$ is always close to zero (cf. Fig. 8.5c, right axis) due to the short life time γ_0^{-1} of the cavity mode, the correlation time decreases with increasing bias because of the larger pumping rate.

8.3 Deterministic Fabrication Technologies

As outlined in the introduction of this chapter, it is by far not trivial to integrate a nanometer-sized, randomly placed and 'hidden' (buried below a capping layer) object like a self-assembled QD into a nanophotonic structure with a precision on a nanometer scale (One workaround on the material's side is the application of site-controlled growth approaches as discussed in Chap. 3.). Without precisely knowing the positions and emission wavelengths of suitable emitters, device integration is like hide-and-seek: A large number of nanophotonic structures has to be fabricated at different positions to finally end up with a functional device that has by chance

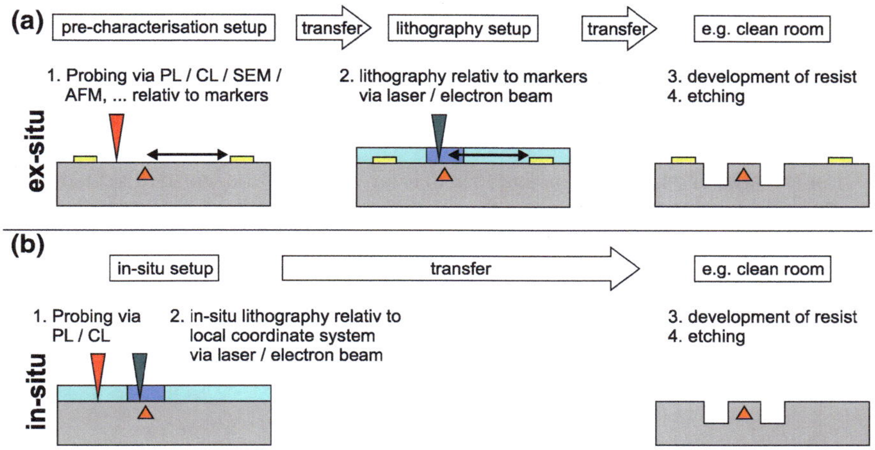

Fig. 8.6 Deterministic processing schemes for the integration of a single QD into a (nano-) photonic device/environment. **a** Ex-situ schemes utilize different setups for probing and lithography with the aid of markers. **b** In-situ schemes perform probing and lithography in the same setup without the need for mapping coordinate systems

a suitable QD at the desired position and wavelength. Depending on the nature of the nanophotonic structure, the required number of attempts might be greater than 100 or even 10,000 in case of QD-microcavities, which makes an up-scaling to large quantum networks based on entanglement distribution between identical quantum-light sources practically impossible. (An estimation of the successful fabrication of *N* 'nodes' (nanophotonic structures) with a single energy-matched QD inside is presented for a multi-node waveguide circuit in Sect. 8.5). Moreover, time-consuming testing and probing is required to check all those devices while at the same time wasting valuable wafer material.

At this point it is quite clear that random device fabrication is not suitable when it comes to devices based on single self-assembled QDs. So, different strategies were developed to gain sufficient information in advance to perform the processing of devices at the right positions and with suitable QDs. Even if the actual combination of pre-characterization and subsequent processing differs, all those schemes follow the idea as sketched in Fig. 8.6: First, spectroscopy or microscopy is performed to identify locations of QDs with respect to a coordinate system and then the actual processing is conducted with respect to those coordinates. Depending on the characterization method applied, also the full spectroscopic information on the QDs is acquired to also select them with respect to application-relevant spectral matching. The most important requirements include precise energetic conformity with the mode of a (micro) cavity [56], a vanishing fine-structure splitting (FSS) for the generation of polarization-entangled photon pairs by utilizing the biexciton-exciton recombination cascade [57], and indistinguishable photons [58] for entanglement-swapping in complex quantum networks.

Existing deterministic methods can be divided into two classes: in-situ and ex-situ. 'In-situ' means that the pre-characterization and the actual processing are performed in the same setup, at the same temperature, and without moving the sample between the two process steps (e.g. [59, 60]). Those are the most sophisticated and attractive schemes that will be discussed later on.

For the sake of completeness and to also look outside of the world of semiconductor QDs, it should also be remarked here, that there exist ways to implement color centers in diamond and silicon carbide in a deterministic way. Prominent examples can be found in [61–64].

8.3.1 Ex-situ Schemes

Ex-situ schemes are the most commonly used approaches since they do not require a dedicated all-in-one system to perform all steps. The first step (pre-characterization) is usually performed in one system with respect to marker structures on the sample's surface (Fig. 8.6a). All the spectral and/or structural information is acquired with respect to those markers that define a local coordinate system. The second step (lithography/processing) is done in another setup (e.g. a standard electron-beam lithography (EBL) setup) with respect to the markers. Table 8.1 lists a number of methods/publications with the respective pre-characterization and processing techniques as well as their lateral (alignment) accuracy.

Another class of ex-situ approaches deals with the precise fixation of an (preselected) emitter on or into a nanophotonic structure. This way a nanowire with a QD inside was integrated into a SiN-based photonic circuit by PECVD [76] or a QD inside a photonic wire was placed onto the core of a fiber pigtail [77]. Another prominent example is the placement of a diamond nanocrystal with a single NV center inside on a photonic crystal cavity for precise optical coupling to its mode maximum [78]. An external-mirror microcavity above a planar QD sample with a lower epitaxial distributed Bragg reflector (DBR) section was realized by bonding the external mirror to a fiber and optimizing its position above the sample by x/y/z piezo scanning [79]. Although these methods result in good device performances, they are not compatible with a fully monolithic or on-chip device concept and electrical operation is hardly possible in most cases.

8.3.2 In-situ Schemes

The in-situ methods with spatial and spectral pre-characterization give the largest versatility. They allow for the target-oriented selection of single QDs and the successive patterning of the resist without switching the setup and without moving the sample (cf. Fig. 8.6b). This speeds up the overall process, requires a single coordinate system only, and there are no markers on the sample that might be obstacles for widespread

Table 8.1 A compilation of deterministic ex-situ processing schemes for the integration of single semiconductor QDs into (nano-) photonic structures or environments

Pre-characterization/ determination of QD position	Processing technique	Lateral accuracy (l: locating a QD, a: aligning a structure, f: full process)	Structures fabricated	References
SEM	EBL	f: 25 nm	Photonic crystals	[65]
AFM	EBL	f: 30 nm	Photonic crystals	[66]
μPL and in-situ marker writing	EBL	l: 50 nm, f: 150 nm	Photonic crystal	[67]
μPL	EBL	l: <10 nm, f: <30 nm	Photonic crystal	[68]
AFM (no markers)	Movement of Au particles via AFM tip	f: <30 nm	Plasmonic nano-antennas	[69]
μPL	EBL	f: <50 nm	Photonic crystals	[70]
CL	EBL	f: <50 nm	Plasmonic nano-antennas	[71]
SEM	EBL	f: 11 nm	Plasmonic nano-antennas	[72]
μPL	EBL	l: <30 nm (+SIL: <10 nm)	Ring resonators	[23]
CL and SEM	EBL	f: 50 nm	Plasmonic nano-antennas	[73]
μPL	EBL	≈ l: <10 nm, f: 25 nm	Micropillars	[74]
μPL and in-situ marker writing	3D laser writing	l: <2 nm, a: >220 nm	Immersion lenses	[75]

SEM Scanning electron microscope/microscopy; μPL Micro photoluminescence spectroscopy; AFM Atomic force microscope/microscopy; CL Cathodoluminescence spectroscopy

structures like, e.g., on-chip waveguide circuits. Contrary to the ex-situ approaches there is resist on the sample from the very beginning. In the case of QDs, as most of the spectral pre-characterization is done at Helium temperatures, the resist is also exposed at such temperatures which requires the selection of suitable resists and a completely new exploration and understanding of their low-temperature properties (see e.g. [80–82] for investigations of resists for low-temperature EBL).

So far, three different in-situ techniques have been reported (overall process accuracy in parenthesis), namely two-color μPL in combination with in-situ optical lithography [59] (50 nm), [17] (accuracy not given), [83] (50 nm), single-color μPL with in-situ optical lithography [84] (50 nm), and CL with in-situ EBL [85] (accuracy not given), [22] (34 nm). All these techniques have different advantages and disadvantages. The optical methods use a 'low-energy' or 'low-power' laser for optical spectroscopy that does not interact with the resist too much. The spectroscopic part is therefore completely independent of the lithography step but this technique is limited to the lateral resolution of optical spectroscopy and lithography. The CL/EBL technique on the other hand affects the resist already during the CL step via the electron beam irradiation. This renders extra requirements on the resist properties and processing as it must not become insensitive for the actual EBL step. A big advantage is the high lateral resolution of spectroscopy and resist patterning [82, 86] and the ability for three-dimensional structuring [81] as provided by high-performance gray-scale EBL.

8.3.2.1 In-situ Electron Beam Lithography

During in-situ EBL, also dubbed CL lithography, both the excitation of the sample and the patterning of the resist are done by the electron beam of a scanning electron microscope. Figure 8.7 displays a schematics of the system that was realized in the framework of SFB 296/787 at the Technische Universität Berlin. The main frame consists of a JEOL JSM 840 scanning electron microscope (SEM). The customized CL extension consists of an He-flow cryostat, an elliptical mirror, and a 500 mm monochromator with a CCD camera for the 300–1000 nm wavelength range and an InGaAs diode-line detector for extended wavelengths of up to 1700 nm. Hardware EBL-functionality is enabled by integrating a beam blanker. The fully integrated software for CL spectroscopy and in-situ EBL has been customized and is completely self-made.

An apparent question is, if an EBL resist is still usable for the patterning step when it was already exposed with a finite electron dose during the spectral mapping of the sample during the pre-characterization step. Usually the spectroscopic dwell time per mapping point and spectrum cannot be shorter than 30 ms which already introduces an electron dose of approximately 6 mC/cm^2 per mapping voxel for a beam current of 0.5 nA. Resists in their 'normal' regime of operation will be completely exposed by such a high dose. Interestingly, there is a number of EBL resists that are usually operated in a positive-tone regime (resist molecules are cracked upon exposure) but that are re-concatenating for larger electron doses [80, 82]. Figure 8.8 displays this feature for the resists PMMA 950k and CSAR 62. The figure depicts the remaining resist thickness after its development as a function of the initially applied electron dose.

It is quite obvious that both resists exhibit a positive-tone behavior for small electron doses whereas they enter a negative-tone regime for larger doses. The critical doses are the clearing dose C_1 when the resist becomes fully soluble and the onset dose C_O when the process of re-concatenating starts. The dose window in between gives the possible doses and hence the possible dwelling times for the spectral mapping. In other words, the mapping drives the resist in its positive regime. The pattering of the resist (as required for the envisaged nanophotonic structures) is performed in the negative-tone regime (cf. marked regions in Fig. 8.8). This leads to a rather simple fabrication process since the inverted resist can directly be used as an etch mask for the final plasma etching process. Noteworthy, the critical doses are highly

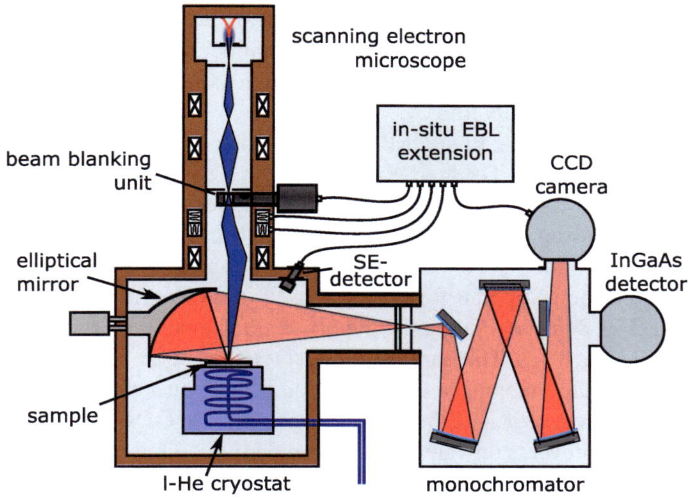

Fig. 8.7 Schematics of the in-situ EBL setup at the Institute of Solid State Physics at TU Berlin. The JEOL JSM 840 SEM is equipped with in-situ spectroscopy and EBL extensions as described in the text. It offers low-temperature operation below 5 K for spectroscopy and lithography. Reprinted by permission from Springer Nature from [87]

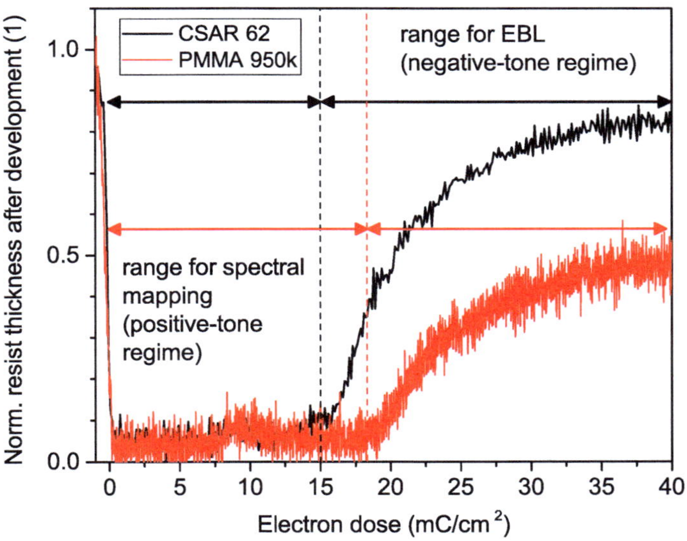

Fig. 8.8 Resist thickness after development as a function of the initially applied electron dose at $T = 7\,\mathrm{K}$ for PMMA 950k and CSAR 62. The regimes for pre-characterization (spectral mapping) and the successive in-situ EBL step are marked by arrows. Due to the different molecular structure of the resists there is a difference in the onset of the negative-tone regime

temperature sensitive [80, 82]. The lower the temperature the larger are the doses C_1 and C_O. This is very beneficial for the low-temperature operation of the resist during mapping and patterning as it allows for increased luminescence dwell times as compared to room temperature.

A sketch of the complete in-situ EBL process using the example of mesa fabrication is depicted in Fig. 8.9: The resist-coated sample it put into the in-situ EBL setup, cooled down to $T = (4 - 10)$ K and an area is mapped by spatially and spectrally resolved CL spectroscopy (Fig. 8.9a). Within the CL map suitable QDs are selected and the actual EBL patterns are written at those locations (Fig. 8.9b). Now the sample is removed from the setup and further processed in a clean room. There the resist is developed (Fig. 8.9c) and the resist pattern is transferred into the semiconductor by dry-etching (Fig. 8.9d). The dual-tone characteristics of the resist provides anther feature: Resist in the mapping area without additional structuring dose is washed away during development and the underlying spectator QDs are removed by the etching step. This prevents unwanted background luminescence from those QDs and increases the performance of the resulting single-QD device.

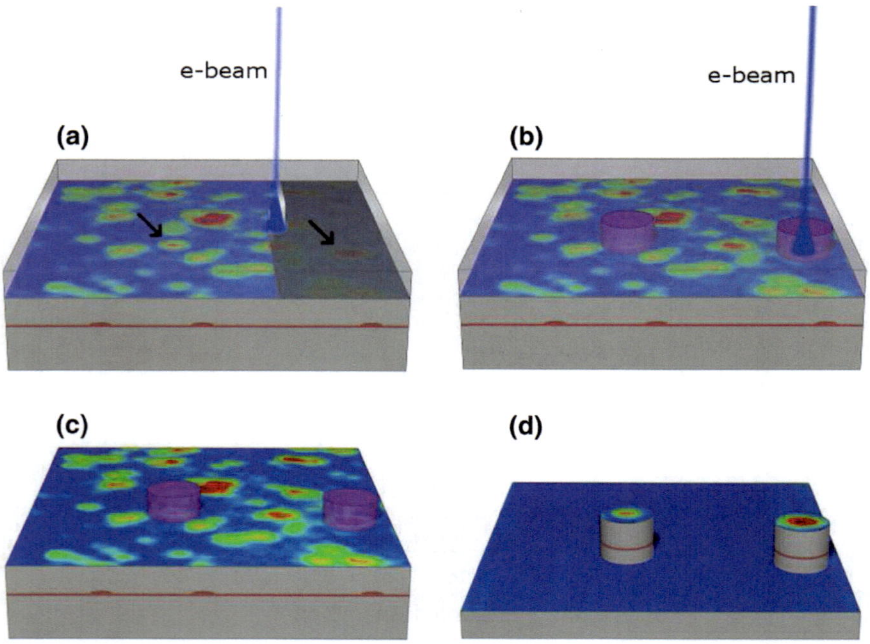

Fig. 8.9 Sketch of the full in-situ EBL process as developed at the TU Berlin. **a** Spectral mapping of QD luminescence, **b** writing of EBL patterns at chosen positions, **c** development of the resist, and **d** dry etching for imprinting the resist pattern. Reprinted from [60], with the permission of AIP Publishing

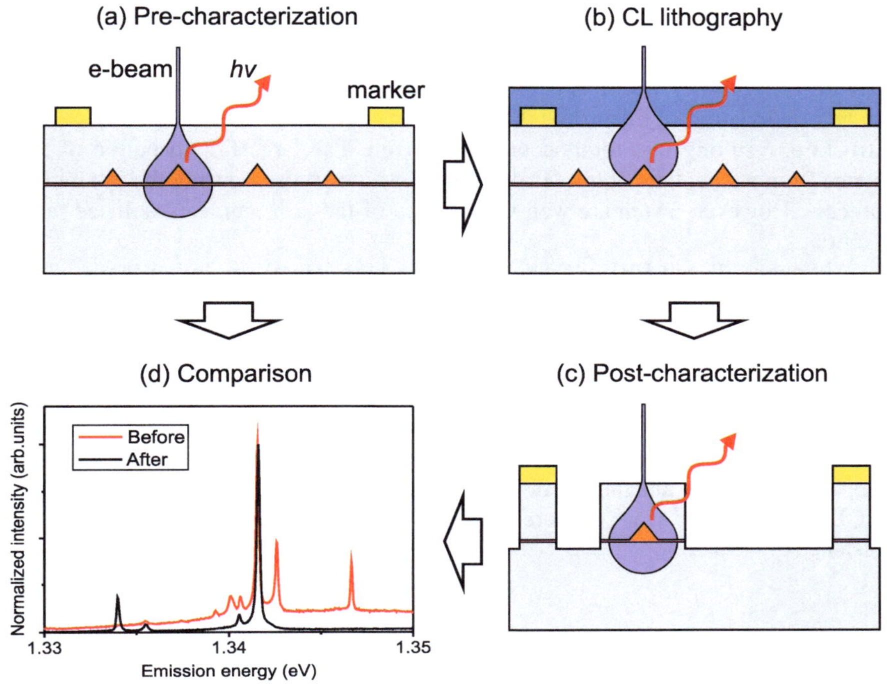

Fig. 8.10 Sketch of the advanced in-situ EBL process. **a** Markers help to roughly pinpoint the position where extensive pre-characterization is performed without resist on the sample. **b** After the pre-characterization the actual in-situ EBL as sketched in Fig. 8.9 is performed. **c** After fully-processing the sample, post-characterization can be performed to check the influence of the processing on the sample by a direct comparison of 'before' and 'after' results (**d**). Reprinted from [88], with the permission of AIP Publishing

If larger dwell times are required for the pre-characterization, a marker-aided extension can be applied to in-situ-EBL, too [88]. Figure 8.10 sketches this approach. Extensive pre-characterization can be performed as the resist will be introduced later on when it comes to the actual in-situ EBL. This is useful for the investigation of QD features such as the luminescence decay time and polarization properties as well as for a 'before and after evaluation' of QDs that are integrated into nanophotonic structures to evaluate a possible influence on the QDs' properties by the device processing. Please note that relocating the QDs is done in-situ—so the markers are only used for a 'rough' navigation on the sample's surface.

By using this method we were able to prove, that the QDs were not deteriorated by the processing [88] which is a very important finding for the utilization of deterministic processing schemes. Otherwise the pre-characterization wouldn't be applicable for the properties of the final device.

Despite the low-temperature and negative-tone operation of the EBL resist, lateral features of acceptable small size are still feasible. With PMMA we were able to demonstrate continuous ridge structures with a width of only 65 nm [86]. When using CSAR 62 even features below 50 nm could be obtained [82]. Evaluation of our lateral accuracy with respect to the integration of single emitters resulted in a standard deviation of 25 nm for the pre-selection and 24 nm for aligning the EBL structure, resulting in a total accuracy of 34 nm. The largest contribution of error stems from a slow temperature drift of the He-flow cryostat during the in-situ EBL process. However, 34 nm are well within state-of-the art accuracies as listed in this section.

Although CL and EBL provide for a very high lateral resolution the areal QD density is of concern. Typical densities of self-organized QDs are on the order of 10^8–10^{10} cm^{-2}, resulting in inter-QD distances from 1 μm to 100 nm. Typically, a QD density $<10^9$ cm^{-2} is sufficient to locate single QDs and perform in-situ EBL on top of them. Finally, the envisaged devices put restrictions on the QD's density. If a single QD is to be integrated into a microlens or mesa with a base width of e.g. 2 μm, a QD density of 10^8 cm^{-2} could be sufficient due to the random areal distribution of self-organized QDs and their known positions due to the CL mapping step. However, a QD density of 10^7 cm^{-2} is more appropriate as it corresponds to a mean inter-QD distance of ≈3 μm.

8.4 Quantum Light Sources Based on Deterministic Quantum Dot Microlenses

So far two different wavelength regimes were exploited with our in-situ EBL: QDs emitting around 930 and 1300 nm. In the following, results for the mature 930 nm regime will be presented. First results for 1300 nm suitable for fiber-based quantum communication in the telecom O-band will follow towards the end of this chapter.

8.4.1 Microlenses for Enhanced Photon Extraction

Semiconductors provide a great platform for the realization of optoelectronic devices as the material processing is well understood and diode structures for electrical operation can be realized. However, the large difference in refractive indexes between a semiconductor and air/vacuum (typically around 2) results in small critical angles where total internal reflection sets in that limits the efficiency of photon extraction to <2% (Fig. 8.11a). Different strategies were developed to get rid of this limiting scenario (see Sect. 8.4.3 for additional examples). A very simple and highly attractive approach is the modification of the surface topology by e.g. attaching lens-like structures (Fig. 8.11b) [14, 20, 22, 75, 89, 90]. Besides an increased light-extraction efficiency, this also allows for tailoring the angular far-field profile to match the

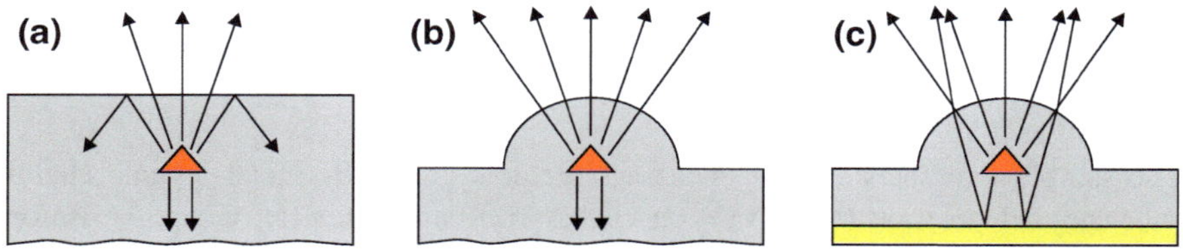

Fig. 8.11 Sketch of the situation of photon extraction with and w/o a lens structure on top of the semiconductor surface. **a** For a planar semiconductor/air interface the photon-extraction efficiency is drastically limited by total internal reflection (TIR). **b** A lens structure reduces TIR. **c** A backside mirror additionally enhances the extraction efficiency towards the upper half-sphere

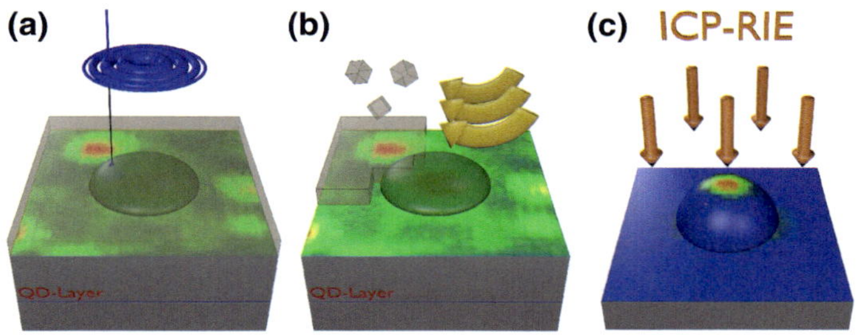

Fig. 8.12 The process of fabricating in-situ EBL lenses is quite similar to that in Fig. 8.9, except that the lens profile is defined by writing concentric circles with varying electron dose (**a**). This results in a 3D lens-shaped resist profile (**b**) that is transferred into the semiconductor by anisotropic dry etching (**c**). Adapted with permission from [22]

acceptance angle and mode profile of adjacent optical elements. One very prominent example would be the coupling into optical fibers [31] (cf. Sect. 8.2).

The most common techniques to add a lens to a semiconductor surface include attaching a separately fabricated dielectrical lens [89, 91, 92], performing (on-site) 3D laser writing [75], or integrating the lens shape monolithically into the semiconductor [20, 22]. Our in-situ EBL process utilizes the latter one [22]. In contrast to Fig. 8.9 the patterning of the resist is performed by writing concentric circles with a decreasing electron dose from the inner to the outer ones. By adjusting the electron dose for each circle correctly, this results in the envisaged 3D lens shape in the resist after its development (Fig. 8.12).

While the lens improves the photon-extraction efficiency from the top, there are still major losses due to luminescence that is emitted towards the bottom side/substrate of the sample. This can be overcome by adding a bottom mirror (Fig. 8.11c). Such mirrors can be constituted by epitaxially grown or evaporated distributed Bragg reflectors (DBRs) and metal layers like gold, silver, or aluminum. To further enhance the light-extraction efficiency from microlenses an anti-reflection coating can be applied. In the simplest case, the lens is coated with a $\lambda/4$-thick dielectric layer whose refractive index is close to $\sqrt{n_{\text{semiconductor}} \cdot n_{\text{air}}}$. This was shown to provide an additional light-extraction enhancement by a factor of ≈ 2 [93].

8.4.2 Description of Sample Templates and Spectroscopic Techniques

The sample templates consist of layer structures grown by metal-organic chemical vapour deposition (MOCVD) on GaAs-(100) substrates (cf. Chap. 3). Before growth, the layer thicknesses were optimized via numerical modeling in the framework described in Sect. 8.2. The following layer structure provides an example that was used for microlenses in combination with a backside DBR for emission around 930 nm. (Sample layouts for different applications are listed in the respective sections of this chapter.) Starting with the substrate, a 300 nm thick GaAs buffer layer and a DBR consisting of 23 pairs of $\lambda/4$-thick $Al_{0.9}Ga_{0.1}As/GaAs$ were deposited. Such a DBR provides a reflectivity of close 100% within the $\approx$100 nm wide stop band centered at 930 nm. Next, 65 nm of GaAs, a single layer of self-organized InGaAs QDs and a 400–420 nm thick GaAs capping layer were grown. The capping layer provides the matrix for the deterministic processing of microlenses via in-situ EBL as described previously in this chapter. The used growth parameters result in a gradient of the QD density along the radius of the wafer. In a certain radius range that is identified by CL mapping, there is a QD density of about 10^7–10^8 cm^{-2} that is very suitable for in-situ EBL processing.

The structures are studied via micro-photoluminescence (μPL) spectroscopy with high spatial, spectral and temporal resolution. The samples are mounted in a liquid-Helium flow-cryostat for operation at cryogenic temperatures (4–40 K) and excited optically via a continuous wave (cw) diode-pumped solid-state laser (660 nm), via a mode-locked Ti:Sapphire laser operating in picosecond mode at a repetition rate of 80 MHz (tunable: 750–1050 nm), or by other cw or pulsed laser sources as specified in the respective subsections. The photoluminescence is collected using a 20 $\times$ microscope objective with a numerical aperture (NA) of 0.4 (if not stated otherwise). The collimated photoluminescence is focused onto the entrance slit of an reflection-grating (double) monochromator with an attached charge-coupled device camera (CCD camera) providing a spectral resolution of 0.017 nm (25 μeV, 6 GHz) or better. Polarization optics (linear polarizers and $\lambda/2$-waveplates) in front of the monochromators enable polarization resolved measurements. Some measurement extensions utilize multi-mode or single-mode fibers that are carefully integrated into the setup to maintain a high detection efficiency of the setup.

Attached to a second output of the monochromator, a fiber-coupled Hanbury-Brown and Twiss (HBT) setup allows for measurements of the second-order photon-autocorrelation $g^{(2)}_{\mathrm{HBT}}(\tau)$. The HBT is constituted of a 50:50 multimode beamsplitter and two single-photon counting modules (SPCMs) based on Si-APDs with a timing resolution of 350 ps in combination with time-correlated single-photon counting electronics.

Alternatively, a Hong-Ou-Mandel (HOM) two-photon interference setup can be connected instead of the HBT setup. It utilizes an asymmetric Mach-Zehnder interferometer with polarization-maintaining (PM) single-mode fibers where the photon stream is split by a 50:50 beam splitter into two paths. One of them is delayed by the temporal pulse distance of the exciting laser in order to facilitate the interference of two consecutively emitted photons at the second beam-splitter. In order to make the photons distinguishable for reference measurements, a $\lambda/2$-waveplate allows for switching the polarization of photons in one arm of the interferometer with respect to the other arm .

Deviations and additions to above listed spectroscopic techniques are given in the respective chapters as well as schematics if necessary for better understanding.

8.4.3 Device Yield and Photon-Extraction Efficiency

As mentioned in Sect. 8.3, deterministic processing technologies should result in much higher device yields as compared to random processing schemes where the positions of suitable targets (e.g. QDs) are unknown. A statistical analysis of a large number of deterministically fabricated QD-microlenses resulted in a device yield larger than 90% when using in-situ EBL. Another appealing feature is the possibility to fabricate a larger number of microlenses within one mapping field in order to speed up the overall process. Figure 8.13 displays such a multi-lens array (a) together with the corresponding single-QD spectrum (b) of the marked lens in panel (a). We observe well-resolved emission lines from different excitonic complexes that were identified by polarization- and power-dependent measurements [22]. Figures 8.13c, d show exemplary lens shapes for a hemispherical lens and a Gaussian-shaped lens, respectively.

Another key point is a high photon-extraction efficiency η_{ext} for a maximized photon flux into the external collection optics. As discussed before, planar and non-structured semiconductor-air interfaces suffer from the large ratio of refractive indexes (e.g. $n_{\text{GaAs}}/n_{\text{air}} \approx 3.5$). This typically limits η_{ext} to $\approx$ 1–2% [14].

So far a number of nanophotonic structures was realized to gain an increased η_{ext} (listed in parenthesis) as compared to an unstructured (planar) semiconductor surface: tapered nanowires (0.72) [18], nano-trumpets (0.75) [94], nano-obelisks (0.36) [95], micropillars (0.36–0.79) [17, 25, 41, 96–99], bulls-eye cavities (0.48) [23], photonic-crystal cavities (0.44) [100], self-aligned defect-DBR structures (0.42) [101], and macroscopic dielectric immersion lenses (0.65) [92].

However, microlenses are very versatile as compared to many of the other structures. They enable an increased η_{ext}, allow for broad-band operation within a typical spectral window of 10 nm, and they can be electrically contacted in an easy and almost planar way. Another appealing feature is the fact, that lens structures also enhance the light *injection* into the QD [102]. This is beneficial for strict resonant excitation schemes as it reduces the required incident laser power and by thus the amount of resonant stray light.

The determination of the photon-extraction efficiency for our microlenses is based on single-photon counting of single-excitonic emission as triggered by pulsed optical excitation of the QD inside. The single-excitonic emission is driven into saturation by a laser with a repetition rate of $f_{\text{laser}} = 80$ MHz. The luminescence is then col-

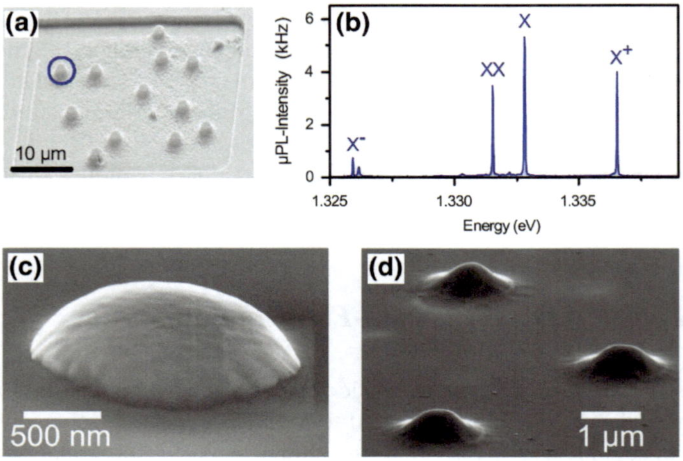

Fig. 8.13 **a** An array of 12 deterministically fabricated microlenses within a single mapping area and an exemplary single-QD spectrum, **b** from the microlens that is marked by the blue circle in (**a**). (X: exciton, $X^{\pm}$: positively/negatively charged exciton, XX: biexciton.) In **c**, **d** examples for a hemispherical and a Gaussian-shaped lens are given. Adapted by permission from Springer Nature from [87] and adapted with permission from [22]

lected by the microscope objective with a given NA (e.g. NA = 0.4) of the μPL setup and detected by a fiber-coupled SPCM that is attached to the second exit port of the monochromator. For liquid-He temperatures an internal quantum efficiency of unity of the optically excited QDs is assumed. Now, the μPL-setup's detection efficiency is needed to deduce η_{ext} into the first lens. Therefore, a tunable diode laser is tuned to the wavelength of the X emission line and focused onto a gold mirror inside the cryostat. The laser is attenuated using neutral-density filters in front of the monochromator to achieve count-rates at the SPCMs comparable to those observed for QD emission. Taking into account the laser power, the reflectance of the gold mirror, the transmission of the cryostat window, the attenuation of the density filters, and the maximal count-rates on the SPCMs a setup efficiency of $\eta_{\text{Setup}} = 0.8\%$ could be deduced. Next, one has to consider that besides an exciton there might be spectator charge carriers (e.g. an extra electron or hole) inside the QD. This determines whether emission from a neutral or charged exciton is observed. As both scenarios stem from a directly triggered emission event in the 'single-exciton regime' their SPCM count rates have to be summed up to obtain the real single-photon emission rate for single-excitonic emission. E.g.

$$\dot{n}^X_{\text{total}} = \dot{n}^X + \dot{n}^{X^+} + \dot{n}^{X^-} \tag{8.23}$$

With $\dot{n}^{X/-/+}$ being the detection rates of emission from neutral, negatively or positively charged excitons.

Now η_{ext} can be deduced by putting it all together:

$$\eta_{\text{ext}} = \frac{r_{\text{detection}}}{\eta_{\text{Setup}}} \quad \text{and} \quad r_{\text{detection}} = \frac{\dot{n}^{X}_{\text{total}}}{f_{\text{laser}}} \tag{8.24}$$

Here, $r_{\text{detection}}$ is the fraction of detected photons with respect to the exciting laser pulses with laser frequency f_{laser}. By following this procedure we were able to demonstrate η_{ext} of $(29 \pm 3)\%$ for a QD-microlens-DBR combination [103] and $(20 \pm 2)\%$ for a QD-microlens above a gold mirror [104]-each for a NA = 0.4 of the external collection optics. An additional boost of η_{ext} could be obtained by combining a DBR-microlens system with a micro-objective on top that was fabricated by 3D femtosecond direct laser writing. Here a η_{ext} of $(40 \pm 4)\%$ was measured in a proof-of-principle experiment [102].

The above demonstrated achievements with respect to the device yield and the photon-extraction efficiency constitute an important starting point for the application of such lens structures in quantum optical settings. A further assessment of such structures with respect to the quantum nature of their emission is given in the following sections.

8.4.4 Verification of Single-Photon Emission

First we investigate the emission statistic from our devices. A crucial benchmark is the purity of single-photon emission, e.g., the suppression of multi-photon events per excitation cycle.

A typical micro-photoluminescence (μPL) spectrum of a deterministic QD microlens chosen for quantum optics experiments is depicted in Fig. 8.14a. The exciting laser was chosen to be resonant with the QD's *p*-shell at a wavelength of 909 nm. Here, the emission is predominantly generated by the positively charged trion state X^+, most probably due to intrinsic carbon doping during growth [60]. To investigate the photon statistics, the emission of the X^+-state is spectrally selected and coupled to the HBT setup (cf. Sect. 8.4.2). Figure 8.14b shows the corresponding raw (blue) and dark-count corrected (red) measurement data of the second-order photon auto-correlation $g^{(2)}_{\text{HBT}}(\tau)$ at an excitation power corresponding to $0.5 \times P_{\text{sat}}$, where P_{sat} denotes the excitation power at which the emission line saturates. Triggered single-photon emission of the deterministic QD microlens is reflected in the almost absent coincidences at zero time delay ($\tau = 0$ ns). At finite τ, coincidence maxima occur at a periodicity corresponding to the pulse-repetition rate of 80 MHz. To achieve highest photon-extraction efficiencies a quantum emitter has to be excited at (or close to) saturation power, e.g. the working point at which the rate of charge-carrier generation equals the inverse radiative lifetime. Hence, we investigated the influence of the excitation power on the single-photon purity. Figure 8.14c, d display the corresponding $g^{(2)}(\tau)$ histograms of the X^+-state for excitation powers P of $1.0 \times P_{\text{sat}}$ and $2.0 \times P_{\text{sat}}$. We observe almost excitation-power-independent values of $g^{(2)}(0) \leq 0.01$ for up to two times the saturation power. This important feature of

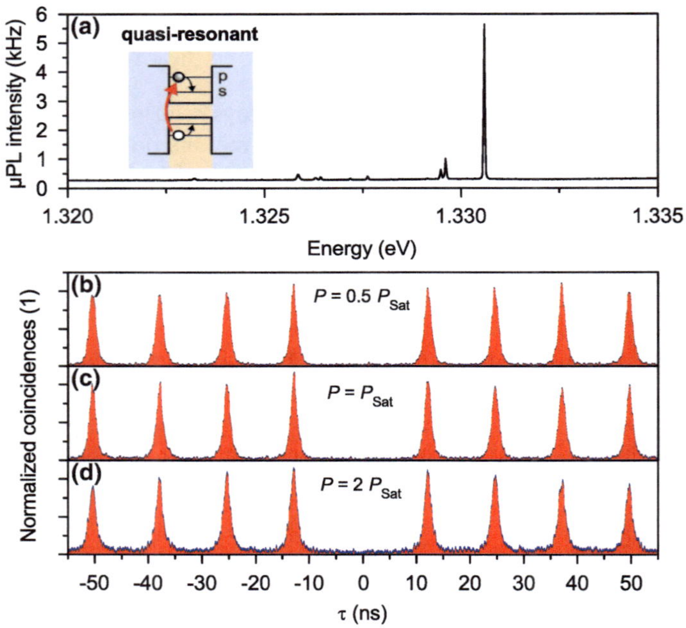

Fig. 8.14 **a** μPL spectrum of a QD microlens under p-shell excitation, showing bright emission of the positively charged trion state X^+ at 1.3305 eV. **b–d** Photon auto-correlation measurements on the X^+ state for excitation powers P of $0.5\times$, $1.0\times$ and $2.0\times$ the saturation power P_{sat}. Adapted with permission from [22]

deterministic microlenses proves that our structures can act as bright quantum light sources with close-to-ideal quantum properties - even at maximum photon flux.

While such excitation-power-independence of $g^{(2)}(0)$ has for instance also been observed in photonic nanowires [18], we go a significant step further in Sect. 8.4.5. There, we demonstrate that our structures not only show a close-to-ideal single-photon emission but at the same time also a high degree of photon indistinguishability. The latter is of crucial importance for advanced quantum communication concepts such as the quantum repeater, which relies on entanglement distribution via Bell-state measurements [105–107].

Having shown, that QD microlenses are well suited for single-photon generation, we will present a method to increase the achievable single-photon flux in the following. To overcome the limitations of standard 80 MHz pulsed laser systems for optical excitation, we use a frequency-doubled mode-locked vertical-external-cavity surface-emitting laser (ML-VECSEL) operating at a pulse repetition rate of close to 500 MHz to excite a deterministically integrated single-QD microlens. For this purpose, a compact laser system based on a ML-VECSEL [108, 109] (see Fig. 8.15a) emitting 508-nm ps-pulses is integrated into a μPL spectroscopy setup (view Fig. 8.15c).

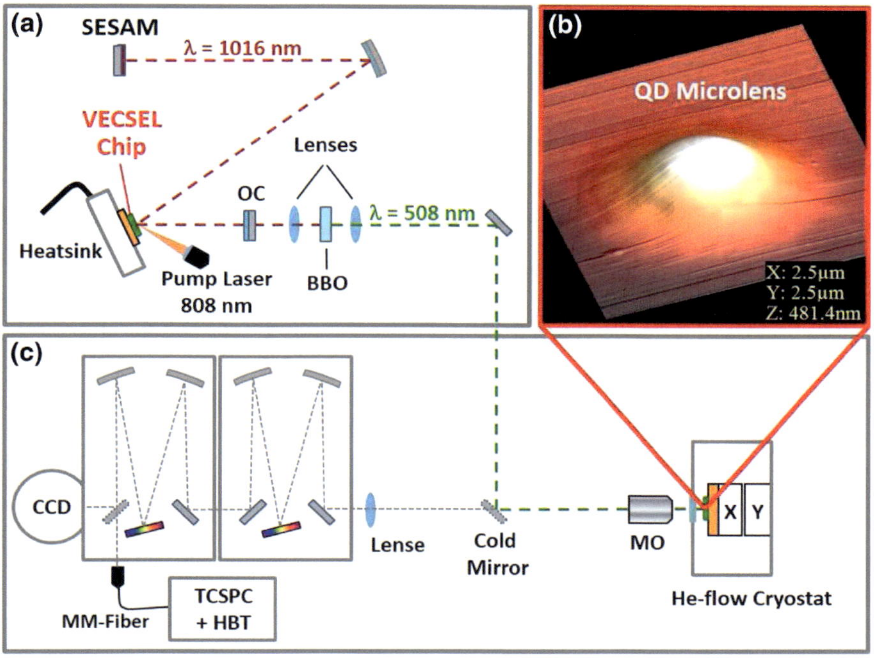

Fig. 8.15 Illustration of the experimental setting used to trigger a quantum dot (QD) single-photon source at a repetition rate of 494 MHz using a mode-locked vertical-external-cavity surface-emitting laser (ML-VECSEL). **a** Arrangement of the compact laser system based on a ML-VECSEL, which is optically coupled to the micro-photoluminescence setup in (**c**) to trigger the QD microlens. **b** Atomic force microscopy image of a deterministically fabricated QD microlens. Reprinted from [110], with the permission of AIP Publishing

Figure 8.16a displays a corresponding μPL spectrum for an excitation power of 1.3 μW with pronounced emission from a positively charged exciton X^+. The integrated emission intensity from this transition is shown in Fig. 8.16b as a function of the ML-VECSEL excitation power on a logarithmic scale.

The emission starts to saturate at an excitation power of $P_{\text{exc}} = 1.3\,\mu\text{W}$ (indicated by red arrow in Fig. 8.16b). The photon emission statistics of this device was investigated by HBT measurements at $P_{\text{exc}} = 1.3\,\mu\text{W}$ (upper panel of Fig. 8.16c). The combined detection rate of both SPCMs was 70 kHz. A clear antibunching with a measured value of 0.22 can be observed at zero time delay ($\tau = 0$). The coincidence maxima with a periodicity of 2.025 ns, correspond to the pulse repetition rate of the ML-VECSEL. Those side peaks are not clearly resolved as their time difference is close to the decay time of the QD ($\approx$1.5 ns) and near the timing resolution of the SPCMs. To quantitatively extract the obscured $g^{(2)}(0)$ value, we modeled the data with equidistant photon pulses represented by Lorentzian profiles with 2.3 ns FWHM and a pulse separation according to the 494 MHz repetition rate of the exciting laser (see [103] for details). The fitted model function (cf. Fig. 8.16c) nicely resembles

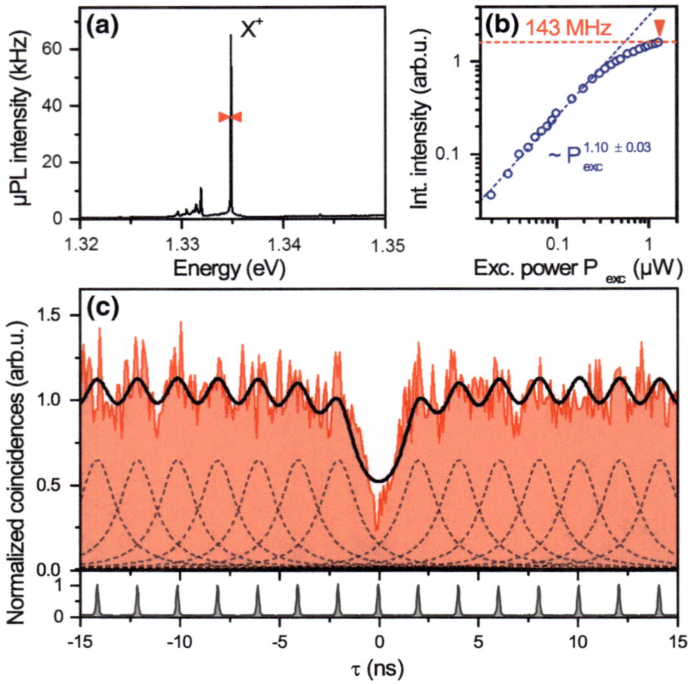

Fig. 8.16 Emission characteristic of a QD microlens triggered by a frequency-doubled ML-VECSEL at a repetition rate of 494 MHz. **a** μPL spectrum revealing bright emission from a positively charged exciton state (X^+). **b** Excitation-power dependence of the X^+ emission intensity. Saturation is reached for 1.3 μW (dashed horizontal line). Here, the single-photon flux emitted by the QD into the first lens of the setup amounts to 143 MHz. **c** Photon-autocorrelation histogram measured on the X^+ emission at saturation. The model curve (solid black line) reveals an upper bound to the antibunching value of $g^{(2)}(0) < 0.03$, clearly proving single-photon emission. For comparison, the lower panel shows the autocorrelation measurement of the ML-VECSEL. Reprinted from [103], with the permission of AIP Publishing

the experimental data and gives an upper bound of $g^{(2)}(0) < 0.03$. Again (cf. beginning of this section), this proves the triggered emission of single-photons from the deterministic QD microlens even at saturation of the QD state.

To quantitatively analyze the emission of the QD microlens in terms of the single-photon flux, we experimentally determined the detection efficiency of our experimental setup following the procedure described in [22]. Exciting the X^+ state at saturation we obtain a single-photon flux F_{SPS} emitted into the microscope objective of (143 ± 16) MHz. Taking into account the repetition rate of 494 MHz, this corresponds to a photon-extraction efficiency of the deterministic single QD microlens of $(29 \pm 3)\%$. The demonstrated single-photon flux of 143 MHz represents a significant improvement compared to previous reports on SPSs based on tapered nanowires [18] as well as optically triggered QD micropillar cavities [96]. To date only electrically pumped QD micropillar cavities operated at GHz repetition rates [44] or gated QD microlenses [110] achieve higher single-photon fluxes.

This proof of principle experiment carried out under above-band excitation points out the high potential of our approach, which can be extended in the future towards resonant excitation of QD-microlenses or other QD based nanophotonic devices. This could be accomplished by deterministically matching the wavelength of respective QD transitions to the emission of a mode-locked VECSEL, or by exploiting wavelength-tunable mode-locked VECSELs in the future. In this light, even the generation of indistinguishable photons at unprecedented emission rates seems to be feasible.

8.4.5 Generation of Indistinguishable Photons

Photon indistinguishability plays a key role in the realization of long-distance QKD schemes based on entanglement distribution via Bell-state measurements [4]. In experiments, the photon-indistinguishability, e.g. the mean two-photon wave-packet overlap, can be evaluated by using a Hong-Ou-Mandel (HOM)-type two-photon interference (TPI) setup [111] (cf. illustration in Fig. 8.17a and description in Sect. 8.4.2). In case of ideal indistinguishability, both photons coalesce at the second beam-splitter in a two-photon Fock state $|2\rangle$. This so called HOM-effect leads to a vanishing number of coincidences at zero time delay ($\tau = 0$ ns) in the histogram $g^{(2)}_{\text{HOM}}(\tau)$ obtained from a coincidence measurement at the two interferometer outputs.

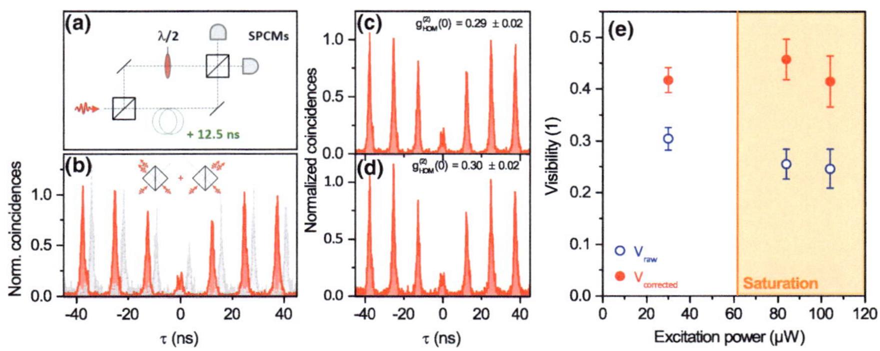

Fig. 8.17 **a** Schematics of a Hong-Ou-Mandel (HOM)-type two-photon interference (TPI) experiment. MO: microscope objective, SPCM: single-photon counting module. **b** TPI histogram $g^{(2)}_{\text{HOM}}(\tau)$ measured with a HOM-type setup for co- (red solid line) and cross- (grey dashed line) polarization of the interfering photons. The cross-polarized data were shifted by +3 ns for clarity. **c–d** TPI histograms for co-polarized photons and increasing optical pump powers P. **e** Visibility values plotted versus the excitation power P. Shaded in orange is the range for which the emission is saturated. Adapted with permission from [22]

Excitation-power-dependent measurements of $g^{(2)}_{\mathrm{HOM}}(\tau)$ were performed on the X^+ emission line of a QD under quasi-resonant excitation. The resulting $g^{(2)}_{\mathrm{HOM}}(\tau)$ histograms are shown in Fig. 8.17b–d, exemplarily also for the cross-polarized (grey dashed line) measurement configuration.

The excitation power has only little impact on the $g^{(2)}_{\mathrm{HOM}}(0)$ values which was also observed for the 'bare' HBT-$g^{(2)}(0)$ values (cf. Fig. 8.14). For a realistic evaluation of TPI visibilities V_{HOM} one has to take into account the finite probability of multi-photon emission events and the imperfections of the HOM setup (see [112] for details). The resulting analysis of the data is presented in Fig. 8.17e and reveals the TPI visibility. Well above saturation a corrected (raw) TPI visibility of $V_{\mathrm{HOM}} = (46 \pm 4)\%$ ($(26 \pm 3)\%$) is obtained.

In contrast to the purely intensity-dependent HBT experiment, the photon-indistinguishability is particular sensitive to spectral jitter and dephasing processes. The dephasing rate of a quantum emitter is described by its coherence time T_2 and the radiative lifetime $T_1 = \Gamma^{-1}$ via $T_2^{-1} = (2T_1)^{-1} + (T_2^*)^{-1}$ [113], where $(T_2^*)^{-1} = \Gamma' + \gamma$ includes pure dephasing due to, e.g., spectral diffusion (Γ') and phonon coupling (γ), which can be tested by time- and temperature dependent TPI experiments. This allows one to modify the interaction time with disturbances between consecutive QD phonons and to vary the phonon number occupation in the system.

For this, the HOM-setup of Fig. 8.17a was modified to enable two-pulse sequences with variable pulse-separation δt as depicted in Fig. 8.18a. Here, the periodic excitation pulses of the mode-locked Ti:Sa laser (80 MHz repetition rate, $\lambda = 909$ nm) are split into a sequence of double-pulses by utilizing a fiber-based asymmetric Mach-Zehnder interferometer (not shown) in the excitation path. Different fiber-delays within one arm lead to corresponding pulse-separations δt from 2.0 to 12.5 ns. Consequently, the variable delay of the HOM interferometer has to match the 'new' pulse

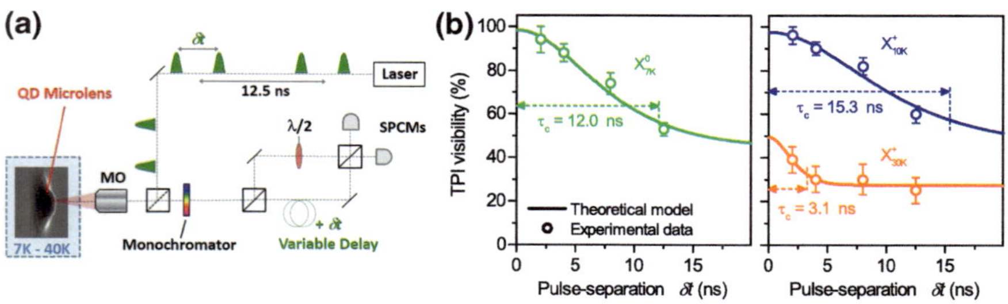

Fig. 8.18 **a** Schematic of the Hong-Ou-Mandel-type two-photon interference setup used for probing the indistinguishability of consecutively emitted photons with variable temporal separation δt (MO: microscope objective, SPCM: single-photon counting module). **b** Measured two-photon interference visibilities of consecutively emitted single photons versus δt for the X^0- and the X^+-state. Experimental data are described quantitatively by a theoretical model assuming a non-Markovian noise correlation resulting from spectral diffusion at ns-timescale. We observe a characteristic temperature-dependent correlation-time τ_c. Adapted with permission from [26]. Copyrighted by the American Physical Society

separation of the excitation pulses. The sample temperature was varied by heating up the cold finger of the cryostat.

Figure 8.18b, left panel presents the obtained raw TPI visibilities as a function of the pulse-separation δt for the neutral exciton X^0. For small δt a plateau-like behavior is observed, at which the visibility remains almost constant with values of $V_{2.0\text{ns}} = (94 \pm 6)\%$ and $V_{4.0\text{ns}} = (88 \pm 4)\%$. For excitation-pulse separations larger than 4 ns, the visibility decreases from $V_{8.0\text{ns}} = (74 \pm 5)\%$ to $V_{12.5\text{ns}} = (53 \pm 3)\%$. Accordingly, the characteristic time scale of spectral diffusion is on the order of several nanoseconds. In addition, a time-dependent analysis of Γ' has been carried out for the charged exciton state X^+ of the same QD at 10 and 30 K (cf. Fig. 8.18b right panel). Again a characteristic correlation time is observed, which decreases at higher temperature.

To support our experimental findings numerical modeling was performed (cf. Chap. 6). The system was described by a Hamiltonian, where the QD was approximated as a two-level system. Dephasing is included by employing the working horse of the phenomenological dephasing description by including a general stochastic force $F(t) = P(t) + D(t)$ with a phonon-induced dephasing (δ-correlated white noise) $P(t)$ and a spectral diffusion $D(t)$ component (colored noise), both shifting the transition energy of the QD. The specific noise correlations depend on the coupling mechanism between the QD and its environment. For example, in case of spectral diffusion random electric fields due to charge fluctuations induce dephasing [114, 115] (see [26] for details). The experimental data was nicely reproduced by the numerical results (solid lines in Fig. 8.18b), thus confirming the influence of electric fields and phonons. The impact of spectral diffusion reveals a strong dependence on the pulse distance while the interaction with phonons almost solely depends on temperature.

For the implementation of MDI-QKD and the quantum repeater concept remote sources of indistinguishable photons are required. So far, only few experiments reported two-photon interference (TPI) from spatially separated QD-based single-photon sources, e.g. [116–120]. Deterministic sources have been applied very recently [121], exploiting high-Q microcavities. Due to the narrow-band characteristics of the microcavity this approach is limited to the light extraction from a single excitonic state and spectral tuning is hardly achievable. Here again, microlenses are beneficial as they offer significantly enhanced photon-extraction efficiency (compared to bulk material), large spectral bandwidth as well as high single-photon purity in terms of the photon-indistinguishability [22]. Additionally they can be combined with piezo actuator for spectral tuning [104] and planar contacts for electrical access [110]. At the same time the in-situ EBL approach allows for a pre-selection of emission energies [60] to realize matching single-photon emitters for scalable quantum networks.

In the following, we demonstrate TPI from two spatially separated, deterministically fabricated single-photon sources based on QDs embedded in photonic microlenses. For this purpose, two cryostats with the microlenses inside were utilized in a confocal microphotoluminescence (μPL) setup (see schematic in Fig. 8.19a). The integrated QDs (labeled QD 1 and QD 2) were chosen such that energetic splitting between s- and p-shell is similar to enable efficient optical excitation via their excited

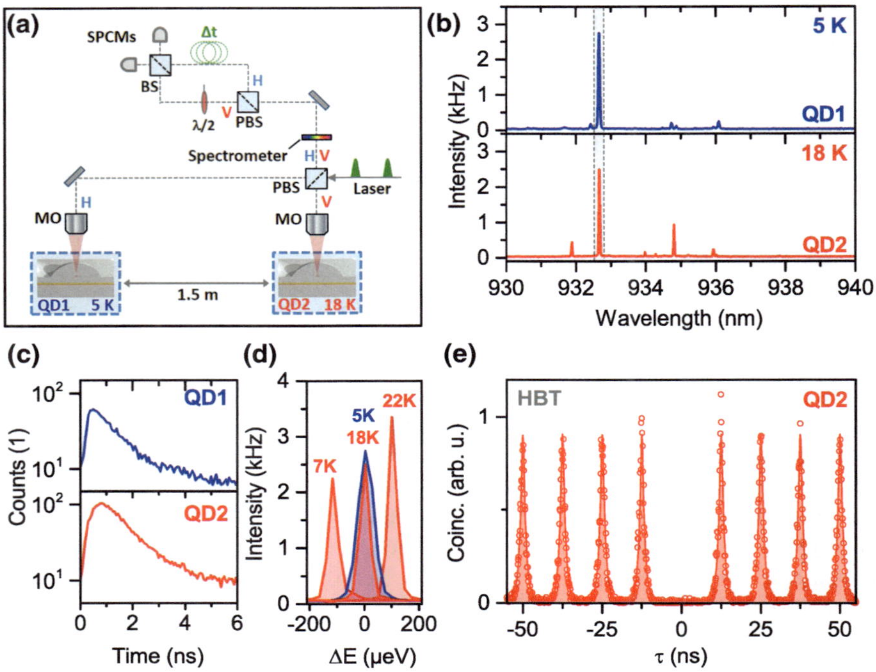

Fig. 8.19 **a** Schematic of the experimental setup used to analyze the TPI from remote QD microlenses. (MO: microscope objective, BS: beam splitter, PBS: polarizing beam splitter, SPCM: single-photon counting module, H/V: polarization direction) **b** μPL spectra of QD 1 and QD 2 showing excellent wavelength matching. **c** Measurement of the lifetimes for QD 1 and QD 2. Temperature tuning is used to precisely match QD 1 and QD 2. **d** HBT autocorrelation measurement for QD 2. Reprinted from [122], with the permission of AIP Publishing

states at 909.5 nm using a mode-locked Ti:sapphire laser operating in picosecond-mode at a repetition rate of 80 MHz.

The photon streams emitted by both sources are orthogonally polarized (using two $\lambda/2$-waveplates), spatially superimposed at a polarizing beam splitter (PBS), energetically filtered in a grating spectrometer and coupled into a Mach-Zehnder-Interferometer (MZI) made out of polarization-maintaining single-mode fiber. The first beam splitter in the MZI is a PBS, and hence spatially splits the orthogonally polarized photons of QD 1 and QD 2 into both interferometer arms. Here, one arm of the MZI includes a fiber-based variable optical delay line for a precise control of the relative temporal delay, while the other arm contains a $\lambda/2$-waveplate for a controlled switching of the polarization from cross- to co-parallel polarization configuration. Next, the photons from QD 1 and QD 2 impinge on a non-polarizing beam splitter (BS) and are finally detected at the two output ports using silicon-based single-photon counting modules (SPCM). Additionally, a filter wheel introduced within one beamline allows for a careful balancing of the intensities of both photon streams

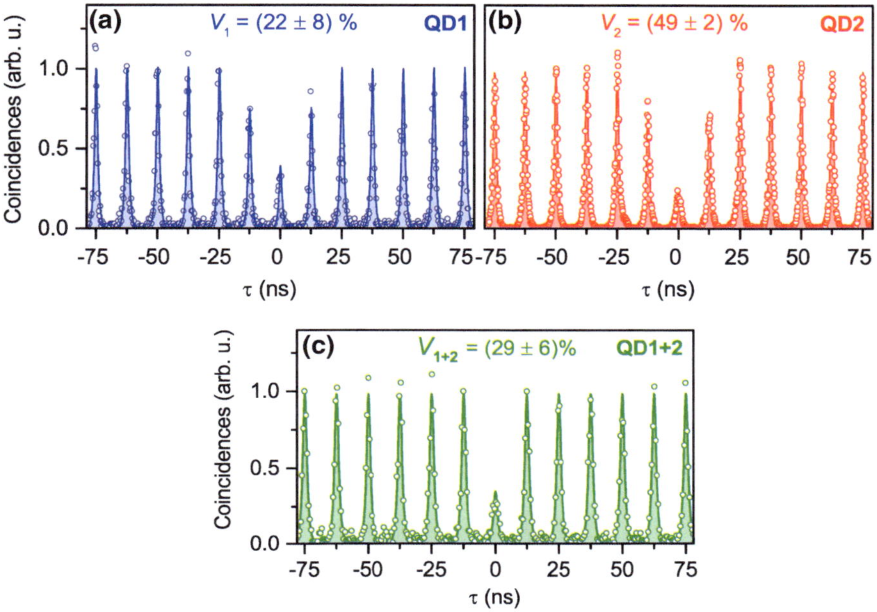

Fig. 8.20 **a, b** Two-photon interference coincidence histograms $g^{(2)}_{\mathrm{HOM}}(\tau)$ and model fit for QD 1 and QD 2 at $T = 10\,\mathrm{K}$. **c** Coincidence histogram $g^{(2)}_{\mathrm{HOM}}(\tau)$ resulting from mutual two-photon interference of the remote single-photon sources from (**a**) and (**b**). A raw two-photon interference visibility of $V_{1+2} = (29 \pm 6)\%$ is observed. Reprinted from [122], with the permission of AIP Publishing

at the second beam splitter inside the MZI, which is crucial for the observation of the TPI effect.

The emission spectra of both QDs are shown in Fig. 8.19b. The emission of QD 2 was fine-tuned to that of QD 1 by setting the temperature of its cryostat accordingly (Fig. 8.19c). Matching was obtained at temperatures of $T_1 = 5\,\mathrm{K}$ and $T_2 = 18\,\mathrm{K}$. We determined the radiative lifetimes of both excitonic complexes via time-resolved μPL measurements (see Fig. 8.19c) to be $\tau_1 = (0.98 \pm 0.02)\,\mathrm{ns}$ and $\tau_2 = (1.03 \pm 0.03)\,\mathrm{ns}$. Measuring the autocorrelation function $g^{(2)}_{\mathrm{HBT}}(\tau)$ of microlenses in the HBT-type setup, excellent single-photon emission with an uncorrected antibunching of $g^{(2)}_{\mathrm{HBT}}(0) = 0.01 \pm 0.01$ is revealed, as displayed exemplarily for QD 2 in Fig. 8.19d.

Next, the photon-indistinguishability of both QD-microlenses was investigated separately in a HOM measurement with an excitation pulse separation of 12.5 ns. The corresponding coincidence histograms for parallel polarized photons in Fig. 8.20a, b demonstrate strongly reduced coincidences at zero time delay, proofing quantum mechanical interference of consecutively emitted photons.

The rather moderate visibilities observed for both microlenses ($V_1 = (22 \pm 8)\%$ and $V_2 = (49 \pm 2)\%$) are mainly attributed to stochastic fluctuating electric fields in the QD's vicinity as was discussed before (see also [26]).

Now we demonstrate the HOM results for the remote source experiment. Hence the photons from both QDs were injected into the HOM setup. Figure 8.20c presents the coincidence histogram $g^{(2)}_{\mathrm{HOM}}(\tau)$ from which an uncorrected TPI visibility of $V_{1+2} = (29 \pm 6)\%$ is extracted by fitting.

This visibility value is competitive with state-of-the-art results reported for deterministic devices without post-selection ($(40 \pm 4)\%$) [121]. Applying resonant excitation schemes, we anticipate significantly improved performance on the level demonstrated for non-deterministic devices ($(91 \pm 6)\%$) [116]. In order to compare the experimental result with the theoretical expectation, we analyzed the radiative lifetimes τ_i and pure dephasing rates τ_i^* for both QD-microlenses to obtain a remote TPI visibility of $V_{1+2}^{\mathrm{theo}} = (24 \pm 13)\%$, in agreement with our experimentally determined value V_{1+2} (see [122] for details).

8.4.6 *Demonstration of a Twin-Photon Source*

Compared to the huge progress achieved in the development of single-photon sources, the on-demand generation of more complex photonic states is still in its infancy. An interesting example for multipartite quantum light states are photon twins, that is pairs of temporally correlated photons with identical properties. It is well known, that QDs allow for the generation of correlated photon pairs by exploiting the biexciton-exciton radiative cascade [24]. Related experiments in the InGaAs/GaAs material system, however, in most cases rely on pairs of photons with different energies [28, 124], excluding the possibility to directly generate photon twins as was demonstrated before in the GaN/AlN material system [125].

In this section, we demonstrate an integrated source of photon twins, that is pairs of photons with identical energy and polarization, highly correlated in time. For this purpose, we use a QD exhibiting an energy degenerate biexciton-exciton radiative cascade that was integrated deterministically within a monolithic microlens (Fig. 8.21b).

While a single electron-hole pair in a QD is called exciton, two electron-hole pairs constitute a biexciton. As a result of Coulomb and exchange interactions between the confined charge-carriers, this state typically shows a finite binding energy $E_{\mathrm{bin}}^{\mathrm{XX}}$ in the range of some meV with respect to the case of two separated and uncorrelated excitons [126]. Hence, this binding energy is directly reflected by the energy separation between the exciton's and biexciton's recombination in a spectrum. If the biexciton's recombination is of higher energy, this configuration is called 'anti-binding'. Another noteworthy feature of the biexciton$\rightsquigarrow$exction$\rightsquigarrow$vacuum-state radiative cascade is a possible fine-structure splitting of the single excitonic state that is inherited by the biexciton transition [127] (cf. Sect. 8.4.7). The resulting radiative cascade emits pairs of photons in two possible decay channels, one being linear-horizontally (H) and the other one linear-vertically (V) -polarized. If a QD with $E_{\mathrm{X}}^{\mathrm{H}} = E_{\mathrm{XX}}^{\mathrm{H}}$ ($E_{\mathrm{FSS}} = |E_{\mathrm{bin}}^{\mathrm{XX}}|$) (Fig. 8.21a) is selected, we end up with a real two-photon source of H-polarized emission.

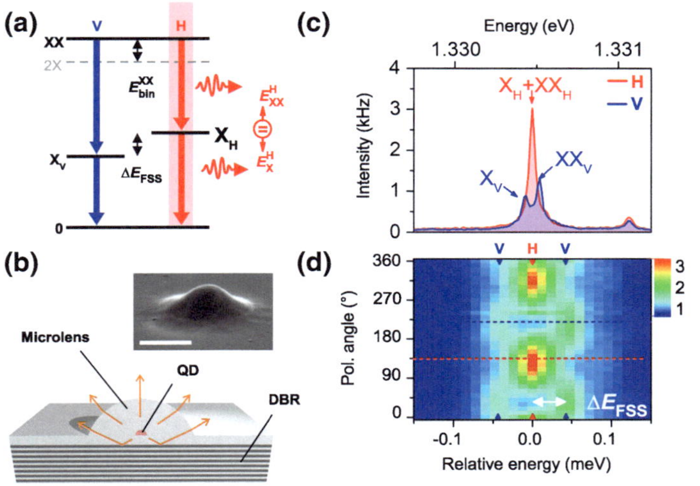

Fig. 8.21 **a** Energy level scheme of the radiative biexciton-(XX) exciton-(X) cascade. If $\Delta E_{\mathrm{FSS}} = E_{\mathrm{bin}}^{\mathrm{XX}}$, the photons within the H-polarized decay channel have identical energy and polarization. **b** Illustration of a deterministically fabricated QD microlens and SEM picture (scale bar: 1 μm). **c** Spectrally resolved μPL: for H-polarization the superimposed emission of exciton and biexciton leads to an increased emission intensity compared to V-polarization. **d** A quantitative analysis of polarized emission reveals $\Delta E_{\mathrm{FSS}} = |E_{\mathrm{bin}}^{\mathrm{XX}}| = (51 \pm 6)$ μeV. Reprinted with permission from [128]

The polarized emission features of this cascade are shown in Fig. 8.21c under above-bandgap (λ = 850 nm) continuous-wave excitation. While the V-doublet is well resolved, the H-transitions of XX and X are degenerate which is also evident in Fig. 8.21d a polarization-resolved map of photoluminescence spectra is depicted. A quantitative analysis of the spectra from Fig. 8.21d yields $E_{\mathrm{FSS}} = E_{\mathrm{bin}}^{\mathrm{XX}} = (51 \pm 6)$ μeV. The observation of an antibinding biexciton state ($E_{\mathrm{XX}}^{\mathrm{V}} > E_{\mathrm{X}}^{\mathrm{V}}$) can be related to a relatively small QD size [129].

Next, polarization-resolved photon-correlation measurements [130] were performed to investigate the temporal correlations. In Fig. 8.22a the cross-correlation coincidence histogram $g_{\mathrm{cross}}^{(2)}(\tau)$ is displayed, where biexciton (exciton) photons trigger the start (stop) events. An asymmetric bunching effect is observed for positive delay times τ, owing to the cascaded emission of photon-pairs within the V-decay-channel [131]. For the H-polarized decay channel exciton and biexciton photons are energetically degenerate and temporal correlations can be probed via photon autocorrelation measurements (Fig. 8.22b). The coincidence histogram is dominated by a pronounced bunching signature at zero delay and, due to the absence of time-ordering of the detected photons, a symmetric behavior in τ is observed. This reveals a high degree of two-photon correlations, which proves that this unique biexciton-exciton radiative cascade serves as a source of true photon twins. At finite delay times ($\tau = \pm 2$ ns) there are clear anti-bunching signatures, stating the non-classicality of the emitted light state. These observations agree quantitatively with a theoretical

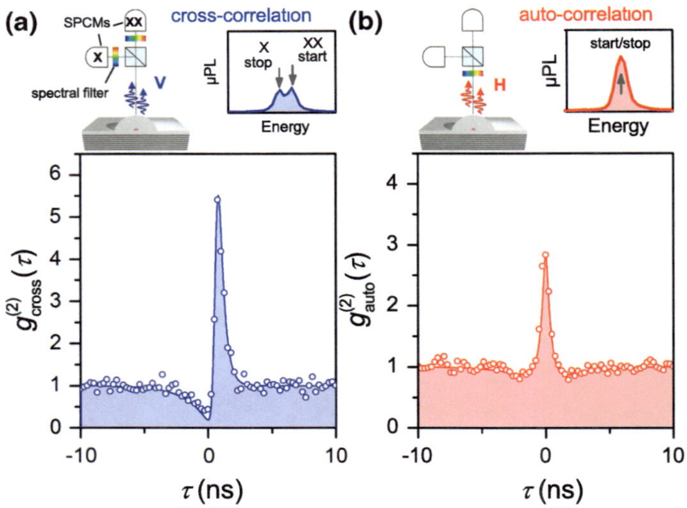

Fig. 8.22 **a** Photon cross-correlation histogram $g^{(2)}_{\mathrm{cross}}(\tau)$ for the V-polarized decay channel. The antibunching-bunching signature proves the cascaded emission of XX-X photon pairs. **b** Photon auto-correlation histogram $g^{(2)}_{\mathrm{auto}}(\tau)$ for the H-polarized decay channel where biexciton and exciton emission is superimposed. The pronounced symmetric bunching indicates a high degree of two-photon correlations, due to the emission of photon twins. Solid curves correspond to theoretical simulations based on a four-level master equation approach. Reprinted with permission from [128]

model (solid curves) (cf. Chap. 6) based on a four-level rate equation approach (see [128] for details).

The magnitude of the bunching in Fig. 8.22b itself, however, does not carry information about the probability of having a photon pair per excitation. In fact the bunching value depends mainly on the occupation of the exciton level (inversely proportional), rather than on the biexciton occupation. For this reason and to quantify the efficiency of our source, we introduce the parameter $\alpha = g^{(2)}_{\mathrm{auto}}(0)/g^{(2)}_{\mathrm{cross}}(0)$, i.e. the ratio of the bunching values in auto- and cross-correlation, respectively. The parameter α thereby corresponds to the fraction of two-photon correlations due to twin-photon emission, which naturally follows if one considers that the observable of our photon auto-correlation measurement on the H-polarized decay channel results from the superposition of a total of four different photon correlations (see [87] for details). Hence, by comparing the measured cross- and auto-correlation traces, one obtains information about the efficiency of the twin-photon cascade. With increasing excitation, the cascade efficiency steadily increases and reaches a maximum value of $\alpha = (39 \pm 3)\%$. From this, we can deduce the twin-photon emission rate (TPR) collected via the microscope objective to be (234 ± 4) kHz. This represents a significant improvement (x5) compared to photon twins generated with atoms [132].

Further investigations are performed in triggered-mode operation. Figure 8.23a shows the auto-correlation histogram of the twin-photon H-channel for pulsed excitation at a repetition rate of $f = 80$ MHz and an excitation wavelength of $\lambda = 850$ nm.

The strong bunching is a clear signature for pulsed twin-photon emission. Next, the indistinguishability of both photons within the H-channel is investigated via a Hong-Ou-Mandel (HOM) experiment. Here, the QD is excited via its p-shell ($\lambda = 904.5$ nm) and a symmetric Mach-Zehnder interferometer is utilized with a half-wave plate within one interferometer arm for switching between co- and cross-polarized measurement configurations. Figure 8.23b presents the $g^{(2)}_{\mathrm{HOM}}(\tau)$ results for both configurations. A clear decrease of coincidences is observed at $\tau = 0$ for co-polarized measurement configuration in contrast to the cross-polarized setting. The corresponding visibility of two-photon interference equals $V = \frac{g^{(2)}_{\mathrm{HOM},\perp}(0) - g^{(2)}_{\mathrm{HOM},||}(0)}{g^{(2)}_{\mathrm{HOM},\perp}(0)\times 1/2} = (56 \pm 9)\%$. Here, $g^{(2)}_{\mathrm{HOM},\perp/||}(0)$ are the HOM values for zero time delay for perpendicular and parallel polarization configurations of the Mach-Zehnder interferometer. Limiting factors for the observed indistinguishability are discussed in [128]. A higher visibil-

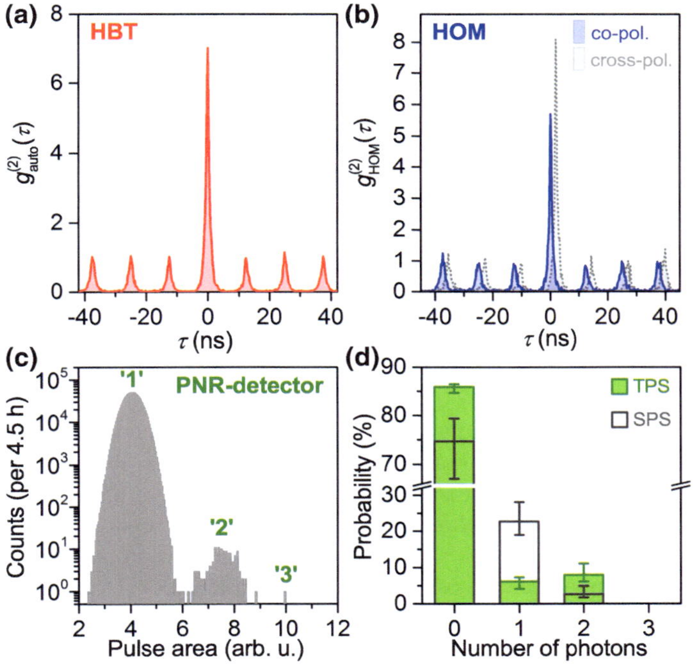

Fig. 8.23 **a** Photon auto-correlation histogram $g^{(2)}_{\mathrm{auto}}(\tau)$ of the twin-photon source (TPS) operated in triggered mode. **b** Hong-Ou-Mandel two-photon interference experiment using photon twins under pulsed p-shell excitation (data in cross-polarization shifted by +2 ns for clarity). **c** Pulse-area-distribution histogram from a photon-number-resolving (PNR) detection system. The peaks correspond to one ('1'), two ('2') or three ('3') photons. **d** Reconstructed photon number distribution deduced from the PNR measurement in (**c**) taking into account the losses of the experimental setup. For comparison, results obtained for a quantum dot single-photon source (SPS) are displayed. Reprinted with permission from [128]

ity could be gained by applying strictly resonant excitation schemes via simultaneous one- and two-photon excitation or by exploiting cavity effects in future works.

For a direct detection of the twin-photon states, a superconductor-based photon-number-resolving (PNR) transition-edge sensor is utilized [133]. Here the twin-photon source was driven by a pulsed diode laser at a repetition rate of 1 MHz ($\lambda = 661$ nm). Twin-photon events (labeled '2' in Fig. 8.23c) are clearly identified and the photon number distribution is derived by taking into account the total losses of the experimental setup (Fig. 8.23d). The probability for twin-photon emission ($p_{\text{twin}} = 8.0^{+3.1}_{-1.8}\%$) is noticeably larger than the probability for single-photon emission ($6.2^{+1.2}_{-2.0}\%$). These relationship completely changes when replacing the twin-photon source by a single-photon source ($22.7^{+5.4}_{-3.8}\%$ single-photon emission and only $2.6^{+2.3}_{-0.9}\%$ photon pair emission). The relatively large contribution of the '0' photon state in both experiments is partly related to the less efficient excitation by the pulsed diode laser (661 nm) and to the fact that photons from the V-polarized decay channel are not detected in case of the twin-photon source.

Very recently, another group investigated twin-photon cascades in pyramidal QDs grown in a site-controlled fashion [134]. In the work reported by Moroni et al., small biexciton binding energies $E_{\text{bin}}^{\text{XX}} \approx (0.28 \pm 0.43)$ meV and almost negligible fine-structure splittings ($E_{\text{FSS}} \approx 6\,\mu$ev) were realized. Due to the significant overlap of exciton and biexciton emission in this case, similar signatures were observed in photon auto-correlation measurements as presented in Fig. 8.22, showing promise for the realization of twin-photon sources also in scalable bottom-up fabrication techniques.

8.4.7 *Generation of Polarization-Entangled Photon Pairs*

Entanglement is often referred to as the essence of the quantum world since it lies at the heart of the most remarkable purely quantum effects. Photons in entangled states allow us to address fundamental questions regarding profound scientific and philosophical consequences about the quantum nature of light. They enabled the experimental validation of Bell's inequalities [135–138], as proposed by Einstein et al. [139], which concluded the debate about the completeness of the quantum theory. The possibility of using QDs as emitters of entangled photons raised a great interest in the community since these artificial atoms can act as on-demand emitters [28]. QDs can potentially outperform non-linear sources based on parametric down-conversion which suffer from the Poissonian statistics of their emission characteristics. The latter property is of crucial importance for the efficient realization of numerous quantum communication schemes and ultimately for the realization of quantum repeaters [105].

The use of QDs as entangled photon emitters was originally proposed by Benson et al. [24]. The idea behind the theory is the same as for an atom: the radiative cascade constituted by the ground state 0, the excitonic state X and the biexcitonic XX state (see Fig. 8.24a) forms two energy-indistinguishable recombination paths (from the biexciton via the exciton to the ground state) if the intermediate excitonic fine structure is degenerate. The photons constituting the entangled pairs have different energies because of the biexciton binding energy as shown on the typical spectrum plotted on Fig. 8.24b. Their state is one of the four maximally entangled states and can be written in the linear basis as follows (H/V are horizontal/vertical polarizations):

$$|\psi\rangle = \frac{1}{\sqrt{2}}(|HH\rangle + |VV\rangle),$$

The proposed scheme was confirmed by a number of publications which demonstrated entangled photon emission from InGaAs/GaAs QDs [124, 140, 141]. The experimental technique used for the evaluation of the emitted entangled state is called 'quantum tomography' and consists of the reconstruction of the two-photon state in the density matrix picture. In this representation, a non-pure state resulting from a non-perfect emitter can be described, as shown in Fig. 8.25. Lots of efforts were made for the reduction of the FSS since the latter removes the exciton's degeneracy that is needed for the entanglement. Approaches as diverse as epitaxial growth in (111) direction [142, 143], growth of highly symmetric GaAs QDs [144–146],

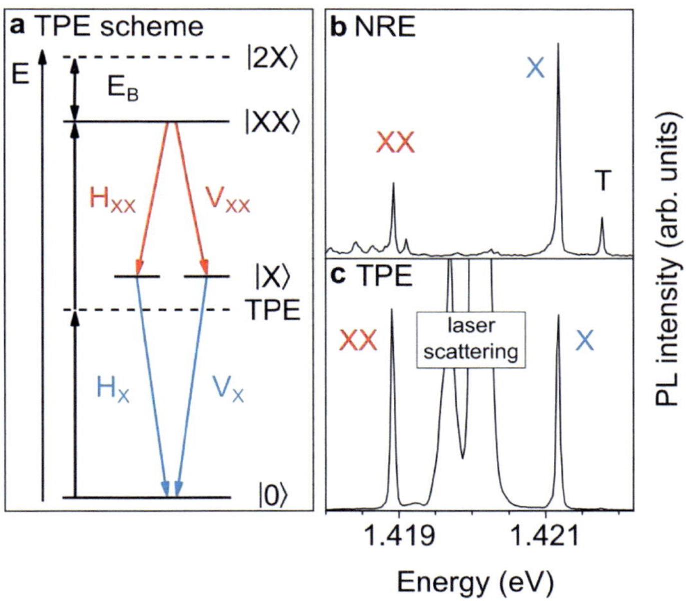

Fig. 8.24 **a** Scheme of the relevant states in a neutral quantum dot with a description of the two-photon excitation scheme. **b** Typical spectrum of a quantum dot under non-resonant excitation. **c** Typical spectrum of a quantum dot resonantly excited through the two-photon excitation of the virtual state. Adapted by permission from Springer Nature from [28]

external piezo-applied stress [147] or electric [148] and magnetic field tuning [149] were explored. Some of them showed convincing evidences of reduction and even suppression of the FSS below the homogeneous emission linewidth of the exciton which is the key requirement.

An alternative to these technological methods is based on a post-selection in time of the excitonic wavepackets. One can take advantage of Heisenberg's relation, expressing that higher temporal resolution in determining the dynamics of the XX-X decay implies larger uncertainty in energy. If the latter is larger than the related excitonic fine structure of the QD, the "which-path information" is lost and quantum entanglement of the paired photons can be measured even in the presence of a FSS. The observed degree of entanglement is then only limited by the detectors temporal resolution which has to be compared with the inverse precession frequency of the excitonic phase defined by the FSS.

In the presence of a non-zero FSS, the electron spin up and electron spin down excitonic states are non-degenerate and are no eigenstates of the system. The exciton state will therefore evolve with time. Two-photon state can be written as follows (H/V horizontal, D/A diagonal, R/L circular polarization states) [150]:

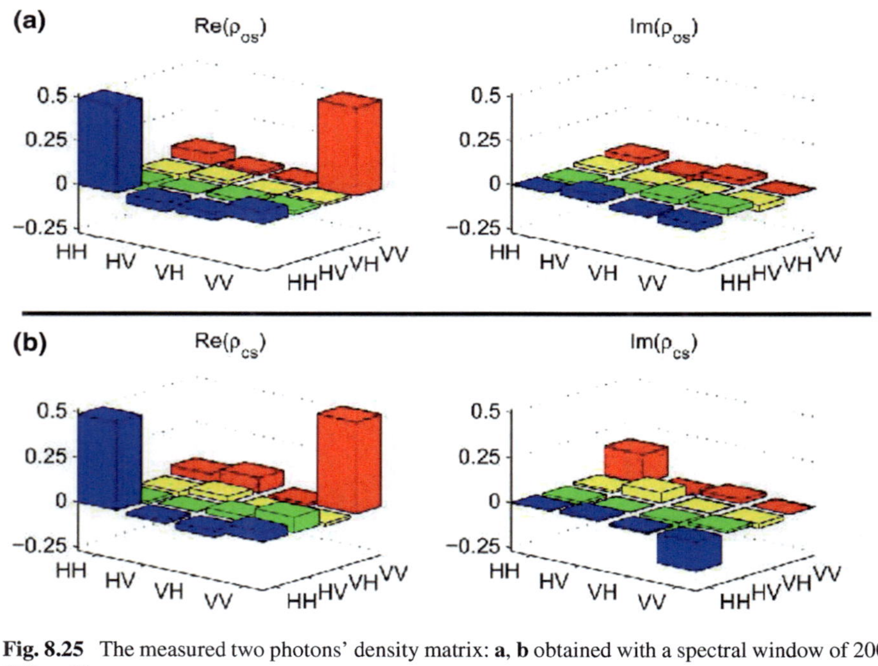

Fig. 8.25 The measured two photons' density matrix: **a**, **b** obtained with a spectral window of 200 (25) μeV

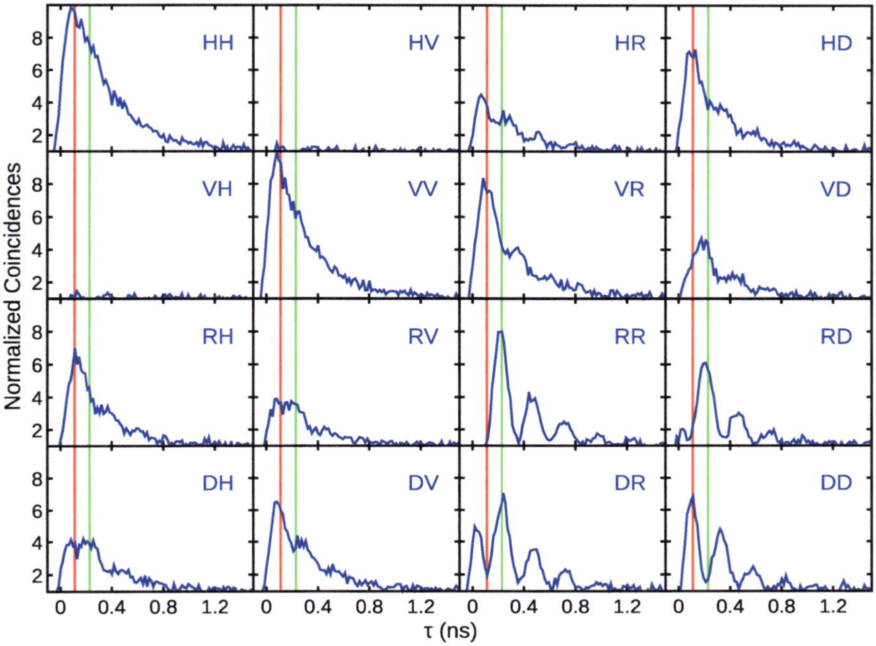

Fig. 8.26 16 time resolved polarization-dependent correlation measurements used for the quantum tomography for the exemplary sample QDM1. Reprinted from [151], with the permission of AIP Publishing

$$|\psi(\tau)\rangle = \frac{1}{\sqrt{2}}(|HH\rangle + e^{i\Delta\tau}|VV\rangle),$$

with Δ the frequency imposed by the FSS energy and τ the time delay between the excitonic and biexcitonic photon emissions. By rewriting this state in the diagonal basis (D/A), one can show that it oscillates between $|\phi^+\rangle = \frac{1}{\sqrt{2}}(|DD\rangle + |AA\rangle)$ and $|\phi^-\rangle = \frac{1}{\sqrt{2}}(|DA\rangle + |AD\rangle)$. One obtains a similar result in the circular basis (R/L). This excitonic phase evolution, expressed by this oscillation, can be tracked by correlating the photons when they are projected in these two bases.

Figure 8.26 shows the 16 correlation measurements performed in the three complementary bases (H/V horizontal, D/A diagonal, R/L circular), used for the reconstruction of the two-photon density matrix. While a bunching is measured in the HH basis (and no bunching in the HV basis), expressing the classical correlations between the excitonic and the biexcitonic photons, quantum correlations revealed in the circular basis (RR measurement) are showing oscillations due to the precession of the excitonic phase at the frequency of the FSS.

Figure 8.27 shows the results provided by the method after deconvolution from the experimental resolution for two different quantum dots (marked as QDM1 and

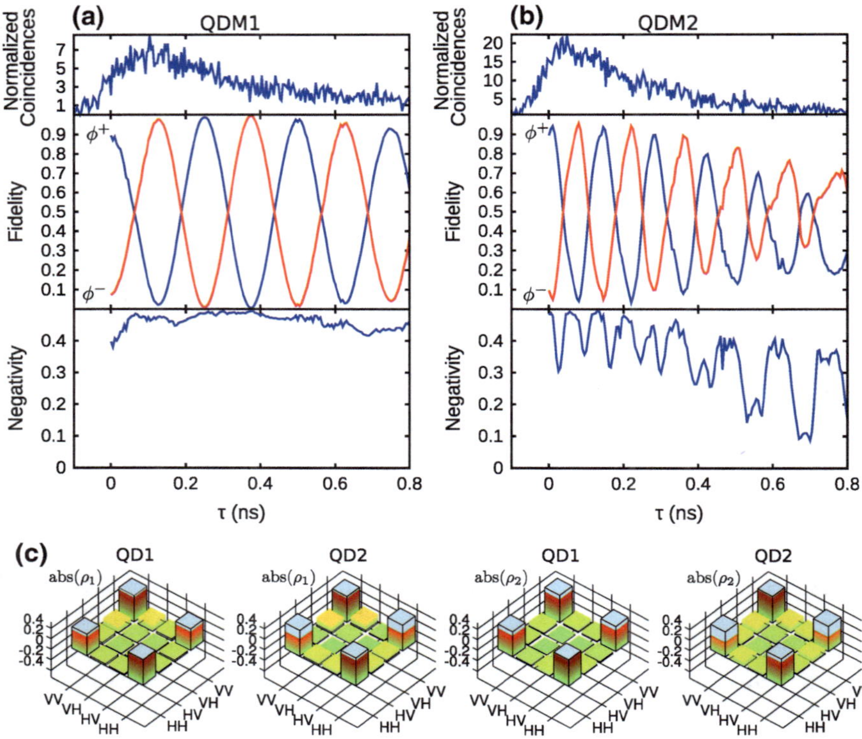

Fig. 8.27 **a** Deconvoluted data obtained for QD QDM1. Upper panel: exciton decay curve. Intermediate panel: measured fidelities to the maximally entangled Bell states (blue curve for $|\phi^+\rangle$, and red curve for $|\phi^-\rangle$). Lower panel: negativity of the two-photon state as a function of the delay after deconvolution, **b** deconvoluted data obtained with QDM2, **c** superposed norms of the density matrices reconstructed before convolution (in color), and corrected after deconvolution (in blue) for QDs QDM1 and QDM2. Reprinted from [151], with the permission of AIP Publishing

QDM2). Figure 8.27a, b depict the decay curves (upper panel), the fidelity to the Bell states (middle panel) and the negativity (lower panel) for QDM1 and QDM2. The fidelities are showing oscillations with an amplitude very close to unity without damping (for QDM1) along the full wavepacket, demonstrating that QDM1 emits nearly perfectly entangled photons. This is confirmed by the negativity (lower panel), which quantifies the separability of the density matrix. A value close to 0.5 (maximal entanglement) is found up to a delay of 0.8 ns. Figure 8.27c shows the final density matrix after deconvolution (blue part) for the two QDs.

This work has combined two important technologies for the development of structures approaching the on-demand regime. They were obtained using deterministically fabricated QD-microlenses, crucial in this case for the broadband extraction of the generated photon pairs, and the coherent control of the biexcitonic state (discussed in the next section).

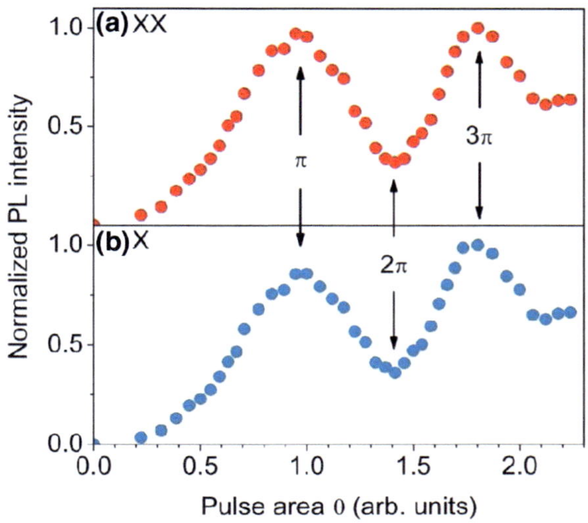

Fig. 8.28 Integrated intensities, under resonant excitation of **a** the biexcitonic emission line, and of **b** the excitonic emission line versus the square root of the pump power (proportional to the excitation pulse area). The abscissa is renormalized in π-units such that the first biexciton intensity maximum, which corresponds to the first rotation from 0 to XX in the Bloch sphere, is reached for a pulse area equal to π. Adapted by permission from Springer Nature from [28]

8.4.7.1 Coherent Control of QD States by Two-Photon Excitation

The interaction of a laser field with an atomic system (for example a two-level system) has opened the way for the observation of some of the most striking features in quantum physics, such as the Mollow Triplet [27], Rabi oscillations [152], Ramsey interferences [153] and Autler-Townes splittings [154]. These phenomena have been extensively studied and demonstrated on the two-level systems constituted of the excitonic transition (or the charged exciton) in QDs. In the pulsed regime, the phase coherence of the laser can be transmitted to the excitonic state and the final state can be controlled by the pulse area [155]. For considerations of momentum conservation, the biexcitonic three level system cannot be addressed by a single photon excitation protocol. A few decades ago, the field of atom physics took benefit of the two-photon absorption theory developed in 1931 by Goeppert-Mayer, who considered the possibility to excite a two-photon transition by placing the laser energy at half the transition energy, resonant with a virtual state [156]. In a QD biexcitonic system, this virtual state is situated between the excitonic and biexcitonic transitions. This excitation protocol is very attractive for applications because it allows for the resonant excitation of the biexciton (initial state of the entangled photon emitting cascade) with the laser energy being detuned from the exciton and biexciton transitions.

I. Pulsed Two-Photon Excitation

The quantum dot is addressed by tuning a linearly polarized pulsed laser into resonance to the virtual state of the biexciton two-photon excitation (see Fig. 8.24a). Figure 8.24c shows a QD spectrum under this two-photon excitation.

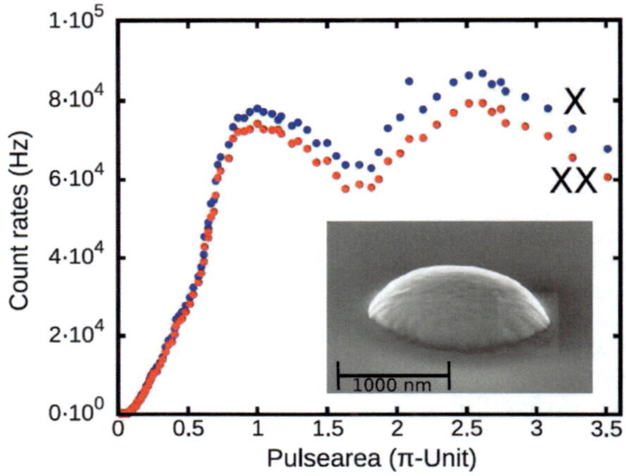

Fig. 8.29 μPL intensity of the XX transition (red) and of the X transition (blue) as a function of the two-photon resonant pulse area. Inset: scanning electron microscopy picture of a microlens. Reprinted from [151], with the permission of AIP Publishing

The intensities of the two lines are similar since only a biexciton can be prepared after each pulse, and the exciton is populated through the radiative cascade after the recombination of the biexciton. Figure 8.28 shows the excitation power dependence of the integrated intensity of the exciton and biexciton lines, respectively. Well pronounced Rabi oscillations are observed for the biexciton emission intensities (Fig. 8.28a). They reflect the coherent nature of the two-photon excitation process together with a high biexciton occupation probability. The observed damping is due to decoherence processes which might be caused by an enhanced coupling to phonons when the quantum dot state rotates faster between 0 and XX (i.e. when the pulse area is increased). The two distinctive signatures of the two-photon excitation process are the slow increase of the intensity at low power and the change of the Rabi oscillations frequency with the pump power. The first maximum of the XX intensity oscillation is considered as the most ideal working point for applications. With this scheme, the probability of generating an entangled pair after a pulse is evaluated to be close to the on-demand regime (0.9 probability) and the photon indistinguishability visibility is measured to be 0.86 [28].

Figure 8.29 shows similar results obtained with deterministic microlenses. The experiment was reproduced for two different QD microlenses. Sample QDM 1 shows a single photon-extraction efficiency of $(8.4 \pm 1.0)\%$ which leads to a photon pair efficiency (X and XX) of $(0.71 \pm 0.02)\%$. These values are comparable to the actual state of the art in the field of efficient entanglement production with semiconductor devices [143]. The achieved value is about two orders of magnitude higher than values achieved for standard self-assembled quantum dots [144, 146]. QDM 2 shows a single photon-extraction efficiency of $(6 \pm 1)\%$ (photon pair efficiency of $(0.36 \pm 0.02)\%$).

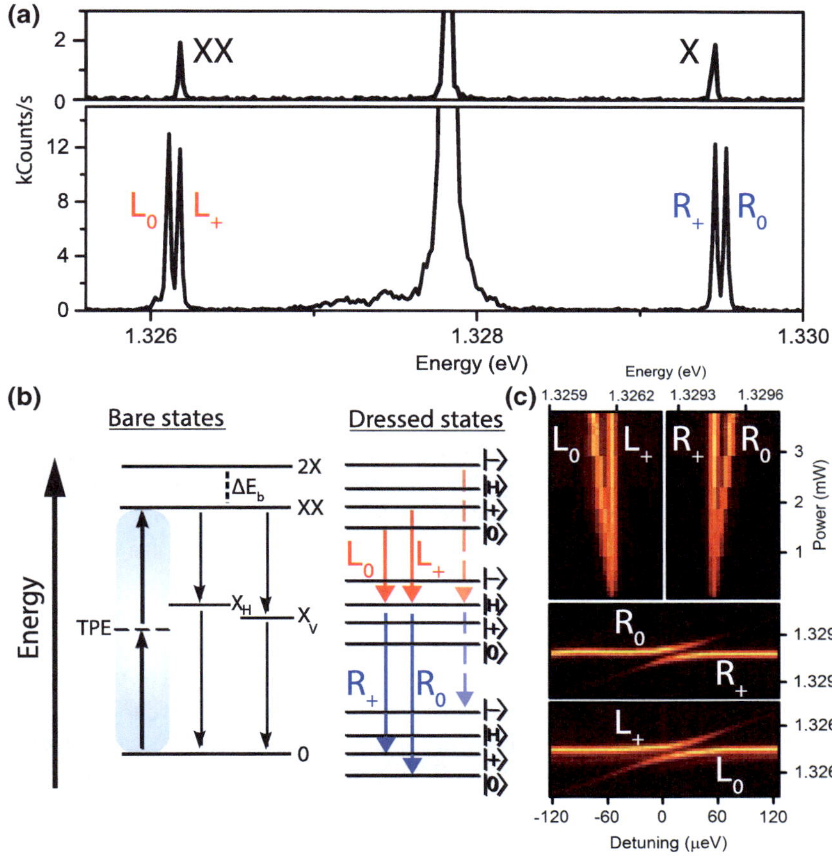

Fig. 8.30 **a** Typical emission spectra of the two-photon coherently driven QD: for an excitation power $P_o = 100\,\mu\text{W}$ (upper spectrum), for an excitation power of $P = 38\,P_o$ (lower spectrum). **b** Left: the relevant QD excitonic bare states. Right: the dressed states emerging after the two-photon laser coupling to the excitonic states. **c** Upper graph: spectra of the lines L_+, L_0, R_+ and R_0 as a function of the power. Lower graph: spectra of these lines for different laser detunings to the two-photon virtual state. Adapted with permission from [157]. Copyrighted by the American Physical Society

II. Continuous-Wave Two-Photon Excitation

The two-photon continuous wave regime is experimentally more demanding: the probability that two photons excite simultaneously the virtual state is smaller and at the same time results in a different physics for the explanation of the observed dressed steady-states. In analogy to the two-level dressing of the Mollow triplet, the dressing presented now deals with a three level "ladder" radiative cascade, giving rise to a "two-photon" ladder of three dressed states [28].

Figure 8.30a shows typical spectra of the QD $|G\rangle$-$|X\rangle$-$|XX\rangle$ under coherent two-photon CW laser driving. The spectra are displayed for two different excitation powers (upper spectrum for $P_o = 100\,\mu$W and lower spectrum for $P = 38\,P_o = 3.8$ mW respectively) (i.e. $E_l = E_X - \Delta E_b/2 = E_{XX} + \Delta E_b/2$, with E_l the laser energy, E_X the exciton energy and ΔE_b the biexciton binding energy). The laser is vertically polarized and excites exclusively the vertical excitonic state $|V\rangle$, leaving the horizontally polarized exciton state $|H\rangle$ unexcited. The resulting emission lines are pictured in the schematics of Fig. 8.30b. The two lower energy emission lines L_0 and L_+ correspond to two transitions from the dressed states to the bare biexcitonic state $|H\rangle$ (L_0 from the dressed state $|0\rangle$ and L_+ from the dressed state $|+\rangle$). $|0\rangle$ and $|+\rangle$ are constituted by a coherent superposition of the original biexciton $|XX\rangle$, exciton $|X\rangle$ and ground state $|G\rangle$. The transitions from $|H\rangle$ to the dressed states (R_0 is the transition towards $|0\rangle$ and $|+\rangle$) correspond to the two lines R_0 and R_+ on the high energy side. These transitions are shown in Fig. 8.30b. Figure 8.30c (upper panel) shows the splitting of the dressed states as a function of the cw laser excitation power. As expected for a resonantly driven system, the splitting increases with the excitation power. The energy of the dressed state $|+\rangle$ (E_+) stays unaffected, whereas the state $|0\rangle$ (E_0) is shifted to higher energies as follows: $\Delta_+ = E_+ - E_{XX} = 0$, $\Delta_0 = E_0 - E_{XX} = -\frac{1}{4}(\sqrt{\Delta E_b^2 + 32(\hbar\Omega)^2} - \Delta E_b)$.

Figure 8.30c (lower panel) shows the spectra of the lines around the original $|X\rangle$ and $|XX\rangle$ energies as the laser is swept through the two-photon resonance energy, at constant excitation power. The anticrossing observed for the two doublets is a signature of the coherent interaction between the laser and the excitonic system. By setting the laser detuning to an energy larger than 80 μeV, one can prepare a particular dressed state with a near-unity fidelity, and force the cascaded emission of the photon pairs to take a chosen path through the dressed states ladder. If the laser is detuned on the higher energy side of the two-photon excitation, the lines L_+ and R_0 are the most intense whereas L_0 and R_+ are suppressed. If the laser is detuned to the energy side, L_0 and R_+ are the privileged and L_+ and R_0 are suppressed.

By tracking the emission intensity of the dressed states as a function of the time, one can observe time domain two-photon Rabi oscillations. Figure 8.31a presents the measured emission probability of a photon L_+ or L_0 with respect to the time. These oscillations are different from the one presented in the pulsed regime case in the sense that they are a live tracking of the driven dressed states in the quantum dot. In the pulsed regime, the pulse is already passed as the state is read out and the dressed states have already relaxed towards the bare states. The lines R_+ and R_0 are not showing oscillation (Fig. 8.31b), since the coherence was already lost in the cascade. The two-photon signature consists of the linear dependence of the Rabi frequency with respect to the laser power (Fig. 8.31c).

One interesting feature of the dressed radiative cascade is the reordering of the photons through the strong coupling. The 0-X-XX cascade has a natural time ordering since the initial biexciton state always recombines before an excitonic photon can be emitted. In the case of the dressed states ladder presented on Fig. 8.30b, the photons emitted by a dressed state and the bare exciton during its transition towards the same

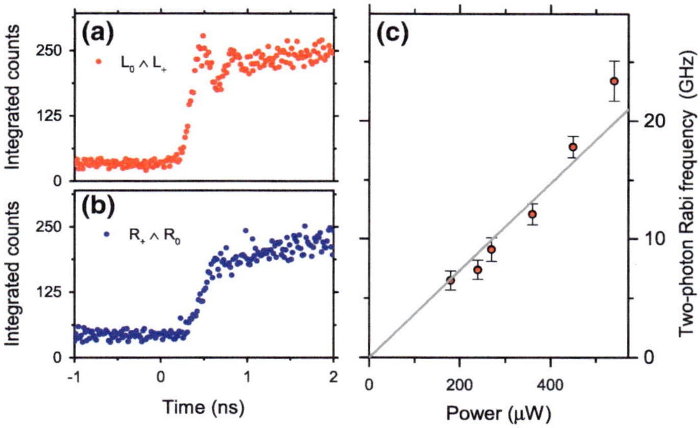

Fig. 8.31 **a** Intensity of the dressed states ($L_+ + L_0$) after the rise of the 20 ns long driving pulse, **b** intensity of the excitonic emission ($R_+ + R_0$) after the rise of the same 20 ns driving pulse, **c** Rabi frequency as a function of the laser power. Adapted with permission from [157]. Copyrighted by the American Physical Society

dressed state have no time ordering. Figure 8.32a shows the correlation between the L_0 and R_0 photons. The observed bunchings for the positive and the negative delays reveals a loss of the natural time ordering of the bare XX-X-0 ladder. On the other hand the correlations between L_0 and R_+ photons (Fig. 8.32c), involving different dressed states, show an anti-bunching for the negative delays. The system has to be repumped between the emission of the two photons and the time ordering is restored. Further control over the time reordering can be reached by tuning the driving laser energy. For example, the time reordering lost between the R_+ and L_+ photons is recovered when the laser is tuned to an energy slightly smaller than the virtual state for a shift of 60 μeV.

8.4.8 *Strain Tuning of the Emission Energy*

In-situ EBL is a very powerful technique to integrate pre-selected single QDs into nanophotonic structures as it is demonstrated and outlined in this chapter for mesas, microlenses and waveguides. Although the pre-selection happens with respect to the emission energy also, the accuracy of several 10 μeV is not sufficient to enable indistinguishable photons from different sources. Tuning mechanisms via temperature [158] (also cf. Sect. 8.4.5) and electric fields [159, 160] were successfully applied but they might be unfavorable due to increased coupling to phonons or complex doping and contacting schemes that may degrade the performance of the device. Here, strain tuning via piezo actuators has proven to be very flexible while maintaining a high optical quality [161–163].

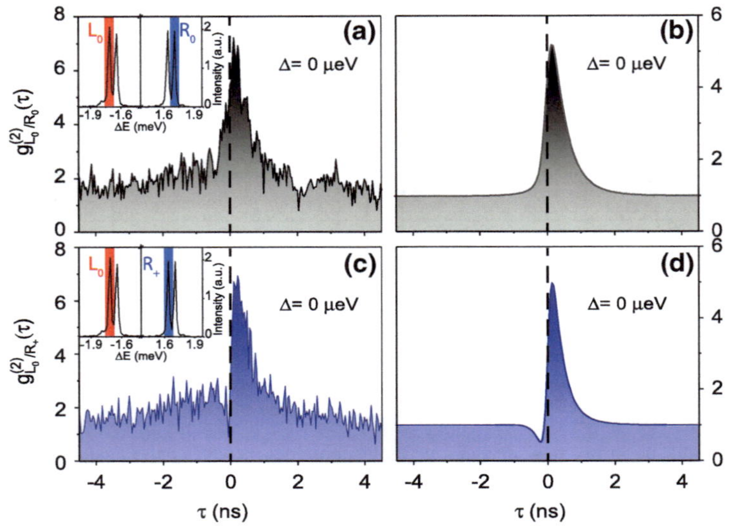

Fig. 8.32 Correlations between the dressed states. **a** Measured correlation between L_0 and R_0 for an excitation power of $P = 4.8$ mW, **b** calculated cross-correlation between L_0 and R_0 for $\Omega = \Delta E_b/13$, **c** measured cross-correlation between L_0 and R_+ for $P = 4.8$ mW, **d** calculated cross-correlation between L_0 and R_+ for $\Omega = \Delta E_b/13$. The theoretical curves have been convoluted with the experimental resolution of 140 ps (standard deviation). For the numerical model, measured lifetimes of the exciton and the biexciton ($\tau_{XX} = 314$ ps and $\tau_X = 742$ ps) were taken as fixed parameters. Adapted with permission from [157]. Copyrighted by the American Physical Society

We combine the concept of strain tuning with our deterministic in-situ EBL approach by performing a flip-chip and gold-bonding process before the actual deterministic integration of single QDs [104]. The gold layer also acts as a backside mirror [164] and replaces in this context the afore utilized AlGaAs/GaAs DBR used for the previous devices. Figure 8.33 sketches the work flow: It starts with a sample whose layer structure was optimized by performing numerical calculations with respect to the envisaged microlens structure. It also contains an AlGaAs etchstop/sacrificial layer for the subsequent flip-chip process. The sample is coated with gold and bonded onto a gold-coated piezo actuator while a temperature of 600 K and a pressure of 6 MPa are applied. Now the substrate is removed by wet chemical etching using a stirred solution of hydrogen peroxide and ammonium hydroxide and the AlGaAs layer is detached by etching with hydrochloric acid. Finally, a thin membrane with the QD layer inside remains on the piezo actuator. For the given example the membrane has a thickness of 570 nm and the QDs reside 200 nm above the gold-bond layer. Now the in-situ EBL processing is performed as described earlier in Sect. 8.3.2.

As material for the piezo actuator we have chosen PIN-PMN-PT ($Pb(In_{1/2}Nb_{1/2})O_3$-$Pb(Mg_{1/3}Nb_{2/3})O_3$-$PbTiO_3$) due to a higher coercive field strength of 6 kV/cm as compared to standard PMN-PT. The strain-tunable device is tested via μPL spectroscopy at cryogenic temperature. First of all it is of interest if the fab-

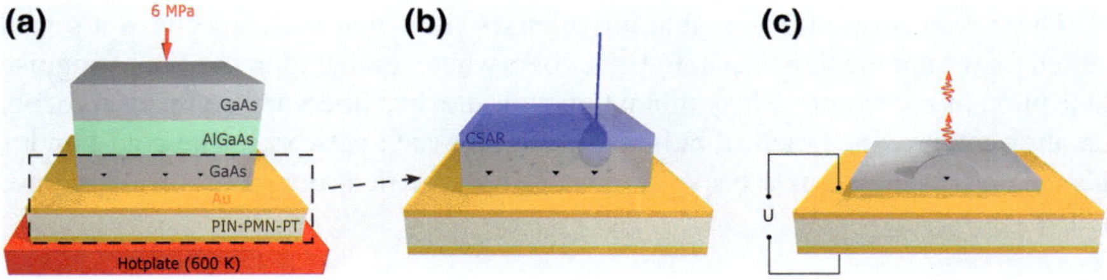

Fig. 8.33 Schematics of the fabrication process for deterministic microlenses on piezo actuators. **a** The gold-coated sample is bonded onto the piezo actuator by applying temperature and pressure. **b** After removal of the substrate, in-situ EBL is performed to fabricate microlenses with pre-selected QDs inside. **c** The finished device is ready for piezo tuning by applying a voltage to the piezo actuator

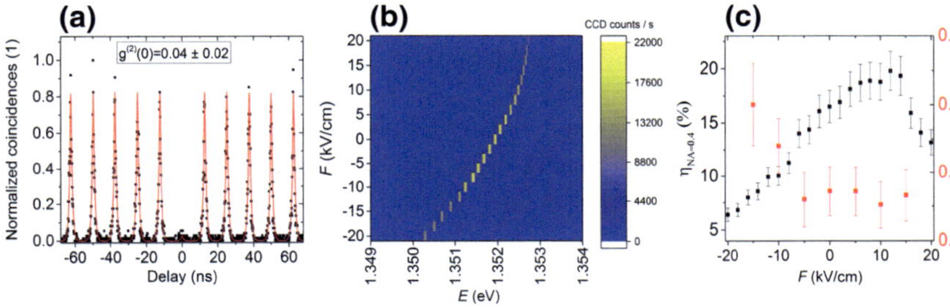

Fig. 8.34 **a** Second-order photon-autocorrelation histogram for pulsed optical excitation. **b** Demonstration of the spectral tunability which amounts to 2.5 meV. **c** Photon-extraction efficiency (η) and $g^{(2)}(\tau = 0)$ as a function of the voltage applied to the piezo actuator

rication process touches the quantum-optical emission properties of the integrated QD as it is located inside a thin membrane that has experienced several etching steps and a gold layer nearby that was heated up to 600 K. Pulsed HBT-measurements reveal a second-order photon-autocorrelation value of $g^{(2)}(\tau = 0\,\text{ns}) = 0.04 \pm 0.02$ (Fig. 8.34a) that clearly states the emission of single photons. This value remains small over a larger tuning range as shown in Fig. 8.34c, right axis. Seemingly, the device is suitable for single-photon emission. The full tuning range was determined to amount to 2.5 meV, such exceeding the QD pre-selection accuracy by far to enable multi-source energy matching. The photon-extraction efficiency was also investigated as a function of the voltage applied to the piezo actuator. It reaches a maximum value of $\eta = (20 \pm 2)\%$ and decreases for larger applied voltages or electric fields, respectively. This trend is tentatively attributed to strain-induced effects that reduce the internal oscillator strength of the integrated QD.

The comparison of the experimental results with numerical calculation in the framework of continuum elasticity, eight-band $k \cdot p$-theory and the configuration interaction method revealed that $\approx$40% of the strain from the piezo actuator is transferred to the QD.

These results on piezo-tunable microlenses show that such structures are well-suited for the application in quantum networks where remote sources of indistinguishable photons are required. In combination with attached fibers and compact cryocoolers they can provide practical building blocks for such networks. The gold-bonding layer is also suitable as a backside contact for electrical operation of microlenses [110].

8.4.9 Quantum Dot Single-Photon Sources Emitting at Telecom Wavelength

So far, results for InGaAs/GaAs-based quantum light sources in the 930 nm wavelength range were presented. This regime benefits from well-established growth techniques and the applicability of highly-efficient Si-based detectors. On the other hand side this wavelength range is not very well suited for fiber-based communication due to large absorption losses. While free-space solutions are of interest for metropolitan point-to-point links [165] or very long-distance quantum communication via satellites [166], they are hardly scalable and suffer from data losses if the line of sight is blocked. In contrast, fiber-based quantum communication is very robust and can in principle use the existing fiber-infrastructure of classical optical data communication by combining the quantum channel and the classical channel in one and the same system. In order to enable fiber-based quantum communication the sources need to match typical telecom wavelengths, i.e. the O-band around 1.3 μm with minimum dispersion or the C-band around 1.55 μm with minimum absorption losses. For long-distance (>100 km) quantum communication the C-band with the lowest absorption losses would be certainly preferable. In contrast, local area networks can also be realized using the O-band with potentially higher transmission rates due to the lower dispersion of the optical fibers in this communication band. With this background it is highly interesting to develop triggered single-photon sources emitting in the 1.3 μm wavelength range and first experiments have already demonstrated fiber-based quantum key distribution over 35 km [167].

Compared to defect related single-photon emitters in the solid state, such as NV-centers in diamond, self-assembled In(Ga)As QDs have the great advantage that the emission wavelength can be tailored to a wide extend by the material composition and process parameters during the epitaxial growth process. For instance, using the partial capping and annealing technique the emission wavelength can be blue shifted by reducing the QD height and thus increasing the quantum confinement [168]. On the other hand, the emission can be red-shifted by increasing the In content in In(Ga)As QDs. Unfortunately, this leads to a built-up of strain in the lattice mismatched InAs/GaAs material system. For emission beyond 1.1–1.2 μm strain-engineering of the low-dimensional semiconductor system is required. This can be achieved by introducing a thin strain-reducing layer (SRL) of InGaAs or InAlAs above or around InAs QDs grown on GaAs substrates in order to shift the emission wavelength of high-quality In(Ga)As QDs to 1.3 μm [169–171]. Here, the ad-layer reduces the strain in the QD system which, otherwise, could lead to a high density of non-radiative defect states limiting the quantum efficiency of the QDs. More recently, the emission wavelength of high-quality single In(Ga)As/GaAs QDs has been extended even to the 1.55 μm range by depositing the InAs QDs on InGaAs metamorphic buffer layers [172]. It should be noted that in an alternative approach researchers have succeeded in growing high-quality 1.55 μm InAs QDs on InP substrate by a applying a QD ripening technique [173, 174].

In the following we will focus our attention towards In(Ga)As QDs emitting at 1.3 μm and their deterministic integration into microstructures with enhanced photon-extraction efficiency. The layer and device design aims at future direct fiber-coupling of triggered single-photon sources at 1.3 μm. Based on numerical simulations (cf. Sect. 8.2) it turned out that simple mesa structures promise higher extraction efficiencies into low-NA optical fibers than standard microlenses, especially when combined with lateral Bragg reflectors. Therefore, as an intermediate step the micromesa geometry was prioritized and optimized for maximum extraction efficiency into free space based on a sample layout with a lower DBR section.

The respective semiconductor heterostructures were grown by MOCVD on GaAs(100) substrates. They include a lower DBR with 23 pairs of $\lambda/4$ thick $Al_{0.9}Ga_{0.1}As$/GaAs layers and $In_{0.8}Ga_{0.2}As$ QDs acting as active medium. The QDs were capped by a nominally 4 nm thick $In_{0.2}Ga_{0.8}As$ strain reducing layer followed by a 630 nm thick GaAs layer which finalizes a $\approx 2\lambda$ cavity.

As described before, in-situ EBL processing is performed to fabricate deterministic nanostructures. Figure 8.35 shows results from the fabrication process.

Figure 8.35c, d compare CL and μPL emission of one selected QD before and after integration into the micromesa. Dashed vertical lines support the assignment of the corresponding emission lines. Even though the spectral resolution during the CL measurement was rather low, it is possible to identify the corresponding patterns of emission lines.

Taking the detection efficiency of the setup into account a photon-extraction efficiency of $(10 \pm 2)\%$ was determined for the QD-micromesa under study [30]. This compares well to a numerically calculated value of about 9% for such nanostructures and is among the highest numbers reported for telecom-wavelength single-photon sources. Cavity-based devices with narrow-band enhancement showed a photon-extraction efficiency of 3.3% in case of micropillar cavities [175] and 36% for InP-based QDs in photonic crystal cavities (into NA = 0.7) [176]. Broadband approaches have resulted so far in a maximum photon-extraction efficiency of 6% using a tapered-mesa design [177]—all of which were fabricated non-deterministically. Recently, also optical in-situ lithography has been applied to realize deterministic QD-based single-photon sources with enhanced photon-extraction efficiency which, in this case, were fabricated using wet chemical etching [83].

Single-photon emission was demonstrated using a HBT setup equipped with superconducting nanowire single-photon detectors (SNSPDs). The fiber-coupled SNSPDs have a quantum efficiency of 20%, 10 dark counts/s and a temporal resolution of about 100 ps at 1.3 μm. Figure 8.36a shows a photon autocorrelation

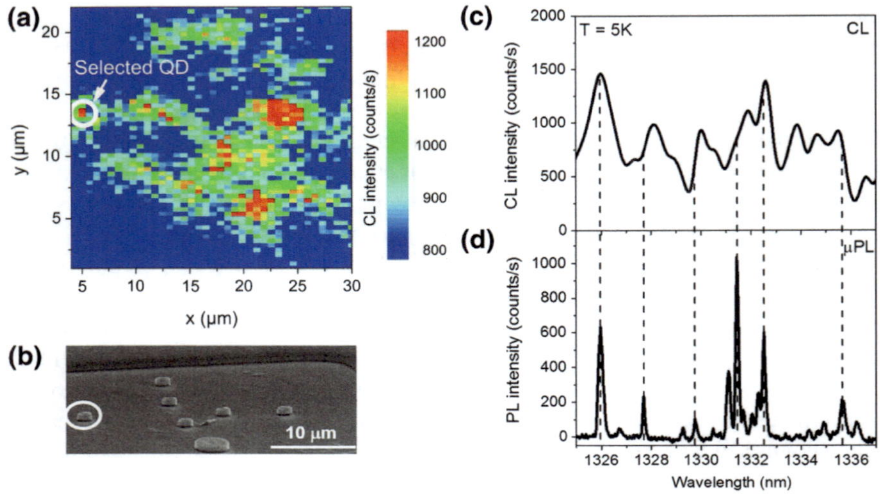

Fig. 8.35 Deterministic nanofabrication of QD-micromesas emitting in the telecom O-band. **a** Low-temperature CL intensity map showing the spatially resolved emission of InGaAs QDs and a selected QD marked by a white circle. **b** SEM image of deterministically fabricated QD-micromesas. The QD marked in (**a**) is integrated into the mesa highlighted by the white circle. **c** Low-resolution CL spectrum at the position of the selected QD in (**a**). **d** μPL emission spectrum of the fabricated QD-micromesa marked in (**b**). Reprinted with permission from [30]

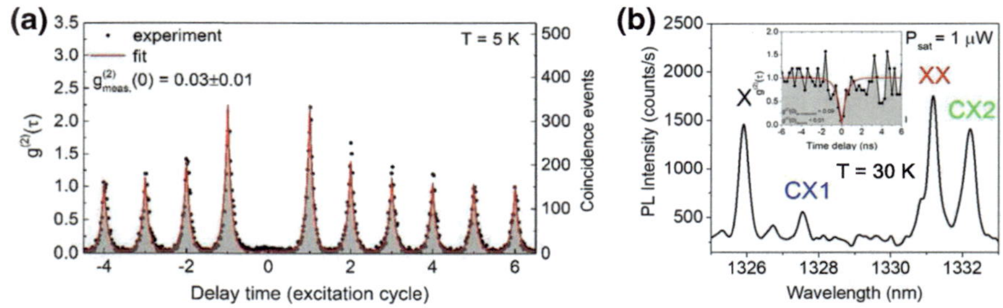

Fig. 8.36 Photon autocorrelation studies of InGaAs-QD-based single photon sources emitting at 1.3 µm. **a** Autocorrelation function of a non-deterministically fabricated QD-microlens under pulsed p-shell excitation at 5 K. **b** μPL emission spectrum and autocorrelation function of a deterministically fabricated QD-micromesa under non-resonant cw excitation at 30 K. Adapted with permission from [30] and [178]

histogram of a non-deterministically fabricated QD-microlens which is excited in its p-shell via pulsed optical excitation by a $\approx$1250 nm laser running at 76 MHz and at a sample temperature of 5 K. Multi-photon emission is strongly suppressed as indicated by the pronounced antibunching behavior with $g^{(2)}(0) = 0.03 \pm 0.01$ [178]. The strong bunching observed for the side-peaks is explained in terms of a blinking-behavior or emission-wavelength fluctuations [178]. Panel (b) (inset) demonstrates single-photon emission of a deterministically fabricated QD-micromesa under non-resonant CW excitation at 30 K. It results in a measured $g^{(2)}(0)$-value of 0.09 which corresponds to $g^{(2)}(0) = 0.01$ when taking the temporal resolution of the HBT setup into account [30]. Here, the higher temperature was chosen to demonstrate the application potential for stand-alone single-photon sources based on a compact Stirling cryocooler with a base temperature of about 30 K [123, 179] (see also Sect. 8.4.9.1). The corresponding emission spectrum shows X and XX emission of the target QD and two additional lines of non-intentionally integrated QDs. This observation is explained by the fact that the QD density in the present sample is higher than 10^8 cm^{-2} so that in addition to the pre-selected QD further QDs are unintentionally integrated within the micromesa. Further growth optimization will aim at QD densities $<10^8$ cm^{-2} to ensure that each micromesa (or microlens) contains only one pre-selected QD.

The technological and experimental progress presented in this section illustrates the great potential of InGaAs QDs to act as high-quality single-photon emitters in the telecom O-band. Further growth optimization is needed to reduce the QD density and to improve the optical and quantum optical quality in terms of quantum efficiency, multiphoton suppression and photon indistinguishability of these quantum emitters to enable the realization of high-performance triggered single-photon sources at telecom wavelengths. In parallel, the device design and fabrication technology needs to be refined to maximize the broadband photon-extraction efficiency and to pave the way for direct fiber-coupling which will be crucial for the development of stand-alone single-photon sources for practical QKD solutions.

8.4.9.1 A Plug-and-Play Single-Photon Source

While self-assembled InGaAs/GaAs QDs have a great potential for the application in quantum communication systems, their efficient utilization still requires low-temperature operation to make them superior when compared to quantum emitters operated at room temperature [180–184]. Consequently there is a high demand for a compact and easy-to-operate QD-emitters including the required cooling.

Here we present the first realization of a stand-alone and easy-to-operate quantum light source based on a fiber-coupled QD integrated within a compact Stirling cryocooler. The QD is deterministically embedded within a monolithic microlens via in-situ electron-beam lithography and precisely coupled to an optical fiber by employing an optical alignment-process and epoxide adhesive bonding at room-temperature.

For the optical alignment procedure of the fiber core to a QD-microstructure, a 80 nm-thick gold mask containing 10 μm $\times$ 10 μm apertures with a pitch of 150 μm and corresponding markers (for unambiguous identification in consecutive experiments) is defined via optical (UV) lithography and the microlenses are processed within these apertures via in-situ EBL (cf. Fig. 8.37a–c).

The positioning and gluing of the fiber is aided by active optical feedback (Fig. 8.37 d): When a target aperture is approached by the fiber, the sample is excited

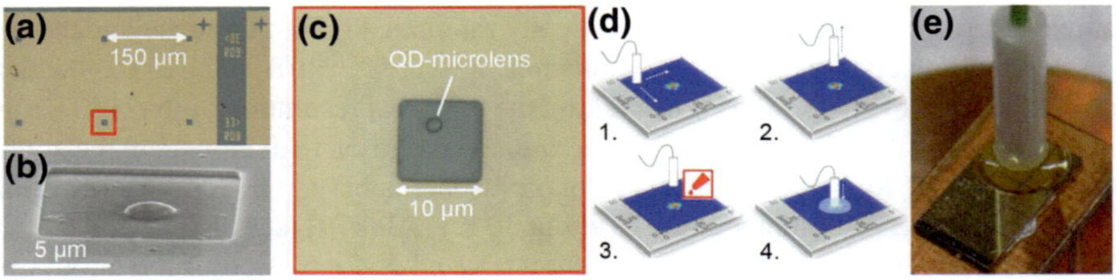

Fig. 8.37 Illustration of the process for active fiber-coupling of QD microlenses at room temperature. **a**, **b**, **c** A gold mask on the sample surface contains apertures (pitch: 150 μm, width: 10 μm) for the fabrication of microlenses via in-situ EBL. Additionally there are marker patches for easy re-allocation. **d** The fiber positioning with active optical feedback includes: locating an aperture by scanning the fiber and monitoring the GaAs luminescence (1), precise lateral and vertical positioning of the fiber above a target aperture (2), applying glue (3), bringing the fiber to the desired position and monitoring/correcting that position while the glue hardens. **e** Microscopy image of a finished device. Reprinted with permission from [123]

optically via the fiber and the fiber also collects the sample's luminescence. By observing the luminescence intensity the optimal position (lateral and vertical) can be determined. Finally the fiber is glued onto the sample while its position is monitored via optical feedback and corrected if necessary.

To test the accuracy of our procedure, we performed this fiber coupling for fiber-core diameters of 25 and 9 μm. In all cases we were able to precisely locate the positions of the apertures, suggesting an alignment accuracy better than 9 μm.

We use a commercially available and quickly reactive epoxide for gluing, which is developed for fiber connector preparation and is thus index matched to optical fibers. The index matching between fiber-core and adhesive further reduces reflection losses at the interface which will enhance the extraction efficiency. Moreover, our approach can easily be adopted for other solid-state based SPSs, such as micropillar cavities [41] and bullseye resonators [185], as well as for SPSs emitting at telecom wavelengths.

Next, we integrate the fiber-coupled QD-microlens into a compact and economic Stirling cryocooler [186] within a 19" chassis to build a stand-alone fiber-coupled SPS (see Fig. 8.38a), called 'QSource' in the following.

The Stirling cryocooler (Model: Cryotel-GT from Sunpower) provides a reasonable cooldown-time of ≈25 min, an operating temperature of ≈40 K, a temperature stability of ±5 mK, and an active vibration cancellation extension for stable operation. This makes this system very suitable for single-QD applications. The overall system performance was evaluated by measuring the second-order photon-autocorrelation function $g^{(2)}(\tau)$ for a sample temperature of $T = 40$ K. The fiber-coupled single-photon source (device 1) showed bright emission from the positively charged exciton X^+. The output fiber is coupled into a spectrometer from where the emission is directed into a fiber-based HBT setup for coincidence measurements. Figure 8.39a presents the histogram of $g^{(2)}(\tau)$ (time-bin width: 275 ps) under cw excitation at 651 nm resulting in $g^{(2)}(0) = 0.07 \pm 0.05$.

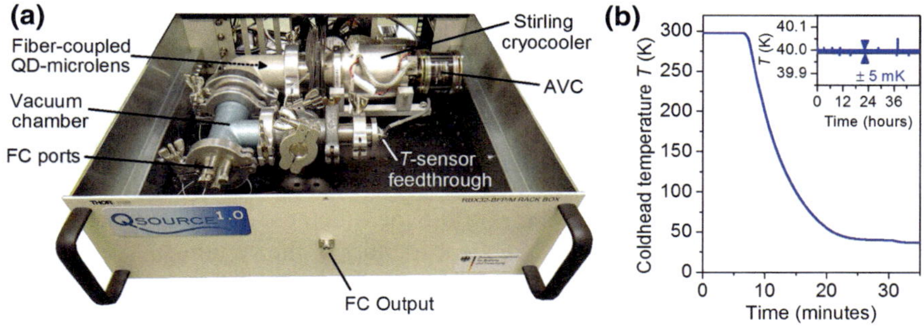

Fig. 8.38 **a** The 19" QSource module comprising a Stirling cryocooler with vacuum-feed-throughs for optical fibers and electronic connections. An active vibration cancellation (AVC) extension stabilized the system. **b** Monitoring of the cool-down cycle and of temperature stability (inset) of the cryostat's cold-head. Reprinted with permission from [123]

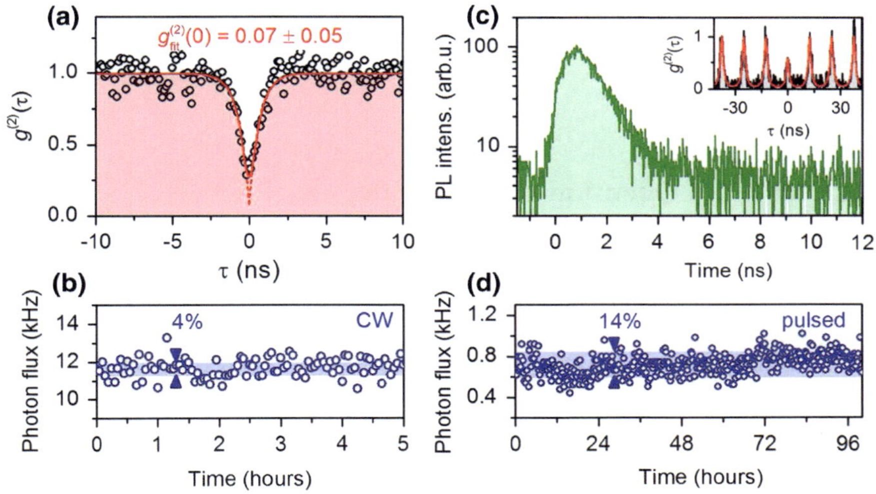

Fig. 8.39 Evaluation of the QSource system. **a** cw HBT measurement of device 1 with $g^{(2)}(0) = 0.07 \pm 0.05$ as determined by a fit to the data (dashed line). **c** Photon flux as recorded by the single-photon counting modules during the $g^{(2)}(\tau)$-measurement shown in (**a**). **b** Time-resolved measurements of device 2 under pulsed excitation at 80 MHz. Inset: $g^{(2)}(\tau)$-histogram revealing triggered non-classical light emission. **d** Photon flux of device 2 under pulsed excitation during a user-intervention-free test run over a period of 100 h. Reprinted with permission from [123]

For device 1 a mean single-photon flux at the HBT setup of 11.7 kHz was recorded. Pulsed operation of the QSource was tested at 855 nm with an excitation repetition rate of 80 MHz (pulse-width: 80 ps) using another fiber-coupled microlens—device 2 (Fig. 8.39b). The transient decay of the QD's luminescence follows a mono-exponential trace with a lifetime of (1.0 ± 0.1) ns. The corresponding $g^{(2)}(\tau)$-histogram under pulsed excitation is depicted in the inset of Fig. 8.39b, which also proves triggered emission of non-classical light. Fitting the experimental data with a model according to [44] yields antibunching with $g^{(2)}(0) = 0.57 \pm 0.05$. The non-ideal suppression of multi-photon events is attributed to uncorrelated background emission of the nearby wetting layer for this particular device. The long-term stability of the QSource system is shown in Fig. 8.39c, d. A photon flux at the pulsed HBT setup of 720 Hz (including 40 Hz dark counts) is detected with a standard deviation of 14% during the entire 100-h measurement period without any user intervention, again confirming the excellent stability of the system.

Taking into account the overall transmission of our experimental setup, the photon detection rates at the single-photon counting modules (SPCMs) yield photon-extraction efficiencies of our present demonstrator in the range of 0.28–0.39%. For device 1, we independently measured a photon-extraction efficiency of 14.4% into $NA = 0.4$ (in a free-space experiment), which is consistent with device simulations based on finite element calculations (see Supplementary of [22]). Hence, we estimate a fiber-coupling efficiency of 2.7% for device 1 used for the measurements in Fig. 8.39a, b. We anticipate significantly improved coupling efficiencies into optical fibers, by employing recently developed on-chip micro-optics [102] in future QSource modules.

8.5 On-Chip Quantum Circuits with Deterministically-Integrated Quantum Dots

The process of miniaturization and monolithic integration also enters the field of quantum circuitry for the routing and manipulation of single photons [187–190]. Bulky and extensive optical lab components will be replaced by compact on-chip circuits. Si-based on-chip photonics [191] is by far the most mature technology platform but the realization of on-chip single-photon emitters that are based on Si-technology is a greater obstacle. One way out is to fabricate Si-based low-loss waveguides and couple them in a hybrid way to e.g. InGaAs-based nano emitters [192]. Another strategy is to completely move to e.g. InGaAs/GaAs-based material. The latter approach allows for a real monolithic integration of emitters and waveguide-structures [188] (see [189] for a review).

8.5.1 Fabrication of Monolithic Waveguide Structures and an On-Chip HBT Circuit

The fabrication of complex waveguide structures is a clear advancement with respect to the previously discussed microlenses. The waveguide structures extend over a much larger distance, require much more time to be written, and surface roughness with related losses by light scattering is of greater importance than for lens structures.

Hence, the self-made in-situ EBL software was completely reworked to meet the new requirements. This way we were able to realize building blocks for on-chip optical networks and also fabricate an on-chip HBT structure with a deterministically integrated QD as single-photon emitter. The importance of a deterministic fabrication technique for waveguide circuits becomes obvious when more than one QD needs to be connected to the circuitry for e.g. quantum gate operations. An estimation of the successful fabrication of N 'nodes' with a single energy-matched QD inside was made for a multi-node waveguide circuit (see [39], supporting info for details). In short: Assuming that we have a QD density of $10^8/\mathrm{cm}^2$, a node/target area of $100\,\mathrm{nm} \times 10\,\mu\mathrm{m}$ (the QD must be located within this patch), a spectral tolerance of 1 meV, and $N = 3$ nodes, we would end up with a yield of $<10^{-3}$ for non-deterministic processing and ≈ 0.94 for our deterministic in-situ EBL technique. In other words, the ratio of success between deterministic and non-deterministic processing is ≈ 1000! Please note, if the target area was even smaller, the ratio would be a great deal larger.

For realizing the beam splitter, the multimode-interference (MMI) approach [193] is chosen. It is more robust against processing deviations than the 'classical' evanescent field coupler [194] and includes no bended sections that would increase the transmission losses [195].

The GaAs-based sample was grown by molecular beam epitaxy (MBE) on GaAs(100) substrate and consists of a GaAs waveguide region above an AlGaAs cladding layer. A low density layer ($<10^7$ cm^{-2}) of InAs QDs is grown at a vertical position corresponding to the maximum of the expected fundamental TE mode. Waveguides with a width of $w = 450$ nm for single-mode operation are fabricated by etching $d = 360$ nm deep into the GaAs matrix. By calculations it is estimated that $\approx 7\%$ of the QD emission is coupled into the waveguide mode for each of the two propagation directions at a wavelength of 900 nm. Figure 8.40 sketches the actual

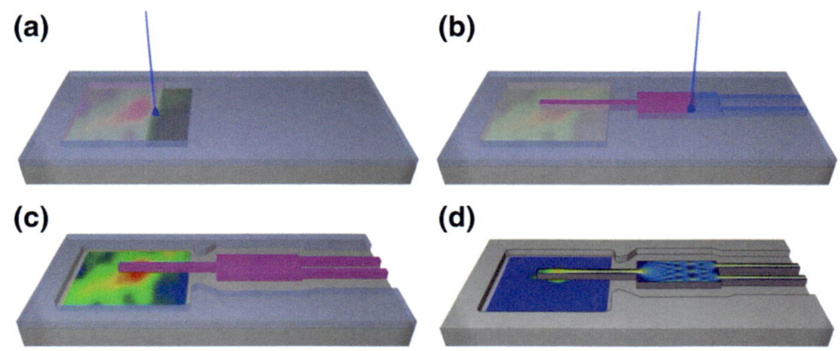

Fig. 8.40 Schematics of the deterministic in-situ EBL process for an on-chip waveguide structure with a multimode-interference (MMI) beam splitter. **a** Within a larger field-of-view a smaller area is mapped via CL. **b** The waveguide structure is written into the resist with a single pre-selected QD in its incoming terminal. **c** The resist is developed and the pattern is transferred into the semiconductor by dry etching (**d**). The MMI section in (**d**) is overlain with a finite-element simulation of the internal field distribution. Adapted with permission from [39]

in-situ EBL fabrication process of a waveguide structure including a 1×2 MMI section in analogy to the lens fabrication process (see [39] for details). Please note that special care has to be taken to get equal patterning doses inside and outside of the mapping region as that region already gained some electron dose during the QD-pre-selection step. The dimensions of the MMI sections were simulated beforehand to realize a 50:50 splitting. A width of 6.05 μm and a length of 68.88 μm were deduced as optimal parameters. The calculated transmission efficiency should amount to 97%.

A false-color SEM micrograph of the final on-chip structure is shown in Fig. 8.41a including information on the position of the QD and the mapping area. The incoming/outgoing ports of the MMI section are tapered for an efficient in/out-coupling. For measuring the routed photons, two output ports (P1 and P2) are ending at the cleaved facet of the chip.

An important property of interest is the photon loss as introduced by straight and bend waveguide sections. Here a novel idea was put into practice by fabricating U-shaped structures as depicted in Fig. 8.41b, c. The underlying concept is that a QD will couple into both transmission directions with an equal strength whereas one direction has a longer distance to its output port as compared to the other direction or additional bends, respectively.

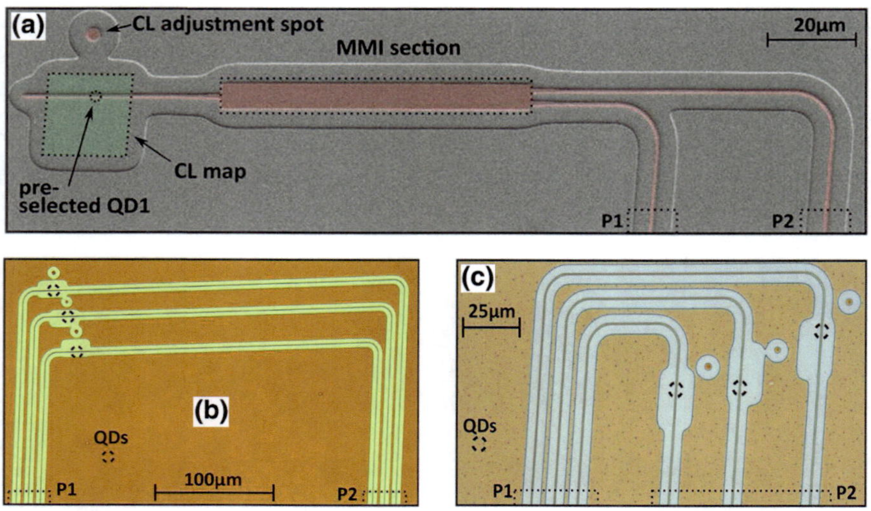

Fig. 8.41 **a** False-color SEM micrograph of the full structure with the pre-selected QD, the MMI beam splitter, and the two output port P1 and P2. The area of CL mapping and an adjustment spot for setting up the CL/SEM optics are marked, too. U-shaped structures for determining transmission losses (**b**) and bend losses (**c**). Adapted with permission from [39]

8.5.1.1 Basic Structural and Optical Characterization

For the optical characterization of the waveguide structures a μPL setup was extended to allow for excitation/detection from the top (standard configuration for all other experiments in this chapter) and simultaneous excitation/detection from the side [196]. For this the cryostat was equipped with an additional side window and the additional optical paths were carefully adjusted to allow for an optimal efficient and simultaneous operation of top and side paths. A 20x optical microscope objective as well as a 50x objective are used and a 785 nm cw diode laser was attached for basic characterization (see [39] for details). A polarization dependent measurement of the QD's luminescence from the waveguide's facet is presented in Fig. 8.42a. It clearly reveals a 98% TE polarization, e.g., monomode waveguiding and at the same time an efficient coupling of the QD to the TE mode. Next, the transmission losses are of interest. By investigating a number of U-shaped structures as presented in Fig. 8.41b the propagation loss can be determined as (12.1 ± 6.2)dB/mm. This value is larger than reported by others for this material system [195, 197]. This deviation is mainly attributed to a larger sidewall roughness due to the liquid-He EBL step and a not optimal dry etching recipe. The bend loss measured by utilizing the structures as shown in Fig. 8.41c equals (1.11 ± 0.69)dB/bend which is comparable to values reported in the works cited above.

The splitting ratio of the MMI beam splitter is determined by measuring absolute emission intensities from the two output ports P1 and P2. The obtained data is presented in Fig. 8.42b. Within the accuracy of the measurement a 50:50 splitting is confirmed. The transmission through the MMI was measured to amount to $61 \pm 15\%$. The deviation from the calculated optimal value of 97% is attributed to a measured deviation in length of $\Delta L = 0.4$ μm (0.6%).

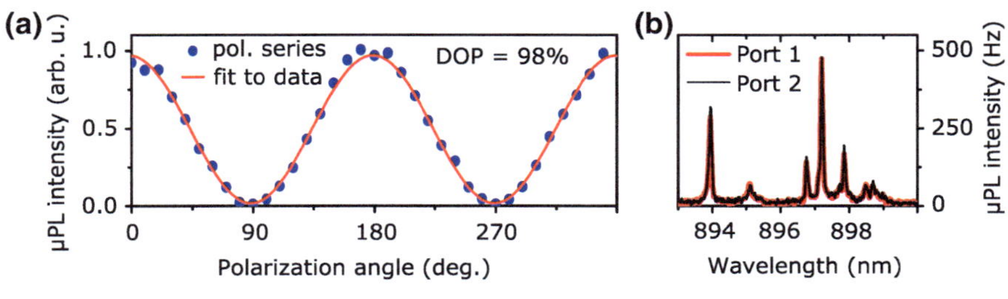

Fig. 8.42 **a** Polarization resolved emission from the output ports that demonstrates that there is single-mode TE transmission only (DOP: degree of polarization). **b** As-measured spectra from both output ports that clearly reveal an optimal 50:50 beam splitting within the MMI section. Adapted with permission from [39]

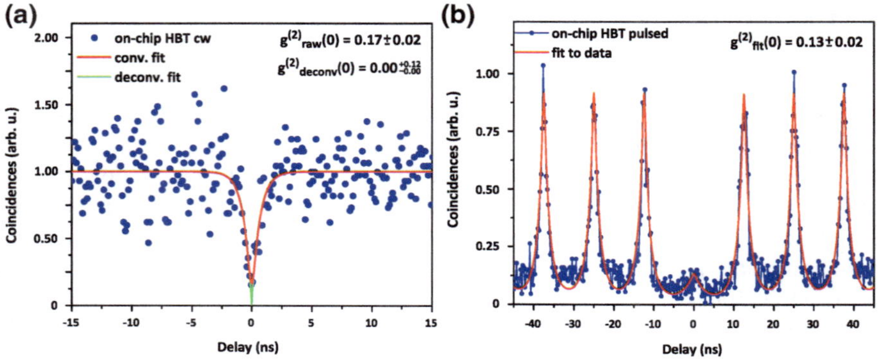

Fig. 8.43 HBT measurements on both output ports for **a** cw and **b** pulsed optical excitation together with fit functions to reveal the actual $g^{(2)}(\tau = 0)$ values. Adapted with permission from [39]

8.5.1.2 Verification of HBT Functionality

We now focus on whether the fabricated structure preserves the single-photon emission capability of the integrated QD and works as an on-chip HBT circuit. For this the QD was excited by a cw and pulsed laser, respectively (cf. Sect. 8.4.4) and the respective auto-correlation histograms are shown in Fig. 8.43. In case of cw excitation a $g^{(2)}(0)$ value of $0.00^{+0.12}_{-0.00}$ is extracted by fitting the data. For pulsed excitation it amounts to $0.13^{+0.02}_{-0.02}$. Thus the functionality of our device is confirmed. The non-vanishing counts at $\tau = 0$ are attributed to carrier recapture processes [102, 198] and/or uncorrelated spectator emitters somewhere in the waveguide.

These results nicely demonstrate that deterministic in-situ EBL is also very suitable for the realization of larger (networks of) on-chip structures with pre-selected QDs as emitters of non-classical light. Future work will target towards the integration of logical on-chip gates and on-chip photon detectors.

8.6 Conclusion and Outlook

This chapter introduced a cutting-edge approach for the realization of nanophotonic structures for quantum communication and on-chip photonic circuits. For this a novel and deterministic device fabrication technology was developed based on a combination of low-temperature cathodoluminescence spectroscopy (CL) and in-situ electron-beam lithography (EBL). The first fabricated devices include microlens structures with single quantum dots (QDs) inside and on-chip waveguide structures based on the InGaAs/GaAs(100) material system. Most experiments were performed in the 930 nm wavelength regime while first results also showed a successful technology transfer into the telecom O-band around 1300 nm. The low-temperature characteristics of EBL resists were thoroughly investigated and suitable ones were identified that operate both in a positive- and negative-tone regime. The quantum-optical properties of the fabricated devices were comprehensively analyzed by micro-photoluminescence (μPL) techniques including Hanbury-Brown and Twiss (HBT) and Hong-Ou-Mandel (HOM) two-photon interference experiments to test for the purity of single-photon emission and the degree of indistinguishability of the emitted photons, respectively. The emission of polarization-entangled photon pairs from the biexciton-exciton recombination cascade was proven by quantum state tomography. The presented results compare very well with state-of-the-art achievements world-wide and demonstrate that e.g. microlenses present a versatile solution with respect to photon-extraction efficiency, single-photon emission, indistinguishability of emitted photons, broad-band operation, fiber coupling, strain tuning, and the ability to be electrically operational. Last but not least an estimation of the device yield clearly favors such a deterministic processing technology over a non-deterministic one (cf. Sect. 8.5).

Still those structures and results are capable of further development. Numerical calculations in the frame-work of a full 3D finite-element approach have shown (cf. Sect. 8.2) that the photon-extraction efficiency into free space and, more important, into optical fibers can be improved. The epitaxial source material can further be optimized to minimize detrimental influences of fluctuating electric fields to enhance the photon indistinguishability. Besides a mere optimization of the crystal quality also internally applied electric fields via delta doping [199] could facilitate even better quantum properties. Another appealing approach is the utilization of QDs grown on GaAs(111) substrates to predictably reduce the fine-structure splitting of the biexciton/exciton recombination cascade for the assured generation of polarization-entangled photon pairs [200].

Telecom wavelength regimes can be made accessible further as well (e.g. telecom C-band around 1550 nm) by direct fiber coupling with single-mode optical fibers [31]. Both are crucial for the practical realization and application of earthbound long-distance quantum communication schemes [1]. Multi-node quantum repeater networks [2] for the secure distribution of crypto keys is on top of the list to fight cybercrime and theft of data. Here, also the integration of quantum memories is crucial for proper node functionality [201, 202].

So far we dealt with the integration of single QDs as emitters of non-classical light. When it comes to on-chip circuitry for optical data processing and quantum computing [190, 203], also the realization of logic elements and the deterministic integration of on-chip detectors is highly desirable [189]. Our approach is also suitable for this. It only requires the integration of more advances EBL hardware like a specimen stage with laser interferometry and stichless writing.

Another appealing approach is the combination of our technology platform with atomic systems [202, 204] and dielectric circuits [192, 205, 206] to build up hybrid systems that benefit from the best of all worlds. Also the transfer to other (semiconductor-based) material systems like nitrides, selenides, phosphides, and single-photon emitting (2D) transition metal dichalcogenides [207] is on the agenda.

Acknowledgements Many people contributed to this work including (in alphabetical order) A. Carmele, S. Fischbach, L. Gantz, D. Gershoni, J. Große, M. Gschrey, C. Heine, L. Heindel, M. von Helversen, T. Heuser, T. Höhne, C. Hopfmann, A. Kaganskiy, A. Knorr, A. Musiał, D. Quandt, J. Schall, C. Scharfenorth, A. Schlehahn, E. Schmidgall, F. Schmidt, M. Schmidt, R. Schmidt, P. Schnauber, J.-H. Schulze, M. Seifried, G. Sek, N. Srocka, M. Strauß, A. Strittmatter, A. Thoma, B. Wohlfeil, U. Woggon, and J. Wolters. We are grateful for their support and skills. The research leading to these results has received funding from the German Research Foundation via CRC 787, Re2974/8-1 and Re2974/12-1, from the European Research Council under the European Union's Seventh Framework ERC Grant Agreement No. 615613, the German Federal Ministry of Education and Research (BMBF) through the VIP-project QSOURCE (Grant No. 03V0630), from the European Regional Development Fund (EFRE) of the European Union in the framework of the programme to promote research, innovation, and technologies (*Pro* FIT) via the project FI-SEQUR (Grant No. 10160387), from the German-Israeli-Foundation for Scientific Research and Development, Grant-No.: 1148-77.14/2011, and from the project EMPIR 14IND05 MIQC2 (the EMPIR initiative is co-funded by the European Union's Horizon 2020 research and innovation programme and the EMPIR Participating States).

References

1. N. Sangouard, C. Simon, J. Minář, H. Zbinden, H. de Riedmatten, N. Gisin, Phys. Rev. A **76**, 050301 (2007). https://doi.org/10.1103/PhysRevA.76.050301
2. N. Gisin, R. Thew, Nat. Photon. **1**, 165 (2007). https://doi.org/10.1038/nphoton.2007.22
3. H.J. Kimble, Nature **453**, 1023 (2008). https://doi.org/10.1038/nature07127
4. N. Gisin, G. Ribordy, W. Tittel, H. Zbinden, Rev. Mod. Phys. **74**, 145 (2002). https://doi.org/10.1103/RevModPhys.74.145
5. W.K. Wootters, W.H. Zurek, Nature **299**, 802 (1982). https://doi.org/10.1038/299802a0
6. C.H. Bennett, G. Brassard, in *Proceedings of IEEE International Conference on Computers, Systems and Signal Processing* (Bangalore, India, 1984), pp. 175–179. https://doi.org/10.1016/j.tcs.2014.05.025
7. M. Sasaki, M. Fujiwara, H. Ishizuka, W. Klaus, K. Wakui, M. Takeoka, S. Miki, T. Yamashita, Z. Wang, A. Tanaka, K. Yoshino, Y. Nambu, S. Takahashi, A. Tajima, A. Tomita, T. Domeki, T. Hasegawa, Y. Sakai, H. Kobayashi, T. Asai, K. Shimizu, T. Tokura, T. Tsurumaru, M. Matsui, T. Honjo, K. Tamaki, H. Takesue, Y. Tokura, J.F. Dynes, A.R. Dixon, A.W. Sharpe, Z.L. Yuan, A.J. Shields, S. Uchikoga, M. Legré, S. Robyr, P. Trinkler, L. Monat, J.B. Page, G. Ribordy, A. Poppe, A. Allacher, O. Maurhart, T. Länger, M. Peev, A. Zeilinger, Opt. Express **19**, 10387 (2011). https://doi.org/10.1364/OE.19.010387
8. H.K. Lo, M. Curty, K. Tamaki, Nat. Photon. **8**, 595 (2014). https://doi.org/10.1038/nphoton.2014.149
9. H.K. Lo, M. Curty, B. Qi, Phys. Rev. Lett. **108**, 130503 (2012). https://doi.org/10.1103/PhysRevLett.108.130503
10. Z.S. Yuan, Y.A. Chen, B. Zhao, S. Chen, J. Schmiedmayer, J.W. Pan, Nature **454**, 1098 (2008). https://doi.org/10.1038/nature07241
11. A.J. Shields, Nat. Photon. **1**, 215 (2007). https://doi.org/10.1038/nphoton.2007.46
12. I. Aharonovich, D. Englund, M. Toth, Nat. Photon. **10**, 631 (2016). https://doi.org/10.1038/nphoton.2016.186
13. P. Michler, A. Kiraz, C. Becher, W.V. Schoenfeld, P.M. Petroff, L. Zhang, E. Hu, A. Imamoğlu, Science **290**, 2282 (2000). https://doi.org/10.1126/science.290.5500.2282
14. W. Barnes, G. Björk, J. Gérard, P. Jonsson, J. Wasey, P. Worthing, V. Zwiller, Eur. Phys. J. D **18**, 197 (2002). https://doi.org/10.1140/epjd/e20020024
15. V. Zwiller, T. Aichele, O. Benson, New J. Phys. **6**, 96 (2004). https://doi.org/10.1088/1367-2630/6/1/096

16. N. Somaschi, V. Giesz, L. De Santis, J.C. Loredo, M.P. Almeida, G. Hornecker, S.L. Portalupi, T. Grange, C. Antón, J. Demory, C. Gómez, I. Sagnes, N.D. Lanzillotti-Kimura, A. Lemaítre, A. Auffeves, A.G. White, L. Lanco, P. Senellart, Nat. Photon. (2016). https://doi.org/10.1038/NPHOTON.2016.23
17. S. Unsleber, Y.M. He, S. Gerhardt, S. Maier, C.Y. Lu, J.W. Pan, N. Gregersen, M. Kamp, C. Schneider, S. Höfling, Opt. Express **24**, 8539 (2016). https://doi.org/10.1364/oe.24.008539
18. J. Claudon, J. Bleuse, N.S. Malik, M. Bazin, P. Jaffrennou, N. Gregersen, C. Sauvan, P. Lalanne, J.M. Gérard, Nat. Photon. **4**, 174 (2010). https://doi.org/10.1038/NPHOTON.2009.287x
19. M.E. Reimer, G. Bulgarini, N. Akopian, M. Hocevar, M.B. Bavinck, M.A. Verheijen, E.P. Bakkers, L.P. Kouwenhoven, V. Zwiller, Nat. Commun. **3**, 737 (2012). https://doi.org/10.1038/ncomms1746
20. F.W. Ostermayer, Appl. Phys. Lett. **43**, 642 (1983). https://doi.org/10.1063/1.94461
21. Y.S. Kim, J. Kim, J.S. Choe, Y.G. Rob, H. Jeon, J. Woo, IEEE Photon Technol. Lett. **12**, 507 (2000). https://doi.org/10.1109/68.841268
22. M. Gschrey, A. Thoma, P. Schnauber, M. Seifried, R. Schmidt, B. Wohlfeil, L. Krüger, J.H. Schulze, T. Heindel, S. Burger, F. Schmidt, A. Strittmatter, S. Rodt, S. Reitzenstein, Nat. Commun. **6**, 7662 (2015). https://doi.org/10.1038/ncomms8662
23. L. Sapienza, M. Davanço, A. Badolato, K. Srinivasan, Nat. Commun. **6**, 7833 (2015). https://doi.org/10.1038/ncomms8833
24. O. Benson, C. Santori, M. Pelton, Y. Yamamoto, Phys. Rev. Lett. **84**, 2513 (2000). https://doi.org/10.1103/PhysRevLett.84.2513
25. S. Strauf, N.G. Stoltz, M.T. Rakher, L.A. Coldren, P.M. Petroff, D. Bouwmeester, Nat. Photon. **1**, 704 (2007). https://doi.org/10.1038/nphoton.2007.227
26. A. Thoma, P. Schnauber, M. Gschrey, M. Seifried, J. Wolters, J.H. Schulze, A. Strittmatter, S. Rodt, A. Carmele, A. Knorr, T. Heindel, S. Reitzenstein, Phys. Rev. Lett. **116**, (2016). https://doi.org/10.1103/PhysRevLett.116.033601
27. A. Muller, E.B. Flagg, P. Bianucci, X.Y. Wang, D.G. Deppe, W. Ma, J. Zhang, G.J. Salamo, M. Xiao, C.K. Shih, Phys. Rev. Lett. **99**, 187402 (2007). https://doi.org/10.1103/PhysRevLett.99.187402
28. M. Müller, S. Bounouar, K.D. Jöns, M. Glässl, P. Michler, Nat. Photon. **8**, 224 (2014). https://doi.org/10.1038/nphoton.2013.377
29. K. Takemoto, Y. Nambu, T. Miyazawa, Y. Sakuma, T. Yamamoto, S. Yorozu, Y. Arakawa, Sci. Rep. **5**, 14383 (2015). https://doi.org/10.1038/srep14383
30. N. Srocka, A. Musiał, P.I. Schneider, P. Mrowiński, P. Holewa, S. Burger, D. Quandt, A. Strittmatter, S. Rodt, S. Reitzenstein, G. Sęk, AIP Adv. **8**, 085205 (2018). https://doi.org/10.1063/1.5038137
31. P.I. Schneider, N. Srocka, S. Rodt, L. Zschiedrich, S. Reitzenstein, S. Burger, Opt. Express **26**, 8479 (2018). https://doi.org/10.1364/OE.26.008479
32. P.I. Schneider, X.G. Santiago, V. Soltwisch, M. Hammerschmidt, S. Burger, C. Rockstuhl, ACS Photonics **6**, 2726 (2019). https://doi.org/10.1021/acsphotonics.9b00706
33. L. Zschiedrich, H. Greiner, S. Burger, F. Schmidt, Proc. SPIE **8641**, 86410B (2013). https://doi.org/10.1117/12.2001132
34. J.A. Nelder, R. Mead, Comput. J. **7**, 308 (1965). https://doi.org/10.1093/comjnl/7.4.308
35. Y. Zhang, S. Wang, G. Ji, Math. Probl. Eng. **2015**, 931256 (2015). https://doi.org/10.1155/2015/931256
36. S. Das, P.N. Suganthan, I.E.E.E. Trans, Evol. Comput. **15**, 4 (2011). https://doi.org/10.1109/TEVC.2010.2059031
37. B. Shahriari, K. Swersky, Z. Wang, R.P. Adams, N. De Freitas, Proc. IEEE **104**, 148 (2016). https://doi.org/10.1109/JPROC.2015.2494218
38. C.K. Williams, in *Learning in Graphical Models* (Springer, 1998), pp. 599–621. https://doi.org/10.1007/978-94-011-5014-9_23
39. P. Schnauber, J. Schall, S. Bounouar, T. Höhne, S.I. Park, G.H. Ryu, T. Heindel, S. Burger, J.D. Song, S. Rodt, S. Reitzenstein, Nano Lett. **18**, 2336 (2018). https://doi.org/10.1021/acs.nanolett.7b05218

40. Z. Yuan, B.E. Kardynal, R.M. Stevenson, A.J. Shields, C.J. Lobo, K. Cooper, N.S. Beattie, D.A. Ritchie, M. Pepper, Science **295**, 102 (2002). https://doi.org/10.1126/science.1066790
41. T. Heindel, C. Schneider, M. Lermer, S.H. Kwon, T. Braun, S. Reitzenstein, S. Höfling, M. Kamp, A. Forchel, Appl. Phys. Lett. **96**, 011107 (2010). https://doi.org/10.1063/1.3284514
42. F. Hargart, C.A. Kessler, T. Schwarzbäck, E. Koroknay, S. Weidenfeld, M. Jetter, P. Michler, Appl. Phys. Lett. **102**, 011126 (2013). https://doi.org/10.1063/1.4774392
43. A. Boretti, L. Rosa, A. Mackie, S. Castelletto, Adv. Opt. Mat. **3**, 1012 (2015). https://doi.org/10.1002/adom.201500022
44. A. Schlehahn, A. Thoma, P. Munnelly, M. Kamp, S. Höfling, T. Heindel, C. Schneider, S. Reitzenstein, APL Photon. **1**, 011301 (2016). https://doi.org/10.1063/1.4939831
45. M. Kantner, M. Mittnenzweig, T. Koprucki, Phys. Rev. B **96**, 205301 (2017). https://doi.org/10.1103/PhysRevB.96.205301
46. M. Kantner, M. Mittnenzweig, T. Koprucki, Proc. SPIE **10526**, 1052603 (2018). https://doi.org/10.1117/12.2289185
47. M. Kantner, Proc. SPIE **10912**, 109120U (2019). https://doi.org/10.1117/12.2515209
48. H.P. Breuer, F. Petruccione, in *The Theory of Open Quantum Systems* (Oxford University Press, Oxford, 2002). https://doi.org/10.1093/acprof:oso/9780199213900.001.0001
49. S. Selberherr, in *Analysis and Simulation of Semiconductor Devices* (Springer, Vienna, 1984). https://doi.org/10.1007/978-3-7091-8752-4
50. A. Wojs, P. Hawrylak, S. Fafard, L. Jacak, Phys. Rev. B **54**, 5604 (1996). https://doi.org/10.1103/physrevb.54.5604
51. H. Haug, S.W. Koch, in *Quantum Theory of the Optical and Electronic Properties of Semiconductors*, 4th edn. (World Scientific, Singapore, 2004). https://doi.org/10.1142/5394
52. M. Kantner, U. Bandelow, T. Koprucki, J.H. Schulze, A. Strittmatter, H.J. Wünsche, IEEE Trans. Electron. Dev. **63**, 2036 (2016). https://doi.org/10.1109/ted.2016.2538561
53. R. Ferreira, G. Bastard, in *Capture and Relaxation in Self-Assembled Semiconductor Quantum Dots* (Morgan & Claypool Publishers, San Rafael, CA, 2015), pp. 2053–2571. https://doi.org/10.1088/978-1-6817-4089-8
54. V. Palankovski, R. Quay, in *Analysis and Simulation of Heterostructure Devices*. Series in Computational Microelectronics (Springer, Vienna, 2004). https://doi.org/10.1007/978-3-7091-0560-3
55. M. Sotoodeh, A.H. Khalid, A.A. Rezazadeh, J. Appl. Phys. **87**, 2890 (2000). https://doi.org/10.1063/1.372274
56. S. Reitzenstein, IEEE, J. Sel. Top. Quantum Electron. **18**, 1733 (2012). https://doi.org/10.1109/JSTQE.2012.2195159
57. M. Bayer, G. Ortner, O. Stern, A. Kuther, A.A. Gorbunov, A. Forchel, P. Hawrylak, S. Fafard, K. Hinzer, T.L. Reinecke, S.N. Walck, J.P. Reithmaier, F. Klopf, F. Schäfer, Phys. Rev. B **65**, 195315 (2002). https://doi.org/10.1103/PhysRevB.65.195315
58. Y.M. He, Y. He, Y.J. Wei, D. Wu, M. Atatüre, C. Schneider, S. Höfling, M. Kamp, C.Y. Lu, J.W. Pan, Nat. Nanotechnol. **8**, 213 (2013). https://doi.org/10.1038/nnano.2012.262
59. A. Dousse, L. Lanco, J. Suffczyński, E. Semenova, A. Miard, A. Lemaître, I. Sagnes, C. Roblin, J. Bloch, P. Senellart, Phys. Rev. Lett. **101**, 267404 (2008). https://doi.org/10.1103/PhysRevLett.101.267404
60. M. Gschrey, F. Gericke, A. Schüßler, R. Schmidt, J.H. Schulze, T. Heindel, S. Rodt, A. Strittmatter, S. Reitzenstein, Appl. Phys. Lett. **102**, 251113 (2013). https://doi.org/10.1063/1.4812343
61. S. Tamura, G. Koike, A. Komatsubara, T. Teraji, S. Onoda, L.P. McGuinness, L. Rogers, B. Naydenov, E. Wu, L. Yan, F. Jelezko, T. Ohshima, J. Isoya, T. Shinada, T. Tanii, Appl. Phys. Expr. **7**(11), 115201 (2014). https://doi.org/10.7567/APEX.7.115201
62. T. Schröder, M.E. Trusheim, M. Walsh, L. Li, J. Zheng, M. Schukraft, A. Sipahigil, R.E. Evans, D.D. Sukachev, C.T. Nguyen, J.L. Pacheco, R.M. Camacho, E.S. Bielejec, M.D. Lukin, D. Englund, Nat. Commun. **8**, 15376 (2017). https://doi.org/10.1038/ncomms15376
63. M. Radulaski, M. Widmann, M. Niethammer, J.L. Zhang, S.Y. Lee, T. Rendler, K.G. Lagoudakis, N.T. Son, E. Janzén, T. Ohshima, J. Wrachtrup, J. Vučković, Nano Lett. **17**, 1782 (2017). https://doi.org/10.1021/acs.nanolett.6b05102

64. Y. Zhou, Z. Mu, G. Adamo, S. Bauerdick, A. Rudzinski, I. Aharonovich, W.b. Gao, New J. Phys. **20**(12), 125004 (2018). https://doi.org/10.1088/1367-2630/aaf2ac
65. A. Badolato, K. Hennessy, M. Atatüre, J. Dreiser, E. Hu, P.M. Petroff, A. Imamoğlu, Science **308**, 1158 (2005). https://doi.org/10.1126/science.1109815
66. K. Hennessy, A. Badolato, M. Winger, D. Gerace, M. Atatüre, S. Gulde, S. Fält, E.L. Hu, A. Imamoğlu, Nature **445**, 896 (2007). https://doi.org/10.1038/nature05586
67. K.H. Lee, F.S.F. Brossard, M. Hadjipanayi, X. Xu, F. Waldermann, A.M. Green, D.N. Sharp, A.J. Turberfield, D.A. Williams, R.A. Taylor, Nanotechnology **19**, 455307 (2008). https://doi.org/10.1088/0957-4484/19/45/455307
68. S.M. Thon, M.T. Rakher, H. Kim, J. Gudat, W.T.M. Irvine, P.M. Petroff, D. Bouwmeester, Appl. Phys. Lett. **94**, 111115 (2009). https://doi.org/10.1063/1.3103885
69. M. Pfeiffer, K. Lindfors, P. Atkinson, A. Rastelli, O.G. Schmidt, H. Giessen, M. Lippitz, Phys. Stat. Sol. b **249**, 678 (2012). https://doi.org/10.1002/pssb.201100788
70. T. Kojima, K. Kojima, T. Asano, S. Noda, Appl. Phys. Lett. **102**, 011110 (2013). https://doi.org/10.1063/1.4773882
71. G. Nogues, Q. Merotto, G. Bachelier, E. Hye Lee, J. Dong Song, Appl. Phys. Lett. **102**, 231112 (2013). https://doi.org/10.1063/1.4809831
72. M. Pfeiffer, K. Lindfors, H. Zhang, B. Fenk, F. Phillipp, P. Atkinson, A. Rastelli, O.G. Schmidt, H. Giessen, M. Lippitz, Nano Lett. **14**, 197 (2014). https://doi.org/10.1021/nl403730q
73. M. Jeannin, P. Rueda-Fonseca, E. Bellet-Amalric, K. Kheng, G. Nogues, Nanotechnology **27**, 185201 (2016). https://doi.org/10.1088/0957-4484/27/18/185201
74. Y.M. He, J. Liu, S. Maier, M. Emmerling, S. Gerhardt, M. Davanco, K. Srinivasan, C. Schneider, S. Höfling, Optica **4**, 802 (2017). https://doi.org/10.1364/OPTICA.4.000802
75. M. Sartison, S.L. Portalupi, T. Gissibl, M. Jetter, H. Giessen, P. Michler, Sci. Rep. **7**, 39916 (2017). https://doi.org/10.1038/srep39916
76. I.E. Zadeh, A.W. Elshaari, K.D. Jöns, A. Fognini, D. Dalacu, P.J. Poole, M.E. Reimer, V. Zwiller, Nano Lett. **16**, 2289 (2016). https://doi.org/10.1021/acs.nanolett.5b04709
77. D. Cadeddu, J. Teissier, F.R. Braakman, N. Gregersen, P. Stepanov, J.M. Gérard, J. Claudon, R.J. Warburton, M. Poggio, M. Munsch, Appl. Phys. Lett. **108**, 011112 (2016). https://doi.org/10.1063/1.4939264
78. T. van der Sar, J. Hagemeier, W. Pfaff, E.C. Heeres, S.M. Thon, H. Kim, P.M. Petroff, T.H. Oosterkamp, D. Bouwmeester, R. Hanson, Appl. Phys. Lett. **98**, 193103 (2011). https://doi.org/10.1063/1.3571437
79. A. Muller, E.B. Flagg, M. Metcalfe, J. Lawall, G.S. Solomon, Appl. Phys. Lett. **95**, 173101 (2009). https://doi.org/10.1063/1.3245311
80. M. Gschrey, R. Schmidt, A. Kaganskiy, S. Rodt, S. Reitzenstein, J. Vac. Sci. Technol. B **32**, 061601 (2014). https://doi.org/10.1116/1.4896671
81. P. Schnauber, R. Schmidt, A. Kaganskiy, T. Heuser, M. Gschrey, S. Rodt, S. Reitzenstein, Nanotechnology **27**, 195301 (2016). https://doi.org/10.1088/0957-4484/27/19/195301
82. A. Kaganskiy, T. Heuser, R. Schmidt, S. Rodt, S. Reitzenstein, J. Vac. Sci. Technol. B **34**, 061603 (2016). https://doi.org/10.1116/1.4965883
83. M. Sartison, L. Engel, S. Kolatschek, F. Olbrich, C. Nawrath, S. Hepp, M. Jetter, P. Michler, S.L. Portalupi, Appl. Phys. Lett. **113**, 032103 (2018). https://doi.org/10.1063/1.5038271
84. K. Sawicki, F.K. Malinowski, K. Gałkowski, T. Jakubczyk, P. Kossacki, W. Pacuski, J. Suffczyński, Appl. Phys. Lett. **106**, 012101 (2015). https://doi.org/10.1063/1.4905306
85. F. Donatini, L.S. Dang, Nanotechnology **21**, 375303 (2010). https://doi.org/10.1088/0957-4484/21/37/375303
86. M. Gschrey, R. Schmidt, J.H. Schulze, A. Strittmatter, S. Rodt, S. Reitzenstein, J. Vac. Sci. Technol. B **33**, 021603 (2015). https://doi.org/10.1116/1.4914914
87. T. Heindel, S. Rodt, S. Reitzenstein, Single-photon sources based on deterministic quantum-dot microlenses (Springer International Publishing, Cham, 2017), pp. 199–232. https://doi.org/10.1007/978-3-319-56378-7_6
88. A. Kaganskiy, M. Gschrey, A. Schlehahn, R. Schmidt, J.H. Schulze, T. Heindel, A. Strittmatter, S. Rodt, S. Reitzenstein, Rev. Sci. Instr. **86**, 073903 (2015). https://doi.org/10.1063/1.4926995

89. V. Zwiller, G. Björk, J. of Appl. Phys. **92**, 660 (2002). https://doi.org/10.1063/1.1487913
90. Y. Ma, G. Ballesteros, J.M. Zajac, J. Sun, B.D. Gerardot, Opt. Lett. **40**, 2373 (2015). https://doi.org/10.1364/OL.40.002373
91. S. Moehl, H. Zhao, B.D. Don, S. Wachter, H. Kalt, J. Appl. Phys. **93**, 6265 (2003). https://doi.org/10.1063/1.1567035
92. Y. Chen, M. Zopf, R. Keil, F. Ding, O.G. Schmidt, Nat. Commun. **9**, 2994 (2018). https://doi.org/10.1038/s41467-018-05456-2
93. P. Schnauber, A. Thoma, C.V. Heine, A. Schlehahn, L. Gantz, M. Gschrey, R. Schmidt, C. Hopfmann, B. Wohlfeil, J.H. Schulze, A. Strittmatter, T. Heindel, S. Rodt, U. Woggon, D. Gershoni, S. Reitzenstein, Technologies **4**, 1 (2016). https://doi.org/10.3390/technologies4010001
94. M. Munsch, N.S. Malik, E. Dupuy, A. Delga, J. Bleuse, J.M. Gérard, J. Claudon, N. Gregersen, J. Mørk, Phys. Rev. Lett. **110**, 177402 (2013). https://doi.org/10.1103/PhysRevLett.110.177402
95. J.H. Kim, Y.H. Ko, S.H. Gong, S.M. Ko, Y.H. Cho, Sci. Rep. **3**, (2013). https://doi.org/10.1038/srep02150
96. O. Gazzano, S. Michaelis de Vasconcellos, C. Arnold, A. Nowak, E. Galopin, I. Sagnes, L. Lanco, A. Lemaître, P. Senellart, Nat. Commun. **4**, 1425 (2013). https://doi.org/10.1038/ncomms2434
97. J.H. Kim, T. Cai, C.J.K. Richardson, R.P. Leavitt, E. Waks, Optica **3**, 577 (2016). https://doi.org/10.1364/OPTICA.3.000577
98. X. Ding, Y. He, Z.C. Duan, N. Gregersen, M.C. Chen, S. Unsleber, S. Maier, C. Schneider, M. Kamp, S. Höfling, C.Y. Lu, J.W. Pan, Phys. Rev. Lett. **116**, (2016). https://doi.org/10.1103/PhysRevLett.116.020401
99. M. Pelton, C. Santori, J. Vučković, B. Zhang, G.S. Solomon, J. Plant, Y. Yamamoto, Phys. Rev. Lett. **89**, 233602 (2002). https://doi.org/10.1103/PhysRevLett.89.233602
100. K. Madsen, S. Ates, J. Liu, A. Javadi, S. Albrecht, I. Yeo, S. Stobbe, P. Lodahl, Phys. Rev. B **90**, 155303 (2014). https://doi.org/10.1103/PhysRevB.90.155303
101. S. Maier, P. Gold, A. Forchel, N. Gregersen, J. Mørk, S. Höfling, C. Schneider, M. Kamp, Opt. Express **22**, 8136 (2014). https://doi.org/10.1364/OE.22.008136
102. S. Fischbach, A. Schlehahn, A. Thoma, N. Srocka, T. Gissibl, S. Ristok, S. Thiele, A. Kaganskiy, A. Strittmatter, T. Heindel, S. Rodt, A. Herkommer, H. Giessen, S. Reitzenstein, ACS Photon. **4**, 1327 (2017). https://doi.org/10.1021/acsphotonics.7b00253
103. A. Schlehahn, M. Gaafar, M. Vaupel, M. Gschrey, P. Schnauber, J.H. Schulze, S. Rodt, A. Strittmatter, W. Stolz, A. Rahimi-Iman, T. Heindel, M. Koch, S. Reitzenstein, Appl. Phys. Lett. **107**, 041105 (2015). https://doi.org/10.1063/1.4927429
104. M. Schmidt, M.V. Helversen, S. Fischbach, A. Kaganskiy, R. Schmidt, A. Schliwa, T. Heindel, S. Rodt, S. Reitzenstein, Opt. Mater. Express **10**, 76 (2020). https://doi.org/10.1364/OME.10.000076
105. H.J. Briegel, W. Dür, J.I. Cirac, P. Zoller, Phys. Rev. Lett. **81**, 5932 (1998). https://doi.org/10.1103/PhysRevLett.81.5932
106. J. Hofmann, M. Krug, N. Ortegel, L. Gérard, M. Weber, W. Rosenfeld, H. Weinfurter, Science **337**, 72 (2012). https://doi.org/10.1126/science.1221856
107. Z.L. Yuan, A.R. Dixon, J.F. Dynes, A.W. Sharpe, A.J. Shields, Appl. Phys. Lett. **92**, 201104 (2008). https://doi.org/10.1063/1.2931070
108. B. Heinen, T. Wang, M. Sparenberg, A. Weber, B. Kunert, J. Hader, S.W. Koch, J.V. Moloney, M. Koch, W. Stolz, Electron. Lett. **48**, 516 (2012). https://doi.org/10.1049/el.2012.0531
109. M. Gaafar, C. Moller, M. Wichmann, B. Heinen, B. Kunert, A. Rahimi-Iman, W. Stolz, M. Koch, Electron. Lett. **50**, 542 (2014). https://doi.org/10.1049/el.2014.0157
110. A. Schlehahn, R. Schmidt, C. Hopfmann, J.H. Schulze, A. Strittmatter, T. Heindel, L. Gantz, E.R. Schmidgall, D. Gershoni, S. Reitzenstein, Appl. Phys. Lett. **108**, 021104 (2016). https://doi.org/10.1063/1.4939658
111. C.K. Hong, Z.Y. Ou, L. Mandel, Phys. Rev. Lett. **59**, 2044 (1987). https://doi.org/10.1103/PhysRevLett.59.2044

112. C. Santori, D. Fattal, J. Vučković, G.S. Solomon, Y. Yamamoto, Nature **419**, 594 (2002). https://doi.org/10.1038/nature01086
113. J. Bylander, I. Robert-Philip, I. Abram, Eur. Phys. J. D **22**, 295 (2003). https://doi.org/10.1140/epjd/e2002-00236-6
114. V. Türck, S. Rodt, O. Stier, R. Heitz, R. Engelhardt, U.W. Pohl, D. Bimberg, R. Steingrüber, Phys. Rev. B **61**, 9944 (2000). https://doi.org/10.1103/physrevb.61.9944
115. Y.M. Galperin, B.L. Altshuler, J. Bergli, D.V. Shantsev, Phys. Rev. Lett. **96**, 097009 (2006). https://doi.org/10.1103/physrevlett.96.097009
116. A. Delteil, Z. Sun, W.b. Gao, E. Togan, S. Faelt, A. Imamoglu, Nat. Phys. **12**, 218 (2015). https://doi.org/10.1038/nphys3605
117. R.B. Patel, A.J. Bennett, I. Farrer, C.A. Nicoll, D.A. Ritchie, A.J. Shields, Nat Photon **4**, 632 (2010). https://doi.org/10.1038/nphoton.2010.161
118. E.B. Flagg, A. Muller, S.V. Polyakov, A. Ling, A. Migdall, G.S. Solomon, Phys. Rev. Lett. **104**, 137401 (2010). https://doi.org/10.1103/PhysRevLett.104.137401
119. W. Gao, P. Fallahi, E. Togan, A. Delteil, Y. Chin, J. Miguel-Sanchez, A. Imamoğlu, Nat. Commun. **4**, (2013). https://doi.org/10.1038/ncomms3744
120. P. Gold, A. Thoma, S. Maier, S. Reitzenstein, C. Schneider, S. Höfling, M. Kamp, Phys. Rev. B **89**, 035313 (2014). https://doi.org/10.1103/PhysRevB.89.035313
121. V. Giesz, S.L. Portalupi, T. Grange, C. Antón, L. De Santis, J. Demory, N. Somaschi, I. Sagnes, A. Lemaître, L. Lanco, A. Auffèves, P. Senellart, Phys. Rev. B **92**, (2015). https://doi.org/10.1103/physrevb.92.161302
122. A. Thoma, P. Schnauber, J. Böhm, M. Gschrey, J.H. Schulze, A. Strittmatter, S. Rodt, T. Heindel, S. Reitzenstein, Appl. Phys. Lett. **110**, 011104 (2017). https://doi.org/10.1063/1.4973504
123. A. Schlehahn, S. Fischbach, R. Schmidt, A. Kaganskiy, A. Strittmatter, S. Rodt, T. Heindel, S. Reitzenstein, Sci. Rep. **8**, 1340 (2018). https://doi.org/10.1038/s41598-017-19049-4
124. N. Akopian, N.H. Lindner, E. Poem, Y. Berlatzky, J. Avron, D. Gershoni, B.D. Gerardot, P.M. Petroff, Phys. Rev. Lett. **96**, 130501 (2006). https://doi.org/10.1103/PhysRevLett.96.130501
125. G. Callsen, A. Carmele, G. Hönig, C. Kindel, J. Brunnmeier, M.R. Wagner, E. Stock, J.S. Reparaz, A. Schliwa, S. Reitzenstein, A. Knorr, A. Hoffmann, S. Kako, Y. Arakawa, Phys. Rev. B **87**(24), 245314 (2013). https://doi.org/10.1103/PhysRevB.87.245314
126. S. Rodt, R. Heitz, A. Schliwa, R.L. Sellin, F. Guffarth, D. Bimberg, Phys. Rev. B **68**, 035331 (2003). https://doi.org/10.1103/PhysRevB.68.035331
127. R. Seguin, A. Schliwa, S. Rodt, K. Pötschke, U.W. Pohl, D. Bimberg, Phys. Rev. Lett. **95**, 257402 (2005). https://doi.org/10.1103/PhysRevLett.95.257402
128. T. Heindel, A. Thoma, M. von Helversen, M. Schmidt, A. Schlehahn, M. Gschrey, P. Schnauber, J.H. Schulze, A. Strittmatter, J. Beyer, S. Rodt, A. Carmele, A. Knorr, S. Reitzenstein, Nat. Commun. **8**, 14870 (2017). https://doi.org/10.1038/ncomms14870
129. S. Rodt, A. Schliwa, K. Pötschke, F. Guffarth, D. Bimberg, Phys. Rev. B **71**, 155325 (2005). https://doi.org/10.1103/PhysRevB.71.155325
130. D.V. Regelman, U. Mizrahi, D. Gershoni, E. Ehrenfreund, W.V. Schoenfeld, P.M. Petroff, Phys. Rev. Lett. **87**, 257401 (2001). https://doi.org/10.1103/PhysRevLett.87.257401
131. A. Kiraz, S. Fälth, C. Becher, B. Gayral, W.V. Schoenfeld, P.M. Petroff, L. Zhang, E. Hu, A. Imamoğlu, Phys. Rev. B **65**, 161303 (2002). https://doi.org/10.1103/PhysRevB.65.161303
132. J.K. Thompson, Science **313**, 74 (2006). https://doi.org/10.1126/science.1127676
133. M. Schmidt, M. von Helversen, M. López, F. Gericke, E. Schlottmann, T. Heindel, S. Kück, S. Reitzenstein, J. Beyer, J. Low Temp. Phys. (2018). https://doi.org/10.1007/s10909-018-1932-1
134. S. Moroni, S. Varo, G. Juska, T. Chung, A. Gocalinska, E. Pelucchi, J. Cryst. Growth **506**, 36 (2019). https://doi.org/10.1016/j.jcrysgro.2018.10.005
135. J.S. Bell, Physics **1**, 195 (1964). https://doi.org/10.1103/PhysicsPhysiqueFizika.1.195
136. B. Hensen, H. Bernien, A.E. Dréau, A. Reiserer, N. Kalb, M.S. Blok, J. Ruitenberg, R.F.L. Vermeulen, R.N. Schouten, C. Abellán, W. Amaya, V. Pruneri, M.W. Mitchell, M. Markham, D.J. Twitchen, D. Elkouss, S. Wehner, T.H. Taminiau, R. Hanson, Nature **526**, 682 (2015). https://doi.org/10.1038/nature15759

137. M. Giustina, M.A.M. Versteegh, S. Wengerowsky, J. Handsteiner, A. Hochrainer, K. Phelan, F. Steinlechner, J. Kofler, J.Å. Larsson, C. Abellán, W. Amaya, V. Pruneri, M.W. Mitchell, J. Beyer, T. Gerrits, A.E. Lita, L.K. Shalm, S.W. Nam, T. Scheidl, R. Ursin, B. Wittmann, A. Zeilinger, Phys. Rev. Lett. **115**, 250401 (2015). https://doi.org/10.1103/PhysRevLett.115.250401
138. L.K. Shalm, E. Meyer-Scott, B.G. Christensen, P. Bierhorst, M.A. Wayne, M.J. Stevens, T. Gerrits, S. Glancy, D.R. Hamel, M.S. Allman, K.J. Coakley, S.D. Dyer, C. Hodge, A.E. Lita, V.B. Verma, C. Lambrocco, E. Tortorici, A.L. Migdall, Y. Zhang, D.R. Kumor, W.H. Farr, F. Marsili, M.D. Shaw, J.A. Stern, C. Abellán, W. Amaya, V. Pruneri, T. Jennewein, M.W. Mitchell, P.G. Kwiat, J.C. Bienfang, R.P. Mirin, E. Knill, S.W. Nam, Phys. Rev. Lett. **115**, 250402 (2015). https://doi.org/10.1103/PhysRevLett.115.250402
139. A. Einstein, B. Podolsky, N. Rosen, Phys. Rev. **47**, 777 (1935). https://doi.org/10.1103/PhysRev.47.777
140. R.M. Stevenson, R.J. Young, P. Atkinson, K. Cooper, D.A. Ritchie, A.J. Shields, Nature **439**, 179 (2006). https://doi.org/10.1038/nature04446
141. R. Hafenbrak, S.M. Ulrich, P. Michler, L. Wang, A. Rastelli, O.G. Schmidt, New J. Phys. **9**, 315 (2007). https://doi.org/10.1088/1367-2630/9/9/315
142. E. Stock, T. Warming, I. Ostapenko, S. Rodt, A. Schliwa, J.A. Töfflinger, A. Lochmann, A.I. Toropov, S.A. Moshchenko, D.V. Dmitriev, V.A. Haisler, D. Bimberg, Appl. Phys. Lett. **96**, 093112 (2010). https://doi.org/10.1063/1.3337097
143. M.A.M. Versteegh, M.E. Reimer, K.D. Jöns, D. Dalacu, P.J. Poole, A. Gulinatti, A. Giudice, V. Zwiller, Nat. Commun. **5**, 5298 (2014). https://doi.org/10.1038/ncomms6298
144. D. Huber, M. Reindl, Y. Huo, H. Huang, J.S. Wildmann, O.G. Schmidt, A. Rastelli, R. Trotta, Nat. Commun. **8**, 15506 (2017). https://doi.org/10.1038/ncomms15506
145. T. Kuroda, T. Mano, N. Ha, H. Nakajima, H. Kumano, B. Urbaszek, M. Jo, M. Abbarchi, Y. Sakuma, K. Sakoda, I. Suemune, X. Marie, T. Amand, Phys. Rev. B **88**, 041306 (2013). https://doi.org/10.1103/PhysRevB.88.041306
146. G. Juska, V. Dimastrodonato, L.O. Mereni, A. Gocalinska, E. Pelucchi, Nat. Photon. **7**, 527 (2013). https://doi.org/10.1038/nphoton.2013.128
147. J. Zhang, J.S. Wildmann, F. Ding, R. Trotta, Y. Huo, E. Zallo, D. Huber, A. Rastelli, and O. G. Schmidt, Nat. Commun. **6**, 10067 (2015). https://doi.org/10.1038/ncomms10067
148. A.J. Bennett, M.A. Pooley, R.M. Stevenson, M.B. Ward, R.B. Patel, A.B. de la Giroday, N. Skold, I. Farrer, C.A. Nicoll, D.A. Ritchie, A.J. Shields, Nat. Phys. **6**, 947 (2010). https://doi.org/10.1038/nphys1780
149. R.M. Stevenson, R.J. Young, P. See, D.G. Gevaux, K. Cooper, P. Atkinson, I. Farrer, D.A. Ritchie, A.J. Shields, Phys. Rev. B **73**, 033306 (2006). https://doi.org/10.1103/PhysRevB.73.033306
150. R.M. Stevenson, A.J. Hudson, A.J. Bennett, R.J. Young, C.A. Nicoll, D.A. Ritchie, A.J. Shields, Phys. Rev. Lett. **101**, 170501 (2008). https://doi.org/10.1103/PhysRevLett.101.170501
151. S. Bounouar, C. de la Haye, M. Strauß, P. Schnauber, A. Thoma, M. Gschrey, J.H. Schulze, A. Strittmatter, S. Rodt, S. Reitzenstein, Appl. Phys. Lett. **112**, 153107 (2018). https://doi.org/10.1063/1.5020242
152. H. Kamada, H. Gotoh, J. Temmyo, T. Takagahara, H. Ando, Phys. Rev. Lett. **87**, 246401 (2001). https://doi.org/10.1103/PhysRevLett.87.246401
153. H. Jayakumar, A. Predojević, T. Huber, T. Kauten, G.S. Solomon, G. Weihs, Phys. Rev. Lett. **110**, 135505 (2013). https://doi.org/10.1103/physrevlett.110.135505
154. X. Xu, B. Sun, P.R. Berman, D.G. Steel, A.S. Bracker, D. Gammon, L.J. Sham, Science **317**, 929 (2007). https://doi.org/10.1126/science.1142979
155. N.H. Bonadeo, Science **282**, 1473 (1998). https://doi.org/10.1126/science.282.5393.1473
156. M. Goeppert-Mayer, Ann. Phys. 273–295 (1931). https://doi.org/10.1002/andp.19314010303
157. S. Bounouar, M. Strauß, A. Carmele, P. Schnauber, A. Thoma, M. Gschrey, J.H. Schulze, A. Strittmatter, S. Rodt, A. Knorr, S. Reitzenstein, Phys. Rev. Lett. **118**, 233601 (2017). https://doi.org/10.1103/PhysRevLett.118.233601

158. T. Farrow, P. See, A.J. Bennett, M.B. Ward, P. Atkinson, K. Cooper, D.J.P. Ellis, D.C. Unitt, D.A. Ritchie, A.J. Shields, Nanotechnology **19**, 345401 (2008). https://doi.org/10.1088/0957-4484/19/34/345401
159. W. Heller, U. Bockelmann, G. Abstreiter, Phys. Rev. B **57**, 6270 (1998). https://doi.org/10.1103/PhysRevB.57.6270
160. C. Kistner, T. Heindel, C. Schneider, A. Rahimi-Iman, S. Reitzenstein, S. Höfling, A. Forchel, Opt. Express **16**, 15006 (2008). https://doi.org/10.1364/OE.16.015006
161. F. Ding, R. Singh, J.D. Plumhof, T. Zander, V. Křápek, Y.H. Chen, M. Benyoucef, V. Zwiller, K. Dörr, G. Bester, A. Rastelli, O.G. Schmidt, Phys. Rev. Lett. **104**, 067405 (2010). https://doi.org/10.1103/PhysRevLett.104.067405
162. R. Trotta, P. Atkinson, J.D. Plumhof, E. Zallo, R.O. Rezaev, S. Kumar, S. Baunack, J.R. Schröter, A. Rastelli, O.G. Schmidt, Adv. Mat. **24**, 2668 (2012). https://doi.org/10.1002/adma.201200537
163. R. Trotta, J. Martń-Sánchez, J.S. Wildmann, G. Piredda, M. Reindl, C. Schimpf, E. Zallo, S. Stroj, J. Edlinger, A. Rastelli, Nat. Commun. **7**, 10375 (2016). https://doi.org/10.1038/ncomms10375
164. S. Fischbach, A. Kaganskiy, E.B.Y. Tauscher, F. Gericke, A. Thoma, R. Schmidt, A. Strittmatter, T. Heindel, S. Rodt, S. Reitzenstein, Appl. Phys. Lett. **111**, 011106 (2017). https://doi.org/10.1063/1.4991389
165. T. Heindel, C.A. Kessler, M. Rau, C. Schneider, M. Fürst, F. Hargart, W.M. Schulz, M. Eichfelder, R. Roßbach, S. Nauerth, M. Lermer, H. Weier, M. Jetter, M. Kamp, S. Reitzenstein, S. Höfling, P. Michler, H. Weinfurter, A. Forchel, New J. Phys. **14**, 083001 (2012). https://doi.org/10.1088/1367-2630/14/8/083001
166. S.K. Liao, W.Q. Cai, W.Y. Liu, L. Zhang, Y. Li, J.G. Ren, J. Yin, Q. Shen, Y. Cao, Z.P. Li, F.Z. Li, X.W. Chen, L.H. Sun, J.J. Jia, J.C. Wu, X.J. Jiang, J.F. Wang, Y.M. Huang, Q. Wang, Y.L. Zhou, L. Deng, T. Xi, L. Ma, T. Hu, Q. Zhang, Y.A. Chen, N.L. Liu, X.B. Wang, Z.C. Zhu, C.Y. Lu, R. Shu, C.Z. Peng, J.Y. Wang, J.W. Pan, Nature **549**, 43 (2017). https://doi.org/10.1038/nature23655
167. P.M. Intallura, M.B. Ward, O.Z. Karimov, Z.L. Yuan, P. See, A.J. Shields, P. Atkinson, D.A. Ritchie, Appl. Phys. Lett. **91**, 161103 (2007). https://doi.org/10.1063/1.2799756
168. L. Wang, A. Rastelli, O.G. Schmidt, J. Appl. Phys. **100**, 064313 (2006). https://doi.org/10.1063/1.2349432
169. K. Nishi, H. Saito, S. Sugou, J.S. Lee, Appl. Phys. Lett. **74**, 1111 (1999). https://doi.org/10.1063/1.123459
170. J. Bloch, J. Shah, W.S. Hobson, J. Lopata, S.N.G. Chu, Appl. Phys. Lett. **75**, 2199 (1999). https://doi.org/10.1063/1.124963
171. V.M. Ustinov, N.A. Maleev, A.E. Zhukov, A.R. Kovsh, A.Y. Egorov, A.V. Lunev, B.V. Volovik, I.L. Krestnikov, Y.G. Musikhin, N.A. Bert, P.S. Kop'ev, Z.I. Alferov, N.N. Ledentsov, D. Bimberg, Appl. Phys. Lett. **74**, 2815 (1999). https://doi.org/10.1063/1.124023
172. M. Paul, F. Olbrich, J. Höschele, S. Schreier, J. Kettler, S.L. Portalupi, M. Jetter, P. Michler, Appl. Phys. Lett. **111**, 033102 (2017). https://doi.org/10.1063/1.4993935
173. M. Yacob, J.P. Reithmaier, M. Benyoucef, Appl. Phys. Lett. **104**, 022113 (2014). https://doi.org/10.1063/1.4861940
174. A. Kors, J.P. Reithmaier, M. Benyoucef, Appl. Phys. Lett. **112**, 172102 (2018). https://doi.org/10.1063/1.5023184
175. Z.S. Chen, B. Ma, X.J. Shang, H.Q. Ni, J.L. Wang, Z.C. Niu, Nanoscale Res. Lett. **12**, 378 (2017). https://doi.org/10.1186/s11671-017-2153-2
176. J.-H. Kim, T. Cai, C. J. K. Richardson, R. P. Leavitt, and E. Waks, Optica. **3**, 577 (2016). https://doi.org/10.1364/OPTICA.3.000577
177. T. Usuki, Y. Sakuma, S. Hirose, K. Takemoto, N. Yokoyama, T. Miyazawa, M. Takatsu, Y. Arakawa, J. Phys.: Conf. Ser. **38**, 140 (2006). https://doi.org/10.1088/1742-6596/38/1/034
178. Ł. Dusanowski, P. Holewa, A. Maryński, A. Musiał, T. Heuser, N. Srocka, D. Quandt, A. Strittmatter, S. Rodt, J. Misiewicz, S. Reitzenstein, G. Sęk, Opt. Express **25**, 31122 (2017). https://doi.org/10.1364/OE.25.031122

179. A. Schlehahn, L. Krüger, M. Gschrey, J.H. Schulze, S. Rodt, A. Strittmatter, T. Heindel, S. Reitzenstein, Rev. Sci. Instr. **86**, 013113 (2015). https://doi.org/10.1063/1.4906548
180. P. Michler, A. Imamoğlu, M.D. Mason, P.J. Carson, G.F. Strouse, S.K. Buratto, Nature **406**, 968 (2000). https://doi.org/10.1038/35023100
181. C. Kurtsiefer, S. Mayer, P. Zarda, H. Weinfurter, Phys. Rev. Lett. **85**, 290 (2000). https://doi.org/10.1103/PhysRevLett.85.290
182. B. Lounis, W.E. Moerner, Nature **407**, 491 (2000). https://doi.org/10.1038/35035032
183. M.J. Holmes, K. Choi, S. Kako, M. Arita, Y. Arakawa, Nano Lett. **14**, 982 (2014). https://doi.org/10.1021/nl404400d
184. S. Deshpande, T. Frost, A. Hazari, P. Bhattacharya, Appl. Phys. Lett. **105**, 141109 (2014). https://doi.org/10.1063/1.4897640
185. S. Ates, L. Sapienza, M. Davanco, A. Badolato, K. Srinivasan, IEEE, J. Sel. Top. Quantum Electron. **18**, 1711 (2012). https://doi.org/10.1109/JSTQE.2012.2193877
186. A. Veprik, S. Riabzev, G. Vilenchik, N. Pundak, Cryogenics **45**, 117 (2005). https://doi.org/10.1016/j.cryogenics.2004.05.007
187. J. Carolan, C. Harrold, C. Sparrow, E. Martín-López, N.J. Russell, J.W. Silverstone, P.J. Shadbolt, N. Matsuda, M. Oguma, M. Itoh, G.D. Marshall, M.G. Thompson, J.C.F. Matthews, T. Hashimoto, J.L. O'Brien, A. Laing, Science **349**, 711 (2015). https://doi.org/10.1126/science.aab3642
188. G. Reithmaier, M. Kaniber, F. Flassig, S. Lichtmannecker, K. Müller, A. Andrejew, J. Vuckovic, R. Gross, J.J. Finley, Nano Lett. **15**, 5208 (2015). https://doi.org/10.1021/acs.nanolett.5b01444
189. C.P. Dietrich, A. Fiore, M.G. Thompson, M. Kamp, S. Höfling, Laser Photon. Rev. **10**, 870 (2016). https://doi.org/10.1002/lpor.201500321
190. P. Lodahl, Quantum Science and Technology **3**, 013001 (2017). https://doi.org/10.1088/2058-9565/aa91bb
191. A. Politi, M.J. Cryan, J.G. Rarity, S. Yu, J.L. O'Brien, Science **320**, 646 (2008). https://doi.org/10.1126/science.1155441
192. E. Murray, D.J.P. Ellis, T. Meany, F.F. Floether, J.P. Lee, J.P. Griffiths, G.A.C. Jones, I. Farrer, D.A. Ritchie, A.J. Bennett, A.J. Shields, Appl. Phys. Lett. **107**, 171108 (2015). https://doi.org/10.1063/1.4935029
193. L.B. Soldano, E.C.M. Pennings, J. Light. Technol. **13**, 615 (1995). https://doi.org/10.1109/50.372474
194. N. Prtljaga, R.J. Coles, J. O'Hara, B. Royall, E. Clarke, A.M. Fox, M.S. Skolnick, Appl. Phys. Lett. **104**, 231107 (2014). https://doi.org/10.1063/1.4883374
195. K.D. Jöns, U. Rengstl, M. Oster, F. Hargart, M. Heldmaier, S. Bounouar, S.M. Ulrich, M. Jetter, P. Michler, J. Phys. D **48**, 085101 (2015). https://doi.org/10.1088/0022-3727/48/8/085101
196. A. Musiał, C. Hopfmann, T. Heindel, C. Gies, M. Florian, H.A.M. Leymann, A. Foerster, C. Schneider, F. Jahnke, S. Höfling, M. Kamp, S. Reitzenstein, Phys. Rev. B **91**(20), 205310 (2015). https://doi.org/10.1103/PhysRevB.91.205310
197. U. Rengstl, M. Schwartz, T. Herzog, F. Hargart, M. Paul, S.L. Portalupi, M. Jetter, P. Michler, Appl. Phys. Lett. **107**, 021101 (2015). https://doi.org/10.1063/1.4926729
198. T. Aichele, V. Zwiller, O. Benson, New J. Phys. **6**, 90 (2004). https://doi.org/10.1088/1367-2630/6/1/090
199. M.J. Stanley, C. Matthiesen, J. Hansom, C. Le Gall, C.H.H. Schulte, E. Clarke, M. Atatüre, Phys. Rev. B **90**, 195305 (2014). https://doi.org/10.1103/physrevb.90.195305
200. A. Schliwa, M. Winkelnkemper, A. Lochmann, E. Stock, D. Bimberg, Phys. Rev. B **80**, 161307 (2009). https://doi.org/10.1103/PhysRevB.80.161307
201. C. Simon, H. de Riedmatten, M. Afzelius, N. Sangouard, H. Zbinden, N. Gisin, Phys. Rev. Lett. **98**, (2007). https://doi.org/10.1103/physrevlett.98.190503
202. A.I. Lvovsky, B.C. Sanders, W. Tittel, Nat. Photon. **3**, 706 (2009). https://doi.org/10.1038/nphoton.2009.231

203. P. Kok, W.J. Munro, K. Nemoto, T.C. Ralph, J.P. Dowling, G.J. Milburn, Rev. Mod. Phys. **79**, 135 (2007). https://doi.org/10.1103/RevModPhys.79.135
204. N. Akopian, L. Wang, A. Rastelli, O.G. Schmidt, V. Zwiller, Nat. Photon. **5**, 230 (2011). https://doi.org/10.1038/nphoton.2011.16
205. M. Davanco, J. Liu, L. Sapienza, C.Z. Zhang, J.V. De Miranda Cardoso, V. Verma, R. Mirin, S.W. Nam, L. Liu, K. Srinivasan, Nat. Commun. **8**, 889 (2017). https://doi.org/10.1038/s41467-017-00987-6
206. A.W. Elshaari, I.E. Zadeh, A. Fognini, M.E. Reimer, D. Dalacu, P.J. Poole, V. Zwiller, K.D. Jöns, Nat. Commun. **8**, 379 (2017). https://doi.org/10.1038/s41467-017-00486-8
207. P. Tonndorf, R. Schmidt, R. Schneider, J. Kern, M. Buscema, G.A. Steele, A. Castellanos-Gomez, H.S.J. van der Zant, S. Michaelis de Vasconcellos, R. Bratschitsch, Optica **2**, 347 (2015). https://doi.org/10.1364/optica.2.000347

Chapter 9
Quantum Networks Based on Single Photons

Oliver Benson, Tim Kroh, Chris Müller, Jasper Rödiger, Nicolas Perlot and Ronald Freund

Abstract Quantum networks rely on the transfer of quantum information between stationary quantum nodes. Physically the nodes are connected by single photons. In a first part of this chapter we address different components of a quantum network. We start with a discussion of semiconductor photon sources with an emission wavelength near 900 nm. In order to make them suitable for fiber-networks a conversion to the telecom band is required. We describe how such converters can be realized with the help of nonlinear optics. Next we address photon storage devices as crucial components of quantum repeaters, which are necessary to establish quantum key distribution (QKD) over long distances. We concentrate on the approach of room-temperature gas cells filled with alkali atoms and outline first promising result. In a second part we address a special QKD protocol, the so-called time-frequency (TF) protocol. It can mostly be realized with off-the-shelf components and its encoding of quantum bits in frequency and time suggests a straightforward way to utilize multiplexing. We analyze the TF-protocol numerically before we report on an actual free-space link over 100 m as a testbed for a quantum network in a realistic environment.

9.1 Introduction

Ten years ago Jeff Kimble formulated the vision of a "quantum internet" [1]. Based on the laws of quantum physics, it would allow secure communication and provide remote access to quantum computers with unprecedented computation power. While nowadays the best-known and most advanced application of a quantum internet is quantum key distribution (QKD), other useful applications will likely be discovered in the next decade [2].

O. Benson (✉) · T. Kroh · C. Müller
Institute of Physics, Humboldt-Universität zu Berlin, Berlin, Germany
e-mail: oliver.benson@physik.hu-berlin.de

J. Rödiger · N. Perlot · R. Freund
Heinrich Hertz Institute, Fraunhofer Institute for Telecommunications, Berlin, Germany

M. Kneissl et al. (eds.), *Semiconductor Nanophotonics*, Springer Series in Solid-State Sciences 194, https://doi.org/10.1007/978-3-030-35656-9_9

A closer look to a real implementation of a quantum network—the physical backbone of the quantum internet—reveals its fundamental unit cell: a stationary system with long coherence times entangled with a flying quantum bit, i.e. a photon (see Fig. 9.1a). Such a fundamental cell can be connected to another one with the help of a joint Bell-state measurement on the two photons as shown in Fig. 9.1b. Bridging a longer distance via optical fibers requires photons in the telecom band. This is no fundamental limitation since the photon wavelength can be coherently converted to other wavelength using nonlinear optical effects in quantum frequency converters [3]. We report on this in Sect. 9.3. Alternatively, intermediate bi-color entangled photon sources can be utilized (see Fig. 9.1b).

A first step towards connecting two unit cells of a quantum network has been taken recently by entangling two solid-state qubits [4]. The latter pioneering experiment used defect centers in diamond. Despite this important success it is still unclear what

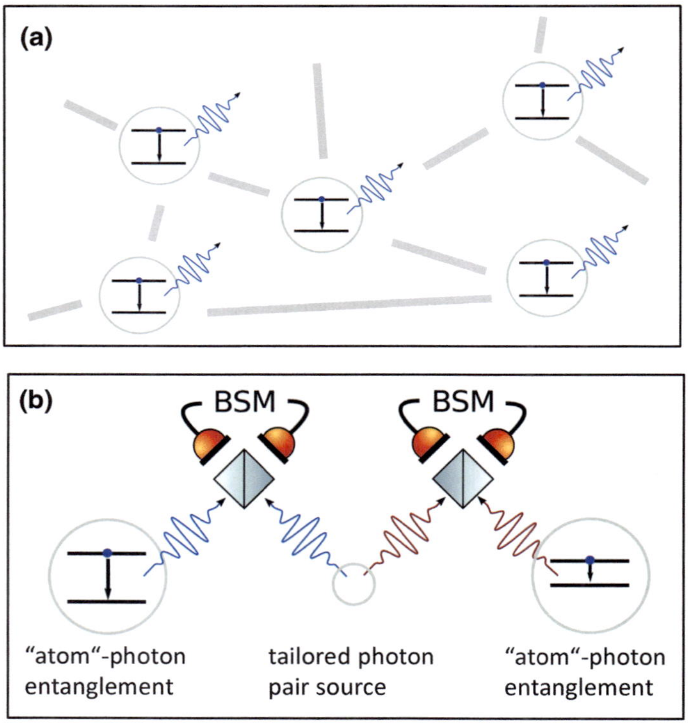

Fig. 9.1 **a** Illustration of several unit cells integrated in a larger quantum network as the physical backbone of the quantum internet. **b** Procedure to connect two fundamental unit cells of a quantum network via Bell state measurements (BSM). An "atom-like" system hosting a stationary quantum bit with long decoherence time is entangled with a photon. BSM provides entanglement swapping. Even if the two unit cells are dissimilar, e.g. two different physical implementations, they can still be connected/entangled via BSM with the help of an intermediate bi-color entangled photon source

technology platform is best suited for a full-fledged quantum network. Semiconductors would perfectly complement an established classical communication technology, whereas atoms or ions would provide very long coherence times and excellent coupling to photons.

In a first part of this chapter we address several aspects of the realization of a quantum network based on semiconductor quantum dots (QD). They are excellent sources for single and even entangled photons (Sect. 9.2). Since they already emit in the near infrared, a frequency conversion to the telecom band has to bridge only a moderate energy gap (Sect. 9.3). First steps to combining semiconductor sources with atomic memories have also been taken recently (Sect. 9.4).

A second part of this chapter concerns quantum key distribution (QKD). Ultra-bright true single photons may render useful for such an application. However, todays QKD protocols provide unconditional security even with classical attenuated sources (Sect. 9.5). This is a huge advantage for commercial applications. In the last section, we finally report on the state-of-the-art of our own efforts to realize a quantum communication network in the Berlin metropolitan area.

9.2 Single-Photon Generation and Manipulation

Ultra-bright single photon sources are useful tools for quantum communication or remote quantum computing [1]. The different properties required of the single-photon sources are discussed in this section.

9.2.1 Properties of Single Photons in Quantum Networks

A single-photon source is a non-classical light emitter, which deterministically emits only one photon at a time. This property can be verified with a Hanbury Brown and Twiss (HBT) setup, consisting of a 50/50 beam splitter and two single-photon detectors and correlation electronics [5] as shown in Fig. 9.2a.

The two detectors behind the 50/50 beam splitter should never detect photons simultaneously since the source emits photons one-by-one (see Fig. 9.2b). An ideal single-photon source has a dip in the second order intensity correlation function going down to 0, but background light increases this value. For this reason, a source is considered to be a single-photon source if the normalized dip is 0.5 or lower. Possible single photon sources are color defect centers [6], super conducting qubits [7], atoms [8], ions [9], molecules [10], or semiconductor quantum dots [11].

A quantum network can in principle use an existing fiber network. However, quantum states, i.e. single photons, cannot simply be sent between two arbitrarily distant nodes due to losses in the fibers [12]. The achievable distance between nodes is largest if the wavelength of the transmitted photons is at telecom wavelength, so that the losses in a fiber are minimized. In classical communication this is not an

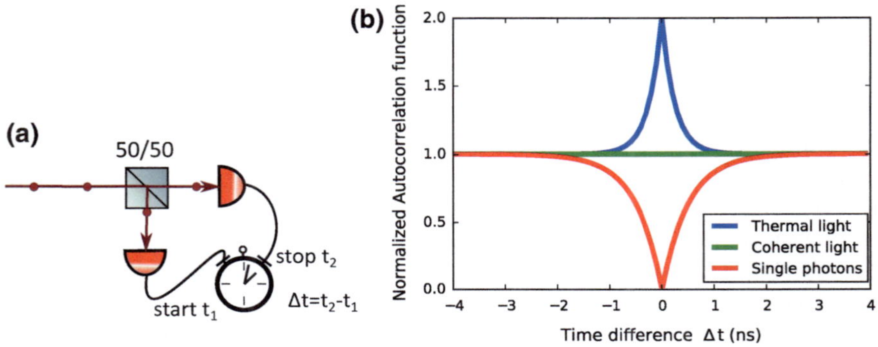

Fig. 9.2 **a** Hanbury Brown and Twiss setup to measure an intensity autocorrelation function. **b** Theoretical second-order correlation function for thermal light (blue), coherent light (green) and single photons (red). The intensity of thermal light is fluctuating with a higher probability of emitting multiple photons at once, leading to a peak in the correlation function for $\Delta t = 0$. A coherent light source emits uncorrelated photons, so that the correlation function is constant. For an ideal single-photon source, only one photon is emitted at once. This results in a dip because the two detectors should never click at the same time. The characteristic value reached at $\Delta t = 0$ is also called $g^{(2)}(0)$ value

issue because the signal is simply amplified after a certain distance. It is not possible with single photons since an unknown quantum state cannot be copied [13].

Quantum repeaters overcome this limitation without measuring or copying the quantum state by using entangled photon pairs and two-photon interference in Bell-state measurements [14]. The photons involved in the interference must be indistinguishable except for maximally one property, which is usually the time of photon detection (time-bin) or polarization. That is why single-photon sources must emit photons with stable frequency and bandwidth. Moreover, two distinct single photon sources have to emit photons that are indistinguishable with respect to each other. Tunable photon sources are particularly useful since the sources can be matched to a common frequency. The emission wavelength can be stabilized to an atomic transition line as a frequency standard. Each node can be locked to that transition line separately, even over large distances. Although single-photon sources exist already in the telecommunication wavelength range of 1.3 and 1.55 μm [15–18], the atomic transitions, which can serve as frequency standard and quantum memory in the near infrared (NIR), are not suited for long distance communication in optical fiber. However, this problem can be circumvented since the wavelength of a single photon can be wavelength-converted. We discuss this in Sect. 9.3. In short, the photon source should be wavelength tunable so that it can be locked to the atomic standard of the quantum network and emit entangled photon pairs.

Additionally, the single-photon source should be as bright as possible, so that higher communication rates are feasible. It would also be advantageous if the photons can be emitted on demand so that the emission time can be controlled.

9.2.2 Semiconductor Single-Photon Sources

Epitaxial semiconductor quantum dots are among the most prominent candidates for light sources in quantum networks [19]. They can emit a single photon or generate pairs of entangled photons.

A quantum dot is a nanostructure that consists of about 10^5 atoms of a semiconductor material with a smaller energy bandgap $E_g^{(A)}$ which is situated within another semiconductor of a comparatively larger bandgap $E_g^{(B)}$ (Fig. 9.3a). According to its small size of only a few nanometers—below the electron's and hole's de Broglie wavelengths—it acts as a finite potential well with only a few discrete bound states for both charges in the conduction band and the valence band, respectively. It is thanks to those discrete states, and their associated narrow emission spectra upon recombination of an electron-hole pair, that quantum dots are also called artificial atoms.

The quantum dot can be filled with various combinations of charges which correspond to different quasiparticles, each. The exciton (X) is a bound state between an electron and a hole. The electron and hole recombine at a rate Γ_X under emission of one photon. According to the Pauli exclusion principle, another exciton may populate the quantum dot at the same time with opposite electron and hole spins, forming a biexciton state (XX). There are also charged excitons—an electron-hole pair with an additional charge—or excited excitonic states. However, here we will focus on the exciton and biexciton ground states.

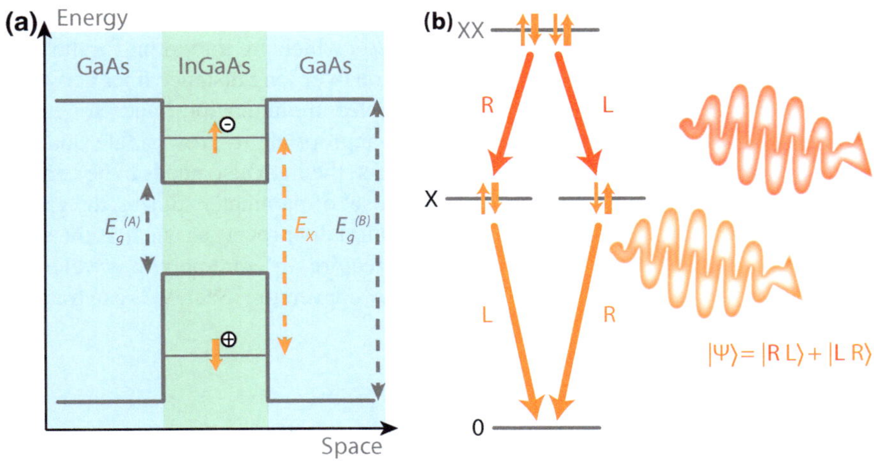

Fig. 9.3 Semiconductor quantum dot as a single and entangled photon source. **a** Electrons and holes are trapped in discrete states of the quasi zero dimensional potential well from different band gap energies which form the quantum dot. Only one electron-hole pair can recombine at a time and emit exactly one photon. **b** The biexciton (XX) cascade can take two recombination paths via two different spin configurations of the excitons (X). The polarization of the two emitted photons is an entangled state if both paths are otherwise indistinguishable

The Coulomb interaction between multiple charges being confined in the quantum dot changes the energy E_{XX} of the first exciton's recombination with respect to the second one, E_X, in the so called biexciton cascade (Fig. 9.3b). Both photons are therefore emitted at different wavelengths and can be separated by spectral filters. Since any of these two decays may only happen once at a time, both can generate single photons.

The biexciton cascade may also produce pairs of entangled photons [20]. There are two paths along which the biexciton can decay: (1) emitting a right circular polarized biexciton photon with $h\nu_{XX} = E_{XX}$ and a left circular polarized exciton photon, $h\nu_X = E_X$, or (2) the other way round (Fig. 9.3b). The paths are quantum mechanically indistinguishable as long as both exciton states have the same energy and the photon pair can be described by $|\psi\rangle = |R, L\rangle + |L, R\rangle$, one of the maximally entangled Bell states which can be used for quantum repeater based communication.

Quantum dots are outstanding in many points of the general wish list of requirements for a quantum network [21]: (i) Deterministic photon generation is achieved by on-demand population (98%) of the biexciton state via two-photon resonant excitation [22] and world record single photon purity $g^{(2)}(0) = 7.5\times10^{-5}$, as-measured and without background correction [23]. (ii) Entanglement fidelity reaches up to 97.8% [24] and (iii) photon indistinguishability, measured by Hong-Ou-Mandel interference, yields near-unity (99.6%) results [25]. (iv) By contrast, collection efficiencies close to one have still to be demonstrated with quantum dots, but are tackled by our colleagues who utilize resonators and monolithic lenses to crank up the extraction efficiency (Chap. 7). Unfortunately, until today it has not been possible to realize all these properties in one quantum dot at the same time. Such a dot would represent an ultimate single-photon source.

At this point, we would like to add two criteria which are important for the technical implementation of quantum communication over long distance fiber networks. The photons need to be compatible with the telecommunication bands for lowest absorption in optical fibers. Thus, it is highly appropriate to grow single quantum dots with emission at 1.3 or 1.55 μm [15–18]. Yet, the light also needs to be compatible for quantum memories. Picking the proper set of parameters during the growth allows for rather accurate control over the quantum dot properties, such as the emission wavelength. But for a network that might require different photon wavelengths for transmission and storage, coherent frequency conversion (Sect. 9.3) can translate between those demands.

9.3 Frequency Conversion of Quantum Light

Nonlinear optics describes the interaction between light and a nonlinear material for high intensity incident light fields. The focus of this part is a second order nonlinear process, involving three frequencies. This can be exploited to convert photons from NIR to the telecom band. Finally, conversion of single photons generated from a QD is presented.

9.3.1 Nonlinear Quantum-Optics

Nonlinear processes can be described by classical, driven dipole oscillators. Incident light causes the dipoles of the nonlinear material to oscillate [26]. Those oscillating dipoles emit electromagnetic fields, and the phase of the oscillating dipole depends on the phase of the incident light. The radiated fields will add constructively if there are multiple dipoles oscillating in phase or destructively if they are out of phase. The radiated frequency does not have to be the same as the incident frequency. Furthermore, any frequency can be radiated as long as the conditions of energy conservation

$$\omega_1 = \omega_2 + \omega_3$$

and conservation of momentum, referred to as phase matching,

$$k_1 = k_2 + k_3$$

are fulfilled. The phase matching condition is material-dependent and can be tuned by changing the material properties. This can be done, for example, via heating. Temperature tuning is quite limited and is only used as a fine-tuning mechanism. A commonly used technique to tailor the possible output frequencies is periodic poling of the material [27]. Therefore, the orientation of the dipoles in the material is changed periodically when the radiated light fields would start to interfere destructively. With that, the phase matching condition is expanded to

$$k_1 = k_2 + k_3 + k_{\text{quasi}}$$

where k_{quasi} depends on the length of the poling period. This permits a large range of wavelengths to be generated by a nonlinear process.

The most frequently used nonlinear process is second-harmonic generation (SHG), where the generated field is exactly twice the frequency of the input field [28]. However, these processes are not limited to just one input field. Two input frequencies can lead to sum-frequency generation (SFG) or difference-frequency generation (DFG), where the sum or the difference of the two input frequencies is generated, respectively [26]. The phase matching condition determines which of these processes dominates. SFG and DFG can be used to alter the frequency of incoming light fields, even of single photons [3]. This can be exploited to convert single photons from the VIS–NIR range to telecom wavelength as needed in quantum networks.

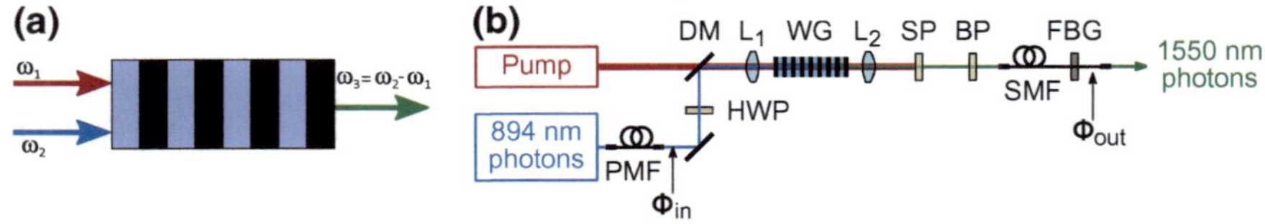

Fig. 9.4 Schematic of a DFG setup. **a** Light beams with two input frequencies ω_1 and ω_2 are combined and interact with the periodically poled nonlinear crystal. Light with a third frequency ω_3, which is the difference of the two input frequencies, is emitted by the crystal. **b** Setup for DFG of 894 nm photons to 1550 nm. The intense 2.1 μm pump laser and the 894 nm photons are superimposed at a dichroic mirror (DM). Lens L1 focuses the light into the nonlinear material waveguide (WG). The converted photons are collimated by Lens L2 and are guided to a filter system: two dichroic mirrors are used as a shortpass at 1800 nm (SP), a 25 nm bandpass (BP) filter at 1550 nm, and a single mode fiber (SMF) with a 3.5 nm fiber Bragg-grating (FBG). Image **b** is taken from [29] (CC BY 3.0)

9.3.2 Conversion of Photons in the Telecom Band

In quantum networks with the Cs D1 transition line at 894 nm as an atomic frequency standard, 894 nm photons could be converted to a wavelength in the telecom C-band (1530–1565 nm) via DFG. Figure 9.4 shows a schematic of the experiment [29]. Photons from a strong 2100 nm pump laser are mixed with the 894 nm photons to generate 1557 nm photons via DFG. The nonlinear material is a 48 mm long waveguide of periodically poled lithium niobate doped with 7 mol% ZnO (NTT Electronics) on a lithium tantalate substrate. The internal conversion efficiency in the waveguide depends strongly on the pump power of the 2100 nm laser and was measured to be 70% with a pump power of 300 mW. The overall conversion efficiency is defined as the 1557 nm photon flux in the fiber after the filtering system Φ_{out} divided by the 894 nm photon flux Φ_{in} directly after the input fiber [3]. The end-to-end conversion efficiency is reduced by the internal conversion efficiency (70%), the coupling efficiency into the waveguide (90%), losses at the lenses and the filter system (60%) and the coupling efficiency into the single mode fiber (35%). This leads to a total conversion efficiency of $(13.1 \pm 0.2)\%$.

9.3.3 Conversion of Photons from a Single Quantum Dot

The setup from Fig. 9.4 can be used to convert the wavelength of single photons from a quantum dot, without losing the single photon character (see Fig. 9.5). Moreover, the $g^{(2)}(0)$ value can even be improved via a frequency conversion because it acts like an additional frequency and polarization filter of the emitted quantum dot photons. This can be seen in Fig. 9.5, where the $g^{(2)}(0)$ value before conversion is 0.26 ± 0.03 and after conversion $0.01 \pm^{0.05}_{0.01}$.

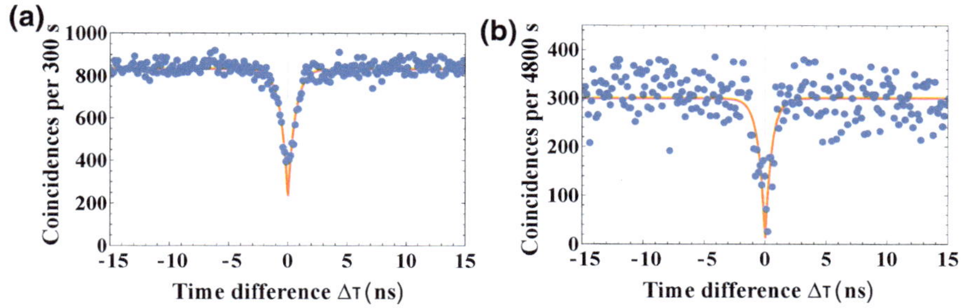

Fig. 9.5 Coincidence counts in an HBT experiment of photons from a single quantum dot under continuous wave excitation. **a** The coincidence measurement at the QD emission wavelength of 894 nm clearly shows a dip at time difference of zero. The orange line is a theoretical curve. **b** Background corrected coincidence measurement after the conversion of the quantum dot photons to 1550 nm. The dip is significantly deeper compared to the unconverted case, since the conversion acts as a frequency filter

Quantum networks need not only single photons but also entangled photon pair sources at telecom wavelengths. Quantum dots can emit entangled photon pairs from a biexciton cascade as discussed in Sect. 9.2.2 [20, 30]. For this reason, we measured the temporal correlation of a QD biexciton cascade with and without the conversion of the biexciton photon (see Fig. 9.6). The temporal correlation of the biexciton remains unchanged after the frequency conversion, showing that the frequency conversion

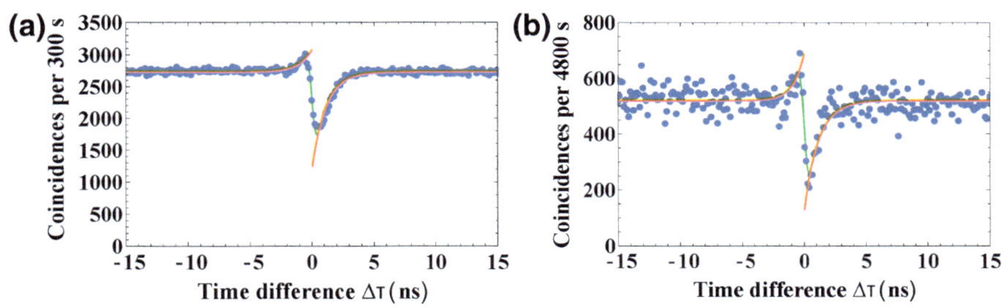

Fig. 9.6 Coincidence measurement between the biexciton and the exciton photons of a QD biexciton cascade. **a** Correlation measurement with unconverted photon from the biexciton cascade at 894 nm. Coincidence counts are increased for negative Δt since in this case a photon is first detected from the biexciton-to-exciton decay. This projects the QD to the exciton state, which can decay immediately. For positive Δt coincidence counts are reduced because the QD is projected to the ground state by the first detected photon and cannot emit another photon immediately again. The orange line indicates a theoretical curve and the green line is the convolution of the theoretical curve with the detector's temporal resolution. **b** Background corrected correlation measurement after conversion of the biexciton photon to 1550 nm. The temporal shape after the conversion is identical to the correlation of (**a**) but the dip is deeper showing an enhanced single photon character caused by the conversion

can be applied to just one photon from a pair without losing the correlation between them. The depth of the coincidence dip is improved here as well.

In summary, it was found that the wavelength of a single photon can be manipulated without disturbing its single photon character or the temporal cross-correlation of the biexciton cascade. This is of great interest for entanglement generation, especially for time-bin entanglement [31]. Moreover, it is a crucial technique for building quantum networks, since it allows combining dissimilar quantum systems, where each can contribute with its advantages. Furthermore, the losses in fibers can be reduced drastically, decreasing the number of required repeater nodes in a quantum network.

9.4 Single-Photon Storage

9.4.1 Concepts of Photon Storage

Quantum memories play a vital role to synchronize the workflow in different fields of quantum information processing [32]. Be it either to temporally coincide two independent photons for a joint Bell state measurement in quantum repeaters [14, 33] or storing qubits for quantum error correction in a quantum computer [34]. In this section, we focus on photon storage in optical quantum networks (Fig. 9.7a).

The emergence of the quantum repeater concept allowed for arbitrarily large transmission distances of quantum states [14]. This is of highest interest for fiber-based quantum communication, which is limited to about a hundred kilometers due to absorption losses in the fibers in simple point-to-point QKD protocols. Subsequently,

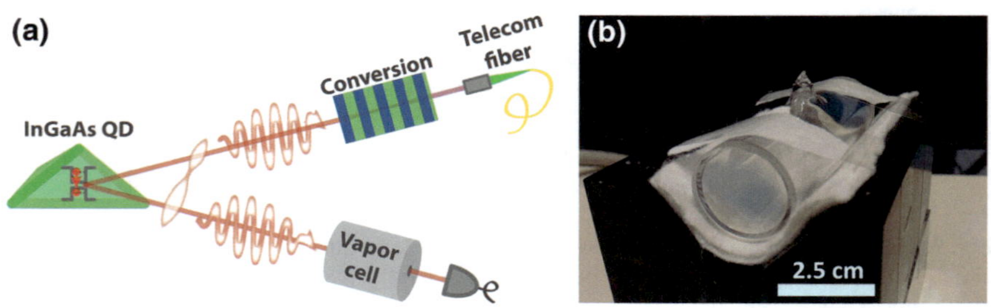

Fig. 9.7 **a** Atomic vapor memory in a quantum network. Entanglement can be distributed over long distances in telecom fiber. One photon of the entangled pair from the quantum dot biexciton cascade is stored in an atomic vapor cell while the other is frequency-converted to 1560 nm and sent to a remote network node. **b** Warm atomic vapors (here cesium) are kept in glass cells. The windows are anti-reflection coated. They are usually shielded by multiple layers of μ-metal, to suppress external electromagnetic fields, and are heated in order to control the vapor density

in the past two decades, various efforts have been made to investigate possible candidates of a physical system that is able to store a photonic state for reasonably long times without notable degradation of the quantum state.

Storage time and integrity of the stored photon are only two out of a set of physical properties that have to be fulfilled by the hybrid system of photon and memory [35]:

- *Fidelity*: The quantum memory is supposed to store a photon for a certain amount of time without altering its quantum state. The integrity of the quantum state during the storage process is described by means of its fidelity. It is defined as the square root of the overlap integral between a pure input state Ψ_{in} and the arbitrary output state, denoted by the density matrix $\rho_{\text{out}} : F = \sqrt{\langle\Psi_{\text{in}}|\rho_{\text{out}}|\Psi_{\text{in}}\rangle}$ [36]. In the ideal case of no change of the quantum state upon storage, the fidelity would be $F = 1$. For a source of indistinguishable photons, it can be measured by Hong-Ou-Mandel interference [37] between a stored photon and another one from the same source.
- *Efficiency*: For low loss operation, each photon that is sent into a storage device should be retrievable afterwards. Therefore, the efficiency of the storage process is defined as the probability η to re-emit a stored photon into a specific optical mode, e.g. of the fiber optical network. It is measured as the ratio between the input single photon rate R_{in} and the recovered photon rate $R_{\text{out}} : \eta = R_{\text{out}}/R_{\text{in}}$. This definition also contains losses from other non-ideal optical elements or coupling into single mode fibers, aside from losses that are caused by the physical storage process itself.
- *Noise*: Detection of noise photons increases errors in quantum networks. One figure of merit is the mean photon number $\mu_1 = p_n/\eta$ in the input state, which is necessary to have a signal to noise ratio SNR ≥ 1, with the memory efficiency η and the unconditional noise probability p_n per attempt [38]. For a higher probability to detect a stored single photon ($\mu_1 \leq 1$) than a noise detection, the overall efficiency needs to exceed the noise probability, $\eta > p_n$.
- *Storage time*: Achievable storage times are limited by decoherence of the physical memory systems. It plays an important role for long-distance communication using quantum repeaters, because entanglement has to be established over the whole distance. A photon in a fiber would take about 0.2 s to go round earth's equator, in case it is not absorbed and without any delay from quantum repeater nodes. Therefore, for the ultimate goal of a worldwide quantum communication network, storage times at the order of seconds would be required for reasonable entangle rates, even for smaller networks [12]. On the way, a storage time of about 500 μs represents a milestone for repeater-less quantum networks at an urban scale, as it corresponds to a photon travelling in a 100 km fiber.
- *Frequency*: Quantum optical networks require a frequency standard, on the one hand to enable fundamental quantum interference between two indistinguishable photons from different sources, but on the other hand the photon wavelength should lie in the low loss O- (1310 nm) or C- (1550 nm) telecommunication bands. *Frequency Conversion of Quantum Light* (Sect. 9.3) makes interconversion possible, if a quantum memory is not available within this range.

- *Bandwidth*: The acceptance bandwidth of the memory should be larger than, or at least at the order of the spectral width of, the quantum light sources in the network. Smaller bandwidth would act as a frequency filter and, conclusively, change the stored quantum state after re-emission, reducing the fidelity.
- *Multiplexing*: Storing many photons in a memory at the same time will be important to further increase transmission rates. Much larger memory bandwidth than that of the photons, for example, would allow for multiplexing by storing many photons at slightly different frequencies around a standard. As a subsidiary condition, the independent readout of each of these states is mandatory.

9.4.2 Atomic Gas Cells

There exists a broad variety of physical systems that are able to store quantum states of light, such as rare earth ions in solids [39], NV centers in diamond [40], spin state of semiconductor QDs [41], single trapped ions [42], cold trapped atomic ensembles [43], Raman-type memories [44] or room temperature alkali gases [45].

Alkali atoms are especially interesting since they only have one bright valence electron. Its spin can be used to map quantum states of light to either a single trapped atom or as a spin wave in a large ensemble of atoms. In our labs we work with room temperature cesium vapor cells [46].

There are different approaches to use them as a memory, e.g. Raman transitions or electromagnetically induced transparency (EIT, cf. Sect. 9.4.4) can be exploited. Raman-memories often suffer from four wave mixing [47], making more complicated cavity-based schemes necessary [48]. EIT memories [45], on the other hand, require only a coupling laser and an atomic vapor cell a few tens of Kelvins above room temperature.

The cell is a glass cylinder that is capped with anti-reflection coated optical windows (Fig. 9.7b). The alkali atoms are filled into the cell under vacuum and remain inside at a low vapor pressure. An oven usually stabilizes the temperature and heats up the vapor for larger optical density of the gas. This increases the interaction strength between light and matter.

In order to utilize the atomic transitions as a frequency standard, the (hyper-) finestructure splitting of the atoms in each cell of the network has to be identical. Multi-layered μ-metal housings reduce external electric and magnetic fields by more than 5 orders of magnitude, down to effectively zero, establishing identical transitions in each network node.

The main processes that limit storage times in vapor cells are collisions of atoms amongst each other, or with the vapor cell walls. In both cases the collision may cause spin flips and consequently reduce the coherence of the overall state. To counter this problem, it is possible to reduce the spin flip rate at the walls by coating the inside of the cell with paraffin [49]. An inert buffer gas can be added to decrease the inter-atom scattering rate of the alkali atoms [50].

9.4.3 Interfacing Quantum Dots and Atomic Vapors

The wavelength of alkali atom based vapor memories ranges around 780 nm (Rb D2), 795 nm (Rb D1), 852 nm (Cs D2) up to 894 nm (Cs D1). Single quantum dots can be grown as InAs or InGaAs inclusions in GaAs to emit at 894 nm (1.3868 eV) very close to the GaAs bandgap of 1.424 eV; or in the shorter wavelength range as GaAs quantum dots in an AlGaAs environment. This made it possible to demonstrate first hybrid atom-QD quantum system.

In a first experiment by Nika Akopian et al. a magnetic field tunes the quantum dot exciton emission to the Rb D2 transitions via Zeeman shift [51]. The quantum dot single photon is then slowed down and partially delayed for about 8 ns due to group velocity reduction between the two 6.8 GHz split hyperfine transitions of the electronic ground state. This gave the first impetus for the quantum dot community to demonstrate hybrid interaction between natural atoms and quantum dots—as their artificial siblings.

However, tuning the quantum dot emission to atomic resonances turned out to be difficult. On one side because magnetic or temperature tuning of excitonic states are limited to the order of 0.3 nm each. A fundamentally more serious point is that both tuning methods may cause a degradation of the quantum state of light. The increase of temperature in the quantum dot environment makes photon–phonon coupling in the event of exciton recombination more probable, which would add to the line broadening and reduce photon indistinguishability or, more general, quantum state fidelity. Magnetic field tuning results in Zeeman splitting of the two exciton states [51]. Nonetheless, this degeneracy is crucial to generate polarization entangled photon pairs from a quantum dot [20].

Accordingly, other methods of frequency tuning had to be investigated. Resonance fluorescence of a strong light field at a two-level system shows the phenomenon of the Mollow triplet line splitting [52]. The splitting is proportional to the Rabi frequency and scales with the square root of the laser power. These side bands also show temporal antibunching of scattered photons [53], proving single-photon emission. Mollow side bands of a quantum dot were used for spectroscopy by tuning them over the Cs D1 transitions [54] and a Cs cell was exploited as a Faraday anomalous dispersion optical filter (FADOF) to selectively switch single components of the Mollow triplet on and off [55]. Unfortunately, the Rabi tuning of quantum dot resonance fluorescence is limited to some 10 GHz, an order less than e.g. temperature tuning.

A major breakthrough has been achieved when strain tuning of the semiconductor lattice became applicable [56, 57]. Experiments quickly followed performing spectroscopy if the Rb D2 transitions with quantum dot single photons [58] or delaying single photons between the 9.192 GHz split Cs ground state [59]. No degradation of the quantum states has been observed so far under strain tuning. It is even possible, with full control over strain in all crystal axes, to generate entangled photons at atomic transitions [60].

In a recent experiment [61], quantum dot single photons have been delayed in a Cs cell and the Hong-Ou-Mandel interference of two consecutive photons was measured for three different cases: without a Cs cell, both photons traversing the cell, or only one of the two being delayed in a cell. The preservation of the quantum interference demonstrates a high fidelity of the delaying process which is a precursor of storing quantum dot photons in a memory.

9.4.4 Single-Photon Storage

EIT-based quantum memories in atomic vapors rely on a Λ-type level scheme [32, 45, 46]. A strong coupling laser field ω_c (Fig. 9.8a) is sent through the atomic vapor, close or on the transition between the excited $|e\rangle$ and the memory state $|m\rangle$. The coupling between light field and matter generates new "dressed" eigenstates (dashed red). Scanning a weak probe field ω_p, e.g. single photons, around the $|e\rangle - |g\rangle$ transition will not only reveal a line splitting in the absorption (Fig. 9.8b), but also a steep increase of the refractive index in the absorption window (Fig. 9.8c).

Similar to the group velocity reduction between resonances that were mentioned before [51, 59, 61], this causes slow light propagation in the so called EIT window around $\omega_p = 0$. In addition, the coupling field can be adiabatically turned off while a single photon is delayed in the cell. This will cause its absorption and the quantum state remains as one excitation to the memory state $|m\rangle$ in a superposition over all atoms in the optical path. The photon is finally emitted again when the coupling field is switched back on.

In 2017 an EIT-based photon storage experiment with weak laser pulses in Rb vapor demonstrated the compatibility of such a quantum memory with quantum dots: In order to store a single photon with a high efficiency η, the acceptance bandwidth has to be larger than the photon's spectrum (cf. Sect. 9.4.1). Laser light with an average photon number of 1 per trial and a FWHM bandwidth of 0.66 GHz was stored for 50 ns with an intrinsic efficiency of $\eta_{\text{int}} = 17\%$ [45]. This already suits

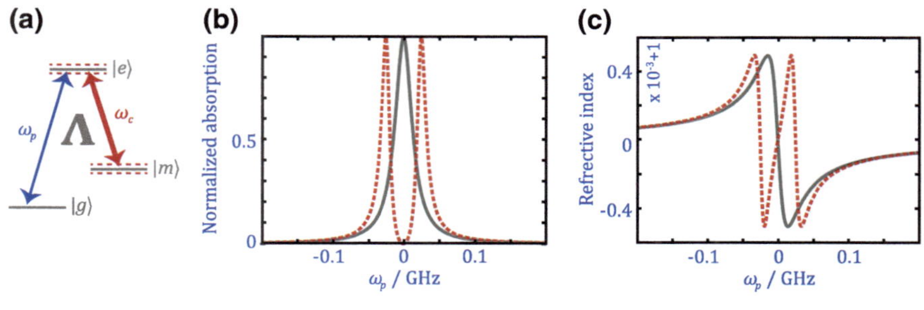

Fig. 9.8 Electromagnetically induced transparency (EIT) in an atomic vapor. **a** Lambda scheme of a three-level system with strong coupling field (red) and probe light (blue). **b** Probe field absorption with (dashed red) and without (gray) coupling laser. **c** Respective refractive indices

well the Fourier-limited linewidth of quantum dot emission, which usually lies at about 0.2 GHz.

New ladder-type schemes, in contrast to the Λ-type level systems, use a higher excited electronic state as a memory [62, 63]. The involved atomic levels, coupling laser, and single photon wavelength can be chosen in a way that the Raman scattering noise in this process can be further reduced. Being able to use these higher excitations also broadens the "spectrum" of quantum light sources that can be combined with one of these states.

With the precise tunability of quantum dots and high collection efficiencies (Chap. 7) on the one hand and more control over the atomic states in vapor cells, as well as their coherence on the other, the hybrid combination of atomic memories and single quantum dots will surely grow to their full potential in coming years.

9.5 Quantum Communication

In the first part of this section, we introduce quantum key distribution (QKD) and explain its strengths and weaknesses compared to classical cryptography. To give an overview of all the existing QKD protocols would be beyond the scope of this book chapter. However, we give an overview of the different types of QKD protocols. In the second part of this section, we introduce the time-frequency (TF) protocol followed by the third part where the TF protocol is further discussed in terms of higher alphabets and security.

9.5.1 Quantum Key Distribution (QKD) Protocols

In the modern interconnected world, secure communication becomes increasingly important. Nowadays, secure communication is mostly ensured by cryptographic algorithms relying on certain problems which are very hard to solve by known algorithms running on classical computers. However, new algorithms, faster computers or quantum computers are an imminent threat to those cryptographic algorithms.

A provable secure method to communicate, even in the presence of those threats, is quantum key distribution, arguably the most advanced quantum technology. Here, two parties, commonly called Alice and Bob share a secret key, which can be used for one-time pad encryption.

What makes QKD secure is that an eavesdropping attempt by an adversary, commonly called Eve, can be detected. Alice and Bob use single quantum states (usually carried by photons) to produce a raw key. Eavesdropping is essentially a quantum mechanical measurement, which leaves traces on the quantum states and can thus be detected. The no-cloning theorem forbids copying the quantum states without distorting them. By knowing the amount of leaked information to Eve, Alice and

Bob can either produce a secret key, totally unknown by Eve, or, if the amount of leaked information is too large, discard the raw key.

The BB84 protocol was the first QKD protocol, proposed in 1984 by Bennett and Brassard [64]. Here the polarization of single photons is used to encode the raw key qubits. Two bases, each with two bits, are defined in the polarization space (e.g. $\{0°, 90°\}$ and $\{-45°, 45°\}$). Alice and Bob perform the following steps:

i. Alice randomly prepares photons in one of the four states and sends them to Bob.
ii. Bob measures the photons randomly in one of the two described bases. A raw key is then generated.
iii. After the raw-key transmission, Alice and Bob communicate publicly about which bases they have used and only keep the qubits where their bases are identical.
iv. Alice and Bob publicly compare a random subset of the remaining qubits from which they deduce the amount of information leaked to Eve. If Eve's information of the key is larger than the mutual information of Alice and Bob, QKD is aborted, otherwise it is continued.
v. Alice and Bob use error correction and privacy amplification algorithms to create a mutual and secret key that can be used e.g. for one-time pad encryption.

Despite being the first, BB84 is not the only QKD protocol. QKD protocols are generally divided into two classes of protocols. The two classes differ in which properties of light are used or alternatively how the qubits are measured. The BB84 protocol is a so-called discrete variable (DV) protocol, but there are also DV protocols using other degrees of freedom than polarization, some of which are shown in Table 9.1. Here light is regarded as a stream of particles that need to be measured with single-photon detectors. In continuous variable (CV) protocols [65–67], light is regarded as a wave, and coherently detected.

Table 9.1 Different QKD protocols are compared in terms of their detection technique, the used dimensions, the capability of photons to carry more than one bit, recent world records of key rate and their ability to be implemented according to standard classical communication techniques ("quantum & classical")

	Discrete variable (DV)-QKD			Continuous variable (CV)-QKD
Detection technique	Single-photon detection			Coherent detection
Dimension	Energy-time uncertainty	Polarization	Differential phase	Quadrature-phase shift keying
>1 bit/photon	Yes	No	No	Yes
World records	Yes [68–70]	No	Yes [71]	No
Quantum and classical	Yes	No	No	Yes

DV protocols can be further distinguished. In prepare-and-measure protocols (e.g. BB84) Alice sends single photons to Bob. In entanglement-based protocols, a source not necessary in possession of Alice or Bob sends out entangled photons to Alice and Bob, which both measure them [72] such that Alice and Bob's measurements are correlated. A third class of DC protocols are measurement-device independent protocols [73], where Alice and Bob send out indistinguishable photons to a third party that performs a bell-test measurement with them.

The before-mentioned no-cloning theorem, which ensures QKD's security, hinders amplification of quantum signals without distorting them. Due to exponential losses in optical fibers, QKD is thus usually restricted to a few hundred kilometers, with the distance record being slightly above 400 km [73], while the key rate decreases exponentially with distance.

One way of overcoming this distance barrier is to use trusted nodes, which act as receiving and sending units. However, the secret key needs to be stored at such trusted nodes. This means one has to trust the provider's integrity as well as its ability to protect the nodes. Trusted node networks are already implemented as research networks [74–76], the longest being the 2000 km network in china [77].

To overcome this distance barrier without trusted nodes, quantum repeaters [14] are currently being developed. By entanglement swapping in combination with quantum memories, quantum repeaters could adopt the role classical repeaters have in classical communication networks. Another approach to extend the transmission distance is QKD over satellites [78, 79]. The quadratic losses of free-space transmission have a significant advantage over the exponential losses of fibers and could thus ensure global QKD coverage. Quantum repeaters and satellite QKD are very promising with both being in the early stages of development.

9.5.2 The Time-Frequency (TF-) Protocol

The time-frequency (TF-) protocol is a BB84 like DV QKD protocol using the energy-time uncertainty relation to ensure security. Like in BB84, two bases—here one in the time and the other in the frequency domain—are used to encode the qubits on single photons. Pulse position modulation (PPM) is used in the time domain and frequency shift keying (FSK) in the frequency domain. Compared to other protocols, like BB84, those bases allow each photon to carry more than one raw-key qubit. The TF protocol can mainly be built out of standard telecom components. The only non-standard components are the Geiger mode avalanche photon detectors (APD), for single-photon measurement. An implementation of the TF protocol was reported in [80]. A simple sketch of a TF protocol implementation can be seen in Fig. 9.9.

In [81] we have discussed the modulations, especially the width of the pulses and their influence on the secret key rate. The results of this paper will be summarized in Sect. 9.5.3.

An example of the two modulations in the time and frequency domain are shown in Fig. 9.10 for $M = 4$ symbols per basis. Here Gaussian pulse shapes are chosen for

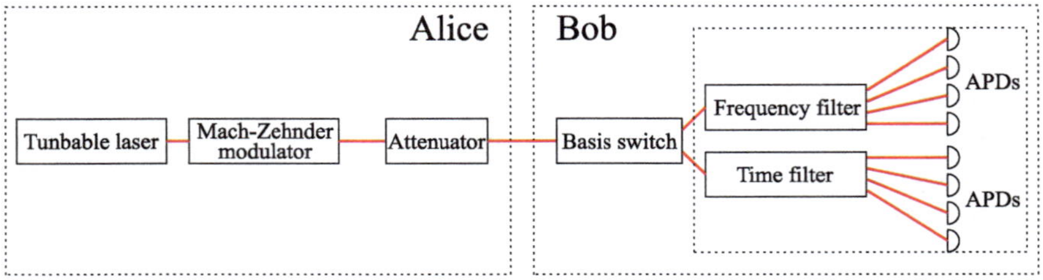

Fig. 9.9 Proposed setting for the time-frequency (TF) protocol. Mostly standard telecom components can be used. The only non-classical devices are the single-photon detectors [avalanche photon detectors (APD)]

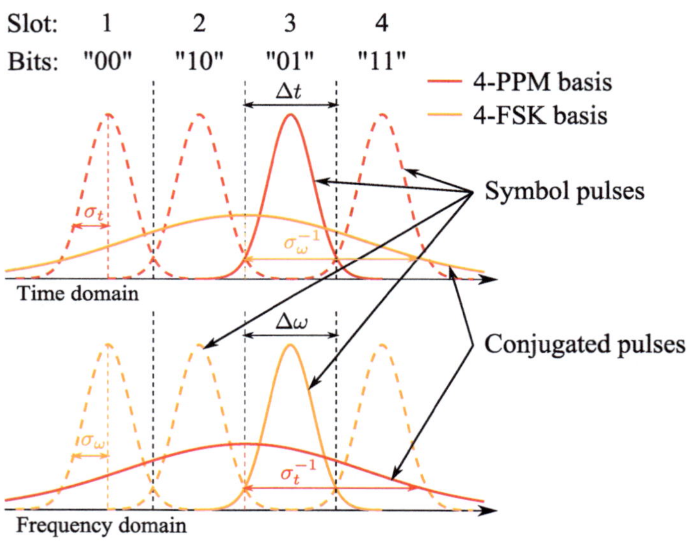

Fig. 9.10 Time-frequency (TF) protocol bases. M-PPM in the time domain and M-FSK in the frequency domain for $M = 4$. Arbitrary values for the number of pulses M are possible. The pulses can be interpreted as probability distributions for photon detection. The symbol pulses (either in time or in frequency) carry the qubits, while in the other basis the conjugated pulse does not carry any information and instead will lead to random measured symbol (from [81]. Copyright 2017, The American Physical Society)

both bases. Depending on the width of the symbol pulses, they spill in the neighboring channels causing additional errors.

With regard to Sect. 9.5.3, it is expedient to define the width of the symbol and conjugated pulses at this point. Normalized and assuming symmetry between the time and the frequency basis, pulse widths normalized to pulse separations are defined by

$$\alpha \equiv \frac{2\sigma_t}{\Delta t} = \frac{2\sigma_\omega}{\Delta\omega}$$

and

$$\beta \equiv \frac{2\sigma_\omega^{-1}}{M\Delta t} = \frac{2\sigma_t^{-1}}{M\Delta\omega}.$$

M is the number of symbols per basis. σ_t and σ_ω are half the $1/e$ width of the energy density of the symbol pulses and Δt and $\Delta\omega$ are the spacing between the symbol pulses, in either the time or the frequency domain (as defined in Fig. 9.10). The widths of a symbol pulse and the conjugated pulses in the opposite basis are reciprocal.

9.5.3 Numerical Studies, Higher Alphabets and Security Issues

In the TF protocol, the energy-time uncertainty relation ensures that, if a photon is measured in the respectively other basis it was prepared in, its information can mostly not be accessed by Eve. Herein lies one of the most notable difference to the BB84 protocol giving Eve a potential backdoor to getting additional information.

We have investigated an intercept/resend attack where Eve exploits this. Here she gains an advantage, if she uses subsequent filters, first determining the time of arrival of a photon, then redirecting it to a frequency-resolving detector or vice versa. Hence, she gets information in both bases with one measurement. As usual in intercept/resend attacks, she would then resend photons in a state dependent on what she measured to Bob. However, the filtering process will distort the filtered photons, thus decreasing the information Eve can get by the subsequent measurement process. Calculating the mutual information for Alice and Bob and comparing it with Eve's information on the key, we could find an optimal value for α.

However, the above-described eavesdropping strategy does not utilize every weakness. Eve could perform an attack where she utilizes the imperfect overlap of symbol and conjugated pulses, as described in [82]. We assume that maximizing the overlap of both bases leads to an optimum for β. Both, the optimal value for α and β are displayed depending on M and the fraction ε of bits Eve is eavesdropping on in Fig. 9.11 (top). Further, one can calculate the amount of information in bits per sent photon, i.e. the capacity C, which is displayed in Fig. 9.11 (bottom).

In conclusion, for secure communication with the TF protocol, the pulse widths are very important. Counterintuitively, broader pulses can lead to a higher capacity although more leakage into neighboring symbol-channels occur. The ability to send more qubits per photon is especially valuable, when the detectors saturate (e.g. due to high repetition rates or in low loss scenarios), because the key rate can be increased without increasing the number of detected photons. Similarly, when the system repetition-rate is limited (e.g. for InGaAs APDs, which usually are operated in gating mode to minimize noise [83]) the key rate can also be increased.

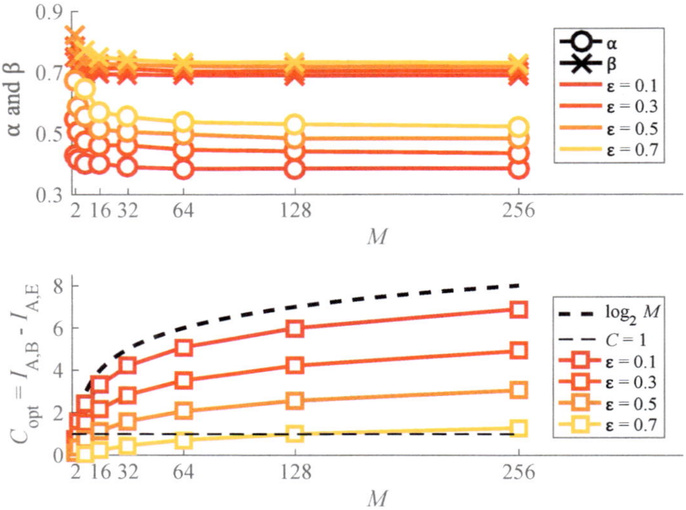

Fig. 9.11 Top: optimal values for the width of the symbol pulses α and the conjugated pulses β depending on the number of symbols per basis. Bottom: capacity C depending on the number of symbol pulses per basis (from [81]. Copyright 2017, The American Physical Society)

9.6 Free-Space Quantum Link

In the first part of this section, free-space QKD is introduced, followed by the second part where we introduce the free-space testbed in the Berlin city center. In the third part, the testbed are evaluated and possible future advancements are discussed.

9.6.1 Free Space QKD Transmission

Whether on ground or over satellites, free-space QKD has raised interests in the recent years. With quadratic instead of exponential losses over the distance, free-space QKD it is the only alternative to quantum repeaters to enable global coverage of quantum communication. Different satellite QKD scenarios are depicted in Table 9.2. Like for fiber-based networks, networks with trusted and untrusted satellites exist.

Beside satellite QKD, free-space QKD links have additional important applications. Firstly, there are scenarios, where fiber-links are not possible, like e.g. with vehicles, vessels, planes or drones. Otherwise at certain scenarios it might be cost effective to build a free-space link instead of laying new fibers, e.g. in last-mile scenarios.

One of the most important experiments of the last years is the quantum communication satellite Micius, a low earth orbit (LEO) satellite, which performed prepare-and-measure [78] and later an entanglement based QKD between two cities 1200 km

Table 9.2 Different satellite QKD scenarios

Must satellite be trusted?	Yes	No		
Satellite role	One-way distributor	Entangled photon source	Bell-measurement station	Relay mirror
Depiction	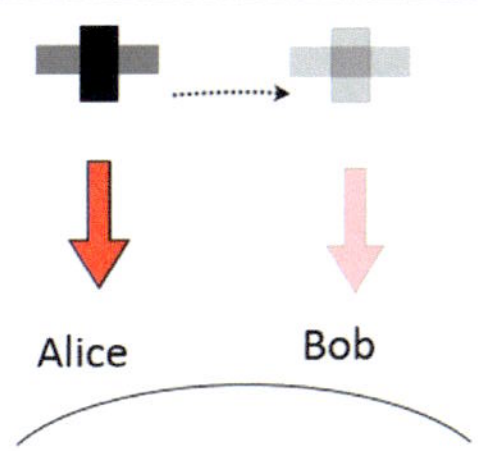	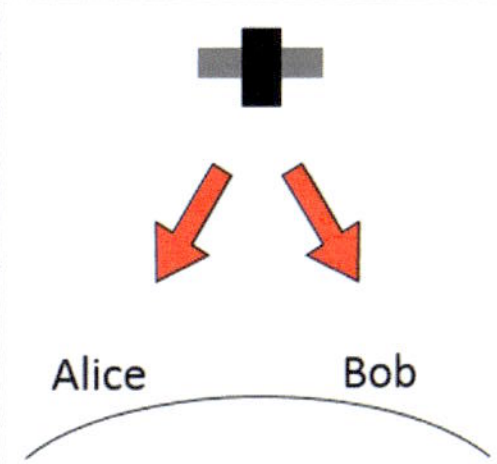	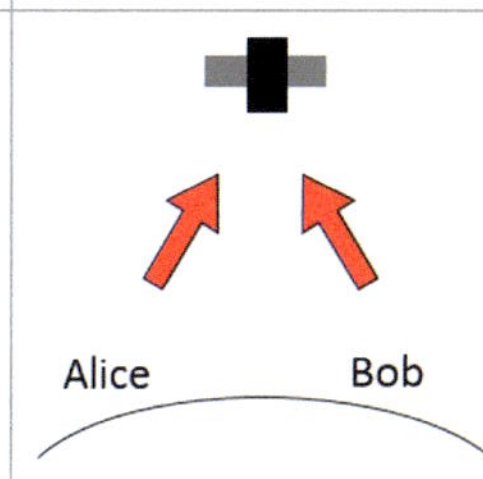	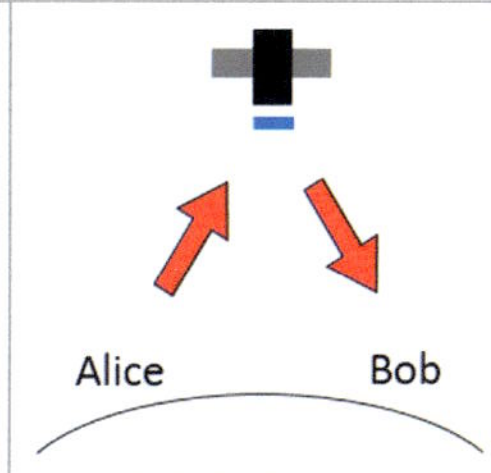
QKD principle	Prepare and measure	Entanglement based	MDI	Prepare and measure

apart [79]. The longest terrestrial QKD link had a distance of 144 km [84, 85]. The highest-loss free-space QKD transmission was performed over 53 km with 48 dB of loss [86].

9.6.2 Experimental Implementation of a Quantum Testbed

As mentioned above, free-space QKD hold advantages for implementing QKD networks. However, free-space QKD links are also challenging to implement. Therefore, it is urgently required to have a test environment for testing QKD components and systems over free-space channels. At Fraunhofer Heinrich Hertz Institute, together with the Humboldt-Universität zu Berlin we have established various testbeds with free-space distances up to 1 km connected to various buildings on the campus of the Technical University of Berlin, to be seen in Fig. 9.12.

Similar to the TF protocol described above, a lot of QKD systems use advanced components developed for classical communication which are often optically interfaced over single mode fibers. Therefore, one of our requirements for the optical

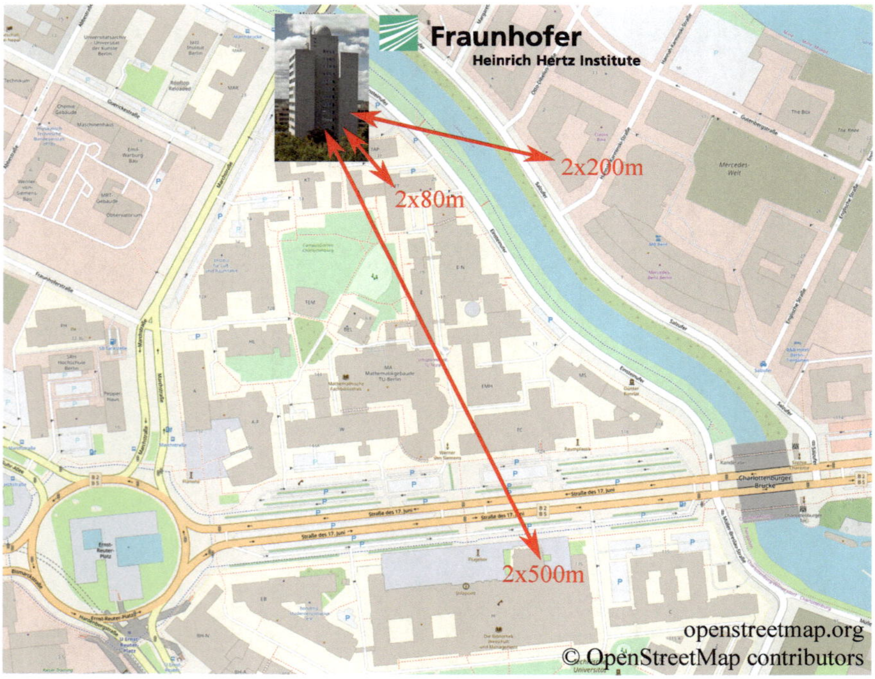

Fig. 9.12 Free-space testbeds at the TU campus in Berlin with distances up to 1 km for benchmarking QKD systems and components

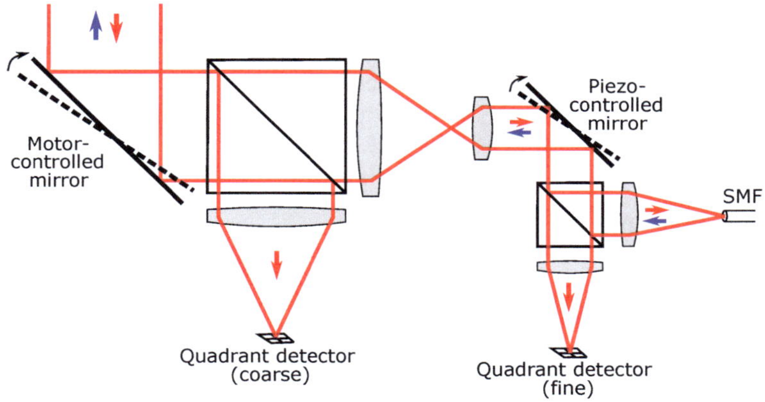

Fig. 9.13 Antenna design with coarse and fine tracking for single mode coupling. From [87]. Copyright 2018, VDE Verlag GmbH

antennas was to have single-mode fiber interfaces. For establishing and keeping the link, a pointing mechanism is the core part of the free-space antennas, minimizing the losses. The principle of the tracking mechanism can be seen in Fig. 9.13. More information on the tracking antennas can be found in [87].

In the antennas the optical signal is divided into two beams, where one is coupled into the fiber and the other is focused on a quadrant detector. Usually quantum beams are very weak and cannot be used for tracking. In those cases, a beacon beam is necessary. The position of the beam on the quadrant detector is processed over a feedback loop, which controls a fine steering Piezo mirror. It will correct tip and tilt distortions of the incoming beam in real time. To cover coarse, large angles on the one hand and precise, small angles on the other, the beam can be divided twice. For coarse-steering, the mirror is moved by motors whereas, for fine-steering, a more precise piezo actuator is used.

Beam deviations could be caused by turbulence or vibrations/movements of the terminals. The tracking works on a millisecond time scale. If necessary, a remote controlled mirror helps to double the link distance and allows the sender and receiver terminal to be side-by-side in just one building. With these terminals, stable QKD transmission with the TF-QKD protocol is feasible over the free-space testbed with minimal losses due to beam deviations.

9.6.3 *Evaluation and Improvements*

To extend the coverage of test scenarios, an additional fiber testbed would be necessary. To be as close to real life conditions, it should consist of different lengths of buried fiber. On the other hand, the logical next step for free-space QKD would be to develop satellite QKD links. Here a ground link with equivalent properties as a ground to satellite link might be beneficial. One possibility is a moving terminal at a distance in the order of a few hundred meters, which moves with the same angular velocity and has the same amount of losses (by scaling the antenna size) as a link with a LEO satellite.

9.7 Conclusion and Outlook

The last two sections of this chapter addressed the most advanced applications of quantum networks, quantum key distribution. Although classical attenuated sources are easy to implement and still provide quantum security, true single-photon sources may after all render advantageous. This is the case if the single photon emission rate can be drastically enhanced and if the on-demand emission probability of a single photon approaches unity. Semiconductor single-photon sources have seen a great improvement in this direction in the last years [88]. Also an on-demand emission of entangled photon pairs is so far only possible with sources based on quantum dots [89].

The necessary cooling of quantum dot sources to nearly liquid Helium temperature has been regarded as major drawback. However, due to the advance of cryotechnology, stand-alone closed-cycle systems are available [90] (see also Chap. 7). This allows for the realization of modular quantum light sources, which may be easily integrated in quantum networks.

Quantum networks have been discussed with respect to secure information exchange over large distances. Obviously, on a global scale this is only feasible with free-space satellite links. On a medium-distance scale fiber-networks are the natural implementation, since classical communication is already established there. An integration of classical and quantum communication in a joint fiber-network is the major challenge in the near future.

As pointed out, a key requirement for an acceptance of non-classical light sources is to demonstrate a high application readiness level. Integration in maintenance-free and (possibly) miniaturized modules is required. With availability of such modules a quantum network on an even shorter distance scale can be envisioned. Figure 9.14 speculates about an exchange of information between several users and (mobile) objects.

For a future internet of things further advancement in technology is needed. Moreover, novel and improved protocols have to be developed, where a quantum advantage in multi-partite secure communication can be fully exploited.

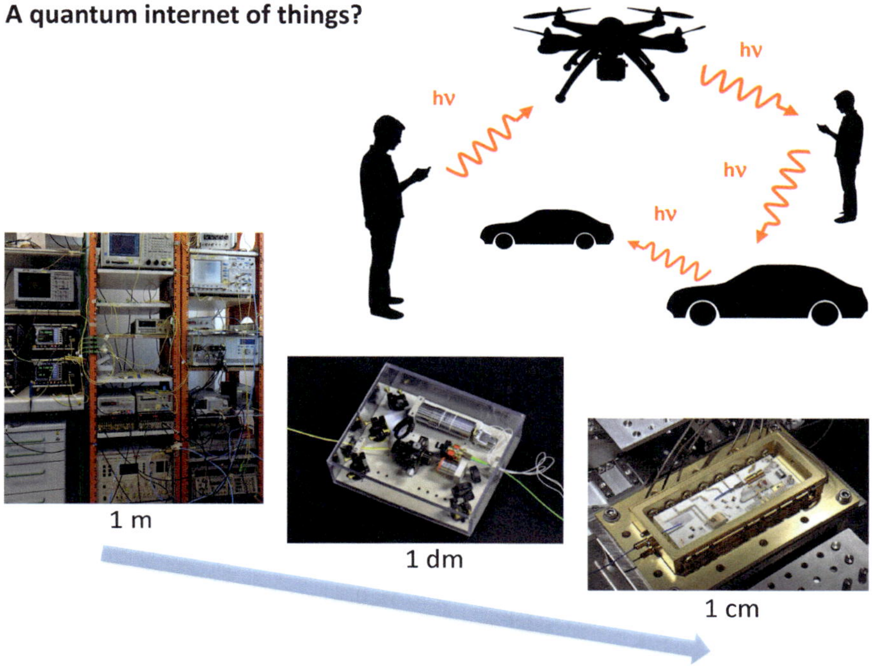

Fig. 9.14 Speculation on a possible application of miniaturized quantum modules in a quantum network, where information is exchanged securely between (mobile) objects

References

1. H.J. Kimble, The quantum internet. Nature **453**, 1023–1030 (2008). https://doi.org/10.1038/nature07127
2. S. Wehner, D. Elkouss, R. Hanson, Quantum internet: a vision for the road ahead. Science **362**, 1–9 (2018). https://doi.org/10.1126/science.aam9288
3. S. Zaske, A. Lenhard, C.A. Kessler, J. Kettler, C. Hepp, C. Arend, R. Albrecht, W.-M. Schulz, M. Jetter, P. Michler, C. Becher, Visible-to-telecom quantum frequency conversion of light from a single quantum emitter. Phys. Rev. Lett. **109**, 147404 (2012). https://doi.org/10.1103/PhysRevLett.109.147404
4. H. Bernien, B. Hensen, W. Pfaff, G. Koolstra, M.S. Blok, L. Robledo, T.H. Taminiau, M. Markham, D.J. Twitchen, L. Childress, R. Hanson, Heralded entanglement between solid-state qubits separated by three metres. Nature **497**, 86–90 (2013). https://doi.org/10.1038/nature12016
5. V. Zwiller, H. Blom, P. Jonsson, N. Panev, S. Jeppesen, T. Tsegaye, E. Goobar, M.-E. Pistol, L. Samuelson, G. Björk, Single quantum dots emit single photons at a time: antibunching experiments. Appl. Phys. Lett. **78**, 2476–2478 (2001). https://doi.org/10.1063/1.1366367
6. E. Togan, Y. Chu, A.S. Trifonov, L. Jiang, J. Maze, L. Childress, M.V. Dutt, A.S. Sørensen, P.R. Hemmer, A.S. Zibrov, M.D. Lukin, Quantum entanglement between an optical photon and a solid-state spin qubit. Nature **466**, 730–734 (2010). https://doi.org/10.1038/nature09256
7. A. Wallraff, D.I. Schuster, A. Blais, L. Frunzio, R. Huang, J. Majer, S. Kumar, S.M. Girvin, R.J. Schoelkopf, Strong coupling of a single photon to a superconducting qubit using circuit

quantum electrodynamics. Nature **431**, 162–167 (2004). https://doi.org/10.1038/nature02851
8. J. Volz, W. Weber, D. Schlenk, W. Rosenfeld, V. Vrana, K. Saucke, C. Kurtsiefer, H. Weinfurter, Observation of entanglement of a single photon with a trapped atom. Phys. Rev. Lett. **96**, 030404 (2006). https://doi.org/10.1103/physrevlett.96.030404
9. B.B. Blinov, D.L. Moehring, L. Duan, C. Monroe, Observation of entanglement between a single trapped atom and a single photon. Nature **428**, 153–157 (2004). https://doi.org/10.1038/nature02377
10. B. Lounis, W.E. Moerner, Single photons on demand from a single molecule at room temperature. Nature **407**, 491–493 (2000). https://doi.org/10.1038/35035032
11. W.B. Gao, P. Fallahi, E. Togan, J. Miguel-Sanchez, A. Imamoglu, Observation of entanglement between a quantum dot spin and a single photon. Nature **491**, 426–430 (2012). https://doi.org/10.1038/nature11573
12. N. Sangouard, C. Simon, H. Riedmatten, N. Gisin, Quantum repeaters based on atomic ensembles and linear optics. Rev. Mod. Phys. **83**, 33 (2011). https://doi.org/10.1103/RevModPhys.83.33
13. W.K. Wootters, W.H. Zurek, A single quantum cannot be cloned. Nature **299**, 802–803 (1982). https://doi.org/10.1038/299802a0
14. H.-J. Briegel, W. Dür, J.I. Cirac, P. Zoller, Quantum repeaters: the role of imperfect local operations in quantum communication. Phys. Rev. Lett. **81**, 5932 (1998). https://doi.org/10.1103/PhysRevLett.81.5932
15. M. Benyoucef, M. Yacob, J.P. Reithmaier, J. Kettler, P. Michler, Telecom-wavelength (1.5 μm) single-photon emission from InP-based quantum dots. Appl. Phys. Lett. **103**, 162101 (2013). https://doi.org/10.1063/1.4825106
16. A. Kors, K. Fuchs, M. Yacob, J.P. Reithmaier, M. Benyoucef, Telecom wavelength emitting single quantum dots coupled to InP-based photonic crystal microcavities. Appl. Phys. Lett. **110**, 031101 (2017). https://doi.org/10.1063/1.4974207
17. N. Srocka, A. Musiał, P.-I. Schneider, P. Mrowiński, P. Holewa, S. Burger, D. Quandt, A. Strittmatter, S. Rodt, S. Reitzenstein, G. Sęk, Enhanced photon-extraction efficiency from InGaAs/GaAs quantum dots in deterministic photonic structures at 1.3 μm fabricated by in-situ electron-beam lithography. AIP Adv. **8**, 085205 (2018). https://doi.org/10.1063/1.5038137
18. C.-Y. Lu, J.-W. Pan, Structural and optical properties of InAs/(In)GaAs/GaAs quantum dots with single-photon emission in the telecom C-band up to 77 K. Phys. Rev. B **98**, 125407 (2018). https://doi.org/10.1103/PhysRevB.98.125407
19. I. Aharonovic, D. Englund, M. Toth, Solid-state single-photon emitters. Nat. Photonics **10**, 631–641 (2016). https://doi.org/10.1038/NPHOTON.2016.186
20. O. Benson, C. Santori, M. Pleton, Y. Yamamoto, Regulated and entangled photons from a single quantum dot. Phys. Rev. Lett. **84**, 2513–2516 (2000). https://doi.org/10.1103/PhysRevLett.84.2513
21. C.-Y. Lu, J.-W. Pan, Push-button photon entanglement. Nat. Photonics **8**, 174–176 (2014). https://doi.org/10.1038/nphoton.2014.29
22. M. Müller, S. Bounouar, K.D. Jöns, M. Glässl, P. Michler, On-demand generation of indistinguishable polarization-entangled photon pairs. Nat. Photonics **8**, 224–228 (2014). https://doi.org/10.1038/NPHOTON.2013.377
23. L. Schweickert, K.D. Jöns, K.D. Zeuner, S.F. Covre da Silva, H. Huang, T. Lettner, M. Reindl, J. Zichi, R. Trotta, A. Rastelli, V. Zwiller, On-demand generation of background-free single photons from a solid-state source. Appl. Phys. Lett. **112**, 093106 (2018). https://doi.org/10.1063/1.5020038
24. D. Huber, M. Reindl, S.F. Covre da Silva, C. Schimpf, J. Martín-Sánchez, H. Huang, G. Piredda, J. Edlinger, A. Rastelli, R. Trotta, Strain-tunable GaAs quantum dot: a nearly dephasing-free source of entangled photon pairs on demand. Phys. Rev. Lett. **121**, 033902 (2018). https://doi.org/10.1103/physrevlett.121.033902
25. N. Somaschi, V. Giesz, L. De Santis, J.C. Loredo, M.P. Almeida, G. Hornecker, S.L. Portalupi, T. Grange, C. Antón, J. Demory, C. Gómez, I. Sagnes, N.D. Lanzillotti-Kimura, A. Lemaítre, A. Auffeves, A.G. White, L. Lanco, P. Senellart, Near-optimal single-photon sources in the solid state. Nat. Photonics **10**, 340–345 (2016). https://doi.org/10.1038/nphoton.2016.23

26. R.W. Boyd, in *Nonlinear Optics*. (Academic Press, 2013)
27. M. Yamada, N. Nada, M. Saitoh, K. Watanabe, First-order quasi-phase matched LiNbO3 waveguide periodically poled by applying an external field for efficient blue second-harmonic generation. Appl. Phys. Lett. **62**, 435 (1993). https://doi.org/10.1063/1.108925
28. P.A. Franken, A.E. Hill, C.W. Peters, G. Weinreich, Generation of optical harmonics. Phys. Rev. Lett. **7**, 118 (1961). https://doi.org/10.1103/PhysRevLett.7.118
29. T. Kroh, A. Ahlrichs, B. Sprenger, O. Benson, Heralded wave packet manipulation and storage of a frequency-converted pair photon at telecom wavelength. Quantum Sci. Technol. **2**, 034007 (2017). https://doi.org/10.1088/2058-9565/aa736c
30. J. Zhang, J.S. Wildmann, F. Ding, R. Trotta, Y. Huo, E. Zallo, D. Huber, A. Rastelli, O.G. Schmidt, High yield and ultrafast sources of electrically triggered entangled-photon pairs based on strain-tunable quantum dots. Nat. Commun. **6**, 10067 (2015). https://doi.org/10.1038/ncomms10067
31. H. Jayakumar, A. Predojević, T. Kauten, T. Huber, G.S. Solomon, G. Weihs, Time-bin entangled photons from a quantum dot. Nat. Commun. **5**, 4521 (2014). https://doi.org/10.1038/ncomms5251
32. A.I. Lvovsky, B.C. Sanders, W. Tittel, Optical quantum memory. Nat. Photonics **3**, 706–714 (2009). https://doi.org/10.1038/nphoton.2009.231
33. L.-M. Duan, M.D. Lukin, J.I. Cirac, P. Zoller, Long-distance quantum communication with atomic ensembles and linear optics. Nature **414**, 413–418 (2001). https://doi.org/10.1038/35106500
34. P.W. Shor, Scheme for reducing decoherence in quantum computer memory. Phys. Rev. A **52**, R2493(R) (1995). https://doi.org/10.1103/PhysRevA.52.R2493
35. C. Simon, M. Afzelius, J. Appel, A. Boyer de la Giroday, S.J. Dewhurst, N. Gisin, C.Y. Hu, F. Jelezko, S. Kröll, J.H. Müller, J. Nunn, E.S. Polzik, J.G. Rarity, H. De Riedmatten, W. Rosenfeld, A.J. Shields, N. Sköld, R.M. Stevenson, R. Thew, I.A. Walmsley, M.C. Weber, H. Weinfurter, J. Wrachtrup, R.J. Young, Quantum memories. Eur. Phys. J. D **58**, 1 (2010). https://doi.org/10.1140/epjd/e2010-00103-y
36. M.A. Nielsen, I.L. Chuang, *Quantum Computation and Quantum Information* (Cambridge University Press, Cambridge, 2000)
37. Z.J. Ou, *Multi-Photon Quantum Interference* (Springer, Berlin, 2007)
38. P. Jobez, C. Laplane, N. Timoney, N. Gisin, A. Ferrier, P. Goldner, M. Afzelius, Coherent Spin Control at the Quantum Level in an Ensemble-Based Optical Memory. Phys. Rev. Lett. **114**, 230502 (2015). https://doi.org/10.1103/PhysRevLett.114.230502
39. T. Zhong, J.M. Kindem, J.G. Bartholomew, J. Rochman, I. Craiciu, E. Miyazono, M. Bettinelli, E. Cavalli, V. Verma, S. Woo Nam, F. Marsili, M.D. Shaw, A.D. Beyer, A. Faraon, Nanophotonic rare-earth quantum memory with optically controlled retrieval. Science **357**, 1392–1395 (2017). https://doi.org/10.1126/science.aan5959
40. D.D. Sukachev, A. Sipahigil, C.T. Nguyen, M.K. Bhaskar, R.E. Evans, F. Jelezko, M.D. Lukin, Silicon-vacancy spin qubit in diamond: a quantum memory exceeding 10 ms with single-shot state readout. Phys. Rev. Lett. **119**, 223602 (2017). https://doi.org/10.1103/PhysRevLett.119.223602
41. M. Kroutvar, Y. Ducommun, D. Heiss, M. Bichler, D. Schuh, G. Abstreiter, J.J. Finley, Optically programmable electron spin memory using semiconductor quantum dots. Nature **432**, 81–84 (2004). https://doi.org/10.1038/nature03008
42. Y. Wang, M. Um, J. Zhang, S. An, M. Lyu, J.-N. Zhang, L.-M. Duan, D. Yum, K. Kim, Single-qubit quantum memory exceeding ten-minute coherence time. Nat. Photonics **11**, 646–650 (2017). https://doi.org/10.1038/s41566-017-0007-1
43. P. Vernaz-Gris, K. Huang, M. Cao, A.S. Sheremet, J. Laurat, Highly-efficient quantum memory for polarization qubits in a spatially-multiplexed cold atomic ensemble. Nat. Commun. **9**, 363 (2018). https://doi.org/10.1038/s41467-017-02775-8
44. D.-S. Ding, W. Zhang, Z.-Y. Zhou, S. Shi, B.-S. Shi, G.-C. Guo, Raman quantum memory of photonic polarized entanglement. Nat. Photonics **9**, 332–338 (2015). https://doi.org/10.1038/NPHOTON.2015.43

45. J. Wolters, G. Buser, A. Horsley, L. Béguin, A. Jöckel, J.-P. Jahn, R.J. Warburton, P. Treutlein, Simple atomic quantum memory suitable for semiconductor quantum dot single photons. Phys. Rev. Lett. **119**, 060502 (2017). https://doi.org/10.1103/PhysRevLett.119.060502
46. D. Höckel, O. Benson, Electromagnetically induced transparency in cesium vapor with probe pulses on the single-photon level. Phys. Rev. Lett. **105**, 153605 (2010). https://doi.org/10.1103/PhysRevLett.105.153605
47. P.S. Michelberger, T.F.M. Champion, M.R. Sprague, K.T. Kaczmarek, M. Barbieri, X.M. Jin, D.G. England, W.S. Kolthammer, D.J. Saunders, J. Nunn, Interfacing GHz-bandwidth heralded single photons with a warm vapour Raman memory. New J. Phys. **17**, 043006 (2015). https://doi.org/10.1088/1367-2630/17/4/043006
48. D.J. Saunders, J.H.D. Munns, T.F.M. Champion, C. Qiu, K.T. Kaczmarek, E. Poem, P.M. Ledingham, I.A. Walmsley, J. Nunn, Cavity-enhanced room-temperature broadband raman memory. Phys. Rev. Lett. **116**, 090501 (2016). https://doi.org/10.1103/PhysRevLett.116.090501
49. M.T. Graf, D.F. Kimball, S.M. Rochester, K. Kerner, C. Wong, D. Budker, E.B. Alexandrov, M.V. Balabas, V.V. Yashchuk, Relaxation of atomic polarization in paraffin-coated cesium vapor cells. Phys. Rev. A **72**, 023401 (2005). https://doi.org/10.1103/PhysRevA.72.023401
50. A. Sargsyan, D. Sarkisyan, U. Krohn, J. Keaveney, C. Adams, Effect of buffer gas on an electromagnetically induced transparency in a ladder system using thermal rubidium vapor. Phys. Rev. A **82**, 045806 (2010). https://doi.org/10.1103/PhysRevA.82.045806
51. N. Akopian, L. Wang, A. Rastelli, O.G. Schmidt, V. Zwiller, Hybrid semiconductor-atomic interface: slowing down single photons from a quantum dot. Nat. Photonics **5**, 230–233 (2011). https://doi.org/10.1038/nphoton.2011.16
52. B.R. Mollow, Power spectrum of light scattered by two-level systems. Phys. Rev. **1969**, 188 (1969). https://doi.org/10.1103/PhysRev.188.1969
53. H.J. Kimble, M. Dagenais, L. Mandel, Photon antibunching in resonance fluorescence. Phys. Rev. Lett. **39**, 691 (1977). https://doi.org/10.1103/PhysRevLett.39.691
54. S.M. Ulrich, S. Weiler, M. Oster, M. Jetter, A. Urvoy, R. Löw, P. Michler, Spectroscopy of the D1 transition of cesium by dressed-state resonance fluorescence from a single (In, Ga)As/GaAs quantum dot. Phys. Rev. B **90**, 125310 (2014). https://doi.org/10.1103/PhysRevB.90.125310
55. S.L. Portalupi, M. Widmann, C. Nawrath, M. Jetter, P. Michler, J. Wrachtrup, I. Gerhardt, Simultaneous Faraday filtering of the Mollow triplet sidebands with the Cs-D1 clock transition. Nat. Commun. **7**, 13632 (2016). https://doi.org/10.1038/ncomms13632
56. F. Ding, R. Singh, J.D. Plumhof, T. Zander, V. Křápek, Y.H. Chen, M. Benyoucef, V. Zwiller, K. Dörr, G. Bester, A. Rastelli, O.G. Schmidt, Tuning the exciton binding energies in single self-assembled InGaAs/GaAs quantum dots by piezoelectric-induced biaxial stress. Phys. Rev. Lett. **104**, 067405 (2010). https://doi.org/10.1103/PhysRevLett.104.067405
57. A. Rastelli, F. Ding, J.D. Plumhof, S. Kumar, R. Trotta, C. Deneke, A. Malachias, P. Atkinson, E. Zallo, T. Zander, A. Herklotz, R. Singh, V. Křápek, J.R. Schröter, S. Kiravittaya, M. Benyoucef, R. Hafenbrak, K.D. Jöns, D.J. Thurmer, D. Grimm, G. Bester, K. Dörr, P. Michler, O.G. Schmidt, Controlling quantum dot emission by integration of semiconductor nanomembranes onto piezoelectric actuators. Phys. Status Solidi B **249**, 687–696 (2012). https://doi.org/10.1002/pssb.201100775
58. J.-P. Jahn, M. Munsch, L. Béguin, A.V. Kuhlmann, M. Renggli, Y. Huo, F. Ding, R. Trotta, M. Reindl, O.G. Schmidt, A. Rastelli, P. Treutlein, R.J. Warburton, An artificial Rb atom in a semiconductor with lifetime-limited linewidth. Phys. Rev. B **92**, 245439 (2015). https://doi.org/10.1103/PhysRevB.92.245439
59. J.S. Wildmann, R. Trotta, J. Martín-Sánchez, E. Zallo, M. O'Steen, O.G. Schmidt, A. Rastelli, Atomic clouds as spectrally selective and tunable delay lines for single photons from quantum dots. Phys. Rev. B **92**, 235306 (2015). https://doi.org/10.1103/PhysRevB.92.235306
60. R. Trotta, J. Martín-Sánchez, J.S. Wildmann, G. Piredda, M. Reindl, C. Schimpf, E. Zallo, S. Stroj, J. Edlinger, A. Rastelli, Wavelength-tunable sources of entangled photons interfaced with atomic vapours. Nat. Commun. **7**, 10375 (2016). https://doi.org/10.1038/ncomms10375

61. H. Vural, S.L. Portalupi, J. Maisch, S. Kern, J.H. Weber, M. Jetter, J. Wrachtrup, R. Löw, I. Gerhardt, P. Michler, Two-photon interference in an atom–quantum dot hybrid system. Optica **5**, 367–373 (2018). https://doi.org/10.1364/OPTICA.5.000367
62. K.T. Kaczmarek, P.M. Ledingham, B. Brecht, S.E. Thomas, G.S. Thekkadath, O. Lazo-Arjona, J.H.D. Munns, E. Poem, A. Feizpour, D.J. Saunders, J. Nunn, I.A. Walmsley, High-speed noise-free optical quantum memory. Phys. Rev. A **97**, 042316 (2018). https://doi.org/10.1103/PhysRevA.97.042316
63. R. Finkelstein, E. Poem, O. Michel, O. Lahad, O. Firstenberg, Fast, noise-free memory for photon synchronization at room temperature. Sci. Adv. **4**, eaap8598 (2018). https://doi.org/10.1126/sciadv.aap8598
64. C. H. Bennett, G. Brassard, in *Quantum Cryptography: Public Key Distribution and Coin Tossing*, Conference on Computers, Systems and Signal Processing (Bangalore, India, Dec 1984), pp. 175–179
65. M.D. Reid, Quantum cryptography with a predetermined key, using continuous-variable einstein-podolsky-rosen correlations. Phys. Rev. A **62**, 062308 (2000). https://doi.org/10.1103/PhysRevA.62.062308
66. M. Hillery, Quantum cryptography with squeezed states. Phys. Rev. A **61**, 022309 (2000). https://doi.org/10.1103/PhysRevA.61.022309
67. T.C. Ralph, Continuous variable quantum cryptography. Phys. Rev. A **61**, 010303 (1999). https://doi.org/10.1103/PhysRevA.61.010303
68. N.T. Islam, C. Cahall, A. Aragoneses, A. Lezama, J. Kim, D.J. Gauthier, Robust and stable delay interferometers with application to d-dimensional time-frequency quantum key distribution. Phys. Rev. Appl. **7**, 044010 (2017). https://doi.org/10.1103/PhysRevApplied.7.044010
69. C. Lee, D. Bunandar, Z. Zhang, G.R. Steinbrecher, P.B. Dixon, F.N. Wong, J.H. Shapiro, S.A. Hamilton, D. Englund, in High-rate field demonstration of large-alphabet quantum key distribution, preprint arXiv:1611.01139 (2016)
70. T. Zhong, H. Zhou, R.D. Horansky, C. Lee, V.B. Verma, A.E. Lita, A. Restelli, J.C. Bienfang, R.P. Mirin, T. Gerrits, Photon-efficient quantum key distribution using time–energy entanglement with high-dimensional encoding. New J. Phys. **17**, 022002 (2015). https://doi.org/10.1088/1367-2630/17/2/022002
71. https://www.toshiba.co.jp/about/press/2017_09/pr1501.htm
72. A.K. Ekert, Quantum cryptography based on Bell's theorem. Phys. Rev. Lett. **67**, 661 (1991). https://doi.org/10.1103/PhysRevLett.67.661
73. H.-L. Yin, T.-Y. Chen, Z.-W. Yu, H. Liu, L.-X. You, Y.-H. Zhou, S.-J. Chen, Y. Mao, M.-Q. Huang, W.-J. Zhang et al., Measurement-device-independent quantum key distribution over a 404 km optical fiber. Phys. Rev. Lett. **117**, 190501 (2016). https://doi.org/10.1103/PhysRevLett.117.190501
74. C. Elliott, A. Colvin, D. Pearson, O. Pikalo, J. Schlafer, H. Yeh, Current status of the DARPA quantum network. Quantum Inf. Comput. III Int. Soc. Opt. Photonics **5815**, 138–150 (2005). https://doi.org/10.1117/12.606489
75. M. Sasaki, M. Fujiwara, H. Ishizuka, W. Klaus, K. Wakui, M. Takeoka, S. Miki, T. Yamashita, Z. Wang, A. Tanaka et al., Field test of quantum key distribution in the Tokyo QKD network. Opt. Express **19**, 10387–10409 (2011). https://doi.org/10.1364/OE.19.010387
76. T.P. Spiller, Quantum Communications hub EPSRC. Impact **2018**(5), 12 (2018). https://doi.org/10.21820/23987073.2018.5.12, https://www.quantumcommshub.net/
77. http://english.cas.cn/newsroom/news/201703/t20170324_175288.shtml
78. S.-K. Liao, W.-Q. Cai, J. Handsteiner, B. Liu, J. Yin, L. Zhang, D. Rauch, M. Fink, J.-G. Ren, W.-Y. Liu et al., Satellite-relayed intercontinental quantum network. Phys. Rev. Lett. **120**, 030501 (2018). https://doi.org/10.1103/physrevlett.120.030501
79. J. Yin, Y. Cao, Y.-H. Li, S.-K. Liao, L. Zhang, J.-G. Ren, W.-Q. Cai, W.-Y. Liu, B. Li, H. Dai et al., Satellite-based entanglement distribution over 1200 km. Science **356**, 1140 (2017). https://doi.org/10.1126/science.aan3211
80. M. Leifgen, R. Elschner, N. Perlot, C. Weinert, C. Schubert, O. Benson, Practical implementation and evaluation of a quantum-key-distribution scheme based on the time-frequency uncertainty. Phys. Rev. A **92**, 042311 (2015). https://doi.org/10.1103/PhysRevA.92.042311

81. J. Rödiger, N. Perlot, R. Mottola, R. Elschner, C.-M. Weinert, O. Benson, R. Freund, Numerical assessment and optimization of discrete-variable time-frequency quantum key distribution. Phys. Rev. A **95**, 052312 (2017). https://doi.org/10.1103/PhysRevA.95.052312
82. Y. Zhang, I.B. Djordjevic, M.A. Neifeld, Weak-coherent-state-based time-frequency quantum key distribution. J. Mod. Opt. **62**, 1713–1721 (2015). https://doi.org/10.1080/09500340.2015.1075616
83. N. Namekata, S. Adachi, S. Inoue, 1.5 ghz single-photon detection at telecommunication wavelengths using sinusoidally gated ingaas/inp avalanche photodiode. Opt. Express **17**, 6275 (2009). https://doi.org/10.1364/oe.17.006275
84. R. Ursin, F. Tiefenbacher, T. Schmitt-Manderbach, H. Weier, T. Scheidl, M. Lindenthal, B. Blauensteiner, T. Jennewein, J. Perdigues, P. Trojek et al., Entanglement-based quantum communication over 144 km. Nat. Physics **3**, 481–486 (2007). https://doi.org/10.1038/nphys629
85. T. Schmitt-Manderbach, H. Weier, M. Fürst, R. Ursin, F. Tiefenbacher, T. Scheidl, J. Perdigues, Z. Sodnik, C. Kurtsiefer, J.G. Rarity et al., Experimental demonstration of free-space decoy-state quantum key distribution over 144 km. Phys. Rev. Lett. **98**, 010504 (2007). https://doi.org/10.1103/physrevlett.98.010504
86. S.-K. Liao, H.-L. Yong, C. Liu, G.-L. Shentu, D.-D. Li, J. Lin, H. Dai, S.-Q. Zhao, B. Li, J.-Y. Guan et al., Long-distance free-space quantum key distribution in daylight towards inter-satellite communication. Nat. Photonics **11**, 509 (2017). https://doi.org/10.1038/nphoton.2017.116
87. N. Perlot, J. Roediger, R. Freund, Single-mode optical antenna for high-speed and quantum communications, in *Photonic Networks*, 19th ITG-Symposium. VDE (2018), pp. 1–4
88. P. Senellart, G. Solomon, A. White, High-performance semiconductor quantum-dot single-photon sources. Nat. Nanotechnol. **12**, 1026–1039 (2017). https://doi.org/10.1038/nnano.2017.218
89. T. Müller, J. Skiba-Szymanska, A.B. Krysa, J. Huwer, M. Felle, M. Anderson, R.M. Stevenson, J. Heffernan, D.A. Ritchie, A. Shields, A quantum light-emitting diode for the standard telecom window around 1550 nm. Nat. Commun. **9**, 862 (2018). https://doi.org/10.1038/s41467-018-03251-7
90. A. Schlehahn, S. Fischbach, R. Schmidt, A. Kaganskiy, A. Strittmatter, S. Rodt, T. Heindel, S. Reitzenstein, A stand-alone fiber-coupled single-photon source. Sci. Rep. **8**, 1340 (2018). https://doi.org/10.1038/s41598-017-19049-4

Chapter 10
Vertical Cavity Surface Emitting Laser Diodes for Communication, Sensing, and Integration

J. A. Lott

Abstract I review my research group's work to date on the design, processing, performance, and key physics of state-of-the-art vertical cavity surface emitting lasers (VCSELs) for modern and emerging applications in optical data communication systems, as low to moderate power optical sources for sensing systems, and as very small to low optical power light sources for photonic-electronic integrated circuits. Via reduced complexity epitaxial designs that potentially lead to lower manufacturing and life cycle costs and via novel device geometries and processing methods we demonstrate record small-signal modulation bandwidth, the highest for all VCSELs at any wavelength to date, and record combinations of bandwidth and optical output power which is vital for emerging free space data communication, tracking, and sensing systems. We further demonstrate arrays of VCSELs of various sizes for a plethora of emerging applications supporting the Internet of Things and all manner of energy sustainable interactive gadgets.

10.1 Introduction

The energy used to transmit information as encoded optical and electrical data bits within and between photonic and electronic integrated circuits, within and between computer servers, within and between data centers, across free-space, and ultimately nearly instantly across the earth from any one point to another should be minimized. This conservation of energy follows naturally from our desire to use resources wisely while advancing our technology. The energy consumption for data transmission spans between typically on the order of femtojoules-per-bit for micrometer on-chip distances to picojoules-per-bit for short-reach to medium-reach distances (across optical fiber for example) to well over tens or hundreds of millijoules-per-bit for the transmission of digital data over intercontinental distances (via combinations of optical fiber or free-space microwave links for example). We seek to meet the exploding

J. A. Lott (✉)
Center of Nanophotonics, Institute of Solid State Physics, Technical University Berlin, EW 5-5
Hardenbergstrasse 36, Berlin, Germany
e-mail: james.lott@tu-berlin.de

M. Kneissl et al. (eds.), *Semiconductor Nanophotonics*, Springer Series
in Solid-State Sciences 194, https://doi.org/10.1007/978-3-030-35656-9_10

demand for information and an ability to sense and understand our surroundings within the terrestrial resources available but more importantly to reduce costs and give all people the ability to connect, communicate, learn, navigate, and live well in comfort and safety with access to stored data, commerce, and to each other. The goal of instant communication, information, and interconnected *things*—that is, photonic-electronic gadgets of all kinds—for enhanced environmental awareness and situational awareness implies a collection of highly energy-efficient, independent, and ubiquitous information, computing, and sensing systems operating with minimal impact on the environment via natural or sustainable energy sources.

A key enabling photonic component for next generation communication, sensing, medical diagnostic, and integrated photonic-electronic systems is the vertical cavity surface emitting laser (VCSEL) [1]. The VCSEL evolved from the pioneering contributions and seminal ideas of many different research groups. Soon after the first reports of the first lasers in 1960 and the first semiconductor laser diodes in 1962 Melngailis in 1965 reported an ~220 μm-long *p–n* junction laser diode emitting perpendicular to the device's InSb wafer surface at a peak wavelength of about 5200 nm with pulsed excitation and operating at about 10 K [2]. In 1974 Dingle et al. demonstrated emission from GaAs quantum wells (QWs) surrounded by AlGaAs barrier layers [3], in 1975 van der Ziel demonstrated lasing from multiple GaAs/AlGaAs structures [4], and in 1975 van der Ziel and Ilegems first reported the seminal idea of using semiconductor distributed Bragg reflectors (DBRs) as mirrors [5]. The idea of using semiconductor DBRs in multiple layer optical interference filters for high frequency optical modulation soon helped to inspire the now obvious idea of using DBRs as mirrors for vertical cavity laser diodes. Also circa 1975 Scifres et al. employed a grating on the surface of a horizontal cavity laser diode to achieve surface normal lasing thus forming a distributed feedback semiconductor laser [6], and in 1976 Scifres and Burnham proposed a heterojunction surface emitting laser composed of half-lambda periods of alternating AlGaAs/GaAs layers [7]. Soda et al. in 1979 reported a GaAsInP surface emitting laser diode on an InP substrate with an ~90 μm-long optical cavity and planar metal mirrors, with one thin-film gold semi-transparent mirror for light emission, that operated pulsed at 77 K at a peak emission wavelength of about 1180 nm [8]. This work is frequently cited as the inspiration for what is now called a VCSEL. Since these early publications on surface emitting laser diodes numerous research groups have contributed to the evolution and improvement of the device we now call a VCSEL, most importantly the brilliant ideas and demonstrations by Ogura et al. wherein semiconductor DBRs are used as mirrors to form surface emitting lasers [9–13], the use of QWs placed at optical field intensity antinodes in the optical cavity [14–18], and the use of oxide apertures [19–22]. Many other significant and pioneering contributions on VCSELs, VCSEL arrays, and VCSEL applications were made by many research groups from the 1980s to the present and these contributions are far too numerous to properly credit here. Most notably, critical advances in epitaxial growth and device processing schemes from about the 1980s to the present hugely contributed to the advancement of VCSELs resulting in our modern state-of-the-art, small (micrometer scale chips), low cost, very practical, and highly efficient laser light sources.

As a quick overview of a modern VCSEL structure in Fig. 10.1 we show a few simplified VCSEL schematic diagrams and some basic simulation plots of a generic oxide aperture 980 nm VCSEL. We present here only a superficial first look at the general physical structure of modern VCSELs. Modern VCSELs are grown by metal-organic vapor phase epitaxy (MOVPE) or by molecular beam epitaxy on GaAs substrates. Optically the VCSEL is an asymmetric multiple-layer Fabry-Pérot etalon. Thus, the VCSEL is a resonant cavity with planar mirrors. However, rather than abrupt mirrors such as with semiconductor-to-air interfaces with a power reflectance $R \sim 0.30$ the VCSEL employs top and bottom highly reflecting DBR mirrors with $R \sim 0.999$–0.9999 (~100%) on the highly reflecting side and $R \sim 0.99$ (~99%) to 0.996 on the output coupling side. On resonance the optical field intensity peaks within the lasing cavity decrease roughly exponentially as the optical energy traverses a given DBR mirror. The small fraction of the resonant optical field intensity that escapes from the coupling mirror (by design, the laser emission) becomes the useful VCSEL optical output power.

The photons are generated via the injection of carriers (electrons and holes). The photons create a standing wave within the cavity at the natural optical cavity longitudinal resonance frequency due to the constructive interference of the electromagnetic planewaves as they reflect and transmit across each of the DBR interfaces with a roundtrip phase of an integer multiple of 2π. The injected carriers recombine is the QWs which are placed in the optical cavity at an optical field antinode between the doped *p*-side and *n*-side of the VCSEL. The optical thickness of the optical cavity between the two DBRs is a positive integer multiple of one-half lambda ($\lambda/2$) and most typically either $\lambda/2$, λ (as shown in Fig. 10.1), or 1.5λ. The injected current is funneled into a confined cylindrical volume within the QW active region via oxide apertures. The selectively thermally oxidized apertures are $Al_{0.98}Ga_{0.02}As$ layers that have been converted into dielectric insulating material. These oxide aperture layers serve to block the current flow near the etched mesa sidewalls of the VCSEL so that the current must flow through the center of the VCSEL structure.

The use of roughly circular oxide apertures compared to proton implanted high resistance regions to form current apertures results in a much larger power conversion efficiency for the VCSELs. The use of oxide apertures also allows us to experimentally vary the diameter of the oxide aperture on each wafer piece to study the VCSEL physics and compare the performance of VCSELs as a function of oxide aperture diameter. The oxide apertures have a lower real refractive index compared to the crystalline semiconductor layers and thus these oxide layers additionally may serve to guide the lateral optical modes of the VCSEL. Oxide layers provide an additional degree of freedom in the VCSEL design leading to higher modulation bandwidths and data transmission rates, but potentially also leading to undesired reliability issues, increased differential series resistance, and increased thermal resistance.

In this Chapter we review our applied physics research work on energy-efficient, high bandwidth, and high optical output power VCSELs and VCSEL arrays for applications as light sources for optical communication, sensing, and very physically small and inexpensive integrated photonic-electronic systems. We outline our VCSEL designs and key performance metrics that enable high bit rate communication

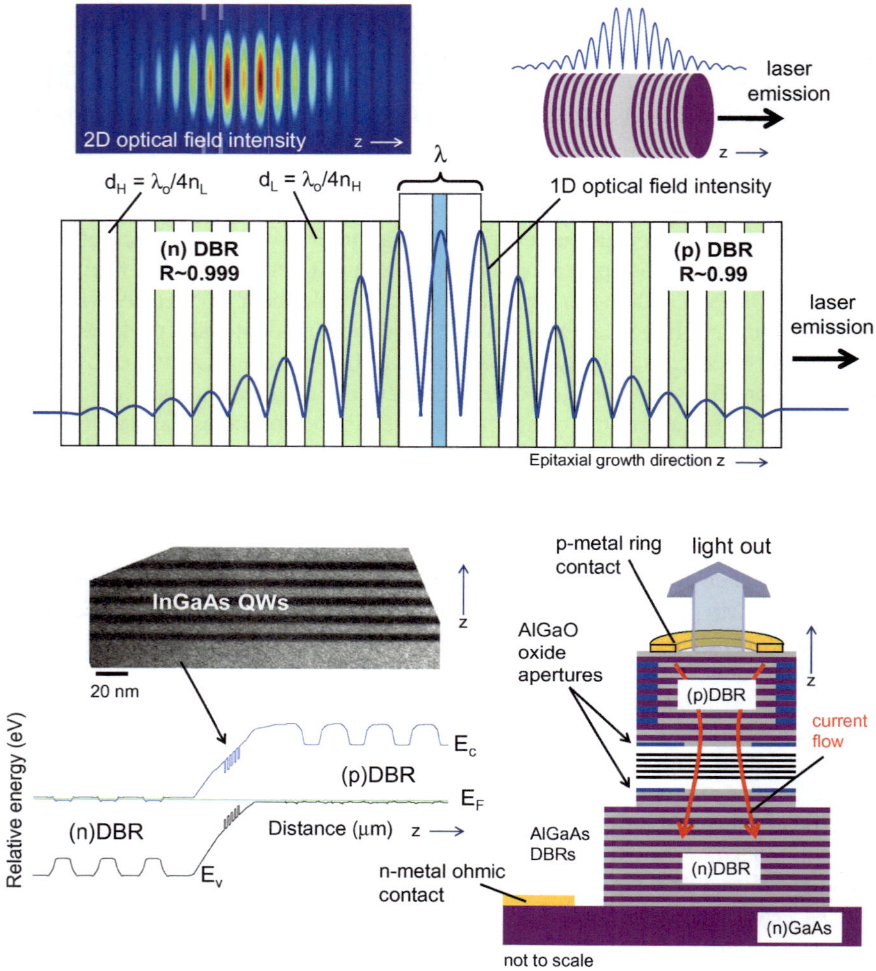

Fig. 10.1 Simplified diagrams and basic simulation plots of a generic oxide aperture 980 nm VCSEL. Optically the VCSEL is an asymmetric multiple-layer Fabry-Pérot etalon that is a resonant cavity in the vertical *z*-axis direction (in the direction perpendicular to the starting GaAs substrate surface). The top and bottom planar mirrors are composed of alternating layers of low index of refraction and high index of refraction AlGaAs, grown epitaxially on a GaAs substrate. The optical cavity active region consists of multiple quantum well (QW) layers (five layers are shown here via a transmission electron microscopy cross-section image). Electrically the VCSEL is a *pn* junction diode in the form of one or more stacked cylinders with some unintentionally doped material in the optical cavity layer. Example one-dimensional (1D) and 2D simulations of the optical field intensity distributions on resonance are plotted, and an example 1D real-space energy band diagram at 0 mA applied forward bias current with about three DBR periods on each side of the optical cavity is included, where E_F is the Fermi level, E_C is the conduction band, and E_v is the valence band

and data systems capable of operation at bit rates from 25 to over 50 gigabits-per-second (Gbps or Gb/s) with energy efficiencies approaching 100 femtojoules-per-bit. These VCSELs are the light sources in modern data centers and supercomputers. In addition to these applications, huge new markets related to the Internet-of-Things and other consumer areas based on both optical fiber links and on free-space optical links will soon dominate and expand the VCSEL market by an estimated compound annual growth rate exceeding 31% from an estimated market value of $738M in 2018 to an estimated market value of $3,775M in 2024 [23]. We further present our VCSEL research which we anticipate will impact a plethora of new and emerging photonic application areas including sensing systems such as light detection and ranging (LiDAR), and 3D object tracking and recognition. This includes new device and packaging designs that optimize both VCSEL bandwidth and optical output power leading to VCSEL arrays capable of sending data or tracking objects at distances of 100s–1000s of meters. Finally we explore VCSEL designs suitable for integration with silicon and other technologies for use in photonic-electronic integrated circuits. We investigate new designs such as the use of subwavelength gratings to reduce the vertical dimensions of VCSELs and new designs that allow the transfer of VCSELs onto silicon host wafers to form complete micrometer-scale or nanometer-scale electronic-photonic integrated systems. This includes highly advanced processing and design concepts, and studies of sub-micrometer oxide aperture diameter VCSEL structures.

10.2 VCSEL Experimental Structures

In Fig. 10.2 we show our experimental VCSEL structure. We use a planarized high frequency (HF) ground-signal-ground (GSG) pad layout for ease of on wafer probing using a standard HF GSG probe head with pin-to-pin spacing of 150 μm [24]. The VCSELs are grown on 76.2 mm-diameter GaAs wafers and processed into double mesa structures using standard cleanroom processing techniques developed at the Technical University of Berlin including ultraviolet contact mask lithography, selective thermal oxidation, thin-film metal deposition via thermal evaporation and electron beam deposition, plasma enhanced chemical vapor deposition of thin SiO and SiN dielectric films, and reactive ion etching of III–V materials such as GaAs, AlGaAs, and InGaAs. We typically process quarter wafer pieces as illustrated in Fig. 10.3, where we employ an experimental VCSEL mask set containing repeating unit cells with variations in VCSEL geometry [25]. The variations include different mesa diameters and different shapes and locations of the n-metal contacts. In Fig. 10.3 we show a colorized wafer map indicating the maximum optical output power of each of the VCSELs on a processed quarter wafer piece.

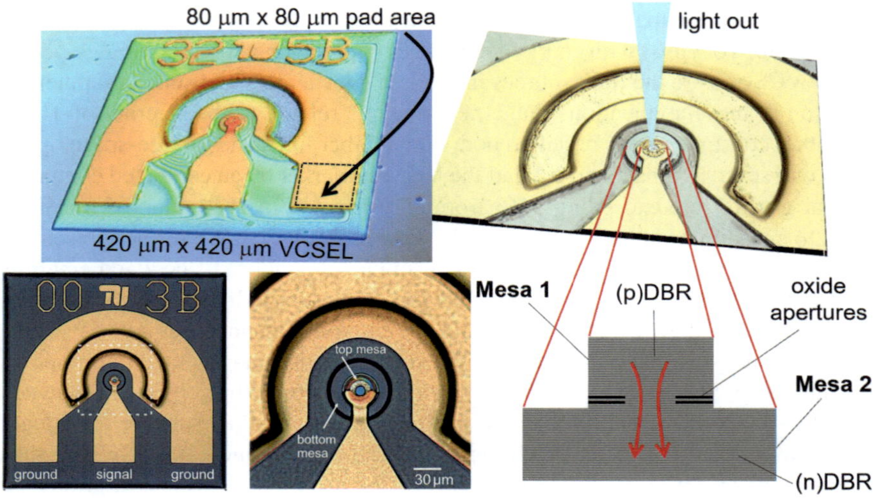

Fig. 10.2 Montage of images and a schematic cross section of the VCSEL structure

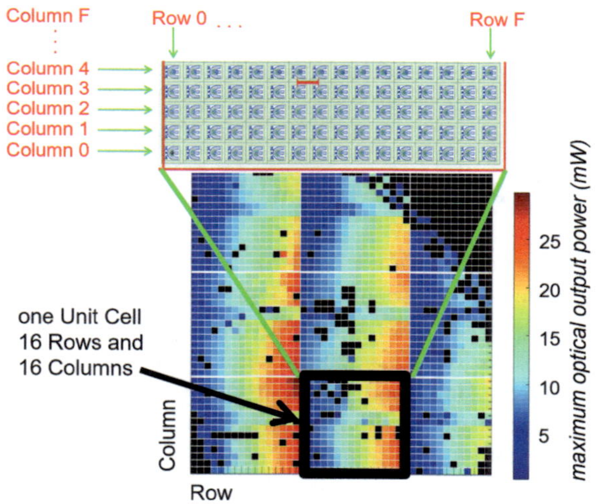

Fig. 10.3 Colorized mapping of the measured continuous wave maximum optical output power for a quarter of a processed 3-inch (76.2 mm) diameter VCSEL wafer, and description of one unit cell in the VCSEL mask set. The colors of each pixel represent the maximum optical output power from one VCSEL as determined from the measured light optical output power (*L*) versus forward bias current (*I*) characteristic at room temperature. A black pixel represents a non-lasing VCSEL or a position off the quarter wafer piece

10.3 VCSEL Processing, Geometric Variations, and Characterization

For a first study of VCSEL processing geometries we investigate 980 nm VCSELs with AlGaAs-based *p*-doped top and *n*-doped bottom DBRs as depicted in Fig. 10.4 [26]. The top and bottom DBRs contain optically inactive GaAs-based current spreading regions. The current spreading regions have an optical thickness of 1.75λ and are highly doped at the position of the electric-field intensity nodes in order to facilitate current spreading. The top *p*-type DBR contains two separate current spreading regions 15 and 4 DBR pairs above the active region which consists of 5 compressively strained InGaAs quantum wells (QWs) placed within a λ/2-thick optical cavity. Transverse optical and current confinement is realized by three thin tapered oxide apertures formed by the selective wet oxidation of three 20 nm-thick $Al_{0.98}Ga_{0.02}As$ layers. Two of these three oxide aperture layers are above the active region in the *p*-DBR whereas the *n*-DBR contains one single oxide aperture layer. In between these two top and one bottom oxide aperture layers, located in the first low-index layers on both the *p*-doped and the *n*-doped sides surrounding the optical cavity are two 10 nm-thick $Al_{0.98}Ga_{0.02}As$ layers that serve as shallow oxide layers (that is, the length of these oxide layers in from the edge of the etched mesa sidewall is smaller than the length of the 20 nm-thick oxide aperture layers in from the same mesa sidewall). In order to further reduce the parasitic capacitance, we employ five additional 20-nm-thick $Al_{0.96}Ga_{0.04}As$ shallow oxide layers positioned directly above the second current spreading layer counted from the top of the *p*-doped DBR. Also, we place one 10-nm-thick $Al_{0.98}Ga_{0.02}As$ deep oxide layer directly above the two *p*-type 20-nm-thick oxide aperture layers. The number and position of the oxide aperture and shallow oxide layers as well as the current spreading layers in the *p*- and *n*-DBRs are illustrated via the scanning electron microscope montage image in Fig. 10.4.

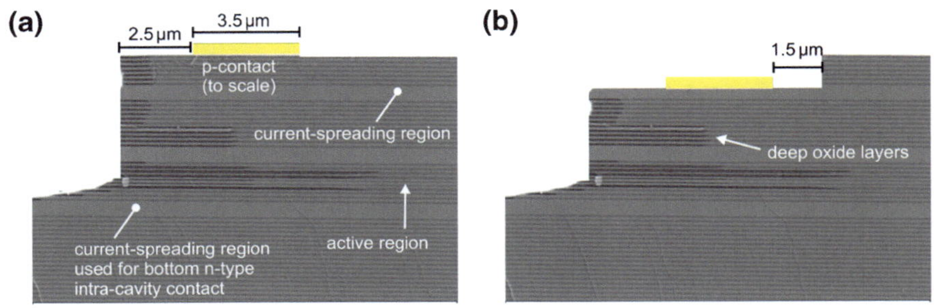

Fig. 10.4 Montage using a scanning-electron microscope (SEM) image of processed epitaxial VCSEL material and a hand-drawn metal contact to illustrate the different positions of the *p*-type top Ti/Pt/Au metal contact relative to the selectively oxidized deep-oxidation and oxide aperture layers for two different processing variations: **a** extra-cavity *p*-type top contact; and **b** intra-cavity top contact made to the first *p*-type current spreading layer from the top. The left edges (mesa sidewalls) of the VCSEL structure in the SEM image are artificially made perfectly vertical. Copyright © 2017 SPIE. Reprinted, with permission, from [26]

The epitaxial VCSEL design allows processing of the material into six different variations of extra-cavity and intra-cavity contacts. Both *n*-type and *p*-type contacts can be realized as extra-cavity (outside of the optical cavity) contacts. Since the *p*-doped top DBR contains two current spreading regions, intra-cavity (within the optical cavity of the asymmetric multilayer Fabry-Pérot etalon—for example within a DBR) *p*-type contacts can be realized at two different positions. The first current spreading region, counting from the topmost epitaxial layer, is 15 DBR pairs away from the optical cavity active region, while the second current spreading region is only 4 DBR pairs above the optical cavity active region. We use the first current spreading region, which is further away from the active region, to realize a *p*-type intra-cavity contact. By using the extra-cavity contact layers and one current spreading region in each of the DBRs for realizing electrical *p*- and *n*-contacts, we realize four different sample processing variations of the same epitaxial wafer material as shown in Fig. 10.5a through Fig. 10.5d. Sample A has extra-cavity top and bottom contacts, sample B has an extra-cavity top-contact and an intra-cavity bottom contact, sample C has an intra-cavity top-contact and an extra-cavity bottom contact, and sample D has top and bottom intra-cavity contacts. One 3-inch (76.2 mm) diameter wafer is cleaved into four quarter pieces and each of these four pieces is processed into one of the four processing samples A, B, C, and D.

For samples A and B, with extra-cavity *p*-type top contacts, the first processing step is the evaporation of the Ti/Pt/Au *p*-contacts with an electron beam metal evaporation system, followed by mesa-etching with an in situ-controlled reactive

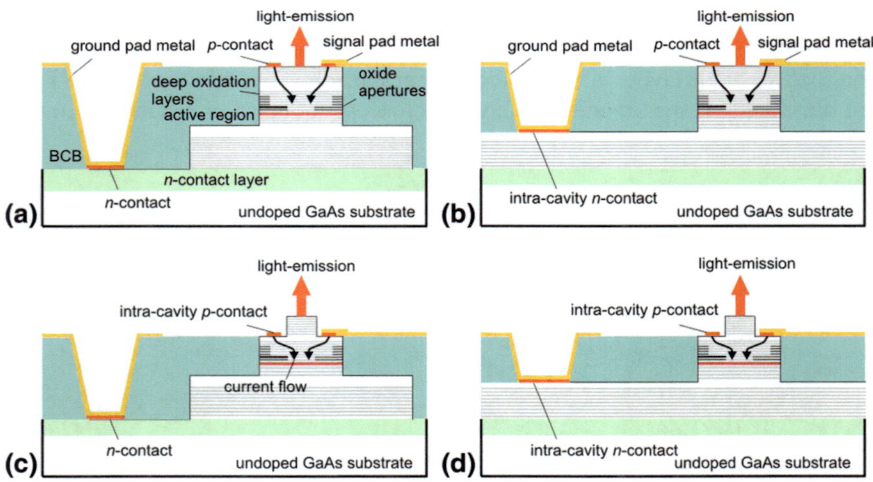

Fig. 10.5 Schematics of the four different realized processing sample variations of the top-emitting oxide-confined 980 nm VCSELs: **a** sample A, extra-cavity top and bottom contacts; **b** sample B, extra-cavity top contact and intra-cavity bottom contact; **c** sample C, intra-cavity top contact and extra-cavity bottom contact; and **d** sample D, intra-cavity top and intra-cavity bottom contacts. The schematics are not to scale. Copyright © 2017 SPIE. Reprinted, with permission, from [26]

ion etching system using a BCl_3/Cl_2 plasma. The first mesa etch is stopped inside the optically 1.75λ-thick current spreading region of the *n*-type bottom DBR. For a better comparison, the etch-time for the first mesa of both samples A and B is chosen to be exactly the same. Since both samples are of an equal size, it can be estimated that the resulting etch-depth will be the same for both samples. In the following step, the oxide layers are formed in an oxidation furnace. Both samples are oxidized together in the same process run. Sample A, with an extra-cavity bottom n-contact, is then etched for a second time with the same process used for the definition of the first mesa, allowing us to contact the GaAs n-contact layer below the bottom DBR, resulting in a double mesa structure. Then the *n*-type Ni/AuGe/Au bottom contact is evaporated using a thermal evaporator. After planarization with photo-active bisbenzocyclobutene (BCB), the final top pad metallization with Cr/Pt/Au is performed using the electron beam evaporator again, resulting in coplanar high frequency contacts in a ground-signal-ground configuration. Samples C and D are processed similarly to samples A and B. The only difference is that a first mesa etch is performed prior to the *p*-type top contact metallization. This mesa etch is stopped after removing the topmost 7 *p*-doped DBR pairs plus roughly up to about a λ/2-optical thickness of the (*p*)GaAs within the top current spreading region.

We perform static LIV measurements of all processed VCSELs from all four samples A, B, C, and D at 25 °C (taken as room temperature) by an automatic wafer mapping system. The measurement data are then automatically analyzed using a script written in the Python programing language. In total, 1145 working VCSELs are included. The rollover current and the maximum optical output power of all investigated VCSELs are plotted versus the differential resistance in Fig. 10.6 (top graph) and Fig. 10.6 (bottom graph), respectively. The VCSELs of sample B have a slightly smaller rollover current and smaller maximum optical output power at a given differential resistance than the VCSELs of the other three samples A, C, and D, while the devices from sample C have the largest rollover currents and largest maximum optical output powers at a given differential resistance.

To compare all 1145 VCSELs and determine the impact of the process variations on the VCSEL properties, we use the differential resistance of the VCSELs, determined by a linear fit of the IV curve between five times the threshold current up to the current at thermal rollover, for comparison. To determine the impact of the different contacting schemes on the dynamic properties of the VCSELs, small-signal and large-signal modulation experiments are performed with VCSELs of particular differential resistance values. We use an HP 8722C vector network analyzer and a New Focus 1434-50-M photodetector module for measuring the small-signal modulation responses in a wide range of bias currents from the threshold current until the thermal rollover current. For each measured VCSEL 40–60 small signal modulation response curves are fitted automatically using a Python script. The D-factors (D), obtained from a linear fit of the resonance relaxation oscillation frequency (f_r) versus the square-root of the bias current (I) above the threshold current (I_{th}) thus $D = f_r/(I - I_{th})^{1/2}$, is plotted versus the differential resistance in Fig. 10.7.

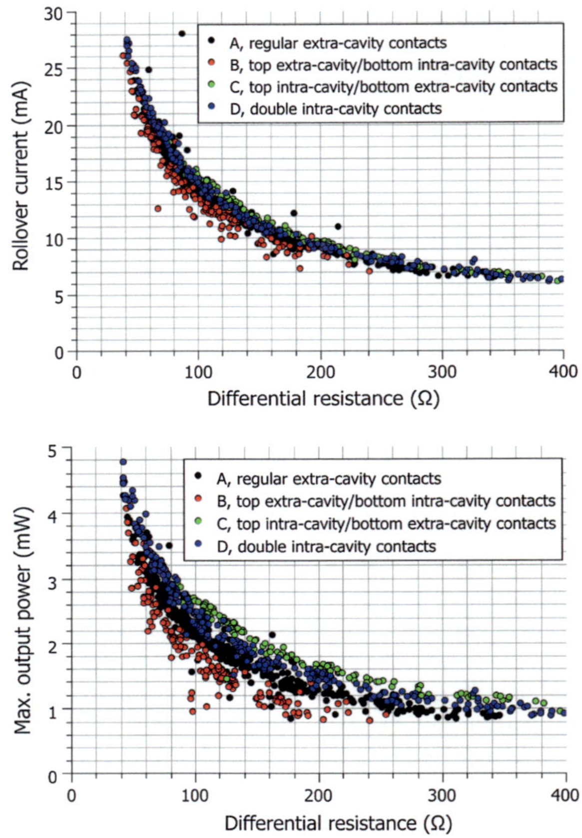

Fig. 10.6 Rollover current (top graph) at 25 °C versus the differential resistance determined for all VCSELs in the four different quarter wafer processing samples A (black circles), B (red circles), C (green circles), and D (blue circles). Maximum optical output power (bottom graph) at 25 °C versus the differential resistance for the same devices as in the top graph. Copyright © 2017 SPIE. Reprinted, with permission, from [26]

While the process variations do not have a significant influence on the D-factor for VCSELs with differential resistance values of roughly between 60 and 80 Ω, that is for VCSELs with larger oxide aperture diameters, there is a significant increase in the D-factor for VCSELs from sample B for differential resistances exceeding 80 Ω. For a given differential resistance, the impact of the process variations on the D-factor is the inverse of the impact of the process variations on the optical output power. While the order from largest to smallest optical output power at a given differential resistance is C, D, A, B, the order from the largest to the smallest D-factor is B, A, C, D. This means that VCSELs from sample B have smaller maximum optical output powers but simultaneously significantly larger D-factors for a given differential resistance larger than 80 Ω as compared to the other samples.

We conclude that when comparing different VCSELs from samples A, B, C, and D with the same differential resistance, we in fact compare VCSELs with different mode-volumes and probably different oxide aperture diameters. For a given differential resistance, the VCSELs from sample B, with a top extra-cavity contact and a bottom intra-cavity contact, have smaller optical output powers and smaller rollover

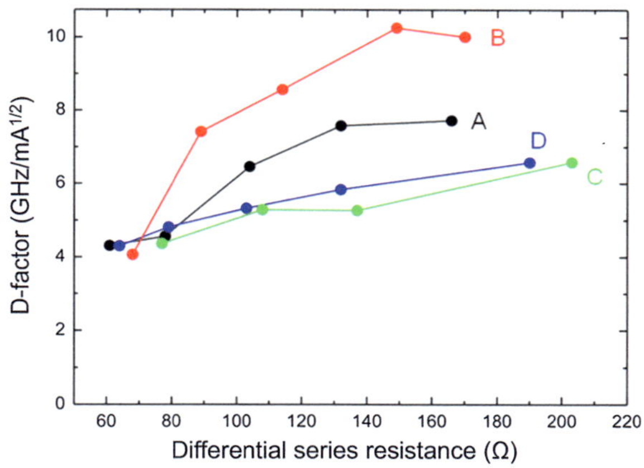

Fig. 10.7 The D-factor versus the differential series resistance for VCSEL samples A, B, C, and D. For differential series resistance values of approximately 85–170 Ω the VCSELs of processing sample B (with the extra-cavity top contact and the intra-cavity bottom contact), have a significantly larger D-factor than the VCSELs of the other processing variations. Copyright © 2017 SPIE. Reprinted, with permission, from [26]

currents but larger D-factors, as if the oxide-aperture diameter is smaller than for the other devices. This likely means that the processing variation for sample B successfully reduced the differential resistance for a given oxide aperture diameter compared to the standard processing of sample A with two extra-cavity contacts.

To determine the potential for data transmission of the processed extra-cavity and intra-cavity contacted VCSELs we perform large-signal modulation experiments. A non-return-to zero (NRZ) bit pattern with a word length of $2^{23} - 1$ bits is used for all measurements. Devices with a differential resistance of 80 and 70 Ω from sample A and sample B, respectively, achieve open optical eyes at 30 and 40 Gbps. For both VCSELs the signal-to-noise ratio exceeds 4.0 at 30 Gbps. We achieve error-free operation at 30 Gbps with both VCSELs, while a slight error-floor is seen for the VCSEL from sample A operating at 40 Gbps. The device from sample B operates error-free at 40 Gbps. The bit error ratio (BER) versus the received optical power is shown for the VCSEL from sample A in Fig. 10.8 (on the left) and for the VCSEL from sample B in Fig. 10.8 (on the right).

A comparison of the static LIV data for our different process variations versus the differential resistance of the VCSELs shows that those VCSELs with a bottom intra-cavity contact and an extra-cavity top-contact achieve smaller maximum optical output power than the VCSELs of the other three contact variations. Simultaneously those VCSELs have a significantly larger D-factor for differential resistance values exceeding approximately 80 Ω as compared to the other process variations. At room temperature our VCSELs operate error-free at up to 40 Gbps with a $2^{23} - 1$ NRZ pseudorandom binary sequence (PRBS) modulation scheme in a back-to-back testing configuration.

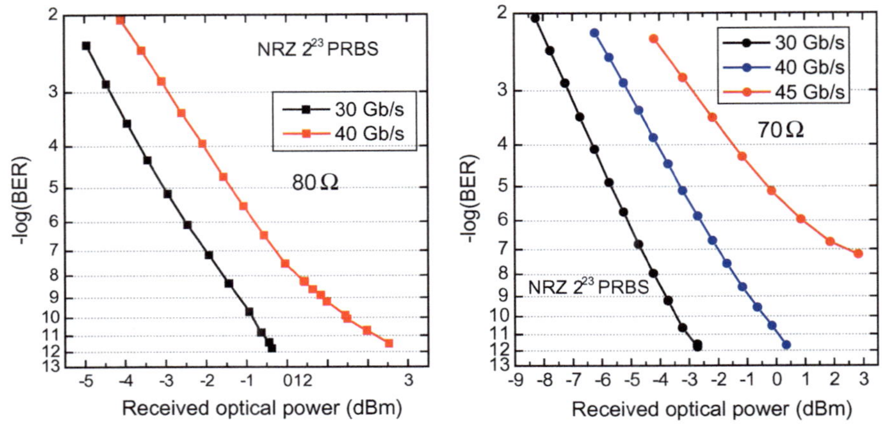

Fig. 10.8 Room temperature bit error ratio (BER) versus received optical power for (on the left) a VCSEL from sample A with a differential resistance of 80 Ω modulated at 30 and 40 Gbps with a $2^{23} - 1$ non-return-to-zero (NRZ) pseudorandom binary sequence (PRBS); and (on the right) the BER versus received optical power for a VCSEL from sample B with a differential resistance of 70 Ω. Copyright © 2017 SPIE. Reprinted, with permission, from [26]

10.4 Reduced Vertical Dimension VCSELs

One method to reduce the vertical thickness of VCSELs, which in turn reduces the epitaxial growth time and thus the epitaxial wafer cost, is to replace the DBR mirrors with subwavelength high index contrast gratings (HCGs). Thinner VCSELs may additionally prove easier to integrate with silicon integrated circuits since they would require less chip area and better match the height of metal interconnecting layers. The physics behind the very high optical reflectivity of the standard HCG requires that the grating must be made of high refractive index material that is surrounded by low refractive index material. This requirement is likely the key reason why only a few HCG designs have been realized for use in VCSELs, of which a high refractive index grating membrane suspended in air is a common design. The suspended membrane containing the HCG mirror may be electrostatically or thermally actuated (caused to move vertically small distances) to produce a micro-electro-mechanical tunable VCSEL with a range of peak emission wavelengths. If an HCG is used as a coupling mirror for an electrically driven VCSEL the low refractive index material—an air gap or a dielectric—must be incorporated in the VCSEL optical cavity. This HCG VCSEL design has practical problems with current spreading, heat dissipation, and mechanical stability.

The subwavelength monolithic high index contrast grating (MHCG) is a special case of an HCG [27]. The high power reflectance R of an MHCG is independent of the refractive index contrast between the grating and the material which lies beneath the grating. A special case of such a situation is when the grating is made of the same material as the cladding. An MHCG may be monolithically integrated with

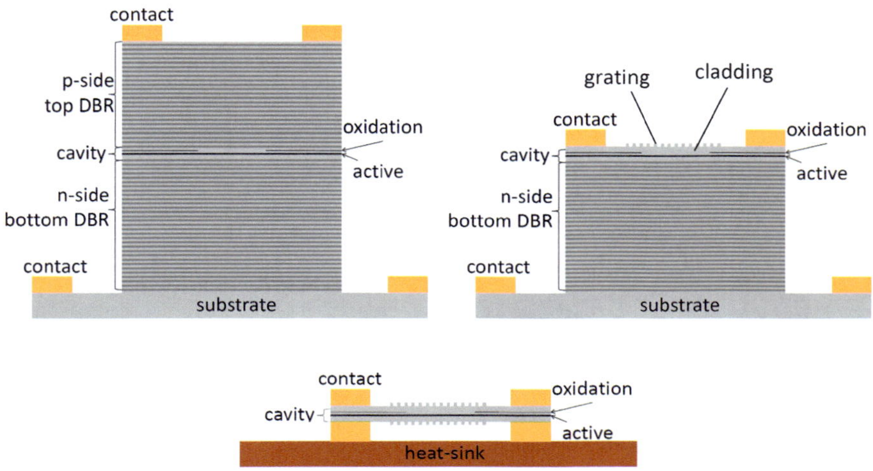

Fig. 10.9 Schematic diagram of conventional all-semiconductor top-emitting VCSEL, a VCSEL where the top DBR is replaced with a monolithic subwavelength high index contrast grating (MHCG) mirror, and a double MHCG VCSEL with top and bottom MHCG mirrors. Copyright © 2016 IEEE. Reprinted, with permission, from [28]

its substrate and may thus be composed of a single material layer such as gallium arsenide. The general idea to use an MHCG to replace standard semiconductor DBRs is depicted in Fig. 10.9. In Fig. 10.10 we show simulations comparing the optical field intensities on resonance of an all-semiconductor DBR VCSEL [28] and a double MHCG VCSEL [29]. The MHCG VCSEL is roughly 10% the thickness of the DBR VCSEL.

We investigate 980 nm VCSELs that employ a composite top mirror consisting of a 5.5 period AlGaAs distributed Bragg reflector (DBR) and a topmost MHCG which replaces the majority of what would be a typical DBR mirror [30]. Static measurements show continuous wave operation of the VCSELs from room temperature up to 75 °C. The VCSEL with a 13.5 μm current oxide aperture diameter operates with quasi-single mode emission from threshold to rollover. Our results open a way to produce all-semiconductor surface emitting lasers capable of emitting at wavelengths from ultraviolet and visible (GaN-based) to infrared (InP- and GaSb-based) with the vertical thickness and thus the manufacturing costs potentially reduced by up to roughly 90% in comparison to the typical DBR VCSELs. These vertically-thin VCSELs have potential applications as light sources in photonic-electronic integrated circuits.

A schematic cross-section and top down view of our DBR MHCG VCSEL is shown in Fig. 10.11. This DBR MHCG VCSEL is a top-emitting, gallium arsenide-based single cavity VCSEL with two mesas and two oxide current confinement apertures. Electrical connections are realized by a top metal ring contact and a bottom half to three-quarter metal-ring contact which are connected to high frequency ground-signal-ground pads deposited on planarizing BCB.

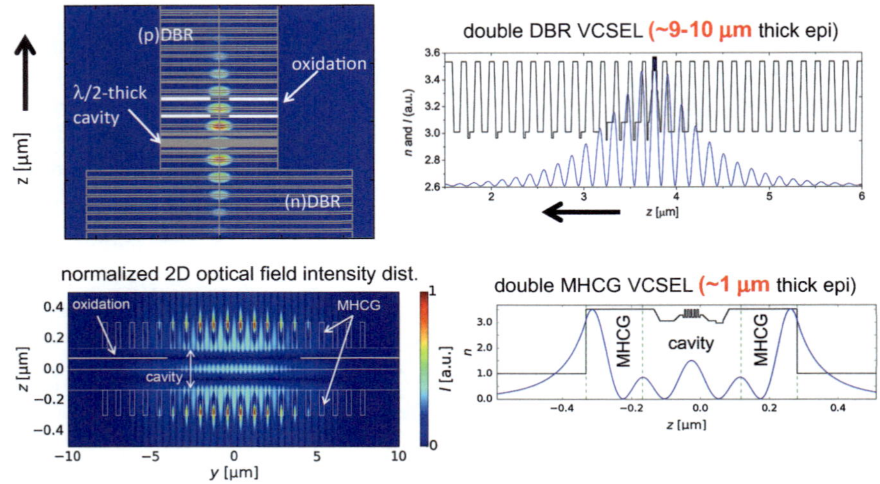

Fig. 10.10 Simulated two-dimensional (left) and one-dimensional (right) fundamental mode optical field intensity distributions on resonance for double DBR and double MHCG 980 nm VCSELs. The black lines (on the right) are the real refractive index profiles and the blue lines are the one-dimensional optical field intensity distributions. Copyright © 2015 and 2016 IEEE. Reprinted, with permission, from [28, 29]

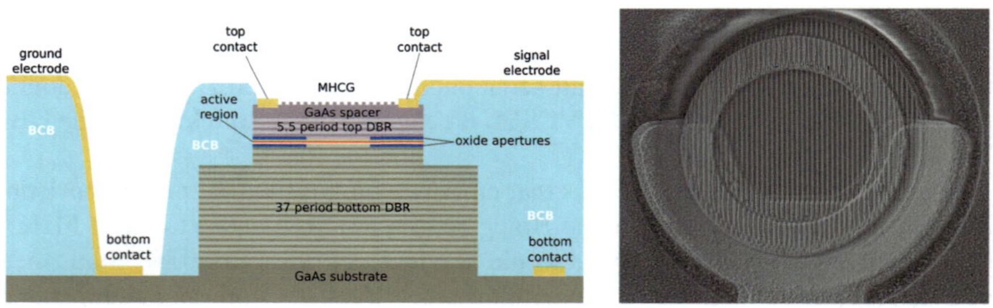

Fig. 10.11 Schematic cross-section of a VCSEL with a top composite DBR and MHCG mirror (left), and a top down view scanning electron microscope (SEM) image of a DBR MHCG VCSEL after device processing (right). Reprinted with permission from [30] copyright © The Optical Society

The epitaxial structure is based on our state-of-the-art communication 980 nm VCSEL design. On the surface of an 450 μm *n*-doped gallium arsenide substrate an ~1.58 μm thick *n*-doped GaAs buffer layer is grown. Following the buffer layer we grow 37 pairs of *n*-doped GaAs/$Al_{0.9}Ga_{0.1}As$ DBRs with simple linear compositionally-graded interfaces. The multi-quantum-well active region consisting of 5 compressively-strained InGaAs quantum wells (QWs) and 6 GaAsP barrier layers is embedded in the center of a half-lambda optically-thick cavity. The interfaces between the active zone and the top and bottom DBRs are realized by AlGaAs graded layers of various compositions. The top coupling mirror is composed of a 5.5 period DBR of *p*-doped GaAs/$Al_{0.9}Ga_{0.1}As$ again with simple linear compositionally-graded interfaces capped with an ~425 nm *p*-doped GaAs phase-matching layer. The last 30 nm of the phase-matching layer is highly doped (to ~10^{20} cm^{-3}) to form a low resistivity electrical contact with a top-contact Au/Zn/Au metal ring. The top-most part of the epitaxial structure is composed of a 341 nm-thick undoped GaAs layer for the MHCG mirror patterning separated from the rest of the structure with a 20 nm-thick GaInP etch-stop layer that is lattice-matched to GaAs. Two Al-rich 20 nm-thick layers for oxide aperture formation are in the first nodes of the optical field intensity standing wave closest to the active region, one each at the interfaces of the top and bottom DBRs and the half-lambda-thick optical cavity. The calculated optical field intensity profile and real refractive index distribution in the top part of the VCSEL structure is presented in Fig. 10.12.

The mask set used in the processing is the same as previously described in Figs. 10.2 and 10.3. The mask set includes several identical unit cells, in the form of a two-dimensional matrix. Every unit cell is divided into 16 rows and 15 columns which results in 240 VCSELs in each unit cell. There is an additional 16th column containing 16 rows of test VCSELs. The mesa diameters are varied within every column from 18 μm (row 0) to 31 μm (row F) with a 0.5 μm step between rows 0 to 4 and with a 1.0 μm step between rows 4 to F. Nine different MHCG mirror geometries were patterned on the sample. All the MHCG types have, by design, identical period $L = 0.500$ μm and thickness $H = 0.298$ μm, while we varied the fill factor (defined as the ratio of the grating stripe width to the grating period) from 0.47 to 0.55 with 0.01 steps. Due to the inaccuracy of the MHCG processing, all the resulting MHCG patterns have a fill factor and thickness smaller than the design value by about 0.1 and 60 nm, respectively.

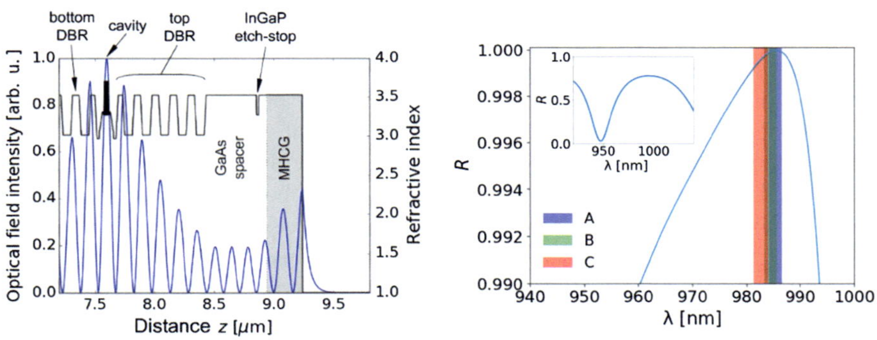

Fig. 10.12 Simulated optical field intensity profile (left, blue line) and real refractive index at 980 nm (left, black line) both versus distance z in the top part of the DBR MHCG VCSEL; and simulated power reflectance (R) of the top DBR MHCG outcoupling mirror (right) and 5.5 top DBR periods (inset) as seen from the optical cavity at normal incidence for TM polarized light and ranges of fundamental mode wavelengths for devices A (blue), B (green) and C (red) from threshold to rollover at 25 °C. In the simulations we use the real measured geometry of the MHCG mirrors as determined from cross-section SEM imaging. Reprinted with permission from [30] copyright © The Optical Society

Our experimental grating parameters result in a nominal power reflectance of the entire outcoupling mirror of 0.9999 for TM-polarized light (where by definition the only non-zero component of the electrical field vector is perpendicular to the stripes) at 980 nm. The shapes of the grating cross-sections are not perfectly rectangular which suggests a reasonably high manufacturing tolerance of the MHCG mirror as predicted in our theoretical work.

The processing of our VCSELs involved planar processing developed and used routinely at the Technical University Berlin for standard DBR VCSEL processing. Additionally, the processing included the patterning of the MHCG gratings by electron beam lithography (EBL). In the first step the wafer sample was covered with an EBL resist and the MHCG fields, 100 μm $\times$ 100 μm in area, were patterned using an EBL tool. After development the patterns were etched using $Cl_2/BCl_3/Ar$ plasma in an ICP-RIE dry-etching reactor. Once the MHCGs were made, standard planar techniques for processing VCSELs were used. Annular patterns in photoresist to define the areas where we will next place top p-metal ring contacts were lain down on top of MHCG fields using standard contact UV-lithography. Then the exposed parts of the MHCGs together with the undoped InGaP etch-stop layer were removed using a mixture of wet- and dry-etching techniques. Then, Au/Zn/Au top metal contact rings were deposited using thermal evaporation.

In the following step, top mesas were patterned and dry-etched using a Cl_2/BCl_3 plasma in an ICP-RIE reactor. The etching was stopped after passing through the active region at the fifth bottom DBR pair to open access to both the top and bottom Al-rich layers, which after selective thermal wet-oxidation for 130 min at 420 °C and 50 mbar formed the top and the bottom oxide apertures. A second dry-etching step was then performed in order to form the second mesa and reach the (n+)GaAs buffer layer that lies just under the bottom *n*-doped DBR. The sample was then planarized using a photo-sensitive BCB spin-on film, which was selectively removed to expose the top mesa together with the top *p*-metal contact rings and create openings for the bottom *n*-metal contact deposition. The sample was covered with an ~300 nm thick PECVD-deposited SiN layer to increase the mechanical stability of the BCB and to improve the adhesion of metal to the BCB surface. The SiN was then patterned and etched by RIE to once again expose the top mesas and bottom contact openings. In the final step, Ni/AuGe/Au GSG pads were deposited using thermal evaporation in a standard photoresist lift-off process. For this processing we use the ground part of the GSG pad (ground-signal-ground, see Fig. 10.2) as the bottom *n*-metal contact. Thanks to this approach the processing was simplified by removing one lithography step. We thus did not perform our standard bottom *n*-metal contact lithography and deposition step since the GSG pad metal served also as *n*-contact metal without any noticeable deterioration of the device performance.

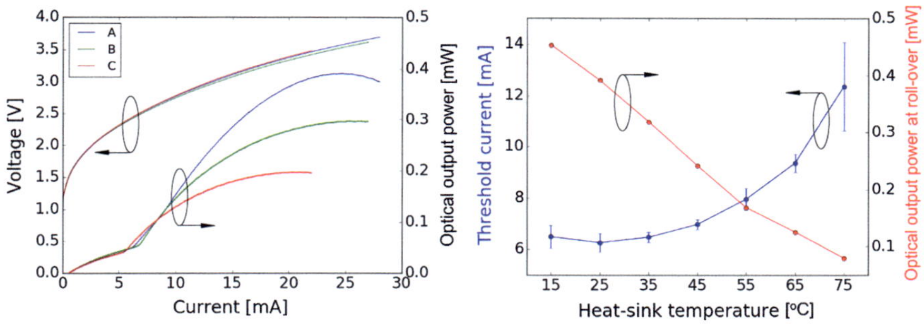

Fig. 10.13 Light output power-voltage-current (LIV) characteristics for devices A, B, and C taken at 25 °C (on the left); and threshold current (blue) and optical output power at rollover (red) vs. heat-sink temperature for device A (on the right). The set of curves enclosed by ovals are directed by the arrows toward the left or right vertical axes. Reprinted with permission from [30] copyright © The Optical Society

We designate our processed DBR MHCG VCSELs device A, B, and C. Devices A and B have a top mesa diameter of 31 μm while device C has a top mesa diameter of 28 μm. We measure the oxide aperture diameters by focused ion beam trenching to be 16.5 μm for device A and B and 13.5 μm for device C. As seen in Fig. 10.13 (left) the three VCSELs have typical diode IV curves. One can also see well pronounced threshold and rollover currents in the LIV curves taken at 25 °C. The threshold currents are equal to 6.18 mA, 6.75 mA, and 5.48 mA, which correspond to threshold current densities equal to ~2.89 kA/cm^2, ~3.16 kA/cm^2 and ~3.83 kA/cm^2 for devices A, B and C, respectively. The maximum optical output power at 25 °C is equal to 0.39 mW, 0.30 mW, and 0.20 mW for devices A, B and C, respectively. In Fig. 10.13 (right) we show the threshold current and optical output power at rollover as a function of the heat-sink temperature for device A. The data is extracted from the LIV curves measured at temperatures from 15 °C to 75 °C with a 10 °C step. The VCSEL exhibits a maximum optical output power of 0.45 mW at 15 °C and a minimum threshold current of 6.18 mA at 25 °C.

Emission spectra a little below the VCSEL threshold (at a forward bias of 5 mA) and above the threshold taken at 25 °C for devices A and C are presented in Fig. 10.14 (right) and 10.14 (left), respectively. Device A as well as device B whose spectra is not shown emit in a double lateral mode over the full range of forward bias currents from threshold to rollover. Device C in contrast emits in a quasi-single mode over the full range of forward bias currents from threshold to rollover with a maximal side mode suppression ratio (SMSR) equal to 36 dB at 11 mA. Emission spectra for a device identical to devices A and B but without the MHCG mirror indicates that lasing is not possible for VCSELs with the 5.5 period top DBR but without the MHCG. Reference VCSELs with a 15.5 period top AlGaAs DBR and without the MHCG show strong multimode operation for oxide aperture diameters of ~5 μm and larger. Hence, the single and double-mode operation of the DBR MHCG VCSELs with such large oxide aperture diameters (ϕ ~ 13.5 and 16.5 μm) must result directly from the influence of the MHCG portion of the top coupling mirror.

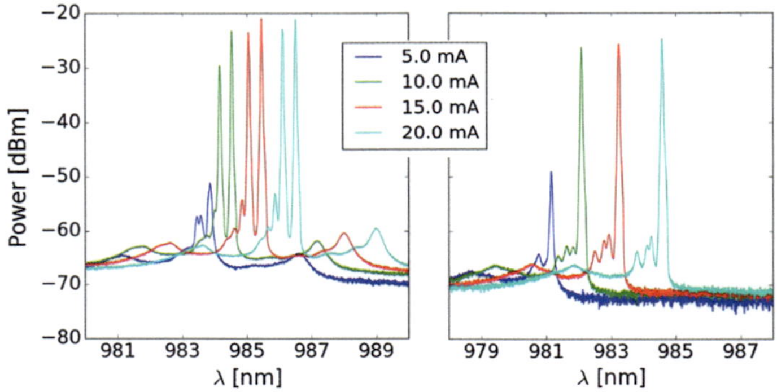

Fig. 10.14 Emission spectra: (on the left) of the double mode VCSEL A; and (on the right) of the single mode VCSEL C both for various forward bias currents taken at a 25 °C ambient temperature. For both Devices A and C the emission spectrum taken at 5 mA is below the laser threshold. Reprinted with permission from [30] copyright © The Optical Society

10.5 High Modulation Bandwidth VCSELs

The small-signal modulation bandwidth (f_{3dB}) of VCSELs is a key parameter for VCSEL applications in high bit rate communication systems. We seek to increase the VCSEL f_{3dB} to its ultimate limit while simultaneously maintaining reliable VCSEL operation and low-cost manufacturing. While the limit of the f_{3dB} of conventional single cavity vertical cavity surface emitting lasers (VCSELs) remains to be determined, it is well recognized that further improvements in VCSEL performance will likely proceed together with improvements in thermal heat transfer, thus with the elimination of oxide apertures. In 2016 we reported a room temperature (RT) f_{3dB} bandwidth for conventional single cavity VCSELs of 30 GHz [31]. It is well established that the VCSEL f_{3dB} may be increased by reducing the VCSEL mode volume by, for example reducing the VCSEL oxide aperture diameter and the optical cavity thickness [32]. The consequence is reduced optical output power, increased threshold current density and thus potential reliability issues, and more. We investigate the f_{3dB} of one of our new 980 nm VCSEL designs as illustrated in Fig. 10.15. We use simple distributed Bragg reflector (DBR) linear grading, near constant doping, and two thin and opposing oxide apertures [25]. We use oxide apertures as a research convenience, as these allow us to produce high performance VCSELs on a single wafer with oxide aperture diameters (ϕ) of ~ 20 μm to <1 μm. To counter a reduction in ϕ we decrease the number of top coupling DBR periods to 14.5 (so $R < 0.990$) to increase the optical output power. With ϕ ~ 3 μm we consistently achieve a f_{3dB} exceeding 30 GHz at room temperature.

Figure 10.16 shows the results of an oxidation study on several cleaved pieces of one of our *simplicity* VCSEL wafers. We perform selective thermal oxidations at 420 °C and 50 mbar in a deionized water vapor environment. We thus can accurately estimate our ϕ based on: 1) the oxidation rate of 0.0658 μm/min from Fig. 10.16; 2)

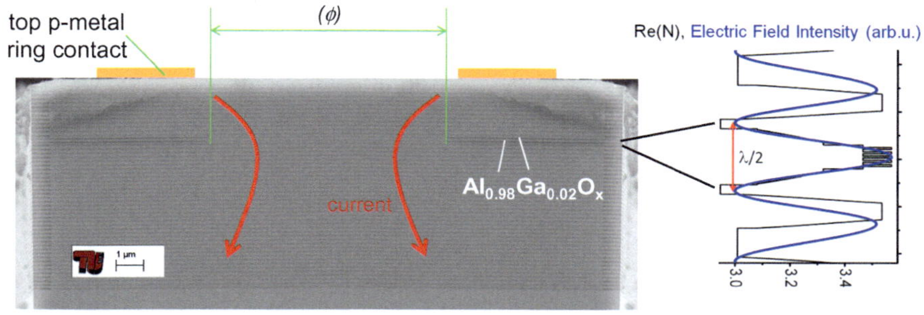

Fig. 10.15 Montage of a scanning electron microscope image and added labels and representative top metal contacts forming a cross-section (on the left) of a 980 nm VCSEL indicating the flow of current from a top anode metal contact toward a bottom cathode contact (not shown) within an area defined by the oxide aperture diameter (ϕ). Real refractive index profile (black line) and electric field intensity (blue curve) on resonance (on the right) in and around a VCSEL's half-lambda-thick optical cavity active region

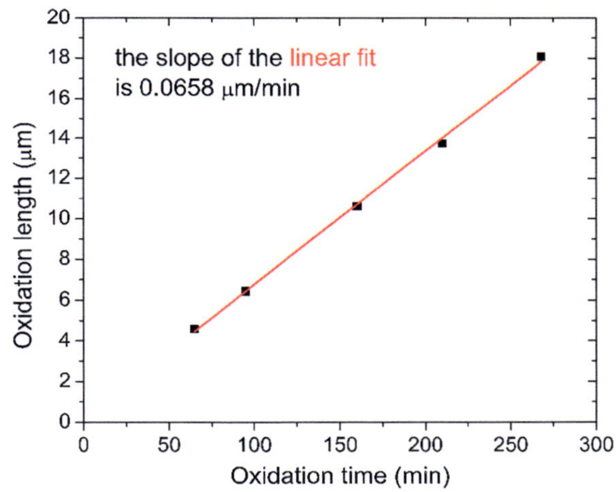

Fig. 10.16 Oxidation length versus time for a double oxide aperture 980 nm VCSEL, revealing a linear oxidation rate of 0.0658 μm/min. Copyright © 2019 IEEE. Reprinted, with permission, from [1]

the measured emission spectra; and 3) our experience gained from device fabrication cycles. In Fig. 10.17 (left) we show the LIV characteristics of a fabricated set of our VCSELs at 25 °C (see [1] and [33]). We oxidized this set of VCSELs for 140 min, so the oxidation length in from the mesa 1 edge is expected to be ~9.2 μm. In Fig. 10.17 (left) the LI characteristic labelled $\phi \sim 0 + \delta$ μm is for a VCSEL that is nominally expected to have $\phi \sim 0.1$ μm (that is, an 18.5 μm mesa 1 diameter minus 2 times the oxidation length of ~9.2 μm). At 25 °C the maximum LI slope efficiency is 1.04 W/A and the maximum optical output power at rollover is 5.6 mW for the VCSEL with $\phi \sim 3$ μm. The maximum external differential quantum efficiencies $\eta_{qe} = (q\lambda_o/hc) \cdot (\Delta L/\Delta I)$ for the VCSELs in Fig. 10.17 are 0.78–0.86, where q is the electron charge, h is Planck's constant, and c is the speed of light. In Fig. 10.17 (right) we see the

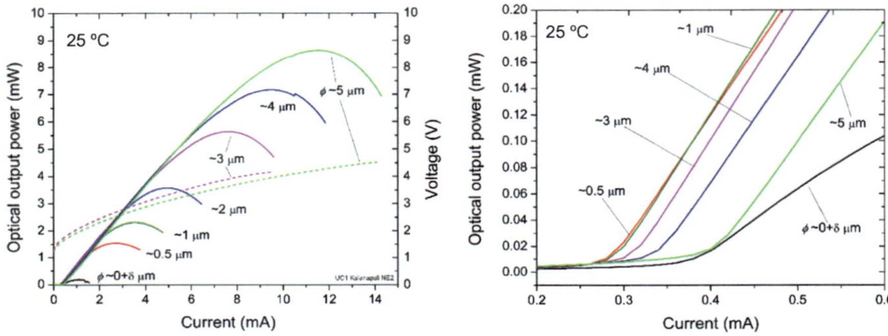

Fig. 10.17 Light-output-power versus current (LI, solid lines) and voltage-current (VI, dashed lines) characteristics for VCSELs with estimated oxide aperture diameters from near zero to $\phi \sim 5$ μm at 25 °C (left); and an expanded view of the LI characteristics (right). Copyright © 2019 IEEE. Reprinted, with permission, from [1]

minimum threshold is for the VCSEL with $\phi \sim 0.5$ μm, indicating that the VCSEL with $\phi \sim 0.1$ μm suffers from scattering and other optical losses as expected for such a small aperture.

Using the finite element method [34] we may numerically model the 980 nm VCSELs as a function of ϕ and as a function of the exact position of the two oxide aperture layers within the VCSEL structure [35]. In Fig. 10.18 (top images) we plot the electric field intensity on resonance and the real refractive index profile along the vertical (epitaxial growth) z-axis. The VCSEL and immediate surrounding material is treated as a cylinder surrounded by air with rotational symmetry about the vertical z-axis. We model the VCSEL first with the two oxide aperture layers centered on nodes of the electric field intensity on resonance, and second with the oxide aperture layers slightly offset from these same electric field intensity nodes away from the optical cavity active region. In Fig. 10.18 (bottom images) we plot the resonant wavelength as a function of ϕ, where we see the expected blue shift of the LP01 and LP11 optical mode resonances as ϕ decreases. The blue shift is stronger when the oxide aperture layers are offset from the electric field intensity nodes since in this case the optical field more strongly interacts with the oxide layers and thus is more strongly optically guided and confined. This result is reinforced by the graphs in Fig. 10.19 where we show the simulated normalized electric field intensity versus distance across the horizontal axis of the VCSEL. The mode confinement remains nearly constant for ϕ smaller than ~6 um, whereas when the oxide apertures are offset from the electric field intensity nodes the mode confinement for 980 nm VCSELs remains strong down to ~1 μm. This correlates well with our LIV results given in Fig. 10.17.

In Fig. 10.20 (left) we show the emission spectra for VCSELs with ϕ near zero and with $\phi \sim 3$ μm. The spectra for the two VCSELs are offset by 10 dB for clarity. The emission is almost purely single-mode for the smallest VCSEL with a side mode suppression ratio (SMSR) > 40 dB. For the VCSEL with $\phi \sim 3$ μm the emission is

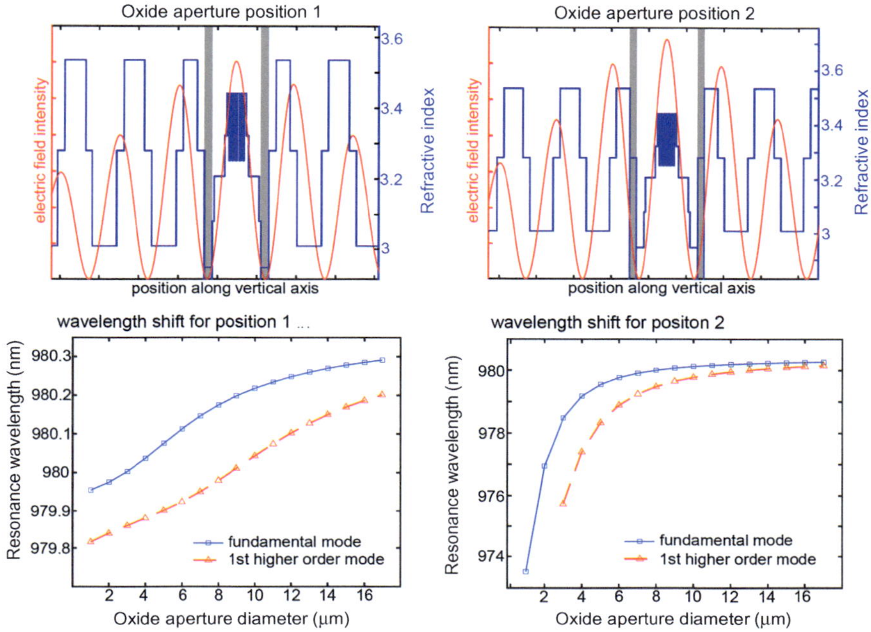

Fig. 10.18 Simulated electric field intensity on resonance and real refractive index profile (top graphs), and resonance wavelength for the LP01 fundamental and LP11 first higher order modes versus oxide aperture diameter (bottom graphs) for the VCSELs in Fig. 10.17. On the left the oxide aperture layers are centered on nodes of the electric field intensity, whereas on the right the oxide aperture layers are slightly offset from the same nodes moving away from the optical cavity. Copyright © 2018 SPIE. Reprinted, with permission, from [35]

quasi-single mode with a SMSR of about 35 dB that decreases with increasing bias current. In Fig. 10.20 (right) we show the small-signal modulation response for one of our 980 nm VCSELs with ϕ ~ 3 μm at bias currents of 1.0, 1.5, and 3.75 mA. We achieve a record maximum bandwidth of 35.5 GHz.

In Fig. 10.21 we show the LIV characteristics of 980 nm VCSELs that are identical to the 980 nm reduced complexity VCSELs in Figs. 10.17 and 10.20 except these VCSELs have 15.5 top DBR periods (rather than 14.5 top DBR periods). As expected the LIV characteristics are similar showing lower threshold current and lower optical output power for a given oxide aperture diameter. From Fig. 10.22 we see the maximum f_{3dB} values are slightly lower but still reach above 25 GHz at 25 °C and above 20 GHz at 85 °C for VCSELs with $\phi \le 4$ μm [36].

In Fig. 10.23 we plot the estimated energy to data ratio (EDR) for the same 980 nm VCSELs as in Fig. 10.22. The EDR is a figure-of-merit used to compare the energy efficiency of VCSELs. The EDR is given by $EDR = I \cdot V/BR$, where I is the static bias current and V is the static bias voltage used to obtain a given f_{3dBmax}. In this EDR calculation the modulation energy (power = energy/time) that is added to the VCSEL bias is neglected. The bit rate (BR) is determined from the bandwidth (f_{3dB})

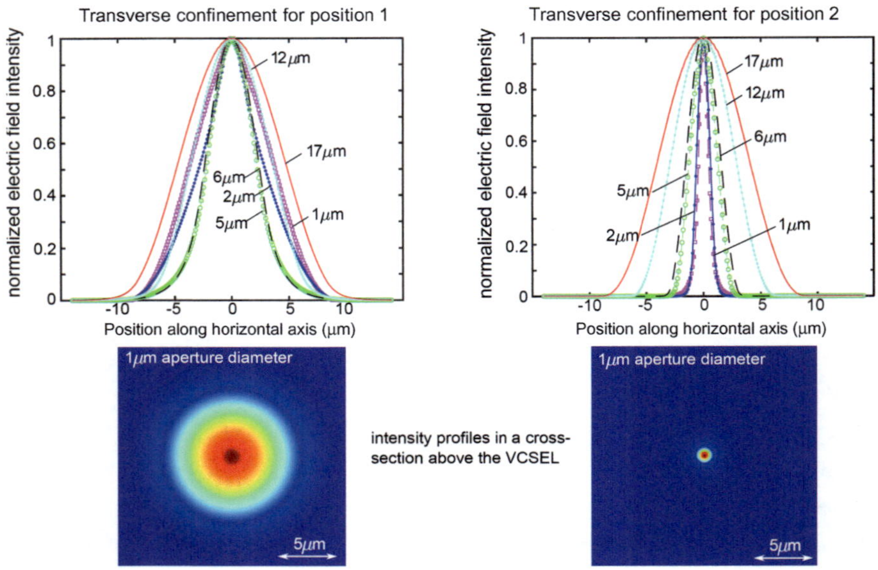

Fig. 10.19 Normalized electric field intensity of the LP01 mode for two different oxide aperture positions and for different oxide aperture diameters (top graphs). Electric field intensity along the horizontal (*x*-axis and *y*-axis) at the maximum of the electric field intensity in the vertical direction along the centerline of the VCSEL (bottom graphs). For oxide aperture positions corresponding to those in Fig. 10.18 (left) the mode confinement vanishes for smaller oxide aperture diameters. For a position of the oxide aperture layers inside the same VCSEL device depicted in Fig. 10.18 (right) the mode is transversely confined over the entire given oxide aperture diameter range. Copyright © 2018 SPIE. Reprinted, with permission, from [35]

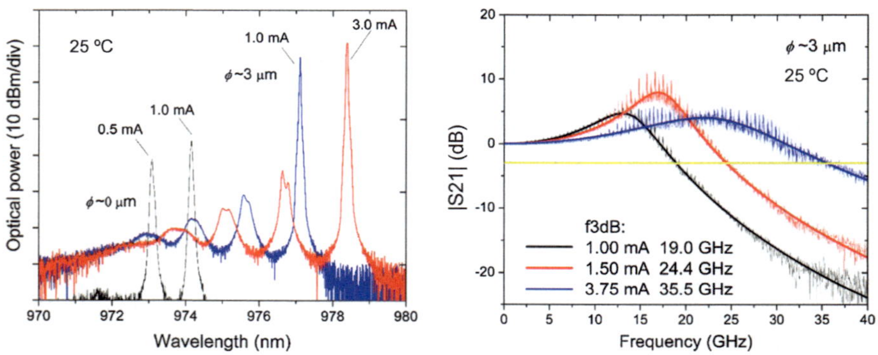

Fig. 10.20 Spectral emission of VCSELs with estimated oxide aperture diameters at $\phi \sim 0$ and ~ 3 μm at 25 °C (left); and small-signal modulation frequency response |S21| versus frequency for VCSELs with an $\phi \sim 3$ μm VCSEL at 25 °C for three different bias currents (right). Copyright © 2019 IEEE. Reprinted, with permission, from [1]

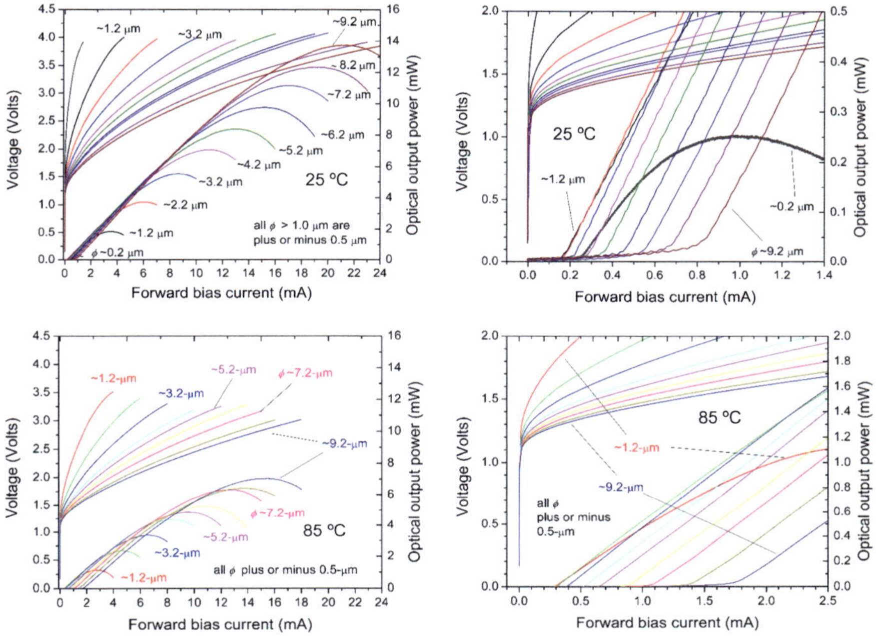

Fig. 10.21 Measured LIV characteristics of 980 nm VCSELs with oxide aperture diameters from ~1.2 to 9.2 μm at 25 and 85 °C (left) and zoomed in to see the threshold currents (right). Copyright © 2019 IEEE. Reprinted, with permission, from [36]

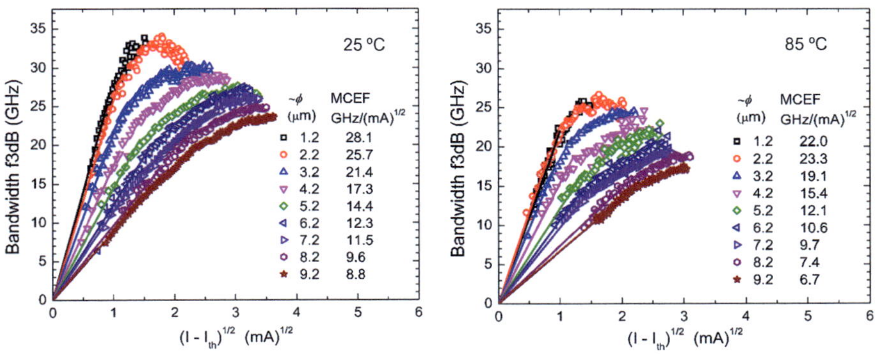

Fig. 10.22 Small-signal modulation bandwidth (f_{3dB}) versus the square root of the quantity the bias current minus the threshold current for 980 nm VCSELs with oxide aperture diameters from ϕ ~ 1.2 to 9.2 μm at 25 and 85 °C. The linear slopes at smaller bias currents before saturation are the modulation current efficiency factors (MCEFs). The maximum f_{3dB} exceeds 30 GHz when ϕ is below ~4 μm. Copyright © 2019 IEEE. Reprinted, with permission, from [36]

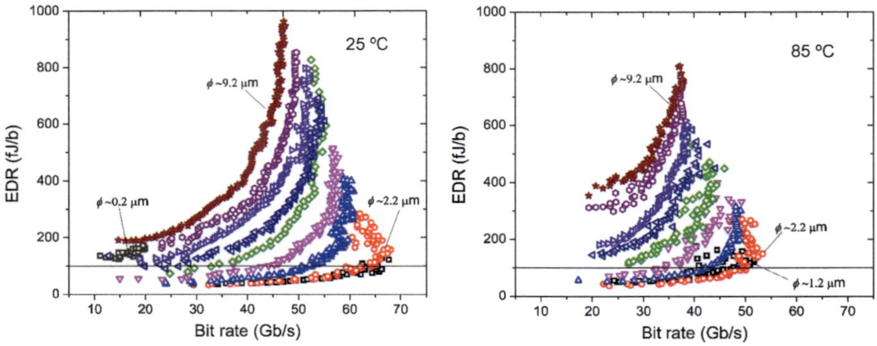

Fig. 10.23 Energy to data ratio versus bit rate for the VCSELs in Fig. 10.22 at 25 and 85 °C. This data is generated by letting the spectral efficiency equal 2 bits (an estimated value), such that the error free bit rate (BR) is taken to be twice the f_{3dB}. We anticipate we may possibly achieve energy efficiencies at or below 100 fJ/b for the VCSELs with ϕ below ~4 μm. Copyright © 2019 IEEE. Reprinted, with permission, from [36]

from BR $= \eta \cdot f_{3dB}$. The proportionality constant is the spectral efficiency η which is typically ~2 (with units of bits), although η may be less than 2 or close to 3. With this formalism which is based on the Shannon-Hartley theorem for data communication with noise [37] we roughly estimate our VCSELs are capable of operating at bit rates of 60 Gbps with EDRs below 100 fJ/b.

10.6 VCSELs for Higher Power

Emerging applications for VCSELs include time-of-flight sensing for focusing, light-detection and ranging for automobile and drone proximity sensing, three-dimensional imaging and tracking for smartphones, and free-space optical interconnections for data centers, mesh networks, and backhaul networks. We envision low-cost and highly flexible VCSEL-based free-space optical links with data capacities of 10–100 Gbps across distances of meters to several kilometers. We thus require light sources with moderate to large optical output power and simultaneously moderate to large bandwidth. We increase the optical output power of VCSELs by increasing their aperture diameters (ϕ), by using arrays of VCSELs, or both. In contrast we increase the small-signal modulation bandwidth of VCSELs by decreasing their ϕ. We investigate the optical power-bandwidth trade-off in VCSELs using a simplified 980 nm VCSEL design [38]. We use a bottom distributed Bragg reflector (DBR) with predominantly binary (GaAs/AlAs) periods and we use 15.5 periods of $Al_xGa_{1-x}As$ ($x = 0.0$ and 0.9) in our top coupling DBR to boost our optical output power. With this epitaxial design and our high frequency GSG pad layout configuration we achieve a record room temperature small-signal modulation bandwidth (f_{3dB}) of 23 GHz for our ϕ ~ 16 μm VCSELs as follows.

In Fig. 10.24 (on the left) we show the static light-output-power vs. current (LI) and current-voltage (IV) characteristics for our VCSELs with ϕ ~ 10, 12, 14, and ~16 μm at ~25 °C (room temperature—RT). We estimate ϕ based on: (1) selective oxidation tests as in Fig. 10.16 performed prior to VCSEL fabrication and imaging using scanning electron microscopy; (2) the measured emission spectra and LIV characteristics; and (3) our experience gained from device fabrication cycles. We characterize groups of 16 VCSELs, where each group forms one column on our processed quarter-wafer piece as illustrated in Fig. 10.3. The top mesa 1 diameter is 18, 18.5, 19, 19.5, 20, 21, 22, 23, …, 31 μm in each column, and the device-to-device pitch is 600 μm. We oxidize all VCSELs simultaneously such that the VCSEL with an 18 μm mesa 1 diameter has ϕ ~ 2.5 μm. For the ϕ ~ 16 μm VCSEL with a 31 μm mesa 1 diameter at 25 °C the maximum LI slope efficiency is ~ 0.95 W/A and the maximum optical output power at rollover is about 26 mW. The peak optical output power of the ϕ ~ 10 μm VCSEL with a 25 μm mesa 1 diameter is about 15 mW at 25 °C. The maximum external differential quantum efficiency η_{qe} for all 16 VCSELs in the given column is within 0.67–0.75 (unitless).

In Fig. 10.24 (right) we show the measured emission spectra from a VCSEL with ϕ ~ 14 μm. As expected the VCSEL emission is multiple-mode due to the relatively large oxide aperture diameter. In fact we expect single mode lasing from our 980 nm VCSELs (as in Fig. 10.20) when ϕ is at or below ~3 μm and quasi-single mode laser when ϕ ~ 4 μm as in Fig. 10.25 (left; blue line). For all larger oxide apertures we expect multiple mode lasing. In Fig. 10.25 (right) we show the small-signal modulation frequency response for the 980 nm VCSEL with ϕ ~ 16 μm at three bias currents. We achieve a record bandwidth of ~23 GHz for the relatively large oxide aperture diameter VCSEL at a bias of 33.5 mA with a corresponding optical output power at f_{3dBmax} of about 23 mW. This is the continuous wave (CW) optical output power from Fig. 10.24 (left).

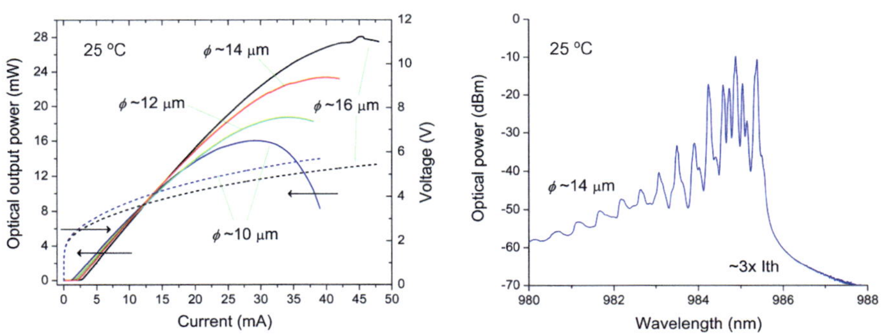

Fig. 10.24 Optical output power (solid lines) and voltage (dashed lines) versus current (on the left) for VCSELs with ϕ ~ 10, 12, 14, and 16 μm at 25 °C (blue, green, red, and black lines, respectively). The ϕ are estimated to be ±1 μm. Emission spectra at 25 °C for a 980 nm VCSEL with ϕ ~ 14 μm (blue line on the right) (see [38])

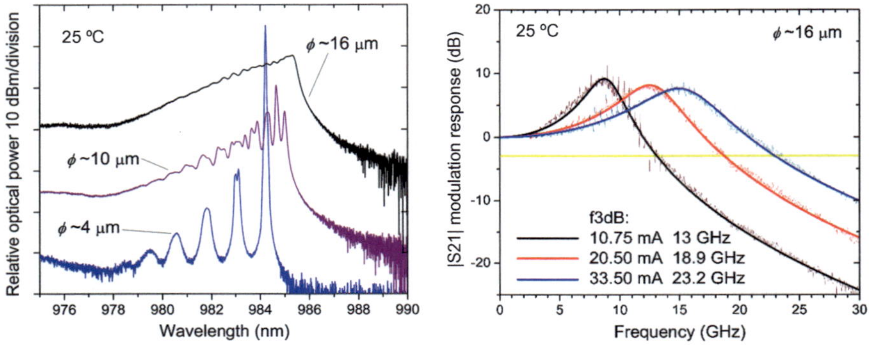

Fig. 10.25 Emission spectra at 25 °C for 980 nm VCSELs: (on the left) above threshold for ϕ ~ 4 μm (blue line); at threshold for ϕ ~ 10 μm (purple line); and just below threshold for ϕ ~ 16 μm (black line). Small-signal modulation response (magnitude |S21|) at three increasing static bias currents for a 980 nm VCSEL with ϕ ~ 16 μm at 25 °C (see [38])

Next we measure the maximum f_{3dB} for several columns of processed VCSELs and plot f_{3dBmax} and the corresponding optical output power at the CW bias that gives us f_{3dBmax}, both as a function of ϕ. An example result is the graph shown in Fig. 10.26. For our 980 nm VCSELs with the dominant binary (AlAs/GaAs) bottom DBR we achieve a record 23 GHz bandwidth for a VCSEL emitting an average of about 23 mW when ϕ ~16 μm (as shown in Fig. 2.24 (right)).

As previously mentioned, VCSELs are key light sources for low power data communication optical links due to their low cost, high energy efficiency, large optical fiber coupling alignment tolerance, and large modulation bandwidths. Large bandwidths enable transmission rates up to about 50 gigabit-per-second (Gbps) at room temperature using non-return-to-zero (NRZ) on-off keying modulation with 980 nm VCSELs. A transmission rate of 25 Gbps or higher usually requires small aperture

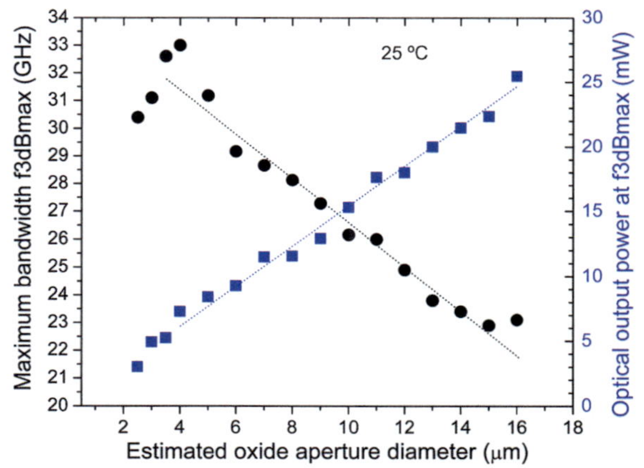

Fig. 10.26 Maximum –3 dB small-signal modulation bandwidth f_{3dBmax} (circles) versus ϕ (±1 μm) from ϕ ~ 2.5 to ~16 μm for 980 nm VCSELs at 25 °C and static optical output power at the bias current giving f_{3dBmax} (squares) (see [38])

diameter (<10 μm) VCSELs to provide the required modulation bandwidth. This (prior to the result we obtained in Fig. 10.26) restricts the output power to ~10 mW per VCSEL, resulting in a limited power budget for long-reach links. We next present a single VCSEL which enables the use of relatively large oxide aperture diameters of ϕ ~ 12 and 17 μm while still enabling record high data rates of 30 and 25 Gbps at corresponding average optical output powers of 17 and 23 mW, respectively. The large oxide aperture diameter also reduces the differential resistance, minimizing the typically large impedance mismatch between the VCSEL and the driver, leading to a more efficient direct modulation.

In contrast to contemporary low power VCSEL applications, emerging applications for moderate to high optical output power VCSELs include time-of-flight sensing, light detection and ranging (LiDAR), object tracking and illumination, and free-space communication. We may scale-up the optical output power of VCSELs by increasing their aperture diameters, by using electrically-parallel arrays of VCSELs, or both. For free-space data communication via VCSEL optical sources we require both moderate-to-high optical output power (10s–100s of mW) and moderate-to-high bandwidth suitable for data bit rates of about 25 Gbps or higher. Compared to previously published results for VCSELs with large oxide aperture diameters we achieve record bandwidths of ~22–24 GHz and record "error free" back-to-back data transmission rates of 25 Gbps as shown in Fig. 10.26. Next we perform a data transmission test using large oxide aperture diameter VCSELs to verify these devices are capable of transmitting data at high bit rates while simultaneously emitting large average optical output powers.

In this experiment as before our VCSEL structure is grown on {001}-oriented *n*-doped GaAs substrates by metal-organic vapor phase epitaxy. The distributed Bragg reflector (DBR) mirrors are composed of $Al_xGa_{1-x}As$ (with $x = 0.0$ and 0.9 with compositionally-graded interfaces). The bottom DBR contains 37 AlGaAs pairs while the top DBR uses 14.5 DBR pairs to enable high optical output power as for the VCSELs in Fig. 10.17. In Fig. 10.27 we plot the room temperature (RT ~ 25 °C) light-optical-output-power and voltage versus current (LIV) characteristics for VCSELs with oxide aperture diameters ϕ ~ 12 μm and ϕ ~ 17 μm at 25 °C [39]. The threshold voltage is ~1.6 V and the maximum external differential quantum efficiency is ~80% for both devices. The differential resistance is about 30 Ω and 40 Ω at 28 mA and 20 mA for the 17 μm and the 12 μm VCSEL, respectively.

The small-signal modulation response is obtained by coupling the output light into an OM1 multimode optical fiber connected to a New Focus Model 1434 25 GHz bandwidth photodetector (PD) module attached to a calibrated Hewlett-Packard 8722C 40 GHz Vector Network Analyzer. In Fig. 10.28 the relaxation oscillation resonance frequency (f_r) and the −3 dB frequency (f_{3dB}) are plotted, both as a function of the square root of the quantity—the applied current minus the threshold current. From the slopes of the linear portion of the plots, both the D-factor and the modulation current efficiency factor (MCEF) are determined. We perform back-to-back nonreturn to zero (NRZ) on-off keying (OOK) large signal modulation experiments using pseudorandom binary sequences (PRBS) of word length $2^7 - 1$ bits. Error-free data transmissions (defined here as a bit error ratio <1E-12) at 25 and 30 Gbps with open

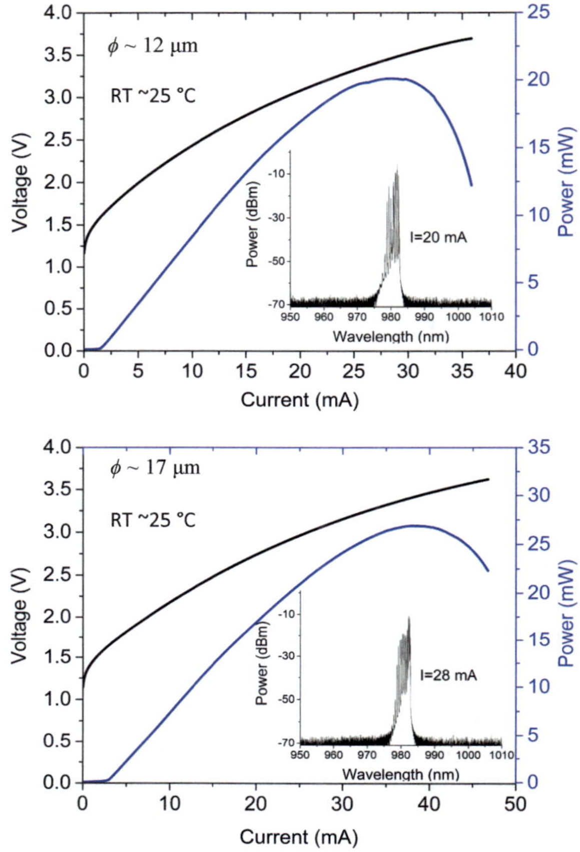

Fig. 10.27 Room temperature LIV curves for 12 and 17 μm oxide aperture diameter 980 nm VCSELs. Insets: optical emission spectra. Reprinted with permission from [39] copyright © The Optical Society

eyes are achieved for the 17 μm and 12 μm VCSELs, respectively, as shown in Fig. 10.29. The VCSELs are biased at $I = 28$ mA and $I = 20$ mA, for corresponding average (operating) optical output powers of ~23 and 17 mW. We achieve record transmission rates of 25 and 30 Gbps with bit error ratios below 1E-12 for these large oxide aperture VCSELs. We anticipate the key to a simultaneous high optical power and high modulation bandwidth VCSEL technology for free space optical communication and general optical links lies in optimized epitaxial designs and VCSEL and VCSEL array packaging geometries for the given application.

10.7 VCSEL Arrays

We investigate small arrays of electrically-parallel VCSELs with 3 and 7 VCSELs organized in triangular (3) and hexagonal (7) patterns [40]. Each VCSEL in each array has approximately the same top mesa 1 diameter and oxide aperture diameter. Our

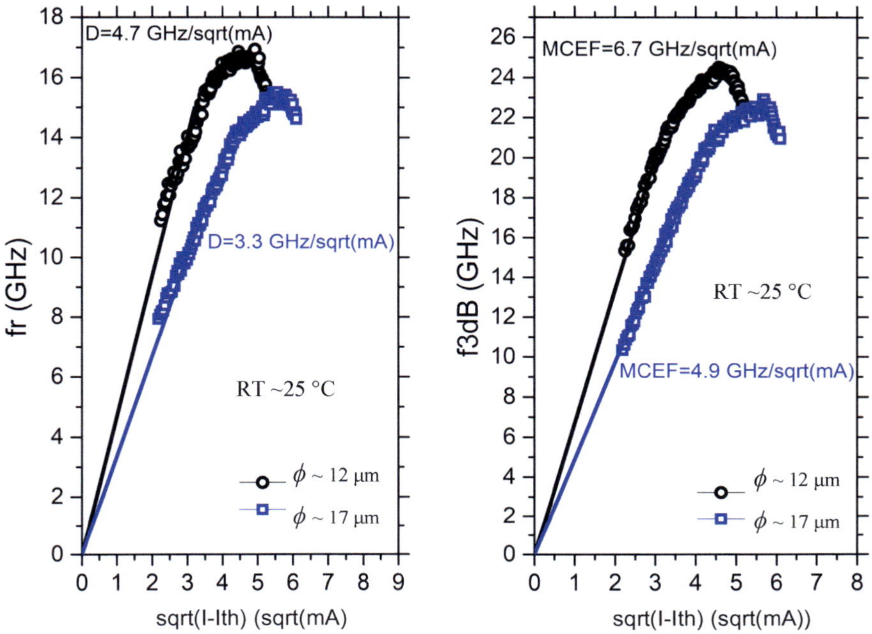

Fig. 10.28 Relaxation oscillation resonance frequency f_r (left) and small signal modulation bandwidth f_{3dB} (right) for 980 nm VCSELs at room temperature (RT ~ 25 °C). The solid lines are linear fits to the data at bias currents below frequency saturation. Reprinted with permission from [39] copyright © The Optical Society

arrays have emerging applications in short-distance (a few meters or less) to medium-distance (up to about a kilometer across free-space) sensing and in free-space optical communication systems. This includes backhaul communication systems in urban, suburban, and rural areas that support 5G and beyond telecommunication networks operating at 10 Gbps or faster. Furthermore, global network infrastructure can enrich underprivileged communities and connect the last families of humans in inaccessible, remote, and other hard to reach areas. New markets and enriched lives result from no cost, instantaneous Internet access. The biggest advancements of humans have been tied to advancements in communications for example clay tablets, the alphabet, the printing press, telephony, radio, and television. A key challenge now is to send data as fast as possible over varied distances affordably, thus energy efficiently and safely to have more information working for or entertaining you. We foresee VCSELs and VCSEL arrays as the key enabling photonics technology for this near-term vision.

In addition to communication systems as pointed out in the introduction, VCSEL arrays are needed to enable low-cost ubiquitous LiDAR systems, human safety systems as in autonomous automobiles, and Internet-of-Things gadgets. We investigate the trade-offs in optical output power, small-signal modulation bandwidth (f_{3dB}), power conversion efficiency, geometric layout (number of VCSELs in the array,

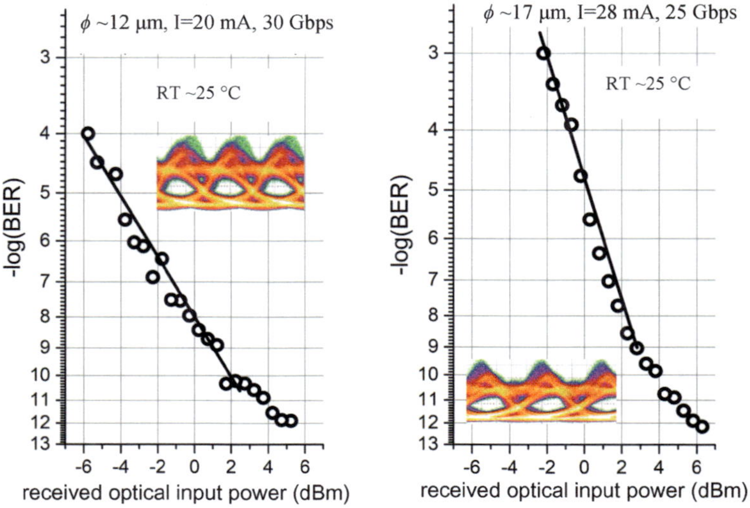

Fig. 10.29 Back-to-back (across ~1 m of OM4 MMF) bit error ratio (BER) versus received optical input power at 30 Gbps for the 12 μm VCSEL (left) and at 25 Gbps for the 17 μm VCSEL (right) at RT. Insets: corresponding eye diagrams at the lowest BER. Reprinted with permission from [39] copyright © The Optical Society

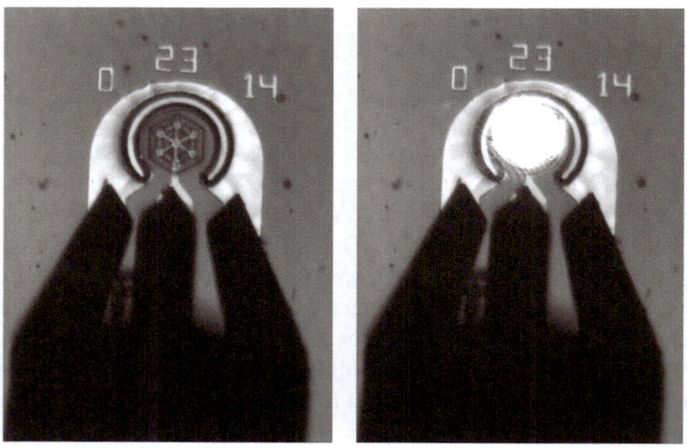

Fig. 10.30 Images of example 7-element VCSEL arrays below and above threshold

VCSEL-to-VCSEL placement, physical geometry), far-field emission pattern, and more. From our initial studies to date we report here example static LIV characteristics and f_{3dB} for a septuple VCSEL array. We have achieved analogous results with triple (3) and novemdecuple (19) VCSEL arrays.

In Fig. 10.30 we show images of an example 7-element 980 nm VCSEL array undergoing on wafer, LIV testing. The room temperature LIV characteristics are given in Fig. 10.31 (left). The optical output power at rollover is about 60 mW indicating the peak optical power of each of the 7 VCSELs is ~8.6 mW. Higher optical output power is possible by, for example simply increasing the oxide aperture diameters of each of the 7 VCSELs. In Fig. 10.31 (right) we plot the LI slope efficiency and the differential series resistance (R_{diff}). We note that for electrically-parallel arrays the R_{diff} (for identical sets of VCSELs) is roughly the resistance of a single VCSEL divided by the number of VCSELs in the array. In Fig. 10.31 (on the right) the differential resistance varies from about 50 Ω to below 20 Ω as the bias current increases, while the maximum slope efficiency is above 0.8 up to a bias current of about 50 mA.

In Fig. 10.32 (left) we show example emission spectra for the 7-element VCSEL array below and well above the ~3 mA threshold current. The emission is fully

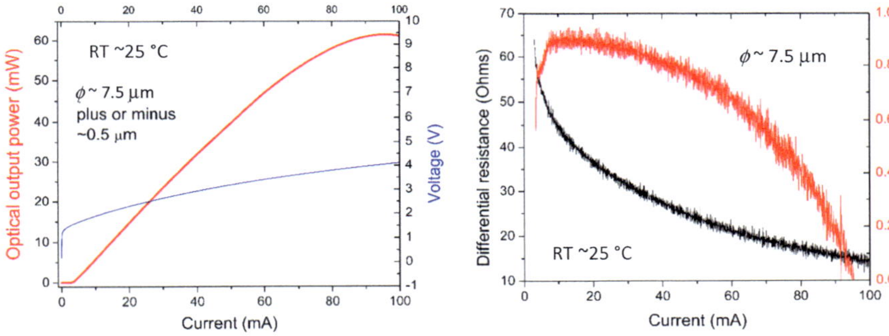

Fig. 10.31 Light output power-voltage-current (LIV) characteristic for a 7-element electrically-parallel 980 nm VCSEL array taken at ~25 °C (left); and LI slope and differential series resistance (right). Copyright © 2019 IEEE. Reprinted, with permission, from [1]

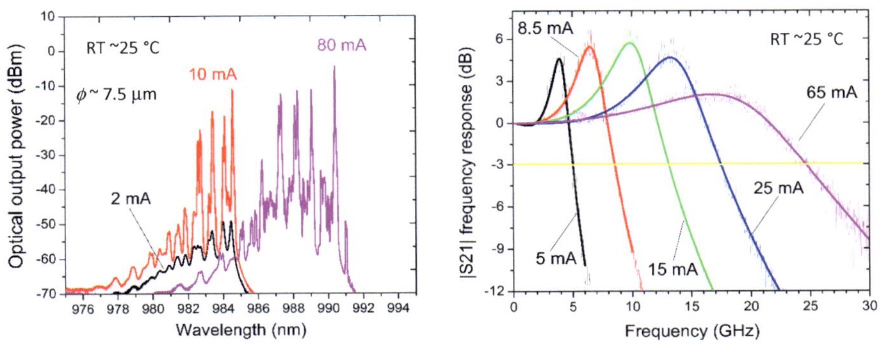

Fig. 10.32 Room temperature emission spectra for a 7-element (septuple) electrically-parallel VCSEL array (left); and small-signal modulation response at various bias currents indicating a maximum −3 dB bandwidth at about a bias current of 65 mA of ~25 GHz (right). Copyright © 2019 IEEE. Reprinted, with permission, from [1]

multiple-mode and is a mix of the 7 approximately identical VCSELs each emitting in a multiple-mode of a few peaks at rollover. In Fig. 10.32 (right) we show the small-signal modulation frequency response of the 7-element VCSEL array at various bias currents. We achieve an f_{3dB} as high as a record 25 GHz (for 7-element VCSEL arrays). We achieve similar results at lower optical output powers for our triple arrays.

In Fig. 10.33 (left) we plot f_r and f_{3dB} versus the square root of the quantity the bias current minus the threshold current. In Fig. 10.33 (right) we plot f_{3dB} versus the bias current and note f_{3dB} remains about 15 GHz (which should be suitable for error free data transmission at 25 Gbps) from a bias of about 20 mA to well over 70 mA.

Finally, in Fig. 10.34 we plot the threshold current versus the emission area for 980 nm single VCSELs and small VCSEL arrays. The data is very close to linear indicating the threshold current scales with the total active VCSEL area. As the electrically-parallel arrays increase in the number of elements we expect some small added resistive and thermal losses due to current spreading and joule heating will tend to slightly degrade the array performance. We see a slight increase in threshold current above the linear increase as defined by the single VCSELs for VCSEL arrays which increases as the number of VCSEL elements increases. In Fig. 10.35 we plot the optical output power versus f_{3dB} for 980 nm single VCSELs and small VCSEL arrays where for all VCSELs ϕ ~ 7.5 μm. The single VCSELs achieve higher bandwidths while the VCSEL arrays achieve very comparable bandwidth but with much higher optical output power. One may consider employing many small oxide aperture diameter VCSELs to achieve both high bandwidth and high total optical output power.

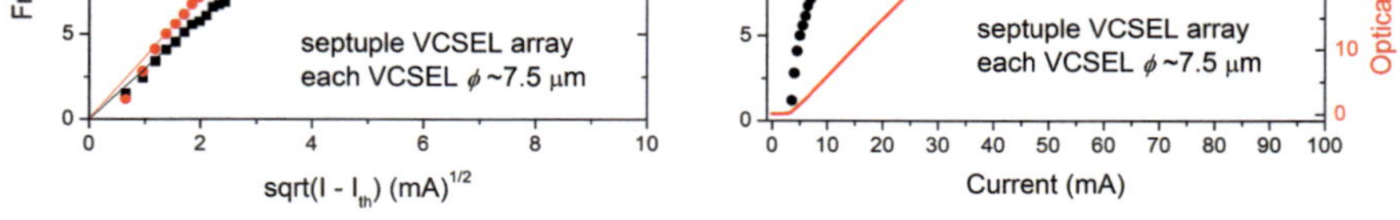

Fig. 10.33 Relaxation resonance frequency (f_r; black squares) and (f_{3dB}; red circles) versus the square root of the applied current minus the threshold current (left); and small-signal modulation bandwidth (f_{3dB}) versus applied bias current (right) for 980 nm electrically-parallel septuple VCSEL arrays. Each VCSEL in a given array has an oxide aperture diameter ϕ ~ 7.5 μm. Copyright © 2019 IEEE. Reprinted, with permission, from [1]

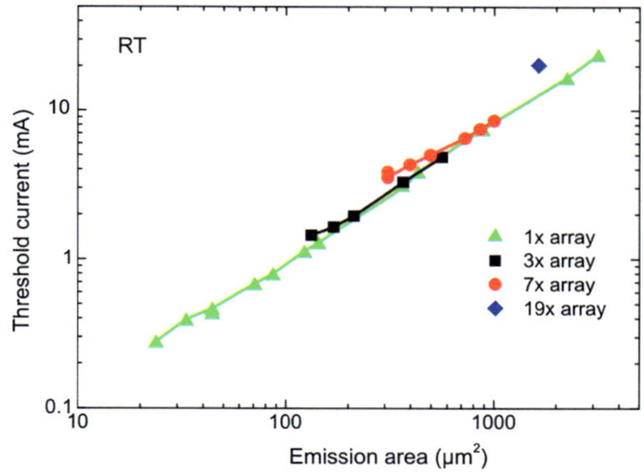

Fig. 10.34 Threshold current versus emission area at room temperature for 980 nm single VCSELs, and for triple, septuple, and novemdecuple (19-element) 980 nm VCSELs. Copyright © 2019 IEEE. Reprinted, with permission, from [1]

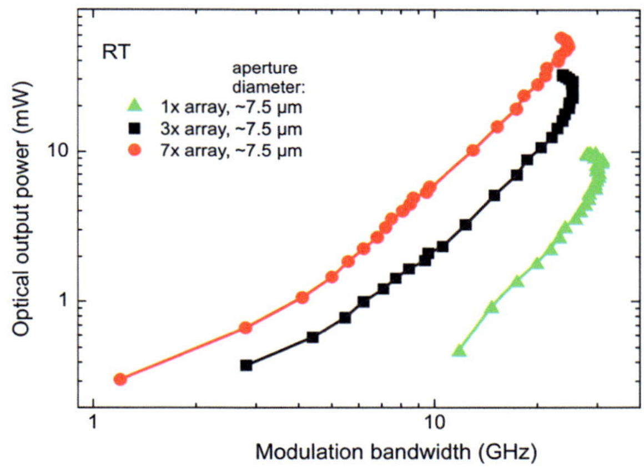

Fig. 10.35 Optical output power versus f_{3dB} for 980 nm single VCSELs, and for triple and septuple VCSEL arrays. Copyright © 2019 IEEE. Reprinted, with permission, from [1]

10.8 Conclusion and Outlook

Via our reduced complexity VCSEL epitaxial designs we demonstrate single VCSELs with four variations of metal contact geometry and find the performance of all designs is potentially well suited for hybrid integration with silicon electronic-photonic integrated circuits or as light sources for short-reach or medium-reach data communication. We achieve the first VCSELs employing a monolithic high index contrast subwavelength grating which enables a drastic vertical thickness reduction and the potential for reduced manufacturing cost. We further demonstrate a record small-signal modulation bandwidth (f_{3dB}) of 35 GHz suitable for data communication through standard multiple-mode optical fiber, and we show the potential for optical data transfer above 50 Gbps with an energy consumption of only 100 femto-joule per bit. We investigate the trade off in optical output power and f_{3dB} for individual

VCSELs as a function of oxide aperture diameter and demonstrate record combinations of average optical output power and f_{3dB}, for example a VCSEL with a 23 GHz bandwidth that can emit ~25 mW. We further investigate electrically-parallel VCSEL arrays and demonstrate f_{3dB} of 25 GHz at 60 mW.

In our most recent work using single 980 nm VCSELs from the set of VCSELs described in Fig. 10.17 (but with $\phi \sim 7$ μm) we have achieved a record 55 Gbps error-free (BER < 1E-12) nonreturn-to-zero (NRZ; i.e. using standard 2-level pulse amplitude modulation, PAM-2, without error correction and without the use of pre-emphasis in the driving signal) data transmission across 100 m of standard OM4 optical fiber [41]. We have also achieved back-to-back error-free NRZ data transmission at 40 Gbps with our triple and septuple VCSEL arrays [42], and 18 GHz bandwidth, 25 Gbps data transmission, and over 200 mW of operating CW optical output power with our novemdecuple (19-element) VCSEL arrays, demonstrating the great potential of the small VCSEL array technology in a variety of new applications. With refined device packaging we anticipate the small VCSEL array technology will swiftly become ubiquitous in consumer products. While high bandwidth VCSELs continue to enable advances in data center communication and high performance computing, a plethora of new life-enhancing applications can be expected in sensing, LiDAR, human interaction and safety systems, in free-space communication for the Internet of Things, 5G (fifth generation) optical wireless backhaul systems, rural, suburban, and urban last mile connections, and also in integrated systems for low cost medical diagnostics, optical computing, additive manufacturing, and much more.

Acknowledgements This work is supported by the German Research Foundation via the Collaborative Research Center 787. I gratefully acknowledge the brilliant work of my research group members Nasibeh Haghighi, Dr. Marcin Gębski, and Dr. Philip Moser, and our collaborations with Prof. Tomasz Czyszanowski (Lodz University of Technology, Poland) and Dr. Martin Zorn (JENOPTIK Optical Systems GmbH, Germany).

References

1. N. Haghighi, P. Moser, J.A. Lott, Power, bandwidth, and efficiency of single VCSELs and small VCSEL arrays. IEEE J. Sel. Topics Quant. Electr. **25**(6), 1700615, 1–15 (2019). https://doi.org/10.1109/jstqe.2019.2922843
2. I. Melngailis, Longitudinal injection plasma laser of InSb. Appl. Phys. Lett. **6**, 59–60 (1965). https://doi.org/10.1063/1.1754164
3. R. Dingle, W. Wiegmann, C.H. Henry, Quantum states of confined carriers in very thin $Al_xGa_{1-x}As$-GaAs–$AlxGa_{1-x}As$ heterostructures. Phys. Rev. Lett. **33**(14), 827–830 (1974). https://doi.org/10.1103/PhysRevLett.33.827
4. J.P. van der Ziel, R. Dingle, R.C. Miller, W. Wiegmann, W.A. Nordland Jr., Laser oscillation from quantum states in very thin GaAs–$Al_{0.2}Ga_{0.8}As$ multilayer structures. Appl. Phys. Lett. **26**(8), 463–465 (1975). https://doi.org/10.1063/1.88211
5. J.P. van der Ziel, M. Ilegems, Multilayer GaAs–$A1_{0.3}Ga_{0.7}As$ dielectric quarter wave stacks grown by molecular beam epitaxy. Appl. Opt. **14**, 2627–2630 (1975). https://doi.org/10.1364/AO.14.002627

6. D.R. Scifres, R.D. Burnham, W. Streifer, Highly collimated laser beams from electrically pumped SH GaAs/GaAlAs distributed—feedback lasers. J. Appl. Phys. **26**(48), 48–50 (1975). https://doi.org/10.1063/1.88068
7. D. Scifres, R.D. Burnham, Distributed feedback diode laser. US Patent US 3983509, 28 Sep 1976
8. H. Soda, K. Iga, C. Kitahara, Y. Suematsu, GaInAsP/InP surface emitting injection lasers. Jpn. J. Appl. Phys. **18**, 2329–2330 (1979). https://doi.org/10.1143/JJAP.18.2329
9. M. Ogura, T. Hata, N.J. Kawai, T. Yao, GaAs/Al_xGa_{1-x} As multilayer reflector for surface emitting laser diode. Jpn. J. Appl. Phys. **22**, L112–L114 (1983). https://doi.org/10.1143/JJAP.22.L112
10. M. Ogura, T. Hata, T. Yao, Distributed feed back surface emitting laser diode with multilayered heterostructure. Jpn. J. Appl. Phys. **23**, L512–L514 (1984). https://doi.org/10.1143/JJAP.23.L512
11. M. Ogura, T. Yao, Surface emitting laser diode with $Al_xGa_{1-x}As$/GaAs multilayered heterostructure. J. Vac. Sci. Technol., B **3**, 784–787 (1985). https://doi.org/10.1116/1.583099
12. K. Iga, S. Kinoshita, F. Koyama, Microcavity GaAlAs/GaAs surface-emitting laser with $I_{th} =$ 6 mA. Electron. Lett. **23**, 134–136 (1987). https://doi.org/10.1049/el:19870095
13. T. Sakaguchi, F. Koyama, K. Iga, Vertical cavity surface-emitting laser with an AlGaAs/AlAs Bragg reflector. Electron. Lett. **24**, 928–929 (1988). https://doi.org/10.1049/el:19880632
14. P.L. Gourley, T.J. Drummond, Visible, room temperature, surface emitting laser using an epitaxial Fabry–Perot resonator with AlGaAs/AlAs quarterwave high reflectors and AlGaAs/GaAs multiple. Appl. Phys. Lett. **50**, 1225–1227 (1987). https://doi.org/10.1063/1.97916
15. J.L. Jewell, A. Scherer, S.L. McCall, Y.H. Lee, S. Walker, J.P. Harbison, L.T. Florez, Low-threshold electrically pumped vertical-cavity surface-emitting microlasers. Electron. Lett. **25**, 1123–1134 (1989). https://doi.org/10.1049/el:19890754
16. Y.H. Lee, J.L. Jewell, A. Scherer, S.L. McCall, J.P. Harbison, L.T. Florez, Room-temperature continuous-wave vertical-cavity single-quantum-well microlaser diodes. Electron. Lett. **25**, 1377–1378 (1989). https://doi.org/10.1049/el:19890921
17. Y.H. Lee, B. Tell, K. Brown-Goebeler, J.L. Jewell, J.V. Hove, Top-surface-emitting GaAs four-quantum-well lasers emitting at 0.85 μm. Electron. Lett. **26**, 710–711 (1990). https://doi.org/10.1049/el:19900463
18. R.S. Geels, S.W. Corzine, J.W. Scott, D.B. Young, L.A. Coldren, Low threshold planarized vertical-cavity surface-emitting lasers. IEEE Photonics Technol. Lett. **2**, 234–236 (1990). https://doi.org/10.1109/68.53246
19. J.M. Dallesasse, N. Holonyak Jr., A.R. Sugg, T.A. Richard, N. El-Zein, Hydrolyzation oxidation of Al_xGa_{1-x}As–AlAs–GaAs quantum well heterostructures and superlattices. Appl. Phys. Lett. **57**, 2844–2846 (1990). https://doi.org/10.1063/1.103759
20. D.L. Huffaker, D.G. Deppe, K. Kumar, T.J. Rogers, Native-oxide defined ring contact for low threshold vertical-cavity lasers. Appl. Phys. Lett. **65**, 97–99 (1994). https://doi.org/10.1063/1.113087
21. K.D. Choquette, K.M. Geib, C.I.H. Ashby, R.D. Twesten, O. Blum, H.Q. Hou, D.M. Follstaedt, B.E. Hammons, D. Mathes, R. Hull, Advances in selective wet oxidation of AlGaAs alloys. IEEE J. Sel. Top. Quantum Electron. **3**, 916–926 (1997). https://doi.org/10.1109/2944.640645
22. M. Dallesasse, N. Holonyak Jr., Oxidation of Al-bearing III–V materials: a review of key progress. J. Appl. Phys. **113**, 051101 (2013). https://doi.org/10.1063/1.4769968
23. Yole Développement 2019, online: http://www.yole.fr/2014-galery-LED.aspx#I00093392
24. N. Haghighi, G. Larisch, M. Gębski, L. Frasunkiewicz, T. Czyszanowski, J.A. Lott, Simplicity VCSELs, in *Proceedings of SPIE 10552, Vertical-Cavity Surface-Emitting Lasers XXII*, 105520 N (2018)
25. N. Haghighi, G. Larisch, M. Zorn, J.A. Lott, High bandwidth versus high optical output power in 980 nm VCSELs. HL 36.7, German Physical Society Spring Meeting, Berlin, 11–16 Mar 2018

26. P. Moser, H. Schmeckebier, M. Gębski, P. Śpiewak, R. Rosales, M. Wasiak, J.A. Lott, Intracavity and extracavity-contacted 980-nm oxide-confined VCSELs for optical interconnects and integration, Invited, in *Proceedings SPIE 101220 J, Vertical-Cavity Surface-Emitting Lasers XXI* (2017). https://doi.org/10.1117/12.2256177
27. M. Marciniak, M. Gębski, M. Dems, E. Haglund, A. Larsson, M. Riaziat, J.A. Lott, T. Czyszanowski, Optimal parameters of monolithic high-contrast grating mirrors. Opt. Lett. **41**(15), 3495–3498 (01 Aug 2016). https://doi.org/10.1364/ol.41.003495
28. J.A. Lott, P. Moser, M. Gębski, M. Dems, M. Wasiak, T. Czyszanowski, Energy-efficient VCSELs for integrated optoelectronic and photonic systems, Invited, in *Proceedings ICP-2016, Kuching, Sarawak* (Borneo, Malaysia, 2016), pp. 14–16. https://doi.org/10.1109/ICP.2016.7510052
29. M. Gębski, M. Dems, J.A. Lott, T. Czyszanowski, Monolithic subwavelength high-index-contrast grating VCSEL. IEEE Photonics Technol. Lett. **27**, 1953–1956 (2015). https://doi.org/10.1109/LPT.2015.2447932
30. M. Gębski, J.A. Lott, T. Czyszanowski, Electrically-injected VCSEL with a single-layer monolithic subwavelength high index contrast grating mirror. Opt. Express **27**(3), 7139–7146 (04 Feb 2019). https://doi.org/10.1364/oe.27.007139
31. R. Rosales, M. Zorn, J.A. Lott, 30-GHz bandwidth with directly current modulated 980-nm oxide-aperture VCSELs. IEEE Photonics Technol. Lett. **29**(23), 2107–2110 (2017). https://doi.org/10.1109/LPT.2017.2764626
32. N. Haghighi, G. Larisch, M. Gębski, M. Marciniak, J.A. Lott, Bandwidth versus oxide aperture diameter for 980 nm Simplicity VCSELs, in *Proceedings of 7th Workshop on Physics and Technology of Semiconductor Lasers*, Krakow, Poland, 15–18 Oct 2017
33. N. Haghighi, G. Larisch, R. Rosales, M. Zorn, J.A. Lott, 35 GHz bandwidth with directly current modulated 980 nm oxide aperture single cavity VCSELs, WD4, in *Proceedings of IEEE International Semiconductor Laser Conference*, Santa Fe, NM, USA, 16–19 Sep 2018
34. M. Noble, J.A. Lott, J.P. Loehr, Quasi-exact optical analysis of oxide-apertured microcavity VCSELs using vector finite elements. IEEE J. Quantum Electron. **34**(12), 2327–2339 (1998). https://doi.org/10.1109/3.736102
35. T. Höhne, L. Zschiedrich, N. Haghighi, J.A. Lott, S. Burger, Numerical computation of resonance modes and of constant-flux modes in VCSELs, in *Proceedings of SPIE Photonics Europe 106821-65*, Strasbourg, France, 22–26 Apr 2018
36. J.A. Lott, N. Haghighi, G. Larisch, M. Zorn, High bandwidth simplicity VCSELs, invited, paper 6, in *Proceedings ICP-2018* (Langkawi, Malaysia, 2018), pp. 09–11. https://doi.org/10.1109/icp.2018.8533202
37. C.E. Shannon, A mathematical theory of communication. Bell Syst. Tech. J. **27**, 379–423, 623–656, July–Oct 1948. https://doi.org/10.1002/j.1538-7305.1948.tb01338.x
38. N. Haghighi, G. Larisch, R. Rosales, J.A. Lott, 23 GHz bandwidth and 25 mW peak optical output power with 980 nm oxide aperture VCSELs, MC2.4, in *Proceedings of IEEE Photonics Conference*, Reston, VA, USA, 30 Sep–04 Oct 2018
39. J.A. Lott, R. Rosales, G. Larisch, N. Haghighi, 25–30 Gbps error-free data transmission with large oxide aperture diameter 980 nm VCSELs, W3A.3, in *Proceedings Optical Fiber Conference (OFC)*, San Diego, CA, 03–07 Mar 2019
40. N. Haghighi, P. Moser, J.A. Lott, Bandwidth and optical output power of VCSELs and VCSEL arrays, in *Proceedings of SPIE 10938*, Vertical-Cavity Surface-Emitting Lasers XXIII, Photonics West 2019, San Francisco, CA, 02–07 Feb 2019. https://doi.org/10.1117/12.2508720
41. N. Haghighi, J. Lavrencik, S.E. Ralph, J.A. Lott, 55 Gbps error free data transmission with 980 nm VCSELs across 100 m of multiple-mode optical fiber, TuE3-3, in *Proceedings of 24th OptoElectronic and Communications Conference (OECC)*, Fukuoka, Japan, 07–11 July 2019
42. N. Haghighi, P. Moser, J.A. Lott, 40 Gbps with electrically parallel triple and septuple 980 nm VCSEL arrays. IEEE/OSA Journal of Lightwave Technology (24 Dec 2019). https://doi.org/10.1109/JLT.2019.2961931

Chapter 11
VCSEL-Based Silicon Photonic Interconnect Technologies

Pascal M. Seiler, Bernd Tillack and Lars Zimmermann

Abstract This chapter motivates the development of silicon photonic interconnect technologies based on vertical-cavity surface-emitting lasers (VCSELs) for future datacom applications, demanding cost effective, spectrally efficient and high-speed transmission links. We shall first present a classification of current interconnect technologies, used transmission techniques and requirements for the next generation of communication links. Furthermore, we shall discuss recent research activities on the integration of VCSELs in silicon photonics and the potential advantages of VCSEL-based silicon photonic transceiver solutions. In the subsequent sections, we introduce commercially available device structures emitting at tele- and datacom wavelengths. Based on experimentally determined characteristics, we model the system behavior and performance for high-speed coherent transmission links using quadrature phase-shift keying (QPSK). In addition, a brief introduction in digital signal processing is given. Following, incoherent, as well as coherent VCSEL-based transmission links are presented and analyzed. System performance and behavior are investigated with long-wavelength VCSELs as transmitter. Dependencies on the data rate and the bias point are further discussed. Finally, we conclude this chapter and give an outlook on the future VCSEL-based interconnects using spectrally efficient transmission techniques.

11.1 Modern Interconnect Technologies and Requirements

The annual global internet protocol (IP) traffic is expected to threefold between 2016 and 2021, reaching 3.3 Zettabytes (ZB) by the end of 2021 [1]. In the same period, the annual global IP traffic in conjunction with data centers (DC) is assumed to increase threefold as well, reaching 20.6 ZB by the end of 2021 (see Fig. 11.1). Following this trend, the amount of hyperscale data centers is expected to nearly double to 628,

P. M. Seiler (✉) · B. Tillack · L. Zimmermann
Technische Universität Berlin, Institut für HF- und HL-Systemtechnologien, Berlin, Germany
e-mail: seiler@tu-berlin.de

B. Tillack · L. Zimmermann
IHP, Leibniz Institut für innovative Mikroelektronik, Frankfurt, Germany

M. Kneissl et al. (eds.), *Semiconductor Nanophotonics*, Springer Series in Solid-State Sciences 194, https://doi.org/10.1007/978-3-030-35656-9_11

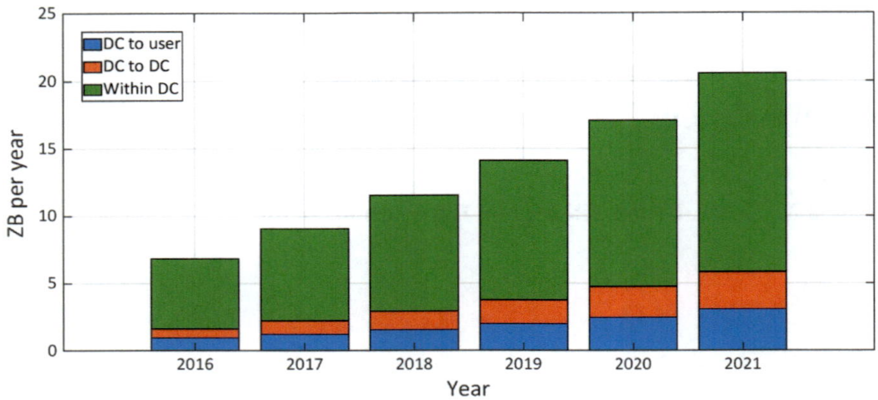

Fig. 11.1 Forecast for the global data center IP traffic growth between 2016 and 2021. Data from [2]

accounting for more than 50% of all data centers [2]. To support this growth in the long term, an evolution of data center interconnect (DCI) technologies, enabling a vast amount of internet-based applications, e.g. streaming services, social media or cloud computing, is necessary. In the past 15 years, Google's intra-data center architecture evolved first from copper interconnects towards optical interconnects in 2007. Five years later Google integrated more complex wavelength division multiplexing (WDM) based interconnects to achieve even higher data rates [3]. One of the main challenges for future DCIs will be to support the immense growth of data center traffic with increasing data rates, while employing efficient, low power consuming and cost effective solutions. The following chapter shall introduce a classification of used interconnect technologies and the typically used transmission techniques. Furthermore, potential future interconnect technologies for transmissions beyond 1 Tbps in conjunction with silicon photonic based interconnects will be addressed [3, 4].

11.1.1 Classification of Interconnects

The requirements and, thus, the interconnect technology vary based on the link distance and the need for high aggregated data rates in conjunction with a low power consumption per bit. Therefore, many very specific transceiver standards have been developed to meet these requirements (see Table 11.1). In metro- and submarine transmission links, the only viable solutions are based on coherent modulation formats, operating in the C-band (around 1550 nm) and are enabled by erbium-doped fiber amplifiers (EDFA) and highly specialized digital signal processing (DSP). For

Table 11.1 Optical transceiver classification based on reach distance and modulation format [5]

Intra-data center			Inter-data center			Metro long-Haul
100 m	500 m	2 km	10 km	40 km	100 km	>100 km
100G SR4	100G PSM4	100G CWDM4	100G LR4	100G ER4	100G PAM4/Coherent	100G Coherent

Table 11.2 Comparison of characteristic features for direct detect and coherent interconnect technologies for data rates >50 GBd

Feature	Direct detect	Coherent detection
DSP	No	Yes
FEC	Yes	Yes
Sensitivity	Low	High
Spectral efficiency	Low	High
Tunable laser	No	Yes
Optical MUX/DEMUX	Yes	No

data centers, a general distinction of link domains can be made into intra-data center (up to 2 km) and inter-data center (up to 100 km) distances. Long reach inter-data center interconnects (more than 40 km) can still profit from the advantages of coherent modulation and detection schemes, due to the increased sensitivity of such transceivers and, hence, greater resistance against the decreasing OSNR in long fiber links (compare Table 11.2). One possible technique is for example to employ discrete multi-tone modulation, which applies different modulation formats (i.e. coherent or incoherent) to different channels in order to transmit at the highest applicable data rate. For shorter link distances (2–40 km), wavelength-division multiplexing (WDM) systems centered around 1300 nm (O-band) employing on-off keying are favored due to the zero dispersion point of standard singlemode fibers (SMFs) in that wavelength range. The less complex modulation and dispersion-free wavelength region allows for a reduction of DSP and, hence, power consumption. The shortest optical intra-data center interconnects are realized by using on-off keying at a wavelength of 850 nm and multimode fibers (MMFs) for up to a couple hundred meters. While normally no DSP is used in these short reach interconnects, there are also commercial transceivers nowadays that allow for forward error correction (FEC) to improve the system performance at the expense of an increased power consumption [3, 5–7].

11.1.2 Road to Coherent Data Center Interconnects

The investigation of more efficient transceivers arises from the need to cope with the drastically growing traffic inside data centers. A further increase in symbol rate of commonly used pulsed-amplitude modulation (PAM) schemes is not feasible, since electronic devices supporting 100 GBd or more are presently and in the near future unavailable [5]. Therefore, complex modulation formats with an increased efficiency are required for future high speed interconnects. A possible strategy to cope with the fast growth of intra DC traffic would be to replace current direct detect systems by coherent ones. Until recently, obstacles to the employment of coherent DCIs for shorter reaches were the use of power consuming DSP and the complexity of coherent transceivers. However, there have been advances in terms of power consumption and density of DSP, increasing the potential use of coherent detection schemes. Between 2012 and 2017, the power consumption per 100 Gbps of real-time transceivers reduced from 125 to 10 W. This was achieved by more spectrally efficient modulation formats and the transition from 40 nm CMOS technology to 16 nm, which is expected to further reduce to 3.75 W in conjunction with 7 nm CMOS technology in the near future [3, 8]. Furthermore, since link distances in the intra- and inter-data center domain are relatively short compared to long-haul transmission links, DSP functions to compensate for polarization mode dispersion (PMD) may be reduced. Additionally, the already employed O-band enables transmissions with reduced dispersion and hence significantly reduced computational heavy DSP. Furthermore, coherent transceivers are far more attractive in terms of flexibility than conventional direct-detection systems. For example, an increase in bit rate while maintaining the symbol rate can be achieved in coherent transceivers by increasing the number of symbols mapped to the signals amplitude and phase. Hence, more complex modulation formats could be used for shorter reaches while less complex ones may be used in scenarios involving larger link distances. However, a technology that supports the requirement to those transceivers, i.e. singlemode waveguiding and polarization depending properties while maintaining a small footprint is needed. Silicon photonics is an attractive technology for these applications due to the inherent singlemode and polarization depending properties of silicon nano-waveguides, perfectly suited for coherent communication links. Furthermore, low-loss and highly dense structures can be achieved on the silicon-on-insulator (SOI) platform. Co-integration with microelectronics and IC fabrication technology make silicon photonics a highly attractive approach for DCI applications with large volume [4, 9–11]. However, silicon does not exhibit a direct bandgap. Therefore, the integration of III–V based lasers sources onto the silicon-based chip is necessary and has been under development for decades [12–18]. Current coherent data links normally use distributed-feedback (DFB) lasers due to their well-defined emission wavelength, high output power and small phase noise, which is desirable for higher modulation formats. However, given by the need to decrease power consumption and reduce the cost, in conjunction with the improvements in terms of density and power consumption of DSP, lasers with lower threshold and, hence, power consumption could be an interesting alternative for these applications. The efficient integration of VCSELs with low threshold currents in coherent systems has been shown as local oscillator on the receiver side [19], as well as externally modulated laser in a coherent transmission link based on quadrature-phase shift-keying [20]. The potential use of VCSELs in coherent systems for link distances below 100 km is subject of ongoing research and a main concern of this chapter.

11.1.3 On the Importance of Quantum Dot Lasers for Silicon Photonics

The heterogeneous combination of silicon photonics with III-V based mature laser sources has been under investigation for many decades. There have been numerous approaches to achieve an efficient integration, including heterogeneous integration via wafer-bonding, flip-chip bonding or heteroepitaxical integration [12–18, 21]. In terms of scalability to larger volume fabrication, the most attractive integration approaches are heteroepitaxical growth or wafer bonding. However, a persistent obstacle for the efficient implementation of large-scale wafer bonding is the mismatch in wafer-size typically used in each material domain. While silicon is currently mostly based on 300 mm wafers, typical III–V substrates are significantly smaller, hence prohibiting an economic large-scale deployment. An alternative approach to potential large-scale manufacturing would therefore be the use of heteroepitaxy. It is well known, that heteroepitaxical approaches suffer from the inherent lattice mismatch between silicon and III–V materials, resulting in bending, cracking and high defect densities. However, recent advances show the potential of monolithically grown, inexpensive laser sources with a high thermal stability. To overcome the mismatch issue between silicon and III-V materials, quantum dot (QD) active layers are typically used for such devices. The highly confined energy states allow for an increased resistance against defects due to the direct growth on silicon, compared to quantum-well (QW) structures [21]. Furthermore, QD lasers have been reported to be significantly more resistant to optical feedback than QW lasers, which is particularly important [17]. While monolithically grown QD lasers on silicon are not yet mature and commercially available, they do show a clear potential to be an efficient and well performing integrated light source for silicon photonics in the near future. Therefore, the investigation of QD-based lasers for the application in silicon photonics is of high relevance.

11.2 Long-Wavelength VCSELs

While VCSELs typically used in data center applications operate around 850 nm, there has also been considerate effort into VCSELs emitting in the O- and C-band [22, 23]. Two main approaches towards long-wavelength VCSELs emitting in the O-band are GaAs-based and Indium-Phosphide-based (InP), with the latter being the material system used for experiments later this chapter. The use of InP-based VCSELs is owed to the present dominance of 1550 nm coherent systems and therefore the starting point of our work. Advantages of GaAs-based VCSELs are high contrast DBRs, allowing for high reflectivity even at a small number of layers and a variety of active media. GaAs is an attractive and well established material platform for QD-based lasers, which are particularly interesting for communication applications, since they can be more defect tolerant and less temperature-sensitive compared to

QWs [22]. 20 Gbps error-free transmission has been shown for short wavelength GaAs-based VCSELs containing submonolayer quantum dots [25]. However, high performing GaAs-based VCSELs emitting in the O-band are still under investigation (see Chap. 10). InP-based VCSELs on the other hand suffer from a lack of lattice-matched materials suitable for high index contrast, resulting in thicker mirror stacks and a smaller thermal conductivity. Regardless, high performing VCSELs emitting in the O- and C-band have been achieved by the implementation of a buried tunnel junction (BTJ) [22].

11.2.1 Device Structure

The device structure of an InP-based long-wavelength VCSEL is shown in Fig. 11.2. In order to achieve a large bandwidth and low threshold current, the active region consists typically of multiple QW, providing a high gain and large relaxation oscillation frequencies [23, 24]. The *n*-doped top distributed Bragg reflector (DBR) is epitaxially grown by alternating InGa(Al)As/InAlAs layer pairs, where the exact composition is designed for the specific emission wavelength. On the other side, a hybrid mirror made of a few CaF_2/ZnS layers capped with gold is being used as bottom mirror. The hybrid mirror, even though the used materials show a smaller thermal conductivity compared to epitaxially grown layers, reduces the thermal resistance of the device, due to the small number of layers required for sufficiently high reflection. Furthermore, the Au finish provides an efficient heat dissipation. However, the

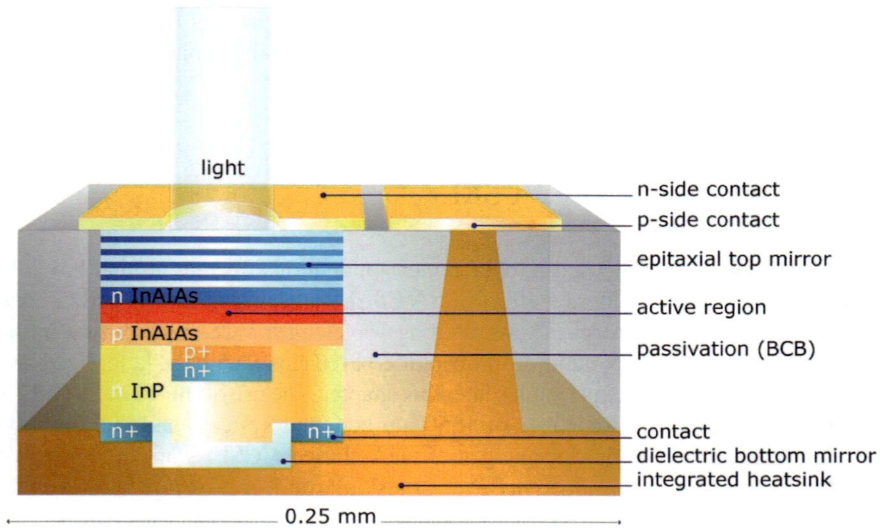

Fig. 11.2 Schematic of an InP-based long-wavelength VCSEL. Courtesy of VERTILAS GmbH

Fig. 11.3 Picture of an InP-based long-wavelength BTJ VCSEL used in this chapter. Courtesy of VERTILAS GmbH

peculiarity of this design is the implementation of a BTJ (compare Fig. 11.2). The BTJ improves the thermal, electrical and optical behavior at the same time, allowing for high performance at long wavelengths. The basic principle is to convert the majority carriers from holes to electrons by introducing one heavily *p*- and *n*-doped InGa(Al)As layer. The shift from holes to electrons as charge carriers allows for smaller optical losses and electrical resistances and hence improved thermal management, given the smaller mobility of holes compared to electrons. Furthermore, this allows for *n*-type contacts on *p*- and *n*-side of the device, effectively improving the electrical performance even further. For the optimal optical performance, the low-bandgap BTJ is placed in a node of the optical field inside the cavity. Due to the unavailability of oxidation techniques, providing efficient current- and photon confinement in GaAs-based devices, the confinement in the InP-based devices is achieved by etching the heavily *n*-doped layer of the BTJ, resulting in a blocking p^+n diode and a low resistance p^+n^+ tunnel diode in the BTJ area, providing high performance current confinement. Additionally, the device is surrounded by low-ε benzocyclobutene (BCB) for passivation [22–24]. A picture of an InP-based BTJ VCSEL with a RF layout used later in this chapter is given in Fig. 11.3.

11.2.2 Operation Characteristics

Figure 11.4 shows the light-current-voltage (LIV) characteristic for a singlemode BTJ VCSEL emitting at 1.55 μm. The device is placed on a ceramic mount which in turn is temperature stabilized by a temperature controller. The threshold current is around 1.3 mA. The VCSEL shows a maximum output power of approximately 2.5 dBm at 20 °C for 14 mA bias current. As can be seen from the power spectral density (PSD) in Fig. 11.5, the VCSEL has a side-mode suppression ratio that is

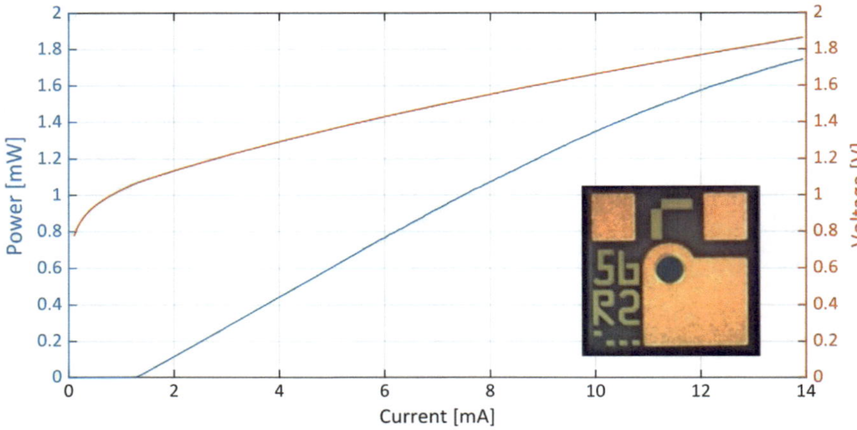

Fig. 11.4 Light-current-voltage characteristic of an 1.55 μm BTJ VCSEL emitting in the C-band at a fixed temperature of 20 °C. The inset shows a picture of a structure used in this chapter. Courtesy of VERTILAS GmbH

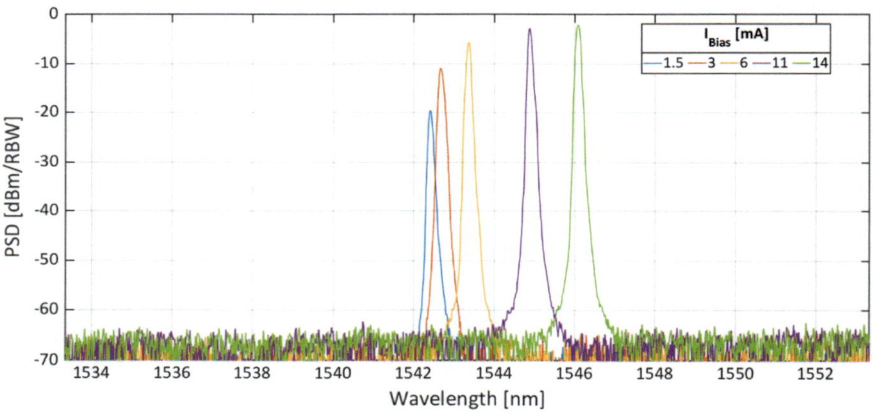

Fig. 11.5 Measured power spectral density of an 1.55 μm BTJ VCSEL at different bias points at a fixed temperature of 20 °C and a resolution bandwidth (RBW) of 0.1 nm

greater than 40 dB. Increasing the temperature from 20 to 35 °C at 6 mA bias current shifts the wavelength by 1.524 nm (comp. Figure 11.6). Increasing it further to 50 °C redshifts the wavelength by additionally 1.611 nm, creating a full wavelength shift of 3.135 nm over the investigated temperature range of 30 °C. The normalized small-signal response S_{21} of the VCSEL under different bias currents is shown in Fig. 11.7. The device is kept at constant 20 °C and is being contacted with a high-frequency microprobe and driven with a small-signal modulation amplitude of −20 dBm. Figure 11.7 shows a 3-dB bandwidth of approximately 8.5 GHz for a bias current of 10 mA.

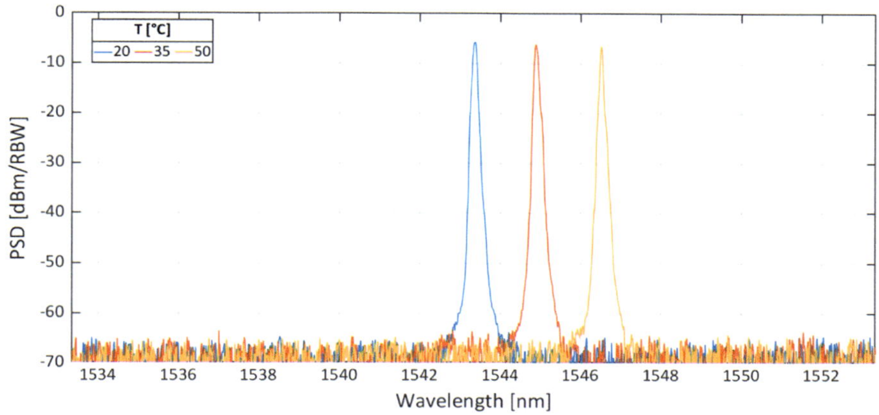

Fig. 11.6 Wavelength shift of an 1.55 μm BTJ VCSEL for 20, 35 and 50 °C at a fixed bias point of 6 mA and a RBW of 0.1 nm

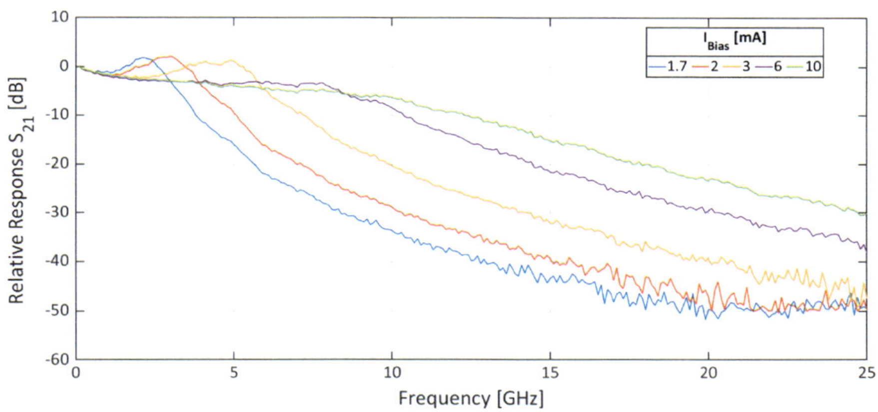

Fig. 11.7 Normalized small-signal response of an 1.55 μm BTJ VCSEL at different bias currents for a fixed temperature of 20 °C

11.3 Characterization of 1.33 μm and 1.55 μm InP VCSELs for Coherent Interconnects

As stated in the preceding sections, the use of long-wavelength VCSELs in coherent transmission systems has been of increasing interest in recent years [20, 26]. However, the use of VCSEL structures in data centers so far chiefly relies on pulse-amplitude modulation. One of the reasons is the fundamentally finite laser linewidth and phase noise, caused by spontaneous emission events. These introduce a phase and intensity change in the laser's field and, hence, a change in refractive index, which in turn causes the finite linewidth, typically expressed as the full-width half-maximum (FWHM) of the lineshape [27]. The resulting linewidth can be expressed by [22]:

$$\Delta\nu = \frac{\beta_{sp}\Gamma \underline{n}}{4\pi\tau_{sp}N}\left(1+\alpha^2\right) \quad (11.1)$$

where β_{sp} is the spontaneous emission factor, Γ the confinement factor, τ_{sp} the spontaneous recombination lifetime, $\underline{n}$ and N the carrier and photon density, respectively. The linewidth enhancement factor α describes the change of the complex refractive index's n real (n') and imaginary (n'') part in dependency of the carrier density [27]:

$$\alpha = \frac{\Delta n'}{\Delta n''} \quad (11.2)$$

In general, lasers with a longer cavity exhibit a smaller linewidth, due to their larger round trip delays and hence larger photon lifetimes and photon number [28, 29]. Due to the small size of VCSELs, linewidths in the area of a couple megahertz have been reported [30]. Furthermore, (11.1) shows a decrease of linewidth linearly proportional to an increase of photon number and, thus, optical power. While this proportionality remains constant for lower bias currents and, hence, photon densities, it tends to saturate in a finite linewidth at higher bias points [29, 31]. Although intrinsic mechanisms, i.e. mode competition, may cause a finite linewidth, the saturation is generally associated with extrinsic influences, i.e. current noise and temperature fluctuations, affecting the distributed Bragg reflectors [31]. These are normally power independent and thus create a finite, measurable linewidth-floor. Therefore, the laser's natural linewidth (often referred to as the intrinsic linewidth) may decrease below this extrinsic linewidth undetected [32]. Due to spontaneous emission essentially being a white noise source, the intrinsic linewidth can be described by a Lorentzian lineshape, while the technical noise, generally showing a $1/f$ dependency, is expressed by a Gaussian lineshape [28, 32]. Therefore, considering intrinsic- and extrinsic contributions, the overall lineshape consists of a convolution of a Lorentzian- and a Gaussian lineshape, resulting in a Voigt-profile [31–33]:

$$V(f,\sigma,\gamma) = \int_{-\infty}^{\infty} G\left(f',\sigma\right)L\left(f-f',\gamma\right)\mathrm{d}f' \quad (11.3)$$

with γ and σ in (11.3) denoting the half-width half-maximum (HWHM) of the Lorentzian- and Gaussian lineshape, respectively. This relation will be used to numerically extract the Lorentzian- and Gaussian linewidth contribution from a measured spectrum by least-square fitting methods.

11.3.1 Intrinsic Linewidth

The delayed self-heterodyne measurement setup used to obtain the intrinsic linewidth of a 1.55 μm singlemode VCSEL is shown in Fig. 11.8. The basic principle aims to mix the optical signal with a delayed copy of itself in order to create a beating within the bandwidth of the photodiode, which doubles the measured intrinsic linewidth [32]. While this can be achieved with self-homodyne techniques, such a method would create a beating of 0 Hz, which is generally to be avoided using electrical spectrum analyzers. Therefore, the optical signal is shifted by a fixed frequency in order to create a stable intermediate frequency >0 Hz, thus allowing for more accurate results. The VCSEL is kept at constant 20 °C and is biased by different currents using a low noise current source. Subsequent to isolation, the signal is split equally into two paths using a standard 3-dB coupler. One output arm is inserted into an acousto-optic modulator (AOM), which outputs a first-order diffraction beam, shifted by 110 MHz. The other arm is amplified with subsequent isolation and optical filtering to compensate for the losses of the AOM and the following 20 km delay-line, which decorrelates the two optical signals [34]. Afterwards, the combined signals impinge on a photodiode with 50 GHz bandwidth and create a beating at a stable intermediate frequency of 110 MHz, due to the AOM's frequency shift. The detection was done with an electrical spectrum analyzer. Afterwards, the relation from (11.3) is used to fit a Voigt-profile to the measured spectra and determine the intrinsic linewidths. This fitting procedure was used in Fig. 11.9. Lineshapes were measured for different bias currents and the intrinsic linewidth is plotted against the inverse output power, to better investigate the proportionality from (11.1). The average of five consecutive measurements were used for each datapoint in Figs. 11.9 and 11.10. Following the linearly inverse proportionality to the number of photons, a linear dependency on the inverse output power can be expected [30]. The extracted intrinsic linewidths in Fig. 11.9 are fitted with a linear curve fit, which shows a good agreement with the measured data. The resulting linewidths are as low as 7 MHz for a bias current of 11 mA with a respective output power of approximately 0.3 mW. A better understanding of the underlying mechanisms can be gained from Fig. 11.10, where the intrinsic Voigt contribution is shown in conjunction with the extrinsic Gaussian

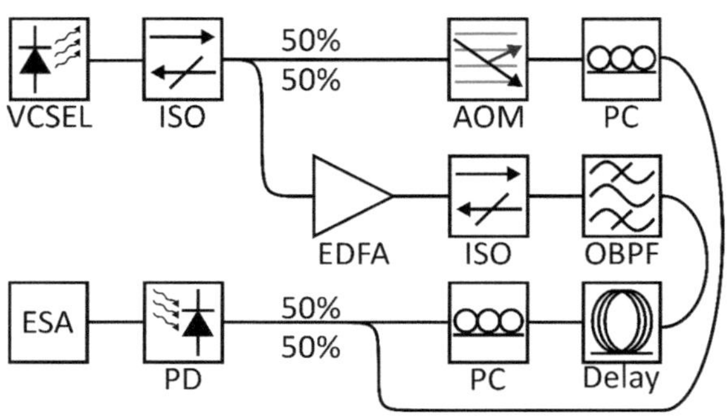

Fig. 11.8 Delayed self-heterodyne measurement setup. ISO: isolator, AOM: acousto-optic modulator, PC: polarization controller, OBPF: optical bandpass filter, PD: photodiode, ESA: electrical spectrum analyser

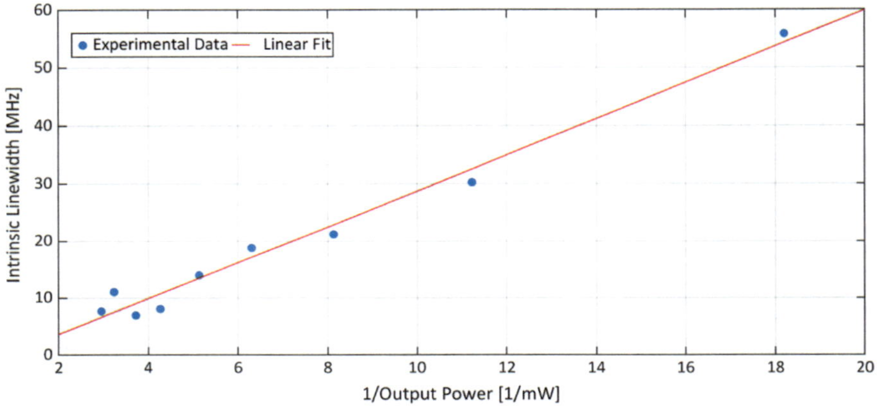

Fig. 11.9 Experimentally determined intrinsic linewidth (FWHM) over inverse output power of an 1.55 μm VCSEL according to (11.3). The data is fitted with a linear curve fit. The determined minimal linewidth is approximately 7 MHz

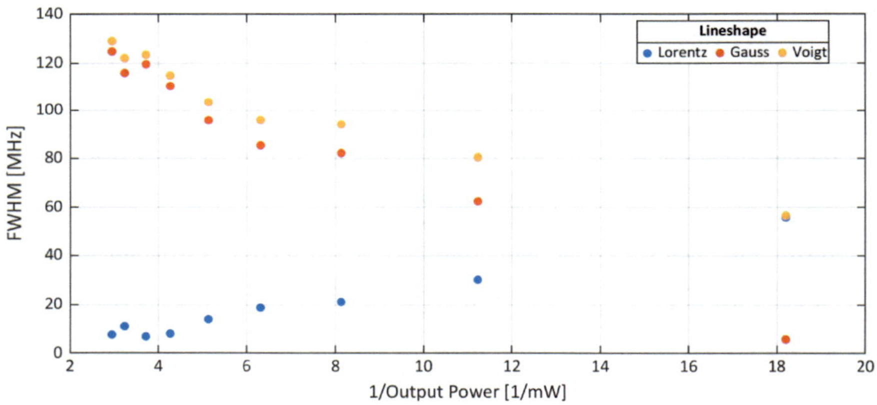

Fig. 11.10 Experimentally determined Gauss and Lorentz contributions (FWHM), as well as overall Voigt FWHM over inverse output power of an 1.55 μm VCSEL according to (11.3). For rising currents, the extrinsic linewidth contribution is increasing and the overall lineshape is further broadened (for the VCSEL in Fig. 11.8)

linewidths and the overall Voigt widths, which is an approximation of the general width of the lineshape. It can be found, that, while the intrinsic linewidth is decreasing for increasing currents and hence output power, the extrinsic linewidth is increasing. However, that increase is accompanied with a rising Voigt width, which can be attributed to a general broadening of the lineshape for increasing currents, possibly related to internal heating. Additional measurements were done for VCSELs emitting at 1330 nm using designated acousto-optic modulator and an optical amplifier for the O-band. However, due to the VCSEL's wavelength being outside the bandwidth of the amplifier, only higher injection currents could be investigated. The determined linewidth is below 10 MHz and thus comparable to that of the 1550 nm devices.

11.4 Modeling of VCSEL-Based Coherent Interconnects

In order to evaluate the system performance of potential VCSEL-based coherent interconnects based on the determined intrinsic linewidth, the expected behavior of a QPSK-based transmission link is simulated. Critical parameters, like phase noise, ASE-noise and frequency offsets between signal and local oscillator, are modeled. Therefore, a short introduction in coherent transmission techniques and the impact of these critical parameters concerning VCSEL-based coherent interconnects is given.

11.4.1 Coherent Transmission Techniques

Transmission techniques like PAM-4 increase the spectral efficiency of the communication link by increasing the number of bits encoded in the amplitude of the optical signal, leaving the phase unutilized. Coherent modulation formats on the other hand employ the modulation also on the signal's phase state, by interfering two amplitude-modulated signals, with one being shifted by 90°. The resulting optical signal maps symbols in the complex plane according to the number of phase/amplitude states of the two optical signals (i.e. QPSK, QAM-16, QAM-64). The real part of the complex signal is called in-phase component and the imaginary part is the quadrature component. Traditional direct-detection schemes can no longer be used to detect these symbols and, hence, coherent receivers are required. A common way to detect the optical signal's phase is to impinge the modulated signal together with a reference laser (also called local oscillator) on a 90°-hybrid. With such a configuration the phase of the modulated laser can be compared with that of the local oscillator (LO), hence, extracting in-phase and quadrature of the combined signal [35]:

$$I_I(t) = R \cdot \sqrt{P_S(t) P_{\mathrm{LO}}} \cdot \cos\big(\omega_{\mathrm{IF}} t + \theta_{\mathrm{Sig}}(t) + \theta_{\mathrm{sn}}(t) - \theta_{\mathrm{LO}}(t)\big), \tag{11.4}$$

$$I_Q(t) = R \cdot \sqrt{P_S(t) P_{\mathrm{LO}}} \cdot \sin\big(\omega_{\mathrm{IF}} t + \theta_{\mathrm{Sig}}(t) + \theta_{\mathrm{sn}}(t) - \theta_{\mathrm{LO}}(t)\big). \tag{11.5}$$

In (11.4) and (11.5), R represents the photodiodes responsivity and P_S and P_{LO} denotes the signal's and local oscillator's power, respectively. While P_{LO} is normally constant, P_S can also vary depending on the modulation format. In heterodyne and intradyne receivers, signal and LO have different frequencies, and, hence, a distorting beating frequency ω_{IF} is exhibited, constantly rotating the constellation. An exception are homodyne receivers, which require an intermediate frequency of 0 Hz. θ_{Sig} represents the phase modulation on the transmitter laser and θ_{sn} and θ_{LO} the signal's and LO's phase noise, respectively [35]. Assuming a constant modulus modulation (i.e. QPSK), the signal power can be assumed to be constant as well. Therefore, critical system impairments are the beating between signal and local oscillator as well as the phase noise, exhibited by transmitter and local oscillator likewise.

11.4.2 Digital Signal Processing

In order to simulate the performance of a basic transmission link employing a VCSEL and higher modulation formats like QPSK, previously mentioned system impairments are modeled. These include the phase noise caused by the finite linewidth of the transmitter and reference laser, which is necessary in coherent receivers. Furthermore, amplitude noise due to amplified spontaneous emission from the EDFA and timing jitter, present in all sampling processes, degrade the signal-to-noise ratio (SNR) and need to be considered as well [29, 36].

11.4.3 Performance of VCSEL-Based Transmission Links for QPSK

Since the most critical parameter about a VCSEL-based coherent interconnect is the phase noise, the following simulation considers solely the VCSEL's intrinsic linewidth and additive white gaussian noise as system impairments. Figure 11.11 shows the impact of different intrinsic linewidths on the modeled transmission link. The investigated linewidths are 10 MHz (a), 15 MHz (b), 25 MHz (c) and 40 MHz (d), which are within the set of determined values in Sect. 11.3. The simulation contains 600 k symbols at an SNR of 14 dB, with a symbol rate of 10 GBd, which is equivalent to a timeframe of 60 μs. The blue samples in Fig. 11.11 represent the received symbols, distorted by amplitude- and phase noise. The red samples show the symbols after digital signal processing, where a second-order decision-directed phase locked loop (DD-PLL) was deployed [20, 37]. The DD-PLL is assumed to be locked prior to transmission. From the recovered symbols in Fig. 11.11, it can be seen that an increasing linewidth causes severe system and thus bit-error rate (BER) degradation. Whilst reducing the laser's linewidth would improve the system performance, a limited finite laser linewidth can be compensated by increasing the symbol rate, thus making the transmission link more resistible to phase noise by reducing the phase error variance [32]. The increased symbol rate reduces the symbol duration, which effectively reduces the impact of phase noise between two adjacent symbols and thus improves the performance of the recovery loop over the transmitted symbol stream. This relation is shown in Fig. 11.12, where different symbol rates are simulated by assuming a constant intrinsic linewidth of 25 MHz. Figure 11.12a represents the same constellation as in Fig. 11.11c, which is 25 MHz intrinsic linewidth at 10 GBd. With

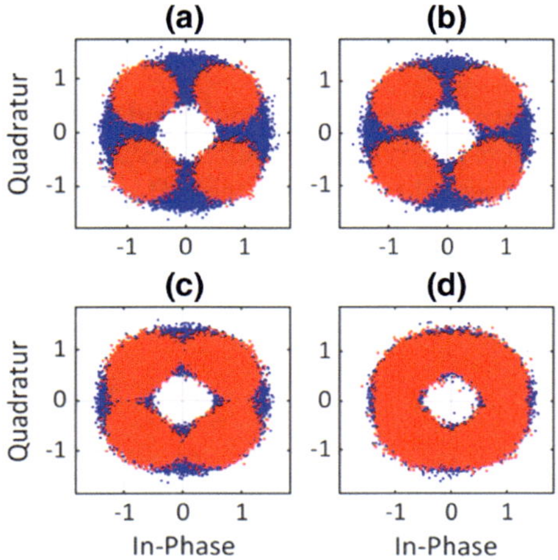

Simulation Parameters	
Symbol Rate	10 GBd
AWGN (SNR)	14 dB
Intrinsic Linewidth (FWHM)	[10, 15, 25, 40] MHz

Fig. 11.11 Modeled constellation diagrams at a fixed symbol rate of 10 GBd and a SNR of 14 dB for different intrinsic linewidths, showing the symbols prior (blue) and after DSP (red). **a** 10 MHz, **b** 15 MHz, **c** 25 MHz, **d** 40 MHz. The table shows the used simulation parameters

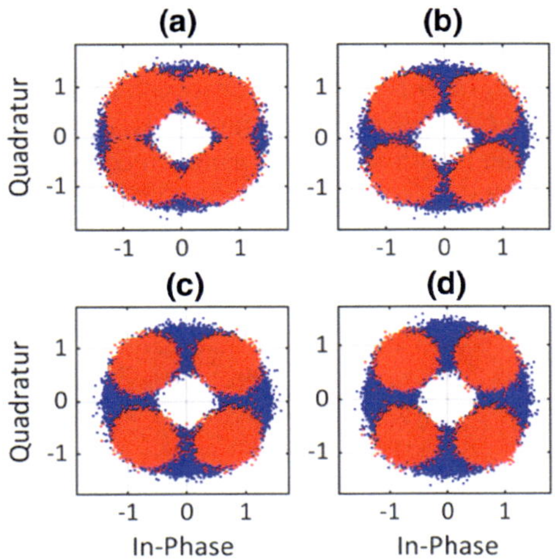

Simulation Parameters	
Symbol Rate	[10, 15, 20, 25] GBd
AWGN (SNR)	14 dB
Intrinsic Linewidth (FWHM)	25 MHz

Fig. 11.12 Modeled constellation diagrams at a fixed intrinsic linewidth of 25 MHz and a SNR of 14 dB for different symbol rates, showing the received symbols prior (blue) and after DSP (red). **a** 10 GBd, **b** 15 Gbd, **c** 20 GBd, **d** 25 GBd. The table shows the used simulation parameters

increasing symbol rate in Fig. 11.12, it can be found that the DD-PLL performance improves and is hence reducing the BER.

11.5 VCSEL-Based PAM-4 Transmission Link

As stated previously, commonly used interconnects are based on on-off-keying or PAM-4 modulation formats. Therefore, future interconnect technologies employing long-wavelength VCSELs need to compare with these present standards. In the following section, a typical PAM-4 measurement setup is presented and evaluated in terms of its performance.

11.5.1 Setup

The measurement setup used for the PAM-4 transmission link is shown in Fig. 11.13. The 1.55 μm singlemode VCSEL is biased with 7 mA at approximately 1.49 V and kept at constant 20 °C. As pattern generator, a digital-to-analog converter with 80 GSps and analog bandwidth of approximately 15 GHz is used to generate the NRZ coded, Nyquist pulse-shaped 10 GBd PAM-4 Signal. A roll-off factor of one is used to limit the bandwidth to 10 GHz. To provide a sufficient peak-to-peak voltage of approximately 580 mV_{pp}, the DAC's signal is amplified by a high-frequency amplifier with 40 GHz bandwidth and 17 dB gain and subsequently attenuated to provide the proper signal voltage. A bias-tee with 67 GHz bandwidth combines bias and high frequency signal prior to injection on the VCSEL. Subsequent, the light directly impinges on a photodiode with 50 GHz bandwidth. Afterwards, the signal is amplified by an amplifier with 11 GHz bandwidth and 12 dB gain. An equivalent-time oscilloscope (ETO) is used to sample the electrical signal, with a synchronized clock signal supplied by the DAC.

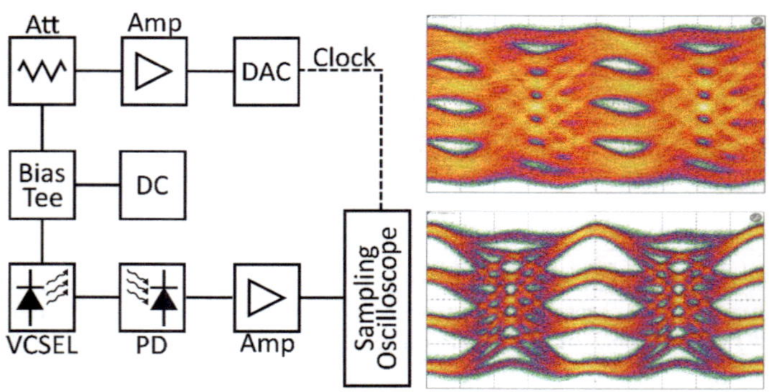

Fig. 11.13 Experimental PAM-4 transmission link (left). DAC: digital-to-analog converter, DC: direct current source, Amp: Amplifier, Att: Attenuator, PD: photodiode. Electrical eye diagram without pre-emphasis (top right) and with pre-emphasis (bottom right)

11.5.2 System Performance

The resulting eye diagram is shown in Fig. 11.14. The received average optical power is approximately −1.5 dBm. The eye diagram shows no clear eye opening, due to linear distortions from the electrical channel and the VCSEL as well as nonlinear effects originating from the VCSEL. Fiber effects are assumed to be of little significance in a back-to-back configuration. To improve the system performance, a pre-emphasis based on a digital filter is used. The filter is based on a small signal analysis of the full setup, including electrical and optical channels, as well as the VCSEL. The pre-distorted eye diagram is presented in Fig. 11.15, showing clear eye openings. 64 k Symbols were oversampled by a factor of eight using the ETO and subsequently resampled to the symbol stream, with the resulting symbols depicted in Fig. 11.16a. The clear distinctions between the four amplitude levels show good agreement with the eye diagram. In order to provide sufficient statistical data, Fig. 11.16b shows the accumulated absolute frequency of sampled symbols based on four chunks of 64 k symbols, resulting in 512 k symbols.

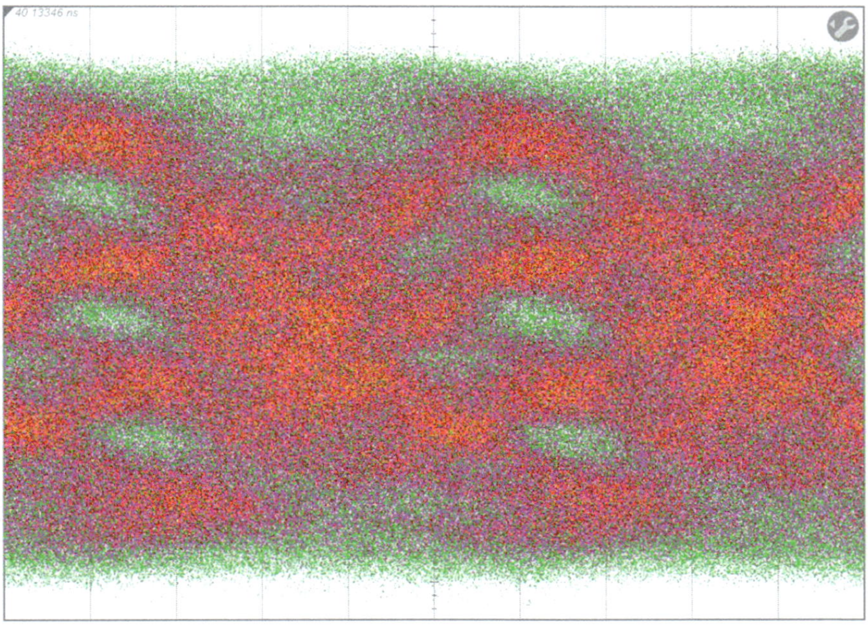

Fig. 11.14 PAM-4 eye diagram without pre-emphasis sampled by the ETO. Vertical resolution is 14 mV/div and horizontal 20 ps/div. The 1550 nm VCSEL is kept at 20 °C

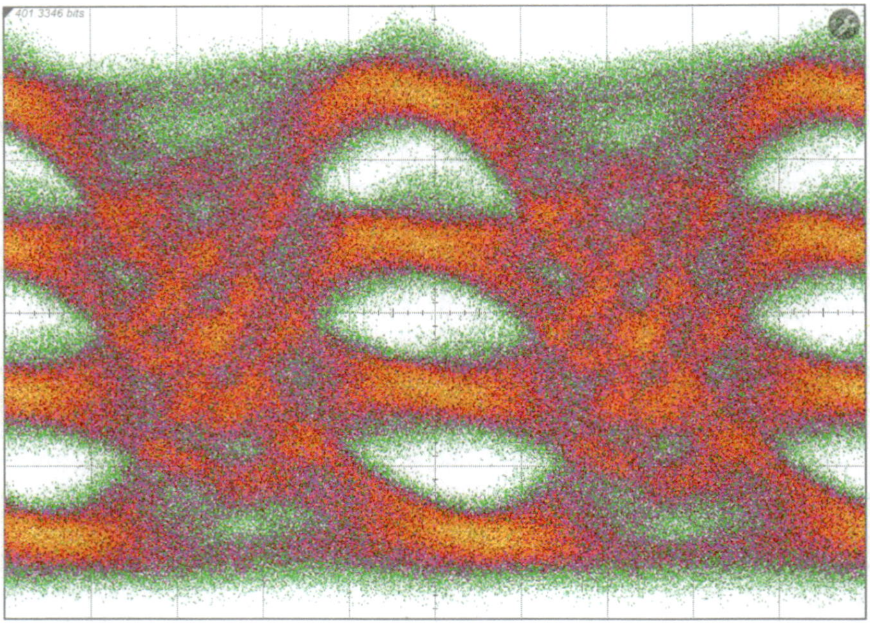

Fig. 11.15 PAM-4 eye diagram with pre-emphasis sampled by the ETO. Vertical resolution is 13 mV/div and horizontal 20 ps/div. The 1550 nm VCSEL is kept at 20 °C

11.6 VCSEL-Based QPSK Transmission Link

With respect to the analysis of the intrinsic linewidth of 1.55 μm singlemode VCSELs in Sect. 11.3 and the simulation of transmission links with respective linewidths, a potential transmission link using QPSK is presented in this section. The DSP algorithms used in Sect. 11.4 will be employed to compensate the most critical system impairments, i.e. the phase noise and the frequency offset due to the beating of signal and local oscillator.

11.6.1 Setup

The measurement setup is shown in Fig. 11.17. The VCSEL is biased at 12 mA in order to provide a sufficiently low linewidth (compare Sect. 11.3). A 40 GHz IQ-modulator in conjunction with a DAC is used to create the in-phase and quadrature signal employing the same NRZ coding and pulse-shaping as in Sect. 11.5 to generate a 10 GBd QPSK signal. Subsequent, the optical signal passes through an amplifier to compensate for the insertion loss due to the modulator and to provide approximately 10 dBm input power on the coherent receiver. After optical bandpass-filtering the

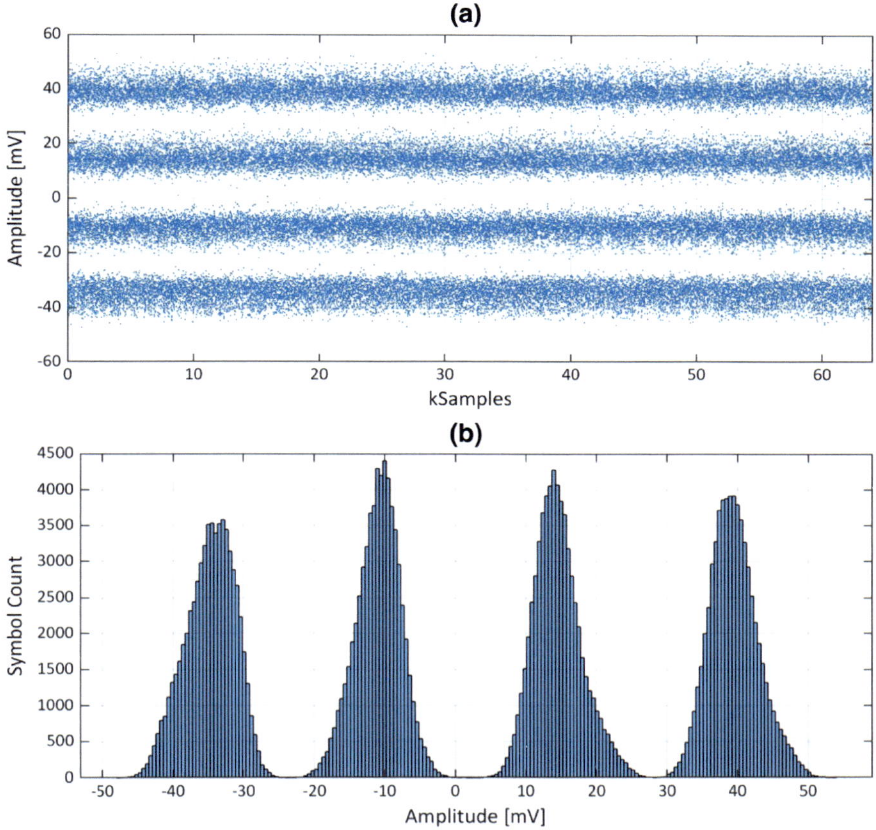

Fig. 11.16 Resampled symbols of the VCSEL-based PAM-4 setup. **a** Resampled symbols for a random pattern containing 64 k symbols. **b** Histogram showing the absolute symbol count for 512 k accumulated symbols

Fig. 11.17 Experimental coherent transmission setup with an VCSEL as transmitter. PC: polarization controller, OBPF: optical bandpass filter, ISO: isolator

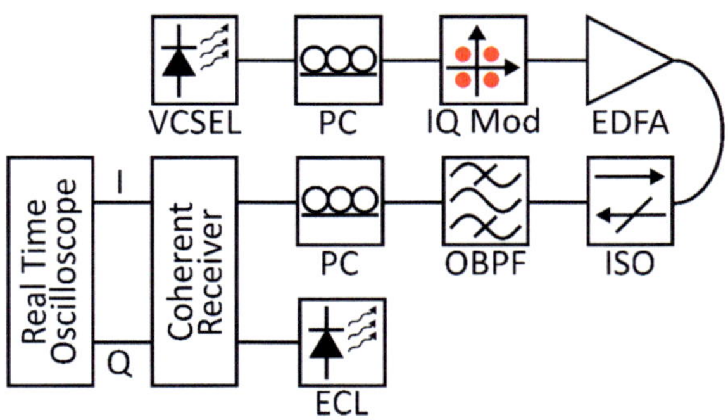

signal impinges on the coherent receiver. The local oscillator in this experiment consists of an external-cavity laser (ECL) with an output power of 13 dBm and an intrinsic linewidth significantly smaller than the VCSEL's. The LO is directly connected to the receiver via a polarization maintaining fiber. Due to the polarization beam splitter (PBS) inside the coherent receiver, 3 dB additional loss can be expected on the LO path. The ECL is carefully fine-tuned in its frequency in order to keep the frequency beating between signal and LO as small as possible. Concluding, the converted electrical signal is sampled by a 33 GHz bandwidth real time oscilloscope with 100 GSps, extracting 1 MS per shot, which is equivalent to chunks of 10 μs.

11.6.2 System Performance

A flowchart of the employed offline signal processing is depicted in Fig. 11.18. The offline-processing was performed in MATLAB with functions from the communication toolbox. The A/D converted signals are oversampled by a factor of ten and furthermore interpolated for the deskewing. The deskew is realized using reference data obtained from two unmodulated ECLs impinging on the coherent receiver, therefore generating sines on the photodiodes. Subsequent to deskewing, the symbols are resampled to the original sample rate and normalized using their root-mean-square.

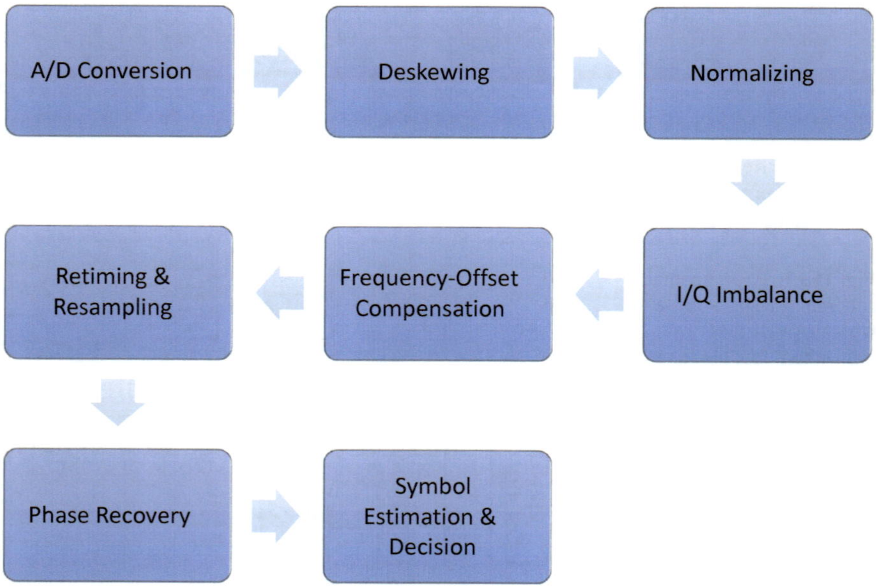

Fig. 11.18 Flowchart of the applied offline digital signal processing. A/D conversion: analog-to-digital conversion

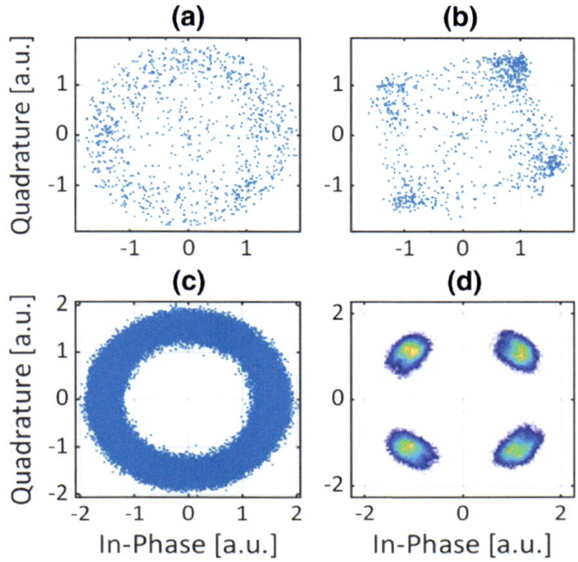

Fig. 11.19 Constellation diagrams for 10 GBd QPSK at various stages of the DSP algorithm: **a** Segment of 100 symbols after I/Q imbalance compensation. **b** Segment of 100 symbols after frequency offset compensation. **c** 64 k symbols after retiming and -sampling. **d** 3D plot of 64 k symbols after phase recovery

During the normalization, IQ imbalances are also compensated for. Subsequently, the frequency offset originating from the beating between the signal and LO are reverted using a coarse FFT-based frequency estimator. The symbols are then retimed and -sampled to the symbol stream. Afterwards, the carrier phase recovery is performed using the DD-PLL introduced in Sect. 11.4. Finally, symbol estimation and -decision is done. Constellations of various stages of this recovery algorithm are depicted in Fig. 11.19, with subplot 11.19a showing a segment of 100 symbols subsequent to deskewing, normalization and IQ imbalance compensation. Afterwards, the frequency offset is compensated, as is depicted in Fig. 11.19b, showing a clearly improved signal constellation with an increased symbol count in the expected reference points. Figure 11.19c shows the retimed and -sampled data for 64 k symbols. Similar to the simulations in Sect. 11.4, the symbols still describe a full circle, due to the presence of significant phase noise and residual rotation from the coarse frequency offset compensation. After passing through the phase recovery algorithm introduced in Sect. 11.4, the 64 k symbols show four clear constellation points, as is depicted in Fig. 11.19d. The data is color-mapped to the symbol count, showing an increased number of symbols in the center of the recovered symbols, as can be expected. After subsequent decoding, the resulting bitstream is error free, reaching the error floor in this experiment due to the 64 k analyzed symbols, which is 7.81×10^{-6}. While this setup was employed as a back to back measurement, the results clearly show the potential of VCSEL-based coherent interconnects. It should be noted, that measurements of other groups on coherent links using 1550 nm VCSELs showed similar BERs in the area of 10^{-5} to 10^{-6} [20]. While the balancing, resampling and phase recovery algorithms are working properly in a 10 GBd QPSK setup employing a VCSEL, problems arose during the measurement concerning the frequency stability of the used VCSEL. Even while the laser is maintained at constant 20 °C during the experiment, the frequency beating is drifting in the domain of multiple GHz, which severely impairs the frequency offset compensation. However, the obstacles imposed by the frequency fluctuation will be overcome by on-chip stabilization techniques, i.e. by employing electrical feedback-loops with integrated photodiodes. For this purpose, we plan to improve the setup in the future. Nonetheless, the demonstrated transmission link with a BER well below the FEC limit shows the high potential of a VCSEL-based coherent interconnect technology. The experimental results indicate, that the intrinsic linewidth of VCSELs is no obstacle to their deployment in coherent transmission systems.

11.7 Conclusion

To cope with the fast growing data center traffic, an evolution from current direct detect interconnects to coherent interconnects is necessary. Due to its polarization depending and singlemode properties, silicon photonics is an ideal candidate for coherent transceiver technology. Furthermore, hetero-epitaxially grown quantum dot lasers on silicon show the potential to be a feedback resistant and temperature insensitive laser source for silicon photonics enabling large scale manufacturing of silicon photonic based interconnects. While the development of such epitaxially grown sources is still under development, hybrid integrated lasers are also relevant for the development of efficient silicon photonic DCIs. VCSELs are currently chiefly used for short optical interconnects and direct detection schemes like PAM. However, the increasing efficiency of DSP in conjunction with less computation hungry algorithms due to the short link distances of intra DCIs may enable their use in low cost coherent interconnects. An integrated VCSEL-based silicon photonic transceiver is conceptualized in Fig. 11.20. The schematic shows a transceiver based on ePIC (electronic-photonic integrated circuit) technology. Due to the cointegration of electronic and photonic circuits on a single chip, a vast amount of functions can be performed efficiently while allowing for a small footprint. As can be found in Fig. 11.20, electrical components like trans-impedance amplifiers (TIAs) and DACs will be realized together with an integrated laser source (i.e. an O-band VCSEL) and an optical IQ modulator. That way, very high performing, efficient and low-cost transmitters are realized. A coherent receiver will be formed by the simultaneous integration of a tunable laser diode (i.e. a distributed feedback laser) as local oscillator, polarization beam splitters and 90° hybrids. With the further deployment of integrated balanced photodiodes, TIAs and analog-to-digital converters, a fully integrated silicon photonic based coherent transceiver is realized. The digital signal processing will be performed in a multi-chip solution on a separate CMOS-based DSP chip, as can be seen in Fig. 11.20. Dedicated CMOS processes are generally better performing at these purely digital tasks compared to ePIC solutions. However, the potential applications for VCSELs in coherent interconnect technologies have to be further investigated due to increasing DSP efficiency and the details concerning the integration of VCSELs

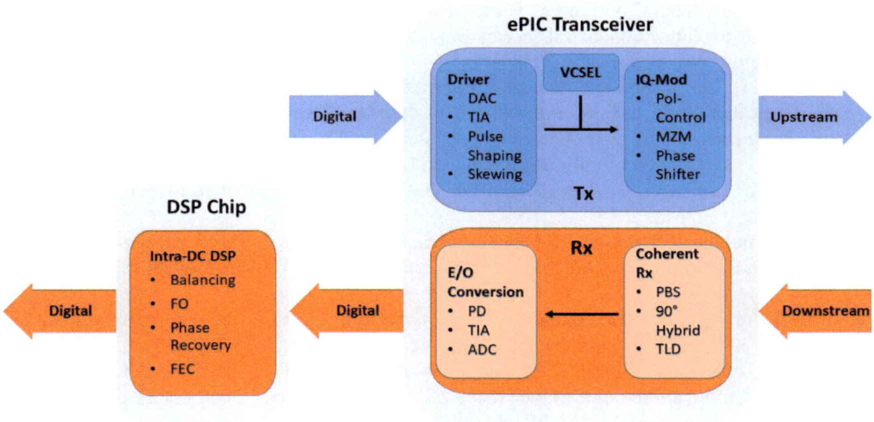

Fig. 11.20 Schematic of an ePIC Transceiver. TIA: trans-impedance amplifier, MZM: Mach-Zehnder-modulator, TLD: tunable laser source, PBS: polarization beam splitter, ADC: Analog-to-digital converter, FO: frequency offset, E/O conversion: electro/optical conversion

in such systems. Within this chapter, we developed and presented a PAM-4 transmission link using a 1.55 μm BTJ VCSEL as a state-of-the-art application. While this is a direct-detection based interconnect, the implementation of a pre-emphasis proved essential in order to achieve an effective transmission. A potential new application for these lasers in coherent interconnects was demonstrated in form of a QPSK transmission link, which achieved a BER below 7.81×10^{-6} and, thus, shows the high potential of VCSEL-enabled coherent interconnects. The significance of inter-DCIs may also further increase in the future with the heteroepitaxical integration of quantum dot lasers on silicon in order to create monolithically integrated coherent transceivers.

Acknowledgements The authors would like to thank VERTILAS GmbH, Munich, Germany, for providing the VCSELs. We'd also like to thank Prof. Dr.-Ing. Dr. H. C. Petermann for his ongoing support and counsel. Furthermore, we want to thank our colleagues at IHP, Innovations for High Performance Microelectronics, Frankfurt (Oder), Germany and at Technische Universität Berlin, Germany—namely Anna Peczek, Marcel Kroh, Georg Winzer and Gregor Ronniger—for their support. Also, we want to thank the group from Prof. James Lott (SFB 787, project C01) for the cooperation and the fruitful discussions.

References

1. Cisco, Cisco Visual Networking Index: Forecast and Methodology, 2016–2021. https://www.cisco.com/c/en/us/solutions/collateral/service-provider/visual-networking-index-vni/complete-white-paper-c11-481360.html. Accessed 5 Nov 2018

2. Cisco, Cisco Global Cloud Index: Forecast and Methodology, 2016–2021. https://www.cisco.com/c/en/us/solutions/collateral/service-provider/global-cloud-index-gci/white-paper-c11-738085.html. Accessed 5 Nov 2018
3. R. Urata, H. Liu, X. Zhou, A. Vahdat, Datacenter interconnect and networking: from evolution to holistic revolution, in *Paper Presented at the 40th Optical Fiber Communications Conference and Exhibition*, Los Angeles Convention Center, Los Angeles, 19–23 Mar 2017
4. C. Doerr et al., Single-chip silicon photonics 100-Gb/s coherent transceiver, in *Paper Presented at the 37th Optical Fiber Communications Conference and Exhibition*, Moscone Center, San Francisco, 9–13 Mar 2014
5. J.C. Rasmussen, T. Drenski, DSP for short reach optical links, in *Paper Presented at the 43rd European Conference on Optical Communication*, Svenska Massan, Gothenburg, 17–21 Sept 2017
6. J.A. Tatum et al., VCSEL-based interconnects for current and future data centers. J. Lightw. Technol. **33**(4), 727–732 (2015). https://doi.org/10.1109/JLT.2014.2370633
7. Y. Li, Y. Zhang, L. Zhang, A. Poon, Silicon and hybrid silicon photonic devices for intra-datacenter applications: state of the art and perspectives [Invited]. Photon. Res. **3**(5), B10–B27 (2015). https://doi.org/10.1364/PRJ.3.000B10
8. H. Zhang et al., Real-time transmission of 16 Tb/s over 1020 km using 200 Gb/s CFP2-DCO. Opt. Expr. **26**(6), 6943–6948 (2018). https://doi.org/10.1364/OE.26.006943
9. M.A. Taubenblatt, Optical interconnects for high-performance computing. J. Lightw. Technol. **30**(4), 448–457 (2012). https://doi.org/10.1109/JLT.2011.2172989
10. G. Winzer et al., Monolithic photonic-electronic QPSK receiver for 28 Gbaud, in *Paper Presented at the 38th Optical Fiber Communications Conference and Exhibition*, Los Angeles Convention Center, Los Angeles, 22–26 Mar 2015
11. B. Wohlfeil et al., First demonstration of fully integrated segmented driver and MZM in 0.25-μm SiGe BiCMOS employing 112 Gb/s PAM4 over 60 km SSMF, in *Paper Presented at the 44th European Conference on Optical Communication*, Fiera Roma, Rome, 23–27 Sept 2018
12. J.P. Reithmaier, M. Benyoucef, III–V integration on Si for photonics, in *Paper Presented at the IEEE Photonics Society Summer Topical Meeting Series*, Newport Beach Marriott Hotel, Newport Beach, 11–13 July 2016
13. A.Y. Liu et al., High performance continuous wave 1.3 μm quantum dot lasers on silicon. App. Phys. Lett. **104**(4), 041104–041104 (2014). https://doi.org/10.1063/1.4863223
14. G.-H. Duan et al., Hybrid III–V silicon photonic integrated circuits for optical communication applications. IEEE J. Sel. Topics Quantum Electron. **22**(6), 379–389 (2016). https://doi.org/10.1109/JSTQE.2016.2614661
15. G.-H. Duan et al., Hybrid III–V on silicon lasers for photonic integrated circuits on silicon. IEEE J. Sel. Topics Quantum Electron. **20**(4), 158–170 (2014). https://doi.org/10.1109/JSTQE.2013.2296752
16. M.J.R. Heck et al., Hybrid silicon photonics for optical interconnects. IEEE J. Sel. Topics Quantum Electron. **17**(2), 333–346 (2011). https://doi.org/10.1109/JSTQE.2010.2051798
17. H. Huang et al., Analysis of the optical feedback dynamics in InAs/GaAs quantum dot lasers directly grown on silicon. J. Opt. Soc. Am. B **35**(11), 2780–2787 (2018). https://doi.org/10.1364/JOSAB.35.002780
18. T. Mitze et al., CWDM transmitter module based on hybrid integration. IEEE J. Sel. Topics Quantum Electron. **12**(5), 983–987 (2006). https://doi.org/10.1109/JSTQE.2006.882643
19. C. Xie et al., All-VCSEL based 100-Gb/s PDM-4PAM coherent system for applications in metro networks, in *Paper Presented at the 40th European Conference on Optical Communication*, Palais des Festivals, Cannes, 21–25 Sept 2014
20. T.N. Huynh et al., Digital coherent communications with a 1550 nm VCSEL, in *Paper Presented at the 38th Optical Fiber Communications Conference and Exhibition*, Los Angeles Convention Center, Los Angeles, 22–26 Mar 2015
21. A.Y. Liu, J. Bowers, Photonic integration with epitaxial III–V on silicon. IEEE J. Sel. Topics Quantum Electron. **24**(6), 1–12 (2018). https://doi.org/10.1109/JSTQE.2018.2854542
22. R. Michalzik (ed.), *VCSELs* (Springer, Berlin Heidelberg, 2013)

23. M.-C. Amann, W. Hofmann, InP-based long-wavelength VCSELs and VCSEL arrays. IEEE J. Sel. Topics Quantum Electron. **15**(3), 861–868 (2009). https://doi.org/10.1109/JSTQE.2009.2013182
24. W. Hofmann, High-speed buried tunnel junction vertical-cavity surface-emitting lasers. IEEE Photon. J. **2**(5), 802–815 (2010). https://doi.org/10.1109/JPHOT.2010.2055554
25. J.A. Lott et al., 20 Gbit/s error free transmission with ~850 nm GaAs-based vertical cavity surface emitting lasers (VCSELs) containing InAs–GaAs submonolayer quantum dot insertions, in *Paper Presented at the SPIE Opto: Integrated Optoelectronic Devices Conference*, San Jose Convention Center, San Jose, 24–29 Jan 2009
26. S.-Y. Kim et al., VCSEL-based coherent detection of 10-Gbit/s QPSK signals using digital phase noise cancellation for future optical access systems, in *Paper Presented at the 33rd Conference On Optical Fiber Communication (OFC/NFOEC), Collocated National Fiber Optic Engineers Conference*, San Diego Convention Center, San Diego, 21–25 Mar 2010
27. C. Henry, Theory of the linewidth of semiconductor lasers. IEEE J. Quantum Electron. **18**(2), 259–264 (1982). https://doi.org/10.1109/JQE.1982.1071522
28. K. Petermann, *Laser Diode Modulation and Noise* (Kluwer Academic Publishers, Dordrecht, 1988)
29. G.P. Agrawal, *Fiber-Optic Communication Systems*, 3rd edn. (Wiley, New York, 2002)
30. F.M. De Sopra et al., Near-infrared vertical-cavity surface-emitting lasers with 3-MHz linewidth. IEEE Photon. Technol. Lett. **11**(12), 1533–1535 (1999). https://doi.org/10.1109/68.806836
31. S. Viciani et al., Lineshape of a vertical cavity surface emitting laser. Opt. Commun. **206**(1–3), 89–97 (2002). https://doi.org/10.1016/S0030-4018(02)01381-0
32. S. Spießberger, Dissertation, Technische Universität Berlin (2012)
33. K.L. Letchworth, D.C. Benner, Rapid and accurate calculation of the Voigt function. J. Quant. Spectrosc. Radiat. Transf. **107**(1), 173–192 (2007). https://doi.org/10.1016/j.jqrst.2007.01.052
34. W. Schmid et al., Delayed self-heterodyne linewidth measurement of VCSELs. IEEE Photon. Technol. Lett. **8**(10), 1288–1290 (1996). https://doi.org/10.1109/68.536630
35. K. Kikuchi, High spectral density optical communication technologies, in *Optical and Fiber Communications Reports*, vol. 6, ed. by M. Nakazawa, K. Kikuchi, T. Miyazaki (Springer, Berlin, Heidelberg, 2010)
36. T. Okoshi, K. Kikuchi, *Coherent Optical Fiber Communications* (Kluwer Academic Publishers, Dordrecht, 1988)
37. T.N. Huynh et al., BER performance of coherent optical communications systems employing monolithic tunable lasers with excess phase noise. J. Lightw. Technol. **32**(10), 1973–1980 (2014). https://doi.org/10.1109/JLT.2014.2319108

Chapter 12
Nitride Microcavities and Single Quantum Dots for Classical and Non-classical Light Emitters

G. Schmidt, C. Berger, A. Dadgar, F. Bertram, P. Veit, S. Metzner, A. Strittmatter, J. Christen, S. T. Jagsch, M. R. Wagner and A. Hoffmann

Abstract Microcavities with InGaN quantum wells or GaN-based quantum dots as active medium are building blocks of electrically-driven, low-threshold surface-emitting lasers or single photon emitters in the visible-to-UV spectral range. In this chapter, we highlight essential developments in epitaxial growth techniques of such nitride-based microcavities and their active regions. Modern analytical techniques for structural and optical characterization of these complex heterostructures as presented in this chapter are essential to solve remaining challenges.

12.1 Introduction

Microcavities confine light to a very small volume and may alter the properties of embedded light emitting objects. Fascinating, fundamental physical effects take place and allow for the realization of challenging semiconductor devices such as surface emitting lasers and non-classical light emitters. Consisting of an optically active region that is surrounded by two highly reflective distributed Bragg reflectors (DBRs), emitted photons travel back and forth inside of the cavity enhancing the interaction between light and matter. The strength of this light-matter coupling can be steered by tailoring the reflectivity of the mirrors and the emission characteristics of the active material. As a consequence of the mode-selection by the cavity the oscillator strength is concentrated into the resonance emission channels of the photoactive region. This leads to an enhanced spontaneous recombination rate and to an enhanced directionality of the emission which is described by the so-called Purcell effect [1]. Employing the Purcell-effect is crucial for single photon sources with high single photon flux and, therefore, single quantum dots (QD) placed inside a microcavity (MC) would be a key element. Vertical-cavity surface emitting laser

G. Schmidt · C. Berger · A. Dadgar (✉) · F. Bertram · P. Veit · S. Metzner · A. Strittmatter · J. Christen
Institut für Physik, Otto-von-Guericke-Universität Magdeburg, Magdeburg, Germany
e-mail: armin.dadgar@ovgu.de

S. T. Jagsch · M. R. Wagner · A. Hoffmann
Institut für Festkörperphysik, Technische Universität Berlin, Berlin, Germany

M. Kneissl et al. (eds.), *Semiconductor Nanophotonics*, Springer Series in Solid-State Sciences 194, https://doi.org/10.1007/978-3-030-35656-9_12

(VCSEL) devices commonly contain quantum wells (QW) as active medium to generate coherent emission [2] with large optical intensity. This goal favors the weak coupling regime, which is characterized by an irreversible interaction between light and matter following Fermis Golden Rule [3]. An electronic transition from an excited state emits a photon, but due to decoherence repeated interaction of both is unlikely. As the thickness of the optically active medium in the direction of the stimulated light emission is very small in a VCSEL only low modal gain is expected. Therefore, the losses through the mirrors usually need to be lower than 1% to achieve stimulated emission, meaning mirror reflectivities above 99% for laser operation. Though, with increasing the mirror reflectivity, the width of the resonance frequency of the microcavity becomes even narrower. The width of this cavity-mode is characterized by the quality-factor (Q-factor) which is also a measure of the cavity-photon lifetime. Microcavities with high Q-factor and an embedded optical emitter with low linewidth broadening can undergo a transition to the strong coupling regime. Here, the coupling strength between the excitonic- and the cavity-mode is large enough forcing interaction between the material oscillator and the electromagnetic field with reversible energy exchange. New quasi-particles, so-called cavity polaritons appear in optical spectra with bosonic properties and are susceptible to Bose-Einstein like condensation [4]. In contrast to conventional lasers, population inversion is no longer a requirement. Instead, these lasers emit coherent light by stimulated relaxation of polaritons towards the lowest polariton energy-state from where they decay spontaneously [5]. Such a transition as fundamental operation principle of a laser would yield threshold-less lasing which has significant merit for nitride-based lasers that typically require a few kA/cm^2 at threshold.

Controlled generation of non-classical states of light is of fundamental interest in quantum optics and a technological challenge in quantum information processing [6–8]. Semiconductor QDs are substantial candidates for the realization of miniaturized single photon emitting devices [9, 10] due to their atom-like energy structure [11, 12] and feasibility of electrically triggered operation [13–16]. In particular, GaN/AlN QDs are promising room temperature single photon emitters [17, 18] because of their high exciton binding energy and sufficient carrier confinement by enormous band offsets between GaN and AlN [19, 20].

12.2 Bragg Mirrors in the Visible to Deep UV Spectral Region

This section illustrates several ways to realize distributed Bragg reflectors (DBRs) in the III-nitride system with different material combinations. The choice of materials mostly depends on the target wavelength, favoring GaN/AlInN for the visible wavelength range, AlGaN/AlInN for the near UV, and AlN/AlGaN Bragg mirrors for wavelengths shorter than 300 nm. DBRs with high average Al-content are mandatory for absorption-free microcavities with GaN QDs emitting in the deep UV (see Sect. 12.4).

Microcavities enforce an optical standing-wave that should have a maximum of the electric field intensity at the optically active region. In principle, the MC can be formed by highly reflective coatings on both ends of the cavity, but simple metal coatings have low reflectivities (<95%) and high absorption losses. Higher reflectivities up to and beyond 99% are enabled by distributed Bragg mirrors, a periodic sequence of two transparent layers with different refractive index and an optical thickness of $\lambda/4$ each. Such Bragg mirrors in the visible to near UV wavelength region in the III-N system are challenging. The refractive index change, even between the binaries AlN and GaN, is rather low and many $\lambda/4$ layer pairs are therefore required to achieve reflectivities above 99%. Furthermore, the large lattice mismatch of the III-Ns inhibits growth of high quality Bragg mirrors out of binary materials, e.g., of AlN/GaN $\lambda/4$ layers [21, 22]. In the III-N material system a lattice-matched material combination, AlInN with either GaN or AlGaN, exists providing a large refractive index contrast (Fig. 12.1) [23–25]. AlInN with an In content of 18% is lattice-matched to GaN, while the AlInN composition can be changed by growth temperature to ensure lattice-matching also to any given AlGaN alloy, e.g. $Al_{0.86}In_{0.14}N$ is lattice-matched to $Al_{0.2}Ga_{0.8}N$. Therefore, even DBRs with large numbers of periods can be grown without issues in strain management.

In heteroepitaxy, strain engineering is required to manage the otherwise steadily increasing tensile stress during epitaxy [27]. This large tensile stress gives rise to cracking of highly reflecting Bragg mirror stacks and VCSEL structures, respectively, with a typical thickness in excess of 7 μm. The growth of such mirror structures is difficult due to the different growth settings of (Al)GaN and AlInN. While (Al)GaN grows best at temperatures above 1000 °C in hydrogen atmosphere, AlInN requires a nitrogen atmosphere at temperatures around 750 °C. Moreover, the temperature controls the In-incorporation into AlInN. Thus the growth of a DBR with high crystalline quality and smooth interfaces requires recurring temperature ramps.

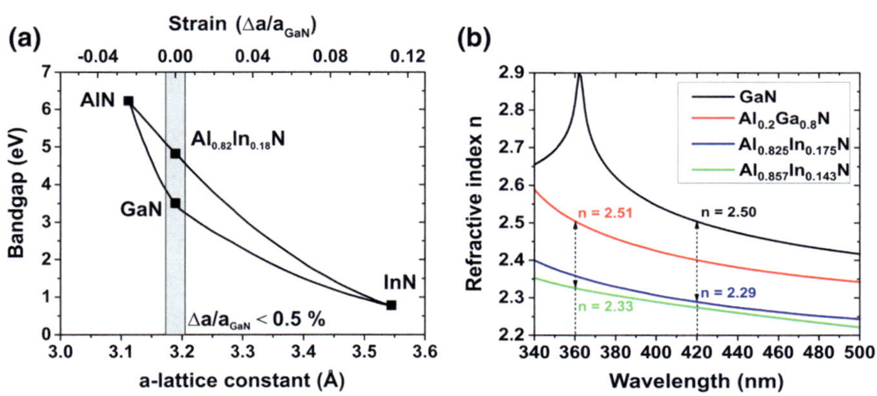

Fig. 12.1 A-lattice constant and lattice mismatch to GaN versus band-edge energy of the III-Ns **a**. Refractive indices for selected III-nitride compounds and possible material combinations for Bragg mirror stacks in the blue to near UV spectral range **b** [26]

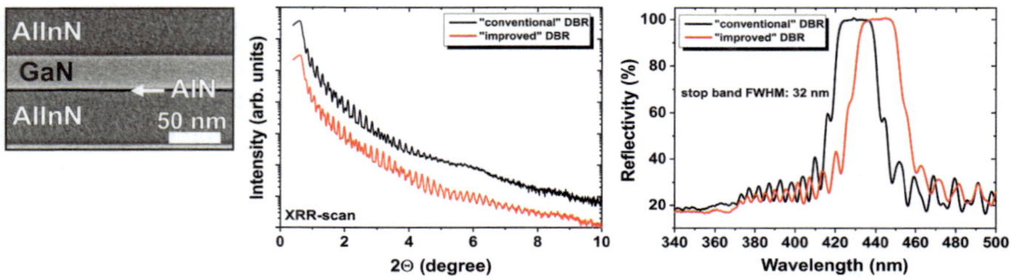

Fig. 12.2 Cross-sectional STEM image of AlInN/GaN/AlInN layers within a DBR (left). A thin AlN layer forms at the upper interface of the AlInN layer by In-loss when heating to GaN growth temperature. With a low-temperature GaN layer grown on top before heating, the interface quality can be improved, which is visible in X-ray reflection measurements (middle). Although the roughness can be reduced, the maximum reflectivity and stop band width is not improved by this approach (right) [28]

During heating, care must be taken to avoid desorption of In from the surface of AlInN (Fig. 12.2). Berger et al. [28] reported a capping layer technique of AlInN by a thin GaN layer that is efficient to preserve the AlInN composition. Bragg mirrors with 35 pairs of AlInN/GaN yield a maximum reflectivity above 99% even if the AlInN is not capped by thin GaN.

Even more demanding are AlGaN/AlInN Bragg mirrors on heterosubstrates, such as sapphire, since the strain engineering is more difficult. The high dislocation density in AlGaN and dislocation climb during growth build up additional tensile strain. For wavelengths around 350 nm, a fully transparent $Al_{0.2}Ga_{0.8}N/Al_{0.86}In_{0.14}N$ heterostructure outperforms the combination of GaN/AlInN, even at a wavelength of 380 nm, only allows a maximum reflectivity of 98.5% [29]. However, a lower refractive-index contrast of 7.5% for AlGaN/AlInN mirrors (at 350 nm) demands for more mirror pairs to achieve a similar reflectivity as for the GaN/AlInN system in the visible spectral region.

As long as growth on foreign substrates is considered, the design of the buffer layer structure needs attention. From a comparison of AlGaN/AlInN mirror stacks either on top of an $AlN/Al_{0.17}Ga_{0.83}N$ buffer or on top of an AlN/GaN/AlGaN buffer layer stack, a significant impact of the strain state is found. The second structure utilizes three AlN interlayers grown at low temperature (LT-AlN) within the AlGaN to modify the strain state, in addition an in situ deposited SiN mask with an intermediate GaN layer to lower the dislocation density [26] (Fig. 12.3, top).

While the latter is not effective to lower the dislocation density [30, 31], the overall quality of the layer is improved (Fig. 12.3, bottom) and the surface shows a reduced roughness with the more complex buffer structure (Fig. 12.4).

Much more important is the change of the initial strain state of the mirror structure. In contrast to the AlN/AlGaN buffer, an AlN/GaN/AlGaN buffer imposes compressive strain (Fig. 12.5). While the DBR grown on the AlN/AlGaN buffer heavily cracks after ~40 periods marked by a sudden drop of the curvature signal, the DBR on the improved buffer structure shows a lower total tension enabling crack-free 45-fold mirror pairs. Such mirrors exhibit maximum reflectivities above 99%, similar to those reported by Feltin et al. [32].

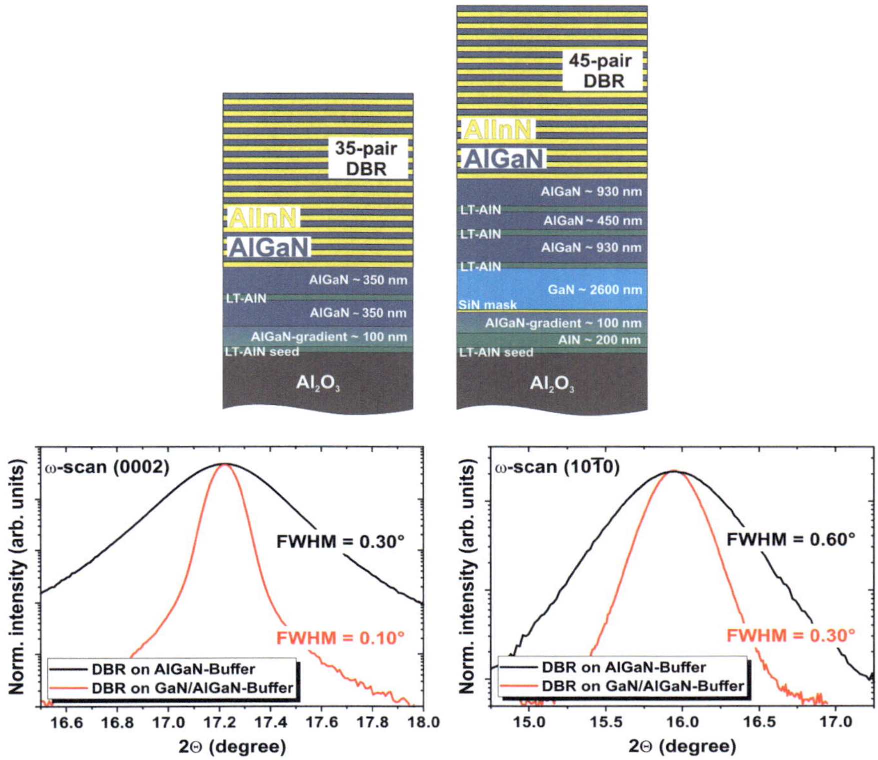

Fig. 12.3 Two different buffers for Bragg mirror stacks were tested (top). A three-fold and two-fold improvement in X-ray ω-scans was detected when using the GaN based buffer [26]

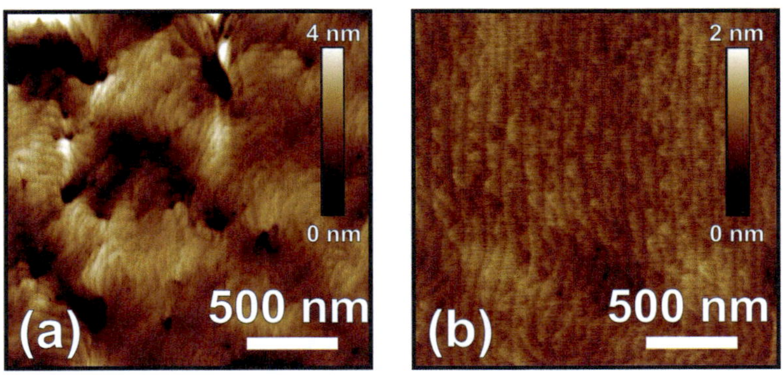

Fig. 12.4 AFM images of a 35-fold and a 45-fold AlGaN/AlInN Bragg mirror on an AlGaN buffer **a** and a GaN based buffer **b**, respectively. The rms roughness decreases from 0.96 to 0.16 nm rms [26]

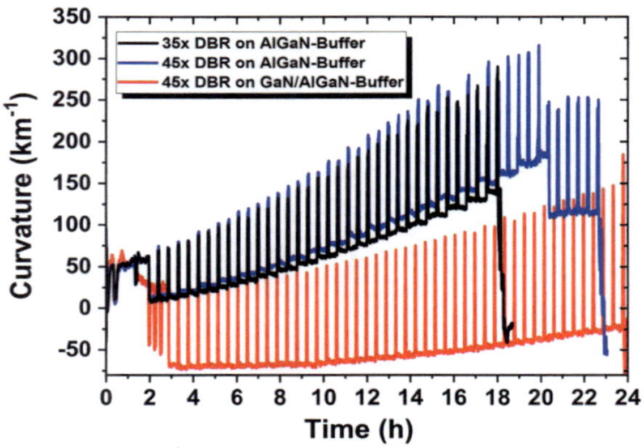

Fig. 12.5 In-situ curvature measurement of AlGaN/AlInN Bragg mirrors on different buffer layers. For the GaN buffer, the initial curvature indicates compressive growth of the mirror stack (negative curvature) while it is under tension for the AlGaN based buffer. The growth of a 45-fold mirror is under increasing tension on an AlGaN buffer, which leads to cracking during growth for the 45-fold mirror (sudden change in curvature after ~20.5 h) [26]

12.3 Microstructure and Emission Properties of Blue/Violet Emitting III-Nitride Microcavities

Microcavities can be accomplished using dielectric mirrors, e.g. with a 10-pair SiO_2/Ta_2O_5 dielectric (amorphous) mirror deposited by electron beam evaporation on top of the otherwise epitaxially grown VCSEL-structures. Figure 12.6 shows a cross-section field emission scanning electron microscope (FE-SEM) image of such

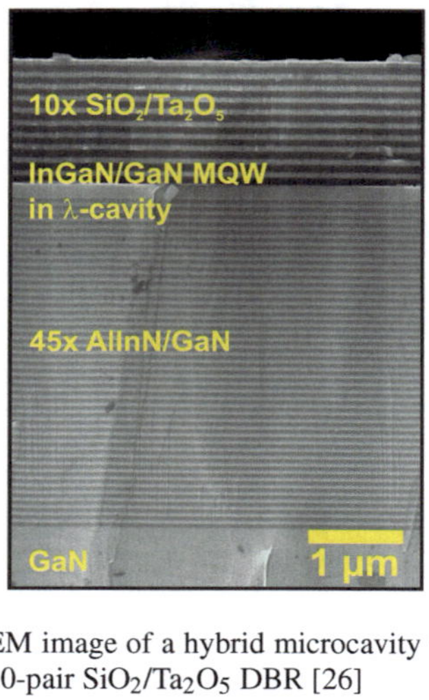

Fig. 12.6 Cross-section FE-SEM image of a hybrid microcavity with 5-fold InGaN/GaN MQW, 45-pair AlInN/GaN DBR and 10-pair SiO_2/Ta_2O_5 DBR [26]

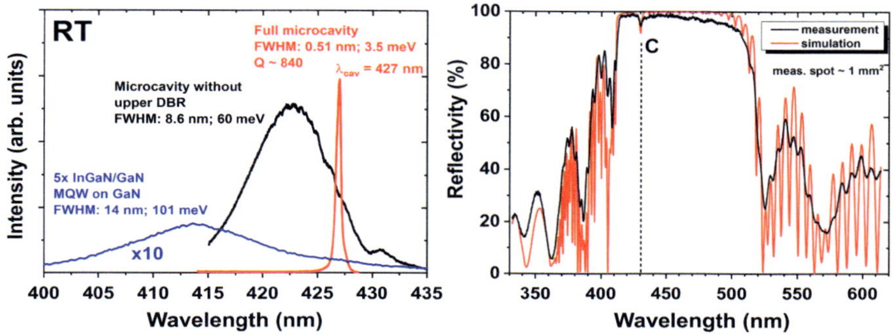

Fig. 12.7 PL-spectra at room temperature (RT) of a hybrid microcavity before and after deposition of an upper SiO_2/Ta_2O_5 DBR compared to a reference InGaN/GaN MQW grown on a GaN/sapphire buffer (left); measured and simulated reflectivity spectrum of the microcavity (right). Simulation parameters are as follows: 45-fold DBR with 46.5 nm AlInN/42 nm GaN, 173 nm GaN cavity with 5-fold 3.2 nm InGaN/4.6 nm GaN MQW, 10-fold DBR with 83 nm SiO_2/53 nm Ta_2O_5 after [26]

a hybrid microcavity with a 5-fold InGaN/GaN MQW as active region. Excellent structual homogeneity in lateral and vertical direction can be realized. Figure 12.7 depicts the corresponding impact on the photoluminescence spectrum before and after deposition of the upper dielectric mirror. The spectrum of the structure without upper DBR is already significantly brighter and narrower compared to a structure with the same active region grown on a GaN-buffer without epitaxial DBR. Thereby, the intensity is increased by a factor of 35 and the linewidth is reduced from 101 to 60 meV. The observed redshift of the luminescence for the MQW grown on top of the DBR is caused by slightly lower temperature during the growth of the InGaN QWs due to a more convex wafer bowing of the microcavity sample. After the deposition of the upper DBR the linewidth becomes as low as 0.5 nm. The luminescence linewidth corresponds to a Q-factor of around 840 which is comparable to other published results [33, 34]. The reflectivity spectrum of the hybrid microcavity is dominated by the broad stop band of the upper mirror. The slight dip at the wavelength of 430 nm marks the spectral position of the cavity mode. The Q-factor could be determined from the width of this dip as well. It should be noted, however, that the photonic disorder of the cavity modes takes place on a length scale of micrometers. Thereby, the measured Q-factor values decrease with increasing size of the probed area. In this particular example, a Q-factor of only $Q = 215$ is found [35, 36]. An evaluation of the photonic disorder can be performed by μ-photoluminescence line scans across the surface of hybrid microcavities. The example shown in Fig. 12.8 was taken on a hybrid microcavity with 13-fold InGaN/GaN MQW embedded in a λ-cavity. Local variations of the emission wavelength of about 5 nm were observed. Furthermore, on a micrometer scale, local spectra revealed strong fluctuations of the emission linewidth with values between 12 and 1.2 meV corresponding to Q-factors between 250 and 2500. Such disorder is mainly caused by thickness fluctuations and the high dislocation density due to heteroepitaxial growth on substrates as sapphire

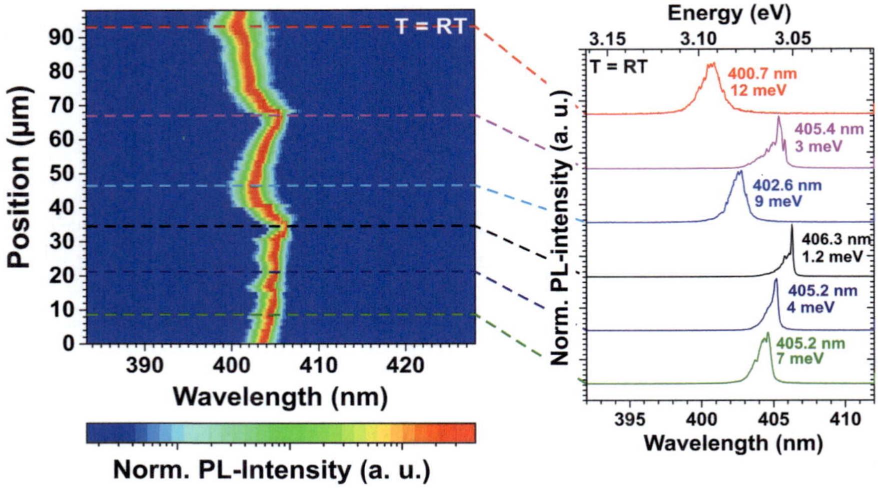

Fig. 12.8 Lateral μ-PL-line scan and local spectra of a hybrid microcavity with a 13-fold InGaN/GaN MQW [26]

or Si. The disorder can be significantly reduced by growing on high-quality free-standing GaN substrates [37]. Optically pumped lasing was demonstrated as shown in Fig. 12.9. Excitation of the microcavity structure was obtained by laser emission of a frequency-doubled Ti-Sapphire-laser. The threshold behavior in the light output power as typically observed for lasing operation is evident upon cranking up the

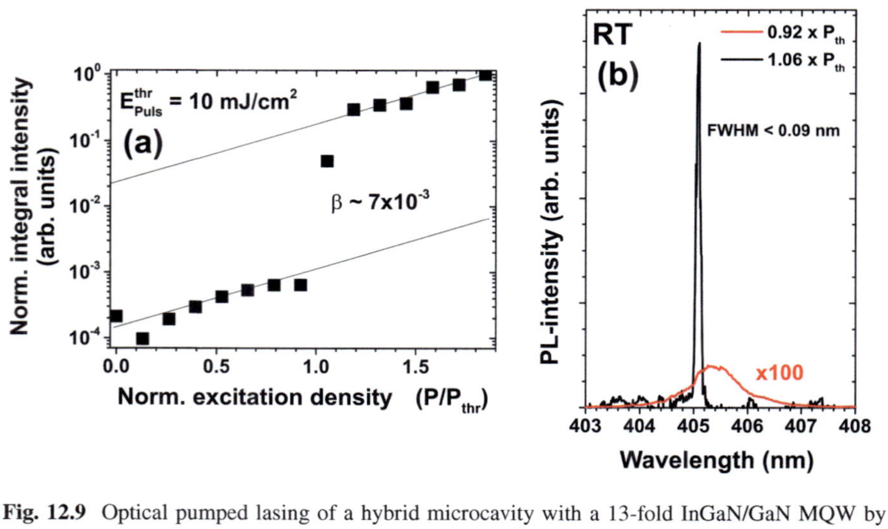

Fig. 12.9 Optical pumped lasing of a hybrid microcavity with a 13-fold InGaN/GaN MQW by excitation dependent PL: normalized integral PL intensity versus normalized excitation density **a** and PL spectra slightly below and above threshold **b** are depicted [26]

excitation energy. More than two orders of magnitude increase in intensity and a linewidth decrease from 1 nm to less than 0.09 nm was seen. Lasing threshold took place at a pulse energy density of 10 mJ/cm^2. The β-factor, which describes the fraction of the spontaneous emission to the stimulated emission, was estimated from the intensity ratio right below and above the lasing threshold. It yields to 7×10^{-3}, which is much higher than for conventional edge emitting lasers [38].

12.3.1 Electric Fields Within AlGaN/AlInN DBRs

For microcavities in the UV spectral range, a lattice-matched combination of AlGaN and AlInN $\lambda/4$-layers enables strain-free epitaxial growth. Both materials exhibit strong spontaneous polarization that affects the band structure significantly. We investigated such an AlInN/AlGaN distributed Bragg reflector grown by metalorganic chemical vapor phase epitaxy (MOVPE). On top of a c-plane sapphire substrate with an AlGaN buffer structure, the DBR is composed of 35 periods of AlInN and AlGaN $\lambda/4$ layers featuring maximum reflectivities of 99% at 360 nm. Lateral homogeneity and sharp interfaces were observed across the whole DBR layer stack. A PL emission signal at 327 nm was identified as emission from the AlInN/AlGaN DBR. In Fig. 12.10a the scanning transmission electron microscope (STEM) high angle annular dark field (HAADF) image from the bottom DBR is shown with the AlGaN buffer layer at the bottom. The simultaneously recorded cathodoluminescence (CL) intensity image with spectral window across the DBR luminescence (see Fig. 12.10b) identifies the AlGaN/AlInN interface as the origin of this emission.

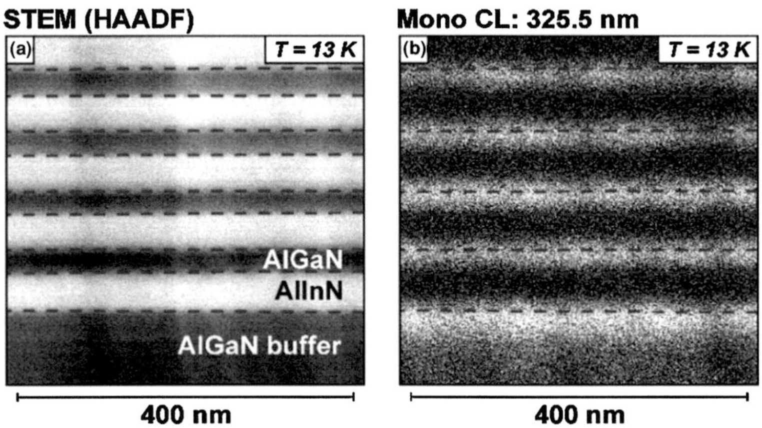

Fig. 12.10 Monochromatic STEM-CL map of AlInN/AlGaN Bragg reflector: **a** STEM (HAADF) image of bottom DBR and AlGaN buffer, **b** CL intensity image of 325.5 nm DBR emission exhibit increased intensity at AlGaN/AlInN interface taken at $T = 13$ K [39]

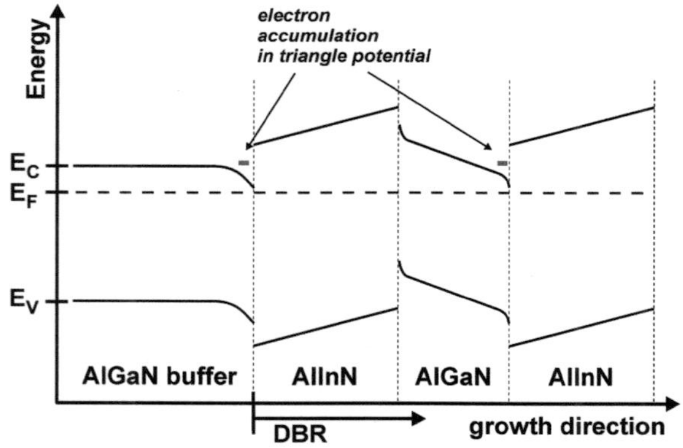

Fig. 12.11 Schematic illustration of the band profile of AlGaN/AlInN DBR structure with the formation of an electron accumulation in the attractive triangle potential at the AlGaN/AlInN interface

For these lattice matched AlGaN/AlInN bilayers a triangularly shaped potential wall is known to be formed at the interface [40] accompanied with a downward bending of the conduction band towards the Fermi level (see Fig. 12.11). This leads to an electron accumulation and an increase of the electron density at this interface. If the conduction band crosses the Fermi level, a two dimensional electron gas will be formed. This results in a high recombination rate of the generated minority carriers, which determine the recombination kinetics in luminescence experiments. Thus, a high intensity at the AlGaN/AlInN interface is to be expected and indeed observed (see Fig. 12.10b). In contrast, at the opposite interface a repulsive potential for electrons exists with no electron accumulation (see Fig. 12.11), leading to a reduced recombination rate of the generated holes. Consequently, a reduced intensity is observed at the AlInN/AlGaN interface compared to the opposite interface. These polarization effects will affect electrical conduction of electrons through such DBR structures as the series resistance will increase.

12.3.2 Plastic Relaxation of 62-Fold InGaN Multiple Quantum Wells in a GaN Cavity

The number of optically active, electric dipoles in a microcavity is an important parameter for the realization of low threshold lasers. For inversion-free operation of polariton lasers, the formation of exciton-polaritons is a prerequisite [41]. The formation of exciton-polaritons takes place, when the square of the light-matter coupling strength g exceeds the mean value of the squared linewidth of exciton and cavity mode [42]. Since g is proportional to the square root of the number of quantum wells located at the antinodes of the light field within the microcavity, the strong coupling regime can be reached by embedding a large number of QWs in the cavity [42]. Beside the absorption probability enhancement of cavity photons, a higher confinement factor emerges [43] along with a threshold reduction for optical amplification. However, there is a variety of limitations for the number of embedded QWs in a microcavity. The most significant constraint is the well-known though undesired phenomena of strain relaxation of the MQW stack. It results from heteroepitaxy of lattice-mismatched materials and, forming additional defects acting as recombination channels, is disadvantageous for optoelectronic devices like VCSELs and polariton lasers. Therefore, a better understanding of the relaxation mechanism and its onset conditions is necessary.

Here, the limitations of the coherent growth of lattice-mismatched InGaN/GaN QW structures on AlInN/GaN DBR layers will be discussed. The investigated sample consists of a 62-fold $In_{0.14}Ga_{0.86}N$/GaN QW active region embedded in a 3λ GaN microcavity on top of an $Al_{0.82}In_{0.18}N$/GaN distributed Bragg reflector. Figure 12.12 is a cross-sectional HAADF image taken in STEM mode of the active region of the microcavity (MC) structure [44]. At the very bottom, the uppermost AlInN/GaN quarter-wavelength layer pair of the DBR can be seen. The DBR layers exhibit laterally homogeneous, sharp interfaces with layer thicknesses of 50 and 37 nm for the AlInN and GaN layers, respectively. The cavity section on top, including the embedded 62 InGaN QWs (brighter HAADF intensity), is 402 nm thick. In addition, the QW thickness increases progressively by more than 1 nm from the bottom (2 nm) to the top of the structure (3.3 nm).

First we refer to the InGaN QW emission at low temperature (16 K) recorded by the CL spectral imaging technique. The spatially averaged spectrum (Fig. 12.13a) exhibits an InGaN QW exciton emission at 404 nm (3.068 eV) with an asymmetrical line shape and a full width at half maximum (FWHM) of 75 meV. In Fig. 12.13b, the STEM (HAADF) image of the cavity region and the upper DBR structure is shown. The AlInN DBR layers exhibit a weaker HAADF intensity than the GaN DBR layers. V-pits run from the surface down through the optically active layers.

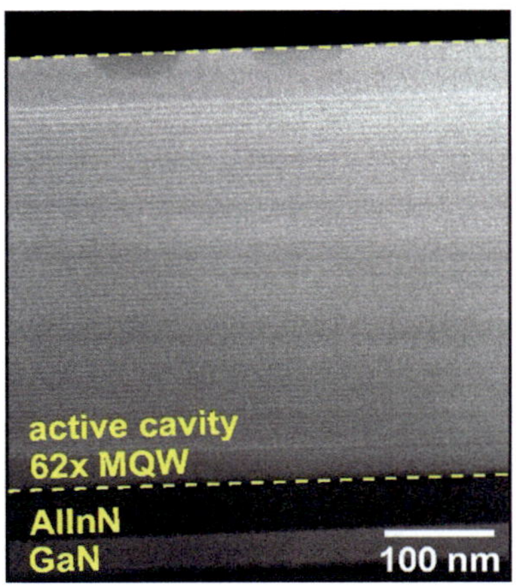

Fig. 12.12 HAADF image of 62 InGaN quantum wells with GaN barriers embedded in a 3λ cavity on top of the AlInN/GaN DBR [44]

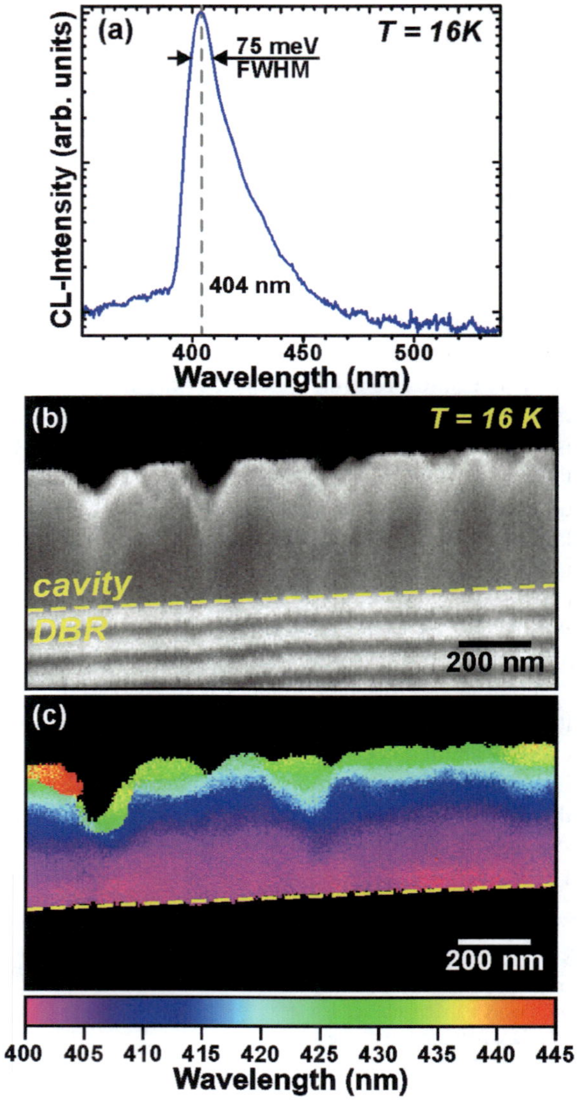

Fig. 12.13 Spectral CL mapping of the MC structure: **a** spatially integrated CL spectrum with **b** HAADF image and, **c** CL peak wavelength image visualizing red-shift towards surface [44]

The HAADF image corresponds to the spatially integrated spectrum displayed in Fig. 12.13a. By following the color-coded peak wavelength in a CL wavelength image (Fig. 12.13c) the strong red-shift of the InGaN emission from 400 nm (3.099 eV) at the bottom to 430 nm (2.883 eV) at the top of the active region ($\Delta E_{QW} = 216$ meV) is quantified. Upon a more detailed evaluation, we find only a small peak wavelength shift in the bottom part of the InGaN-QW structure. Most of the redshift, about $\Delta\lambda = 30$ nm, appears in the upper part within a distance of $\Delta x = 250$ nm. This vertically inhomogeneous emission causes the 75 meV broad asymmetric integral luminescence at 404 nm.

An even better insight into the evolution of the excitonic QW luminescence along the growth direction is possible with a CL spectrum line scan from the top of the DBR section, across the cavity towards the surface. In Fig. 12.14, the color-coded CL intensity is logarithmically plotted versus position in the vertical direction and wavelength. The $x = 0$ nm position marks the DBR/cavity interface. We observe a progressive red-shift in the InGaN emission from 407 nm (3.045 eV) at the bottom of the cavity to 429 nm (2.889 eV) at the surface accompanied by a decrease in intensity. The CL peak wavelengths marked as black dots exhibit a continuous red-shift from the DBR/cavity interface up to a position lying 191 nm above the DBR/cavity interface. This position is marked by the black dashed line in Fig. 12.14 and corresponds to the position of the 32nd QW. Here, the CL peak wavelength jumps abruptly by 3.6 nm ($\Delta E = 26.5$ meV). Tracking the QW emission further along the growth direction, the peak wavelength continuously shifts to longer wavelengths with a simultaneous reduction in the intensity maximum and increasing FWHM. At the position of the 55th QW ($x = 322$ nm in Fig. 12.14), an additional abrupt peak wavelength jump by $\Delta\lambda = 5.9$ nm ($\Delta E = 40.7$ meV) occurs. Above the latter, the peak position continuously shifts to longer wavelengths until reaching the surface. The origin of the continuous red-shift in the InGaN MQW luminescence is most likely due to an increasing QW thickness in the growth direction and/or a higher In incorporation with increasing strain. In contrast, the step-like behavior in the CL peak at the position of the 32nd QW could be primarily attributed to an abrupt change in the strain state of the InGaN/GaN MQW. The coherence of the epitaxially grown layers is therefore lost, leading to a sudden change of the strain dependent band gap. X-ray diffraction measurements confirm a partially relaxed strain-state of the InGaN-QW structure. From these results, one would assign the onset of partial

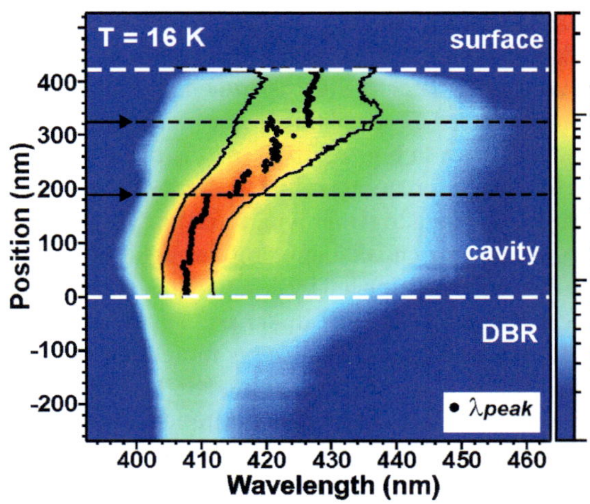

Fig. 12.14 STEM-CL line scan along the cross-section of the cavity region. The black dots and lines indicate peak wavelength and spectral position at half maximum, respectively. The black dashed lines mark the positions of plastic strain relaxation [44]

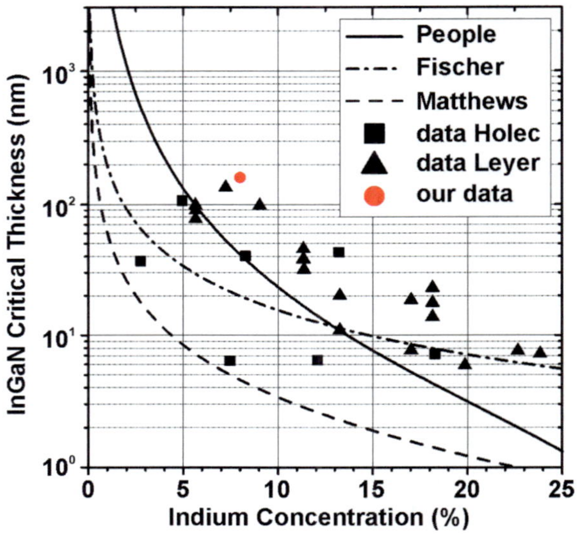

Fig. 12.15 Critical layer thicknesses for InGaN as a function of the indium concentration. Theoretical critical thicknesses after the models of People and Bean [48], Fischer et al. [46] as well as Matthews and Blakeslee [45]. Adapted from [51]

plastic relaxation to the 32nd QW layer corresponding to a 159 nm strain accumulation thickness starting from the 1st QW. Thereafter, the strain accumulation process resumes within the next 191 nm up until the 55th InGaN QW is grown.

Exceeding the critical thickness of a layer results in strain relaxation, after which the coherence to the underlying lattice dissolves. Different models for the calculation of the critical thickness of pseudomorphically grown epitaxial layers exist. Strain relaxation by the formation of misfit dislocations is common to all of the following models. Very popular is a calculation of the critical thickness after Matthews and Blakeslee, based on a force approach that has been originally developed for cubic III–V heterostructures [45]. Expansions of this model include the interaction of dislocations and the wurtzite structure of III-nitride materials [46, 47]. In contrast, the work of People and Bean uses an empirical energy approach, which is controversially discussed, but matches with experimental data [48]. Calculated according to those different models, the critical thicknesses for pseudormorphically grown InGaN on GaN layers is shown in Fig. 12.15. Here, the dislocation glide system $<11\bar{2}3>$, $\{11\bar{2}2\}$,[1] and linear interpolation of material constants is applied. Figure 12.15 also shows experimental data together with modeling [47, 49, 50].

Assuming strain accumulation throughout the whole stack, the MQW structure may also be approximated by a single InGaN layer with a mean indium concentration of 8%. Then, the critical thickness would exceed the theoretical calculations by far. For instance, the difference is more than one order of magnitude as compared to the Matthew and Blakeslee model. The smallest deviation exists to the People and Bean approach. In this concentration range, experimentally determined critical thicknesses of 100 nm for [In] = 8%, 30 nm for [In] = 10% and 35 nm for [In] = 8.5%

[1]The slip directions are denoted by the $<11\bar{2}3>$ directions. This slip occurs on the $\{11\bar{2}2\}$ planes.

are reported. Obviously, a critical thickness of 159 nm for a MQW stack with a mean indium concentration of 8% is significantly larger. Several origins for this discrepancy can be named:

1. The InGaN MQW stack is not strain free due to intrinsic and thermal stress. These effects are not included in the theoretical assumption.
2. The presence of threading dislocations in III-nitride materials grown on foreign substrates by heteroepitaxy influences the strain of subsequently grown layers [52]. Hence, the critical thickness has to be considered in terms of the particular threading dislocation density.
3. All the models as well as the experimental data refer to the relaxation of single thick InGaN layers. The insertion of GaN barriers in a MQW structure leads, however, to compressive forces. An alternate approach to include the essence of the MQW structure is provided by a model of Bykhovski et al. [53]. The upper limit of the critical thickness of GaN/AlN superlattices can be approximated by Bykhovski et al. [53] and Glauser et al. [54] to be 1.45 times the critical thickness of a single layer.

12.3.3 Carrier Localization in a Pseudomorphically Grown InGaN MQW/DBR Structure

With the results of the previous section, not more than 32 InGaN/GaN QWs with an average In content of 8% should be used in MQW structures embedded in monolithically grown MCs to preserve its pseudomorphic strain state.[2] In the following, a relaxation free, MOVPE-grown fully pseudomorphical 29-fold InGaN/GaN MQW stack is analyzed. Structurally homogeneous 1.4 nm thick $In_{0.11}Ga_{0.89}N$/GaN QWs on top of a lattice matched $Al_{0.82}In_{0.18}N$/GaN DBR can be seen in Fig. 12.16b. Asymmetrical reciprocal space maps recorded by XRD around the $(11\bar{2}4)$ diffraction peak prove the pseudomorphic strain state of the whole structure (see Fig. 12.16c). In particular, plastic relaxation of the MQW stack does not occur.

An analysis of the spectral fine structure of luminescence from the individual InGaN-QWs within the stack is possible by a spectral STEM-CL line scan in growth direction on a ~75 nm thin TEM lamella (see Fig. 12.17). The estimated excitation depth δx [see Sect. 4.2.1, (4.1)] is given in Fig. 12.17. Within the normalized line scan, a broad but spectrally constant emission at 390 nm is present across the whole stack which verifies pseudomorphic growth of the structure. However, at each spot in the structure, the emission is a superposition of several sharp emission lines. The local spectrum of the 10th QW as shown in Fig. 12.17c exhibits several overlapping emission lines adding up to a 40 meV broad luminescence band. The recombination labelled A_1 at 381.8 nm wavelength allows for evaluation of the emission properties

[2]For other thicknesses, In concentration or MC designs this number may vary.

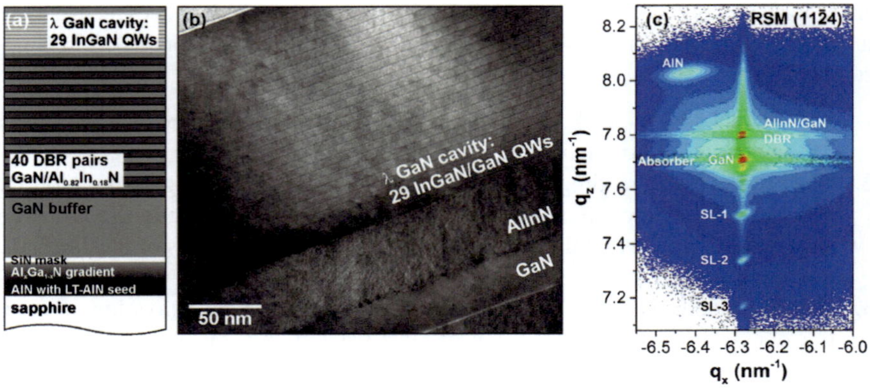

Fig. 12.16 29-fold $In_{0.11}Ga_{0.89}N$ MQW stack on top of an $Al_{0.82}In_{0.18}N/GaN$ DBR: **a** schematic sample setup, **b** bright field TEM image of the MQW region, and **c** XRD reciprocal space map [55]

of a single emission line. Here, the FWHM is only 4.8 meV pointing to strongly localized electron-hole pairs.

Three different pictures of carrier localization and formation of QD like states in InGaN QWs can be taken to explain the sharp emission lines:

1. Random concentration fluctuations in InGaN ternary alloys on the atomic scale: Theoretical calculations show a strong localization of holes [56–58] due to statistical In fluctuations causing an intrinsic hole localization center around In atoms [56] or In–N–In–N ... chains [59, 60]. Chichibu et al. for instance, concluded from monoenergetic positron annihilation spectroscopy a hole (positron) localization at atomic In–N clusters indicating the predicted In–N localization centers [61].
2. Fluctuations of the QW thickness cause changes of the quantization energy [62]: The nanometer length scale of such fluctuations would primarily lead to quantized levels for electrons [57, 58]. Exciton localization due to monolayer fluctuations in InGaN QW layers is argued by several groups [63, 64].
3. Changes of the In concentration on a length scale of several lattice constants: Such fluctuations would also locally alter the band gap as well as spontaneous and piezoelectric polarization fields. Indium-rich regions then form QD-like attractive potential [59, 60].

Although sharp emission lines from InGaN QWs with 4.8 meV FWHM were observed the first time in spatially resolved cathodoluminescence experiments, the spectra are still broadened, most likely by localization on the local atomic and supra atomic range. Carrier localization originating in thickness fluctuations seems to be unlikely because of the very homogenous QW thicknesses measured by TEM.

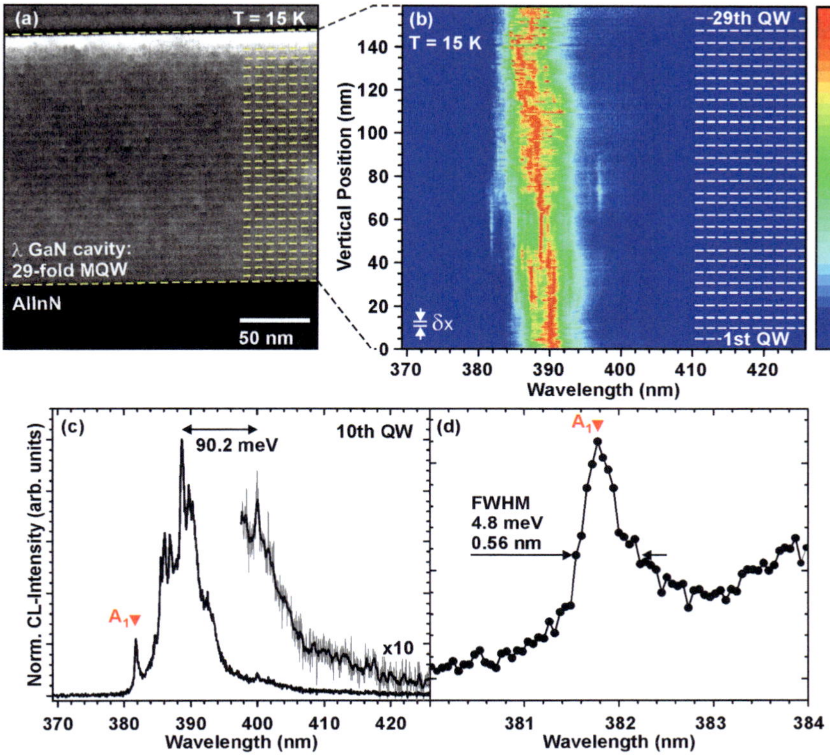

Fig. 12.17 STEM-CL investigation of the $In_{0.11}Ga_{0.89}N$ MQW emission fine structure at low temperature: **a** HAADF image, **b** spectral line scan across MQW stack, **c** local spectrum of 10th QW, **d** magnified view of A_1 emission with 4.8 meV FWHM [55]

12.3.4 Local Properties of Excitonic and Photonic Modes in Violet Emitting Microcavities

In the following section, a discussion of excitonic as well as photonic disorder in a hybrid MC is done. The microcavity structure is described in Sect. 12.3 and exhibits lasing under optical pumping (see Fig. 12.9). First, the vertical inhomogeneity of the $In_{0.136}Ga_{0.864}N$ MQW emission is analyzed by highly spatially resolved STEM-CL analysis. Local STEM-CL spectra from the cavity region from bottom towards growth direction are plotted in Fig. 12.18. Despite slight changes in line shape, the individual spectra show a constant peak wavelength at 407.9 nm. These local spectra, being representative for the whole MQW structure, prove a negligible inhomogeneity in the vertical direction. Nevertheless, single weak emission lines from localized carriers can be observed occasionally (see Sect. 12.3.3).

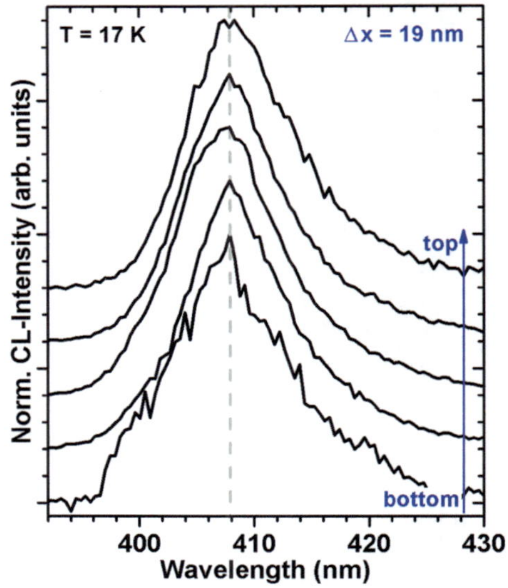

Fig. 12.18 Local low temperature STEM-CL spectra in equidistant steps of 19 nm from bottom to top of the GaN cavity exhibit a constant peak wavelength at 407.9 nm [55]

The lateral inhomogeneity of the excitonic $In_{0.136}Ga_{0.864}N$/GaN MQW luminescence on submicrometer scale was assessed from plan-view CL on a structure without the top dielectric mirror at 5 K and a low acceleration voltage of 5 kV. The resulting spatially averaged spectrum with a luminescence peak at 406.8 nm is depicted in Fig. 12.19. An asymmetric line shape is observed with a spatially averaged FWHM of 91 meV. Figure 12.19 contains also a histogram that allows for a quantification

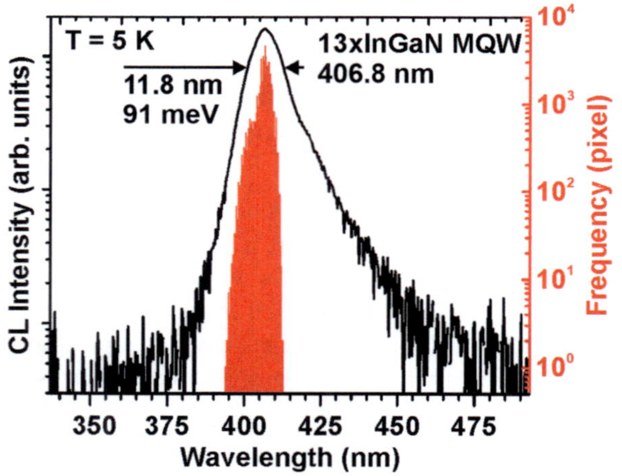

Fig. 12.19 Spatially averaged CL spectrum (black) and peak wavelength histogram (red) of the half microcavity (without top dielectric mirror) exhibits emission of the 13-fold $In_{0.14}Ga_{0.86}N$/GaN MQW at $T = 5$ K [55]

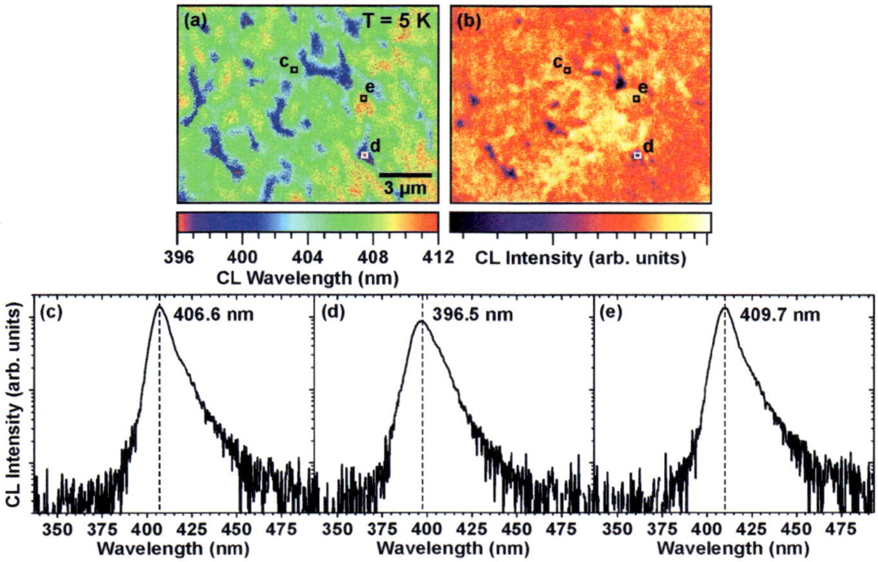

Fig. 12.20 Investigation of the disorder in lateral excitonic emission of the InGaN MQW at 5 K: **a** InGaN peak wavelength image, **b** variation of integral intensity of MQW emission, **c**–**e** local spectra marked in **a**, **b** [55]

of the lateral excitonic disorder together with the CL peak wavelength image in Fig. 12.20a. From these images, the spectral frequency and lateral distribution of the MQW peak wavelength, respectively, is found. In the wavelength image (Fig. 12.20a), regions emitting at different peak wavelength are identified: areas of distinct blue-shifted peak wavelength (400 nm) are surrounded by regions with intensity maxima at wavelengths ranging from 406 to 411 nm. Accordingly, a bimodal distribution of the peak wavelength is found in the histogram (Fig. 12.19). The frequency maximum at 406.6 nm and a shoulder at 401.7 nm are present. The 28 meV full width at half maximum of the frequency distribution reflects the disorder of the lateral excitonic emission of the InGaN MQW. Local spectra in Fig. 12.20c–e resolve respective spectral shifts and only minor intensity changes (see also Fig. 12.20b) of the MQW peaks for each distinct region.

Altogether, the spatially averaged FWHM is a convolution of the disorder of the lateral excitonic emission and local emission linewidth. Data of the local linewidth was obtained from a FWHM mapping as shown in Fig. 12.21e. In most parts of the cavity, the local linewidth accounts to about 69 meV FWHM (see local spectra in Fig. 12.21d, f). Measured linewidths increasing by up to 1.5 times (see local spectra Fig. 12.21a–c) are only locally found.

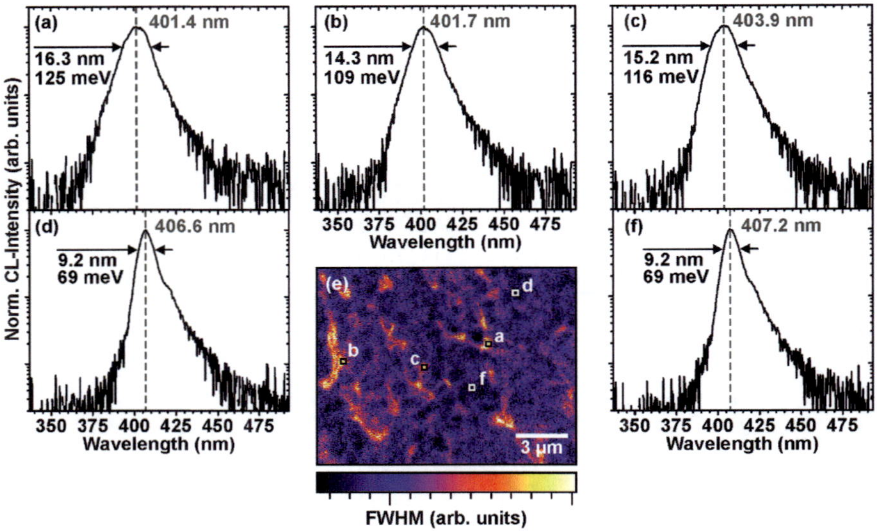

Fig. 12.21 Spatial variation of the MQW emission linewidth: **e** FWHM map, **a**–**c**, **d**, **f** local spectra with inscribed FWHM values [55]

Looking at a sub-micrometer scale, the local InGaN MQW linewidth of 69 meV exceeds the lateral inhomogeneity of 28 meV. Hence, within the excitation volume the dominant contribution of large luminescence FWHMs is attributed to inhomogeneities on the (sub-)nanometer scale. Such effects can be explained with strong carrier localization and formation of quantum dot like states as mentioned in Sect. 12.3.3. The local variations in peak wavelength, integral intensity and linewidth can be taken as a phenomenological description and quantification of the excitonic disorder in InGaN MQWs embedded in microcavities.

Angular resolved PL measurements are required to identify the photon mode based on the angular dispersion of the complete hybrid microcavity (with top dielectric SiO_2/Ta_2O_5 mirror). In such measurements, the in-plane component of the cavity-photon wave vector varies with the detection angle. At small angles, the photon energy dispersion can be approximated by a parabolic dependence on the planar wave vector component (see references [65–69] for a detailed introduction into the physics of microcavities). Due to a roughly 1000-fold larger mass of the exciton compared to the cavity photon, an almost angle independent InGaN MQW emission wavelength is to be expected. In Fig. 12.22b the PL intensity is plotted as a function of spectral position and detection angle. The full microcavity can be optically excited with a non-focused HeCd laser [70]. Under normal incidence, the emission intensity maximum is at 3.093 eV (400.7 nm) and then shifts to higher energies up to 3.279 eV (378.0 nm) at angle of 56°. The energy shift between 0° and 46° can be approximated by a parabola, justifying the assignment to the cavity photon. With a spatially averaged CL spectrum recorded on a half cavity sample (Fig. 12.22a) also the excitonic mode from the

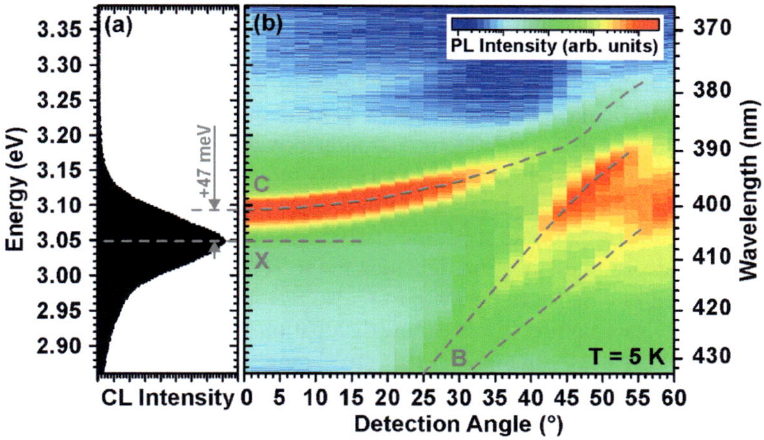

Fig. 12.22 Angle resolved emission of the +47 meV detuned microcavity: **a** excitonic emission (X) of the InGaN MQW from spatially averaged CL spectrum at $T = 5$ K (see Fig. 12.19), **b** angle resolved PL showing angle dispersion of the cavity mode (C) and Bragg modes (B) at $T = 5$ K [55]

same sample region is spectrally identified. By comparison of the energies, a positive detuning of +47 meV between cavity and exciton mode is found. From the absence of anti-crossing of the excitonic and photonic mode in the angular resolved experiments the weak coupling regime can be concluded. A crucial maximum linewidth value for the strong coupling regime of +36 meV has been reported [71] which is exceeded for this InGaN MQW structure.

12.4 GaN Quantum Dots: Formation, Optical and Electronic Properties

In this section we demonstrate the formation of GaN QDs by growth and annealing of a GaN layer on AlN [72] and discuss the formation mechanism in the vicinity of dislocations.

Growth of GaN quantum dots differs significantly from growth in the III-As/P system. In the latter, a Stranski-Krastanov (SK) growth mode is typically used to initiate QD formation after the growth of a thin wetting layer [73]. In the III-As/P system, the QDs usually form after the deposition of a thicker layer during a growth interruption. Then, Ostwald ripening increases larger and shrinks smaller initial QDs [74, 75]. To take place this process requires a sufficiently high surface mobility of the ad-atoms.

Another approach in the III-As/P system is the growth of submonolayers of, e.g., InAs in GaAs where e.g. a superlattice of submonolayers leads to the formation of local compositional fluctuations without forming a 3D surface modulation [76]. While SK QDs have been studied for more than three decades the transfer of this approach to the III-N system is not straightforward. The main difference is the high density of threading dislocations in III–Ns requiring expensive GaN pseudo-substrates with still rather high threading dislocation densities (typical > 10^5 cm^{-2}) when compared to, e.g., GaAs substrates (<10^4 cm^{-2}).

With low threading dislocation densities in the 10^4 cm^{-2} range, AlN wafers are the substrates of choice for the growth of GaN QDs, but are very expensive. Using heterosubstrates as sapphire, the threading dislocation densities in AlN are often well above 10^8 cm^{-3}, typically in the 10^9–10^{10} cm^{-3} range. While in the III-As/P system dislocations strongly inhibit radiative recombination and in the case of QDs lead to the formation of non-radiative clusters, dislocations in III-Ns are not as detrimental for light emission. For example commercial blue LEDs with dislocation densities in the mid 10^7 cm^{-2} to low 10^8 cm^{-2} range exhibit high efficiencies, which can be partly attributed to local In fluctuations in the InGaN QW layers [77].

In our investigations the growth of GaN QDs was initiated with a thin layer (nominally ≈ 2 nm) of GaN deposited at 960 °C and a V/III ratio of 30 on AlN on sapphire followed by a growth interruption and an AlN cap layer (Fig. 12.23). During the growth interruption no TMGa and NH_3 were supplied leading to GaN desorption [72]. In the following section this formation mechanism is described in more detail.

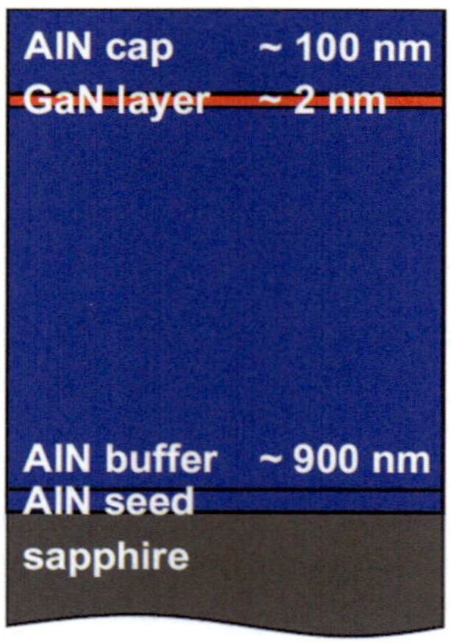

Fig. 12.23 Structure of GaN QD samples on sapphire substrates

12.4.1 GaN Quantum Dot Formation Mechanism

To understand the formation mechanism of GaN-based QDs grown on AlN, different samples with varying desorption times from 0 to 60 s were grown. Cross sectional scanning transmission electron microscopy (STEM) investigations of a sample without growth interruption reveal that after deposition of nominally 2 nm GaN, islands with a thickness of about 19 nm and a width of 100 nm appear. It is observed that they preferentially form close to regions with a high dislocation density (Fig. 12.24). As edge type dislocations lead to an in-plane widening of the AlN lattice, GaN can accumulate at these positions in a more relaxed manner, which is energetically favorable. The same situation occurs when InGaAs QDs on GaAs exceed their critical thickness and strain relaxation via the formation of dislocations occurs and large clusters form [78]. While for GaAs dislocation formation occurs during QD growth, in the case of GaN QDs the AlN buffer layer already contains a high number of dislocations and no SK QDs form but large GaN clusters. In both cases such clusters exhibit no radiative emission as shown in cross sectional STEM cathodoluminescence (CL) in Fig. 12.25. GaN desorption is one possible way to reduce cluster size and split up large clusters in several smaller islands. Here the GaN clusters on AlN were annealed directly after deposition of nominally 2 nm of GaN without TMGa and NH_3 supply in the MOVPE reactor.

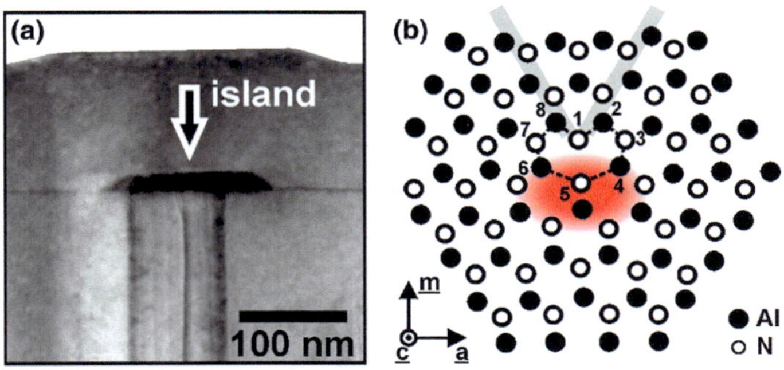

Fig. 12.24 GaN on AlN grown without growth interruption **a** forms clusters on top of dislocations. This can be attributed to the wider AlN in-plane lattice close to edge type dislocations **b**

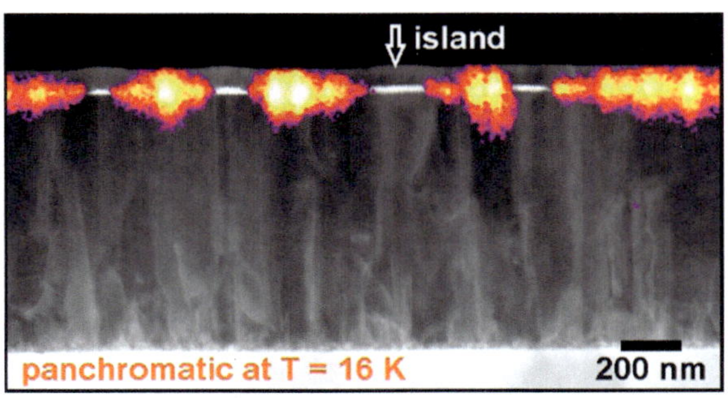

Fig. 12.25 STEM and superimposed panchromatic CL intensity of GaN layer grown without growth interruption on AlN. Luminescence is only visible in the region of the thin GaN QW

Figure 12.26 shows CL line scan spectra of samples with different annealing times. Without desorption step, the integral spectra show a broad luminescence around 250–280 nm which is broadened due to thickness fluctuations. This is visualized by the STEM image in Fig. 12.27. Thereby, the sample was slightly tilted to obtain structural characteristics within the growth plane. Brighter areas show up in the GaN QW layer indicating thickness fluctuations. The broad luminescence blue-shifts and splits up for longer annealing times. At shorter wavelengths the GaN QW luminescence can

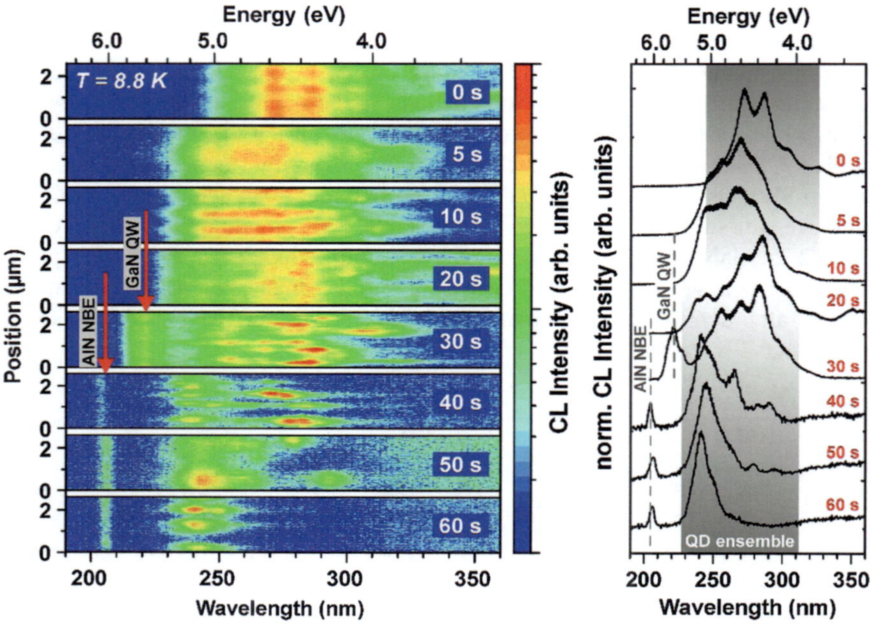

Fig. 12.26 Series of samples grown on AlN with identical GaN deposition time but increasing desorption times. First only a broader emission is observed which splits up for longer desorption times into a GaN QW emission and a locally varying emission which stems from GaN QDs

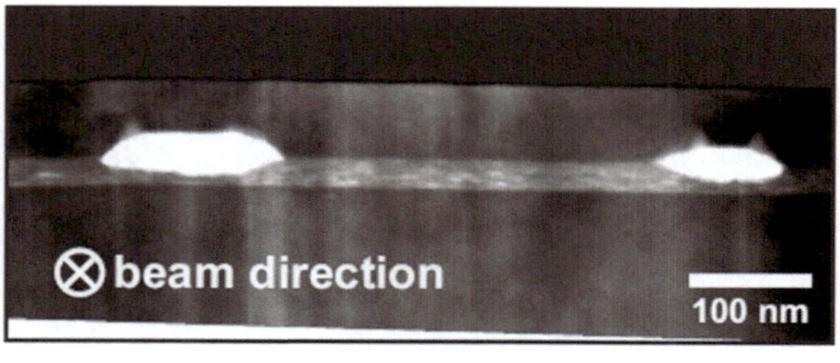

Fig. 12.27 STEM of a GaN cluster (0 s desorption time) on AlN after and the ~2 nm GaN QW showing bright spots due to thickness fluctuations

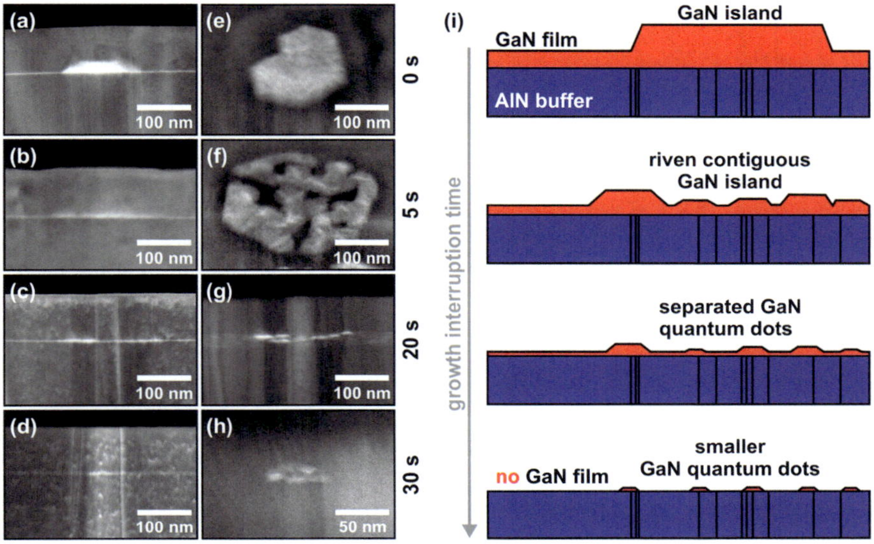

Fig. 12.28 **a–h** 17° tilted cross-sectional STEM of a former GaN cluster after 30 s of growth interruption (top). The compact GaN cluster (see Fig. 12.27) splits up into small islands. The evolution of GaN clusters and QW on AlN is schematically shown in the bottom image **i**

be seen while the broad luminescence at longer wavelengths evolves in distinct sharp local spectra for annealing times $\geq$20 s (Fig. 12.26). This can be attributed to the formation of QDs as also visualized in a cross sectional STEM CL image for annealing times of 30 s in Fig. 12.28. Here, the former large islands show intense sharp luminescence (Fig. 12.26) and in a tilted STEM image the formation of a high number of small islands in the vicinity of threading dislocations is revealed. For long annealing times $\geq$40 s the QW layer and its correlated luminescence disappear.

12.4.2 Quantum Dot Emission from GaN Islands Formed at Threading Dislocations

For the realization of controlled single photon emitters on the basis of semiconductor quantum dots two aspects have to be considered. Firstly, excitation of a single QD only is necessary favoring a low areal density of QDs. Secondly, exciton- and biexciton emission lines have to be energetically separated larger than their spectral line widths to enable the use of single recombination channels. Spectrally overlapping exciton-biexciton emission lines are attractive for the generation of entangled photon pairs.

In the following, the luminescence properties of single self-organized GaN/AlN QD heterostructures are presented and correlated with their real structure. These QDs are clustered around threading dislocations as single GaN islands. The GaN islands resulted from MOVPE growth of a GaN layer on top of an AlN/sapphire template as

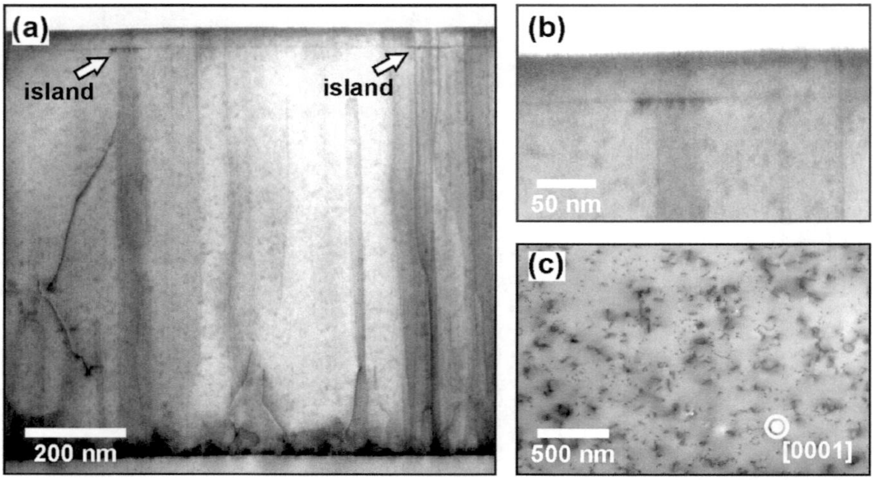

Fig. 12.29 STEM bright field images **a** of the whole structure and **b** magnified view of a single GaN island in cross section as well as **c** of the sample surface showing the threading points of dislocation at the surface [72, 79]. Copyright (2016) The Japan Society of Applied Physics

described in the previous section. After deposition of one to two monolayers GaN at 960 °C without ammonia supply, the growth was interrupted for 30 s and an AlN cap layer was grown on top. In Fig. 12.29, a TEM bright field (BF) image of the whole structure can be seen with the sapphire substrate at the bottom. The AlN buffer layer exhibits a threading dislocation density of 2×10^{10} cm^{-2}. On top of the AlN buffer a continuous GaN film has been formed. Primarily, GaN islands are formed in vicinity of dislocation bundles identifiable due to higher GaN film thickness.

These GaN islands show the most intense emission in panchromatic CL intensity mapping at 15 K revealing the efficient recombination of generated carriers in these structures. The panchromatic CL intensity image was superimposed with the BF image in Fig. 12.30b to show the correlation of high intensity in regions of high dislocation density. Different emission contributions are identified by spectral characterization in STEM-CL. A luminescence peak at 210 nm is attributed to the GaN quantum well layer as directly shown in Fig. 12.32a. Here, the emission energy exceeds calculated values of one monolayer thick GaN/AlN quantum wells. An unintentional incorporation of residual Al cannot be excluded, therefore. Different spectral positions from the broad emission band between 220 and 310 nm exhibit a complementary intensity distribution: exact at the positions of GaN islands high intensities are observed in this spectral range while the QW intensity at 210 nm drops. Simultaneously, but opposite, luminescence from Si-DX centers as well as from O-DX centers are exclusively detected in the AlN layer (Fig. 12.32).

Analyzing the spectral characteristic of individual quantum dots by micro-photoluminescence is beneficial. If the quantum dot density is high, the sample surface should be masked by opaque materials like metals or etched into mesas to

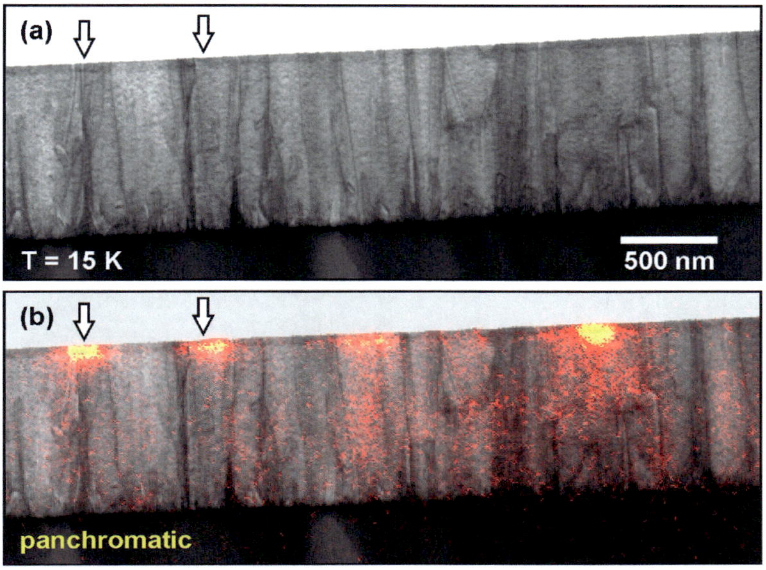

Fig. 12.30 **a** STEM (BF) image of the GaN/AlN heterostructure and **b** panchromatic CL intensity image ($T = 15$ K) in orange-colored contrast superimposed to BF image reveals intense emission in regions of high dislocation density from GaN islands (marked with arrow) [79]. Copyright (2015) American Institute of Physics

allow excitation and detection of only few to single QDs. With the high spatial resolution capability of STEM-CL microscopy, single quantum dots can be excited and their luminescence distribution analyzed even for high QD density ensembles. This is demonstrated in the following, where closely located quantum dots were individually measured in STEM-CL spectroscopy. As structural properties of the QD region are simultaneously resolved by STEM a direct correlation to the optical properties is possible.

First we discuss the properties of single GaN islands as inspected by STEM-CL in plan-view configuration. The bright field image of a region with dislocation bundles is shown in Fig. 12.31a. Due to the low atomic mass thickness contrast in plan-view—islands are very thin compared to the AlN cap and buffer along the primary electron beam irradiation direction—the GaN island structure is not visible. Nevertheless, the panchromatic intensity exhibits a region of high intensity in vicinity of the threading dislocation (TD) bundle (see Fig. 12.31b) and three sharp emission lines in the spectral range of the island luminescence band evolve through local electron excitation in spot mode (see Fig. 12.31c). Hence, a single GaN island is characterized by multiple spectrally sharp lines whose origin is discussed in the following. The main emission lines of the local spectra exhibit peak wavelengths at 250.3, 265.5 and 284.6 nm. The small FWHM of the emission lines of the GaN islands indicates QD emission which is verified by a Hanbury-Brown and Twiss

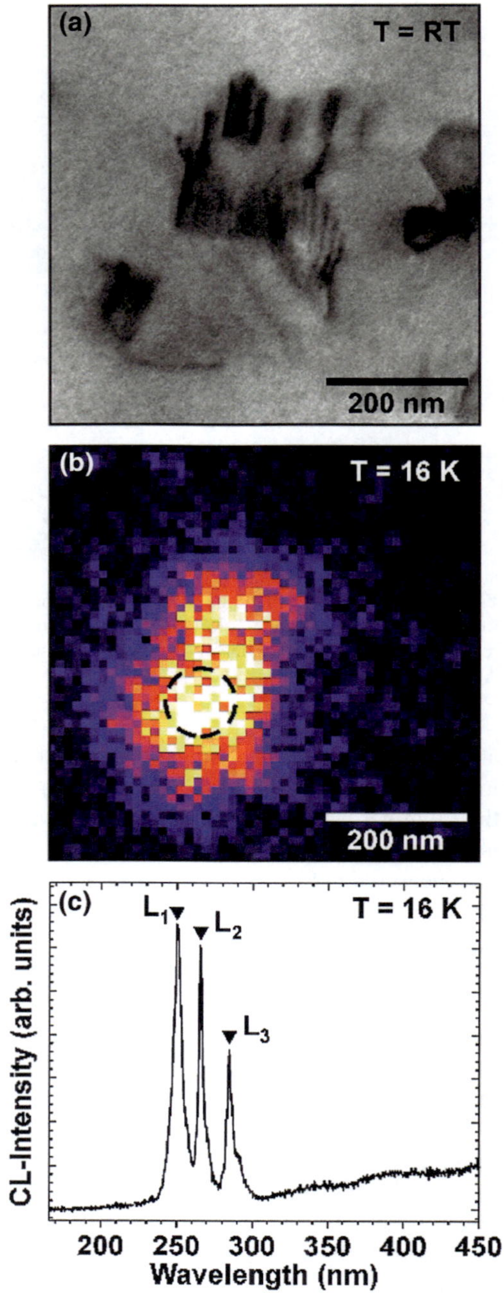

Fig. 12.31 **a** STEM (BF) image of the plan-view sample (slightly tilted) in region of high threading dislocation (TD) density with bright emission in **b** panchromatic CL intensity image. **c** Local electron-beam-excited region/marked in **b** exhibits three sharp main emission peaks at $L_1 = 250.3$ nm (4.951 eV), $L_2 = 265.5$ nm (4.668 eV), and $L_3 = 284.6$ nm (4.355 eV) in CL spectrum [72]. Copyright (2016) The Japan Society of Applied Physics

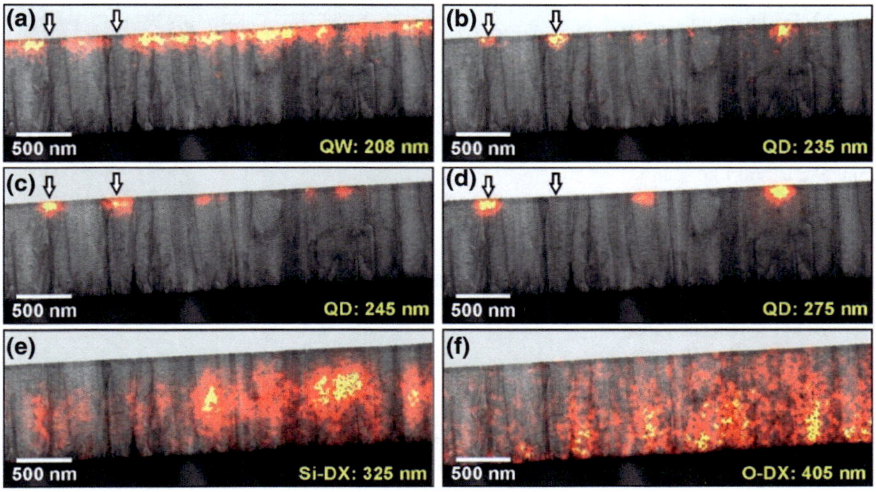

Fig. 12.32 Monochromatic STEM-CL intensity images at 15 K at the spectral positions of **a** GaN QW, **b–d** GaN island and **e–f** AlN-DX centers in orange colored contrast superimposed with simultaneously recorded BF image (same region like in Fig. 12.30) [79] (Copyright (2015) American Institute of Physics)

experiment showing single-photon emission [79]. However, the lateral dimensions of the GaN island of up to 200 nm (see Fig. 12.31b) exceeds the expected exciton Bohr radius of <12 nm in GaN/AlN QDs [80]. Obviously, there is a discrepancy between QD emission from these islands and their lateral dimensions which exceeds the calculated QD regime.

In general, single QD luminescence may exhibit several recombination channels. On one hand, the strong interaction of QD excitons with phonons has been reported to exhibit longitudinal optical phonon replica with a spectral separation of 90–110 meV [81, 82] as well as acoustic phonon sidebands adjacent to the low-energy side of the excitonic recombination signal [81, 83, 84]. On the other hand, multi-excitonic recombination, such as biexciton emission spectrally separated by their binding energies from the exciton emission, has been observed in GaN QDs [80, 85–90]. The highest reported biexciton binding energy of GaN QDs is 52 meV [89]. In our data, the large spectral separation of the main peaks, $\Delta L_{1\text{-}2} = 283$ meV, and $\Delta L_{2\text{-}3} = 313$ meV, significantly exceeds the value of the LO phonon energy as well as that of the biexciton binding energy. Hence, while phonon coupling and multiexcitonic recombination might occur in the QD emission a biexciton–exciton or a phonon-assisted relationship between $L_{1,2,3}$ cannot be concluded. However, the origin of multiple recombination channels in GaN islands is attributed to QD clustering in some GaN islands as verified by a STEM-CL spectral line scan along a GaN island (see Fig. 12.33). Beginning at the lateral position $x = 0$ nm, an intense emission from the GaN QW is observed until $x = 70$ nm. At exactly the position where the QW intensity drops, a multitude of overlapping sharp lines in the emission band

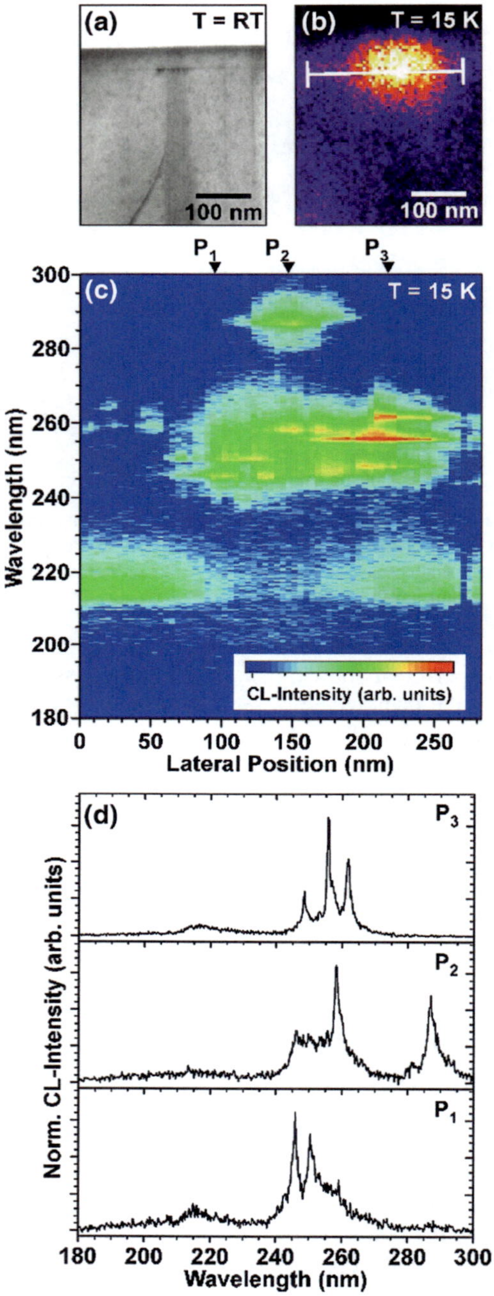

Fig. 12.33 Detailed STEM-CL analysis of a single GaN island in cross section: **a** BF image of island formed at TDs, **b** panchromatic CL intensity image, **c** lateral line scan along the GaN layer [position marked in **b** by white line], and **d** local spectrum at line scan positions $P_{1,2,3;}$ [72]. Copyright (2016) The Japan Society of Applied Physics

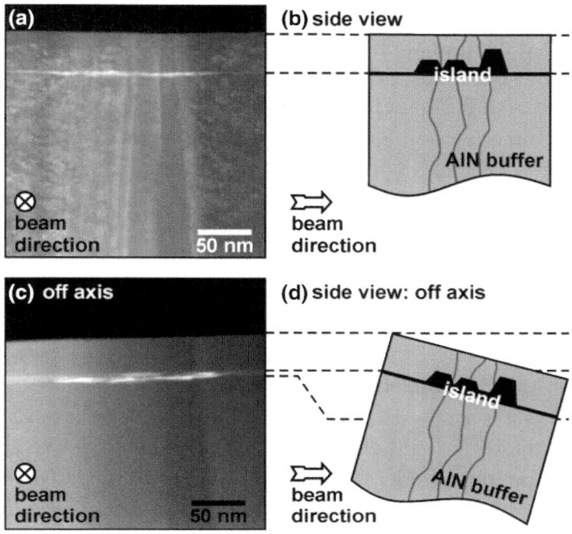

Fig. 12.34 HAADF images of an island for different tilt alignments with the direction of the primary electron beam **a** parallel to the GaN layer and **b** slightly tilted off axis, which shows different atomic mass thickness contrast within the island. The off-axis configuration as well as the thickness fluctuations are shown schematically in **c** [72]. Copyright (2016) The Japan Society of Applied Physics

from 240 to 300 nm appears. Most strikingly, the spectral positions of the emission lines vary along the island cross section, which is equivalent to the existence of spatially separated different QD states within a single island. Figure 12.33d shows local spectra extracted from line scans exhibiting quite disparate emission lines for each line scan position $P_{1,2,3}$. Accordingly, spatially separated and energetically different optical transitions exist within the GaN island.

To investigate the origin of the spatially separated emission lines, a STEM tilt series was performed. In Fig. 12.34a) the STEM (HAADF) image of a single GaN island can be seen. The depicted plane is perpendicular to the basal plane with the result that the primary electron beam passes parallel to the GaN layer. Due to the projection along the electron beam direction, the interpretation of the structural features can be ambiguous because objects at different TEM film depths but the same projected image position overlap and are difficult to distinguish. STEM tilt series enable a three-dimensional tomographic illustration. Figure 12.34b shows the same GaN island as in Fig. 12.34a but tilted off-axis. The projection of the GaN QW is visible as a bright stripe. In contrast, in the region of the GaN island, bright blurs in the HAADF contrast are observed. In off-axis images, regions of thicker GaN layers exhibit more Ga atoms along the primary electron beam direction, resulting in stronger scattering into angles of the HAADF detector. From this atomic mass thickness contrast, the bright blurs are attributed to a locally different thickness of the GaN island visualized by the schematic drawing in Fig. 12.34c. The observed TDs influence the HAADF contrast due to dechanneling effects, but do not fully explain the extended bright blurs in the GaN layer.

The locally varying layer thickness within the islands is assumed to strongly influence the transition energies. In addition, TDs are expected to affect the local strain, resulting in a spatial variation of the band gap as well as piezoelectric polarization fields. For GaN QDs nucleated at single edge-type TDs, Jurczak et al. calculated a strain reduction of 20% compared to an isolated GaN QD in the absence of any TDs [91]. Hence, a complex interplay between both, thickness and strain variations, is assumed for the formation of conglomerates of QD states in an island.

12.4.3 Exciton-Phonon Coupling

In this section, we highlight the optical signatures of exciton-phonon coupling in single wurtzite III-nitride quantum dots, which have been investigated both experimentally [84, 82], by means of SEM-CL and μPL techniques, as well as theoretically, in the framework of 8-band $\mathbf{k} \cdot \mathbf{p}$-theory [92]. Unusually strong exciton-phonon interaction is traced back to the permanent dipole moment of excitonic complexes as a result of the charge carrier separation inherent to wurtzite III-nitride heterostructures.

12.4.3.1 Exciton-Acoustic Phonon Coupling

Elastic and inelastic exciton acoustic-phonon coupling influences the linewidth and—shape of single-QD emission spectra, leading to a broadening of the zero-phonon line (ZPL) and the appearance of phonon sidebands. Both features are exceptionally pronounced in wurtzite III-nitride QD spectra. A fundamental characteristic of III-nitride heterostructures (e.g. quantum dots) is the separation of electrons and holes through large built-in piezo- and pyroelectric fields along the [0001] crystallographic direction (*c*-axis) [93], leading to a significant dipole moment for excitonic complexes in III-nitride QDs. This dipole moment makes excitons sensitive to the local electrostatic environment of the QD, e.g. charged defects, which manifests itself as an inhomogeneous broadening due to spectral diffusion on a nanosecond timescale (see also Sect. 12.4.4) [94, 95].

We studied self-assembled Stranski-Krastanow GaN/AlN QDs which were grown on (0001)-oriented 6H–SiC substrates using low-pressure metal-organic vapor deposition [96]. For the investigation of the exciton acoustic-phonon coupling, we studied the luminescence during cathodoluminescence excitation [84]. The temperature dependent emission shows pronounced asymmetric sidebands on the low-energy side down to a temperature of 5 K (Fig. 12.35a, c). These sidebands stem neither from background luminescence nor from another QD, which can be confirmed by a time dependent investigation of the emission. Instead, the sidebands can be ascribed to inelastic scattering of the charge carriers with acoustic phonons. As the temperature is increased, the temperature dependent occupation of the acoustic phonon modes leads to more symmetric sidebands as the probability for phonon absorption increases. A peculiarity of the nitride material system is the unusual strength of these

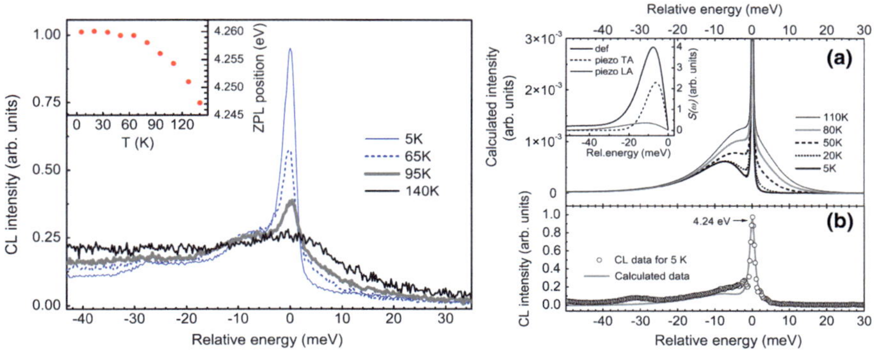

Fig. 12.35 Temperature dependence of a single emission line. (left) Temperature dependence of a single emission line with respect to the zero phonon line (ZPL). *Inset* Temperature dependence of the energetic position of the ZPL. (top right) Calculated temperature dependence. The ZPL peak at 5 K is normalized to 1. *Inset* Spectral densities $S(\omega)$ of phonon coupling mechanisms (deformation potential, piezoelectric coupling) for a model quantum dot. (bottom right) Comparison between experiment and theory for a temperature of 5 K. Taken from [84]. Reprinted figure with permission from [84]. Copyright (2012) American Physical Society

sidebands, when compared to, e.g. arsenide-based QDs. In an accompanying theoretical investigation based on 8-band $\mathbf{k} \cdot \mathbf{p}$-theory, we found that piezoelectric coupling in GaN/AlN QDs is of the same order of magnitude as deformation potential coupling (Fig. 12.35b) (see also Chap. 5 for theoretical background). Strength and position of the acoustic phonon sidebands depend, through the internal field-induced permanent dipole moment, on the height and baselength of the QD. Taller QDs, exhibiting a larger dipole moment, show the most pronounced sidebands, while a decreasing baselength shifts the sidebands to larger relative detunings.

12.4.3.2 Exciton-LO-Phonon Coupling

The permanent dipole moment of excitonic complexes in III-nitride QDs entails strong coupling to the naturally polar longitudinal optical (LO) phonon modes, with exceptionally large Huang-Rhys factors S' of up to ~0.5 [82]. The strong coupling manifests itself in the appearance of pronounced Stokes-shifted phonon replica in the optical spectra, clearly visible even on a linear intensity scale, at characteristic energy differences in the range of 90–110 meV (Fig. 12.36, left) [10189]. By performing a statistical study of single GaN/AlN QDs, in combination with 8-band $\mathbf{k} \cdot \mathbf{p}$ modeling, we have shown that both the Huang-Rhys factor as well as the LO-phonon energy scale with the exciton emission energy of the QD (Fig. 12.36, right). When grown along the c-axis, the emission energy is mainly determined by the QD height via the quantum-confined Stark effect. The data shows that S' decreases with increasing emission energy (smaller QDs) from $S \sim 0.5$ at 3.3 eV to $S \sim 0.06$ at 4.2 eV. A trend that is well reproduced in our model, using a realistic QD structure as determined

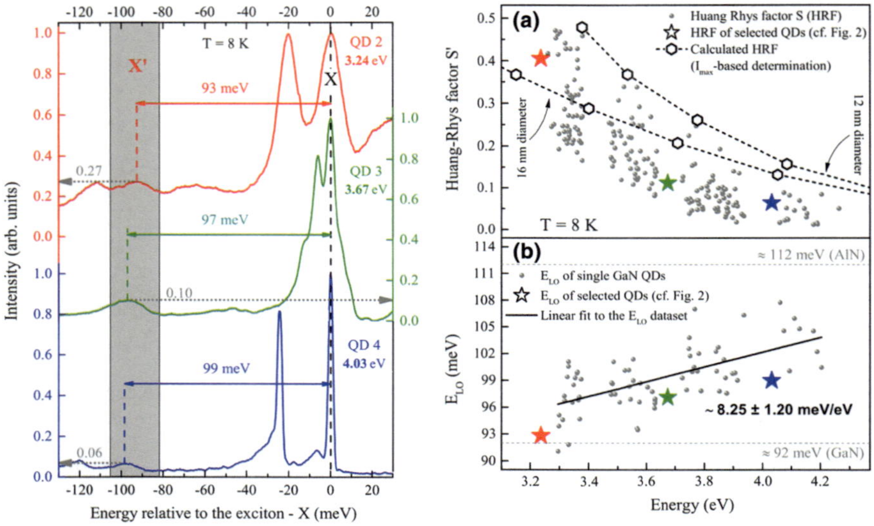

Fig. 12.36 Exciton LO-phonon coupling in single GaN QDs. (left) Normalized μPL spectra of QDs with different transition energies, showing exciton peak X and its first LO-phonon replica X'. (top right) Measured and modelled scaling behavior of the Huang-Rhys factor S' with the X emission energy. (bottom right) Evolution of the LO-Phonon energy with the X transition energy. Colored stars represent the QDs shown on the left. Taken from [82]. Reprinted figure with permission from [82]. Copyright (2015) American Physical Society

by STEM imaging, for varying height of the QDs. The spread in data points at constant emission energy is explained by a non-uniform aspect ratio among the QD ensemble, leading to slightly different scaling of S with the emission energy (cf. dashed lines in Fig. 12.36, top right). For the extracted LO-phonon energy E_{LO} an increase of $\sim$8.25 $\pm$ 1.20 meV/eV with the emission energy of the ZPL is observed. Interestingly, the extracted energies scale between the LO-phonon energies for GaN (~92 meV) and AlN (~112 meV) at cryogenic temperatures.

By investigating possible underlying alloying effects using STEM-HAADF imaging and by modeling the stress induced energy shifts (cf. Fig. 12.37a), we found that the observed shift in LO-phonon energy can best be explained using an effective exciton-LO-phonon interaction volume that extends into the matrix material for smaller QDs. Subsequently, the observed LO-phonon energy scales towards the AlN matrix material as the exciton overlap into the matrix material increases (cf. 11.37b). We found that the observed scaling can be reproduced assuming a spherical interaction volume with a diameter of 2.6 $\pm$ 0.2 nm.

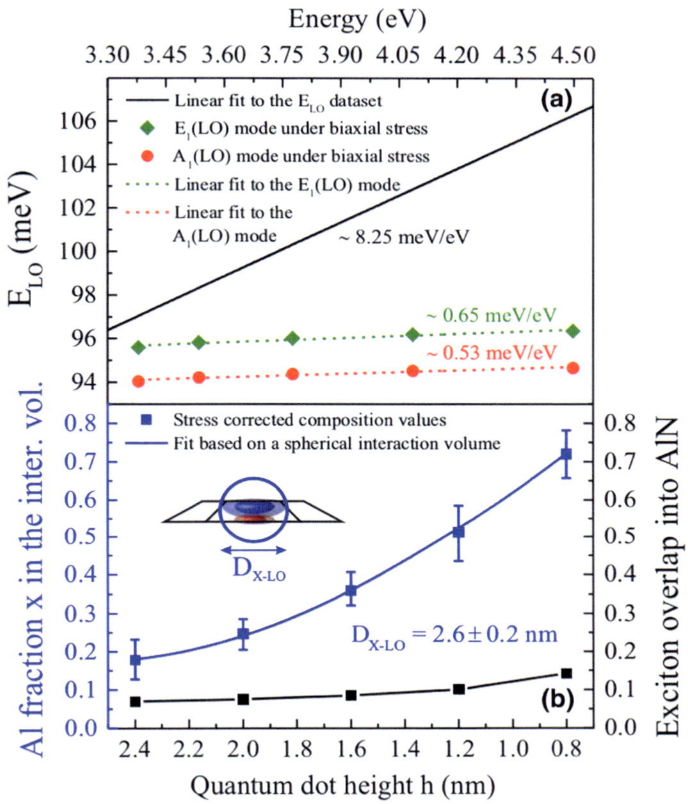

Fig. 12.37 **a** Linear fit to the experimentally observed LO-phonon energies in Fig. 12.36 (black solid line) compared to the modeled shifts of the E_1(LO) (green) and A_1(LO) (red) phonon modes due to biaxial stress. **b** Aluminum fraction in the interaction volume and exciton overlap into the matrix material as a function of the QD height. Taken from [82]. Reprinted figure with permission from [82]. Copyright (2015) American Physical Society

12.4.4 Spectral Diffusion of Excitonic Complexes

The built-in permanent dipole moment of excitonic complexes in single wurtzite III-nitride QDs is known to couple to its electrostatic environment (cf. Fig. 12.38a), resulting in spectral diffusion in the luminescence energy on both fast (~nanoseconds) and slow (~seconds) timescales [94, 95]. The fast variations cannot be resolved in standard luminescence measurements and merely manifest themselves in a substantial broadening of the emission line (when compared to e.g. typical As-based QDs). The slow variations, however, can be readily observed when tracking the luminescence over several minutes.

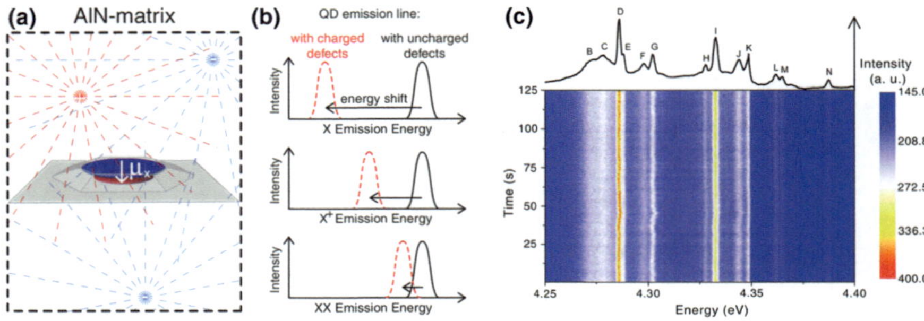

Fig. 12.38 Spectral diffusion of excitonic complexes in III-nitride QDs. **a** Scheme: electrostatic environment in the presence of charged defects. **b** Spectral shift due to the interaction between transition dipole and the electrostatic environment for different excitonic complexes. **c** Intensity-coded time-trace of a typical emission spectrum (top) using CL excitation. Taken from [97]. Reprinted figure with permission from [97]. Copyright (2013) American Physical Society

The fact that the local electrostatic environment of individual QDs is more or less unique is routinely employed to identify emission lines belonging to the same QD. We find that the energy shift of the emission peaks of different excitonic complexes (e.g. exciton, biexciton, etc.)—while temporally correlated—varies in magnitude depending on the total number of charge carriers in the QD and the size of the confinement region through Coulomb interaction between the charge carriers (Fig. 12.38b, c) [97]. The resulting carrier distributions inside the QD will carry different permanent dipole moments, which in turn couple differently to charged defects in the vicinity of the QD. As such, the temporal evolution of the spectrum not only allows the assignment of emission lines belonging to the same QD, but also to assign the emission lines to different excitonic complexes according to our calculations. The method becomes more accurate with increasing Coulomb interaction, e.g. for smaller QDs.

12.4.5 Photon Statistics of the Biexciton Cascade

In this section we present the results of a joined experimental and theoretical study of photon emission from the biexciton cascade in single GaN QDs. In a quantum-optical study, we investigated the influence of biexciton binding energy $E^{\mathrm{B}}_{\mathrm{bind}}$, temperature and excitation density on the photon statistics of the biexciton cascade. We show that the emission process can be changed from cascaded single-photon emission to a bunched two-photon emission process [90]. Consequently, we observe a transition in emission statistics from a sub-Poissonian [17] to a super-Poissonian distribution, indicating efficient photon pair generation in the UV spectral range. GaN/AlN QDs are an ideal material system for studying the biexciton cascade, because the fine-structure splitting of the excitonic bright states, as well as the biexciton binding energy, can be tuned close to or across zero as a function of the QD size [98]. For small absolute biexciton binding energies $\left|E^{\mathrm{B}}_{\mathrm{bind}}\right|$ one can no longer clearly separate exciton and biexciton emission lines and a bunching around zero-delay time is observed in a Hanbury-Brown and Twiss second-order correlation measurement (Fig. 12.39a) (see Chap. 5 for more theoretical information). Such bunching is usually attributed

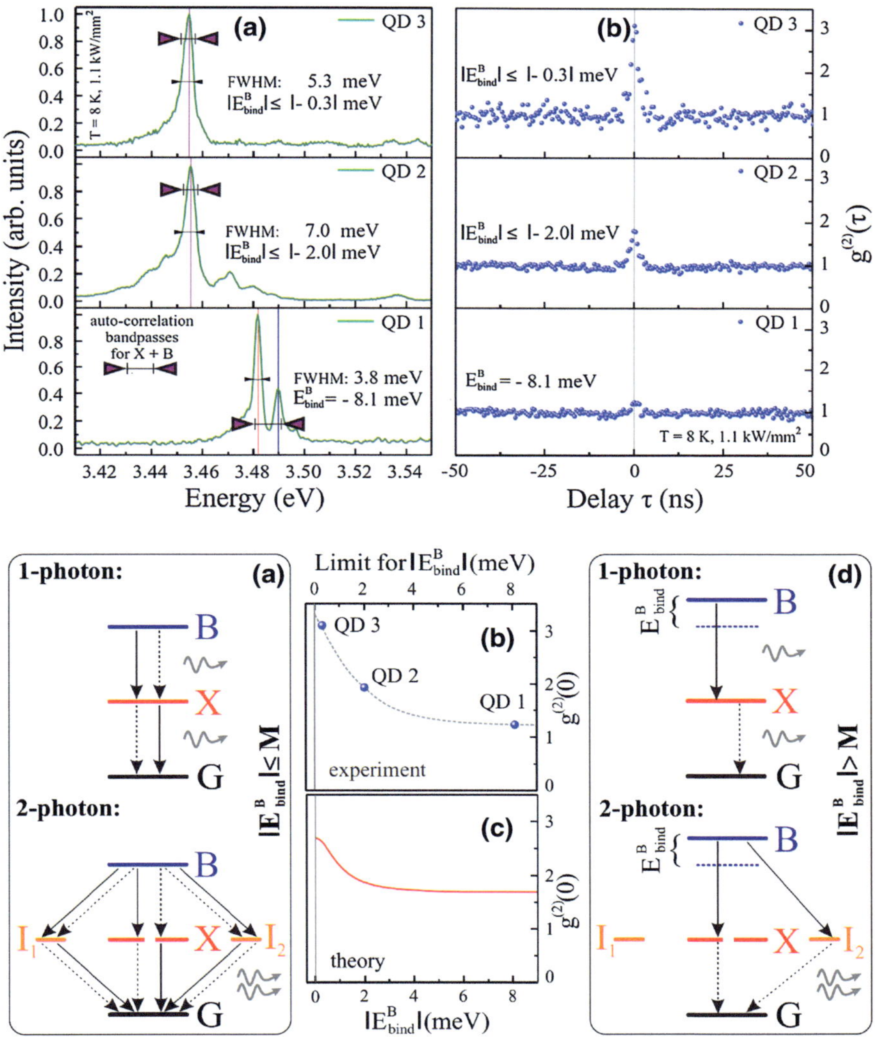

Fig. 12.39 Bunching phenomenon as a function of biexciton binding energy. (**top**) μPL spectra and intensity autocorrelation function for three different GaN QDs with different biexciton binding energies $E^{\mathrm{B}}_{\mathrm{bind}}$. (**bottom**) Level schemes for resonant (left) and off-resonant (right) biexciton cascade. Theory confirms the experimentally observed trend of decreasing bunching with increasing $E^{\mathrm{B}}_{\mathrm{bind}}$ (middle). Taken from [90]. Reprinted figure with permission from [90]. Copyright (2013) American Physical Society

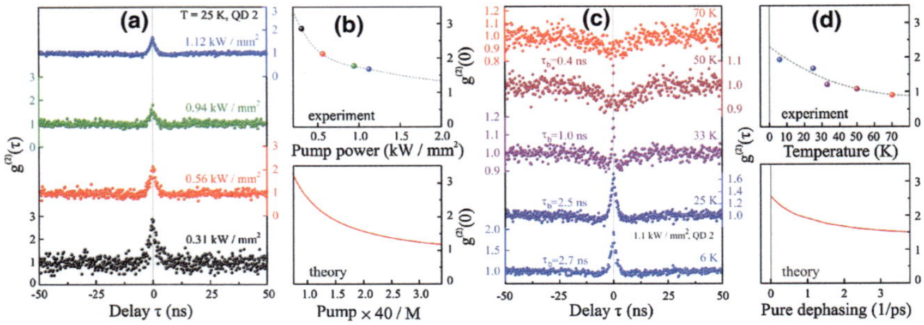

Fig. 12.40 Excitation power and temperature dependence of the bunching. **a**, **b** Excitation power dependence of the second-order correlation function $g^{(2)}(\tau)$ and the zero-delay time value, respectively. **c**, **d** Temperature dependence of the observed bunching and the zero-delay time value. Taken from [90]. Reprinted figure with permission from [90]. Copyright (2013) American Physical Society

to spectral diffusion, which can in theory lead to an artificial bunching due to the emission line 'jumping' in and out of the bandpass window used for the start and stop signal. We are, however, able to exclude such origin by performing bandpass dependent measurements, which do not show any change in emission statistics as a function of window size and position with regard to the emission line.

By numerically solving the quantum state tomography of the biexciton cascade, we reveal the underlying physics of the change in emission statistics. The results are summarized in Fig. 12.39b, showing level scheme and photon modes (indicated by arrows). We find that the number of transition channels for the bunched two-photon emission is increased if the biexciton binding energy is smaller than the electron-photon coupling element M. In the resonant case $\left(\left|E_{\text{bind}}^{\text{B}}\right| \leq M\right)$ there are a total of ten decay processes towards the ground state, whereas only two decay processes remain in the non-resonant case. Consequently, the bunching amplitude at zero-delay time is reduced as bunched two-photon emission becomes less likely with increasing $\left|E_{\text{bind}}^{\text{B}}\right|$, which is observed both experimentally and theoretically.

Figure 12.40 shows excitation density and temperature dependence of the bunching phenomenon for a single quantum dot. In good agreement between experiment and theory we observe a degradation of the bunching with increasing excitation density and temperature. Both trends can be explained within the framework of the theoretical model as consequence of increased excitation-induced and pure dephasing [90]. Decay processes contributing to bunched two-photon emission are found to be twice as sensitive to such dephasings when compared to the cascaded emission and are thus quickly quenched by increasing excitation density or temperature. An experimental indication for the influence of dephasing is the narrowing of the bunching peak around zero-time delay with increasing temperature, indicating a rapid dephasing by acoustic phonons.

12.4.6 Unconventional Biexciton States

In this section, we focus on biexcitons formed in wurtzite GaN/AlN QDs. We identify optical and quantum-optical signatures of an unusual biexciton configuration, which shows a net spin $s = 3$ instead of the expected antiparallel spin configuration for electrons and holes with a net spin of $s = 0$ [80]. As a consequence, radiative decay from this hybrid-biexciton state populates the exciton dark states ($s = 2$) and subsequent emission from the exciton bright states ($s = 1$) requires a phonon-mediated spin-flip. The hybrid-biexciton is predicted within our theoretical model, based on a configuration interaction scheme, for QDs in an intermediate confinement regime, where the attractive exchange interaction between the holes overcompensates their Coulomb repulsion (green shaded area in Fig. 12.41).

In contrast to previous theoretical works [99, 100], our model also reproduces the experimentally observed zero-crossing of the biexciton binding energy in GaN QDs. A biexciton in its ground state usually consists of two electrons and two holes, occupying the energetically lowest electron and hole orbital, respectively, with antiparallel spins (net spin $s = 0$). As we show, this is no longer true if the lateral confinement for holes is sufficiently weakened, reducing the energy spacing of the hole states. This is possible for a range of QD sizes in which electrons are still strongly confined due

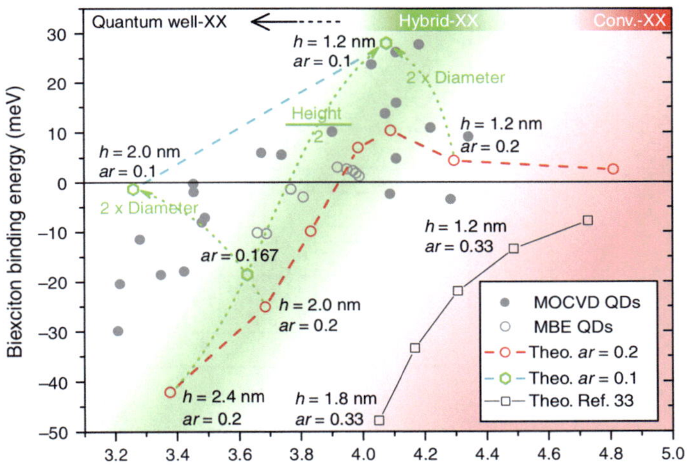

Fig. 12.41 Exotic biexcitons and where to find them. The biexciton binding energy is plotted versus the exciton emission energy. Experimental data points are shown as grey circles. Order of magnitude as well as evolution of the biexciton binding energy are well reproduced within our model (dashed and dotted lines for different aspect ratios of the QDs). Regions where conventional biexcitons or hybrid-biexcitons are expected according to our theory are color-coded in red and green, respectively. The parameters h and ar quantify the height and aspect ratios of the quantum dots, respectively. Taken from [80]. Reprinted figure with permission from [80]. Copyright (2014) Springer Nature

to their bigger effective Bohr radius. Simply put, if the QD is flat and wide enough, the attractive exchange interaction can overcome a weakened Coulomb repulsion. Similar to Hund's rules for filling atomic orbitals, it can become energetically more favorable for the holes to occupy two different orbitals with parallel spins, instead of the same orbital with antiparallel spins, leading to a net spin of 3 (Fig. 12.42).

The different decay paths can be readily observed in a polarization dependent μ-PL investigation. Here, a conventional biexciton would show two linearly polarized biexciton and exciton emission lines, each separated by the fine-structure splitting of the bright states. Instead, the hybrid-biexciton shows only a single, partially linearly polarized, biexciton emission line, since the dark state splitting is too small to be resolved. The hybrid-biexciton also manifests itself in temperature dependent μ-PL and cross-correlation measurements. Since the spin-flip process is phonon mediated, it strongly depends on the sample temperature. In an Arrhenius type investigation of the temperature dependent intensity of the two exciton emission lines we extract two activation energies, representing dark-to-bright state splitting ($E_{\mathrm{DB}} = 0.9 \pm 0.4$ meV) and bright state fine-structure splitting ($E_{\mathrm{FSS}} = 5.1 \pm 0.8$ meV), respectively.

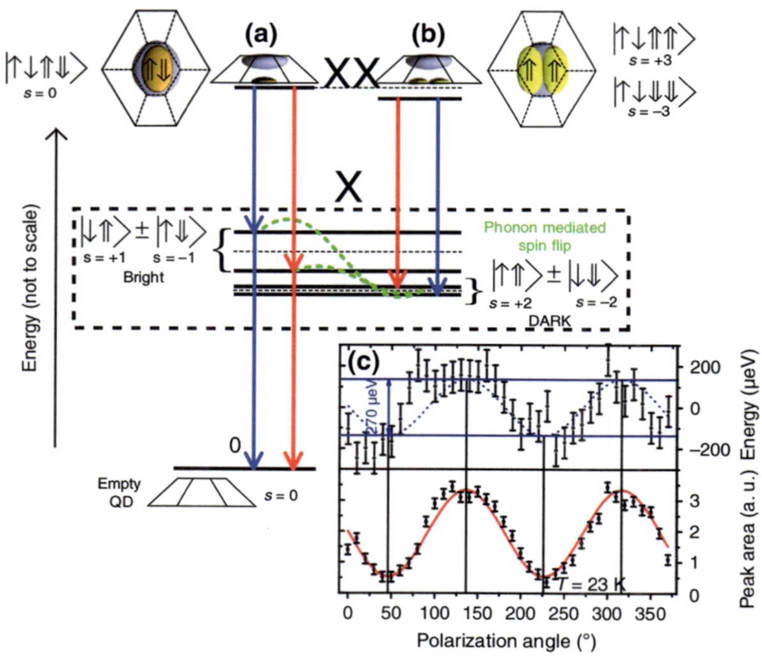

Fig. 12.42 Level scheme of the biexciton cascades. **a** Conventional biexciton ($s = 0$) configuration and its decay via the exciton bright states ($s = 1$). **b** Hybrid-biexciton ($s = 3$) and its decay via the exciton dark states ($s = 2$) and a subsequent phonon-mediated spin-flip into the exciton bright states. **c** Polarization dependent investigation of the hybrid-biexciton emission energy and intensity as an indirect measure of the dark state splitting. Taken from [80]. Reprinted figure with permission from [80]. Copyright (2014) Springer Nature

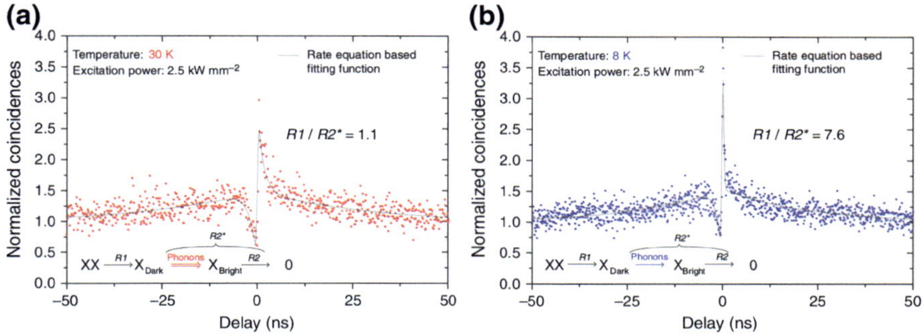

Fig. 12.43 Temperature dependent cross-correlation measurements. Histograms showing the cross-correlation between hybrid-biexciton (start) and bright exciton (stop) emission. By lowering the temperature from 30 K **a** to 8 K **b** at constant excitation power, a transition from the typical biexciton cascade **a** to a spin-blockaded biexciton cascade **b** is observed. The increased ratio between the hybrid biexciton decay rate (R1) and the exciton decay rate R2* at lower temperature demonstrates the influence of the phonon-mediated dark-to-bright exciton transitions. Taken from [80]. Reprinted figure with permission from [80]. Copyright (2014) Springer Nature

Final evidence for a phonon-blockaded cascade process is obtained by temperature dependent cross-correlation measurements (Fig. 12.43). Using the hybrid-biexciton as start and the bright exciton as stop trigger, the transition from a spin-blockaded biexciton cascade at low temperatures to a typical biexciton cascade for temperatures of 30 K and above is observed.

12.4.7 Monolithic Deep UV Bragg Mirrors for GaN QD Microcavities

The realization of GaN QD cavities requires suitable high UV-reflective Bragg mirrors. For microcavities with a monolithically grown bottom mirror layer combinations with lattice constants close to AlN and low absorption, e.g. AlN/AlGaN DBRs, are preferred. Therefore, increasing QD emission energies requires a higher Al-contents in the AlGaN layers which reduces the refractive index contrast of the DBR layers and leads to a higher number of required mirror pairs. E.g. with a combination of AlN/$Al_xGa_{1-x}N$ ($x \geq 0.7$) more than 50 pairs of $\lambda/4$ layers must be grown to achieve a reflectivity >98% at 273 nm with a stop band width of 8.4 nm (Fig. 12.44). Similar values have been reported by Franke et al. [101] for an AlN/$Al_{0.65}Ga_{0.35}N$ DBR near 270 nm exhibiting a reflectivity >97% as well as by Detchprom et al. [102] targeting the wavelength range from 220 to 250 nm.

For AlGaN/AlN DBRs grown on sapphire substrates, strong tensile stress is present during growth originating from the coalescence of initial AlN islands and edge dislocation climb [103]. Thus, thick AlN buffer layers grown to improve the material quality can lead to high tensile stress and relaxation by cracking. This is

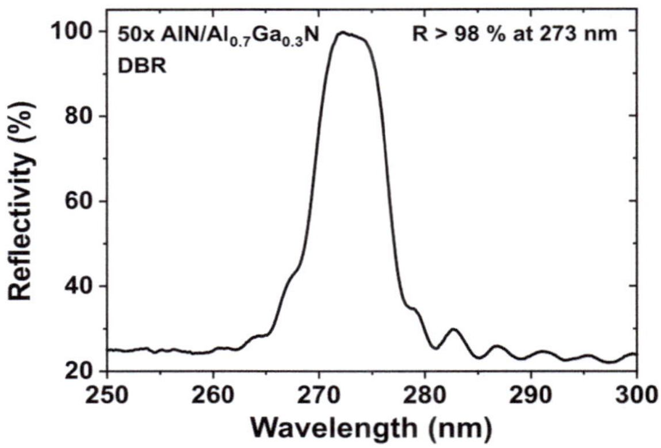

Fig. 12.44 Reflectivity of a 50-fold $AlN/Al_{0.7}Ga_{0.3}N$ Bragg mirror

shown in Fig. 12.45 in the in situ curvature measurement of a 50-fold DBR with two different AlN buffer layer thicknesses. The sample with thick AlN buffer shows a higher initial stress before starting the growth of the DBR layers leading to a heavily cracked sample with spacings of about 50 μm between two adjacent cracks. With a thinner AlN buffer, which is already fully coalesced, e.g., free of pits, the sample shows less than 1 crack per mm. A significant improvement is expected for the growth on AlN substrates which are, however, still very expensive. With AlN substrates growth is likely under compression and DBR stacks with more than 50 periods should be achievable easily while for sapphire substrates tensile stress limits the number of DBR pairs.

During the growth of the Al-rich mirror layer pairs it was observed that instead of obtaining the intended double layers of AlN and AlGaN the layers were actually comprised of three layers as shown in Fig. 12.46. The three layers show sharp interfaces and, because the total thickness of this triple layer stack is about the same as

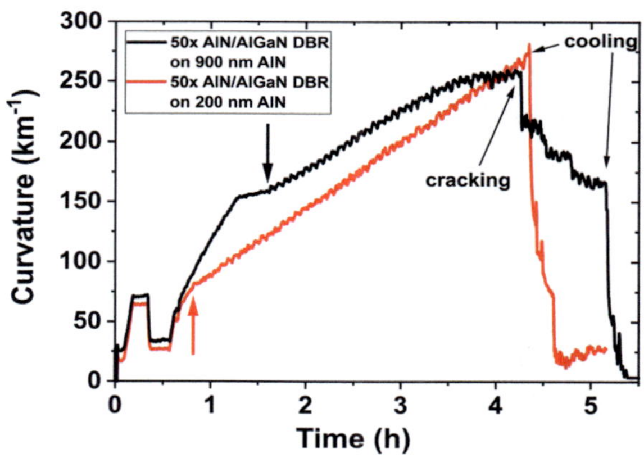

Fig. 12.45 In-situ curvature measurement during the growth of AlGaN/GaN DBRs on differently thick AlN buffer layers indicating tensile stress by the positive curvature value (convex shape). The end of AlN buffer layer growth is marked by an arrow. The surface obtained is smooth with an AFM rms roughness of 0.23 nm (3 × 3 μm^2)

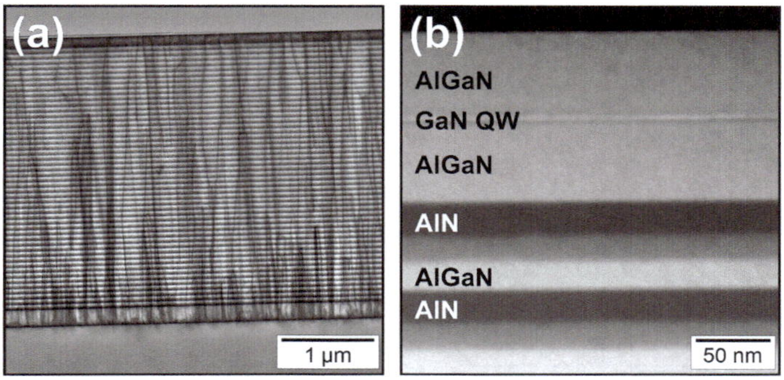

Fig. 12.46 Cross-section TEM image of a 50-pair AlN/AlGaN DBR with an AlGaN cavity with embedded GaN QW and islands **a** and STEM image from the upper DBR periods and the cavity showing a triple stack consisting of 22 nm AlN, 20 nm AlGaN transition layer, 16.5 nm AlGaN instead of a 29 nm/27 nm layer pair for the AlN/AlGaN layer stack **b**

for a double layer stack, the target wavelength of the mirror is still matched but with a slightly lower reflectivity as for the ideal double layer structure. Such a triple layer structure has been also observed by other authors [104, 105]. Liu et al. propose a composition pulling effect which they estimated to reduce overall strain by one order of magnitude [104]. XRD Θ-2Θ scans and reciprocal space maps (rsm) confirm a strictly periodic triple layer structure and coherent growth of the DBR (Fig. 12.47). As the origin of this structure is currently not fully understood avoiding the formation is, however, difficult although reports of two layer structures exist in literature [106].

When growing GaN QDs on a DBR stack a slightly larger lattice constant as compared to the growth on a pure AlN buffer can be expected and, e.g., lead to a

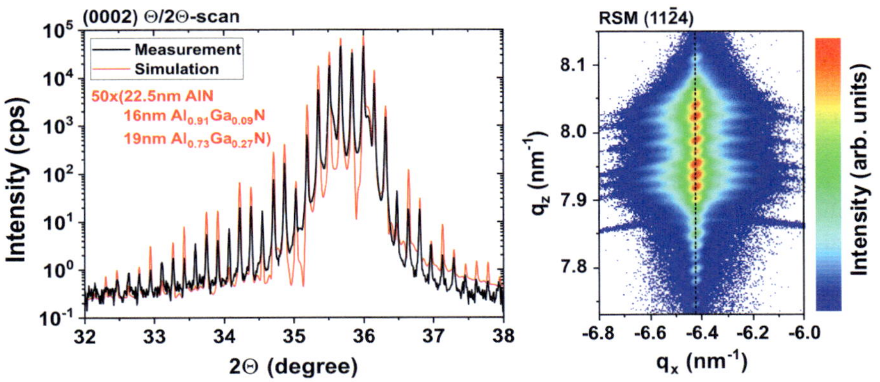

Fig. 12.47 XRD Θ-2Θ scan and simulation of a nominal 50-fold AlN/$Al_{0.8}Ga_{0.2}N$ Bragg mirror (left). The resulting structure consists of 22.5 nm AlN, 16 nm $Al_{0.91}Ga_{0.09}N$ and 19 nm $Al_{0.83}Ga_{0.17}N$. The rsm at the (11$\bar{2}$4) reflection demonstrates coherent growth of the structure (right)

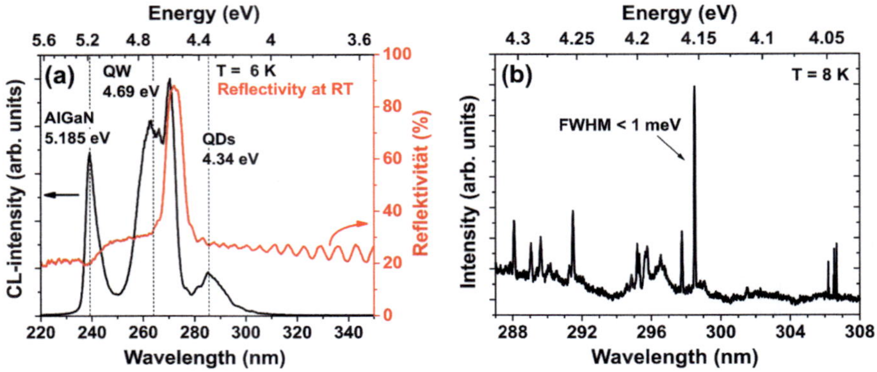

Fig. 12.48 Local CL-spectrum at 6 K and RT reflectivity (red) for a 50-fold AlGaN/AlN mirror with an AlGaN QW cavity but without top mirror. For this sample, the stopband does not match the spectral region of the QDs. Thus, an enhanced emission in the spectral range of the QWs was observed. In μ-PL measurements narrow luminescence was observed in the spectral range of the QDs, which is, however, outside the stop band width

different GaN desorption behavior and GaN QD growth must be adjusted on Bragg mirrors. In direct comparison using X-ray in-plane diffraction measurements, the difference in in-plane mismatch reduces from 2.36% on an AlN buffer to 2.13% on an $Al_{0.73}Ga_{0.27}N$ cavity on top of a DBR structure.

For GaN QDs grown on top of a DBR structure intense luminescence is observed. In Fig. 12.48 the reflectivity at room temperature is shown together with a CL-measurement at 6 K. The emission at 5.185 eV originates from the $Al_{0.73}Ga_{0.27}N$ cavity and the contribution at 4.69 eV is assigned to the GaN quantum well. The high intensity shoulder at 4.58 eV overlaps with the energetic position of the stop band. The small redshift of the stop band with respect to this emission peak in Fig. 12.48 originates in the temperature offset between CL- and reflectivity measurement.

Cross-section CL measurements prove that this luminescence arises from the GaN quantum well and the increased intensity at 4.58 eV is only observed when measuring perpendicular to the surface. An additional luminescence contribution is detected at 4.34 eV and is assigned to the GaN QD ensemble. In μ-PL measurements, sharp emission lines with linewidths below 1 meV are observed in this spectral region. Compared to the QDs grown on AlN buffer structures, these QDs exhibit a red-shifted emission behavior (see Fig. 12.33). Since the DBR and the QDs are strongly detuned, the Bragg reflector has no positive impact on the QD emission characteristics. Most likely, the GaN islands have larger dimensions when grown on the DBR with an AlGaN cavity due to a reduced strain-related desorption rate during the growth interruption. This could be caused by a reduced lattice-mismatch between GaN and the AlGaN cavity (compared to GaN on an AlN buffer) and therefore a higher bonding energy of GaN to the surface.

12.5 Towards Electrically Driven Microcavity Devices

The realization of electrically driven (Al)GaN based microcavity single photon emitters and lasers requires highly conducting layers to enable high current flow through the center of the device. On the *n*-side of the device this can be easily realized by using conducting mirror layers while *p*-type doping is a challenge for the *p*-side of the device. Electrically pumped GaN based VCSEL structures have been achieved previously [107, 108]. Typically an indium-tin-oxide (ITO) layer is deposited on top of the *p*-GaN layer forming a tunneling contact similar to GaN LED technology for solid state lighting applications. However, ITO has a limited transparency in the UV and blue wavelength region and therefore leads to higher losses. An alternative to ITO are highly conductive GaN layers which are fully transparent [25, 109]. While MBE grown GaN can be Si-doped in excess of 10^{19} cm^{-3} Si-doping levels in MOVPE are limited due to the formation of SiN leading to an increase in surface roughness and 3D growth [110]. Thus until now mostly MBE grown tunnel junctions were demonstrated [111], while in MOVPE either very thin Si δ-doped or rough but thicker highly Si-doped GaN layers were applied to achieve a sufficiently high doping level [112]. Alternatively polarization doping introducing, e.g. InGaN in GaN was applied for tunneling junctions, for light emitters though absorption can be an issue. As an alternative to Si-doping or polarization doping germanium doping can be applied. With Ge it is possible to dope GaN in MOVPE in the 10^{20} cm^{-3} range without significant deterioration in surface properties [110]. In direct comparison to ITO GaN:Ge is superior in regard of several properties. While the resistivity can be as low as 4×10^{-4} Ω cm for both compounds, GaN:Ge has a higher transparency and can be epitaxially deposited on the *p*-GaN layer in one growth process. It has been demonstrated that tunneling junctions with excellent current spreading properties can be realized on LEDs [113, 114] and current densities >10 kA/cm^{-2} can be applied through these junctions. In electronics GaN:Ge has been demonstrated to act as an excellent current spreader when applied to quasi-vertical *p-i-n* diodes [115]. Thus for structures with a highly resistive mirror stack it can be also applied as back *n*-type contact layer.

A remaining problem is the formation of the current aperture which can be realized by several methods (Fig. 12.49). One is by selective wet chemical etching of AlInN [116, 117] or the selective oxidation of AlInN [118] forming a highly resistive oxide layer. Another approach is masking the surrounding area of the active region and subsequent overgrowth. In principle one can also use sputtered ITO after deposition of a dielectric aperture mask with the above mentioned limitations in transparency. Comparing the oxidized AlInN layer and the regrowth approach, oxidizing AlInN, as also etching, requires precise control of the process (e.g. pH, temperature). It is also only suited for highly conductive mirror stacks as current flow must be at least through one AlInN/GaN layer pair. For large cavities it only offers a limited

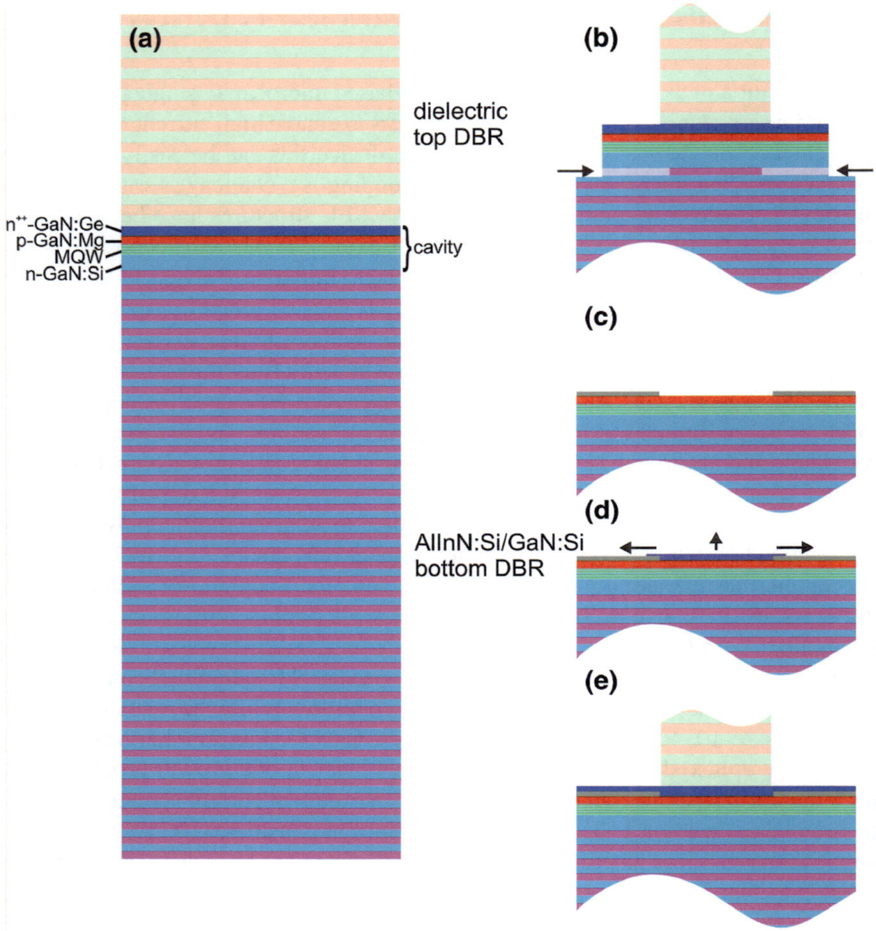

Fig. 12.49 Full VCSEL structure **a** with a 40-fold epitaxial bottom mirror and 3/2 λ-cavity including a 5-fold InGaN MQW a p-type layer (red) and a top n^{++}-GaN:Ge layer (dark blue) forming a tunneling contact to the p-GaN:Mg layer. On top a dielectric mirror, e.g. a 10-fold Si_3N_4/SiO_2 is deposited. For current confinement the structure can be etched down to the first AlInN layer which is then oxidized forming a current aperture **b**. Alternatively shown in **c**–**e** growth can be stopped after p-GaN formation and a dielectric masking **c** can define the current aperture which is then overgrown with a n^{++}-GaN:Ge layer for contacting and current spreading **d** before the upper DBR is deposited **e**, e.g., by sputtering

current confinement due to the higher conductivity and current spreading of the n-type layers. On the other hand overgrowth of p-GaN with highly Ge-doped GaN is not straightforward and must be optimized to achieve a smooth surface.

In principle the top mirror stack can be also an epitaxial structure but several difficulties hamper the growth of such a structure. During heteroepitaxy on sapphire substrates the tensile strain leads to strong tension and cracking of the AlGaN/AlN layers during growth. This may be avoided by improving the material quality as well as growing on low defect density GaN substrates. Nevertheless for GaN/AlInN mirrors, because of the low growth rate of AlInN, often growth times in excess of 36 h are required if a top mirror is included. Also process stability is an issue and modern in situ monitoring techniques as true temperature and growth rate monitoring at several wavelengths are required to adjust growth parameters in order to account for changing conditions. Also processing of such structures is more demanding because the upper mirror layer is typically in excess of 3 μm in thickness. Due to the lack of wet etching techniques and etch stop layers accurate dry etching is required for the contact layers and to define a current aperture. Therefore dielectric top mirrors have significant advantages for device fabrication.

For the investigated GaN QDs in an AlN or Al-rich AlGaN matrix, electrically driven single photon emitters are still difficult to achieve. In this case optical pumping is the best choice as the Al-rich cavity and Bragg mirror layers show very high resistivity.

12.6 Conclusion and Perspectives

Although a few groups demonstrated GaN-based vertical surface emitting lasers much of the involved physics of microcavities remains a subject of ongoing research. This chapter could only address some of these challenges, for most the growth of monolithic DBR structures and optical properties related to InGaN/GaN quantum wells inside such cavities. Epitaxial solutions to growth that require thick DBR structures exist which allow for reflectivities beyond 99%. However, the quest for techniques to achieve even higher reflectivities is still there. Not to speak of all the research that is necessary for electrical operation of microcavity devices. Beside low-resistive mirror structures or efficient intra-cavity injection schemes there is also a strong need for a viable technology to implement current apertures. Newly developed current spreading layers and tunneling contacts using the transparent conductive nitride GaN:Ge, blue and near UV-VCSELs may open up new pathways to a mature technology within the next years.

Single photon emitters based on GaN QDs have been demonstrated and optically pumped operation is in principle possible [79]. Since the GaN QD formation process is driven by desorption kinetics around dislocations rather than by an equilibrium state, control of the QD emission properties is difficult. Novel approaches are demanded providing better epitaxial control by MOVPE of nitride QD density and size, and thereby of their emission wavelength. Beyond that, any single QD-based emitter will rely on a selectively addressable excitation scheme. Therefore, either novel growth methods or post-growth technologies are necessary to isolate suitable single QDs. Both electrical excitation as well as external spectral tuning of single photon emission is an open area in this material system.

Acknowledgements We gratefully acknowledge the German Research Foundation (DFG) for financial support within the Research Instrumentation Program INST 272/148-1 and the Collaborative Research Center SFB 787 "Semiconductor Nanophotonics: Materials, Models, Devices".

Many thanks to Silke Petzold (University of Magdeburg) for her work regarding the specimen preparation.

References

1. E.M. Purcell, Phys. Rev. **69**, 681 (1946)
2. H. Soda, K. Iga, C. Kitahara, Y. Suematsu, Jpn. J. Appl. Phys. **18**, 2329 (1979)
3. P.A.M. Dirac, Proceedings of the royal society of London A: mathematical. Physical and Engineering Sciences **114**, 243 (1927)
4. J. Kasprzak, M. Richard, S. Kundermann, A. Baas, P. Jeambrun, J.M.J. Keeling, F.M. Marchetti, M.H. Szymanska, R. André, J.L. Staehli, V. Savona, P.B. Littlewood, B. Deveaud, L.S. Dang, Nature **443**, 409 (2006)
5. R. Butté, N. Grandjean, Semicond. Sci. Technol. **26**, 014030 (2011)
6. C. Bennett, P. Shor, IEEE Trans. Inf. Theory **44**(6), 2724–2742 (1998)
7. M.A. Nielsen, I.L. Chuang, *Quantum Computation and Quantum Information* (Cambridge University Press, Cambridge and New York, 2000)
8. T.D. Ladd, F. Jelezko, R. Laflamme, Y. Nakamura, C. Monroe, J.L. O'Brien, Nature **464**(7285), 45–53 (2010)
9. D. Bimberg, M. Grundmann, N.N. Ledentsov, *Quantum Dot Heterostructures* (John Wiley, Chichester (England) and New York, 1999)
10. A.J. Shields, Nat. Photonics **1**(4), 215–223 (2007)
11. M. Grundmann, J. Christen, N.N. Ledentsov, J. Böhrer, D. Bimberg, S.S. Ruvimov, P. Werner, U. Richter, U. Gösele, J. Heydenreich, V.M. Ustinov, A.Y. Egorov, A.E. Zhukov, P.S. Kop'ev, Z.I. Alferov, Phys. Rev. Lett. **74** (20), 4043–4046 (1995)
12. O. Stier, M. Grundmann, D. Bimberg, Phys. Rev. B **59**(8), 5688–5701 (1999)
13. P. Michler, A. Imamoglu, M.D. Mason, P.J. Carson, G. Strouse, S.K. Buratto, Nature **406** (6799), 968–970 (2000)
14. P. Michler, A. Kiraz, C. Becher, W.V. Schoenfeld, P.M. Petroff, L. Zhang, E. Hu, A. Imamoglu, Science **290**(5500), 2282–2285 (2000)
15. C. Santori, M. Pelton, G. Solomon, Y. Dale, Y. Yamamoto, Phys. Rev. Lett. **86**(8), 1502–1505 (2001)
16. Z. Yuan, B.E. Kardynal, R.M. Stevenson, A.J. Shields, C.J. Lobo, K. Cooper, N.S. Beattie, D.A. Ritchie, M. Pepper, Science (New York, N.Y.) **295** (5552), 102–105 (2002)
17. S. Kako, C. Santori, K. Hoshino, S. Goetzinger, Y. Yamamoto, Y. Arakawa, Nat. Mater. **5**, 887 (2006)
18. M.J. Holmes, K. Choi, S. Kako, M. Arita, Y. Arakawa, Nano Lett. **14**(2), 982–986 (2014)
19. M. Tchernycheva, L. Nevou, L. Doyennette, F.H. Julien, E. Warde, F. Guillot, E. Monroy, E. Bellet-Amalric, T. Remmele, M. Albrecht, Phys. Rev. B **73** (12) (2006)
20. I. Vurgaftman, J.R. Meyer, L.R. Ram-Mohan, J. Appl. Phys. **89**(11), 5815 (2001)
21. H.P.D. Schenk, E. Feltin, P. Vennéguès, O. Tottereau, M. Laügt, M. Vaille, B. Beaumont, P. de Mierry, P. Gibart, S. Fernández, F. Calle, Phys. Status Solidi (a) **188**, 899 (2001)
22. T. Shirasawa, N. Mochida, A. Inoue, T. Honda, T. Sakaguchi, F. Koyama, K. Iga, J. Cryst. Growth **189–190**, 124 (1998)
23. J.F. Carlin, M. Ilegems, Appl. Phys. Lett. **83**, 668 (2003)

24. J.F. Carlin, C. Zellweger, J. Dorsaz, S. Nicolay, G. Christmann, E. Feltin, R. Butté, N. Grandjean, Phys. Status Solidi (b) **242**, 2326 (2005)
25. C. Berger, A. Dadgar, J. Bläsing, A. Lesnik, P. Veit, G. Schmidt, T. Hempel, J. Christen, A. Krost, A. Strittmatter, J. Cryst. Growth **414**, 105 (2015)
26. C. Berger, Dissertation, Otto-von-Guericke-Universität Magdeburg (2017)
27. A. Dadgar, A. Krost, Epitaxial growth and benefits of GaN on silicon, in *III-Nitride Semiconductors and their Modern Devices*, ed. by B. Gil (Oxford University Press). ISBN 978-0-19-968172-3 (2013)
28. C. Berger, A. Dadgar, J. Bläsing, A. Lesnik, P. Veit , G. Schmidt, T. Hempel, J. Christen, A. Krost, A. Strittmatter, J. Cryst. Growth **414**, 105 (2015)
29. J. Dorsaz, J.F. Carlin, S. Gradecak, M. Ilegems, J. Appl. Phys. **97**, 084505 (2005)
30. H. Amano, M. Iwaya, N. Hayashi, T. Kashima, S. Nitta, C. Wetzel, I. Akasaki, Phys. Stat. Sol. (b) **216**, 683 (1999)
31. J. Bläsing, A. Reiher, A. Dadgar, A. Diez, A. Krost, Appl. Phys. Lett. **81**, 2722 (2002)
32. E. Feltin, J.F. Carlin, J. Dorsaz, G. Christmann, R. Butté, M. Laügt, M. Ilegems, N. Grandjean, Appl. Phys. Lett. **88**, 051108 (2006)
33. T.C. Lu, J.R. Chen, S.W. Chen, H.C. Kuo, C.C. Kuo, C.C. Lee, S.C. Wang, IEEE J. Sel. Top. Quantum Electron. **15**, 850 (2009)
34. J.F. Carlin, J. Dorsaz, E. Feltin, R. Butté, N. Grandjean, M. Ilegems, M. Laügt, Appl. Phys. Lett. **86**, 031107 (2005)
35. G. Christmann, D. Simeonov, R. Butté, E. Feltin, J.F. Carlin, N. Grandjean, Appl. Phys. Lett. **89**, 261101 (2006)
36. Z. Gacevic, G. Rossbach, R. Butté, F. Réveret, M. Glauser, J. Levrat, G. Cosendey, J.F. Carlin, N. Grandjean, E. Calleja, J. Appl. Phys. **114**, 233102 (2013)
37. G. Cosendey, (In,Al)N-based blue microcavity lasers. Dissertation, École Polytechnique Fédérale de Lausanne (2013)
38. S. Kako, T. Someya, Y. Arakawa, Appl. Phys. Lett. **80**, 722 (2002)
39. F. Bertram, J. Christen, G. Schmidt, P. Veit, C. Berger, A. Krost, Microsc. Microanal. **18**, 1874 (2012)
40. M. Akazawa, B. Gao, T. Hashizume, M. Hiroki, S. Yamahata, N. Shigekawa, Appl. Phys. Lett. **98**(14), 142117 (2011)
41. A. Imamoglu, R.J. Ram, S. Pau, Y. Yamamoto, Phys. Rev. A **53**, 4250–4253 (1996)
42. V. Savona, L.C. Andreani, P. Schwendimann, A. Quattropani, Solid State Commun. **93**, 733 (1995)
43. L.A. Coldren, S.W. Corzine, M. Mashanovitch, Diode lasers and photonic integrated circuits. Wiley Series in Microwave and Optical Engineering, vol. 218, 2nd edn. (Wiley, Hoboken and NJ) (2012)
44. G. Schmidt, M. Müller, P. Veit, F. Bertram, J. Christen, M. Glauser, J.-F. Carlin, G. Cosendey, R. Butté, N. Grandjean, Appl. Phys. Lett. **105**(3), 032101 (2014)
45. J. Matthews, A. Blakeslee, J. Cryst. Growth **29**(3), 273–280 (1975)
46. A. Fischer, H. Kühne, H. Richter, Phys. Rev. Lett. **73**, 2712 (1994)
47. D. Holec, Y. Zhang, D.V.S. Rao, M.J. Kappers, C. McAleese, C.J. Humphreys, J. Appl. Phys. **104** (12), 123514 (2008)
48. R. People, J.C. Bean, Appl. Phys. Lett. **47**(3), 322 (1985)
49. M. Leyer, J. Stellmach, C. Meissner, M. Pristovsek, M. Kneissl, J. Cryst. Growth **310**(23), 4913–4915 (2008)
50. C.A. Parker, J.C. Roberts, S.M. Bedair, M.J. Reed, S.X. Liu, N.A. El-Masry, Appl. Phys. Lett. **75**(18), 2776 (1999)
51. C. Hums, *MOVPE Wachstum und Eigenschaften von AlIn(Ga)N Schichten und Schichtsystemen*. Ph.D. thesis, Otto-von-Guericke-Universität Magdeburg, Magdeburg (2012)
52. B.W. Dodson, J.Y. Tsao, Appl. Phys. Lett. **51**(17), 1325 (1987)
53. A.D. Bykhovski, B.L. Gelmont, M.S. Shur, J. Appl. Phys. **81**(9), 6332 (1997)
54. M. Glauser, G. Rossbach, G. Cosendey, J. Levrat, M. Cobet, J.-F. Carlin, J. Besbas, M. Gallart, P. Gilliot, R. Butté, N. Grandjean, Phys. Status Solidi (c) **9**(5), 1325–1329 (2012)

55. G. Schmidt, *Optische Nanocharakterisierung GaN-basierter Quantenstrukturen für Mikrokavitäten*. Doctoral Thesis, Otto-von-Guericke-University Magdeburg, Magdeburg 2017
56. L. Bellaiche, T. Mattila, L.-W. Wang, S.-H. Wei, A. Zunger, Appl. Phys. Lett. **74**(13), 1842 (1999)
57. D. Watson-Parris, M.J. Godfrey, P. Dawson, R.A. Oliver, M.J. Galtrey, M.J. Kappers, C.J. Humphreys, Phys. Rev. B **83** (11) (2011)
58. S. Schulz, M.A. Caro, C. Coughlan, E.P. O'Reilly, Phys. Rev. B **91** (3) (2015)
59. P.R.C. Kent, A. Zunger, Appl. Phys. Lett. **79**(13), 1977 (2001)
60. L.-W. Wang, Phys. Rev. B **63** (24) (2001)
61. S.F. Chichibu, A. Uedono, T. Onuma, B.A. Haskell, A. Chakraborty, T. Koyama, P.T. Fini, S. Keller, S.P. Denbaars, J.S. Speck, U.K. Mishra, S. Nakamura, S. Yamaguchi, S. Kamiyama, H. Amano, I. Akasaki, J. Han, T. Sota, Nat. Mater. **5**(10), 810–816 (2006)
62. S. Chichibu, T. Sota, K. Wada, S. Nakamura, J. Vac. Sci. Technol. B: Microelectron. Nanometer Struct. **16**(4), 2204 (1998)
63. D.M. Graham, A. Soltani-Vala, P. Dawson, M.J. Godfrey, T.M. Smeeton, J.S. Barnard, M.J. Kappers, C.J. Humphreys, E.J. Thrush, J. Appl. Phys. **97**(10), 103508 (2005)
64. M.J. Galtrey, R.A. Oliver, M.J. Kappers, C.J. Humphreys, P.H. Clifton, D. Larson, D.W. Saxey, A. Cerezo, J. Appl. Phys. **104**(1), 013524 (2008)
65. Y. Yamamoto, F. Tassone, H. Cao, *Semiconductor Cavity Quantum Electrodynamics*. Springer Tracts in Modern Physics, vol. 169 (Springer, Berlin and New York) (2000)
66. A. Kavokin, G. Malpuech, *Cavity polaritons*. Thin Films and Nanostructures, vol. 32, 1st edn. (Elsevier, Acad. Press, Amsterdam [u.a.]) (2003)
67. S.M. Dutra, *Cavity Quantum Electrodynamics: The Strange Theory of Light in a Box*. Wiley Series in Lasers and Applications. Wiley, New York (2005)
68. B. Deveaud, *The Physics of Semiconductor Microcavities: From Fundamentals to Nanoscale Devices* (Wiley-VCH, Weinheim, 2007)
69. A. Kavokin, *Microcavities*. Series on Semiconductor Science and Technology, vol. 16 (Oxford University Press, Oxford) (2007)
70. A. Franke, *Licht - Materie -Wechselwirkung in nitridischen Mikrokavitäten*. Doctoral thesis, Otto-von-Guericke-Universität Magdeburg, Magdeburg (2012)
71. M. Glauser, C. Mounir, G. Rossbach, E. Feltin, J.-F. Carlin, R. Butté, N. Grandjean, J. Appl. Phys. **115**(23), 233511 (2014)
72. G. Schmidt, P. Veit, C. Berger, F. Bertram, A. Dadgar, A. Strittmatter, J. Christen, Jpn. J. Appl. Phys. **55**, 05FF04 (2016)
73. I.N. Stranski, L. Krastanow, Sitzungsber. Akad. Wiss. Wien, Math. Naturwiss. Kl., Abt. 2B **146**, 797 (1937)
74. W. Ostwald, Z. Phys, Chem. (Leipzig) **34**, 495 (1990)
75. K. Poetschke, L. Mueller-Kirsch, R. Heitz, R.L. Sellin, U.W. Pohl, D. Bimberg, N. Zakharov, P. Werner, Phys. E (Amsterdam) **21**, 606 (2004)
76. A.E. Zhukov, A.R. Kovsh, S.S. Mikhrin, N.A. Maleev, V.M. Ustinov, D.A. Livshits, I.S. Tarasov, D.A. Bedarev, M.V. Maximov, A.F. Tsatsul'nikov, I.P. Soshnikov, P.S. Kop'ev, Z.I. Alferov, N.N. Ledentsov, D. Bimberg, Electron. Lett. **35**, 1845 (1999)
77. T. Bartel, M. Dworzak, M. Strassburg, A. Hoffmann, A. Strittmatter, D. Bimberg, Appl. Phys. Lett. **85**, 1946 (2004)
78. I.N. Kaiander, R.L. Sellin, T. Kettler, N.N. Ledentsov, D. Bimberg, N.D. Zakharov, P. Werner, Appl. Phys. Lett. **84**, 2992 (2004)
79. G. Schmidt, C. Berger, P. Veit, S. Metzner, F. Bertram, J. Bläsing, A. Dadgar, A. Strittmatter, J. Christen, G. Callsen, S. Kalinowski, A. Hoffmann, Appl. Phys. Lett. **106**, 252101 (2015)
80. G. Hönig, G. Callsen, A. Schliwa, S. Kalinowski, C. Kindel, S. Kako, Y. Arakawa, D. Bimberg, A. Hoffmann, Nat. Commun. **5**, 5721 (2014)
81. G. Callsen, Advanced optical signatures of single, wurtzite GaN quantum dots: From fundamental exciton coupling mechanisms towards tunable photon statistics and hybrid-quasiparticles. Ph.D. thesis, TU Berlin, Berlin (2015)

82. G. Callsen, G.M.O. Pahn, S. Kalinowski, C. Kindel, J. Settke, J. Brunnmeier, C. Nenstiel, T. Kure, F. Nippert, A. Schliwa, A. Hoffmann, T. Markurt, T. Schulz, M. Albrecht, S. Kako, M. Arita, Y. Arakawa, Phys. Rev. B **92**, 235439 (2015)
83. I.A. Ostapenko, Influence of defects, phonons and strain on the luminescence properties of nitride- and arsenide-based quantum dots. Ph.D. thesis, TU Berlin, Berlin (2012)
84. I.A. Ostapenko, G. Hönig, S. Rodt, A. Schliwa, A. Hoffmann, D. Bimberg, M.R. Dachner, M. Richter, A. Knorr, S. Kako, Y. Arakawa, Phys. Rev. B **85**, 081303(R) (2012)
85. G. Callsen, G.M.O. Pahn, Phys. Status Solidi RRL **9**(9), 521–525 (2015)
86. S. Kako, K. Hoshino, S. Iwamoto, S. Ishida, Y. Arakawa, Appl. Phys. Lett. **85**(1), 64–66 (2004)
87. J. Renard, R. Songmuang, C. Bougerol, B. Daudin, B. Gayral, Nano Lett. **8**(7), 2092–2096 (2008)
88. S. Amloy, K.H. Yu, K.F. Karlsson, R. Farivar, T.G. Andersson, P.O. Holtz, Appl. Phys. Lett. **99**, 251903 (2011)
89. K. Choi, S. Kako, M.J. Holmes, M. Arita, Y. Arakawa, Appl. Phys. Lett. **103**(17), 171907 (2013)
90. G. Callsen, A. Carmele, G. Hönig, C. Kindel, J. Brunnmeier, M.R. Wagner, E. Stock, J.S. Reparaz, A. Schliwa, S. Reitzenstein, A. Knorr, A. Hoffmann, S. Kako, Y. Arakawa, Phys. Rev. B **87**, 245314 (2013)
91. G. Jurczak, T.D. Young, P. Dluzewski, Phys. Status Solidi C **10**, 97 (2013)
92. M. Winkelnkemper, A. Schliwa, D. Bimberg, Phys. Rev. B **74**, 155322 (2006)
93. T. Bretagnon, P. Lefebvre, P. Valvin, R. Bardoux, T. Guillet, T. Taliercio, B. Gil, N. Grandjean, F. Semond, B. Damilano, A. Dussaigne, J. Massies, Phys. Rev. B **73**, 113304 (2006)
94. R. Bardoux, T. Guillet, P. Lefebvre, T. Taliercio, T. Bretagnon, S. Rousset, B. Gil, F. Semond, Phys. Rev. B **74**, 195319 (2006)
95. G. Sallen, A. Tribu, T. Aichele, R. Andre, L. Besombes, C. Bougerol, M. Richard, S. Tatarenko, K. Kheng, J.-P. Poizat, Nat. Photonics **4**, 696 (2010)
96. K. Hoshino, S. Kako, Y. Arakawa, Appl. Phys. Lett. **85**, 1262 (2004)
97. G. Hönig, S. Rodt, G. Callsen, I.A. Ostapenko, T. Kure, A. Schliwa, C. Kindel, D. Bimberg, A. Hoffmann, S. Kako, Y. Arakawa, Phys. Rev. B **88**, 045309 (2013)
98. D. Simeonov, A. Dussaigne, R. Butte, N. Grandjean, Phys. Rev. B **77**, 075306 (2008)
99. S. Tomić, N. Vukmirović, Phys. Rev. B **79**, 245330 (2009)
100. S. Tomić, N. Vukmirović, [Phys. Rev. B 79, 245330 (2009)], Phys. Rev. B **86**, 159902(E) (2012)
101. A. Franke, M.P. Hoffmann, L. Hernandez-Balderrama, F. Kaess, I. Bryan, S. Washiyama, M. Bobea, J. Tweedie, R. Kirste, M. Gerhold, R. Collazo, Z. Sitar, Proc. SPIE **9748**, 97481G (2016)
102. T. Detchprohm, Y.S. Liu, K. Mehta, S. Wang, H. Xie, T.T. Kao, S.C. Shen, P.D. Yoder, F.A. Ponce, R.D. Dupuis, Appl. Phys. Lett. **110**(1), 011105 (2017)
103. D.M. Follstaedt, S.R. Lee, A.A. Allerman, J.A. Floro, J. Appl. Phys. **105**, 083507 (2009)
104. B. Liu, R. Zhang, J.G. Zheng, X.L. Ji, D.Y. Fu, Z.L. Xie, D.J. Chen, P. Chen, R.L. Jiang, Y.D. Zheng, Appl. Phys. Lett. **98**, 10 (2011)
105. X.L. Ji, R.L. Jiang, B. Liu, Z.L. Xie, J.J. Zhou, L. Li, P. Han, R. Zhang, Y.D. Zheng, J.G. Zheng, Phys. Status Solidi A **205**, 1572 (2008)
106. L.L. Zhang, Z.-H. Liu, X.-G. Huang, Q.-F. Li, J. Korean Phys. Soc. **65**, 1101 (2014)
107. E. Feltin, G. Christmann, J. Dorsaz, A. Castiglia, J.F. Carlin, R. Butté, N. Grandjean, S. Christopoulos, G. Baldassarri, H.V. Hogersthal, A.J.D. Grundy, P.G. Lagoudakis, J.J. Baumberg, Electron. Lett. **43**, 924 (2007)
108. Y. Higuchi, K. Omae, H. Matsumura, T. Mukai, Appl. Phys. Express **1**, 1211021 (2008)
109. C. Nenstiel, M. Bügler, G. Callsen, F. Nippert, T. Kure, S. Fritze, A. Dadgar, H. Witte, J. Bläsing, A. Krost, Hoffmann, Phys. Status Solidi RRL **9** (12) 716–721 (2015)
110. S. Fritze, A. Dadgar, H. Witte, M. Bügler, A. Rohrbeck, J. Bläsing, A. Hoffmann, A. Krost, Appl. Phys. Lett. **100**, 122104 (2012)

111. C. Skierbiszewski, G. Muziol, K. Nowakowski-Szkudlarek, H. Turski, M. Siekacz, A. Feduniewicz-Zmuda, A. Nowakowska-Szkudlarek, M. Sawicka, P. Perlin, Appl. Phys. Express **11**(3), 034103 (2018)
112. S. Lee, C.A. Forman, C. Lee, J. Kearns, E.C. Young, J.T. Leonard, D.A. Cohen, J.S. Speck, S. Nakamura, S.P. DenBaars, Appl. Phys. Express **11**(6), 062703 (2018)
113. S. Neugebauer, M.P. Hoffmann, H. Witte, J. Bläsing, A. Dadgar, A. Strittmatter, T. Niermann, M. Narodovitch, M. Lehmann, Appl. Phys. Lett. **110**, 102104 (2017)
114. Armin Dadgar, Transparente leitfähige Nitride. Vak. Forsch. Prax. **30**(4), 26 (2018)
115. Yuhao Zhang, Daniel Piedra, Min Sun, Jonas Hennig, Armin Dadgar, Yu. Lili, Tomas Palacios, IEEE Electron Device Lett. **38**, 248 (2017)
116. I.M. Watson, C. Xiong, E. Gu, M.D. Dawson, F. Rizzi, K. Bejtka, P.R. Edwards, R.W. Martin, *Proceedings of SPIE, volume 6993, MEMS, MOEMS, and Micromachining III* (2008), p. 69930E
117. D. Simeonov, E. Feltin, A. Altoukhov, A. Castiglia, J.-F. Carlin, R. Butté, N. Grandjean, Appl. Phys. Lett. **92**, 171102 (2008)
118. J. Dorsaz, H.-J. Bühlmann, J.-F. Carlin, N. Grandjean, M. Ilegems, Appl. Phys. Lett. **87**, 072102 (2005)

Chapter 13
Group III-Nitride-Based UV Laser Diodes

Tim Wernicke, Luca Sulmoni, Christian Kuhn, Günther Tränkle, Markus Weyers and Michael Kneissl

Abstract Recent progress in the development of deep ultraviolet lasers is reviewed as well as challenges for the heterostructure design and epitaxial growth for AlGaN-based laser diodes are discussed. The growth of AlN on sapphire and AlGaN heterostructures is reviewed and its impact on the performance characteristics of lasers in the UVC spectral range is presented. We achieve optically pumped AlGaN multiple quantum well (MQW) lasers near 270 nm with threshold power densities of less than 800 kW/cm^2 and optically pumped AlGaN MQW lasers emitting at record shortest wavelength of 237 nm. We discuss critical stepping stones towards the development of current injection deep UV laser diodes including studies on Si- and Mg-doping of AlGaN with high aluminum mole fractions. n-$Al_{0.8}Ga_{0.2}N$ cladding layers with resistances as low as 0.026 Ω cm are realized and UVC-transparent p-AlGaN cladding layers are developed. Finally, electroluminescence from current injection broad area UV laser diodes is demonstrated with maximum current densities of 4.7 kA/cm^2. At the end, we provide an outlook of future prospect for deep UV laser diodes and discuss alternative approaches, e.g., electron beam pumping.

13.1 Introduction

There is a wide range of applications for UV lasers, e.g., in the area of medical diagnostics, bio-detection, phototherapy, high-throughput drug screening, 3D printing, materials processing, lithography, and optical storage [1]. As can be seen from Fig. 13.1, most of these applications currently employ conventional UV laser sources, including various types of gas lasers like excimer lasers, argon ion lasers, helium-

T. Wernicke (✉) · L. Sulmoni · C. Kuhn · M. Kneissl
Institute of Solid State Physics, Technische Universität Berlin, Berlin, Germany
e-mail: tim.wernicke@physik.tu-berlin.de

G. Tränkle · M. Weyers · M. Kneissl
Ferdinand-Braun-Institut, Leibniz-Institut für Höchstfrequenztechnik, Berlin, Germany

M. Kneissl et al. (eds.), *Semiconductor Nanophotonics*, Springer Series in Solid-State Sciences 194, https://doi.org/10.1007/978-3-030-35656-9_13

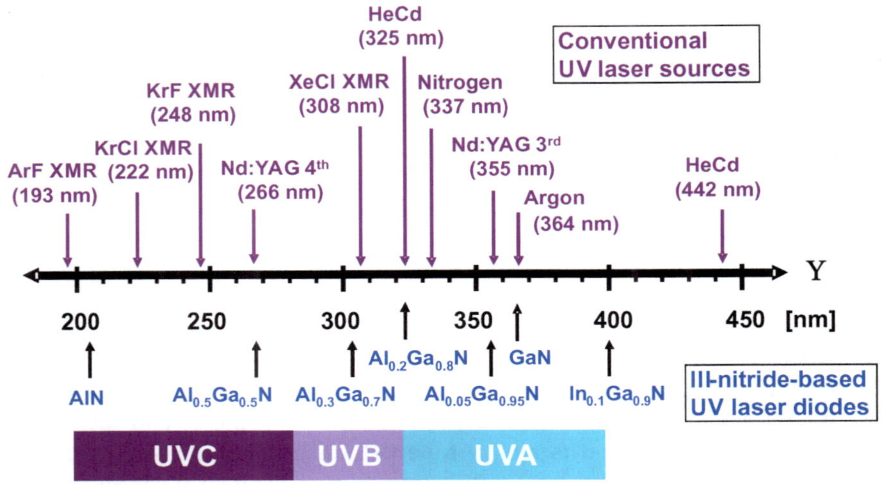

Fig. 13.1 Examples of conventional UV laser sources in the UVA (400–320 nm), UVB (320–280 nm), and UVC (280–200 nm) spectral bands together with the wavelength range that is accessible by AlGaN- and InGaN-based laser diodes

cadmium and nitrogen lasers. In addition, frequency doubled, tripled, and quadrupled solid state lasers are commonly used. For example, the 3rd and 4th harmonic of Nd:YAG lasers emit at 355 nm and 266 nm, respectively. As these solid state lasers can be pumped by lamps or high power IR laser diode stacks, these types of UV laser sources have improved in compactness over the years. However, all the traditional UV laser sources are still very inefficient, often require water cooling, are using toxic gases and/or high voltages, and have very limited lifetimes. Due to their often complex and delicate optical setup, traditional UV laser sources are not really mobile and very sensitive against any kind of vibrations or physical shock. Due to their high efficiency, long lifetimes, small form factors, and robustness, semiconductor-based laser diodes emitting in the infrared and visible spectral range have already found widespread use in many applications, ranging from tele- and data-communication to laser cutting and welding. It is obvious that many applications would also benefit from the development of group III-nitride-based laser diodes emitting in the ultraviolet spectrum. Furthermore, group III-nitride materials provide an excellent foundation for such devices since they enable access to a large portion of the ultraviolet spectrum ranging from the UVA to wavelengths as short as 210 nm in the UVC.

13.2 State-of-the-Art in Group III-Nitride Laser Diode Technologies

13.2.1 Near UV and Blue Laser Diodes

Up to date, various groups have demonstrated current injection laser diodes (LDs) with InGaN, InAlGaN, and AlGaN multiple quantum well (MQW) active regions emitting in the UVA wavelength range [2–13]. As can be seen from Fig. 13.2, the threshold current densities for UV laser diodes rapidly increase with shorter wavelength, and consequently most UV LDs can only be operated under pulsed current injection conditions. In the UVA spectral range, the shortest wavelength current injection laser diode that has been reported is emitting at 327 nm [14]. The $Al_{0.06}Ga_{0.94}N$ multiple quantum well LDs were grown on a low defect density GaN substrate employing an epitaxial lateral overgrowth (ELO) technique to relief the tensile strain of the overgrown n-$Al_{0.3}Ga_{0.7}N$ cladding layer. Due to the very high threshold current density of 35 kA/cm^2, the laser could only be operated under pulsed conditions with extremely short current-pulses of 10 ns and a repetition frequency of 5 kHz. In addition, the measured differential quantum efficiency was only 1.4%, which was partly attributed to the high optical losses of 70 cm^{-1} and poor internal quantum efficiency of merely 17%.[1] Although lower threshold current densities have been reported for

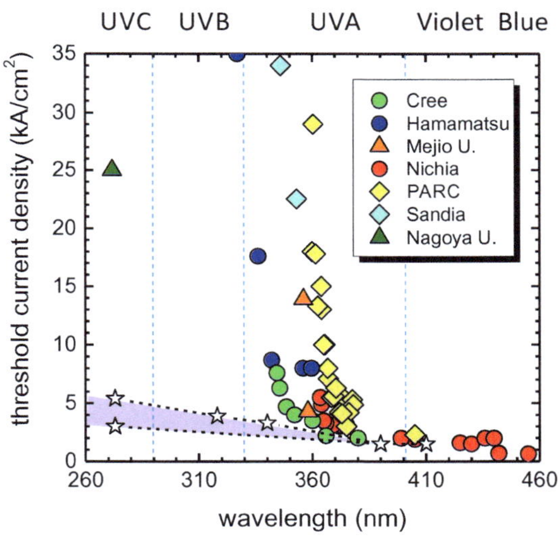

Fig. 13.2 Measured threshold current densities for edge-emitting group III-nitride-based UV laser diodes [2–15]. The grey-shaded area illustrates the simulated threshold current densities based on gain calculations for InGaN and AlGaN multiple quantum well lasers [16, 98]

[1]During proof of this chapter Z. Zhang et al. [15] demonstrated an AlGaN single quantum well laser diode on (0001) bulk AlN substrate emitting at 271.8 nm with a threshold current density of 25 kA/cm^2. The LD was operated under pulsed current conditions with a pulse width of 50 ns and repetition frequency of 2 kHz. This breakthrough was ascribed to a low defect density AlN substrate and incorporation of a polarization doped Mg-free p-AlGaN cladding layer.

longer wavelength UV laser diodes and cw operation has been demonstrated for AlGaN MQW LDs emitting at 348 nm, these lasers only lasted a few seconds under cw operation and exhibited output power levels in the mW range [8]. The performance characteristics of AlGaN MQW laser diodes are in contrast to near UV LDs with InGaN MQW active regions. For near UV lasers, the required aluminum contents in the $Al_xGa_{1-x}N$ waveguide and cladding layers are still quite moderate (typically $x < 0.15$) allowing the coherent growth of the entire laser heterostructure on low defect density GaN without generation of additional defects. This of course also limits the UV emission wavelength that can be obtained, since there is a minimum thickness and aluminum contents requirement for functional cladding or waveguide layers. CW operation of GaN and InGaN QW laser diodes with emission wavelength as short as 369 nm and threshold current densities of 3.5 kA/cm^2 have been reported [2, 3, 5] and InGaN laser diodes emitting in the 370–380 nm wavelength band with cw output power levels of 70 mW are already commercially available [4].

13.2.2 Optically Pumped Deep UV Lasers

Optically pumped AlGaN MQW lasers grown on low defect density bulk AlN substrates with very low threshold power densities of 41 kW/cm^2 for lasers emitting near 266 nm have been demonstrated [18]. More recently, threshold power densities of 6 and 11 kW/cm^2 have been demonstrated for 267 nm AlGaN MQWs lasers with AlN and AlGaN quantum barriers, respectively, by reducing the point defect density in the laser heterostructure [17]. Even shorter wavelengths have been reported for optically pumped InAlGaN and AlGaN MQW lasers [13, 14, 19, 20] on bulk AlN and ELO AlN/sapphire with emission as short as 237 nm [21–23] as also shown in Fig. 13.3. For optical pumping the deep UV lasers, typically ArF excimer lasers (λ_{exc}

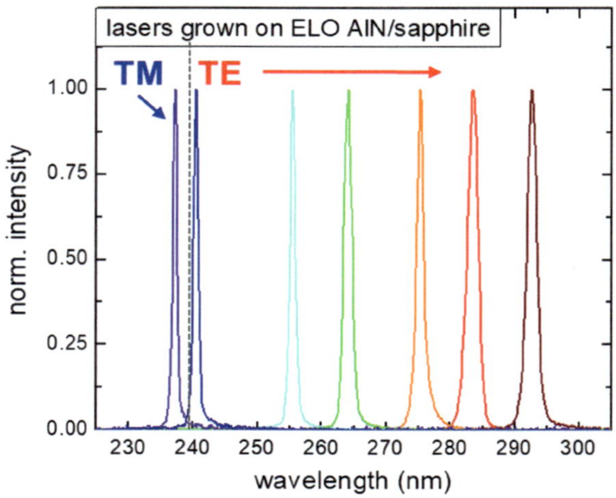

Fig. 13.3 Emission spectra of optically pumped AlGaN MQW laser on ELO AlN/sapphire. TE mode lasing was observed for all wavelength except for the 237 nm laser [25] © 2018 Martin Martens. Reprinted from M. Martens, "Optical gain and modal loss in AlGaN based deep UV lasers", dissertation, TU Berlin

= 193 nm) are employed as excitation source delivering short pulses of 5 ns with repetition frequencies of 100–500 Hz and excitation power densities of up to several tens of MW/cm^2. Since the ArF laser light is strongly absorbed in all layers of the AlGaN heterostructure it is not straightforward to translate these optically pumping results into threshold current densities for current injection laser diodes. In order to do so, one has to accurately estimate the electron and hole carrier densities in each of the AlGaN QWs, which requires detailed understanding of the carrier transport into the QWs and recombination processes in the laser heterostructure. Nevertheless, it is possible to compare these optical pumping results to results from blue and violet lasers where high performance current injection laser diodes have already been demonstrated. Of course, one has to be careful when comparing threshold power densities measured on different experimental setups. For optically pumped AlGaN-based lasers from TU Berlin and FBH, the lowest threshold power densities were around 900 kW/cm^2 for lasers emitting near 270 nm [24]. Photopumping of an InGaN MQW laser with violet emission on GaN substrate measured with the same setup required threshold power densities of about 200 kW/cm^2. On a comparable current injection laser diode emitting at 402 nm, threshold current densities of 2 kA/cm^2 were obtained. If one uses these violet LD results as reference, one would estimate threshold current densities of less than 10 kA/cm^2 for the AlGaN-based deep UV LDs. Similarly, for the AlGaN lasers demonstrated in [21], reference data for current injection in violet LDs exist. In that case, 400 nm optically pumped lasers yielded threshold power densities of 60 kW/cm^2, while the corresponding LDs showed threshold current densities of about 3 kA/cm^2. This would place the expected threshold current densities of these 267 nm AlGaN LDs [21] around 6 kA/cm^2, well within the range that should allow cw operation. These examples illustrate that AlGaN-based laser diodes can in principle reach the UVB (320–280 nm) and even the UVC (280–200 nm) wavelength range and, thus, are capable of replacing many of the existing UV sources like gas and excimer lasers as well solid state lasers.

13.2.3 Electron Beam Pumping of UV Emitters

An alternative option to pump AlGaN-based UV emitters is using a high energy electron beam. E-beam pumped group III-nitride lasers have been first demonstrated in the visible spectral range on InGaN MQW laser emitting near 400 nm [26] and more recently for UVA lasers emitting in the wavelength range from 375 to 385 nm [27]. This pumping scheme has the advantage that it reliefs the challenges of *n*-type or *p*-type doping of AlGaN alloys with high Al contents and also avoids the problem of ohmic contact formation. Since Mg-doped AlGaN layers also add optical losses, e-beam pumped laser heterostructures could also provide lower threshold current densities. Recent research efforts in the DARPA LUSTER program enabled the development of extremely compact electron beam sources with a volume of less than 15 cm^3 [27]. The excitation energies for e-beam pumped UV emitters are typically ranging from 5 to 30 keV allowing the penetration depth of the electron beam to be adjusted between a few 100 nm up to several μm [28]. Even though only about one third of the total power deposited in the semiconductor material is converted into electron-hole pairs, e-beam pumping can be still very efficient even compared to *pn*-junction devices. For example, peak output powers of 230 mW have been demonstrated for $Al_{0.56}Ga_{0.44}N$ MQW heterostructures emitting near 246 nm and pumped by a 12 keV pulsed electron beam with a beam current of 4.4 mA [28]. This corresponds to a wallplug efficiency (WPE) of 0.43%, which actually surpasses the best reported WPE of UV LEDs at that wavelength [29]. However, despite these very encouraging results, no e-beam pumped deep UV laser has been demonstrated yet.

13.2.4 AlGaN-Based Deep UV Laser Diodes

Even though optically pumped AlGaN MQW lasers emitting around 265 nm exhibit extremely low threshold power densities [17, 21], no current injection LD within the UVC wavelength band has been demonstrated so far. This can be attributed to a number of challenges, e.g., the difficulties in the realization of low-loss Mg-doped waveguides for optical confinement as well as efficient carrier injection at high current densities. Figure 13.4 illustrates a typical AlGaN MQW LD heterostructure for emission in the UVC spectral range together with some of the key challenges for the realization of current injection deep UV LDs. One of the main impediments

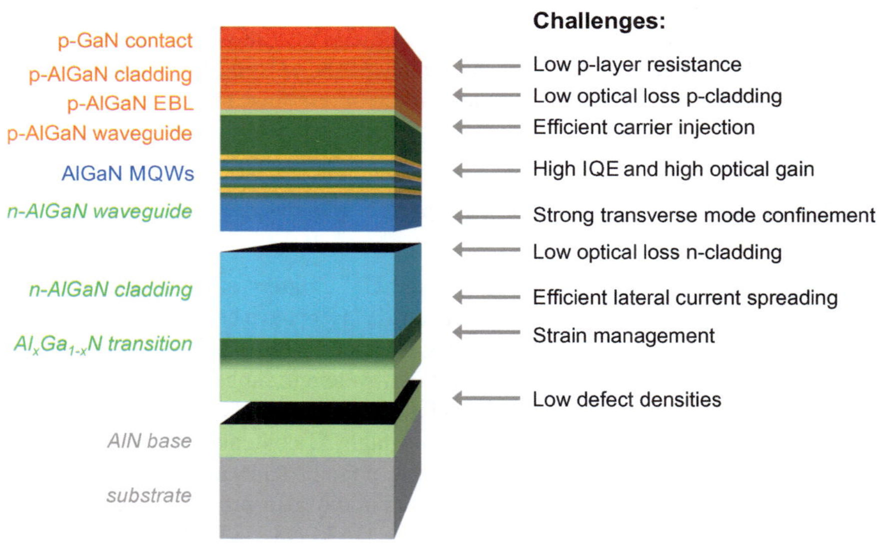

Fig. 13.4 Schematic of an AlGaN-based current-injection UV laser heterostructure together with key challenges that need to be addressed

for deep UV LDs can be attributed to the strong increase of the ionization energy of Mg acceptors in high aluminum mole fraction AlGaN alloys, resulting in highly resistive *p*-doped AlGaN layers. Several studies estimate the Mg acceptor level to be about 500–700 meV above the valence band for AlN [30, 31]. Different approaches to overcome this problem, e.g., by employing Mg-doped short period superlattice (SPSL) AlGaN layers [32, 33], polarization doping [34], Mg-doped hexagonal boron nitride [35], and tunnel junctions [36] have been explored. Several approaches are also investigated in the following section, in particular, Mg-doped AlGaN SPSL, and AlGaN tunnel heterojunctions. In addition, the optical losses from Mg-doped AlGaN layers are investigated as well as approaches to improve the carrier injection efficiency at high current densities.

13.3 Design of AlGaN-Based Deep UV Laser Diodes

The heterostructure design of deep UV LDs is crucial as efficient current injection has to be combined with a separate confinement heterostructure providing strong transverse mode confinement to maximize the overlap with the AlGaN quantum wells, i.e. the confinement factor, and to realize low optical losses. Reducing the optical losses in current injection LDs is particularly challenging due to absorption tails in the highly Mg-doped *p*-AlGaN cladding and waveguide layers as well as strong absorption in the *p*-metal contacts. While these challenges can be mitigated by an optimized laser heterostructure design, additional restrictions are imposed on the range of layer thicknesses and compositions due to the degradation of the material properties, e.g., the formation of defects caused by strain relaxation. Therefore, strain relaxation must be avoided (see Sect. 13.6.1) as conductivity (Sect. 13.6.2) and low non-radiative recombination rates (Sect. 13.6.3)—both crucial for laser operation—can only be achieved for laser heterostructures containing very low threading dislocation densities.

13.3.1 Separate Confinement Heterostructure

A targeted emission wavelength near 270 nm requires AlGaN quantum wells with an Al mole fraction of 42–47%, depending on the QW width, the barrier Al mole fraction and the strain state. In order to provide sufficient electrical confinement for the charge carriers, aluminum mole fractions of more than 55–60% in the quantum barrier are required [37, 38]. Often, even higher Al mole fractions in the AlGaN waveguide and barriers are used to prevent strain relaxation in the AlGaN waveguides as described in Sect. 13.6. The optical confinement of the lasing mode can be achieved by embedding the quantum wells and the waveguide between AlGaN cladding layers with lower refractive index, i.e., higher Al mole fraction. Examples

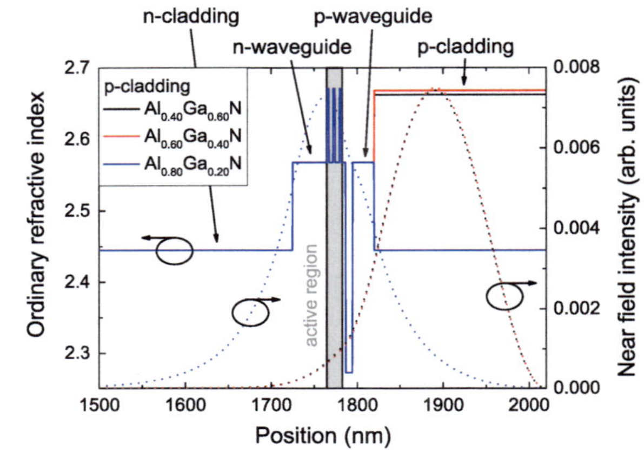

Fig. 13.5 Refractive index profile (at 270 nm) and calculated mode profile of three current injection LD designs, exhibiting a confined optical mode in the waveguide (near position 1770 nm) only in the case of an $Al_{0.8}Ga_{0.2}N$ *p*-cladding layer. Reprinted from [39], with the permission of AIP Publishing

of the refractive index and optical mode profiles for three different heterostructure designs are shown in Fig. 13.5. The structure consists of an AlN base layer, a 1000 nm thick Si-doped *n*-$Al_{0.8}Ga_{0.2}N$ bottom cladding layer, 40 nm $Al_{0.7}Ga_{0.3}N$:Si waveguide, a threefold $Al_{0.47}Ga_{0.53}N/Al_{0.7}Ga_{0.3}N$ MQW, followed by a 6 nm thick AlN electron blocking layer, 25 nm $Al_{0.7}Ga_{0.3}N$:Mg upper waveguide, and 200 nm upper Mg-doped *p*-AlGaN cladding. As can be seen from Fig. 13.5, the simulation of the laser heterostructures with $Al_{0.8}Ga_{0.2}N$, $Al_{0.6}Ga_{0.4}N$, and $Al_{0.4}Ga_{0.6}N$ upper cladding layers shows an optical confinement with a mode centered at the MQW only for the highest Al mole fraction in the *p*-cladding. For lower Al mole fractions in the *p*-AlGaN cladding, the maximum of the transverse optical modes shifts into the *p*-layers, resulting in high optical losses and low confinement factors.

For the *p*-$Al_{0.8}Ga_{0.2}N$ cladding layer a confinement factor Γ, i.e. the overlap integral of the optical mode with the quantum wells, of more than 4% can be achieved [39]. This value is actually higher than for visible wavelength InGaN-based lasers [40] due to the larger refractive index contrast between the different AlGaN layers as well as thinner waveguides (here 93 nm) that can be employed at shorter emission wavelength. Also the formation of substrate modes that is often observed in GaN-based lasers [41] can be avoided since the AlN base layers exhibit a very low refractive index. However, as the whole structure is compressively strained and doping—especially *p*-type doping—is increasingly difficult, the most important challenges are the reduction of optical losses (Sect. 13.7) while achieving high carrier injection efficiencies (Sect. 13.8) and low operation voltages (Sects. 13.4 and 13.8).

13.3.2 Design Rules for Deep UV Laser Heterostructures

In the following, we discuss a number of considerations regarding the laser heterostructure design, in particular the variation of the thicknesses and compositions of the individual layers and its effect on the transverse optical mode profile, optical loss and confinement factor.

1. The Al mole fraction of the waveguide should be as low as strain relaxation allows as a lower Al mole fraction leads to a higher Γ or enables lower Al mole fractions in the cladding layers.
2. Cladding layers should have the same Al mole fraction. A lower Al mole fraction in the *p*-side leads to an undesirable extension of the optical mode towards the Mg doped layers and the contacts (see Sect. 13.7). A lower Al mole fraction in the *n*-side cladding might not be harmful but implies that the Al mole fraction in the *p*-cladding could be reduced to achieve a higher conductivity in this layer. In general, the minimum Al mole fraction in the cladding has to be between 80 and 70%, but certainly above the waveguide Al mole fraction, which is decisively very challenging for *p*-doping.
3. A placement of the electron blocking layer (EBL) (thickness and Al mole fraction determine mainly the injection efficiency) on the upper waveguide would be most beneficial in order to minimize optical absorption in *p*-doped layers and not to disturb the optical mode. However, carrier injection efficiency into the QWs and stability of the device might require a very small distance from the QWs [42]. In this case, the EBL needs to be placed on top of the QWs.
4. The waveguide thicknesses have to be carefully adjusted to find a compromise between optimum carrier injection efficiency and transverse optical mode confinement. For example, increasing the thickness of the *n*-AlGaN waveguide reduces Γ due to the larger modal volume and the shift of the optical mode maximum away from the QWs. A thicker waveguide before as well as after the EBL can be detrimental for the injection efficiency. A slightly asymmetric optical mode with a total waveguide thickness of 93 nm as shown in Fig. 13.5 appears to be a favorable compromise with respect to optical and electrical properties. On the other hand, laser heterostructures with thicker waveguides might exhibit reduced optical losses due to a reduced mode intensity near the metal contacts with the drawback of a decreased injection efficiency and Γ. However, this has not yet been verified by experimental results.
5. The thickness of the *p*-AlGaN cladding should be as thick as possible from an optical point of view. However, thick *p*-AlGaN layers lead to a much higher operation voltage and a reduced maximum injection current (Sect. 13.8.2). A solution might be provided by tunnel junctions allowing to increase the upper cladding layer thickness without high excess voltages since the resistivity of *n*-AlGaN is substantially lower than that of *p*-AlGaN (Sect. 13.8.4).
6. There is not yet enough experimental data to determine the optimum number of QWs and their optimum thickness. The optimum number of QWs depends on the hole transport between the QWs at high current densities. If the holes are mainly injected into the first QWs as suggested by drift diffusion calculations for low current densities [43], a single QW would be optimal. If multiple quantum wells are used in spite of limited hole transport, the QWs will be pumped unevenly and transparency is reached only for the uppermost QW at lower current densities.

In case of unequal pumping, the threshold increases significantly since only one QW exhibits optical gain [44]. Also in comparison to a single quantum well (SQW), the lasing threshold is increased as the additional QWs need to be pumped optically into transparency before lasing can start [45]. However, if the electrical confinement for holes is too weak, multiple QWs should be employed to avoid carrier losses with the additional benefit of a higher confinement factor. The influences of the QW width on gain and threshold is discussed in Sect. 13.7.

13.3.3 Investigated Deep UV Laser Structures

These design rules were applied to develop the growth, fabrication and analysis of UV laser diodes (Sects. 13.3–13.7). Two basic structures were designed:

1. A laser diode heterostructure for current injection investigations (Fig. 13.6, left)
2. A laser heterostructure for optical pumping (Fig. 13.6, right)

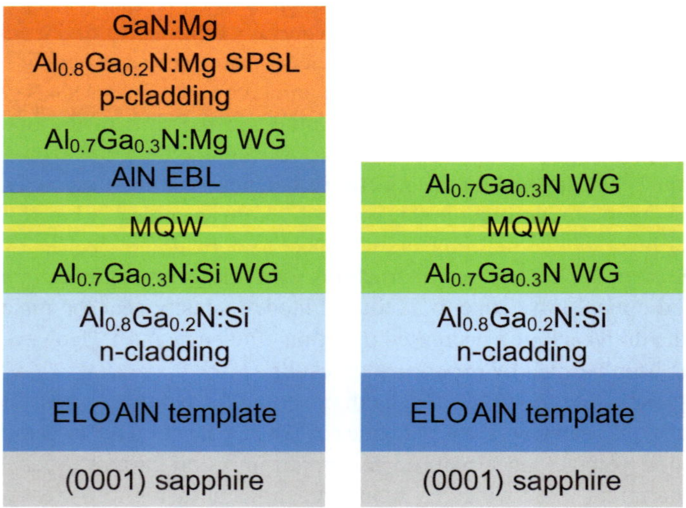

Fig. 13.6 (left) For current injection investigations, the base structure (as described above) consists of an AlN base layer, a 1000 nm thick $Al_{0.8}Ga_{0.2}N$:Si lower *n*-cladding, 40 nm $Al_{0.7}Ga_{0.3}N$:Si lower waveguide, threefold $Al_{0.47}Ga_{0.53}N/Al_{0.7}Ga_{0.3}N$ MQW, 6 nm AlN electron blocking layer, a 25 nm $Al_{0.7}Ga_{0.3}N$:Mg upper waveguide and a 200 nm $Al_{0.8}Ga_{0.2}N$:Mg upper *p*-cladding. (right) As optical pumping experiments were conducted extensively, a structure was defined which allows for pumping at an excitation wavelength of 193 nm (by an ArF* excimer laser), i.e. has a layer thickness <100 nm above the QWs, no blocking layers and has a very similar mode profile as the laser diode heterostructure. To reduce the thickness, the EBL and *p*-cladding are omitted. The mode profile is adjusted by increasing the upper waveguide to a thickness of 85 nm as for thinner upper waveguides the strong refractive index contrast to air pushes the mode towards the *n*-side

13.4 Fabrication of AlGaN-Based UV Laser Diodes

13.4.1 Low Resistance Ohmic Contacts to n-*AlGaN Layer*

Minimizing the contact resistivities on both *p*- and *n*-layers in deep UV LEDs and lasers is essential for improving their WPE due to the high operating voltages and resistive Joule heating. Contact formation on *n*-type III-nitrides is well established [46]. The basic mechanism responsible for ohmic behavior involves the extraction of nitrogen atoms by the contact metal upon thermal annealing and the formation of a heavily *n*-doped thin layer in the proximity of the interface [46, 47]. In fact, N vacancies act as donors in GaN thus enhancing the probability of tunneling for carriers through the thinner Schottky barrier [48]. Due to the need for a transparent current spreading layer, UV light emitters utilize *n*-$Al_xGa_{1-x}N$. Unfortunately, the widely employed Ti-based electrodes form ohmic contacts with *n*-AlGaN only up to an AlN molar fraction of $x = 0.58$ [49]. Our approach consisted in switching the metallization scheme to a V-based electrode, i.e. V/Al/Ni/Au [50]. The generation of a heavily *n*-doped thin layer at the AlGaN interface is then driven by the Al in the metal scheme while the V acts as an indiffusion barrier for Al. Nevertheless, N extraction by Al will become less energetically favorable with increasing Al mole fraction as the AlN bond in the *n*-AlGaN layer is more energetically stable than the GaN one (lower formation enthalpy) [46]. In fact, we were able to obtain ohmic V-based *n*-contacts on plasma etched *n*-AlGaN up to an AlN mole fraction of $x = 0.75$.

In Fig. 13.7 (left) current–voltage (*IV*) characteristics of *n*-contacts on plasma etched *n*-$Al_xGa_{1-x}N$ in the entire alloy composition range are depicted for an electrode distance of 8 μm. By optimizing the thicknesses of the individual layers in a four-metal electrode V/Al/Ni/Au configuration as well as the rapid thermal annealing procedure, we were able to significantly reduce the contact resistivities as well as the Schottky behavior of our *n*-contacts. Linear (ohmic) *IV* curves have been achieved up to an Al mole fraction of $x = 0.75$, above that the rectifying behavior prevails and the *IV*s are more and more non-linear. The highest Al content for which an ohmic contact on *n*-AlGaN has been reported was $x = 0.7$ [51]. In Fig. 13.7 (right), the voltage at 0.1 kA/cm^2, contact and sheet resistivity are shown as a function of the Al mole fraction for *n*-contacts on *n*-$Al_xGa_{1-x}N$ in the entire alloy composition range. These values were measured using the linear transfer length method (TLM) at a current density of 0.1 kA/cm^2. When increasing the Al content in *n*-AlGaN, the contact resistivity increases regularly whereas the voltage at 0.1 kA/cm^2 exhibits a clear kink around $x = 0.75$. Note that a similar behavior has been observed also with the sheet resistivity [52]. It becomes more and difficult above an Al mole fraction of $x = 0.75$ to achieve *n*-layers and *n*-contacts with high conductivity. Our assumption is that this is due to a strong reduction in the amount of activated donors. On the one hand side, the activation energy of donors in the *n*-layers falls deeper into the band gap due to formation of DX centers [52]. On the other hand side, N vacancy formation is hindered as the AlN bond in *n*-AlGaN becomes more and more energetically

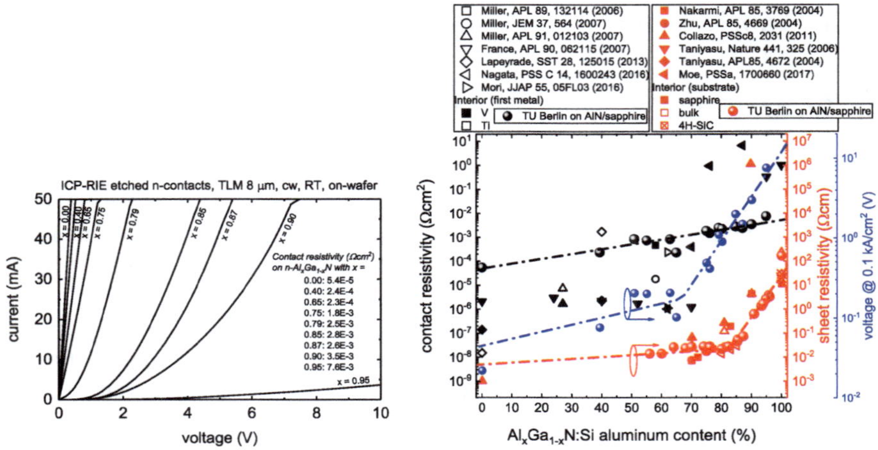

Fig. 13.7 (left) Current–voltage characteristics of n-contacts on plasma etched n-Al$_x$Ga$_{1-x}$N in the entire alloy composition range at an electrode distance of 8 μm. (right) Voltage at 0.1 kA/cm^2, contact and sheet resistivity as a function of the Al mole fraction for n-contacts on n-Al$_x$Ga$_{1-x}$N in the entire alloy composition range. The black scattered points depict the resistivities of V-based (full symbols) and Ti-based (hollow symbols) contacts reported in the literature. The red scattered points depict the sheet resistivities reported in the literature. The full spheres represent our experimental values. L. Sulmoni, TU Berlin, unpublished results

stable with decreasing Ga content which inhibits the formation of a heavily n-doped layer at the interface upon electrode annealing [46]. Nevertheless, albeit still showing Schottky behavior, our n-contacts on AlGaN with Al mole fraction higher than x = 0.75 show contact resistivities orders of magnitude lower than the ones reported in the literature up to now.

Finally, we demonstrated efficient carrier injection in cw operation into deep UV LEDs emitting at 228 nm by implementing the improved V-based n-contacts (Fig. 13.8, left) [53]. At 20 mA, the operating voltage of the LEDs was as low as 10.1 V and the output power was up to 20 μW. The EQE reached a maximum of 0.02% at 15 mA. Thanks to the reduced Schottky barrier seen by the carriers on the n-side due to the optimized n-contact, it was possible to reach a maximum WPE of 0.01% and to push the temperature roll-over to higher currents (i.e. 50 mA). In order to further improve the current uniformity in the emitting area due to the higher resistivity of n-AlGaN layers and of n-contacts in comparison to n-GaN, we implemented an interdigitated n-electrode layout.

The fabrication of low-resistance ohmic n-electrodes to n-AlGaN is not only possible due to a careful adjustment of the metal stack and the related thermal annealing conditions, but also because of optimum dry etching conditions. UV LEDs are typically grown on non-conductive sapphire substrates. This means that plasma etching techniques must be employed in order to expose and subsequently contact the buried n-AlGaN current spreading layer [54]. Plasma-induced damage often results in an etched surface of poor crystalline quality mainly due to problems in maintaining a

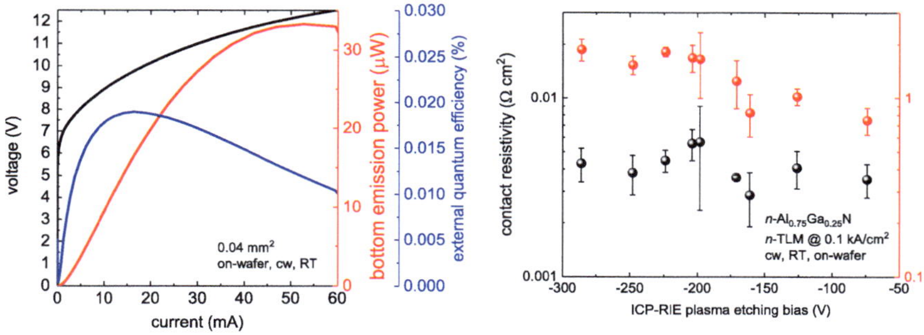

Fig. 13.8 (left) LIV and EQE curves for a 0.04 mm^2 LED emitting at 228 nm under cw operation and at room temperature. (right) Contact resistivity and voltage at 0.1 kA/cm^2 as a function of the applied etching bias in ICP-RIE for *n*-contacts on *n*-AlGaN with an Al content of x = 0.75. L. Sulmoni, TU Berlin, unpublished results

stoichiometric surface because of the different volatilities of the group III and group V etch products thus undermining significantly the optical and electrical properties of *n*-AlGaN layers [55, 56]. Cao et al. reported that when increasing the Al mole fraction beyond x = 0.3 the plasma-induced defects act as deep-level centers, pinning the Fermi level and increasing the Schottky barrier [57]. Moreover, plasma etching can generate dangling bonds at the interface making the semiconductor prone to oxidize [58].

In order to minimize the plasma damages and to assist the *n*-contact formation, we carefully investigated and adjusted the ICP-RIE (inductively coupled plasma reactive ion etching) parameters such as the applied etching bias, the gas species, the ICP plasma power, the HF plasma power and the chamber pressure. According to our findings, the applied etching bias seems to have the largest influence on the *n*-contact conductivity. Figure 13.8 (right) shows the contact resistivity and voltage at 0.1 kA/cm^2 as a function of the applied etching bias for *n*-contacts on *n*-$Al_{0.75}Ga_{0.25}N$. Both the resistivity and voltage clearly improve when decreasing the etching bias. This means that when the chemical component of the etching reaction dominates over the physical component it was possible to minimize the impact of plasma induced damage. In fact, when we increased the physical part in the recipe, e.g., by introducing Ar in the gas mixture (a typical ion species in physical etching systems), we observed an increase in the resistivity (not shown here). We then developed a special plasma etching recipe consisting in 2 steps: a first "fast" step that provides a high etching rate together with steep mesa sidewalls and a second "soft-landing" step that yields low damage for the *n*-contact formation.

As described by Cao et al. [57], it is expected that the formation of plasma-induced compensating defects will become more and more severe when increasing the Al mole fraction in the *n*-AlGaN layers thus making the realization of low resistance *n*-contacts on very high Al mole fraction extremely challenging. Conductive AlN bulk substrates would allow for the usage of *n*-contacts on the backside as used for GaN-based LDs, eliminating the problems in formation of contacts on *n*-AlGaN surfaces prepared by plasma etching. However, the problems in n-doping are most severe for pure AlN and thus it is highly unlikely that the standard approach for laser diodes on n-type substrates can be applied to AlN-based devices in the near future.

13.5 Low Defect Density AlN Templates

13.5.1 Substrates and Templates for AlGaN UV Lasers

Semiconductor lasers have a number of requirements on the substrate or base layer they are grown on. The most prominent is a low density of extended defects that needs to be achieved in the active region of the laser as such extended defects are centers for non-radiative recombination. In case of dominating non-radiative recombination the threshold carrier density for stimulated emission cannot be reached and lasing is prevented. Even if lasing can be achieved, non-radiative recombination leads to strong local heating which can prevent continuous wave lasing and/or lead to rapid device degradation. So the aim has to be to start with a low density of defects, specifically a low threading dislocation density (TDD), in the substrate or base layer and keep this low TDD by pseudomorphic growth of the active device heterostructure. This implies that the crystal structure of the base layer should be the same as that of the active layer structure. Ternary substrates are difficult to make and no viable routes to AlGaN substrates have been reported so far. Also nucleation of ternary layers in the initial phase of heteroepitaxy is challenging, especially if atoms with strongly different bond strength and surface diffusion lengths like Al and Ga are involved. So for UV lasers based on AlGaN the base layer needs to be AlN.

For edge emitting lasers, which are in the focus or our work, facets forming the laser cavity need to be cleaved. This is facilitated if substrate and laser heterostructure feature a common natural cleavage plane. For lasers it usually is desired to have homogeneous spectral gain properties over the cavity length, since otherwise only parts of the cavity can support lasing at a particular wavelength. This implies that the active quantum wells should be homogeneous in thickness and composition. This can only be achieved if step flow growth can be maintained at least up to those QWs. Macro-steps due to the strongly different surface mobility of Ga and Al can lead to accumulation of Ga at the step edges and thus inhomogeneous active regions which may even be more like quantum wires than quantum wells [59]. One additional requirement is that for successful development of epitaxial growth processes for devices the substrates and base layers need to be reproducible with respect to the key properties mentioned above, be available in sufficient quantity and preferably also in diameters that allow processing with standard semiconductor processing equipment. Since UV lasers are still in the research stage, two inch diameter is sufficient.

The usual growth direction is the metal-polar +c direction. While other directions may offer advantages like reduction of the quantum confined Stark effect on semi-

and non-polar planes [60] or different ratio of emission with TE and TM polarization [61] growth on such surfaces is difficult. So we here concentrate on $+c$-plane AlN.

For AlGaN-based UV lasers the choices are limited. AlN single crystals offer the lowest TDD in AlN-based epitaxial structures. However, such substrates are still a research topic themselves and not readily available on the market. UVA lasers have also been realized on GaN substrate by one group [13]. However, the AlGaN heterostructure here is under heavy tensile strain resulting in the risk of cracking. Attempts to move to shorter wavelengths than 335 nm apparently have not been successful up to now. Silicon carbide (SiC) substrates offer a decent lattice matching to AlN and are frequently employed for GaN HFETs. The epitaxial layer stack here starts with an AlN layer, which usually is thin. This application has pushed the development of SiC crystal growth and substrates up to 6" are available. However, the number of SiC manufacturers still is limited. SiC substrates have played role in the early days of GaN blue LEDs and have also been used for AlGaN UV lasers [19, 62, 63]. The substrate of choice for UV LEDs is sapphire. Some requirements on the substrate for lasers and LEDs are different. So for UV LEDs which emit through the substrate, transparency is essential, which is not the case for edge emitting lasers. On the other hand, such lasers need perfect facets and thus cannot be sawn or scribed and broken but need to be cleaved. Nevertheless, the tasks in development of growth processes for the required heterostructures are similar. So most of our work on UV lasers also uses sapphire substrates. The task of obtaining smooth facets has been mastered [64]. To obtain the low TDD necessary for lasing at low threshold growth over patterned substrates is explored.

13.5.2 Bulk AlN Substrates

AlN substrates are grown by sublimation of AlN onto a seed. Details of this process called PVT (physical vapor transport) can be found in a recent review [65]. To obtain a single crystal of high quality requires a single crystal as seed which constitutes a "chicken and egg" problem. Spontaneous nucleation yields crystals of a few mm in diameter at most. Such small crystals can be used as seed for further crystals of some mm in length. Depending on seed orientation and growth conditions the crystals grow with gradually increasing diameter so that a seed with slightly higher diameter can be cut from the end of the grown boule. However, this incremental increase in diameter is slow and it has taken more than 15 years until 2 inch substrates have entered the market [66]. However, such substrates still are not on the open market since they are mostly used in-house. UV lasers on AlN substrate have been reported by a number of groups, for example in [18, 20, 21]. In our own work a limited number of substrates from commercial vendors as well as from the Institute for Crystal Growth (IKZ) in Berlin, a cooperation partner in projects for the development of UVC LEDs could be used [67]. While this did not allow for a systematic development, it highlighted the importance of some substrate properties beyond a low TDD and sufficient diameter.

The first substrate property is related to the surface preparation. Sawing of individual wafers from AlN boules leaves subsurface damage that needs to be removed by chemical-mechanical polishing in combination with chemical etching. Here it is not sufficient to obtain a mirror-like surface since this may contain scratches filled with polycrystalline material which prevents the growth of a single crystalline layer [68]. AlN surfaces form a hydroxide layer at the surface upon contact to humid air. While this layer can be removed in the MOVPE reactor by annealing under NH_3 at high temperature (1250 °C) the reproducibility of this process needs to be verified. Also residues of the surface preparation have to be avoided. Polishing with colloidal SiO_2 for example can leave Si on the surface that is converted to SiN by annealing in NH_3 [69]. SiN is well-known to form an in-situ mask in GaN growth [70] and has a similar effect in AlN [71]. It would be desirable to avoid very high temperatures and have substrates available with a surface that can be used for epitaxy without any special measures. Currently our approach is to subject the substrates to reactive ion etching in an atmosphere containing chlorine and store them under N_2 only over short periods of time in order to ensure reproducible surface properties.

A second issue is related to the crystalline structure of the surface. It has turned out that an only small miscut of the surface (<0.2°) is required in order to maintain step flow and avoid step bunching [72, 73]. This implies that already during crystal growth the lattice planes should not be bowed for example due to a non-optimum lateral temperature profile at the growth front. Only if the lattice planes are not bowed a specified miscut towards a specific crystal direction can be obtained homogeneously over the wafer during the substrate polishing procedure.

13.5.3 SiC Substrates

SiC has already played an important role as substrate in the development of blue LEDs [74]. However, as substrate for LEDs it was too expensive. For UV lasers its relatively good lattice matching and the hexagonal lattice structure are advantageous and edge emitting lasers on SiC have been realized by some groups [19, 62, 63]. Usually the 6H polytype of SiC has been employed. However, SiC has not played a role in our own work on UV emitters and is rarely used worldwide for this application. So it is not discussed further.

13.5.4 Sapphire Substrates

Sapphire is the by far dominant substrate for GaN-based blue-green and also AlN-based UV LEDs. It is available from many vendors in large diameters up to 12 in. (although work in UV lasers still is basic research and 2 in. wafers are sufficient). Nearly all groups working on UV lasers have also activities in UV LEDs and developed their growth procedures for AlN on sapphire. Usually a sequence of growth steps is employed. First AlN is nucleated on the sapphire substrate. However, sapphire is already reacting with the growth atmosphere since it is etched by hydrogen and the surface can partially be converted into AlN by reaction with NH_3. Here the switching sequence of the precursors as well as the nucleation temperature is important to obtain a low TDD in subsequently grown thick layers [75]. Due to the low mobility of Al first small islands nucleate. Depending on the surface structure, these islands can have Al or N polarity. These islands then are coalesced in a second growth step. This also involves conversion of the N-polar grains into Al-polar ones. This is facilitated by formation of AlON [76]. The oxygen can come from partial decomposition of sapphire or from contamination of precursors or the reactor chamber. These sources are difficult to control and thus the intentional addition of O in the initial growth phase has also been investigated [77]. Since the nuclei do not all have identical orientation, dislocations with a density of more than 10^{10} cm^{-2} are formed in the nucleation layer. After islands coalescence AlN is grown preferably at a higher rate to obtain a thick buffer layer. With increasing thickness, threading dislocations can mutually annihilate reducing the TDD. High growth temperatures have been found to be advantageous to obtain AlN layers with low TDD and free of pits [78]. Figure 13.9 shows this reduction in TDD with thickness. However, this process is accompanied by the introduction of tensile strain and finally cracking of the layers limits the thickness that can be obtained and thus also the possible reduction of the TDD.

Others have reported higher thicknesses for planar layers by alternating growth of thin rough layers (for example by pulsed growth [79]) and subsequent smoothing

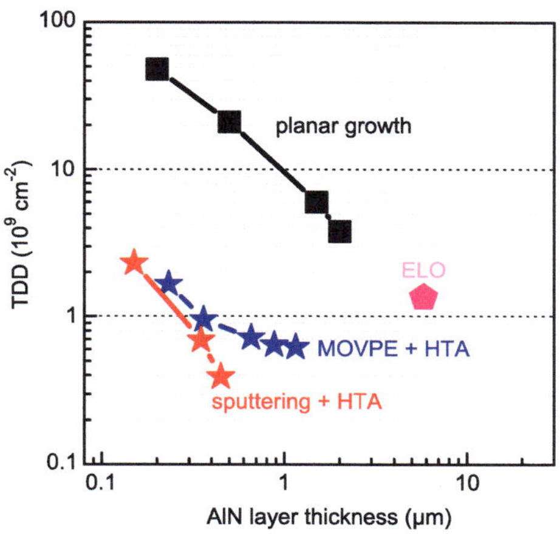

Fig. 13.9 TDD in AlN on sapphire versus layer thickness. For planar growth tensile strain limits the thickness that can be achieved. By patterned growth this thickness can be further extended and TDD can be reduced to 1×10^9 cm^{-2}. Annealing of AlN layers at high temperatures (>1600 °C) is a relatively new approach that can reduce TDD to mid 10^8 cm^{-2}. S. Hagedorn et al., Ferdinand-Braun-Institut, unpublished results

of this rough surface. However, especially smoothing of rough surfaces appears to be very sensitive to growth parameters and growth environments.

Our current route to low defect density is epitaxial lateral overgrowth (ELO) over stripe patterned AlN/sapphire templates [80]. This results in a faster initial reduction of the TDD by bending dislocations towards the sidewalls of the growing ridges. Additionally, it enhances the crack-free thickness that can be grown. In Fig. 13.9 a 7 μm thick layer with a TDD of 1×10^9 cm^{-2} is included as one example. Ridges along the [1−100] orientation of AlN are necessary to obtain a fully coalesced *c*-plane AlN surface [81]. Also the miscut of the substrate is important. With miscut to the *m*-direction, the steps on the ridges flow towards the edges of the ridges facilitating coalescence. If the miscut is towards the *a*-direction, the steps move along the ridges leading to a jagged coalescence front, which is prone to formation of pits [82]. Also the miscut angle is important. For 0.25° offcut to the *m*-direction on the one hand a lower TDD can be achieved by capturing TDs in the inclined coalescence front resulting in areas with low TDD and narrow stripes with higher TDD (Fig. 13.10). However, for such high miscut values macro-steps form at the coalescence boundary. Such macro-steps enhance the incorporation probability for Ga resulting in composition modulation in AlGaN layers [83] (Fig. 13.11) and also inhomogeneous QWs [84]. A lower miscut of 0.1° is slightly less effective in reducing the TDD but avoids macro-steps [72, 85].

This ELO approach was the basis for most of the results presented in the following sections. However, it requires some technological effort and alternative approaches are being sought. One approach that appears to be very promising is to improve the

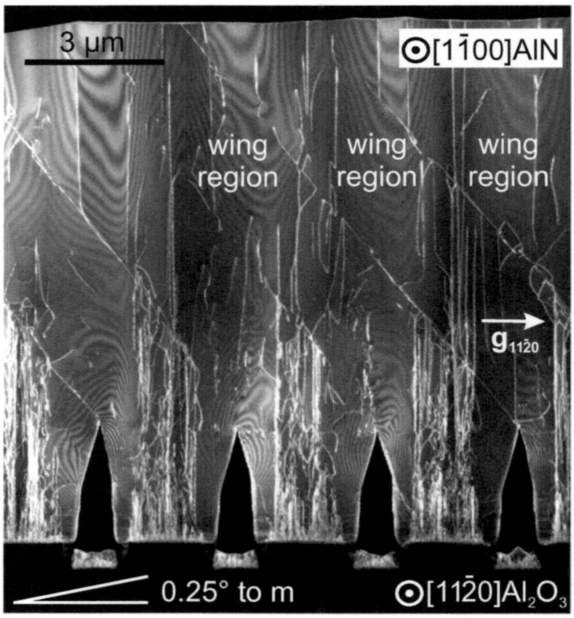

Fig. 13.10 Cross-sectional dark-field image demonstrating TD behavior in ELO-AlN (miscut of the sapphire substrate to its *m*-direction) as well as the presence of inclined grain boundaries starting at the coalescence front. Reprinted from [85], Copyright Elsevier (2014)

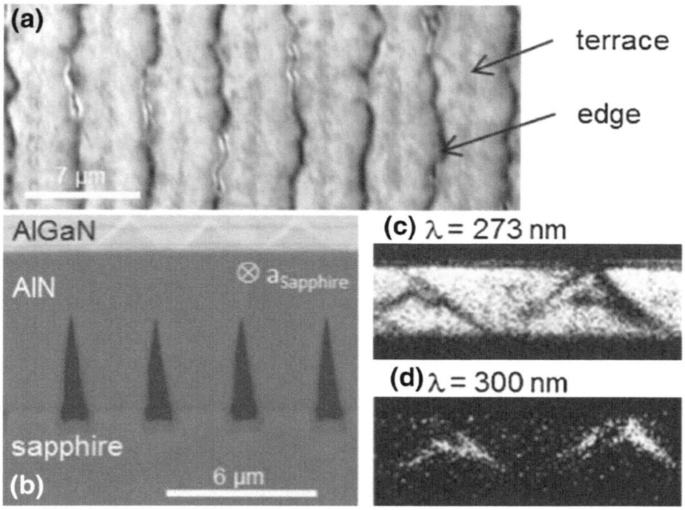

Fig. 13.11 $Al_{0.4}Ga_{0.6}N$ layer on ELO template: plan view panchromatic cathodoluminescence image (**a**), cross-section backscattered electron microscopy image (**b**) and monochromatic cathodoluminescence images from cleaved edge for 273 nm (**c**) and 300 nm (**d**) emission wavelength at 100 K sample temperature. Reprinted from [83], Copyright (2013) The Japan Society of Applied Physics

crystalline quality (i.e. reduce the TDD) by post-growth annealing at high temperatures [86]. Annealing at temperatures above 1600 °C substantially reduces the TDD for example by one order of magnitude for 700 nm thick layers grown by MOVPE from initially 10^{10} cm^{-2} to below 10^9 cm^{-2} (Fig. 13.9). It also works very well for sputtered layers that initially have a columnar structure with a high density of dislocations or rather grain boundaries with *c*-orientation. This approach has successfully been verified for UVC LEDs [87] and also appears very promising for the future work on UV lasers.

13.6 Growth of AlGaN Laser Heterostructures

13.6.1 Pseudomorphic Growth of AlGaN and Critical Layer Thickness

In order to reduce the non-radiative recombination in the active region and to obtain high internal quantum efficiencies, AlGaN laser heterostructures with low dislocation density are required. Given a foreign substrate with a certain threading dislocation density (e.g. ELO AlN/sapphire with TDD $= 1 - 2 \times 10^9$ cm^{-2}), pseudomorphic growth of AlGaN heterostructures provides the highest structural quality without generation of additional dislocations due to strain relaxation [88]. To achieve pseudomorphic growth, the critical layer thickness has to be considered, which is not trivial in laser heterostructures containing AlGaN cladding and waveguide, and barriers and quantum wells with different Al contents. Therefore, we investigated the influence of the $Al_xGa_{1-x}N$ composition of waveguide and barriers on the crystalline properties of an MQW stack with $Al_{0.4}Ga_{0.6}N$ QWs (emission near 280 nm) as depicted in Fig. 13.12. The strain state analysis by X-ray diffraction (Fig. 13.13)

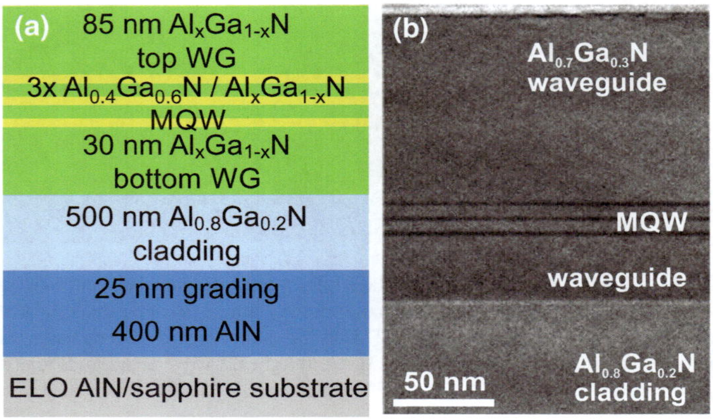

Fig. 13.12 **a** Schematic of AlGaN laser heterostructure and **b** TEM micrograph of UVC lasers composed of MQW active region, AlGaN waveguide and cladding layer on ELO AlN/sapphire substrates. Reprinted from [90], with the permission of AIP Publishing

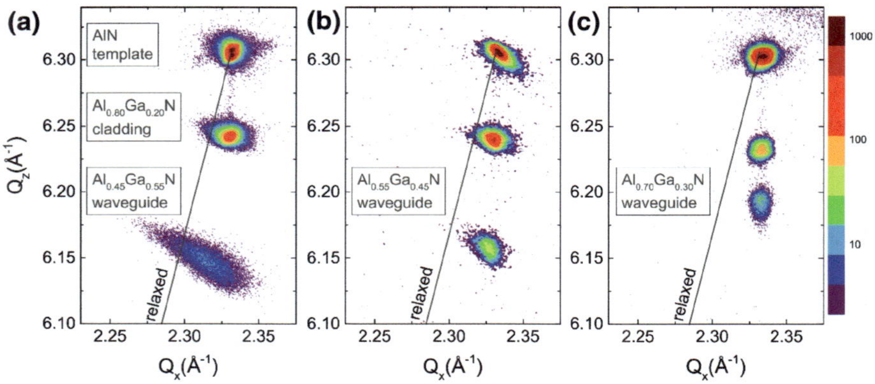

Fig. 13.13 XRD reciprocal space maps of the (10.5) reflection of MQW lasers with different $Al_xGa_{1-x}N$ waveguide composition from $x = 0.45$ (**a**), $x = 0.55$ (**b**) to $x = 0.70$ (**c**) and resulting different strain state. Partial plastic relaxation is observed for waveguides with $x = 0.45$ and $x = 0.55$, pseudomorphic growth for $x = 0.70$. Reprinted from [90], with the permission of AIP Publishing

demonstrates partial plastic relaxation of an $Al_{0.45}Ga_{0.55}N$ waveguide by formation of additional threading dislocations or inclination of existing threading dislocations [85, 89]. Also an $Al_{0.55}Ga_{0.45}N$ waveguide already is slightly relaxed. Only the $Al_{0.7}Ga_{0.3}N$ waveguide is pseudomorphic, enabling low defect density in the MQW and thus low non-radiative recombination rates [90].

13.6.2 Si- and Mg-Doping of AlGaN Materials and Superlattices

Silicon is typically used as dopant in AlGaN materials, however, the AlGaN:Si resistivity strongly depends on the composition. In the bottom AlGaN:Si cladding layer, high Al contents are required to provide sufficient optical confinement and to avoid relaxation. To provide also sufficient current spreading, the influence of the silicon doping level (adjusted by the SiH_4/III gas phase ratio) on the AlGaN:Si layer resistivity was investigated. As Fig. 13.14 indicates, a minimum resistivity can be found for a certain SiH_4/III ratio at a given Al content (0.026 Ωcm for $Al_{0.82}Ga_{0.18}N$:Si). Additionally, for increasing Al contents this minimum shifts towards lower SiH_4/III ratios. Decreasing charge carrier densities due to increasing donor activation energies, DX center formation and impurity incorporation are the main underlying effects [52, 91]. Hence, by careful tuning of the Si doping level, a decent conductivity of AlGaN:Si cladding layers with high Al content can be achieved together with sufficient optical confinement.

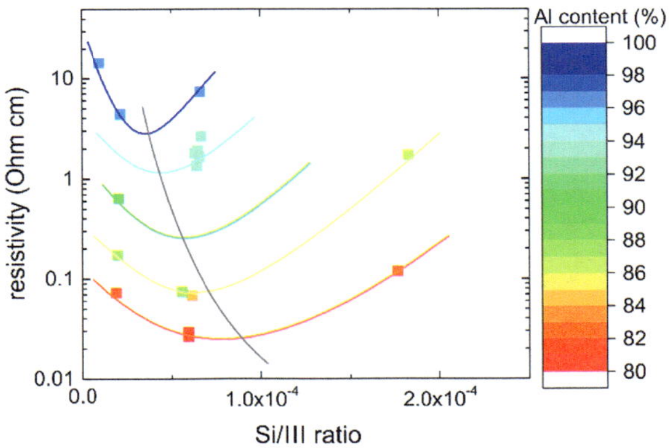

Fig. 13.14 Resistivity of $Al_xGa_{1-x}N$:Si layers with $x > 0.8$ in dependence of Si-concentration and Al-content. Generally, the resistivity strongly increases with increasing Al-content. A minimum resistivity is achieved by an optimized Si-concentration corresponding to 4×10^{18} cm^{-3} for $Al_{0.82}Ga_{0.18}N$:Si. The optimized Si-concentration decreases with increasing Al-content. Reprinted from [91], with the permission of AIP Publishing

In order to achieve low optical losses within the laser resonator, the n-cladding layer should have homogeneous composition without relaxation and provide a smooth surface morphology. As $Al_xGa_{1-x}N$:Si layers with $x \leq 0.8$ can exhibit partial relaxation and growth spirals on the surface (Figs. 13.15a and 13.16a), we utilized an AlGaN:Si-based superlattice (SL) concept with growth interruptions of 10 s after each single superlattice layer [24], enabling 1 μm thick n-cladding layers with homogeneous alloy composition without relaxation as well as a much smoother surface than a bulk cladding AlGaN layer (Figs. 13.15b and 13.16b).

In order to achieve vertical mode confinement, a refractive index contrast to the $Al_{0.7}Ga_{0.3}N$ waveguide requires a top AlGaN:Mg cladding layer with very high Al

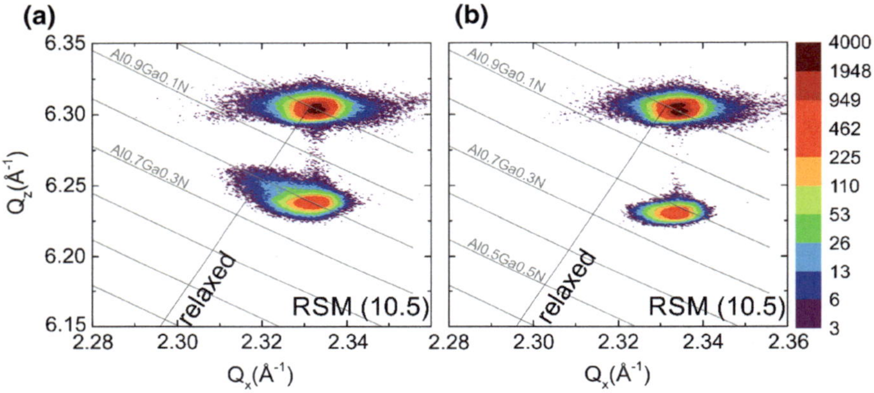

Fig. 13.15 XRD reciprocal space maps of the (10.5) reflection of low resistance (0.025 Ω cm) AlGaN cladding layers, **a** a bulk AlGaN layer and **b** an AlGaN/AlGaN superlattice. The bulk layer exhibits a broadened AlGaN reflection, whereas the SL exhibits a sharp single peak indicating more homogeneous AlGaN. Reprinted from [24], with the permission of John Wiley and Sons

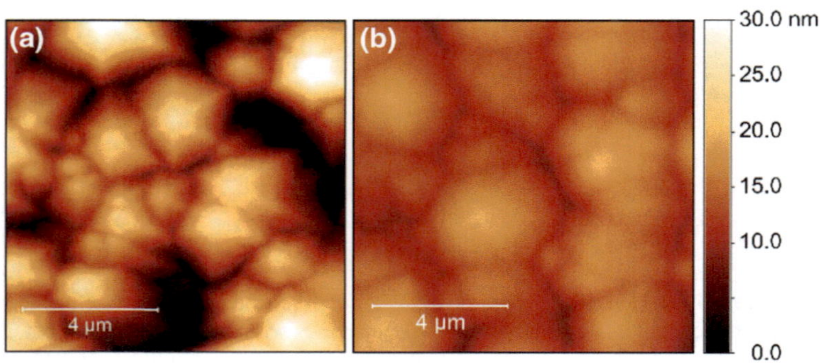

Fig. 13.16 AFM surface topograms of a bulk AlGaN layer (**a**) and an AlGaN/AlGaN SL with growth interruptions of 10 s (**b**), both with 1 μm thickness. With comparable average composition of $x = 0.8$, the SL exhibits reduced spiral height and RMS roughness and thus smoother surface. Reprinted from [24], with the permission of John Wiley and Sons

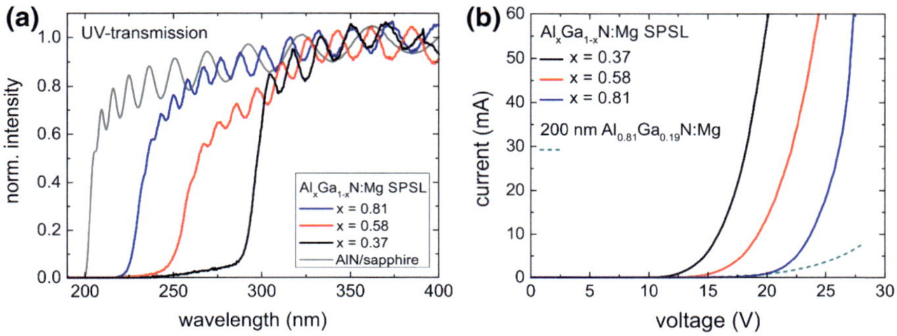

Fig. 13.17 UV transmission spectra of $Al_xGa_{1-x}N$:Mg superlattices with varying Al-content (**a**) and electrical characteristics of LEDs emitting at 270 nm utilizing such superlattices (**b**). Reprinted from [39], with the permission of AIP Publishing

contents exceeding 75%. Figure 13.17a shows normalized UV transmission data of Mg-doped AlGaN short period superlattices (SPSL). With increasing Al mole fraction the band edge shifts to shorter wavelengths, providing quasi-transparency for 270 nm laser emission above 58%. However, full transparency without Mg-related sub-band gap absorption and sufficient refractive index contrast are provided only for $Al_{0.81}Ga_{0.19}N$:Mg. The conductivity of those SPSLs is demonstrated by *IV* characteristics of LED structures (Fig. 13.17b), exhibiting a diode characteristic with increasing operation voltages for increasing average Al contents. Even the LED with $Al_{0.81}Ga_{0.19}N$:Mg superlattice shows 270 nm luminescence at a reasonable voltage. However, bulk $Al_{0.81}Ga_{0.19}N$:Mg layers exhibit much higher resistance and less stability. Therefore, $Al_{0.81}Ga_{0.19}N$:Mg superlattices are the most promising *p*-cladding layers for UVC laser diodes, enabling sufficient vertical mode confinement options with full transparency and reasonable conductivity [39].

The impact of Mg doping level on the properties of $Al_{0.8}Ga_{0.2}N$:Mg superlattices was investigated in a series by varying the Mg/III gas phase ratio between 0% and 5% (Fig. 13.18). Unintentionally-doped samples (0%) exhibit steep absorption edges and smooth surfaces, whereas heavily-doped samples ($\geq$4%) exhibit strong sub-band gap absorption and small surface pits. For Mg concentrations up to 9×10^{19} cm^{-3} (2%) no optical or structural degradation is observed indicating that such doping levels are suited for UVC laser diodes.

13.6.3 *Growth and Optical Properties of AlGaN Quantum Wells*

Three-fold AlGaN/AlGaN MQWs for emission in the UVC spectral range are grown by MOVPE and analyzed by temperature-dependent PL measurements in order

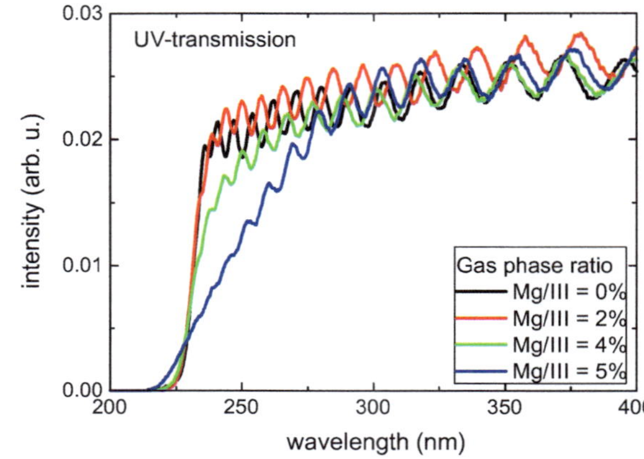

Fig. 13.18 UV transmission spectra of Mg-doped $Al_xGa_{1-x}N$ SPSL with average composition of $x = 0.8$. High Mg-doping results in crystal quality deterioration and increased sub-band gap absorption. For Mg-concentrations up to 9×10^{19} cm^{-3} (2%) no sub-band gap absorption is observed

to determine the spectral emission properties and to estimate the internal quantum efficiency. Room temperature emission spectra are dominated by the AlGaN MQW emission near 270 nm and exhibit minor contributions from the barriers and waveguide layers near 240 nm (Fig. 13.19a). At low temperature (5 K) the total PL intensities are strongly increased due to a reduced non-radiative recombination rate. Furthermore, the emission spectra exhibit comparable intensities of MQWs and waveguide/barriers, which indicates a higher recombination rate within the waveguide/barriers due to a suppression of carrier diffusion into the quantum wells at low temperatures. The temperature-dependent emission of MQWs grown on various types of substrates with different dislocation density is shown in Fig. 13.19b. Here, filled symbols represent the spectrally integrated AlGaN MQW emission in the 4.0–4.7 eV energy range, hollow symbols represent the spectrally integrated MQW and waveguide/barrier emission in the 4.0–5.5 eV energy range. The influence of the different kinds of substrates on the internal quantum efficiency can be estimated

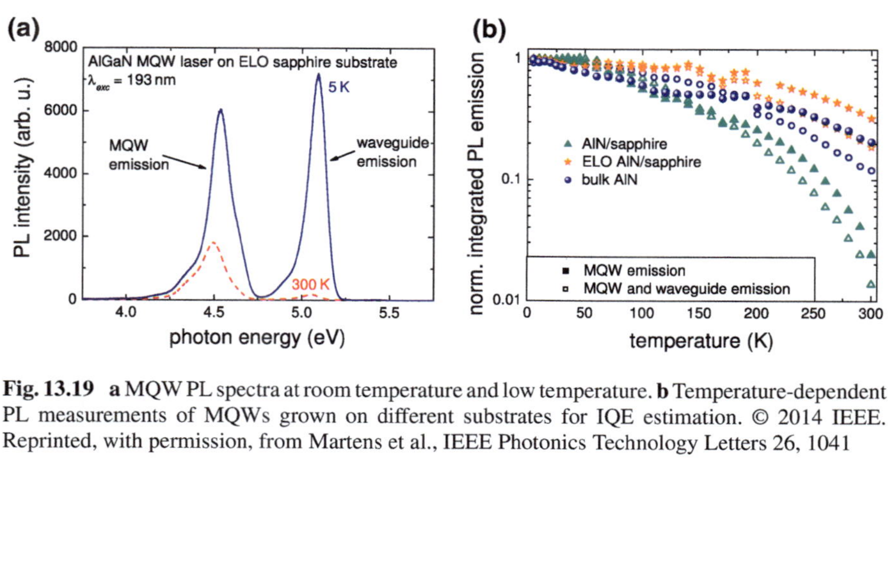

Fig. 13.19 **a** MQW PL spectra at room temperature and low temperature. **b** Temperature-dependent PL measurements of MQWs grown on different substrates for IQE estimation.

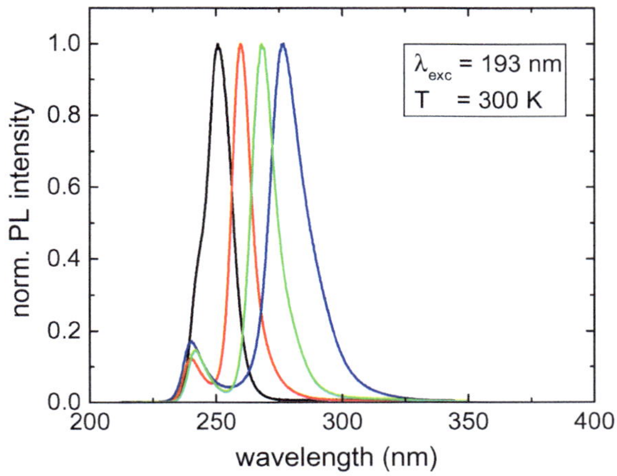

Fig. 13.20 Room temperature PL spectra of AlGaN/AlGaN MQW with varying composition with emission ranging from 250 to 280 nm © 2018 Martin Martens. Reprinted from M. Martens, "Optical gain and modal loss in AlGaN based deep UV lasers", dissertation, TU Berlin

from the normalized plot. MQWs grown on planar AlN/sapphire substrates exhibit the lowest normalized emission intensity in the range of 1–5% at 300 K. Both ELO AlN/sapphire and bulk AlN substrates allow for MQWs with less intensity reduction at 300 K, indicating an estimated internal quantum efficiency in the range of 10–30% [67].

Based on such AlGaN MQWs, heterostructure variations of the barrier thickness or composition, the quantum well thickness or composition as well as the quantum well number can be performed in order to experimentally determine the influence on estimated IQE values, emission polarization or lasing thresholds. Figure 13.20 shows room temperature PL spectra of AlGaN/AlGaN MQWs with various compositions enabling intentional tuning of the UVC emission ranging from 250 to 280 nm. To achieve even shorter emission wavelengths, the heterostructure has to be modified (increasing barrier/waveguide composition) to ensure full transparency for the MQW emission.

13.7 Gain and Losses in Deep UV AlGaN Lasers by Optical Pumping

13.7.1 Optical Pumping for Lasing Threshold and Gain Measurements

Optical pumping of laser heterostructures is an effective technique to investigate the modal gain of $Al_xGa_{1-x}N$ QWs as well as determining the optical losses in the waveguides especially if current injection laser diodes have not yet been realized. For this purpose the excitation laser (e.g., an ArF* excimer laser emitting at 193 nm) was focused to form a narrow stripe (14 μm × 2000 μm) which is also depicted in

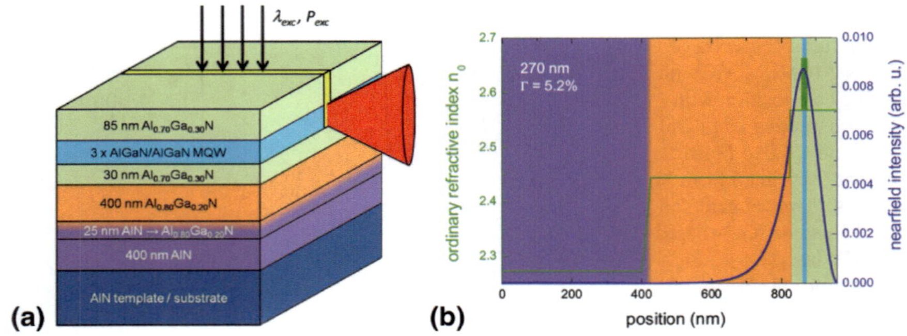

Fig. 13.21 **a** Sketch of the optical pumping experiment. The excitation laser is focused in a stripe (14 μm × 2000 μm) on the surface of the laser bar with cleaved facets exciting the AlGaN MQW active region [67] © 2014 IEEE. Reprinted, with permission, from Martens et al., IEEE Photonics Technology Letters 26, 1041. **b** Refractive index and mode profile of the laser

Fig. 13.21. Since, for non-resonant pumping, the light from the excitation laser is strongly absorbed in all $Al_xGa_{1-x}N$ layers, a special laser heterostructure for optical pumping was employed that includes no *p*-doped layers and no upper cladding layer in order to maximize the carrier density in the QWs. The optically pumped laser heterostructures as shown in Fig. 13.21 consist of an AlN base layer, a 1000 nm thick $Al_{0.8}Ga_{0.2}N$:Si lower *n*-cladding, 40 nm $Al_{0.7}Ga_{0.3}N$:Si lower waveguide, followed by a threefold $Al_{0.47}Ga_{0.53}N/Al_{0.7}Ga_{0.3}N$ MQW and a 85 nm $Al_{0.7}Ga_{0.3}N$:Mg upper waveguide. The mode profile centered near the MQWs exhibits a confinement factor of 5.2% and a near-field intensity distribution in the lower cladding similar to the current injection laser (see Sect. 13.3).

Two kinds of measurements were conducted: (1) Measurement of the lasing threshold power density [67] on laser bars with a cavity length of 0.8 mm or 1 mm and the stripe spanning the entire length of the bar. (2) Measurement of gain spectra by variable stripe length measurements (VSLM) [92] on laser bars of 2 mm length and excitation stripe lengths between 0.05 and 0.5 mm in order to avoid optical feedback. Whereas the laser threshold is relatively easy to acquire and compare between different heterostructures, gain spectroscopy allows for deeper insights and the determination of material gain, transparency excitation power density and internal losses.

13.7.2 Optical Gain in Dependence of the Emission Wavelength

Lasing was demonstrated under optical pumping for emission wavelengths ranging between 237 and 290 nm, as can be seen also in Fig. 13.3, depicting the respective lasing spectra. The lasers with emission wavelengths from 255 to 290 nm all have the

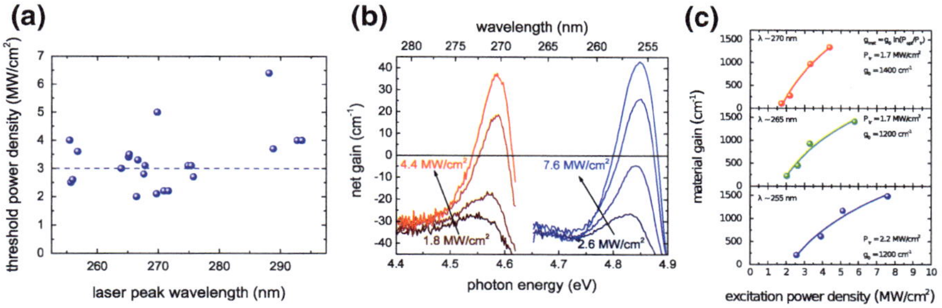

Fig. 13.22 **a** Lasing thresholds for lasers with QW Al mole fraction from 38% (294 nm) to 52% (250 nm). **b** Excitation power density dependent gain spectra of lasers with 255 and 270 nm emission wavelength. **c** Dependence of the peak gain on the excitation power density fitted by a logarithmic gain model, © 2018 Martin Martens. Reprinted from M. Martens, "Optical gain and modal loss in AlGaN based deep UV lasers", dissertation, TU Berlin

same optical confinement heterostructure described above and only the aluminum mole fraction in the AlGaN QWs was varied from 38 to 52%. In Fig. 13.22a the measured lasing thresholds are plotted [25]. It should be noted that the reported threshold power density values are referring to the total power density at the sample surface and do not take into account loss mechanisms like surface reflection, lateral diffusion of carriers and inhomogeneous excitation within the laser heterostructure. Therefore, the reported power densities are certainly overestimating the threshold power density. Nevertheless, as the setup is the same for all lasers, this approach still allows the comparison of different laser heterostructures. The thresholds for optically pumped lasers are varying for the different wavelengths and samples but overall are very similar for all structures around 3 MW/cm^2. This suggests that the optical losses, the carrier injection into the quantum wells as well as the gain behave very similar for all emission wavelengths. Indeed, gain characteristics measured by the VSLM are very similar as shown in Fig. 13.22b for lasers emitting at 255 and 270 nm. Both structures exhibit similar modal losses of 30 cm^{-1} that can be extracted from the low energy side of the spectrum. At excitation power densities around 2 MW cm^{-1} a positive modal gain (gain relative to the internal losses) of 5–10 cm^{-1} is observed for both lasers. At higher pumping power densities similar net gain values of 40 cm^{-1} are generated whereas the slightly higher power density in the 255 nm emitting laser also corresponds to a slightly higher threshold compared to the 270 nm emitting laser. For such high pumping levels the FWHM of the gain spectra are very similar with 70–80 meV indicating similar homogeneity for both emission wavelengths. The gain spectra were further analyzed by calculating the peak material gain [25] and fitting the peak material gain with a logarithmic gain model [25, 93]. In this model, the material gain is described by:

$$g_{\mathrm{mat}} = g_0 \ln\left(\frac{P_{\mathrm{opt}}}{P_{\mathrm{tr}}}\right) \qquad (13.1)$$

Here g_{mat} is the material gain, g_0 is the gain coefficient, P_{opt} the excitation power density and P_{tr} the transparency excitation power density. The results for three lasers emitting at 255, 265 and 270 nm are shown in Fig. 13.22c. The fits result in a P_{tr} of 1.7 MW/cm^2 for the lasers emitting at 265 and 270 nm. The slightly higher P_{tr} of 2.2 MW/cm^2 of the 255 nm laser might be explained by the much smaller band offset from QW to barriers and waveguides. P_{tr} refers to the excitation power that is required to reach population inversion within the quantum well. P_{tr} depends on spontaneous and nonradiative recombination rates, carrier escape from the quantum wells as well as transport of charge carriers from the upper waveguide, where they are mainly generated, into the quantum wells [25]. Here radiative and nonradiative recombination within the upper waveguide lead to loss of carriers and increase of P_{tr} (i.e. decrease the injection efficiency). Within the investigated wavelength range these processes do not depend significantly on the Al mole fraction in the QW. As can be seen, the measured gain coefficient g_0 is independent of the emission wavelength and exhibits a value of 1200–1400 cm^{-1}. g_0 depends mainly on the wavefunction overlap of electrons and holes, the injection efficiency and the inhomogeneous broadening, which appear to be similar for the considered lasers.

13.7.3 Optical Polarization and Valence Band Ordering

In the $Al_xGa_{1-x}N$ material system the valence band structure changes with the emission wavelength. The positive crystal field in GaN and the negative crystal field in AlN [94] are leading to nearly purely polarized emission from each of these valence bands either parallel or perpendicular to the [0001] *c*-axis [95]. This transition is not only influenced by composition but also by compressive strain and electronic confinement allowing dominant TE emission (E-field perpendicular to the *c*-axis) from AlGaN QWs on (0001) *c*-plane AlN templates with emission wavelengths as short as 240 nm [38, 96]. Similarly, optically pumped lasers exhibit TE polarized laser emission with emission wavelength of 255 nm. For a lasing wavelength of 241 nm a switch to TM polarized emission is observed [17, 25, 97] in good agreement with other reports. For the design of lasers a sufficient energetic separation of the valence bands is desirable since close to the band crossing point the transparency carrier density is increased since also the bands not contributing to gain in the desired mode (either TE or TM) need to be populated. However, the optical pumping experiments described above do not hint to an increase of the lasing threshold due to a reduced valence band separation within the wavelength range of 255–290 nm.

The optimum QW thickness depends mainly on the quantum confined Stark effect (QCSE) and the nonradiative recombination rate as well as the internal losses. Increasing the QW thickness increases the confinement factor as well as reduces the spontaneous emission due to the QCSE, both reducing transparency excitation power density [98]. However, the decreased spontaneous emission rate will not reduce the threshold if the effective lifetime at the threshold carrier density is dominated by Auger recombination [99] or nonradiative recombination. However, an increased

QW thickness also reduces the gain coefficient due to the reduced overlap of electron and hole wavefunctions [98] potentially increasing the laser threshold as well as reducing the slope efficiency of lasers. Additionally, the energetic separation of the different valance bands should be considered which also reduces with increasing QW thickness [96]. For thick $Al_{0.48}Ga_{0.52}N$ QWs a small separation of the valence bands is expected thereby increasing the density of states and the transparency carrier density. Experimentally we found a reduction of the threshold by increasing the QW width from 1 nm to 4.4 nm (also changing the emission wavelength from 260 to 270 nm) and ascribe this effect mainly to the increased confinement factor.

13.7.4 Loss Mechanisms in Deep UV Lasers

To achieve low threshold lasing, not only the gain is important but also losses within the waveguide. Losses in semiconductor lasers can be caused by absorption of the optical mode at point defects [100], mode leakage [101] and scattering at morphological features within the waveguide. For deep UV lasers on ELO AlN, scattering can play a significant role as the morphology can exhibit pits and macro-steps as shown in Fig. 13.23. These pits are generated from dislocations with a screw component [90] similar to the growth of InGaN [102]. The relative growth rate of (0001) and {10-11} facets, i.e., the pit formation, can be influenced by the growth conditions [103]. An increased growth temperature leads to a pit-free surface as shown in Fig. 13.23b). With a lasing wavelength of (278 ± 5) nm under optical excitation, the lasing threshold of these samples is strongly influenced by the surface, as the pit-free sample has a much lower lasing threshold of 2.8 MW/cm^2 than the laser exhibiting pits with 6.4 MW/cm^2 [90].

Even though pit-free, the surfaces in Figs. 13.23a and 13.23b exhibit macro-steps with a height of 15 nm (also shown in Fig. 13.24c). These steps are formed during the coalescence of the ELO AlN (see also Sect. 13.5) [72, 104] and are stable in

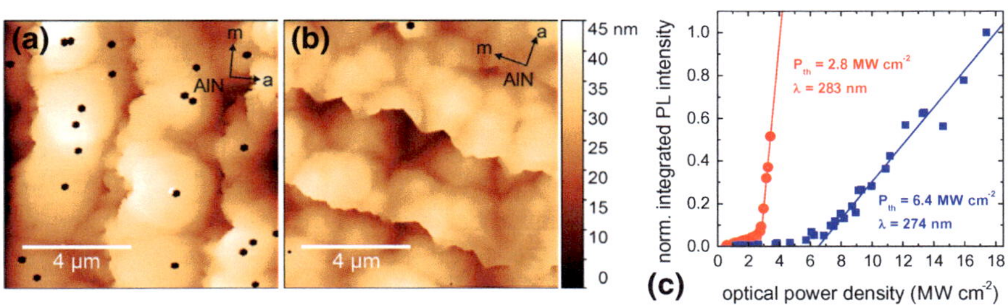

Fig. 13.23 AFM topograms of optically pumped lasers grown on ELO AlN **a** at 900 °C and **b** at 1070 °C, **c** Threshold power density determination of lasers grown at 900 °C (blue) and 1070 °C (red). Reprinted from [90], with the permission of AIP Publishing

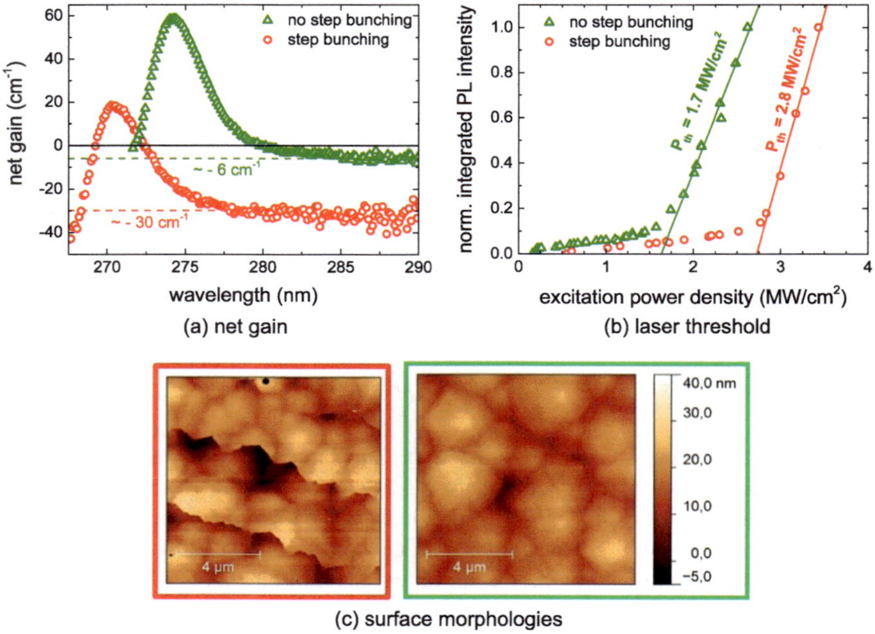

Fig. 13.24 **a** Gain spectra of lasers grown on different templates with **b** corresponding laser thresholds and **c** surface morphologies (red: step bunching, green: no step bunching) [25], © 2018 Martin Martens. Reprinted from M. Martens, "Optical gain and modal loss in AlGaN based deep UV lasers", dissertation, TU Berlin

further overgrowth in spite of growth conditions favoring step flow growth [105, 106]. By reducing the offcut of the sapphire substrate from 0.2° to (0.1 ± 0.05)°, the macro-steps formed during coalescence are smaller and can be smoothed out [72, 104]. The AFM topogram in Fig. 13.24c shows that the resulting surface of the optically pumped laser is now free of macro-steps and dominated by spiral hillocks generated by the high supersaturation necessary for efficient *n*-doping and high conductivity, as described in Sect. 13.6. The laser threshold of the sample without step bunching is reduced to 1.7 MW/cm^2 in comparison to the laser with step bunching (see Fig. 13.24b). Gain spectra of these lasers are shown in Fig. 13.24a [25]. Both samples show a clear gain peak at 270 nm and 275 nm, respectively (the difference in wavelength is due to the variation of the Al content between the different MOVPE runs). However, the internal losses that can be extracted from the spectra on the long wavelength side are very different: 30 cm^{-1} on the sample with step bunching and 6 cm^{-1} on the sample without step bunching, shifting the entire gain spectrum of the sample with step-bunching down by 24 cm^{-1}. Therefore, a much higher pumping power is required to overcome the losses and achieve lasing on the sample with step bunching. The higher internal losses can be explained by more scattering on the sample with step bunching. Another possible reason for the lower threshold without step bunching is the homogeneity of the QWs. Step-bunching can induce local variations of the aluminum mole fraction in the QWs [84], thereby broadening the gain spectrum.

Although the scattering losses are strongly reduced, the spiral hillocks can also influence the local incorporation of gallium changing the local emission wavelength [25, 73, 84] which leads to broadening of the gain peak and a reduction of the peak gain [107]. An $Al_xGa_{1-x}N/Al_yGa_{1-y}N$ superlattice with an average aluminum content of 80% can—without reducing the conductivity—improve the surface roughness as described in Sect. 13.6. Bulk and superlattice lower cladding layer both exhibit surfaces with spiral hillocks. However, with the superlattice the height of the spiral hillocks is reduced leading to a reduced RMS roughness of 2.5 nm in comparison to 5.8 nm for the bulk layer (both on 10 μm × 10 μm). Optical pumping reveals a lower threshold power density of 0.9 MW/cm^2 for the laser with superlattice cladding compared to that with bulk lower cladding with the same average Al mole fraction of 80%, shown in Fig. 13.25. As the optical losses (see Fig. 13.24a) are already quite low for lasers with a bulk AlGaN cladding, we assume that the reduced surface roughness improved the homogeneity of the quantum wells and the peak gain. These results show that for lasers grown on sapphire substrates, it will always be necessary to find a tradeoff between the morphology and conductivity of the *n*-cladding [73, 108]. On bulk AlN a lower surface roughness even for AlGaN grown with a high supersaturation can be achieved due to the absence of screw dislocations allowing for potentially even lower thresholds (see also Sect. 13.2).

As the lasing threshold excitation power density is a rather indirect figure an optically pumped 405 nm laser grown on bulk GaN substrate was measured on the same setup used for optical pumping of deep UV lasers. The laser exhibited a threshold power density of 0.2 MW/cm^2 which relates to a threshold current density of 2.1 kA/cm^2 for the full laser diode heterostructure [25]. The lowest thresholds of UVC lasers of 0.9 MW/cm^2 measured on the setup indicate that there is no major

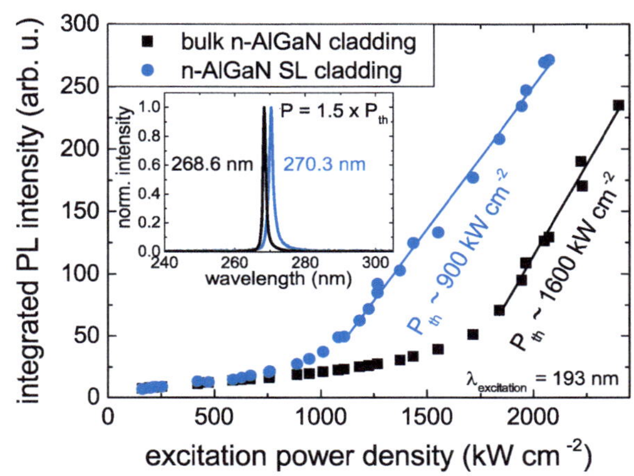

Fig. 13.25 Optical pumping of lasers with bulk $Al_{0.8}Ga_{0.2}N$ cladding and superlattice cladding showing a reduced threshold for the superlattice cladding Reprinted from [24], with the permission of John Wiley and Sons

roadblock for UVC laser diodes from the viewpoint of confinement, gain and scattering losses. However, to achieve current injection lasing, gain and confinement must be further optimized as the current injection, especially the *p*-doping is so challenging (see Sect. 13.8). A further reduction of the threshold is possible by improving the homogeneity, reducing the nonradiative recombination rates by reducing the threading dislocation density and also by optimizing the QW number, QW and barrier width and interface quality. All these challenges can only be solved by growing lasers on bulk AlN as the threading dislocation density on sapphire would dominate over all other effects. However, in current injection laser diodes a critical part of the absorption is related to Mg used for *p*-doping of the waveguide and cladding [100]. Therefore, an experimental determination of the absorption in AlGaN:Mg is very important for the design of current injection UVC laser diodes and the estimation of the expected threshold current density.

In order to determine the losses due to *p*-doping with Mg we performed gain spectroscopy of AlGaN lasers with a partially doped upper $Al_{0.7}Ga_{0.3}N$ waveguide [109]. Figure 13.26a shows the sample design. In order to differentiate Mg induced internal losses from other internal losses, an undoped structure was grown and compared to lasers with different fractions of the upper waveguide doped with Mg. In order to extrapolate the losses of fully doped structure, lasers with a mode overlap with the Mg doped portion of the waveguide of 10, 20 and 40% were grown [109]. From the gain measurements the internal losses were determined by reading the loss on the long wavelength side of the spectrum similar to Fig. 13.24a. In Fig. 13.26b the measured losses are plotted over mode overlap with the Mg doped waveguide. The laser with the undoped waveguide exhibits internal losses of 10 cm^{-1}. The internal losses increase linearly with increasing mode overlap with the Mg doped waveguide to up to 30 cm^{-1} for a mode overlap of 40%. By subtracting the internal losses not related

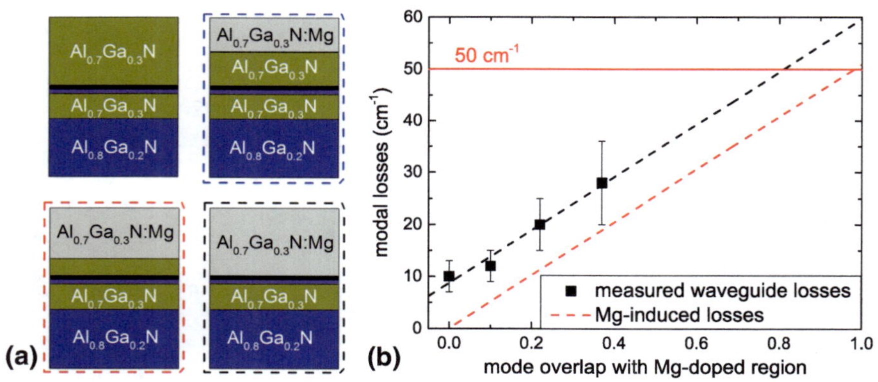

Fig. 13.26 **a** Laser heterostructures for the determination of the Mg induced losses in AlGaN with different portions of the upper waveguide doped with Mg. **b** Absorption losses determined by gain spectroscopy on the samples of (**a**) exhibiting mode overlap with the Mg doped region from 0 to 40%. Reprinted from [109], with the permission of AIP Publishing

to Mg an absorption coefficient of the AlGaN:Mg of 50 cm^{-1} can be consistently extrapolated for all samples which is similar to values determined for GaN:Mg [100]. This shows that Mg induced absorption also plays a significant role for the losses in current injection UVC lasers. However, by optimizing the mode profile acceptable modal losses similar to InGaN based laser diodes can be achieved.

Optical pumping experiments allowed to investigate gain and losses in AlGaN based lasers providing a critical stepping-stone for the realization of current injection laser diodes in the UVC spectral range. We found that high modal gain can be produced under sufficient pumping and threshold as well as gain characteristics do not depend on the emission wavelength between 255 and 290 nm. Lasers with shorter emission wavelength might suffer from a smaller separation of the valance bands. To minimize the threshold, barrier and quantum well width and composition need to be optimized for low threshold considering also the losses within the structure (low losses allow for wide QWs with increased confinement factor and potentially a reduced transparency excitation power density). Smooth surface morphologies are critical in order to reduce scattering losses and decrease the inhomogeneous broadening of the gain spectra. Finally, the absorption losses in AlGaN:Mg were extracted from gain measurements yielding this important parameter for the design of current injection lasers diodes.

13.8 Development of Current Injection Deep UV Laser Diodes

13.8.1 *Low Resistance* n-*AlGaN Current Spreading Layers*

Besides electrical conductivity, compositional homogeneity and surface morphology of AlGaN:Si layers, which is discussed in Sect. 13.6 of this chapter, the AlGaN:Si composition can also influence the injection efficiency in UVC LEDs. Figure 13.27 shows the emission power of 265 nm emitting LEDs with varying AlGaN:Si composition, all of them without strong relaxation or absorption at this wavelength. The emission power is strongly increasing for Al contents above 70%, indicating more efficient carrier injection. This should be considered in UV laser diode design, but fortunately is in line with the high Al content required for pseudomorphic growth that also enhances optical confinement.

13.8.2 *Mg-Doped AlGaN Short Period Superlattices*

Achieving sufficient electrical conductivity of AlGaN:Mg-based *p*-SPSL cladding layers with high Al content $x > 0.6$ is very challenging. The voltages in *IV*-measurements of UVC laser diode heterostructures are strongly determined by these

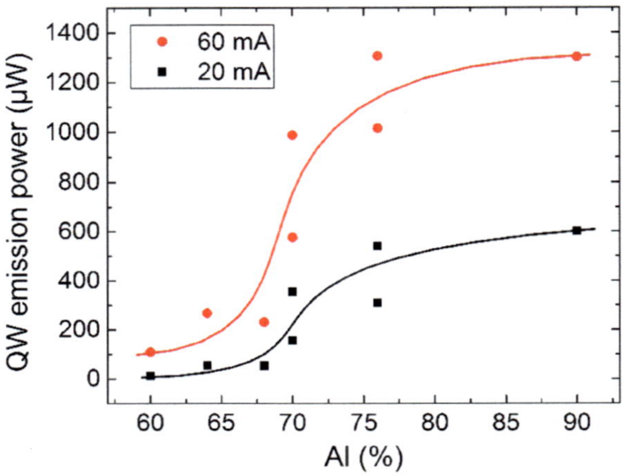

Fig. 13.27 Emission power of 265 nm emitting UVC LEDs grown on AlGaN:Si with different composition

p-layers as they exhibit the highest resistance within these devices. With increasing *p*-SPSL thickness *IV*-curves are expected to reduce in slope as determined by the increasing series resistance R combined with a constant turn-on voltage (see calculation in Fig. 13.28a). Instead, the measured diode structures exhibit increasing turn-on voltages with almost comparable slopes (Fig. 13.28b). Therefore, the *p*-SPSL resistance depends on the operating point of the diode, which implies a correlation with the electric field F in the *p*-AlGaN SPSL. The evaluated electric field in the *p*-SPSL and its resistivity in dependence on current are shown in Fig. 13.29a. In combination, the influence of the electric field on the resistivity is shown in Fig. 13.29b: With increasing electric field, the *p*-SPSL resistivity is decreasing by orders of magnitude, resulting in values of approximately 10 kΩ cm for 0.5–0.7 MV/cm.

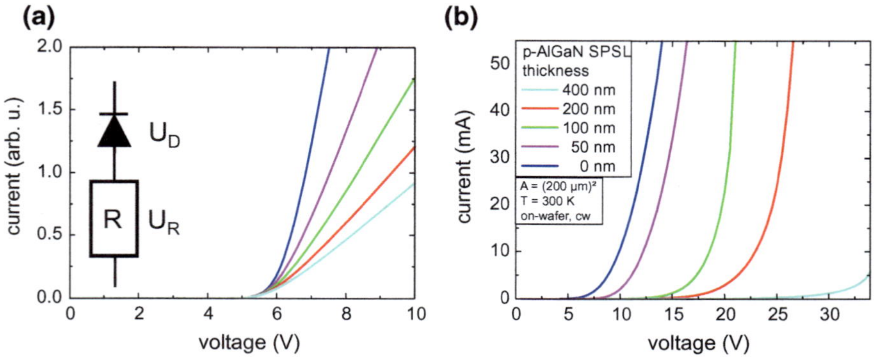

Fig. 13.28 *IV*-characteristics of UVC laser diode heterostructures with increasing $Al_{0.76}Ga_{0.24}N$:Mg *p*-SPSL thickness, **a** calculation with increasing series resistance *R* and **b** *IV*-measurement

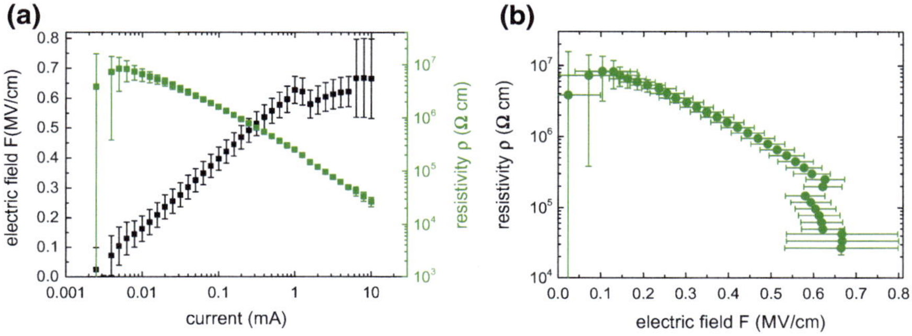

Fig. 13.29 **a** Evaluated electric field and resistivity of the Mg-doped $Al_{0.86}Ga_{0.14}N/Al_{0.66}Ga_{0.34}N$ SPSL at different LED operation points (currents), **b** a strong resistivity reduction by two orders of magnitude with increasing electric field to values around 0.5 MV/cm, indicating field-enhanced acceptor ionization

Such a field dependence of the resistivity could be expected for field-induced ionization processes of magnesium acceptors in AlGaN materials, generally referred to as Poole-Frenkel-effect (PF). It describes the reduction of the effective ionization energy of a localized state by external electric fields [110–113]. The resulting current density is correlated to the square root of the electric field, see (13.2), where the bulk ionization energy Φ_t and the relative optical permittivity $\varepsilon_r^{\text{opt}}$ are materials properties. This correlation of current density and electric field is shown in Fig. 13.30 in a representation easily comparable to the terms of the PF-equation. The measurement data of UVC laser diode heterostructures with $Al_{0.76}Ga_{0.24}N$:Mg p-SPSL is well reproduced by the PF simulation line (green) which assumes a relative optical

Fig. 13.30 Correlation of current density and electric field of $Al_{0.76}Ga_{0.24}N$:Mg containing diode structures in the representation of the equation for a field-induced acceptor ionization model (Poole-Frenkel-effect). The measurement is well reproduced by the PF-simulation assuming a reasonable $\varepsilon = 5$

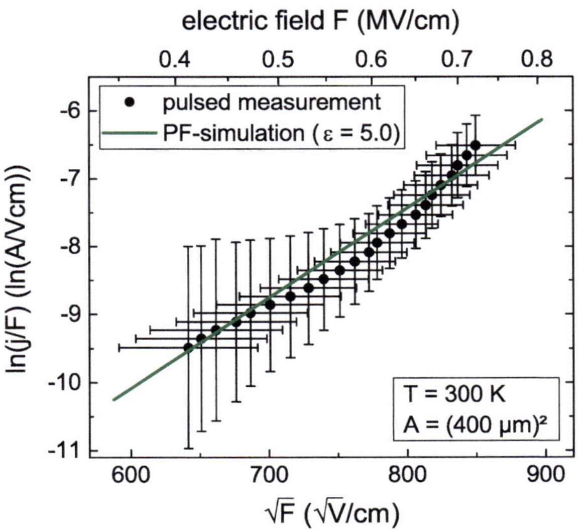

permittivity $\varepsilon = 5$ that appears reasonable compared to literature values ($\varepsilon_r^{\mathrm{opt}}$(AlN) $= 4.6$, $\varepsilon_r^{\mathrm{opt}}$(GaN) $= 5.35$) [114]. This investigation illustrates the potential use of AlGaN:Mg SPSL *p*-cladding layers in UV laser diodes and the beneficial resistivity reduction mechanism in operation due to the Poole-Frenkel effect. However, even lower *p*-AlGaN layer resistivities and operating voltages are needed to enable current injection UVC laser diodes.

$$j_{\mathrm{PF}} \propto F \exp\left\{-\frac{e_0}{k_{\mathrm{B}}T}\left[\Phi_t - \left(\frac{e_0}{\pi \varepsilon_r^{\mathrm{opt}} \varepsilon_0}\right)^{1/2} F^{1/2}\right]\right\} \quad (13.2)$$

Field-dependent current density influenced by the bulk acceptor ionization energy Φ_t as well as a reduction that depends on the relative optical permittivity $\varepsilon_r^{\mathrm{opt}}$.

13.8.3 Efficient Carrier Injection and Carrier Confinement in Deep UV AlGaN LDs by Electron Blocking Heterostructures

One crucial part in the UVC laser diode heterostructure is the electron blocking layer (EBL) that must prevent electron leakage from the active region into the *p*-side while facilitating efficient hole injection into the active region. The investigation of the AlN EBL width (Fig. 13.31a) in UVC LEDs reveals almost zero emission without EBL, whereas an AlN EBL increases the emission power to values near 1 mW that are almost constant for different widths. That demonstrates, that an EBL is necessary and an EBL width of 1.5 nm already appears sufficient. Furthermore, the EBL composition influences the carrier injection (Fig. 13.31b). With a low Al content, the electrons are not sufficiently blocked. Also an AlN EBL is not ideal,

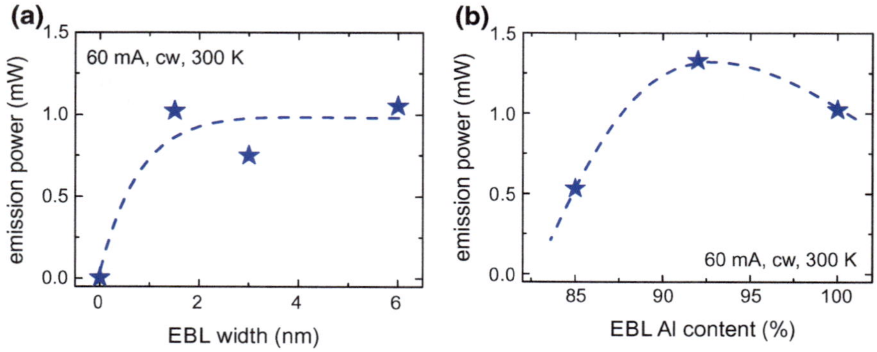

Fig. 13.31 Emission power of UVC laser diode heterostructures for different, **a** thickness and **b** composition of the electron blocking layer, indicating a large impact on carrier injection efficiency

but a maximum of emission power can be found for an $Al_xGa_{1-x}N$ EBL with $x = 0.92$. These examples illustrate the impact of the heterostructure design, in particular the EBL, on the carrier injection efficiency and the emission power of UVC LEDs. These findings also hold for laser diode heterostructures.

13.8.4 Efficient Carrier Injection in Deep UV AlGaN LD by Tunnel Heterojunctions

Light extraction from top-emitting UV LEDs is challenging, since not only all metals typically used as *p*-contacts but also the *p*-GaN contact layer absorb the emission from the active region. Contacts directly on *p*-AlGaN with high Al mole fraction without absorbing GaN:Mg cap layers suffer from a high *p*-contact resistance with a high Schottky barrier, which results in very high operating voltages. To increase the transmittance of the *p*-contact and the current spreading in the *p*-layers, it is therefore necessary to find a suitable transparent conducting layer (TCL) for the UV spectral range with high electrical conductivity. Different approaches have been proposed in order to overcome this problem: metal alloys combining high electrical conductivity with low optical absorption in the UV (such as Rh, Pd/Al, ITO/Al) [115], multi-layer graphene [116] or 1D photonic crystals [117]. Another promising alternative to standard *p*-contacts is to inject holes into the *p*-layers by means of efficient tunnel junctions (TJs). This allows low resistivity *n*-contacts on both sides of the device [118]. This way, a thin transparent top *n*-layer can be used as current spreading layer thus acting as an excellent native TCL. In a TJ configuration, electrons can tunnel from the valence band of the *p*-type material into the conduction band of the *n*-type material when the depletion width of the *p*-*n* junction is narrow enough [119]. MOVPE-grown TJ LEDs (TJ-LEDs) have already been demonstrated in the GaN/GaN and InGaN/GaN material system [120, 121]. In the AlGaN material system, only MBE-grown TJs combined with MOVPE-grown UV LEDs have been reported in literature so far [122]. Low resistivity TJs are even more challenging for UV devices due to the larger band gap and higher activation energies when increasing the Al mole fraction in the AlGaN layers.

We have successfully demonstrated AlGaN-based TJ-LEDs emitting at 270 nm grown entirely by MOVPE (Fig. 13.32a) [123]. UV-sensitive EL microscopy images measured from the top show homogeneous emission intensity with a slight enhancement near the top contact (Fig. 13.32b). The TJ heterostructure consists of a transparent AlGaN:Mg^{++} layer, an (Al)GaN interlayer (thickness <10 nm) and a transparent AlGaN:Si top current spreader. The interlayer was implemented to facilitate carrier tunneling at the interface. The TJ-LEDs have been annealed in-situ to activate the Mg acceptors and V/Al/Ni/Au *n*-contacts were deposited on the top and bottom AlGaN:Si layers. Without an interlayer, current injection and light emission was possible but at extremely high operation voltages and low current levels. LIV curves of a 0.35 mm^2 TJ-LED featuring a thin interlayer are shown in Fig. 13.32c under

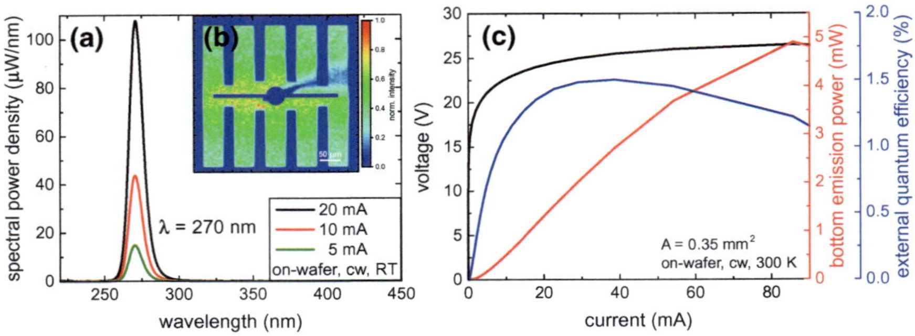

Fig. 13.32 Emission spectrum of TJ-LEDs (**a**) with UV-sensitive EL image (**b**). Reprinted from [123], with the permission of Photonics Research. LIV and EQE curves for a 0.35 mm^2 TJ-LED under cw operation at room temperature (**c**)

cw driving operation. Typically, the operation voltage and the output power are 23 V and 1.3 mW, respectively, measured at 20 mA on wafer. A maximum EQE of 1.5% is reached at 40 mA.

The development of an AlGaN-based UVC TJ-LED shows routes to reduce operating voltage and the optical losses also in LDs in the same wavelength range. In fact, the poor electrical characteristics of *p*-AlGaN layers with high Al mole fraction present even more severe problems when dealing with high-power lasers operating at high-current densities (typically 5–10 kA/cm^2) [124]. Moreover, one has also to deal with an efficient way to confine the optical mode inside the laser cavity. As mentioned before, the more transparent the AlGaN layers (i.e. higher Al mole fraction), the lower their conductivity. To ensure a sufficient refractive index contrast between the waveguide and the cladding layers in order to confine the emitted light, this imposes *p*-AlGaN cladding layers with high Al mole fraction but of poor conductivity thus increasing the device series resistance. In the TJ configuration, the top *n*-layer due to its conductivity provides lateral current injection and the *n*-contact can be moved from the top of the ridge to its side. Thus air could be used as top cladding [125]. This would substantially reduce the optical losses induced by optical mode overlap with the upper metallic contact and the doped AlGaN cladding and at the same time increase the confinement factor. The advantage in implementing a TJ in a standard UVC LD is that in the final lasing structure one will mainly deal with the electrical losses inside the contacts and the layers. In order to achieve lasing in the UVC spectral range, we will investigate a device design consisting in an edge emitting LD where the ridge is formed by a buried TJ as already reported in the GaN/InGaN material system by [125, 126].

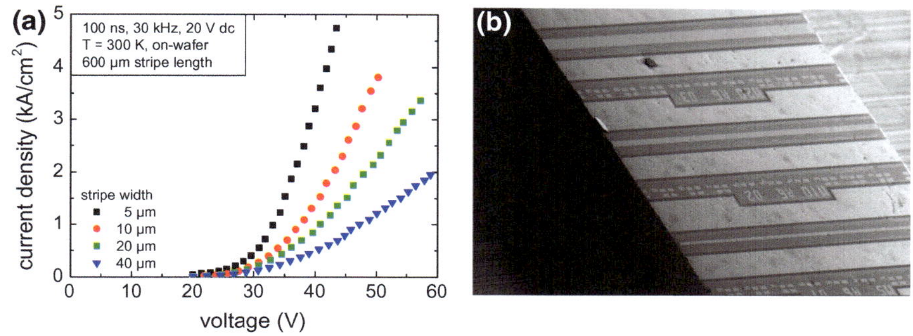

Fig. 13.33 **a** Pulsed *IV*-characteristic of UVC laser diode structures with $Al_{0.81}Ga_{0.19}N$:Mg superlattices exhibiting current injection up to current densities of 4.7 kA/cm^2, SEM image of processed and cleaved laser bar (**b**), © 2018 Martin Martens. Reprinted from M. Martens, "Optical gain and modal loss in AlGaN based deep UV lasers", dissertation, TU Berlin

13.8.5 High Density Pulsed Current Injection in UV Laser Diodes

In order to obtain population inversion in UV laser diodes pulsed current injection with high densities is necessary. This requires conductive AlGaN:Si and AlGaN:Mg layers as well as stable *n*-contacts and *p*-contacts that can sustain current densities in the range of several kA/cm^2, as described in Sect. 13.2 of this chapter. To address this issue with decent mode confinement at the same time, laser diode heterostructures containing $Al_{0.81}Ga_{0.19}N$:Mg superlattices were processed and *IV* characteristics in the high current density range were evaluated (Fig. 13.33a). In pulsed operation, high current densities of more than 1 kA/cm^2 were obtained for laser structures with 600 μm resonator length and different contact stripe widths. For the smallest structure (5 μm × 600 μm), the highest current density of more than 4.7 kA/cm^2 could be obtained. Processed and cleaved laser bars with different stripe widths are depicted in the SEM image Fig. 13.33b.

13.9 Conclusion and Outlook

We have demonstrated optically pumped AlGaN multiple quantum well (MQW) lasers near 270 nm with threshold power densities of less than 900 kW/cm^2 and realized optically pumped AlGaN MQW lasers emitting at record shortest wavelength of 237 nm. A transition from TE mode to TM mode lasing has been observed for UV lasers emitting below 240 nm due to the rearrangement of the valence bands in the AlGaN MQW. Pseudomorphic growth of atomically smooth AlGaN layers for deep UV lasers has been demonstrated including UVC-transparent and low resistance Si- and Mg-doped AlGaN cladding layers with Al mole fractions as high as 80%. Electroluminescence from current injection broad area UV laser diodes was demonstrated with maximum current densities of 4.7 kA/cm^2. These results provide a significant step forward in the demonstration of a first current injection laser diode emitting in the UVC spectral range.

Acknowledgements The authors would like to thank Sven Einfeldt, Carsten Netzel, Ute Zeimer, Jörg Jeschke, Anna Mogilatenko, Sylvia Hagedorn, and Arne Knauer (Ferdinand-Braun-Institut, Leibniz Institut für Höchstfrequenztechnik, Berlin) as well as Martin Martens, Christoph Reich, Norman Susilo, Frank Mehnke, and Martin Guttmann (Institute of Solid State Physics, Technische Universität Berlin) for their contributions to this work. This work was supported by the German Federal Ministry of Education and Research (BMBF) within the "Advanced UV for Life" project and by the Deutsche Forschungsgemeinschaft (DFG) within the Collaborative Research Centre "Semiconductor Nanophotonics" (SFB 787).

References

1. M. Kneissl, J. Rass (eds.), *III-Nitride Ultraviolet Emitters: Technology and Applications*, vol. 227, Series on Material Science (Springer, Cham, 2016)
2. S. Nagahama, T. Yanamoto, M. Sano, T. Mukai, Jap. J. Appl. Phys. **40**, L785 (2001)
3. S. Nagahama, T. Yanamoto, M. Sano, T. Mukai, Jap. J. Appl. Phys. **41**, 5 (2002)
4. S. Masui, Y. Nakatsu, D. Kasahara, S.-i. Nagahama, Proc. SPIE **10104**, 101041H (2017)
5. M. Kneissl, D.W. Treat, M. Teepe, N. Miyashita, N.M. Johnson, Appl. Phys. Lett. **82**, 2386 (2003)
6. M. Kneissl, D.W. Treat, M. Teepe, N. Miyashita, N.M. Johnson, Phys. Stat. Sol. (A) **200**, 118 (2003)
7. M. Kneissl, D.W. Treat, M. Teepe, N. Miyashita, N.M. Johnson, Appl. Phys. Lett. **82**, 4441 (2003)
8. J. Edmond, A. Abare, M. Bergman, J. Bharathan, K.L. Bunker, D. Emerson, K. Haberern, J. Ibbetson, M. Leung, P. Russel, D. Slater, J. Cryst. Growth **272**, 242 (2004)
9. H. Yoshida, Y. Yamashita, M. Kuwabara, H. Kan, Nat. Photonics **2**, 551 (2008)
10. H. Yoshida, Y. Yamashita, M. Kuwabara, H. Kan, Appl. Phys. Lett. **93**, 241106 (2008)
11. Y. Aoki, M. Kuwabara, Y. Yamashita, Y. Takagi, A. Sugiyama, H. Yoshida, Appl. Phys. Lett. **107**, 151103 (2015)
12. M.H. Crawford, A.A. Allerman, A.M. Armstrong, M.L. Smith, K.C. Cross, Appl. Phys. Express **8**, 112702 (2015)
13. H. Taketomi, Y. Aoki, Y. Takagi, A. Sugiyama, M. Kuwabara, H. Yoshida, Jpn. J. Appl. Phys. **55**, 05FJ05 (2016)
14. S. Okawara, Y. Aoki, Y. Yamashita, H. Yoshida, 6th Int. Symp. Growth III-Nitrides (ISGN-6), paper # I-Tu -B3 (2015)
15. Z. Zhang, M. Kushimoto, T. Sakai, N. Sugiyama, L.J. Schowalter, C. Sasaoka, H. Amano, Appl. Phys. Express **12**, 124003 (2019)
16. T. Wunderer, J.E. Northrup, N.M. Johnson, in *III-Nitride Ultraviolet Emitters: Technology and Applications*, ed. by M. Kneissl, J. Rass. Materials Science, vol. 227 (Springer, Cham, 2016)
17. R. Kirste, Q. Guo, J.H. Dycus, A. Franke, S. Mita, B. Sarkar, P. Reddy, J.M. LeBeau, R. Collazo, Z. Sitar, Appl. Phys. Express **11**, 082101 (2018)

18. X.-H. Li, T. Detchprohm, T.-T. Kao, M.M. Satter, S.-C. Shen, P.D. Yoder, R.D. Dupuis, S. Wang, Y.O. Wei, H. Xie, A.M. Fischer, F.A. Ponce, T. Wernicke, C. Reich, M. Martens, M. Kneissl, Appl. Phys. Lett. **105**, 141106 (2014)
19. H. Kawanishi, M. Senuma, T. Nukui, Appl. Phys. Lett. **89**, 041126 (2006)
20. M. Kneissl, Z. Yang, M. Teepe, N.M. Johnson, Proc. SPIE **7230**, 7230-13 (2009)
21. T. Wunderer, C.L. Chua, J.E. Northrup, Z. Yang, N.M. Johnson, M. Kneissl, G.A. Garrett, H. Shen, M. Wraback, B. Moody, H.S. Craft, R. Schlesser, R.F. Dalmau, Z. Sitar, Phys. Stat. Sol. (C) **9**(3–4), 822 (2012)
22. J. Jeschke, M. Martens, A. Knauer, V. Kueller, U. Zeimer, C. Netzel, C. Kuhn, F. Krueger, C. Reich, T. Wernicke, M. Kneissl, M. Weyers, IEEE Photonics Technol. Lett. **27**(18), 1969 (2015)
23. Z. Bryan, I. Bryan, S. Mita, J. Tweedie, Z. Sitar, R. Collazo, Appl. Phys. Lett. **106**, 232101 (2015)
24. C. Kuhn, T. Simoneit, M. Martens, T. Markurt, J. Enslin, F. Mehnke, K. Bellmann, T. Schulz, M. Albrecht, T. Wernicke, M. Kneissl, Phys. Stat. Sol. (A) **215**, 1800005 (2018)
25. M. Martens, Dissertation, Technische Universität Berlin, 2018
26. V.I. Kozlovsky, A.B. Krysa, Y.K. Skyasyrsky, Y.M. Popov, A. Abare, M.P. Mack, S. Keller, U.K. Mishra, L. Coldren, S. DenBaars, M.D. Tiberi, T. George, MRS Internet J. Nitride Semicond. Res. **2**, E38 (1997)
27. T. Wunderer, J. Jeschke, Z. Yang, M. Teepe, M. Batres, B. Vancil, N.M. Johnson, IEEE Photonics Technol. Lett. **29**(16), 1344 (2017)
28. F. Tabataba-Vakili, T. Wunderer, M. Kneissl, Z. Yang, M. Teepe, M. Batres, M. Feneberg, B. Vancil, N.M. Johnson, Appl. Phys. Lett. **109**, 181105 (2016)
29. M. Kneissl, T.-Y. Seong, J. Han, H. Amano, Nat. Photonics **13**, 233 (2019)
30. M.L. Nakarmi, N. Nepal, C. Ugolini, T.M. Altahtamouni, J.Y. Lin, H.X. Jiang, Appl. Phys. Lett. **89**, 152120 (2006)
31. J.L. Lyons, A. Janotti, C.G. Van de Walle, Phys. Rev. Lett. **108**, 156403 (2012)
32. B. Cheng, S. Choi, J.E. Northrup, Z. Yang, C. Knollenberg, M. Teepe, T. Wunderer, C.L. Chua, N.M. Johnson, Appl. Phys. Lett. **102**, 231106 (2013)
33. A.A. Allerman, M.H. Crawford, M.A. Miller, S.R. Lee, J. Cryst. Growth **312**, 756 (2010)
34. J. Simon, V. Protasenko, C. Lian, H. Xing, D. Jena, Science **327**, 60 (2010)
35. R. Dahal, J. Li, S. Majety, B.N. Pantha, X.K. Cao, J.Y. Lin, H.X. Jiang, Appl. Phys. Lett. **98**, 211110 (2011)
36. Y. Zhang, S. Krishnamoorthy, F. Akyol, A.A. Allerman, M.W. Moseley, A.M. Armstrong, S. Rajan, Appl. Phys. Lett. **109**, 121102 (2016)
37. F. Mehnke, M. Guttmann, J. Enslin, C. Kuhn, C. Reich, J. Jordan, S. Kapanke, A. Knauer, M. Lapeyrade, U. Zeimer, H. Krüger, M. Rabe, S. Einfeldt, T. Wernicke, H. Ewald, M. Weyers, M. Kneissl, IEEE J. Sel. Top. Quantum Electron. **23**, 2 (2017)
38. M. Guttmann, F. Mehnke, B. Belde, F. Wolf, C. Reich, L. Sulmoni, T. Wernicke, M. Kneissl, Jap. J. Appl. Phys. **58**, SCCB20 (2019)
39. M. Martens, C. Kuhn, E. Ziffer, T. Simoneit, V. Kueller, A. Knauer, J. Rass, T. Wernicke, S. Einfeldt, M. Weyers, M. Kneissl, Appl. Phys. Lett. **108**, 151108 (2016)
40. M. Röwe, P. Michler, J. Gutowski, S. Bader, G. Brüderl, V. Kümmler, S. Miller, A. Weimar, A. Lell, V. Härle, Phys. Stat. Sol. (A) **194**, 414 (2002)
41. H. Braun, C. Lauterbach, U.T. Schwarz, V. Laino, B. Witzigmann, C. Rumbolz, M.O. Schillgalies, A. Lell, V. Härle, U. Strauß, Phys. Stat. Sol. (C) **4**, 2772 (2007)
42. J. Glaab, C. Ploch, R. Kelz, C. Stölmacker, M. Lapeyrade, N. Lobo Ploch, J. Rass, T. Kolbe, S. Einfeldt, F. Mehnke, C. Kuhn, T. Wernicke, M. Weyers, M. Kneissl, J. Appl. Phys. **118**, 094504 (2015)
43. T. Kolbe, T. Sembdner, A. Knauer, V. Kueller, H. Rodriguez, S. Einfeldt, P. Vogt, M. Weyers, M. Kneissl, Phys. Stat. Sol. (A) **207**, 2198 (2010)
44. S.-N. Lee, H.Y. Ryu, H.S. Paek, J.K. Son, T. Sakong, T. Jang, Y.J. Sung, K.S. Kim, K.H. Ha, O.H. Nam, Y. Park, Phys. Stat. Sol. (C) **4**, 2788 (2007)
45. T. Weig, Dissertation, Albert-Ludwigs-Universität Freiburg im Breisgau, 2015

46. B. Van Daele, G. Van Tendeloo, W. Ruythooren, J. Derluyn, M.R. Leys, M. Germain, Appl. Phys. Lett. **87**, 061905 (2005)
47. S. Ruvimov, Z. Liliental-Weber, J. Washburn, D. Qiao, S.S. Lau, P.K. Chu, Appl. Phys. Lett. **73**, 2582 (1998)
48. A. Motayed, R. Bathe, M.C. Wood, O.S. Diouf, R.D. Vispute, S.N. Mohammad, J. Appl. Phys. **93**, 1087 (2003)
49. M.A. Miller, B.H. Koo, K.H.A. Bogart, S.E. Mohney, J. Electron. Mater. **37**, 564 (2008)
50. M. Lapeyrade, A. Muhin, S. Einfeldt, U. Zeimer, A. Mogilatenko, M. Weyers, M. Kneissl, Semicond. Sci. Technol. **28**, 125015 (2013)
51. N. Nagata, T. Senga, M. Iwaya, T. Takeuchi, S. Kamiyama, I. Akasaki, Phys. Stat. Sol. (C) **14**, 1600243 (2017)
52. F. Mehnke, T. Wernicke, H. Pingel, C. Kuhn, C. Reich, V. Kueller, A. Knauer, M. Lapeyrade, M. Weyers, M. Kneissl, Appl. Phys. Lett. **103**, 212109 (2013)
53. F. Mehnke, L. Sulmoni, M. Guttmann, T. Wernicke, M. Kneissl, Appl. Phys. Express **12**, 012008 (2019)
54. H.W. Jang, C.M. Jeon, J.K. Kim, J.-L. Lee, Appl. Phys. Lett. **78**, 2015 (2001)
55. S.J. Pearton, Appl. Surf. Sci. **117/118**, 597 (1997)
56. Z. Mouffak, A. Bensaoula, L. Trombetta, J. Appl. Phys. **95**, 727 (2004)
57. X.A. Cao, H. Piao, S.F. LeBoeuf, J. Li, J.Y. Lin, H.X. Jiang, Appl. Phys. Lett. **89**, 082109 (2006)
58. M.A. Miller, S.E. Mohney, A. Nikiforov, G.S. Cargill, K.H.A. Bogart, Appl. Phys. Lett. **89**, 132114 (2006)
59. A. Knauer, V. Kueller, U. Zeimer, M. Weyers, C. Reich, M. Kneissl, Phys. Stat. Sol. (A) **210**, 451 (2013)
60. U.T. Schwarz, M. Kneissl, Phys. Stat. Sol. RRL **1**, A44 (2007)
61. T. Kolbe, A. Knauer, J. Stellmach, C. Chua, Z. Yang, S. Einfeldt, P. Vogt, N.M. Johnson, M. Weyers, M. Kneissl, SPIE Proc. **7939**, 79391G (2011)
62. T. Takano, Y. Ohtaki, Y. Narita, H. Kawanishi, Jpn. J. Appl. Phys. **43**(10A), L1258 (2004)
63. H. Sun, E.F. Pecora, J. Woodward, D.J. Smith, L. Dal Negro, T.D. Moustakas, Phys. Stat. Sol. (A) **213**, 1165 (2016)
64. J.R. van Look, S. Einfeldt, O. Krüger, V. Hoffmann, A. Knauer, M. Weyers, P. Vogt, M. Kneissl, IEEE Photonics Technol. Lett. **22**(6), 416 (2010)
65. M. Bickermann, in *III-Nitride Ultraviolet Emitters: Technology and Applications*, ed. by M. Kneissl, J. Rass. Springer Series in Materials Science, vol. 227. (Springer, Cham, 2016)
66. HexaTech Inc., HexaTech Launches 2-Inch AlN Substrates (2018), http://www.hexatechinc.com/news-events.html. Accessed 24 Apr 2018
67. M. Martens, F. Mehnke, C. Kuhn, C. Reich, V. Kueller, A. Knauer, C. Netzel, C. Hartmann, J. Wollweber, J. Rass, T. Wernicke, M. Bickermann, M. Weyers, M. Kneissl, IEEE Photonics Technol. Lett. **26**, 342 (2014)
68. A. Rice, R. Collazo, J. Tweedie, R. Dalmau, S. Mita, J. Xie, Z. Sitar, J. Appl. Phys. **108**, 043510 (2010)
69. A. Mogilatenko, A. Knauer, U. Zeimer, C. Netzel, J. Jeschke, R.-S. Unger, C. Hartmann, J. Wollweber, A. Dittmar, U. Juda, M. Weyers, M. Bickermann, J. Cryst. Growth **505**, 69 (2019)
70. P. Drechsel, P. Stauss, W. Bergbauer, P. Rode, S. Fritze, A. Krost, T. Markurt, T. Schulz, M. Albrecht, H. Riechert, U. Steegmüller, Phys. Stat. Sol. (A) **209**, 427 (2012)
71. A. Mogilatenko, A. Knauer, U. Zeimer, C. Hartmann, H. Oppermann, M. Weyers, J. Cryst. Growth **462**, 18 (2017)
72. A. Knauer, A. Mogilatenko, S. Hagedorn, J. Enslin, T. Wernicke, M. Kneissl, M. Weyers, Phys. Stat. Sol. (B) **253**, 809 (2016) and erratum Phys. Stat. Sol. (B) **253**, 1228 (2016)
73. I. Bryan, Z. Bryan, S. Mita, A. Rice, J. Tweedie, R. Collazo, Z. Sitar, J. Cryst. Growth **438**, 81 (2016)
74. J. Edmond, J. Lagaly, JOM **49**, 24 (1997)
75. O. Reentilä, F. Brunner, A. Knauer, A. Mogilatenko, W. Neumann, H. Protzmann, M. Heuken, M. Kneissl, M. Weyers, G. Tränkle, J. Cryst. Growth **310**, 4932 (2008)

76. S. Mohn, N. Stolyarchuk, T. Markurt, R. Kirste, M.P. Hoffmann, R. Collazo, A. Courville, R. Di Felice, Z. Sitar, P. Vennéguès, M. Albrecht, Phys. Rev. Appl. **5**, 054004 (2016)
77. B. Kuhn, F. Scholz, Phys. Stat. Sol. (A) **188**, 629 (2001)
78. F. Brunner, H. Protzmann, M. Heuken, A. Knauer, M. Weyers, M. Kneissl, Phys. Stat. Sol. (C) **5**, 1799 (2008)
79. H. Hirayama, T. Yatabe, N. Noguchi, T. Ohashi, N. Kamata, Appl. Phys. Lett. **91**, 071901 (2007)
80. M. Imura, K. Nakano, G. Narita, N. Fujimoto, N. Okada, K. Balakrishnan, M. Iwaya, S. Kamiyama, H. Amano, I. Akasaki, T. Noro, T. Takagi, A. Bandoh, J. Cryst. Growth **298**, 257 (2007)
81. V. Kueller, A. Knauer, U. Zeimer, H. Rodriguez, A. Mogilatenko, M. Kneissl, M. Weyers, Phys. Stat. Sol. (C) **8**, 2022 (2011)
82. V. Kueller, A. Knauer, U. Zeimer, M. Kneissl, M. Weyers, J. Cryst. Growth **368**, 83 (2013)
83. A. Knigge, M. Brendel, F. Brunner, S. Einfeldt, A. Knauer, V. Kueller, U. Zeimer, M. Weyers, Jpn. J. Appl. Phys. **52**, 08JF03 (2013)
84. U. Zeimer, J. Jeschke, A. Mogilatenko, A. Knauer, V. Kueller, V. Hoffmann, C. Kuhn, T. Simoneit, M. Martens, T. Wernicke, M. Kneissl, M. Weyers, Semicond. Sci. Technol. **30**, 114008 (2015)
85. A. Mogilatenko, V. Küller, A. Knauer, J. Jeschke, U. Zeimer, M. Weyers, G. Tränkle, J. Cryst. Growth **402**, 222 (2014)
86. H. Miyake, C.-H. Lin, K. Tokoro, K. Hiramatsu, J. Cryst. Growth **456**, 155 (2016)
87. N. Susilo, S. Hagedorn, D. Jaeger, H. Miyake, U. Zeimer, C. Reich, B. Neuschulz, L. Sulmoni, M. Guttmann, F. Mehnke, C. Kuhn, T. Wernicke, M. Weyers, M. Kneissl, Appl. Phys. Lett. **112**, 041110 (2018)
88. F.A. Ponce, J.S. Major, W.E. Plano, D.F. Welch, Appl. Phys. Lett. **65**, 2302 (1994)
89. A. Mogilatenko, J. Enslin, A. Knauer, F. Mehnke, K. Bellmann, T. Wernicke, M. Weyers, M. Kneissl, Semicond. Sci. Technol. **30**, 114010 (2015)
90. C. Kuhn, M. Martens, F. Mehnke, J. Enslin, P. Schneider, C. Reich, F. Krueger, J. Rass, J.B. Park, V. Kueller, A. Knauer, T. Wernicke, M. Weyers, M. Kneissl, J. Phys. D: Appl. Phys. **51**, 415101 (2018)
91. F. Mehnke, X.T. Trinh, H. Pingel, T. Wernicke, E. Janzén, N.T. Son, M. Kneissl, J. Appl. Phys. **120**, 145702 (2016)
92. K.L. Shaklee, R.F. Leheny, Appl. Phys. Lett. **18**, 475 (1971)
93. L.A. Coldren, S.W. Corzine, *Diode Lasers and Photonic Integrated Circuits* (Wiley Interscience, 1995)
94. S.-H. Wei, A. Zunger, Appl. Phys. Lett. **69**, 2719 (1996)
95. K.B. Nam, J. Li, M.L. Nakarmi, J.Y. Lin, H.X. Jiang, Appl. Phys. Lett. **84**, 5264 (2004)
96. C. Reich, M. Guttmann, M. Feneberg, T. Wernicke, F. Mehnke, C. Kuhn, J. Rass, M. Lapeyrade, S. Einfeldt, A. Knauer, V. Kueller, M. Weyers, R. Goldhahn, M. Kneissl, Appl. Phys. Lett. **107**, 142101 (2015)
97. X.-H. Li, T.-T. Kao, M.M. Satter, Y.O. Wei, S. Wang, H. Xie, S.-C. Shen, P.D. Yoder, A.M. Fischer, F.A. Ponce, T. Detchprohm, R.D. Dupuis, Appl. Phys. Lett. **106**, 041115 (2015)
98. W.W. Chow, M. Kneissl, J. Appl. Phys. **98**, 114502 (2005)
99. F. Nippert, M. Tollabi Mazraehno, M.J. Davies, M.P. Hoffmann, H.-J. Lugauer, T. Kure, M. Kneissl, A. Hoffmann, M.R. Wagner, Appl. Phys. Lett. **113**, 071107 (2018)
100. M. Kuramoto, C. Sasaoka, N. Futagawa, M. Nido, A.A. Yamaguchi, Phys. Stat. Sol. (A) **192**, 329 (2002)
101. V. Laino, F. Roemer, B. Witzigmann, C. Lauterbach, U.T. Schwarz, C. Rumbolz, M.O. Schillgalies, M. Furitsch, A. Lell, V. Härle, IEEE J. Quantum Electron. **43**, 16 (2007)
102. Y. Chen, T. Takeuchi, H. Amano, I. Akasaki, N. Yamada, Y. Kaneko, S.Y. Wang, Appl. Phys. Lett. **72**, 710 (1998)
103. Q. Sun, C.D. Yerino, T.S. Ko, Y.S. Cho, I.-H. Lee, J. Han, M.E. Coltrin, J. Appl. Phys. **104**, 093523 (2008)

104. J. Enslin, A. Knauer, A. Mogilatenko, F. Mehnke, M. Martens, C. Kuhn, T. Wernicke, M. Weyers, M. Kneissl, Phys. Stat. Sol. (A) **216**, 1900682 (2019)
105. K. Bellmann, U.W. Pohl, C. Kuhn, T. Wernicke, M. Kneissl, J. Cryst. Growth **478**, 187 (2017)
106. M. Pristovsek, K. Bellmann, F. Mehnke, J. Stellmach, T. Wernicke, M. Kneissl, Phys. Stat. Sol. (B) **254**, 1600711 (2017)
107. J. Jeschke, U. Zeimer, L. Redaelli, S. Einfeldt, M. Kneissl, M. Weyers, Appl. Phys. Lett. **105**, 173501 (2014)
108. P. Reddy, Q. Guo, J. Tweedie, S. Washiyama, F. Kaess, S. Mita, M.H. Breckenridge, R. Kirste, R. Collazo, A. Klump, B. Sakar, Z. Sitar, in *2018 IEEE Research and Applications of Photonics in Defense Conference*, vol. 1 (2018)
109. M. Martens, C. Kuhn, T. Simoneit, S. Hagedorn, A. Knauer, T. Wernicke, M. Weyers, M. Kneissl, Appl. Phys. Lett. **110**, 081103 (2017)
110. H.H. Poole, Philos. Mag. **32**, 112 (1916)
111. J. Frenkel, Phys. Rev. **54**, 647 (1938)
112. J. Piprek, Proc. SPIE **8262**, 82620E-1 (2012)
113. F. Römer, B. Witzigmann, J. Comput. Electron. **14**, 456 (2015)
114. M.E. Levinshtein, S.L. Rumyantsev, M.S. Shur (eds.) *Properties of Advanced Semiconductor Materials GaN, AlN, InN, BN, SiC, SiGe* (J. Wiley and Sons, New York, 2001, pp. 1–47)
115. T. Takano, T. Mino, J. Sakai, N. Noguchi, K. Tsubaki, H. Hirayama, Appl. Phys. Express **10**, 031002 (2017)
116. B.-J. Kim, C. Lee, Y. Jung, K. Hyeon Baik, M.A. Mastro, J.K. Hite, C.R. Eddy Jr., J. Kim, Appl. Phys. Lett. **99**, 143101 (2011)
117. Y. Kashima, N. Maeda, E. Matsuura, M. Jo, T. Iwai, T. Morita, M. Kokubo, T. Tashiro, R. Kamimura, Y. Osada, H. Takagi, H. Hirayama, Appl. Phys. Express **11**, 012101 (2018)
118. M. Malinverni, D. Martin, N. Grandjean, Appl. Phys. Lett. **107**, 051107 (2015)
119. A.G. Chynoweth, W.L. Feldmann, R.A. Logan, Phys. Rev. **121**, 684 (1961)
120. S.-R. Jeon, Y.-H. Song, H.-J. Jang, G.M. Yang, S.W. Hwang, S.J. Son, Appl. Phys. Lett. **78**, 3265 (2001)
121. M. Diagne, Y. He, H. Zhou, E. Makarona, A.V. Nurmikko, J. Han, K.E. Waldrip, J.J. Figiel, T. Takeuchi, M. Krames, Appl. Phys. Lett. **79**, 3720 (2001)
122. Y. Zhang, S. Krishnamoorthy, F. Akyol, S. Bajaj, A.A. Allerman, M.W. Moseley A.M. Armstrong, S. Rajan, Appl. Phys. Lett. **110**, 201102 (2017)
123. C. Kuhn, L. Sulmoni, M. Guttmann, J. Glaab, N. Susilo, T. Wernicke, M. Weyers, M. Kneissl, Photonics Res. **7**, B7 (2019)
124. B.P. Yonkee, E.C. Young, C. Lee, J.T. Leonard, S.P. DenBaars, J.S. Speck, S. Nakamura, Opt. Express **24**, 7816 (2016)
125. M. Malinverni, C. Tardy, M. Rossetti, A. Castiglia, M. Duelk, C. Vélez, D. Martin, N. Grandjean, Appl. Phys. Express **9**, 061004 (2016)
126. C. Skierbiszewski, G. Muziol, K. Nowakowski-Szkudlarek, H. Turski, M. Siekacz, A. Feduniewicz-Zmuda, A. Nowakowska-Szkudlarek, M. Sawicka, P. Perlin, Appl. Phys. Express **11**, 034103 (2018)

Index

M. Kneissl et al. (eds.), *Semiconductor Nanophotonics*, Springer Series in Solid-State Sciences 194, https://doi.org/10.1007/978-3-030-35656-9

T